ADVANCES IN ECOLOGY AND ENVIRONMENTAL SCIENCES

ADVANCES IN ECOLOGY AND ENVIRONMENTAL SCIENCES

P.C. MISHRA,
N. BEHERA
B.K. SENAPATI
B.C. GURU

ASHISH PUBLISHING HOUSE
8/81, Punjabi Bagh
New Delhi-110 026

Published by :

S.B. Nangia
for **ASHISH PUBLISHING HOUSE**
8/81, Punjabi Bagh
New Delhi-110 026

Tel. 50 05 81
541 09 24

Show Room :

5, Ansari Road, Daryaganj
New Delhi-110 002

Tel. 327 40 50

ISBN : 81-7024-676-8

1995

Typesetting by :

Cambridge Printing Works
B-85, Naraina Indl. Area, Phase-II
New Delhi-110 028.

Printed by :

Efficient Offset Printers
New Delhi.

Preface

Study of Ecology as an independent discipline is about 50 years old in India. The ecological theories and principles not only apply to plants and animals in their natural environment but also to human beings on a global scale. This offshoot of ecology i.e. the study of ecological principles and their application to the human situation, a key for sustainable development, has been known as Environmental Sciences. This science not only deals with system concept, resources and pollution, it also encompasses Human Ecology, Population and Control, Environmental Policy and Management, Environmental Law, Environmental Economics and Sustainable Development.

During the last 24 years since the first Earth Day Celebration in 1970, the world lost nearly 200 million hectares of tree cover, deserts expanded by some 120 million hectares, thousands of plant and animal species with which we shared the planet in 1970 no longer exist. Over two decades, some 1.6 billion people were added to the world's population and the world's farmers lost an estimated 480 billion tons of top soil which is roughly equivalent to the amount of India's cropland. Such planetary degradation proceeded despite the environmental protection efforts of the National Governments over the last 20 years.

We are beginning to experience a hotter climate, holes in the protective ozone shield, toxic chemicals in ground water, food contaminated with pesticide residues, depletion of forests, increasing extinction of species and famines. The solution lies with the creation of a sustainable society or alternatively, we have to bear the consequences of rapid deterioration of natural environment. India is still in an advantageous position for being in the early stage of industrial development. It would be possible to take adequate preventive measures to avoid the problems being faced by the developed countries regarding Environmental deterioration. Although National Legislatures passed many laws to protect the environment and thousands of grassroots environmental groups sprang up in response to locally destructive activities, the reason for such failure lies with the lack of willingness to make basic changes needed to reverse it. We have gathered enough information in the changes in the quality of our Land, Water and Air due to industrialization, urbanisation and rapid population growth. What we need is to have implementable policy and laws, to develop low cost technology for waste treatment, to change the human reproductive behaviour and to educate people at grassroots level. A major difficulty in our attempts to solve problems such as over population and pollution is conflict between the short-term welfare of individuals and the long-term welfare of the society.

In order to achieve lasting solutions to such environmental problems Bernard J. Nobel rightly suggests to refocus our attention and priorities on the principles underlying sustainability which include (a) Stabilizing population, (b) attaining a sustainable agriculture (one which does not deplete soil and water resources or contaminate the earth and food supply with pesticides), (c) recycling wastes, (d) developing benign solar energy alternatives, and (e) adopting more energy and resource conservation life style.

Prof. M.C. Dash, the eminent ecologist of India and Professor in the School of Life Sciences, Sambalpur University, played a significant role during the last two decades to deal with various local problems as also such problems existing elsewhere in the country. The earlier ecological work in India mainly centred around the plants and forest ecology but Prof. Dash gave a system approach to the ecological studies highlighting the plant - animal interaction as an important function in the ecological studies. The work on vermitechnology, quantification studies on ecosystem energetics and secondary productivity, characterisation of physico-chemical and Biological characteristics of major rivers and reservoirs of Orissa, conservation of sea Turtle in Orissan coast and Amphibians are some of the pioneering works of Prof. Dash in the state. He has also pioneered the environment education movement in Orissa through the formation of the first non-governmental voluntary organisation called ''Orissa Environmental Consciousness Society'' since 1980.

To commemorate the 55th birth anniversary of Prof. Dash and in recognition of this contribution to ecology, this commemoration volume, ''Advances in Ecology and Environmental Sciences'', has been brought out containing 35 research/review papers contributed by the eminent ecologists of India and abroad in the field of Soil Ecology and Vermitechnology, Eco-Physiology, Forest and Wild Life Ecology and Conservation, Pollution Ecology and Toxicology and Pollution Control Management.

My special thanks and appreciation go to the scientists of India and abroad whose contributions have enriched this volume and to the co-editors who have been most helpful in getting the manuscripts into the final form. I look forward for many more collaboration with them.

I also with to acknowledge the incentive provided by my students, Madhumita, Aditya, Ranjita, Susmita, Julian, Ashish, Smita, Madhuchhanda, Monoranjan, and Minakshi for this work.

P.C. Mishra

Contents

List of Contributors

Siba Prasad Adhikary	P.G. Department of Botany, Utkal University, Bhubaneswar, 751 004, Orissa
M. Alemla Ao	Lecturer, SASRD, NEHU, Medziphema, Nagaland-796 106
Ross A. Alford	Department of Zoology, James Cook University, Townsville, Qld 4811 Australia
N.K. Ambasht	Centre of Advanced Studies in Botany, Banaras Hindu University, Varanasi - 221 005
R.S. Ambasht	Centre of Advanced Studies in Botany, Banaras Hindu University, Varanasi - 221 005
D.M. Backiyavathy	Department of Zoology, Sri Venkateswar University, Tirupati, 517 502, India
S. Basu	School of Life Sciences, Sambalpur University, Jyoti Vihar, 768 019, Orissa
N. Behera	School of Life Sciences, Sambalpur University, Jyoti Vihar - 768 109, Orissa
Tanmay Bhattacharya	Department of Zoology, Vidyasagar University, Midnapore - 721 102, West Bengal
S.K. Bose	Department of Zoology, Ranchi University, Ranchi, India
Prabir Chakraborti	Department of Zoology, Womens' College, Agartala - 799 001, Tripura
Pran K. Chakrabartty	Department of Microbiology, Bose Institute, P-1/12 CIT Scheme VII M, Calcutta - 700 054, India
N.B.K. Chaudhury	Secretary, Orissa Bigyan Academy, Unit-1, Bhubaneswar
H.P. Dani	Lecturer in Botany, Phulbani, Orissa
R.K. Das	School of Life Sciences, Sambalpur University, Jyoti Vihar - 768 019, Orissa
S.K. Das	Fisheries and Limnology Laboratory Department of Zoology, Universtiyi of Kalyani, Kalyani - 741 235, West Bengal

A.K. Dash	Department of Environmental Sciences, Sambalpur University, Jyoti Vihar - 768 019, Orissa, India
P.K. Dash	Department of Botany, Panchayat College, Bargarh, Orissa
Klára Dózsa Farkas	Department of Systematic Zoology and Ecology of the Eötvos Lorånd University, 1088, Budapest, Puskin Utca 3, Hungary
S.S. Dudeja	Department of Microbiology, C.C.S. Haryana Agriculatural University, Hisar - 125 004
B.C. Guru	Depatment of Zoology, Utkal University, Vani Vihar, Bhubaneswar, Orissa
Anup K. Haldar	Department of Microbiology, Bose Institute P-1/12 CIT Scheme VII M, Calcutta- 700 054, India
A.K. Hota	School of Life Sciences, Sambalpur University, Jyoti Vihar - 768 019, Orissa
B.B. Jana	Fisheries and Limnology Laboratory, Department of Zoology, University of Kalyani, Kalyani -741 235, West Bengal
Radha D. Kale	Department of Zoology, University of Agricultural Sciences, (GKVK), Bangalore
K.K. Kapoor	Department of Microbiology, CCS Haryana Agricultural University, Hisar, 125 004
C.S. Kar	Bhitarkanika National Park, Forest Department, Government of Orissa
Ehsan A. Khan	Department of Zoology, Ranchi University, Ranchi, India
Rajee N. Kumar	Department of Botany, University of Delhi, Delhi - 110 007
P.N. Mehrotra	Department of Zoology, Ranchi University, Ranchi, India
A.B. Mishra	School of Life Sciences, Sambalpur University, Jyoti Vihar - 768 019, Orissa
Malay K. Mishra	Department of Botany, Berhampur University, Berhampur - 760 007, Orissa, India
P.C. Mishra	Department of Enviromental Sciences, Sambalpur University, Jyoti Vihar - 768 019 Orissa, India
A.K. Mohanty	Department of Zoology, Utkal University, Vani Vihar, Bhubaneswar

K.G. Mukherjee	Department of Botany, University of Delhi, Delhi - 110 007
U. Nag	Department of Environmental Sciences, Sambalpur University, Jyoti Vihar - 768 019, Orissa, India
M. Dhananjaya Naidu	Department of Zoology, Sri Venkateswar University, Triupati - 517 502, India
N.V. Nandakumar	Department of Zoology, Sri Venkateswar University, Triupati - 517 502, India
Tribhuban Panda	School of Life Sciences, Sambalpur University, Jyoti Vihar - 768 109, Orissa
Ashok K. Pandit	Centre for Research and Development, The University of Kashmir, Naseem Bagh Campus, Hazaratbal, Srinagar - 190 006, India
Rama Pasricha	Department of Botany, University of Delhi, Delhi - 110 007
D.P. Pati	School of Life Sciences, Sambalpur University, Jyoti Vihar - 768 019, Orissa
M. Phillip	Department of Zoology, Ranchi University, Ranchi, India
A.K. Panigrahi	Department of Botany, Berhampur University, Bhanja Vihar - 760 007, Orissa
T. Poukhanmang	Life Science Department, Manipur University, Imphal, 795 003, India
E. Prabhakar	Department of Zoology, Sri Venkateswar University, Tirupati - 517 502, India
Pramila Prasad	School of Life Sciences, Sambalpur University, Jyoti Vihar - 768 019, Orrisa
K.S. Rao	G.B. Pant, Istitute of Himalayan Environment and Development, Kosi Almora 263 643, India
T.V. Rao	School of Life Sciences, Sambalpur University, Jyoti Vihar -768 019, Orissa
Manasi Ray	P.G. Department of Botany, Berhampur University, Berhampur - 760 007, Orissa, India
M. Vikram Reddy	Department of Zoology, Kakatiya University, Warangal, Andhra Pradesh - 506 009
John W. Reynolds	School of Natural Resources, Sir Sandford Fleming College, Lindsay, Ontario, Canada K 9 V 5 E 6

S.K. Sahu	Department of Environmental Science, Sambalpur University, Jyoti Vihar - 768 019, Orissa, India
A.K. Sanya	Zoological Survey of India, 'M' Block, New Alipur, Calcutta - 700 053
G.R. Satpathy	School of Life Sceince, Sambalpur University, Jyoti Vihar - 768 019, Orissa, India
K.G. Saxena	G.B. Pant Institute of Himalayan Environment and Development, Kosi - Almora - 263 643, India
N. Saxena	Department of Zoology, Ranchi University, Ranchi, India
A. Sahu	Department of Botany, Berhampur University, Bhanja Vihar - 760 007, Orissa
M.P. Sinh	Department of Zoology, Ranchi University, Ranchi, India
Ratna Sinha	Department of Zoology, Ranchi University, Ranchi, India
Narendra Singh	772, Stinchomb Drive, Columbus, Chio - 43202, USA
Alok K. Srivastava	Centre of Advance Studies in Botany, Banaras Hindu University, Varanasi - 221 005
G. Subramanian	National Facility of Marine Cyanobacteria, Bharathidasan University, Palkalaiperur, Tiruchirapalli, 620 024, India
N. Thajuddin	National Facility of Marine Cyanobacteria, Bharathidasan University, Palkalaiperur, Tiruchirapalli - 620 024, India
N.S. Thoudam	Life Science Department, Manipur University, Imphal - 795 003, India
P.S. Yadava	Ecology Laboratory, Life Science Department Manipur University, Imphal - 795 003, India
Y.S. Yadava	Advisor (Fisheries), North Eastern Council Secretariat, Ministry of Home Affairs, Shillong - 793 001, Meghalaya.

Professor Madhab Chandra Dash : His Life and Works

Professor Madhab Chandra Dash is well known to the International Scientific Community for his significant contributions to several aspects of ecological sciences. He is better known as Earthworm Scientist throughout the world because of his significant work on terrestrial oligochaetes but he has also made substantial contributions to Herpetology, Ecological energetics, Limnology and Sericulture.

Professor Dash was born to Sri Hare Krushna Dash and late Mrs. Mukta Dash in 1939 in a middle class Brahmin family at Malatipatpur (Tulsichoura) in Puri district of Orissa. After graduating with school certificate from Puri Zila School, he obtained B.Sc (Hons.) from Ravenshaw College, Cuttack and M. Sc. in Zoology from Utkal University in 1963. After serving as a Lecturer in Government Colleges of Orissa for four years (1963-1967) he joined the University of Calgary, Canada (Department of Biological Sciences and Kananaskis Environmental Sciences Centre) as Research Assistant under the famous Ecologist Professor J.B. Cragg to prosecute doctoral work. He obtained Ph. D. degree in May 1970 and continued there as a Post Doctoral Fellow for sometime. He returned to Orissa in September 1970 and after working for a short period at B.J.B. College, Bhubaneswar joined Berhampur University in February, 1971 and started his independent research a career. In April, 1973, Professor Dash joined Sambalpur University and took over the Headship of the Department of Biological Sciences in February 1974. He became a permanent Professor in September 1976. He continued as the Head of the Department till November, 1987 when the Headship became rotational.

Under his able Leadership, the Department of Biological Sciences grew very rapidly (from seven faculty positions and thirteen supporting staff to Seventeen faculty positions and thirty five supporting staff) and became an active centre of research in Ecology, Biochemistry, Genetics and Physiology. The Department of Biological Sciences named as School of Life Science from 1979 celebrates her Silver Jubilee Year in 1994.

The Undergraduate and Postgraduate Biology (Particularly in Zoology) curriculum in Orissan Universities did not have enough components of ecological sciences. Professor Dash was the Pioneer in incorporating the course on ecological sciences in undergraduate and post graduate syllabi in Orissa especially in Berhampur and Sambalpur Universities and in Autonomous Colleges. The newly established Department of Environmental Sciences of Sambalpur University is the brainchild of Professor Dash.

Research Contributions

During his academic career of more than thirty years, he has successfully supervised ten major research projects funded by National and International agencies. Some forty five students have been awarded research degrees (Ph. D. and M. Phil) under his supervision. Independently and in collaboration with his research students, he has published 150 research papers, 3 research monograph (on Oligochaetes, Sea turtle, Brackish water prawn culture), 1 advanced level text book in Ecology (TATA Mc GRAW HILL), 5 research level book chapters on soil biology and energetics (Academic Press, USA, Cambridge University Press, U.K., John Wiley and Sons, USA, Elsevier Netherlands), 2 proceeding volumes (Organic waste Biomass, Verms and Vermitechnology) and three undergraduate level text books. He has organised a number of Summer Institutes, National Seminars, participated and presided over many National and International Seminars on Soil Biology, Secondary Productivity, Ecological Energetics etc. Some scientists have honoured him by naming new species of earthworms on his name (see Fauna of India-Oligochaetes by J.M. Julka, Zoological Survey of India, 1989, Genus *Deshiella*, Julka 1989, Species *Dashiella Khandanesis* Julka, 1989, *Lennogaster Dashi* Julka *et al.*, 1988 *Fridericia Dashi* Lal, 1981).

Soil Ecology

Professor Dash has made significant and pioneering contributions to the understanding of the role of soil oligochaetes in Indian ecosystems by studying their population dynamics, cocoon morphology, emergence pattern, life cycle patterns, metabolic functions, role in soil fertility, interactions with microorganisms and organic waste recycling, energy flow modelling and effect of pesticides on them. These studies could show academic and practical importance of these organisms with regard to the secondary productivity, soil fertility and waste recycling. His work on vermiculture provides a major breakthrough that some Indian Earthworms like *Dichogaster bolui, Drawida willsi, Lampita mauritic* and *Octochaetona surensis* can be effectively utilized for organic waste recycling and for raising tadpoles, chickens, Japanese quails and for the cultivation of edible mushrooms. He is pioneer in India in initiating vermitechnological work. He has also made contributions to the study of population, biomass, metabolic function and interaction of soil nematodes, protozoans and fungi and thier contribution to soil subsystem.

Herpetology and Conservation

The studies made by Professor Dash on the ecological aspects of amphibian metamorphosis, energetics, population biology and conservation of amphibians and of sea turtle (*Lepidochelys olivacea* in the Gahirmatha coast of Orissa) are significant with regard to frog culture and ecological aspects of metamorphosis and bringing new knowledge on ''arribada'', reproductive biology and conservation of sea turtles, which has drawn the attention of conservationists of the world.

Aquatic Ecology

Professor Dash has also studied the water quality and pollution load in Mahanadi water system, fresh water ponds and prawn culture in brackish water confined ponds. His work on limnology has brought new information on community structure, mercury pollution and biomagnification and secondary production in aquatic systems.

Sericulture

His recent work on voltinism, feeding biology, energetics and pathological aspects of Tasar Silk Worm *Antheraea mylitta* also contributes substantially to the sericulture.

Professor Dash is an admired teacher, excellent researcher, lovable person with a high quality analytical and rational mind and an able administrator. It will not be out of place to write few words about his happy family life. He married to Sovita in June 1964 and she is also a great source of inspiration to him. They are the proud parents of three daughters (Sharmila, Dalia and Papiya) and the son, Satya Prakash. Sharmila Dash-Acharya is a biologist, married and settled in USA, Dalia is doctor, Papiya Dash-Mishra is an Engineer married and settled at Bhubaneswar and Satya Prakash is a B.Sc Biology Student

We wish Professor Dash and Madam Dash a long life and good health and their Children a bright and prosperous future.

Editors

Publication of Professor M.C. Dash

Book/Monographs

Dash, M.C., 1993, Fundamentals of Ecology, Tata Mc Graw Hill, New Delhi, 373 P.

Dash, M.C. and P.N. Patnaik, 1993. Brackish Water Prawn Culture in Confined Ponds. Palani Publishers, Palni, India (in press).

Dash, M.C. and C.S. Kar, 1990. The Turtle Paradise Gahirmatha : An Ecological Analysis and Conservation and Strategy. Interprint, New Delhi 295 P.

Dash, M.C., 1983. The Biology of Enchytraeidae (Oilgochaeta) International Book Distributors, Dehradun, India - 171 P.

Dash, M.C. and U.C. Patra, 1993. Higher Secondary Zoology (3rd Ed.) Vidyapuri, Cuttack 778 P.

Dash, M.C., U.C., Patra and B.K. Mohanta, 1993. Higher Secondary Practical Zoology, Vidyapuri, Cuttack (2nd ed.) 193 P.

Dash, M.C., 1988. Evolution of Life, Katab Mahal Cuttack (2nd ed.) 162 P.

Book Chapters

Dash, M.C., 1993. The Enchytraeidae. In Soil Organisms and Decomposition in Tropics (Ed. M.V. Reddy), Springer Verlag (In Press).

Dash, M.C., 1990. The Enchytraeidae (Oligochaeta), In 'Soil Biology Guide' (Ed. D.L. Dindal), John Wiley and Sons, New York, USA, 311-340.

Dash, M.C., 1987. The Other Annelids (Other than Polychaetes) In 'Animal Energetics Vol. I' (Eds. Pandian and Vernberg), Academic Press, U.S.A., 261-299.

Dash, M.C. (Coauthor), 1980. Decomposer Subsystems, In 'Grassland, Systems Analysis, and Man' (Eds. Breymeyer and Van Dyne), Cambridge University Press, U.K., 609-653.

Dash, M.C., 1979. Consumers, In 'Grassland Ecosystems of the World' (Ed. R.T. Coupland), Cambridge University Press, U.K. 219-226.

Seminar Proceedings (Edited)

Dash, M.C., B.K. Senapati and P.C. Mishra, 1985. Organic Waste biomass and their decomposition and utilisation, Proceed. Vol. I., Indian National Seminar on Organic Waste Utilization and Vermicomposting, held December 5-8, 1984, at the School of Life Sciences, Sambalpur Univesity, Burla India 768 019.

Dash, M.C., B.K. Senapati, and P.C. Mishra, 1986. Verms and Vermicomposting Vol. II, Ibid.

Consolidated List of full length Research Publications

Name of Journal		*No. of Papers Published/ Accepted*
Rev. Ecol. Biol. Sol. (Presently Euro. J. Soil Ecol.)	France	11
Alytes	France	04
Sericologia	France	01
Pedobiologia	Germany	10
Oikos	Sewden/Denmark	04
Tropical Ecology	Internal Society (India/USA)	08
Tropical Agril.	Butterworths (Trinidad)	02
Soil Biol. Bicohem	Pergamon (UK)	01
Ecology	USA	01
J. Nematology	USA	01
J. Lepidop. Soc.	USA	02
J. Res. Lepid	USA	02
Buil Env. Cont. Toxico	USA	01
Aquaculture	Netherland	01
INT. J. Inv. Reprod.	Netherland	01
Con. J. Zool.	Canada	01
Megadriologica	Canada	02
Environmentalist	Switzerland	01
Experientia	Switzerland	02
Ann. Zool. Fenn.	Finland	01
Wild Silk Moth	Japan	03
Oxford and IBH (Earthworm Ecology and Land Management)	India/UK	01
Proc. Indian Acd. Sciences	India	04
Ind. Jour Expt. Biol.	India	02

Name of Journal		*No. of Papers Published/ Accepted*
Jour. Bioscience	India	01
Jour. B.N.H-S.	India	01
Pollution Research	India	03
Comp. Physiol. Ecol.	India	08
J. Ecobiol	India	04
J. Soil Biol. & Ecol.	India	02
Current Science	India	01
Science and Cult	India	01
Biol. Bull. India	Inid	04
Oriental Insect	India	01
Journ Zool. Soc.	India	01
Oryza	India	01
Ind. Indust. Eng. Jour.	India	02
Ind. Journ. Agril. Chem.	India	01
Ind. Jour. Poult. Sci.	India	01
Jour. Aquaculture	India	01
Indian Biologist	India	03
Ind. J. Env. Health	India	01
Geobios	India	03
Prakruti	India	01
Sambalpur University Journ.	India	03
Research Papers in Books (published in India)	India	06
Proceedings of Nat Seminars	India	16
Misc. Publ. (India Silk, Sci. Reporter)	India	10
J. Ecotoxicol. Env. Monit.	India	01

Former and Present Doctoral and M. Phil Scholars Associated with Professor Madhab Chandra Dash

Name	Area of research	Present Position and Address
A.V. Thambi	Soil Ecology	Reader in Zoology, Marthama College, Kerala
U.C. Patra	Soil Ecology	Reader in Zoology, Government College, Phulbani, Orissa
B.C. Guru	Soil Ecology	Reader in Zoology, Utkal University, Bhubaneswar
N. Behera	Soil Ecology	Reader in Microbial Ecology School of Life Science Sambalpur University, Orissa
B. Nanda	Soil Ecology	Reader in Zoology, Gangadhar Meher College, Sambalpur (Autonomous) Sambalpur
B.K. Senapati	Soil Ecology	Reader in Ecology, School of Life Sciences Sambalpur University, Orissa
P.C. Mishra	Soil Ecology	Reader and Head, Department of Environmental Science Sambalpur University, Orissa
C.C. Mishra	Soil Ecology	3, Wiltshir Crescent, Wodonga VIC 3690 Australisa (La Taoba University)
Late G.B. Pradhan	Soil Ecology	Ex-Reader in Zoology, Govt. College, Denkanal
A.K. Hota	Herpetology	Senior Lecturer, School of Life Science Sambalpur University, Orissa
Mrs. Binapani Satpathy	Soil Ecology	Senior Lecturer in Botany, Govt. College, Bhawanipatna, Orissa
S.N. Mohanty	Herpetology	Senior Scientist, Central Inland Fresh Water Aquaculture Institute, Bhubaneswar, Orissa
D.P. Pati	Soil Ecology	Research Assistant, School of Life Sciences Sambalpur University, Orissa
P.K. Mishra	Herpetology	Reader in Zoology, Larambha College, Larambha Sambalpur, Orissa

Name	Area of research	Present Position and address
T.V. Rao	Soil Biology and Biochemistry	Reader in Biochemistry, School of Life Sciences Sambalpur University Orissa
B.K. Mohapatra	Herpetology	Soil Analyst, School of Life Sciences Sambalpur University, Orissa
C.S. Kar	Herpetology	Research Officer, Wild Life Wing, Forest Department, Government of Orissa
G.K. Kar	Limnology	Lecturer in Zoology, Panchayat College, Badagaon Sundargarh, Orissa.
Mrs. B.K. Rath	Socio-ecology	Reader in Home Sciences Sambalpur University, Orissa
A.K. Dash	Sericulture	Lecturer in Zoology, Dr. Jadunath College, Balasore, Orissa
Ms. P.N. Patnaik	Limnology and Prawn Culture	Chief Executive Officer, Fish Farmer Development Agency, Dept. of Fishery, Govt. of Orissa, Bhubaneswar
H.K. Patnaik	Soil Ecology	Lecturer in Zoology, Dr. Samantarai Institute of Science and Technology, Gania, Nayagarh, Orissa
G. C. Pandia	Soil Ecology	Senior Lecturer in Zoology, Govt. Women's College, Sambalpur
A.K. Das	Soil Ecology	Senior Lecturer in Zoology, D.A.V. College, Titilagarh, Bolangir, Orissa
B.K. Nayak	Sericulture	Asst. Director (Sericulture) Dept. of Textiles, Govt of Orissa, Bhubaneswar
S.K. Panda	Limnology	Senior Lecturer in Zoology, Govt. M.K.C.G. College, Parlakhemundi, Orissa
Ch. S.K. Mishra	Sericulture	Senior Lecturer in Zoology Orissa University of Agriculture and Technology, Bhubaneswar, Orissa
D.K. Sharma	Limnology	Senior Lecturer in Zoology, Kuchinda College, Kuchinda, Orissa
S.K. Swain	Herpetology	Technical Asst. School of Life Sciences, Sambalpur University, Qrissa

Name	Area of research	Present Position and Address
C.K. Rath	Limnology and Prawn Culture	Senior Lecturer in Zoology, Khallikote (Autonomous) College, Berhampur, Orissa
Mrs. C. Dei	Herpetology	Senior Lecturer in Zoology, Larambha College, Larambha, Orissa
J.K. Mohanta	Herpetology	Research Scholar, School of Life Sciences Sambalpur University, Orissa
A.B. Mishra	Environmental Education	Reader, School of Life Sciences Sambalpur University, Orissa
H.K. Dash	Soil Ecology	Lecturer in Zoology, NAC College, Burla
S.C. Dash	Soil Ecology	Reader in Zoology, B.J.B. College, Bhubaneswar
B.K. Choudhury	Microbial Physiology	Reader in Botany, Government Science College, Rourkela
Mrs. S. L. Mishra	Soil Ecology	Senior Lecturer in Zoology, Government Science College, Rourkela
Mrs. N. Bhatt	Soil Ecology	Senior Lecturer in Zoology S.B. Women's College, Cuttack
Mrs. K. Mishra	Soil Ecology	Senior Lecturer in Zoology, Government Womens' College, Rourkela
B.N. Beura	Soil Ecology	Lecturer in Zoology, MIG Factory College, Sunabeda
K.B. Das	Aquatic Pollution	Superintendent of Fisheries, Government of Orissa, Jujumura, Sambalpur
P.K. Hota	Soil Ecology	Reader in Zoology, G.M. (Autonomous) College, Sambalpur
P.K. Das	Herpetology	Lecturer in Zoology, Gandhi Mahavidyalaya, Rourkela
Mrs. B.P. Behera	Limnology	Lacturer in Zoology, Sarla Mahavidyalaya, Rahama, Cuttack
A.N. Jena	Soil Ecology	Senior Lecturer in Zoology, Muncipal College, Rourkela

ADVANCES
IN
ECOLOGY AND ENVIRONMENTAL SCIENCES

(*Professor M.C. Dash Commemoration Volume*)

Prof. M.C. Dash

Chapter 1

Research Trends in Rhizobial Nitrogen Fixation

Pran K. Chakrabartty AND **Anup K. Halder**

Department of Microbiology, Bose Institute P-1/12 CIT Scheme VII M. Calcutta-700 054, India

Abstract. The biological cycle of nitrogen is fundamental to the persistence of life of this planet. In the early stages there must have adequate chemically fixed nitrogen for emergence of life. By the time oxygen became a permanent component of the atmosphere, selection pressure favouring biological nitrogen fixation arose. However, the exploding human population has largely been supported by the Haber process of fixing molecular nitrogen from atmosphere. This emphasizes the practical importance of the research efforts now being put into the study of biological nitrogen fixation of which plant-rhizobium symbiotic system is the most efficient one. However, much of the studies have been carried out using *Klebsiella pneumoniae* nitrogen fixation system as the model. Nevertheless, the *sym* plasmides as well as *nitf nod, hsp. fix* and related genes of rhizobium symbiosis are now being revealed through molecular genetic studies. The sequential regulation of the expression of *nod* genes for nodulation, the mechanism of regulation of *nif* regulon by oxygen and/or ammonia are under study. The physiological function of uptake hydrogenase (*hup*) gene for energy economy of nitrogen fixation is being carried out. At the same time attempts are being made to increase the efficiency of nitrogen fixing machinery as well as the process of nodulation. Although, root-nodulating symbioses have been the main examples of topics of research, stem-nodulating system of *Sesbania rostrata* and its attendent *Azorhizobium caulinodans* as well as that of *Aeschynomene* are creating excitement and becoming popular for their stem nodules closer to the sites of photosynthesis and relative aerotolerance. For the ultimate aim is to increase the production of crops through the exploitation of nitrogen fixation system, the process of competition for nodulation must be understood and full-proof approaches for nodulation with laboratory improved strains must be realized. Already attempts have been made to nodulate cereals by rhizobium with marginal academic success. Hopefully, someday field grown cereals will get their required nitrogen from the bounty of atmosphere mediated by their own nitrogen-fixing apparatus.

Key words : *Rhizobia, nitrogen fixation, nodulation, energy metabolism, competition*

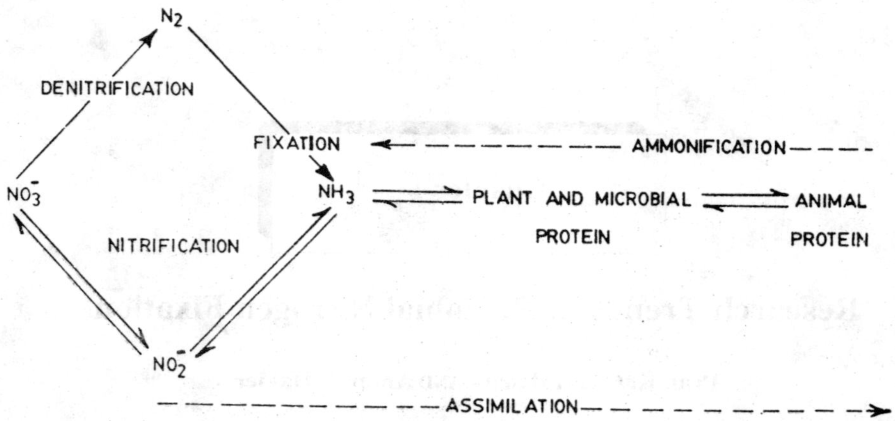

Fig. 1.1. The nitrogen cycle.

Introduction

Nitrogen fixation is the process by which nitrogen gas from air is converted into ammonia. It is essential for life on earth and role of microbes in nitrogen fixation as well as in the process of biological cycle of nitrogen as a whole is crucial. In global nitrogen cycle (Fig. 1.1), it is estimated that every nitrogen atom in the atmosphere cycles once in about 10^6 years, but in a tropical rice field, where nitrogen fixation takes place in the water and denitrification in the mud, the turnover time is reduced to only a few years (Postgate, 1989) emphasizing the contribution of microbes in these processes.

Early dogma based on Oparin/Haldane model (Bernal, 1967) postulated that "primitive soup" must have adequate reserves of chemically fixed nitrogen for the emergence of life. As the primitive atmosphere became less reducing, nitrite and nitrate became the biosphere's major nitrogen reserves and gradually selection pressure favouring the emergence of biological

TABLE 1.1. *Annual Biological Fixation of N_2 and Grain Production on earth*

System	N_2 fixed ($\times$ 10^6 tons/yr)	Area Ha ($\times$ 10^6)	Grain yield ($\times$ 10^6 tons/yr)
Legume,	20	250	130
Non legume,	60	15,000	1300
Cereal plants			
Grass land, forest and wood lands			
Ocean and others	20	--	--

nitrogen fixation arose. As a consequence, nitrogen is no longer a limiting nutrients for most equilibrium or climax ecosystems. In virgin forests, natural savannah and other fertile regions which are in ecological balance, biological fixation provides an adequate input of nitrogen. Without this, nitrogen reserves in the soil would have depleted by the microbial denitrification. The amounts of fixed nitrogen available annually are about 10^8 tons per year from biological nitrogen fixation (Table 1.1) of which a substantial level is contributed by legume-rhizobium symbiotic system.

The world poulation is now about 5×10^9 and estimated to rise to 8×10^9 early in the next century. Already one-third of the world's population depends on industrial nitrogen-fertilizer for its food supply. A sustained, world wide investment in research on biological nitrogen fixation has, therefore, developed over the past two decades. The considerations that the manufacture of ammonia by Haber process is capital intensive with a high energy and raw material consumption emphasize the practical improtance of the research efforts now being put into the study of biological nitrogen fixation.

Classification of root-nodule bacteria

Biological nitrogen fixation is performed by two types of microbes, symbiotic and free-living. Legume-rhizobium symbiosis is agronomically the most important by virtue of its root-nodule system. The tropical plants *Sesbania and Aeschynomene* form nodules on their stems (Subba Rao *et al*, 1980; Yoneyama and Kondo, 1990). *Parasponia* is the only non-legume known to be nodulated by rhizobium (Trinick, 1980). The host infectivity along with certain biochemical tests formed the basis of species recognition of rhizobium (Jordan, 1984). DNA-DNA homology studies (Chakrabarti *et al.* 1990) have also contributed greatly to the current classification of root-nodule bacteria. The root-nodulating, nitrogen-fixing bacteria have been divided into two genera, the fast-growing *Rhizobium* and the slow-growing *Bradyrhizobium*. The stem-nodulating bacteria are placed under the genus *Azorhizobium*. The members of the genus Rhizobium grow fast on yeast extract mannitol agar medium (Vincent, 1970) while the strains of Bradyrhizobium are slow-growing in nature. Current classification merges the former species *R. leguminosarum*, *R. trifolii* and *R. phaseoli* together into a single species and recognizes them as biovars of *R. leguminosarum* due to their broad similarities. *R. meliloti* is still recognized as a distinct separate species. The fast-growing *Rhizobium* of *Lotus* sp., *R. loti* and the fast-growing bacteria nodulating soybean, *R. fredii* have also been validly described. *R. fredii* is also described as *Sinorhizobium fredii* in many recent literatures. Bacteria belonging to the genus *Bradyrhizobium* is represented by a singly species *B. japonicum* and the former species *R. lupini* has been eliminated. *Azorhizobium* nodulating the stem of *Sesbania rostrata* is represented by *A. caulinodans*. For the purpose of this review, members of all the three genera, *Rhizobium, Bradyrhizobium* and *Azorhizobium* will be loosely called as rhizobium (pl. rhizobia).

The development of nodules

The development of nitrogen-fixing nodules is a complex process requiring interaction

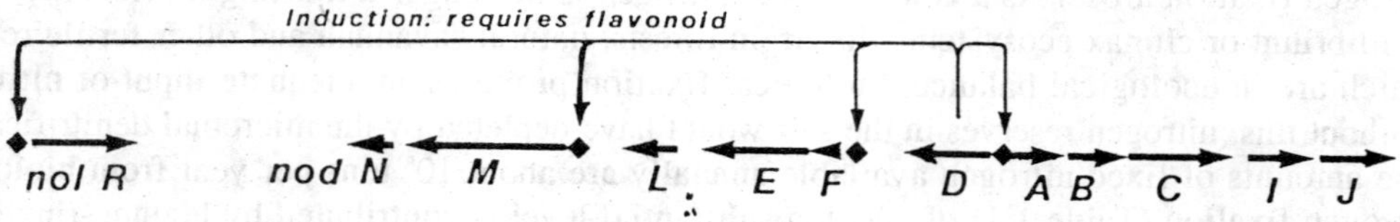

Fig. 1.2. The orientation and relative sizes of the nodulation genes of *Rhizobium leguminosarum* bv. *viciae*. The nodulation genes are indicated by horizontal arrow. The *nod* ABCIJ, *nod* FEL, *nod* MN and *nol* R operons are preceded by a conserved DNA (*nod*-box) sequence represented by '◆', and these operons are indicated by the *nod* D gene product in the presence of a flavonoid such as naringenin.

between the bacterium and the plant host (Kondorosi and Kondorosi, 1986). The first step in the establishment of symbiosis is the recognition and invasion of the appropriate legume by the bacteria. Plant lectins are thought to interact selectively with compatible rhizobium cell surface polysaccharides for initiation of nodulation (Dazzo and Truchet, 1983). Bohlool and Schmidt (1974; 1976) provided first experimental evidence demonstrating that lectins may function as recognition molecules or determinants of host specificity. Host roots also excrete substances which induce physiological processes necessary for nodulation (Halverson and Staoy 1984) and serve as chemoattractants. In soybean exudates dicarboxylic acids and the amino acids glutamate and aspartate are strong attractants of *Bradyrhizobium japonicum* (Barbour *et al*, 1991). As a matter of fact, various stages of nodulation process illustrate a series of cell-to-cell signalling events that involve soluble signal molecules as well as intimate contact between the surfaces of the plant and bacterial cells. Flavonoid molecules are released from legume roots as signal molecules in the initial process. Sensing of flavonoid molecules released from legume roots by the bacterial results in switching on expression of a group of nodulation genes, i.e. several *nod* operons. (*nod* ABCI, *nod* FEL, *nod* MN, *nod* R) (Fig. 1.2) are transcribed at high level (Rossen *et al*, 1985). This activation requires the presence of regulatory gene *nod* D. Flavonoid inducers seem to interact with the product of *nod* D, causing an allosteric shift in the protein that enables it to allow transcription from the inducible nod promoters. *nod* D is transcribed at significant level in free-living rhizobia (Fig. 1.2). The isoflavones, genistein and daidzein present in soybean seed-and root exudates (Porter *et al*, 1985, Smit *et al*, 1990) are the most potent inducers of nod gene expression in *B. japonicum* (Long, 1989 a). Motility and chemotaxis of rhizobium may also be involved in the early stage for more efficient (Caetano-Anolles *et al.*, 1988) and more effective (Ames and Bergman, 1981; Mellor *et al.*, 1987) nodulation of the host.

During nodulation rhizobia attach to the root hairs, penetrate cell wall and form an infection thread which carries the bacteria into the root meristem. Meanwhile, the cortical cells are induced to divide and are invaded by the bacteria released from the branching infection thread. Within the cortical cells the bacteria multiply and differentiate into the morphologically altered bacteroids.

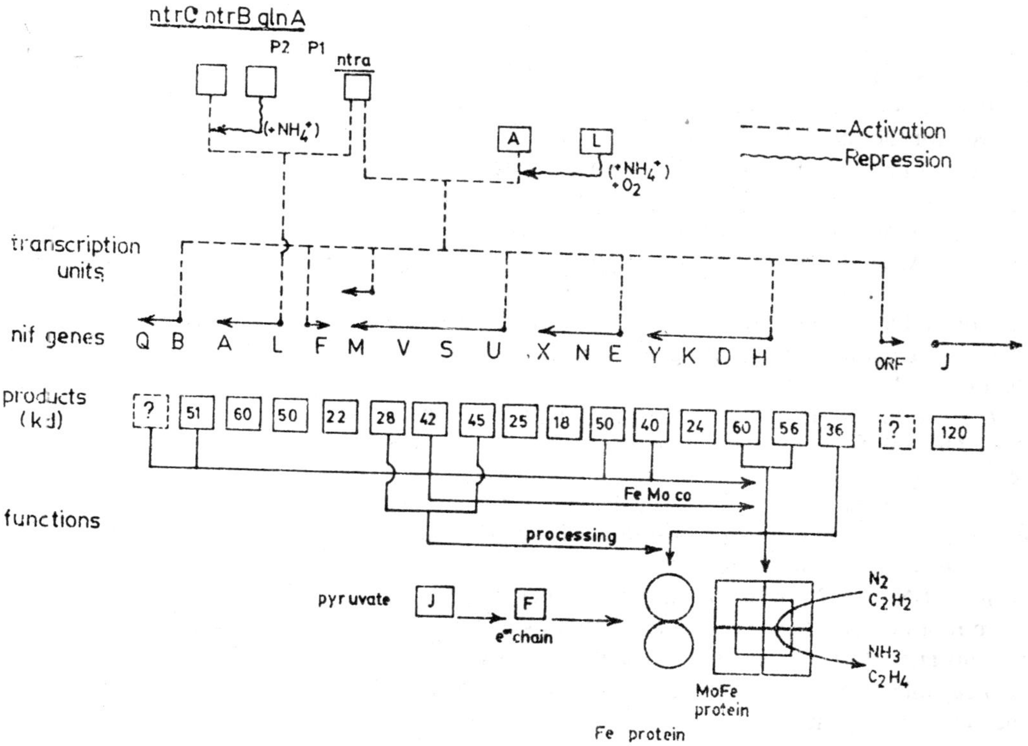

Fig. 1.3. Organization of the *nif* cluster of *Klebsiella pneumoniae.* Model showing the genes and products involved in regulation of the *nif* regulon in response to NH_4^+ and O_2. Regulation of *nif* gene expression at the transcriptional level occurs in a nif specific manner and a global nitrogen regulation (*ntr*) system. The *ntr* A gene product is a sigma factor (σ^{54}) is required for *nif* and *ntr* regulated promoters. The arrows signify directions of transcription of component operons. *Nif* HDK genes code for the structural peptides of nitrogenase; *nif* LA operon specifies proteins which regulate the expression of all other *nif* operons.

(Adapted from D. McCormick, 1988. Bio/Technology, 6, 384.)

Nitrogen fixation and its regulation

The enzyme nitrogenase synthesized in the bacteroids converts atmospheric nitrogen to ammonia which is then assimilated by the plant host. The information enabling rhizobia to fix nitrogen by way of conventional nitrogenase is encoded in *nif* genes which are regulated by a complex set of processes (Fig. 1.3).

If there is adequate ammonia or too much oxygen in the environment the nitrogenase is not made or made inactive. The model system of *nif* regulon is provided in *Klebsiella pneumoniae* where it is a contiguous cluster of 20 genes arranged in 8 operons and occupy about 25 Kb of DNA (Fig. 1.3).

The enzyme nitrogenase is comprised of two proteins, a tetramer with two pairs of non-identical subunits and containing iron-molybdenum cofactor and a dimer with two identical subunits and containing iron. Three distinct structural genes, now known as *nif* H, *nif* D and *nif* K code for the three distinct proteins or peptides of nitrogenase. The remaining 17 genes have different functions (Dixon, 1987). Two *nif* genes in *K. pneumoniae, nif* L and *nif* A. code for proteins which regulate the transcription of other *nif* genes internally (Dixon, 1987) while the nitrogen regulation genes *ntr* A, *ntr* B and *ntr* C 'externally' regulate the transcription of operons. As such transcription of the *nif* operon in *K. pneumoniae* is subject to regulation at two levels. The activity of nif A is modulated by the *nif* L gene product (NIF L) in response to oxygen and low level of fixed nitrogen whereas NTR C activity is regulated by the *ntr* B product (NTR B) according to the nitrogen status of the cell. NTR B is less sensitive to fixed nitrogen than is NIF L.

Klebsiella couple amonia production by nitrogenase to cellular biosynthesis without any detectable export to ammonia into the growth medium, as such, production of nitrogenase is very sensitive to the presence of combined nitrogen (Parejko and Wilson, 1970). In contrast, *nif* genes of rhizobia are normally found to be derepressed within the nodule leading to the export of fixed nitrogen to the host cell cytoplasm. It has been possible to increse the nitrogen fixation capacity of *B. japonicum* in the presence or absence of fixed nitrogen by mutagenizing the bacteria with N-methyl-N-nitro-N'-nitrosoguanidine (Maier and Brill, 1978). Some of these mutants formed more root nodules then the wild type. By inoculation with all the mutant strains, increased dry weights and nitrogen content of inoculated plants were reported. Certain species of rhizobium are also capable of producing nitrogenase in free-living state (Pagan *et al,* 1975) when presence of succinate and a pentose sugar is thought to be necessary (Postgate, 1975).

Energy metabolism

Isolated bacteroids from nodules of snake bean oxidize only C4-dicarboxylic acid such as succinate while bacteroids within the nodule are dependent on the host for the supply of plant photosynthates for their carbon substrate. Sucrose is supposed to be the carbon source that is transported to nodules. However cultured rhizobia metabolize a number of carbon substrates via the Embden-Meyerhof-Parnas (EMP), Entner-Doudoroff (ED), pentose phosphate (PP) pathways and tricarboxylic acid (TCA) cycle (Chakrabarti *et al,* 1987a). Acetate or the storage polymer β-hydroxybutyrate is metabolized by the induction of the enzyme isocitrate lyase (ICL), a key enzyme of glyoxalate pathway, while the enzyme malate synthase is constitutively synthesised in *Rhizobium* and *Bradyrhizobium*. Succinate acts as a catabolite repressor of enzymes of the EMP, ED and PP pathways as well as of ICL activity. The cAMP relieves succinate mediated catabolite repression of ICL activity but not of the enzymes of the EMP, ED and PP pathways (Mandal and Chakrabartty, 1992a; 1993). The marked shift in the metabolism during nitrogen fixation results in a major shift in the pattern of synthesis of cytochromes and other components of the electron transport chain (Chakrabarti *et al.*, 1987b) leading to the formation of new respiratory chain that is functional under low O_2 concentration (Shanmugam *et al,* 1978). Bacteroid electron transport occurs only in low O_2 concentration since higher O_2 concentrations would inactivate the O_2-sensitive nitrogenase activity. In nodule, leghemoglobin is believed to maintain a delicate level of O_2 near the bacteroid membrane (Appleby, 1974), thus allowing the nitrogenase to function. Two additional systems, probably membrane bound, may play an important role in energy metabolism, the hydrogen uptake system (Dixon, 1972) and energy linked nitrate reductase (Rigaud *et al.,* 1973). The reduction of molecular nitrogen to ammonia is an energy demanding process and in vivo energy cost for nitrogen fixation in terms of ATP equivalents may be as many as 30 ATP/N_2 for the nitrogenase reaction. However, nitrogenase also produces hydrogen as a by-product of nitrogen fixation which represents a waste of energy. Efficient diazotrophs including many strains of *Rhizobium* capture this H_2 before it escapes, using an uptake hydrogenase (hup) enzyme, although a functional *hup* system has not so far been detected in any natural strain of *R. meliloti* (Behki *et al,* 1985). In *R. leguminosarum* the plasmid PIJ 1008 determines *hup* ability and the plasmid is transferable. The enzyme is under the control of cyclic AMP (Lim and Shanmugam, 1979) and its expression in free-living cultures is influenced by carbon source (Thimmaiah *et al,* 1986). The *hup* gene from *B. japonicum* has been cloned and its symbiotic expression has been studied (Lambert *et al,* 1987). Since a majority of rhizobium species lacks hup activity, genetic studies have focused on the transfer of hup genes to these species. However, from the view point of agricultural application the most important question is whether the tranferred gense are expressed in nodules. Evidence suggests that expression of hup genes depends on the specific plant host (Garg *et al,* 1985). Nevertheless, if bacteroids derive ATP for dinitrogen reduction primarily from oxidative phosphorylation coupled to electron transport to O_2, a supply of ATP under these conditions of low partial pressure of O_2 in the nodules is likely to be rate limiting for nitrogen fixation. Energy production coupled to anaerobic nitrate reduction can support nitrogen fixation in isolated bacteroids (Rigaud *et al,* 1973). Bhandari *et al.* (1984) presented direct evidence that when nitrate is dissimilated to nitric oxide ATP production was coupled. However, detoxifying role of denitrification, rather than ATP production to support nitrogen fixation, has recently been emphasized (Casella *et al,* 1986). Thus, the evolution of very sophisticated genetic control systems for nitrogen fixation may be understood in the context of a requirement of a large imput of metabolic anergy.

Ammonia assimilation and its regulation

Ammonia produced by nitrogenase is assimilated by rhizobium by glutamine synthetase (GS)-glutamate synthetase (GOGAT) pathway (Bravo & Mora, 1988). No glutamate dehydrogenase activity is detected in this group of bacteria. In contrast to other bacteria, rhizobium contains two forms of GS (Keister *et al,* 1987). GS I has many properties similar to the GS found in enteric bacteria (Somerville and Kahn, 1983) while GS II appears solely in Rhizobiaceae and has properties distinctly different from GS I. Amino acid sequence of GS II more closely resembles to GS of plant than to that of bacteria (Carlson and Chelm, 1986). Ammonia has a strong negative modulation effect on GS II and a mild effect on GS I. GS I is mainly regulated by adenylylation (Ludwig, 1978; Bhandari and Nicholas, 1986). However, using ^{15}N analysis, it has been found that free-living rhizobia excrete as much as 94% of fixed nitrogen presumably as ammonia (O'Gara and Shanmugam, 1976). Thus, a major portion of the fixed nitrogen is exported to the host cell cytoplasm. This explains why the activities of ammonia-assimilatory enzymes leading to biosynthesis of glutamine-glutamate family of amino acids are found to be extremely low in bacteroids (Brown and Dilworth, 1975; Kurz *et al*, 1975). In this context it is worthnoting that plant enzymes rather than the bacteroid enzymes have been proposed to be physiologically important in the assimilation of ammonia produced by nitrogenase (Scott *et al*, 1976). Build up of glutamate by the assimilation of ammonia in plant fraction of nodules may function to repress ammonia assimilation pathway of bacteroids. In free-living dizotrophs this becomes a real problem, the solution of which has been sought in selection of L-methionine-DL-sulfoximine (MSX) resistant mutant strains with altered GS (Lugue *et al*, 1986).

Enhancement of nodulation and nitrogen fixation

If glutamine synthetase activity of nodule plant fraction is selectively inactivated by the delivery of tabtoxinine-β-lactam of *Pseudomonas syringae* pv. *tabaci* by infestation of alfalfa rhizosphere symbiotic nitrogen fixation is enhanced (Knight and Langston-Unkefer, 1988), although normal GS-catalyzed ammonia assimilation is significantly impaired. It must be noted that *Pseudomonas* infestation also resulted in increased nodulation of the host plant in this study.

Increased nodulation of host plant by *Rhizobium* has been noted also when legume seeds are treated with *Pseudomonas fluorescens-putida* group of organism (Kloepper *et al*, 1980), *Bacillus cereus* (Halverson and Handelsman, 1991) or even *Azospirillium* (Plazinski and Rolfe, 1985). Althogh a unified concept of how all of these organisms could independently enhance nodulation of rhizobia is difficult to develop, rapid colonization of the legume rhizoplane with these plant growth promoting rhizobacteria and antagonizing potential deleterious rhizoplane fungi and bacteria which permits favourable environment for root growth may be a likely explanation. Nevertheless, symbiotic associations with Vesicular-arbascular mycorrhizae (VAM) which are common in legumes can caused growth stimulation (Al-Radded Al-Momany, 1991). VAM can increase phosphorus uptake of plants growing in infertile or low phosphate soil (Kessel *et al*, 1985). In *Rizobium*-mycorrhizal fungi-legume association nodulation is better and nitrogen fixation is higher (Smith and Daft, 1977) probably because of higher supply of phosphate. Recently certain strains of *Rhizobium* and *Bradyrhizobium* have been found to be highly capable of solubilizing insoluble phosphates including rock phosphate (Halder *et al.*, 1990). Solubilization was found to be related to pH of the microenvironment of the bacteria. Presence of ammonium ion in the medium caused greater reduction of medium pH by a proton exchange mechanism and higher solubilization of phosphate due to generation of inorganic acids (Halder *et al.*, 1992). It is not yet known if the strains with higher phosphate solubilizing capacity are also better nitrogen fixers.

For obtaining strains of Rhizobium with increased ability to fix nitrogen, mutagenesis followed by screening by rapid effectiveness assay have been successfully tried. Selected strains exhibited higher acetylene reduction per plant in symbiotic association and increased plant dry weight (maier and Brill, 1978). However this method did not gain wide application because of the laborious selection protocol. With the objective to develop high nitrogen fixing strains, methylammonium resistant strains of *Azotobacter* that overproduce nitrogenase have been isolated (Gordon and Jacobson, 1983). As such, it may not be impossible to isolate methylammonium resistant strain of *Rhizobium* under a condition of ex-planta nitrogen fixation. Meanwhile, Agracetus, a Madison (USA) based company, has begun test marketing Gold Coat, its own nitrogen fixing bacterium for soybean - a chemically mutated *B. japonicum* (McCormick, 1988). Successful mutagenesis with transposon Tn5 of *Rhizobium* and *Bradyrhizobium* has also been reported and symbiotically defective mutants as well as auxotrophic mutants have been identified (Rostas *et al*, 1984). Symbiotic mutants showing extended host range was also isolated by Tn5 mutagenesis (Pueppke and Krishnan, 1988). It has been found that glutamine auxotrophs of strain 32H1 (Ludwig and Singer, 1977) and of *R. meliloti* (Kondorosi *et al*, 1977) are NIF⁻ providing evidence for ammonia regulation mediated by GS.

An useful tool for engineering the active site of the enzyme nitrogenase is the product of *nif* V gene which is homocitrate. Homocitrate seems to control the structure of the final FeMoco· and thus the structure of the final cofactor of nitrogenase (Hoover *et al*, 1987). Substituting homocitrate analog will produce aberrant forms of FeMoco- changing the substrate specificity of the resulting nitrogenase and its susceptibility to inhibition.

Enhancement of competition

One of the major problems in the enhancement of symbiotic nitrogen fixation is that when laboratory grown superior strains of nitrogen-fixing root nodule bacteria are applied to field either in the form of seed-inoculum or mixed with soil, they fail to infect and develop nodules in the roots. Majority of the nodules on the roots of the legume plants are occupied by indigenous strains of root nodule bacteria. Thus, it is a problem of competition between the indigenous strains (Berg *et al*, 1988) and the inoculum strain to make nodules on the host roots. The ability to dominate in nodulation of legume roots when other strains are also present in the rhizosphere is called competitiveness. Despite the advancement of knowledge of the *nif* (nitrogen fixation), *nod* (nodulation) and *hsn* (host specific nodulation) genes of rhizobia (Long, 1989b; Hennecke, 1990; Wang and Stacy, 1991), not much is known as to what makes a strain competitive (Dowling and Broughton, 1986; Triplett, 1990), although several characters which may have some bearing on competitive nodulation have been identified; these are motility and chemotaxis (Mellor *et al*, 1987; Caetano-Anolles *et al*, 1988; Liu *et al*, 1989), cell surface polysaccharides (Bhagwatt *et al.*, 1991; Zdor and Pueppke, 1991), bacteriocin production (Triplett, 1990), rate of infection (Hahn and Hennecke, 1988; McDermott and Graham, 1990), substrate responsiveness, and growth rates on soil substrates (Viteri and Schmidt, 1987). A gene controlling the phenomenon of competitive nodulation blocking has also been isolated from *R. leguminosarum* (Davis *et al.*, 1988). Both positive and negative results have been observed for most of these cellular characteristics and phenotypes in a variety of strains and species. Recent work has demonstrated that certain DNA regions of the chromosome of a competitive strain are expressed specifically in response to soil extract and root exudate. The regions can be identified by subtractive hybridization and then cloned. Thus, after screening of 2000 colonies from the gene library of a higher competitive strain USDA 438, two soil extract-activated clones and four root exudate-activated clones were identified. Of the two soil extract-activated clones, one appeared to encode genes expressed in response to both soil extract and root exudate and, when mobilized to USDA 110, a less competitive strain, enhanced the competitive nodulation ability of USDA 110 (Bhagwatt and Keister, 1992).

Rhizobitoxine synthesized by some strains of the legume symbionts is a relatively poor predictor of symbiotic effectiveness (Fuhrmann, 1990). However, insertion of the gene to produce trifolitoxin, an antirhizobial peptide, to *R. leguminosarum* bv. *trifolii* TA1 strain which did not produce it conferred competitiveness for nodulation on clover roots upon it when coinoculated with trifolitoxin sensitive reference strains. Recently trifolitoxin genes have been cloned and transferred to effective strain of *Rhizobium* by conjugation (Triplett, 1988). The know how of transfer of rhizobitoxine gene to a recipient is important to construct strains with both effectiveness and competitiveness.

Although one of the major goals of research on rhizobium is the development of better nitrogen-fixing strains and related biotechnology, selection of improved plant germplasm for high nitrogen fixing genotypes is also equally important. The plant host provides photosynthates as energy source, leghaemoglobin and nodulin, proteins which are important at various phases of the symbiosis. A poor plant phenotype could clearly override the most promising rhizobial genotype. *Sesbania* stem nodules containing *Azorhizobium* bacteroids with bacteriocholorophyll and photosynthetic centres raise the possibility of future transfer of their *nif* genes to plant cells. The nodulation of *Parasponia*, a non-leguminous weed, by diazotrophic *Bradyrhizobium* appears to be an experimentation in the nature's laboratory in the direction of extending the host specificity of rhizobium nodulation. Of the 14 or so not genes *nod* ABC are common among all *Rhizobium* species while *nod* FE, *nod* G, *nod* H and *nod* L are involved in specifying the type of host plant nodulated. Bacteria carrying mutations in such genes often show abnormalities in the root hair curling reactions elicited on their normal hosts and sometimes also affect hosts that would normally be inresponsive to them (Brewin, 1991).

Modern science faces essentially five challanges in the avenue of nitrogen-fixation by rhizobia. It is imperative to construct energy-efficient nitrogenase, to increase the efficiency of existing nitrogen-fixing system, to alter the organisms to the extent that they continue to nodulate and fix nitrogen in presence of nitrates or ammonia, to impart the property of competitiveness and to extend the host range as far as to the level of cereals. Engineering of plants to fix nitrogen directly will then be a dream comes true. At the current rate of advancement of science, it is possible to foresee engineered rhizobium capable of nodulating and fixing nitrogen in presence of nitrate or ammonia in near future. Normally in the presence of soil nitrate legumes do not form nodules. Moreover, when nitrate is applied to a nodulated plant, nitrogen fixation is repressed and the nodules senesce prematurely. In addition to their effects on plant metabolism and nodule morphogenesis, some aspects of bcterial *nod* gene activity are regulated by nitrate or ammonia (Wang and Stacy, 1990). In case of *R. meliloti*, transcription of the *nod* ABC genes are nitrogen regulated (Dusha *et al.*, 1989).

Cocking's team (Al-Mallah *et al.*, 1989) reported an important advance towards nodulated cereals. His team showed that after treatment with cell wall digesting enzymes rice roots became susceptible to infection by rhizobia. After incubation with the bacteria the roots had formed nodule-like structures which contained rhizobia, in about a month. The "rice nodule" had very little nitrogenase activity if at all. This feat has been repeated with wheat (Tchan *et al.*, 1991) and the oil seed rape (Al-Mallah *et al.*, 1990). Obviously the questions now being asked are how to achieve nitrogenase activity of the paranodules. But the biggest hurdles as yet to be faced but not tried are how to stabilize the association, how to make the association a hereditary one, how to regulate the fixation process and how to get the plants to make substances like leghaemoglobin and nodulin. A time table has been offered (McCormick, 1988) for extension of host range of symbiotic bacteria to new host species, 20-30 years; transfer of nitrogen fixation genes directly into a plant genome, 20-30 years, by concerned scientists. The last approach is obviously radical since evolution has so far not produced any nitrogen fixing plant although suggestions of the presence of nitrogen fixing eukaryotes are there (Saha and Sen, 1990).

Conclusions

Rhizobium is primarily a rhizosphere organism. Before the seedling is established in soil, inoculant rhizobia must survive in an atypical environment. Also after nodule senescence when the bacteria are to live in soil as a saprophyte, they are under stress and their number decreases linearly. One of the first stress they suffer is related to the lack of nutrients, as soils are almost invariably deficient in readily metabolizable carbon sources. In such a situation, *Rhizobium* appears to regulate its level of metabolic enzymes for the production of maintenance energy (Mandal and Chakrabarty, 1992b). Problems of survivality of rhizobium in soil, competition from other organisms and the problem of host promiscuity involve the environment, the bacteria and the plant and therefore not easy to regulate. However, with the progress of molecular biology and molecular genetics of both host and bacteria, it is hoped that it will be possible to manipulate the respective genes for better nodulation and nitrogen fixation. This will lead to increased yield of legumes causing economic benefit primarily to the people in the third world. With the advances of research in rhizobial nitrogen fixation, hopefully, malnutrition and starvation will be things of the past someday. Let's keep our fingure crossed.

REFERENCES

Al-Raddad Al-Momany, A. 1991. Response of bean, broad bean and chickpea plants to inoculation with *Glomus* species. *Scientia Horticulturae*, **46** : 195-200.

Al-Mallah, M.K., Davey, M.R. and Cocking, E.C. 1989. A new approach to the nodulation of non-legumes by rhizobia and the transformation of cereals by agrobacteria using enzymatic treatment of root hairs. *Int. J. Gen. Manip. Plants*. **5** : 1-7.

— 1990. Nodulation of oil seed rape (*Brassica napus*) by rhizobia. *J. Plant Physiol.* **137** : 15-19.

Ames, P. and Bergman, K. 1981. Competitive advantage provided by bacterial mobility in the formation of nodules by *Rhizobium meliloti*. *J. Bacteriol.* **148** : 728-729.

Appleby, C.A. 1974. Leghemoglobin. In : *The biology of nitrogen fixation* (Ed. A. Quispel). Amsterdam, North Holland.

Barbour, W.M., Hattermann, D.R. and Stacy, G. 1991. Chemotaxis of *Bradyrhizobium japonicum* to soyabean exudates. *Appl. Environ. Microbiol.* **57** : 2635-2639.

Behki, R.M., Selvaraj, G. and Iyer, V.N. 1985. Derivatives of *Rhizobium meliloti* strains carrying a plasmid of *Rhizobium leguminosarum* specifying hydrogen uptake and pea-specific symbiotic functions. *Arch. Microbiol.* **140** : 352-357.

Berg, R.K., Loynachan, T.K., Zablotowicz, R.M. and Lieberman, M.T. 1988. Nodule occupancy by introduced *Bradyrhizobium japonicum* in lowa soils. *Agron. J.* **80** : 876-881.

Bhagwatt, A.A., Tully, R.E. and Keister, D.L. 1991. Isolation and characterization of a competition defective *Bradyrhizobium japonicum* mutant. *Appl. Environ. Microbiol.* **43** : 800-805.

Bhagwatt, A.A. and Keister, D.L. 1992. Identification and cloning of *Bradyrhizobium japonicum* genes expressed strain selectively in soil and rhizosphere. *Appl. Environ. Microbiol.* **58** : 1490-1495.

Bhandari, B., Naik, M.S. and Nicholas, D.J.D. 1984. ATP production coupled to denitrification of nitrate in *Rhizobium japonicum*, grown in cultures and in bacteroids from *Glycine max*. *FEBS Lett.* **168** : 321-326.

Bhandari, B. and Nicholas, D.J.D. 1986. Properties of glutamine synthetase of bacteroids from root nodules of *Glycine max*. *Plant Sci.* **44** : 89-96.

Bohlool, B.B. and Schmidt, E.L. 1974. Lectins : a possible basis for specificity in the Rhizobium-legume root nodule symbiosis. *Science.* **185** : 269-271.

Bohlool, B.B. and Schmidt, E.L. 1976. Immunofluorescent polar tips of *Rhizobium japonicum*; possible site of attachment or lectin binding. *J. Bacteriol.* **125** : 1188-1194.

Bravo, A. and Mora, J. 1988. Ammonium assimilation in *Rhizobium phaseoli* by the glutamine synthetase-glutamate synthase pathway. *J. Bacteriol.* **170** : 980-984.

Brewin, N.J. 1991. Development of the legume root nodule. *Ann. Rev. Cell Biol.* **7** : 191-226.

Brown, C.M. and Dilworth, M.J. 1975. Ammonia assimilation by Rhizobium cultures and bacteroids. *J. Gen. Microbiol.* **86** : 39-48.

Caetano-Anolles, G., Crist-Ester, D.K. and Bauer, W.D. 1988. Chemotaxis of *Rhizobium meliloti* to the plant flavone luteslin requires functional nodulation genes. *J. bacteriol.* **170** : 3164-3169.

Carlson, T.A. and Chelm, B.K. 1986. Apparent eukaryotic origin of glutamine synthase II from the bacterium *Bradyrhizobium japonicum*. *Nature.* **322** : 568-570.

Casella, S., Shapleigh, J.P. and Payne, W.J. 1986. Nitrite reduction in *Rhizobium ''hedysari''* strain HCNT1. *Arch. Microbiol.* **146** : 233-238.

Chakrabarti, S.K., Mishra, A.K. and Chakrabartty, P.K. 1987a. Metabolism of glucose and gluconate in fast- and slow-growing rhizobia. *Phytochemistry.* **26** : 85-87.

— 1987b. Cytochromes in free-living rhizobia. *Current Microbiol.* **15** : 165-170.

Chakrabarti, S.K., Halder, A.K. and Chakrabartty, P.K. 1990. Relatedness among *Rhizobium species*. *J. Gen. Appl. Microbiol.* **36** : 47-54.

Davis, E.O., Evans, I.J. and Johnston, A.W.B. 1988. Identification of a *nod* X, a gene that allows *Rhizobium leguminosarum* bv *Viceae* strain TOM to nodulate Afganistan peas. *Mol. Gen. Genet.* **212** : 531-535.

Dazzo, F.B. and Truchet, G.L. 1983. Interaction of lectins and their saccharide receptors in the *Rhizobium*-legume symbiosis. *J. Membrane Biol.* **73** : 1-16.

Dixon, R.O.D. 1972. Hydrogenase in legume root nodule bacteroids : occurrence and properties. *Arch. Microbiol.* **85** : 193-201.

Dixon, R. 1987. Genetic regulation of nitrogen fixation. *In : The nitrogen and sulphur cycles,* (Eds. J.A.

Cole and S. Fergusm) Society for General Microbiology Symposium 42. Society for General Microbiology Ltd. Published by Cambridge University Press.

Dowling, D.N. and Broughton. W.J. 1986. Competition for nodulation of legumes. *Ann. Rev. Microbiol.* **40** : 131-157.

Dusha, I., Bakos, A., Kondorosi, A., de Bruijn, F.A. and Schell, J. 1989. The *Rhizobium meliloti* early nodulation genes (*nod* ABC) are nitrogen regulated : isolation of a mutant strain with efficient nodulation capacity on alfalfa in the presence of ammonium. *Mol. Gen. Genet.* **219** : 89-96.

Fuhrmann, J. 1990. Symbiotic effectiveness of indigenous soyabean bradyrhizobia related to serological, morphological, rhizobitoxine and hydrogenase phenotypes. *Appl. Environ. Microbiol.* **56** : 224-229.

Garg, F.C., Garg, R.P., Kukreja, K., Sindhu, S.S. and Tauro, P. 1985. Host-dependent expression of uptake hydrogenase in cowpea rhizobia. *J. Gen. Microbiol.* **131** : 93-96.

Gordon, J.K. and Jacobson, M.R. 1983. Isolation and characterization of *Azotobacter vinelandii* mutant strains with potential as bacterial fertilizer. *Can. J. Microbiol.* **29** : 973-978.

Hahn, M. and Hennecke, H. 1988. Cloning and mapping of a novel nodulation region from *Bradyrhizobium japonicum* by genetic complementation of a deletion mutant. *Appl. Environ. Microbiol.* **54** : 55-61.

Halder, A.K., Mishra, A.K., Bhattacharyya, P. and Chakrabartty, P.K. 1990. Solubilization of rock phosphate by *Rhizobium* and *Bradyrhizobium. J. Gen. Appl. Microbiol.* **36** : 81-92.

Halder, A.K., Banerjee, A. and Chakrabartty, P.K. 1992. Role of NH_4^+ or NO_3^- on release of soluble phosphate from hydroxyapatite by *Rhizobium* and *Bradyrhizobium. J. Basic Microbiol.* **32** : 325-330.

Halverson, L.J. and Stacy, G. 1984. Signal exchange in Plant-microbe interactions. *Plant Physiol.* **74** : 84-89.

Halverson, L.J. and Handelsman, J. 1981. Enhancement of Soyabean nodulation by *Bacillus cereus* UW 85 in the field and in a growth chamber. *Appl. Environ. Microbiol.* **57** : 2767-2770.

Hennecke, H. 1990. Nitrogen fixation genes involved in the *Bradyrhizobium japonicum* - soyabean symbiosis. *FEBS Lett.* **268** : 422-426.

Hoover, T.R., Robertson, A.D., Cerny, R.L., Hayes, R.N., Imperial, J., Shah, V.K. and Ludden, P.W. 1987. Identification of the V factor needed for synthesis of the iron-molybdenum co-factor of nitrogenase as homocitrate. *Nature.* **329** : 855-857.

Jordan, D.C. 1984. Family III Rhizobiaceae Conn 1938, p. 234. *In : Bergey's manual of systematic bacteriology,* Vol. 1 (Eds. N.R. Kreig and J.G. Holt) The Williams and Wilkins Co., Baltimore.

Keister, D., Jones, B.L., Kondorosi, A. and Svab, Z. 1987. Characterization of glutamine synthetase I and II mutants of *Rhizobium. J. Gen. Appl. Microbiol.* **33** : 267-273.

Kessel, C.V., Singleton, P.W. and Hoben, H.J. 1985. Enhanced N-transfer from a soyabean to maize by vesicular arbuscular mycorrhizal (VAM) fungi. *Plant Physiol.* **79** : 562-563.

Kloepper, J.W., Leong, J., Teintze, M. and Schroth, M.N. 1980. Enhancement of plant growth by siderophores produced by plant growth-promoting rhizobacteria. *Nature.* **286** : 885-886.

Knight, T.J. and Langston-Unkefer, P.J. 1988. Enhancement of symbiotic dinitrogen fixation by a toxin-releasing plant pathogen. *Science.* **241** : 951-954.

Kondorosi, A., Svab, Z., Kiss, G.B. and Dixon, R.A. 1977. Ammonia assimilation and nitrogen fixation in *Rhizobium meliloti. Mol. Gen. Genet.* **151** : 221-226.

Kondorosi, E. and Kondorosi, A. 1986. Nodule induction on plant roots by *Rhizobium. Trends Biochem. Sci.* **11** : 296-299.

Kurz, W.G.W., Rokosh, D.A. and LaRue, T.A. 1975. Enzymes of ammonia assimilation in *Rhizobium leguminosarum* bacteroids. *Can J. Microbiol.* **21** : 1009-1012.

Lambert, G.R., Harkar, A.R., Cantrell, M.A., Hanus, F.J., Russel, S.A., Hangland, R.A. and Evans, H.J.

1987. Symbiotic expression of cosmid-borne *Bradyrhizobium japonicum* hydrogenase genes. *Appl. Environ. Microbiol.* **56** : 203-214.

Lim, S.L. and Shanmugam, K.T. 1979. Regulation of hydrogen utilization in *Rhizobium japonicum* by cyclic AMP. *Biochem. Biophys. Acta.* **584** : 479-492.

Liu, R., Tran, M. and Schmidt, E.L. 1989. Nodulating competitiveness of a non-motile Tn7 mutant of *Bradyrhizobium japonicum* in non-sterile soil. *Appl. Environ. Microbiol.* **55** : 1895-1900.

Long, S.R. 1989a. *Rhizobium* genetics. *Ann. Rev. Genet.* **23** : 483-506.

— 1989b. *Rhizobium*-legume nodulation : Life together. *Cell.* **56** : 203-214.

Ludwig, R.A. and Singer, E.R. 1977. Glutamine synthetase and control of nitrogen fixation in *Rhizobium. Nature.* **267** : 245-248.

Ludwig, R.A. 1978. Control of ammonium assimilation in *Rhizobium* 32H1. *J. Bacteriol.* **135** : 4 : 114-123.

Lugue, F., Santero, E., Torolero, M., Medina, J.R. 1986. Genetic alteration of glutamine synthetase in *Azotobacter vinelandii. Microbios. Lett.* **33** : 29-32.

Mandal, N.C. and Chakrabartty, P.K. 1992a. Regulation of enzymes of glyoxylate pathway in root-nodule bacteria. *J. Gen. Appl. Microbiol.* **38** : 417-427.

— 1992b. Carbohydrate metabolic enzymes of *Rhizobium* during carbon starvation. *Indian J. Exp. Biol.* **30** : 804-807.

— 1993. Sccuinate mediated catabolite repression of enzymes of glucose metabolism in root-nodule bacteria. *Current Microbiol.* **26** : 247-251.

Maier, R.J. and Brill, W.J. 1978. Mutant strains of *Rhizobium japonicum* with increased ability to fix nitrogen in soyabean. *Science* **201** : 448-450.

McCormick, D. 1988. How biotech is dealing with its nitrogen fixation. *Biotechnology.* **6** : 383-385.

McDermott, T. and Graham, P.H. 1990. Competitive ability and efficiency in nodule formation of strains of *Bradyrhizobium japonicum. Appl. Environ. Microbiol.* **56** : 3035-3039.

Mellor, H.Y., Glenn, A.R., Arwas, R. and Dilworth, M.J. 1987. Symbiotic and competitive properties of motility mutants of *Rhizobium trifolii* TA1. *Arch Microbiol.* **148** : 34-39.

O'Gara, F. and Shanmugam, K.T. 1976. Regulation of nitrogen fixation by rhizobia; export of fixed N_2 as NH_4^+. *Biochem. Biophys. Acta.* **437** : 313-321.

Pagan, J.D., Child, J.J., Scowcraft, W.R. and Gibson, A.H. 1975. Nitrogen fixation by *Rhizobium* cultured on a defined medium. *Nature.* **256** : 406-407.

Parejko, R.A. and Wilson, P.W. 1970. Regulation of nitrogenase synthesis by *Klebsiella pneumoniae. Can J. Microbiol.* **16** : 681-685.

Plazinski, J. and Rolfe, B.G. 1985. Influence of *Azospirillium* strain on the nodulation of cloves by *Rhizobium* strains. *Appl. Environ. Microbiol.* **49** : 984-989.

Porter, P.M., Banwart. W.L. and Hassett, J.J. 1985. HPLC isolation and GC-MS identification of genistein, daidzen, and coumestrol from unhydrolysed soyabean root extract. *Environ. Exp. Bot.* **25** *:* 229-232.

Postgate, J. 1975. Rhizobium as a free living nitrogen fixer. *Nature.* **256** *: 363.*

Postgate, J. 1989. Trends and Perspectives in Nitrogen Fixation *Research. Adv. Microbial Physiol.* **30** : 1-22.

Pueppke, S.G. and Krishnan, H.B. 1988. Symbiotic mutants of wide host range rhizobia that nodulate soybean. p. 22-23. *In : Scientific Programme and Abstracts, International Symposium and Workshop, Biological nitrogen fixation associated with rice production.* Oxford and IBH Publishing Co. Pvt. Ltd., New Delhi.

Rigaud, J., Bergersen, F.J., Turner, G.L. and Daniel, R.M. 1973. Nitrate dependent anaerobic acetylene reduction and nitrogen fixation by soyabean bacteroids. *J. Gen. Microbiol.* **77** : 137-144.

Rossen, L., Shearman, C., Jhonston, A.W.B., Downie, J.A. 1985. The *nod* D gene of *Rhizobium leguminosarum* is autoregulatory and in the presence of plant exudate induces the *nod* A, B, C genes. *EMBO J.* **4** : 3369-3374.

Rostas, K., Sista, P.R., Stanley, J. and Verma, D.P.S. 1984. Transposon mutagenesis of *Rhizobium japonicum. Mol. Gen. Genet.* **197** : 230-235.

Saha, V. and Sen, M. 1990. Nitrogen fixation by *Candida tropicalis. Proc. Indian Acad. Sc.* (Plant Sci.), **100** : 343-352.

Scott, D.B., Robertson, J.G. and Farnden, K.J.F. 1976. Ammonia assimilation in lupin nodules. *Nature.* **263** : 703-705.

Shanmugam, K.T., O'Gara, F., Anderson, K. and Valentina, R.C. 1978. Biological Nitrogen Fixation. *Ann. Rev. Plant Physiol.* **29** : 263-276.

Smit, G., Puvanesarajah, Carlson, R. and Stacy, G. 1990. *Bradyrhizobium japonicum nod* D can specifically be induced by soybean seed extract compounds which donot induce the *nod* YABC genes. p. 274. *In : Nitrogen Fixation - achievements and objectives* (Eds. P.M. Gresshoff, L.E. Roth, G. Stacy and W.E. Newton) Chapman and Hall Ltd., London.

Smith, S.E. and Daft, M.J. 1977. Interactions between growth, phosphate content and nitrogen fixation in mycorrhizal and non-mycorrhizal *Medicago sativa Aust. J. Plant Physiol.* **4** : 403-413.

Somerville, J.E. and Kahn, M.L. 1983. Cloning of the glutamine synthetase I gene from *Rhizobium meliloti. J. Bacteriol.* **156** : 168-176.

Subba Rao, N.S., Tilak, K.V.B.R. and Singh, C.S. 1980. Root nodulation studies in *Aeschynomene aspera. Plant Soil.* **56** : 491-494.

Tchan, Y.T., Zeman, A.M.M. and Kennedy, I.R. 1991. Nitrogen fixation in paranodules of wheat roots by induced free-living diazotrophs. *Plant Soil.* **137** : 43-47.

Thimmaiah, S.K., Sharma, N.D. and Lodha, M.L. 1986. Regulation of uptâke hydrogenase expression in mungbean (*Vigna radiata*) *Rhizobium* strain S-24, effect of carbon and nitrogen source. *Indian J. Exp. Biol.* **24** : 238-241.

Trinick, M.J. 1980. Growth of *Parasponia* in agar tube culture and symbiotic effectiveness of isolates from *Parasponia* sp. *New Phytologist.* **85** : 37-45.

Triplett, E.W. 1988. Isolation of genes involved in nodulation competitiveness for *Rhizobium leguminosarum bu. trifolii* T24. *Proc. Natl. Acad. Sci.* (USA). **85** : 3810-3814.

Triplett, E.W. 1990. Construction of a symbiotically effective strain of *Rhizobium leguminosarum bv. trifolii* with increased nodulation competitiveness. *Appl. Environ. Microbiol.* **56** : 98-103.

Vincent, J.M. 1970. *A manual for the practical study of the root-nodule bacteria.* IBP Handbook No. 15 Blackwell Scientific Publications. Oxford.

Viteri, S.E. and Schmidt, E.L. 1987. Ecology of indigenous soil rhizobia : response of *Bradyrhizobium japonicum* to readily available substrate. *Appl. Environ. Microbiol.* **53** : 1872-1875.

Wang, S.P. and Stacy, G. 1990. Ammonia regulation of *nod* genes in *Bradyrhizobium. Mol. Gen. Genet.* **223** : 329-331.

— 1991. Studies of the *Bradyrhizobium japonicum nod* D_1 promoter : a repeated structure for the *nod* box. *J. Bacteriol.* **173** : 3356-3365.

Yoneyama, T. and Kondo, M. 1990. *Sesbania* spp., *Aeschynomene indica* and *Crotalaria* spp. are amide exporters. *Soil Sci. Plant Nutr.* **36** : 689-693.

Zdor, R.E. and Pueppke, S.G. 1991. Nodulation competitiveness of Tn 5 induced mutants of *Rhizobium fredii* USDA 208 that are altered in motility and extracellular polysaccharide production. *Can J. Microbiol.* **37** : 52-58.

Chapter 2

Ecology of Legume Root Nodule Bacteria

K.K. Kapoor AND **S.S. Dudeja**

Department of Microbiology, CCS Haryana Agricultural University, Hisar-125 004, India

Abstract. The survival and establishment of root nodule bacteria in the soil is crucial for the performance of legume - *Rhizobium* symbiosis. Many direct and indirect approaches have been tried to investigate the ecology of rhizobia. Some of these include use of antibiotic resistance markers, immunofluorescence, bacteriophage typing, monoclonal antibodies, DNA hybridization and whole cell protein pattern analysis. Large number of abiotic and biotic factors influence the population of rhizobia in the soil and consequently the nodulation and nitrogen fixation of legumes. Among abiotic factors; temperature, moisture, soil pH, soil salinity and available nutrients control the survival of inoculated as well as native rhizobia population. A wide range of biological interactions between rhizobia and other soil microorganisms are possible which have an important bearing on size and composition of rhizobial population in soil. Some of these biotic factors affecting rhizobial population include competition among rhizobial strains as well as competition between rhizobia and other soil microorganisms; antagonism by related strains and other soil microorganisms, and predation by protozoa and bacteriophages. Development of efficient rhizobia strains resistant to these abiotic and biotic stresses can improve nodulation and nitrogen fixation in soil.

Key words : *Ecology, Rhizobia, Nodulation, Biotic Stress, Abiotic Stress*

Introduction

The ability to fix atmospheric nitrogen is restricted to several procaryotic microorganisms, including members of the family Rhizobiaceae, Azotobacteraceae, Enterobacteriaceae, Bacillaceae, Rhodospirallaceae, Actinomycetaceae and Cyanobacteria. The members of the plant family Leguminose are able to enter a beneficial relationship with soil bacteria of the genera *Rhizobium, Bradyrhizobium* and *Azorhizobium* (Long, 1989). In this association the bacteria induce the plant to develop a new plant organ, the root nodule in which they house to fix atmospheric nitrogen. The endosymbiotic association helps the legumes to assimilate atmospheric nitrogen. A sequence of events allows the establishment of a symbiotic state where nitrogen is fixed to ammonia and subsequently assimilated into organic nitrogenous compounds.

The ecology of rhizobia in the soil is crucial to the establishment and performance of the legume-*Rhizobium* symbiosis. The excitement over prospects for genetic improvement of rhizobia has unfortunately over shadowed the need for an understanding of its ecology. It will be a futile exercise if any "super bug" engineered in the laboratory to have an economic impact fails to withstand the stresses of the soil. It must be able to survive and grow in the rhizophere, and compete successfully with well adapted indigenous rhizobia for nodulation of the appropriate host.

The neglect for the study of ecology of *(Brady) Rhizobium* is mainly because of lack of suitable methodology. The understanding of ecology of rhizobia in soil has been mainly based on the indirect approach by accessing the proportion of nodules formed by the inoculant strain. No doubt, such information is of practical importance but it does not directly reflect the changes in soil and rhizosphere rhizobial populations. Presently good number of techniques are available to monitor the rhizobial population directly in the soil. Direct quantitative measurements have been carried out using plant infection method (Brockwell, 1963), fluorescent antibody technique (Reyes and Schmidt, 1979), antibiotic marker technique (Ramirez and Alexander, 1980), genetic markers (Bao *et al.*, 1991; Sharma *et al.*, 1991) and non-radioactive DNA hybridization (Sano and Lindostrom, 1992). Indirect approaches to monitor, the fate of rhizobia in the soil is through the analysis of the nodules. After crushing the nodules, rhizobia are identified by slide agglutination, immunodiffusion, ELISA (enzyme linked immunosorbent assay) and DNA hybridization. In another indirect approach, rhizobia from the nodules are isolated and subsequently identified by antibiotic resistance, bacteriophage typing, polyacrylamide gel electrophoresis and immunoblot techniques.

1. Indirect Approaches to the Ecology of Rhizobia

In contrast to the few techniques that allow a direct approach to the ecology of rhizobia, there are larger number of techniques available for the indirect nodule analysis. In this the rhizobia that has attained nodule niche are identified. Further identification may involve the direct analysis of the nodule or by isolation of the rhizobia and subsequent analysis of the pure isolate.

1.1 *Nodule Analysis*

To identify the *(Brady) Rhizobium* Strains directly from nodule bacteroids, serological techniques have been tried several years ago and were based on whole cell agglutination (Hughes and Vincent, 1942). Subsequently other serological techniques were used that offered greater resolution and reliability. These techniques were refined to discriminate between strains that showed ambiguous cross reacting tendencies. These included gel immunodiffusion assay (Dudman and Brockwell, 1988; Humphrey and Vincent, 1976; Dudeja and Khurana, 1988) enzyme linked immuno sorbent assay-ELISA (Morley and Jones, 1980) and immuno flourescence-IF (Roughly *et al.*, 1976; Moawad *et al.*, 1984; Robert and Schmidt, 1983; May and Bohlool, 1983; Kingsley and Bohlool, 1983). Currenlty *Brady/Rhizobium* strains are identified by IF or by ELISA. Particularly ELISA is capable of serotyping large number of nodules and providing highly specific and sensitive method of strain identification. This technique has successfully been used for both fast and slow growing strains (Rice *et al.* 1984). Recent improvements in the technique include use of a flourescent substrate (Morley and Jones, 1980) and substitution of beta galactosidase for alkaline phosphatase (Martensson *et al.*, 1984). Further use of monoclonal antibodies has greatly enchanced the precision of strain identification.

Indentification of *Rhizobium* strains directly in the legume root nodules by means of DNA colony hybridization was first reported by Hodgson and Roberts (1983). In comparison to immunological and antibiotic marker techniques, this reduces the cross reactions and permits studies in the genetically unmodified strains. DNA colony hybridization has a very high generic and strain specificity. DNA homology was used to differentiate *R. fredii*, *R. leguminosarum* and *Galega* rhizobia from other *Rhizobium* species (Wedlock and Jarvis, 1986). Cooper *et al.* (1987) reported the hybridization of crushed *Lotus pedunculatus* root nodules with P labelled total genomic DNA probes to identify *Rhizobium loti and Bradyrhizobium* sp. (*Lotus* rhizobia).

1.2 *Identification after isolation of rhizobia from nodules*

Serological techniques are reliable for identification of rhizobia in nodules only if there is no cross reaction with other strains, or if the cross reactivity can be removed by adsorption. Numerous alternatives to serological techniques are available, once rhizobia are isolated from the nodules. Some times rhizobia that develop stable resistance to fairly high levels of antibiotics can be used for strain identification for direct ecological studies. Advantage of antibiotic resistant mutants is considered in ecological studies to evaluate the success of an antibiotic marked strain in competition studies (McLough in *et al.*, 1984). Many rhizobia are also naturally resistant to a range of concentrations of various antibiotics and since this resistance varies with the antibiotic and with the strain, a given strain may express a distinctive pattern of antibiotic resistance and such variation may be a useful taxonomic character (Graham, 1963). This intrinsic antibiotic resistance has been utilized for rhizobial ecology (Obaton, 1973; Kingsley and Bohlool, 1983; Khuranal *et al.*, 1991) Antibiotic resistant strains may vary in their effectiveness. Spontaneous antibiotic resistant mutants differed in competitiveness in forming nodules on clover (Bromfield and Jones, 1979; Jones and Bromfield, 1978) but no change was observed for rhizobia nodulating medic (Amarger, 1981) and chickpea (Sharma *et al.*, 1991). Further, such mutants has been reproted to differ in effectiveness (Schwinghamer and Dudman, 1973). Loss of the antibiotic markers further make this method unsuitable for strain identification (Brockwell *et al.*, 1977).

1.2.1 *Bacteriophage Typing*

The rhizobial strain typing by means of bacteriophage sensitivity has been used for many years for both fast and slow growing rhizobia. The rhizophages bind with some surface components of *Rhizobium* and are strain specific. The host specificity of these phages forms the basis of strain typing. Strains of *R. meliloti* has been distinguished from *R. leguminosarum, R. phaseoli or R. trifolii* by bacteriophage typing (Staniewski, 1970a, b). Depending upon this 71 strains of *R. meliloti* have been divided into 10 groups using 7 phages. Similarly Lesly (1982) detected as many as 80 phage groups in a collection of *R. meliloti* on the basis of patterns of reaction obtained with 15 phages. On the basis of host specificity, phage typing system has limited applications because rhizophages for all the strains are not available and some phages has wide host range. Therefore this method has also limited suitability for strain identification.

1.2.2 *Polyacrylamide Gradient Gel Electrophoresis (PAGGE)*

Gel electrophoresis for the differentiation of whole cell proteins of *Rhizobium* has been proposed for detailed characterization of *Rhizobium* strains of ecological interest (Noel and Brill, 1980). This method has been used to describe degree of strain heterogeneity in natural populations (Young, 1985; Bromfield *et al.*, 1986; Thurman and Bromfield, 1988). The degree of heterogeneity depends on the identification method employed and a group of isolates homogenous for one characteristic may vary in respect of another character (Dughri and Bottomly, 1983). To determine the genetic variation within natural populations of *R. leguminosarum* biovar *trifolii* polyacrylamide gel electrophoresis of isoenzymes glucose-6-phosphate dehydrognase, superoxide dismutase and B-galactosidase has been used (Harrison *et al.*,1989). Difference in the degree of strain polymorphism were observed.

1.2.3 *Immunoblot Technique*

Identification of *Rhizobium* strains by enzyme linked immunosorbent assay in colonies picked from plates and suspension of each colony in buffer, adjustment of cell densities, individual transfer of cell suspension to microdilution plate wells and manipulation of microdilution plates with specialized immuno assay equipment. *Rhizobium* extracts and whole cells (Ayanaba *et al.*, 1986) have been immunologically identified at the strain level after electrophoretic or manual transfer to nitrocellulose membranes. In immunoblot technique a replica of rhizobial colonies on nitrocellulose membrane is obtained and then this membrane is processed with a specific primary antibody and enzyme conjugated second antibody, and a precipitating substrate. The result is coloured spots on the membrane corresponding to specific colony-antibody interactions (Olsen and Rice, 1989). This procedure simultaneously tests all colonies and eliminates requirement for specialized enzyme linked immunosorbent assay equipment.

2. Direct approach to the ecology of rhizobia

The ecology of root nodule bacteria in the soil and especially in the rhizosphere is very important in determining the establishment and performance of the *Rhizobium*-legume symbiosis. But it is extremely difficult to study rhizobia directly in the soil. However, different approaches which has been used to study the ecology of rhizobia in the soil are described here.

2.1 *Immuno fluorescence (IF)*

The IF was first applied to *Rhizobium* ecology in 1968 (Bohlool and Schmidt, 1968). This technique has been used to study the autoecology of rhizobia, which elicit a distinctive immuno response. The use of IF for direct study of rhizobia in the soil and rhizosphere has expanded in recent years with the availability of fluorescence microscopy and improvement in the extraction procedures to separate rhizobia from soil particles prior to IF examination (Kingsley and Bohlool, 1981). Specific protocols are not common to all soils. To avoid the problem of nonspecific staining by other bacteria, monoclonal antibodies has been used which have enhanced the strain specificity.

2.2 *Monoclonal Antibodies*

The immunofluorescence technique has been useful as a tool in ecological studies (Bohlool and Schmidt, 1980). Due to high specificity and sensitivity of this technique, recently this has been used for the detection and enumeration of genetically modified microorganisms in soil (Ford and Olson, 1988). The use of monoclonal antisera in the immunofluorescence technique offers the promise of fluorescent antibodies with greater specificity, since even high quality polyclonal antisera usually consist of over 90% immunoglobulins with little or no avidity for the antigen of interest. The remaining specific immunoglobulins are diverse and could react with a wide spectrum of different antigenic determinants (Roitt, 1984). Monoclonal antibodies in contrast possess a high specific titre of homogenous immunoglobulins. Monoclonal antibodies, have only recently been used for the identification of rhizobia for ecological studies (Wright, 1990; Velez *et al.*, 1988). Their high specificity has led to the problem of the stability of antigenic determinants under extremes of environmental conditions because two out of five monoclonal antibodies produced against a culture of *B. japonicum* did not react with the same strain after re-isolation from soybean nodules (Velez *et al.*, 1988). However in another report monoclonal antibodies were found to be stable (Kinkle and Schmidt, 1992).

2.3 *Antibiotic resistance markers*

Strains of root nodule bacteria resistance to antibiotics have been used for ecological studies. Soil dilution plating on medium containing the specific antibiotic alongwith various antibacterial and antifungal agents has been used to enumerate spontaneous antibiotic resistant mutants of *Rhizobium* (Pena-Cabriales and Alexander, 1983). To be useful in ecological studies, antibiotic resistance markers must be stable, unaltered in symbiotic effectiveness and these ecological attributes should not be present in the native rhizobia. Sensitivity of the antibiotics resistance/selective plating was found to be in the range of 30 rhizobia g^{-1} of soil (Bushby, 1981). Combination of selective antibiotic medium with immunoenzymatic assay to detect an antibiotic resistant mutant of *R. japonicum* in soil has also been used.

2.4 *Genetically marked strains for identification*

To study the fate of introduced rhizobia in the soil, methods are required to differentiate between the introduced strain and native rhizobial population. Therefore, rhizobia have been constructed to introduce stable markers so as to identify these from soil. Recently antibiotic markers have been introduced in rhizobia by transposition of *Tn5* and *Tn7* (Bio *et al.*, 1991; Selvaraj and Iyer, 1983; Simon *et al.*, 1983; Sharma *et al.*, 1991). Integration of transposon into chromosome or megaplasmid of *Rhizobium* give rise to stable marker. Transposon tagged plasmid in *Rhizobium* could be used to study the survival and persistence of introduced strains in the soil.

Recently transposon containing promoterless reporter gene(s) like *lac* Z (structural gene for β-galactosidase), *gus* A (β-glucuronidase) and *lux* A (luciferase) are available and these genes are introduced into root nodule bacteria to monitor the fate of these strains in the soil and rhizosphere. Such reconstructed rhizobia are directly identified on plates containing chromogenic substrate like X-gal (5-Bromo-4-chloro-3-indolyl-β-galactopyranoside) or X-gluc (5-Bromo-4-chloro-3-indolyl-β glucuroniside) or by bioluminescence visible to naked eyes or by X-ray films. Dev Vrat (1992) constructed random *lac* Z fusion in *Rhizobium* sp. (*Cicer*) using transposon p SUP-TnB-20 containing promoterless *lac* Z gene. These reconstructed rhizobia were identified directly on plates from soil and nodules. Likewise *gus* A fusion were constructed in *Rhizobium* to use these fusions for survival and persistence studies in the soil (Wener *et al.*, 1992). Gene probes specific to insertion sequence of *R. meliloti* have been used to identify the constructed rhizobia in the soil (Wheatcroft and Watson, 1988). In case of *Bradyrhizobium japonicum* gene probe consisting of *csn* gene, *nod* A have been used to detect the genetically constructed strains in the soil (Sadofasky *et al.*, 1989).

2.5 *Non-radioactive DNA-DNA hybridization*

To monitor the fate of rhizobia or any other organism in soil, a reliable, sensitive, fast, easy to use and inexpensive technique of identification is needed. The DNA-DNA hybridization has already responded positively to identify rhizobia directly from crushed root nodules (Cooper *et al.*, 1987) and to differentiate different strains using Southern blots of total DNA digestion patterns (Wheatcroft and Waston, 1988; Lindstrom *et al.*, 1990). This technique can also be used to directly monitor the number of rhizobia in soil. There are two main approaches to identify bacteria directly from soil by DNA-DNA hybridization : the cell extraction and direct lysis method. Tsai and Olson (1991) detected specific sequence of *Pseudomonas* DNA from soil by DNA-DNA hybridization with a ^{32}P labelled probe. However from the point of view of safety, radioactive labelling of probe DNA is not desirable especially for repetitive and large scale field testing. Therefore, Sano and Lindstrom (1992) applied non radioactive DNA-DNA hybridization to identify *Rhizobium galegae* which forms nitrogen fixing nodules on roots of *Galega* species (Lindstrom, 1989), a perennial forage legume for temperate conditions. Total bacterial DNA from the soil samples was extracted using lysozyme and a freeze thaw procedure and the sample DNA was attached to nitrocellulose or nylon membranes. During a 3h hybridization 10pg of sample DNA could be detected on a nylon filter with digoxigenin-dUTP labelled *R. galegae* total DNA.

3. Factors Affecting the Ecology of Rhizobia

For greater part of their existence, rhizobia survive as free living bacteria in the soil. Both abiotic and biotic factors influence the population of rhizobia in the soil, which in turn is responsible for the successful interaction of rhizobia and the legume. Highly efficient rhizobial strains following their introduction into the soil often do not persist in significant numbers. Nevertheless some attention has been paid on the use of strains which are able to persist in the soil in sufficient numbers to ensure proper nodulation (Graham, 1992). The abiotic factors influencing survival of rhizobia include the effect of soil physical environment and nutrient availability, while the biotic factors comprise interactions with other living organisms.

3.1 *Abiotic factors*

The soil physical environment has a profound effect on the free living rhizobia as well as the symbiotic performance. The abiotic factors affecting rhizobia include; temperature, moisture, soil pH, soil salinity and level of available nutrients.

3.1.1 *Temperature*

Rhizobia are mesophilic organisms and various species and strains differ in their tolerance to the elevated temperature (Hartel) and Alexander, 1984). The elevated temperatures delay nodule initiation, nodule development, and interfere with nodule functioning in the temperate legumes. The strains of rhizobia from the tropical soils are generally more tolerant to temperature stress than those from temperate locations (Wilkins, 1967). There are some basic differences between fast and slow growing rhizobia in their tolerance to heat. Slow growers were found to be relatively resistant to drying and high temperature as compared to fast growers (Marshal, 1964). The strains of cowpea rhizobia grew in mannitol amended nonsterile soil at 29-35°C but not at 40°C. Decline in the population of these bacteria occurred in dry, nonsterile soil incubated at 43°C for 7 days (Osa-Afiana and Alexander, 1982). The survival and persistence of *Bradyrhizobium* sp. (*Cajanus*) in the soil and the rhizosphere of pigeonpea was determined by the most probably number and the fluorescent antibody technique. The survival was poor and the inoculated strain did not persist upto the next cropping season (Dudeja and Khurana, 1988, 1989). In Western Australia, the death of *R. trifolii* is responsible for the poor growth of subterranean clover in the second year after planting. The rhizobia fail to survive in adequate numbers in the hot dry soil at the end of first season (Chatel and Parker,

1973). The invoivement of plasmids in resistance to heat stress in *R. trifolii* was tested by Baldini and Weaver (1992). Some cured derivatives survived better than the parent at 38 or 42°C but for the others the parent survived better than the cured derivatives. Sharma and Lakshminarayana (1989) reported curing of pigeonpea rhizobia for nodulation genes by high temperature resulting in poor nodulation.

Rhizobia are generally tolerant to low temperature. Danso and Alexander (1974) observed good survival of *R. meliloti* at 6°C. The strains isolated from temperate regions are well adopted to survival at low temperature as compared to the strains from warmer regions (Alexander, 1985). Cloutier *et al.* (1992) reported that the temperate rhizobia produced more cold sensitive proteins under freezing conditions and their cold adaptation does not seem to provide them with better survival at freezing temperatures.

3.1.2 *Moisture*

Very low or very high soil moisture contents are harmful for rhizobia. The effect of limiting soil moisture on survival of rhizobia is more pronounced as compared to excess soil moisture. Drying of soil decreases the number of viable rhizobia although the rate of death varies with other environmental conditions. A single exposure of the soil to drying reduced the viable rhizobial population by 99% or more, and several cycles of soil wetting followed by drying reduced the population still further (Pena-Cabriales and Alexander, 1979). Temperature and moisture jointly affect the growth and survival of rhizobia in the soil (Dudeja and Khurana, 1989), Desiccation even at moderate temperature is known to be highly detrimental to the survival of rhizobia Marshall, 1964; Pena-Cabriales and Alexander, 1979). Danso and Alexander (1974) reported that viability of *R. meliloti* decreased from 10^8 per gram to less than 10^1 per gram after drying for four weeks in sand. However, when drying took place in soil with high clay content, the survival rate was much higher with 10^6 cells per gram still viable after 8 weeks. The protective effect of clay and organic matter has been observed by the other workers also (Bushby and Marshall, 1977; Chao and Alexander, 1982). The strains of rhizobia within a species vary considerably in their ability to survive drying. Nearly half of the viable cells of some strains of cowpea rhizobia are not killed in one drying cycle whereas more than 99% of the cells of other strains die under identical circumstances (Osa-Afiana and Alexander, 1982). Death rates seem to be dependent on the rate of drying; rapid drying being more detrimental as compared to slow drying (Van Rensburg and Stridjom, 1980).

Little work has been carried out on the effect of waterlogging on the ecology at rhizobia Osa-Afiana and Alexander (1979) reported rapid decline in the population of *R. trifolii* as compared to that of *B. japonicum* in the soils flooded over a six week period. Periodic waterlogging is not important for the survival of rhizobia as suggested by good nodulation of subterranean clover in the waterlogged soils (Parker *et al.*, 1977).

3.1.3 *Soil pH*

A great deal of variability in the pH tolerance among rhizobia has been observed. Different strains within the same rhizobial species vary considerably in the response to soil pH (Lowendrof and Alexander, 1983). Extremes of pH frequently reduce the nodule formation by reducing the colonization of soil and the legume rhizosphere by rhizobia. Pure culture studies have shown that lowest pH for growth of rhizobia varies from 4 to 6. In general slow growing *R. japonicum* and *R. lupini* and cowpea rhizobia are less sensitive to acidity as compared to the fast growing strains such as *R meliloti*. Maintenance of intracellular pH is important for growth of rhizobia under acidic conditions. Acid tolerant strains of *R. meliloti* are able to maintain an alkaline cytoplasmic pH when the external pH is acidic (O'Hara *et al.*, 1989; Chen *et al.*, 1993). Acid tolerant strains of *R. leguminosarum* are reported to be more efficient.

Acidic soils have low levels of P, Ca and Mo and high concentration of Al and Mn. In addition to low pH, Al and Mn may reach levels toxic to the plants and rhizobia (Keyser and Munns, 1979; Munns and Keyser, 1981; Flis *et al.*, 1993). Graham *et al.* (1982) observed that a strain of *Rhizobium phaseoli* tolerant to low pH and high Mn and Al concentrations survived better than sensitive strains when inoculated in an acid soil.

High pH also adversely affects the growth and survival of rhizobia. Only a few rhizobial strains are capable of growth above pH 9.0 (Graham and Parker, 1964). In the alkali soils the carbonates and bicarbonates of sodium have generally more severe effect on legume growth compared to corresponding rhizobia (Bhardwaj, 1974).

3.1.4 *Soil salinity*

The growth of most strains of rhizobia is sensitive to salt stress, however few reports of occurrence of salt tolerant strains are there (Steinborn and Roughley, 1975; Singleton *et al.*, 1982; Yelton *et al.*, 1983; Elsheikh and Wood, 1990). Many studies have been conducted to establish the effect of salts on the growth of rhizobia. These studies have demonstrated that the inhibition of growth of rhizobia usually requires high salt levels (Ethiraj *et al.*, 1972); Steinborn and Roughley, 1975; Rai, 1983; Kumar *et al.*, 1990). It has been observed that the rhizobia are able to proliferate at the salt levels that do not permit the growth of the host plant (Bhardwaj, 1975). The mechanism of salt tolerance in rhizobia appears similar to that of other bacteria. High external salts induce an increase in the intracellular pool of amino acids and sometimes of K^+ in salt tolerant strains (Hua *et al.*, 1983; Yap and Lim, 1983). High concentration of the intracellular compatible solutes such as glycine betaine is maintained under salt stress in several rhizobia (Fougere and Rudulier, 1990). Fujihara and Yoneyama (1983) reported that glutamate and K^+ contents of salt tolerant strain of *R. fredii* increased in response to sodium chloride stress. Singleton *et al.* (1982) suggested that more attention should be given to the influence of salinity on symbiosis rather than to the survival of rhizobia. This is because the existing strains of *Rhizobium* are able to survive at high salinities.

3.1.5 *Available nutrients*

Rhizobia are dependent on the soil organic matter for supply of carbon and energy.

Correlation between the soil organic matter levels and the size or strain composition of populations of rhizobia have been reported. Bezdicek (1972) showed a positive correlation between the distribution of some serotypes of *B. japonicum* and the soil organic matter. Direct relationship between the rhizobial survivility and decaying root material in soil has been reported by (Robert and Schmidt, 1983). In a laboratory study Pena-Cabriales and Alexander (1983) observed that the growth of a number of species of rhizobia in the soil increased after addition to mannitol and plant residues, although the latter had a perceptible effect only when competition and predation were limited. Fast growing rhizobia have the ability to utilize a more diverse range of carbon sources than the slow growing rhizobia. The latter have the ability to utilize aromatic compounds (Elkan and Kuykendall, 1982), which may be of significance for the survival and persistence of rhizobia particularly in non-rhizosphere soil where the degradation products of lignin may be the most abundant organic material (Barnet *et al.*, 1988). In general levels of available energy substrates are more in the rhizosphere. This lead to the stimulation of growth of rhizobia in the rhizosphere compared to the bulk soil (Robert and Schmidt, 1983).

Adequate amounts of soluble phosphate is also required for successful host-rhizobial symbiosis. *Rhizobium japonicum* grown in a phosphate rich medium accumulates phosphorus as polyphosphate granules which is sufficient for its maintenance for one to two generations in a low P medium (Cassman *et al.*, 1981; Beck and Munns, 1984).

3.2 *Biotic factors*

A wide range of biological interactions that take place between rhizobia and other soil micro-organisms are reported to cause changes in the size and composition of rhizobial population in soil.

3.2.1 *Competition*

Competition for limited nutrients is generally put forward as a major factor for the disappearance of microorganisms inoculated into the soil. The ideal rhizobial strain for inoculation should have the ability to persist and grow in the soil or around plant roots. The highly effective strains of rhizobia generally fail to grow under natural soil condition. Difficulties in the establishment of inoculated strains of rhizobia in the soil with established population of native rhizobia may be due to inter-strain competition (Holland, 1970; Bohlool and Schmidt, 1973). The factors responsible for competition are not yet fully understood. The host plant and environmental factors can influence the relative competitiveness of rhizobial strains (Jones and Morley, 1981). The only means of obtaining the desired strains is to screen them under greenhouse conditions. Even the strains which do well under greenhouse conditions fail to perform under field conditions (Dudeja and Khurana, 1988). Furthermore, a rhizobial strain successful in competition with other strains in the presence of one variety of legume may be unsuccessful or perform poorly with the other varieties of the same plant species (Roughley *et al.*, 1976; Khurana *et al.*, 1991). Successful establishment of introduced *Rhizobium* is also greatly influenced by its poor movement in soil. For example, very few cells of *R. trifolii* are able to move even as short as three cm from the point of introduction into the soil (Hamdi 1971). A *Rhizobium* with poor mobility will not be able to generate nodules at a distance from the point of its introduction into the soil with seed. Under these circumstances vertical and lateral roots get infected by indigenous rhizobia which are usually ineffective.

3.2.2 *Antagonism*

Microbial antagonism by bacteria and fungi has been demonstrated against rhizobia under cultural conditions as well as in the sterile soil (Holland and Parker, 1966; Pugashetti *et al.*, 1982; Habte and Barrion, 1984) although the basis of the antagonism is usually unknown. Biologically formed toxins may also influence the establishment of rhizobia. For example, extracts of soils supporting clovers that nodulated poorly were toxic to *R. trifolii* growing in the laboratory media and it has been postulated that such inhibitors restrict the development of rhizobia in the soil (Chatel and Parker, 1972).

Many strains of rhizobia produce bacteriocins which inhibit the growth of the related strains. Bacteriocins are a diverse group of substances, composed largely of proteins which are produced by many bacteria and act to kill certain related strains. The influence of bacteriocins in the soil has been examined only in a few instances. Schwinghamer and Brockwell (1978) observed that a sensitive strain of *R. trifolii* was supressed when it was added to peat alongwith a bacteriocin producer in a non-sterile soil, the proportion of nodules formed by an effective bacteriocin resistant *R. trifolii* strain was greater when inoculated with a bacteriocin producing strain, than in the absence of a bacteriocin producer (Hodgson *et al.*, 1985). It seems that bacteriocidal function of bacteriocins could be significant in providing the bacteriocinogenic strain with some competitive advantage over other rhizobia.

3.2.3 *Predation and Parasitism*

Predation seems to be a significant factor in reducing bacterial population in the soil. Many protozoa, known to be predatory on rhizobia, appreciably reduce their density without altogether eliminating them (Danso *et al.*, 1975; Habte and Alexander, 1977).

The predatory bacterium *Bdellovibrio bacteriovorus* parasitizes certain rhizobia and cause their lysis. However, they have a negligible role in control of rhizobial population in soil (Keya and Alexander, 1975).

Bacteriophages have also been implicated as agents in control of rhizobial populations in the soil. Bacteriophages specific for rhizobia are of common occurrence in soils. Exposure of sensitive rhizobia to phages leads to selection of variants with reduced sensitivity to phage lysis. Evans *et al.* (1979) reported that phage reduced the rhizoplane population of *R. trifolii* and induced the growth of variant strains which were less susceptible to phage and usually ineffective in nitrogen fixation. Phages are present in some areas in relatively high numbers and under this situation phage resistant mutants have a competitive advantage (Barnet, 1980; Lawson and Barnet, 1987).

Conclusions

The study of ecology of soil microorganisms including rhizobia is fraught with many difficulties. Many indirect and direct approaches have been followed to study ecology of rhizobia. Many of the techniques followed have inherent limitations and do not predict the *in situ* situation. Some of these techniques allow to monitor the population dynamics of inoculated rhizobia in soil, rhizosphere and intact nodules. The refinement of techniques can go a long way in predicting the status of rhizobia in soils and the interactions of various abiotic and biotic stresses on the survival of rhizobia. Accordingly strains resistant to various abiotic and biotic stresses can be obtained which can help to improve nodulation and nitrogen fixation. The phenomenon of strain competition is poorly understood. Highly efficient strains of rhizobia often fail to compete with the native ones for nodulation. The understanding of inter strain competition can help to derive efficient rhizobial strains which are also capable of competing with native strains for nodule occupancy.

REFERENCES

Alexander, M. 1985. Ecological constraints on nitrogen fixation in agricultural ecosystems. *Adv. Microbial Ecol.* **8** : 163-183.

Amarger, N. 1981. Competition for nodule formation between effective and ineffective strains of *Rhizobium meliloti*. *Soil Biol. Biochem.* **14** : 489-493.

Ayanaba, A., Weiland, K.D. and Zablotowicz, R.M. 1986. Evaluation of diverse antisera, conjugates, and support media for detecting *Bradyrhizobium japonicum* by indirect enzyme linked immunosorbent assay. *Appl. Environ. Microbiol.* **52** : 1132-1138.

Baldani, J.I. and Weaver, R.W. 1992. Survival of clover rhizobia and their plasmid cured derivatives in soil under heat and drought stress. *Soil Biol. Biochem.* **24** : 737-742.

Bio, Y., Lies, D.P., Fu, H. Roberts, G. 1991. An improved *Tn7* based system for the single copy insertion of cloned genes into chromosomes of gram negative bacteria. *Gene* **109** : 167-168.

Barnet, Y.M. 1980. The effect of rhizobiophages on populations of *Rhizobium trifolii* in the root zone of clover plants. *Can J. Microbiol.* **26** : 572-576.

Barnet, Y.M., Trinick, M.J., Date, R.A. and Roughley, R.J. 1988. Ecology of the root nodule bacteria. p. 1-22. In : *Microbiology in Action* (Eds W.G. Murrell and I.R. Kennedy), Research Studies Press, Letchworth.

Beck, D.P. and Munns, D.N. 1984. Phosphate nutrition of *Rhizobium* spp. *Appl. Environ. Microbiol.* **44** : 278-282.

Bezdicek, D.F. 1972. Effect of soil factors on the distribution of *Rhizobium japonicum* serogroups. *Soil Sci. Soc. Am. Proc.* **36** : 305-307.

Bhardwaj, K.K.R. 1974. Growth and symbiotic effectiveness of indigenous *rhizobium* species of a saline-alkaline soil. *Proc. Indian Natl. Sci. Acad.* **40** : 540-543.

Bhardwaj, K.K.R. 1975. Survival and symbiotic characteristics of *Rhizobium* in saline alkali soils. *Plant Soil* **43** : 377-385.

Bohlool, B.B. and Schmidt, E.L. 1968. Non specific staining : its control in immuno-fluorescence examination of soil. *Science* **162** : 1012-1014.

Bohlool, B.B. and Schmidt, E.L. 1973. Persistence and competition aspects of *Rhizobium japonicum* observed in soil by immunofluorescence microscopy. *Soil Sci. Soc. Am. Proc.* **37** : 561-564.

Bohlool, B.B. and Schmidt, E.L. 1980. The immunofluorescence approach in microbial ecology. *Adv. Microbial. Ecol.* **4** : 203-241.

Brockwell J. 1963. Accuracy of a plant infection technique for counting populations of *Rhizobium trifolii. Appl. Microbiol.* **11** : 377-383.

Brockwell, J., Schwinghamer, E.A. and Gault, R.R. 1977. Ecological studies of root nodule bacterial introduced into field environment. V. Critical examination of the stability of antigenic and streptomycin resistance markers for strain identification. *Soil Biol. Biochem.* **9** : 19-24.

Bromfield, E.S.P. and Jones, D.J. 1979. The competitive ability and symbiotic effectiveness of doubly labelled antibiotic resistant mutants of *Rhizobium trifolii. Ann Appl. Biol.* **91** : 211-219.

Bromfield, E.S.P., Sinha, I.B. and Wolynetz, M.S. 1986. Influence of location, host cultivar and inoculation on the composition of naturalized populations of *Rhizobium meliloti* in *Medicago sativa* nodules. *Appl. Environ. Microbiol.* **51** : 1077-1084.

Bushby, H.V.A. 1981. Changes in the number of antibiotic resistant rhizobia in the soil and rhizosphere of field grown *Vigna mungo* cv. Regur. *Soil Biol. Biochem.* **13** : 241-245.

Bushby, H.V.A. and Marshall, K.C. 1977. Some factors affecting the survival of root nodule bacteria on desiccation. *Soil Biol. Biochem.* **9** : 143-147.

Chao, W.L. and Alexander, M. 1982. Influence of soil characteristics on the survival of *Rhizobium* in soils undergoing drying. *Soil Sci. Soc. Am. J.* **60** : 949-952.

Cassman, K.G., Munns, D.N. and Beck, D.P. 1981. Phosphorus nutrition of *Rhizobium japonicum* : Strain differences in phosphate storage and utilization. *Soil Sci. Soc. Am. J.* **45** : 517-520.

Chatel, D.L. and Parker, C.A. 1972. Inhibition of rhizobia by toxic soil-water extracts. *Soil Biol. Biochem.* **4** : 289-294.

Chatel, D.L. and Parker, C.A. 1973. Survival of field grown rhizobia over the dry summer period in Western Australia. *Soil Biol. Biochem.* **5** : 415-423.

Chen, H., Gartner, E. and Rolfe, B.G. 1993. Involvement of genes on a megaplasmid in the acid tolerant phenotype of *Rhizobium leguminosarum* biovar *trifolii* : *Appl. Environ. Microbiol.* **59** : 1058-1064.

Cloutier, J., Prevost, D., Nadeau, P. and Antoun, H. 1992. Heat and cold shock protein synthesis in arctic and temperate strains of rhizobia. *Appl. Environ. Microbiol.* **58** : 2045-2853.

Cooper, J.E., Bjourson, A.J. and Thompson, J.A. 1987. Identification of *Lotus* rhizobia by direct DNA hybridization of crushed root nodules. *Appl. Environ. Microbiol.* **53** : 1705-1707.

Danso, S.K.A. and Alexander, M. 1974. Survival of two strains of *Rhizobium* in soil. *Soil Sci. Soc. Am. Proc.* **38** : 86-89.

Danso, S.K.A., Keya, S.O. and Alexander, M. 1975. Protozoa and the decline of *Rhizobium* populations added to soil. *Can. J. Microbiol.* **21** : 884-895.

Dev Vrat. 1992. Construction of genetically marked strains of *Rhizobium* sp. (*Cicer*). M.Sc. Thesis, CCS Haryana Agricultural University, Hisar.

Dudeja, S.S. and Khurana, A.L. 1988. Survival and competitiveness of *Bradyrhizobium* sp. in the rhizosphere of pigeonpea (*Cajanus cajan*). *Biol. Fertil. Soils.* **7** : 63-67.

Dudeja, S.S. and Khurana, A.L. 1989. Persistence of *Bradyrhizobium* sp. (*Cajanus*) in a sandy loam. *Soil Biol. Biochem.* **21** : 709-713.

Dudman, W.F. and Brockwell, J. 1988. Ecological studies of root nodule bacteria introduced into field environments. I. A survey of field performance of clover inoculants by gel immune diffusion serology. *Aust. J. Agric. Res.* **19** : 739-747.

Dughri, M.H. and Bottomley, P.J. 1983. Complementary methodologies to deliniate the composition of *Rhizobium trifolii* population in root nodules. *Soil Sci. Soc. Am. J.* **47** : 939-945.

Elkan, G.H. and Kuykendall, R. 1982. Carbohydrate metabolism. p. 147-166. In : *Nitrogen Fixation* vol. 2 : *Rhizobium* (Ed. W.J. Broughton). Clarendon Press, Oxford.

Elsheikh, E.A.E. and Wood, M. 1990. Salt effects on survival and multiplication of chickpea and soybean rhizobia. *Soil Biol. Biochem.* **22** : 343-347.

Ethiraj, S., Sharma, H.R. and Vyas, S.R. 1972. Studies on salt tolerance of rhizobia. *Indian J. Microbiol.* **12** : 87-92.

Evans, J., Barnet, Y.M. and Vincent, J.M. 1979. Effect of bacteriophages in the colonization and nodulation of clover roots by a strain of *Rhizobium trifolii. Can. J. Microbiol.* **25** : 968-973.

Flis, S.E., Glenn, A.R. and Dilworth, M.J. 1993. The interaction between aluminium and root nodule bacteria - Review. *Soil Biol. Biochem.* **25** : 403-418.

Ford, S. and Olson, B.H. 1988. Methods for detecting genetically engineered microorganisms in the environments. *Adv. Microbial. Ecol.* **10** : 45-79.

Fougere, F. and Rudulier, D.L. 1990. Glycine betaine biosynthesis and catabolism in the bacteriods of *Rhizobium meliloti* : Effect of salt stress. *J. Gen. Microbiol.* **136** : 2503-2510.

Fujihara, S. and Yoneyama, T. 1993. Effects of pH and osmotic stress on cellular polyamine contents in soybean rhizobia *Rhizobium fredii* P200 and *Bradyrhizobium japonicum* A1017. *Appl. Environ. Microbiol.* **59** : 1104-1109.

Graham, P.H. 1963. Antibiotic sensitivities of the root nodule bacteria. *Aust. J. Biol. Sci.* **16** : 557-559.

Graham, P.H. 1992. Stress tolerance in *Rhizobium* and *Bradyrhizobium* and nodulation under adverse soil conditions. *Can J. Microbiol.* **38** : 675-684.

Graham, P.H. and Parker, C.A. 1964. Diagnostic features in the characterization of root nodule bacteria of legumes. *Plant Soil.* **20** : 383-396.

Graham, P.H., Viteri, S.E., Mackie, F., Vangoes, A.T. and Palacios, A. 1982. Variation in acid soil tolerance among strains of *Rhizobium phaseoli. Field Crops Research.* **5** : 121-127.

Habte, M. and Alexander, M. 1977. Further evidence for the regulation of bacterial population by soil protozoa. *Arch. Microbiol.* **113** : 181-183.

Habte, M. and Barrion, M. 1984. Interaction of *Rhizobium* sp. with toxin producing fungus in culture medium and in a tropical soil. *Appl. Environ. Microbiol.* **47** : 1080-1083.

Hamdi, Y.A. 1971. Soil water tension and the movement of rhizobia. *Soil Biol. Biochem.* **3** : 121-126.

Harrison, S.P., Jones, D.G. and Young, J.P.W. 1989. *Rhizobium* population genetics : Genetic variation within and between populations from diverse locations. *J. Gen Microbiol.* **135** : 1061-1069.

Hartel, P.G. and Alexander, M. 1984. Temperature and desiccation tolerance of cowpea rhizobia. *Can. J. Microbiol.* **30** : 820-823.

Hodgson, A.L.M. and Roberts, W.P. 1983. DNA colony hybridization to identify *Rhizobium* strains. *J. Gen. Microbiol.* **129** : 207-212.

Hodgson, A.L.M., Roberts, W.P. and Waid, J.S. 1985. Regulated nodulation of *Trifolium subterraneum* inoculated with bacteriocin producing strains of *Rhizobium trifolii. Soil Biol. Biochem.* **17** : 475-478.

Holland, A.A. 1970. Competition between soil and seed borne *Rhizobium trifolii* in nodulation of introduced *Trifolium subterraneum. Plant Soil.* **32** : 293-302.

Holland, A.A. and Parker, C.A. 1966. Studies on microbial antagonism in the establishment of clover pasture. II. The effect of saprophytic soil fungi upon *Rhizobium trifolii* and the growth of subterranean clover. *Plant Soil.* **25** : 329-340.

Hua, S.T., Tsai, V.Y., Lichens, G.M. and Noma, A.T. 1983. Accumulation of amino acids in *Rhizobium* sp. strain WE 1001 in response to sodium chloride salinity. *Appl. Environ. Microbiol.* **44** : 135-140.

Hughes, D.Q. and Vincent, J.M. 1942. Serological studies of the root nodule bacteria. III. Test of neighbouring strains of the same species. *Proc. Linnean Soc. New South Wales.* **67** : 142-152.

Humphery, B.A. and Vincent, J.M. 1975. Specific and shared antigens in strains of *Rhizobium meliloti. Microbios.* **13** : 71-76.

Jones, D.G. and Bromfield, E.S.P. 1978. A study of the competitive ability of streptomycin and spectinomycin mutants of *Rhizobium trifolii* using various marker techniques. *Ann. Appl. Biol.* **88** : 448-450.

Jones, D.G. and Morley, S.J. 1981. Marker techniques, competitive ability and host strain preference in the clover - *Rhizobium* symbiosis. p. 429-439. In : *Current Perspectives in Nitrogen Fixation* (Eds. A.H. Gibson and W.E. Newton), Australian Academy of Science, Canberra.

Keya, S.O. and Alexander, M. 1975. Factors affecting growth of *Bdellovibrio* on *Rhizobium. Arch. Microbiol.* **103** : 37-43.

Keyser, H.H. and Munns, D.N. 1979. Tolerance of rhizobia to acidity, aluminium, and phosphate. *Soil Sci. Soc. Am. J.* **43** : 519-523.

Khurana, A.L., Sharma, P.K. and Dudeja, S.S. 1991. Influence of host, moisture and native rhizobial population on nodule occupancy in chickpea (*Cicer arietinum*). *Zbl. Mikrobiol.* **146** : 137-141.

Kingsley, M.T. and Bohlool, B.B. 1981. Release of *Rhizobium* spp. from tropical soils and recovery for immunofluorescence enumeration. *Appl. Environ. Microbiol.* **42** : 241-248.

Kingsley, M.T. and Bohlool, B.B. 1983. Characterization of *Rhizobium* sp. (*Cicer arietinum* L.) by immuno fluorescence, immunodiffusion and intrinsic antibiotic resistance. *Can. J. Microbiol.* **29** : 518-526.

Kinkle, B.K. and Schmidt, E.L. 1992. Stability of a monoclonal antibody determinant in soil populations of *Bradyrhizobium japonicum. Soil Biol. Biochem.* **24** : 819-820.

Kumar, V., Bala, R. and Nafees, G. 1990. Effect of salinity, alkalinity and acidic conditions on growth of cowpea *Rhizobium. Acta. Botanica. Indica.* **18** : 290-292.

Lawson, R.A. and Barnet, Y.M. 1987. Environmental factors influencing number of *Rhizobium leguminosarum* biovar. *trifolii* and its bacteriophages in two field soils. *Appl. Environ. Microbiol.* **53** : 1125-1131.

Lesley, S.M. 1982. A bacteriphage typing system for *Rhizobium meliloti. Can. J. Microbiol.* **28** : 180-189.

Lindstrom, K. 1989. *Rhizobium galegae,* a new species of legume root nodule bacteria. *Int. J. Syst. Bacteriol.* **39** : 365-367.

Lindstrom, K., Lipsanen, P. and Kaijalainen, S. 1990. Stability of markers used for identification of two *Rhizobium galegae* inoculant strains after five years in the field. *Appl. Environ. Microbiol.* **56** : 444-450.

Long S.R. 1989. Legume - *Rhizobium* symbiosis : Life together underground. *Cell* **56** : 203-214.

Lowendorf, H.S. and Alexander, M. 1983. Identification of *Rhizobium phaseoli* strains that are tolerant or sensitive to soil acidity. *Appl. Environ. Microbiol.* **45** : 737-742.

Marshall, K.C. 1964. Survival of root-nodule bacteria in dry soils exposed to high temperatures. *Aust J. Agric. Res.* **15** : 273-281.

Martensson, A.M., Gustafsson, J.G. and Ljunggren, H. 1984. A modified, highly sensitive enzyme-linked immunosorbent assay (ELISA) for *Rhizobium meliloti* strain identification. *J. Gen. Microbiol.* **130** : 247-253.

May, S.N. and Bohlool, B.B. 1983. Competition among *Rhizobium leguminosarum* strains for nodulation of lentils (*Lens esculenta*). *Appl. Environ. Microbiol.* **44** : 960-965.

McLoughlin, T.J., Bordeleau, L.M. and Dunican, L.K. 1984. Competition studies with *Rhizobium trifolii* in a field experiment. *J. Appl. Bacteriol.* **56** : 131-135.

Moawad, H.A., Ellis, W.R. and Schmidt, E.L. 1984. Rhizosphere response as factor in competition among three serogroups of indigenous *Rhizobium japonicum* for nodulation of field grown soybeans. *Appl. Environ. Microbiol.* **47** : 607-612.

Morely, S.J. and Jones, D.G. 1980. A note on a highly sensitive modified ELISA technique for *Rhizobium* strain identification. *J. Appl. Bacteriol.* **49** : 103-109.

Munns, D.N. and Keyser, H.H. 1981. Response of *Rhizobium* strains to acid and aluminium stress. *Soil Biol. Biochem.* **13** : 115-118.

Noel, K.D. and Brill, W.J. 1980. Diversity and dynamics of indigenous *Rhizobium japonicum* populations. *Appl. Environ. Microbiol.* **42** : 255-258.

Obaton, M. 1973. The use of spontaneous antibiotic resistant mutants for ecological study of *Rhizobium. Bull. Ecol. Res. Commun.* **17** : 170-171.

Olsen, P.E. and Rice, W.A. 1989. *Rhizobium* strain identification and quantification in commercial inoculants by immunoblot analysis. *Appl. Environ. Microbiol.* **55** : 520-522.

O'Hara, G.W., Gross, T.J., Dilworth, M.J. and Glenn, A.R. 1989. Maintenance of intracellular pH and acid tolerance in *Rhizobium meliloti. Appl. Environ. Microbiol.* **55** : 1870-1876.

Osa-Afiana, L.D. and Alexander, M. 1979. Effect of moisture on survival of *Rhizobium* in soil. *Soil Sci. Soc. Am. J.* **42** : 925-930.

Osa-Afiana, L.D. and Alexander, M. 1982. Clays and the survival of *Rhizobium* in soil during desiccation. *Soil Sci. Soc. Am. J.* **46** : 285-288.

Parker, C.A., Trinick, M.J. and Chatel, D.L. 1977. Rhizobia as soil and Rhizosphere inhabitants p. 311-352. In : *A Treatise on Dinitrogen Fixation* (Eds. R.W.F. Hardy and A.H. Gibson), Wiley-Interscience, New York.

Pena-Cabriales, J.J. and Alexander, M. 1979. Survival of *Rhizobium* in soils undergoing drying. *Soil Sci. Soc. Am. J.* **43** : 962-966.

Pena-Cabriales, J.J. and Alexander, M. 1983. Growth of *Rhizobium* in soil amended with organic matter. *Soil Sci. Soc. Am. J.* **47** : 241-245.

Pugashetti, B.K., Angle, J.S. and Wagner, G.H. 1982. Soil microorganisms antagonistic towards *Rhizobium japonicum. Soil Biol. Biochem.* **14** : 45-49.

Rai, R. 1983. The salt tolerance of *Rhizobium* strains and lentil genotypes and the effects of salinity on aspects of symbiotic nitrogen fixation. *J. Agric. Sci.* **100** : 81-89.

Ramirez, C. and Alexander, M. 1980. Evidence suggesting protozoan predation in *Rhizobium* associated with germinating seeds and in the rhizosphere of beans (*Phaseolus vulgaris* L). *Appl. Environ. Microbiol.* **40** : 492-499.

Reyes, B.G. and Schmidt, E.L. 1979. Population densities of *Rhizobium japonicum* strain 123 estimated directly in soil and rhizosphere. *Appl. Environ. Microbiol.* **37** : 854-858.

Rice, W.A., Olsen, P.E. and Page, W.J. 1984. ELISA evaluation of the competitive abilities of two *Rhizobium meliloti* strains. *Can J. Microbiol.* **30** : 1187-1190.

Robert, F.M. and Schmidt, E.L. 1983. Population changes and persistence of *Rhizobium phaseoli* in soil and rhizospheres. *Appl. Environ. Microbiol.* **45** : 550-556.

Roitt, I.M. 1984. *Essential Immunology.* Blackwell, London.

Roughley, R.J., Blowes, W.M. and Herridge, D.F. 1976. Nodulation of *Trifolium subterraneum* by introduced rhizobia in competition with naturalized strains. *Soil Biol. Biochem.* **8** : 403-407.

Sano, A. and Lindstrom, K. 1992. Detection of *Rhizobium galegae* from one gram soil samples by non-radioactive DNA-DNA hybridization. *Soil Biol. Biochem.* **24** : 969-977.

Sadofasky, M.J., Gregan, P.B. and Keyser, H.H. 1989. DNA hybridization probe for use in determining restricted nodulation among *Bradyrhizobium japonicum* sero-cluster 123 field isolates. *Appl. Environ. Microbiol.* **56** : 1768-1774.

Schwinghamer, E.A. and Dudman, W.F. 1973. Evaluation of spectinomycin resistance as a marker for ecological studies with *Rhizobium* spp. *J. Appl. Bacteriol.* **36** : 263-272.

Schwinghamer, E.A. and Brockwell, J. 1978. Competitive advantage of bacteriocin and phage producing strains of *Rhizobium trifolii* in mixed cluture. *Soil Biol. Biochem.* **10** : 383-387.

Selvaraj, J. and Iyer, V.N. 1983. Suicide plasmid vehicle for insertion mutagenesis in *Rhizobium meliloti* and related bacteria. *J. Bacteriol.* **156** : 1292-1300.

Sharma, P.K., Anand, R.C. and Lakshminarayana, K. 1991. Construction of *Tn5* tagged mutants of *Rhizobium* sp. (*Cicer*) for ecological studies. *Soil Biol. Biochem.* **23** : 881-885.

Simon, R., Pfiefer, U. and Puhler, A. 1983. A broad host range mobilization system for *in vivo* genetic engineering : Transposon mutagenesis in gram negative bacteria. *Biotechnol.* **1** : 784-791.

Singleton, P.W., El-Swaify, S.A. and Bohloo, B.B. 1982. Effect of salinity on *Rhizobium* growth and survival. *Appl. Environ. Microbiol.* **44** : 884-890.

Steinborn, J. and Roughley, R.J. 1975. Toxicity of sodium and chloride ions to *Rhizobium* spp. in broth and peat culture. *J. Appl. Bacteriol.* **39** : 133-138.

Staniewski, R. 1970a. Relationship among different *Rhizobium* strains determined by phage lysis. *Acta Microbiol. Pol. Ser. B.* **2** : 3-12.

Staniewski, R. 1970b. Typing of *Rhizobium* by phages. *Can. J. Microbiol.* **16** : 1003-1009.

Thurman, N.P. and Bromfield, E.S.P. 1988. Effect of variation within and between *Medicago* and *Melilotus* species on the composition and dynamics of indigenous population of *Rhizobium meliloti. Soil Biol. Biochem.* **20** : 31-38.

Tsai, Y.L. and Olson, B.H. 1991. Rapid method for direct extraction of DNA from soil and sediments. *Appl. Environ. Microbiol.* **57** : 1070-1074.

Van Rensburg, J.H. and Stridjom, B.W. 1980. Survival of fast and slow growing *Rhizobium* spp. under conditions of relatively mild desiccation. *Soil Biol. Biochem.* **12** : 353-356.

Velez, D., MacMillan, J.D. and Miller, L. 1988. production and use of monoclonal antibodies for identification of *Bradyrhizobium japonicum* strains. *Can. J. Microbiol.* **34** : 88-92.

Wedlock, D.N. and Jarvis, B.D.W. 1986. DNA homologies between *Rhizobium* and *Bradyrhizobium* species. *Int. J. Syst. Bacteriol.* **36** : 550-558.

Wener, D., Streit, W. and Kaasch, K. 1992. Nodulation competition of *Rhizobium leguminosarum* and *Rhizobium tropici* by gus fusion. *Biol. Fertil. Soils.* **14** : 140-144.

Wheatcroft, R. and Watson, R.J. 1988. A positive strain identification method for *Rhizobium meliloti. Appl. Environ. Microbiol.* **54** : 574-576.

Wilkins, J. 1967. Effects of high temperatures on certain root-nodule bacteria. *Aust. J. Agric. Res.* **18** : 299-304.

Wilson, K.J., Giller, K.E. and Jefferson, R.A. 1990. β-glucosonidase (*Gus*) operon fusions as a tool for studying plant microbe interactions pp. 226-229. In : *Advances in Molecular Genetics of Plant Microbe Interaction.* (Eds. Hennecke, H. and Verma, D.P.S.). Kluwer Academic Publishers, Dordrecht.

Wright, S.F. 1990. Production and epitope analysis of monoclonal antibodies against a *Rhizobium leguminosarum* biovar *trifolii* strain. *Appl. Environ. Microbiol.* **66** : 2262-2264.

Yap, S.F. and Lim, S.T. 1983. Response of Rhizobium sp. UMKL 20 to sodium chloride stress. *Arch. Microbiol.* **135** : 224-228.

Yelton, M.M., Yang, S.S., Edie, S.A. and Lim, S.T. 1983. Characterization of an effective salt tolerant fast growing strain of *Rhizobium japonicum. J. Gen. Microbiol.* **129** : 1537-1547.

Young, J.P.W. 1985. *Rhizobium* population genetics : Enzyme polymorphism in isolates from peas, clover, beans and lucerne grown at the same site. *J. Gen. Microbiol.* **131** : 2394-2408.

Chapter 3

Taxo-Ecological Studies of Coprophilous Fungi – A Review

Rajee N. Kumar, Rama Pasricha, Narendra Singh* AND K.G. Mukerji

Department of Botany, University of Delhi-110 007, India

Abstract. The review deals with taxonomic and ecological aspects of the fungal community development on dung and other cuprous substrates. Inspite of numerous taxonomic reports, compilation of complete coprophilous fungal flora is far from complete. Fungal succession and their substrate utilisation pattern in the light of ''Nutritional hypothesis'' has been discussed. Considering the potentiality of fungi as decomposer, the review suggests the systematic exploration of coprophilous fungal community for solving many ecological problems.

Key words : *Coprophilous Fungi, Taxo-ecology, succession, Decomposition.*

Introduction

Dung of animals forms a unique habitat since it is rich in nutrients and thus supports a definite community by providing a complex of niches to various organisms viz., bacteria, actinomycetes, myxobacteria, fungi, protozoa, molluscs and nematodes. Dung holds a fascination especially for mycologists because of the fact that it has a complex nutrient spectrum by virtue of which it supports a rich mycoflora. Of all the various microorganisms colonizing this substrate, fungi received the attention first yet the originators of its details are difficult to trace. By critical examination of the literature it was found that most of the work has been done on the taxonomy of these fungi and not their ecology. The lack of interest in this field can be attributed to a number of reasons; the complex nature of the substrate, the physical and chemical nature, lack of taxonomic revision, nomenclature problems, wrong determination of fungi colonizing the dung, in addition to other ecological factors (Lodha, 1974).

These studies are helpful in screening out the decomposers which are responsible for active

* 772 Stinchomb Drive, Columbus, Ohio 43202, U.S.A.

turnover of minerals from herbivore faeces in grassland ecosystem (Angel and Wicklow, 1974; 1975; Waksman, 1938). By looking at the increasing fertilizer cost, studies on dung decomposition and release of nutrients from this substrate becomes important with a view that organic farming can be practiced in India which can replace chemical fertilizers to a large extent.

Taxonomical aspects

The earliest record of taxonomic work on dung fungi is available from Berkeley, (1836). Later, the work of Griffth (1901) on North-American Sordariaceae and classical studies of Massee and Salmon (1901, 1902) on English coprophilous fungi set the pace for present century to further explore and elaborate systematic mycology. The work of coprophilous taxonomy progressed further with important contribution of Milliard (1903a, b), Buller (1909), Gibbs (1909), Scchmidt (1912), Brown and Graff (1913) and Ingold (1928). Extensive work has also been done by Cain (1934, 1950, 1956a, b, 1957, 1961a, b, c, 1962), Moreau (1953) and Fernier (1954) on the taxonomy of coprophilous sphaeriales.

During the 1960-1970's lot of work has been reported on coprophilous ascomycets from all over the world (Lundqvist, 1960, 1964a, b, 1973, 1974; Toth, 1963, 1965, 1967; Brummeler, 1967; Eckbald, 1968; Richardson and Wattling, 1968, 1969; Mirza and Cain, 1969; Maloch and Cain, 1970, 1971 and 1972; Furuya and Udagawa, 1972a, b, 1973, 1977a, b; Krug and Cain, 1972; Minoura and Yamada, 1976; Jeng and Cain, 1977; Jeng and Krug, 1977a, b; Liou and Chingchen, 1977; Aas, 1978; Garcia and Noeni, 1978).

There are also many contributions from India in this field (Ajrekar and Dharmarejulu, 1931; Mahju, 1933; Ginai, 1936; Lodha, 1964, 1971a; Subramanian and Lodha, 1964a, b, 1968; Kar and Pal, 1968; Mukerji, 1968a, b, 1969, 1970 1978; Saxena et al., 1969; Saxena and Mukerji, 1970a, b, c, d, 1972, 1973a, b, c, 1974; Kelkar, 1971; Behera et al., 1973; Iyer et al.,1973; Narendra, 1973, 1974; Waraitch, 1976; Pathak and Aggarwal, 1977a, b; Kaur and Chauhan, 1991). Mukerji and Singh (1974), Singh (1978), Von Arx *et al.* (1978a, b), Singh and Mukerji (1979) have described many species of coprophilous Ascomycetes and one new genus *Leucosphaera* from Delhi. Nannenga Bremekamp et al. (1979) have described two new coprophilous Myxomycetes from India.

Recently, much work has been done on Mucorales, Hyphomycetes and other groups (Mehrotra et al., 1971; Richardson and Leadbeater, 1972; Upadhyay, 1973; Hesseltine, 1975; Subramanian and Chandrasekhara, 1979; Rajeevalochana and Mukerji, 1980; Bawcutt, 1983; Menon and Mukerji, 1983; Anson et al., 1985; Buffin and Hennebert, 1985; Pasricha and Mukerji, 1987; Taj-Al-Deen et al., 1990; Ebersohn and Eicker, 1992).

There is ample scope for further exploration of natural fungal flora from dung. Everyday, lot of new reports are coming from various parts of the world (Barrasa and Moreno, 1984; Locquin-Linard, 1984, 1990; Solans, 1984; Kimbrough and Kenneth, 1985; Blackwell and Malloch, 1991; Vallolosera et al., 1991; Dissing, 1992; Jalink, 1992; Lundqvist, 1992). All this has resulted in the discovery of many fungi new to the field of Mycology.

Ecological aspects

During the last few years lot of work has been done on the succession of fungi and their role on decomposition of dung (Watling, 1963; Harper and Webster, 1964; Nicholson et al., 1966; Lodha, 1967, 1974; Mitchell, 1970; Webster, 1970; Wicklow and Moore, 1974; Singh, 1978; Rajeevalochana, 1980; Pasricha, 1986; Ebersohn and Eicker, 1992). Some of the important aspects such as coprophilous substrate, growth characteristics, the occurrence of dung fungi, mode of dispersal and distribution, analysis of various factors that govern their succession and biological interactions among fungal species and with other organisms occurring on dung are discussed.

The presence of a particular coprophilous fungus on faeces depends on (1) the density of spore inoculum on the food given (2) the ability of the fungus to increase its propagule density during passage through the digestive tract, (3) the competitive saprophytic ability of the fungus in faeces of a given nutrient composition, texture and volume, (4) the conditions of incubation (Wicklow and Moore, 1974). The occurrence of mycoflora on a particular dung is more or less regular depending on the nature of food intake of the animal and the macro- and micro-environments where the dung is excreted. The plant material, a herbivore eats is already infested and it is partially decomposed as a result of the microbial activity (Webster, 1956, 1957; Hudson and Webster, 1958; Bonner and Fergus, 1959; Webster and Dix, 1960; Hering, 1965; Hogg and Hudson, 1966; Hudson, 1968). After digestion, the vegetable matter ingested by the animal is finally dropped on the soil in the form of dung or faecal pellets. Fungi are the most important agents of decomposition (Webster, 1970) and these organisms begin their activity simultaneously as the dung is voided by the animal.

Present understanding of the colonization of fungi on dung comes from studies based either on direct incubation of freshly vioded dung (Watling, 1963; Harper and Webster, 1964; Webster, 1970; Richardson, 1972; Angel and Wicklow, 1974, 1975; Lodha, 1974; Wicklow and Moore, 1974; Bell, 1975; Wicklow and Angel, 1980) or from collection of naturally occurring dung at different stages of decay and subsequently incubated in the laboratory for observation (Angel and Wicklow, 1983; Wicklow and Angel, 1983). There are several reports on succession of fungi on undecomposed dung artificially exposed to different natural environments (Nicholson *et al.*, 1966; Wicklow and Hirschfield, 1979; Yocom and Wicklow, 1980; Dickinson et al., 1981).

Coprophilous fungi are represented by large numbers of Mucorales, Gymnoascaceae, Eurotiaceae, Ascobolaceae, Sordariaceae, Chaetomiaceae, Coprinaceae and Bolbitaceae. Of these some are very specific and they rarely occur on other substrates. The true coprophilous fungi, are species of *Pilobolus, Pilairia, Ascobolus* and *Coprinus*.

Coprophilous succession

Succession of fungi on dung represents a model on the basis of which the fungal succession on other substrates has also been explained (Hudson, 1968). It has been observed that there is a regular pattern of succession of certain groups of fungi on dung (Massee and Salmon, 1901, 1902,; Tubaki, 1954; Webster, 1970; Lodha, 1974; Kaur and Chauhan, 1991).

Garrett (1950, 1951 and 1966) introduced the concept of preferential nutrient utilization of the substrate by different fungal groups in relation to succession. Griffin (1960) found that the primary colonizers in succession had high competitive saprophytic ability, rapidly utilizing the less complex nutrients of the substrate. Burges (1939) assigned Mucorales, as a fungal group capable of utilizing readily available sugars. The capacity to utilize cellulose is found amongst Ascomycetes and Deuteromycetes, Whereas, the majority of lignin decomposers are basidiomycetes. In general, succession of fungi has been interpreted as a nutritional sequence based on the known facts of the physiology and nutrition of different groups of fungi and on nutrient content of organic materials at various stages of decay. Based upon these concepts, Garrett (1963) proposed a very generalized scheme of fungal succession which was adapted for coprophilous substrates also and later named as 'Nutritional hypothesis' (Webster, 1970).

According to this hypothesis, the succession of fruiting bodies on dung is attributed to nutritional requirements of these groups of fungi. There is an apparent correlation between the decomposition of progressively complex carbon sources and the taxonomic disposition of different fungal groups. Mucorales which are called as sugar fungi appear in the early phase of the sequence and have a capacity for rapid spore germination and fast growth rate and are not known to decompose cellulose and lignin (Thaysen and Bunker, 1927; Reese and Levinson, 1952). They show a rapid growth on substrate rich in soluble nutrients and disappear as these substances get depleted. At this stage, the Hyphomycetes and Ascomycetes which are generally able to decompose cellulose (Siu and Reese, 1953) begin to dominate the substrate. Lignin is decomposed towards the end of the sequence and therefore, many basidiomycetous species appear late in the succession (Waksman, 1938; Burges, 1939; 1958). Though this succession is well known and widely accepted it has never been experimentally tested and some very important ecological factors have been neglected.

The 'Nutritional hypothesis' stimulated several critical studies on coprophilous succession and its was found to be different from the one proposed earlier (Harper and Webster, 1964; Larsen, 1971; Singh, 1978; Rajeevalochana, 1980; Parsricha, 1986). Mucorales appear within 8 days of incubation of dung. Hyphomycetes appear early in the succession along with some Mucorales, and remain throughout the period of succession. Pezizales follow within 10 days and persist upto 16 days. Ascomycetous fungi belonging to sphaeriales like Chaetomium sp., *Podospora* sp. and Sordaria sp. appear within 8-10 days. Basidiomycetes like *Coprinus* sp. mature within 6 days of incubation (Larsen, 1971). It was observed that Basidiomycetes do not appear late in succession as proposed earlier (Lodha, 1974; Singh, 1978).

There is no clear cut sequence in the appearance of different fungal groups and it is overlapping (Rajeevalochana, 1980; Pasricha, 1986). The sequence of appearance and the species compostion of the mycoflora belonging to the fungal groups that participate in the succession varies from dung to dung, but the general pattern is the same (Fig. 3.1). The variation is because of the difference in time for which the food remains within the gut and the consistency of the dung. The minimum time required by a fungus for the production of mature fruiting bodies must also be an important factor in determining the succession (Harper and Webster, 1964; Lodha, 1974; Nagy and Harrower, 1979)

Succession was explained on the basis of run down of nutrients. It seems that 'Nutritional hypothesis' was based on the biochemical analysis of compost–horse dung manure, during the course of its decomposition (Waksman et al., 1939). However, the relation between the biochemical compostion of dung at various stages of decomposition with sequential appearance of fruiting bodies was never analysed systematically. According to Waksaman (1938) it was observed that the percentage of water soluble organic matter remained high during the entire period of decomposition, whereas, the hemicellulose, cellulose and lignin declined to some extent. It was considered the hemicellulose and cellulose were utilized by Ascomycetes and Hypohomycetes and degraded into simpler water soluble substances from the time the process of decomposition started. The percentage of lignin increased because it was utilized more slowly than other substances during decomposition.

Successional studies have shown that most of the Mucorales continue to grow for the entire period of decomposition (Lodha, 1974; Singh, 1978; Rajeevalochana, 1980). This may be because of either the availability of simple sugars which were broken down from cellulose by the cellulolytic fungi (Ascomycetes and Hyphomycetes) or they can also utilize cellulose and hemicellulose. Studies of Spalding (1963) on *Rhizopus stolonifer* and Rangaswami and Chandrasekaran (1963) on *Cunninghamella elegans* revealed that Mucorales can utilize cellulose also. Later, Harrower and Nagy (1979) have surveyed a number of coprophilous fungi for their ability to grow and sporulate on a wide range of carbon sources. There is no correlation between the observed coprophilous succession and the bility of different species to utilize different nutrients. It was reported that mucorales grew relatively well but sporulated poorly on media containing a single sugar and the carbon source and Ascomycetes showed a poor sporulation on carboxymethyl cellulose. Recently Taj Al-Deen et al. (1990) have worked on cellulolytic activity of some coprophilous fungi. Most of the isolates belonging to different groups were capable of growing and producing clearing zones on cellulose agar media. Fries (1956) reported that Basidiomycetes were capable of growing on substrates other than lignin.

All these reports deviate from the concept of nutritional hypothesis. Similar views were also expressed by Harper and Webster (1964) and Webster (1970). Although the nutritional hypothesis may be inapplicable to the coprophilous succession, it may apply to other fungal successions (Nagy and Harrower, 1979).

Rayner (1978) defined succession as the sequential occupation of the same site by mycelia of different fungi or associations of fungi. But it is very difficult to define succession of fungi on this basis because fungal succession on dung is based on the occurrence of fungal fruiting structures and data on mycelial growth in dung are still lacking in most instances (Dickinson and Underhay, 1977). More information is needed about the vegetative development of fungi on dung. Most of the knowledge concerning succession on dung comes from laboratory studies which may not be

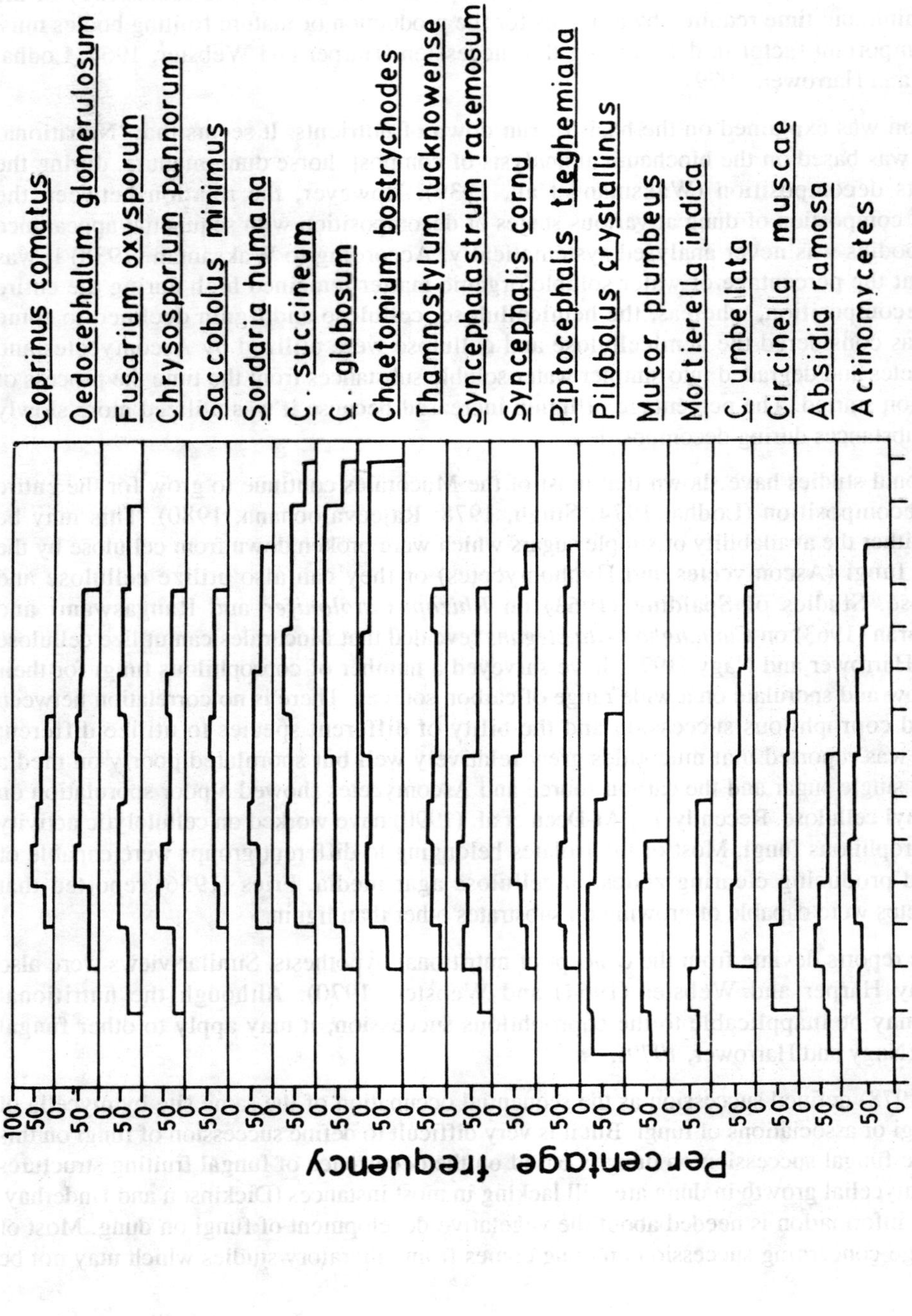

Fig. 3.1 General pattern of succession of some dominant fungi representing different fungal groups.

comparable with those under field conditions. Pattern of fungal species appearing on dung pellets exposed in the filed cannot be duplicated in moist chamber incubation of fresh or field dried faeces because of numerous environmental fluctuations in the field (Angel and Wicklow, 1975, 1983). Therefore, it is important to distinguish between succession occurring during moist chamber incubation and succession pattern observed in nature (Webster, 1970).

Variations were observed in the fungal community composition of dung collected from nature depending on the dung type, age of dung, place of collection, time or interval of exposure and habitat in which dung was incubated (Angel and Wicklow, 1975, 1983; Yocom and Wicklow, 1990). However, communities occurring on same dung type, on similar age of dung and from the same site of collection of dung showed high similarity (Angel and Wicklow, 1983). Yocom and Wicklow (1980) have shown a differential fungal community response to specific habitats associated with dung succession.

The organization of coprophilous fungal communities is influenced by a number of factors like species composition of inoculum (Harper and Webster, 1964), digestive process as it affects survival and germination of spores (Webster, 1970; Wicklow et al., 1980), growth rates and fruiting times of individual species (Harper and Webster, 1964; Wicklow and Hirschfield, 1979) and competitive interactions among fungal species (Ikediugwu and Webster, 1970a, b; Singh and Webster, 1973; Wicklow and Hirschifield, 1979; Wicklow and Yocom, 1981).

Coprous substrate

Faecal substrates provide a complete self contained fungal community. Coprophilous fungal communities have been found very suitable for study of community dynamics (Yocom and Wicklow, 1980; Wicklow, 1981; Wood and Cooke,. 1984). Coprophilous substrates provide a large number of fungal communities having similar fungal inoculum and nutrient composition (Wicklow and Yocom, 1981). Wood and Cooke (1984) designed physically and chemically homogeneous units of faeces (copromes) for studying eco-physiological aspects of community development. Coprophilous fungal communities have several properties which provide an excellent experimental system to measure the impact of biotic and abiotic environmental factors on the composition of fungal community (Wicklow, 1981).

The sequence of organisms involved in decomposition is related to the type of substrate and the environment. The prior presence of associated decomposer organisms in the substrate, presence of nutrient components and the availability of water are important factors in accelerating the process. In this connection, herbivore dung is an unique substrate in combining all the above requisites. There is no other example of naturally colonized substrate of such physical, chemical and microbial uniformity (Wood and Cooke, 1984).

The herbivore dung is the best suited substrate for the coprophilous fungi although these also occur on dung of carnivorous mammals and birds but rarely on the dung of reptiles and amphibians. It indicates that the coprophilous fungi are more adapted to the warm blooded condition and it has been an impoitant facrtor in the evolution of coprophilous habitat (Webster,

1970). Vander Drift and Witkamp (1959) found that the faecal pellets of soil arthropod were more favourable habitat for litter microorganisms than the whole leaf litter.

There are different factors which influence the growth and sporulation of fungi on any dung substrate (i) physical nature of the dung, its consistency and mositure content, (ii) the chemical nature of dung (iii) biological nature of the dung i.e. interactions among the organisms present on it. Among these the nutritional status, water content and humidity are very important. These factors do not however, operate in isolation but conjunction with each other (Webster and Dix, 1960).

Physical nature of dung formed as a result of chewing and mastication during the process of digestion and finally the formation of semi-solid pellets is highly suitable for fungal activity. It provides good aeration for the growth of microorganisms deep in the substratum (Lodha, 1974). Another unsual feature of dung as a substrate for fungal growth is its high pH. Most investigators have given values above 6.5 (Fries, 1956; Harper, 1962). These high values often near neutrality are reflected in the pH optima of growth of coprophilous fungi while many other fungi grow best at low pH (Webster, 1970).

Most other substrates are less favourable for fungal growth than dung, particularly with respect to moisture content (Hudson, 1968). Dung moisture influences the extent of fungal growth and sporulation. Dickinson and Underhay (1977) reported that colonization and sporulation of coprophilous fungi are influenced by moisture content of dung in laboratory or in the field and found that it was also important in controlling the spread of hyphae into the centre of dung pellets. Underhay and Dickinson (1978) observed that in summer due to high temperature the dung pellets lose moisture and thus the rate of growth and sporulation of coprophilous fungi was very slow. There was restricted fungal growth, at water contents, slightly lower than in freshly voided dung (Dickinson et al., 1981). This indicates that moisture content of the dung is a very important factor in controlling the fungal activity. Further, Harrower and Nagy (1979) and Kuthubudeen and Webster (1986) have also reported that moisture content of substrate especially in terms of physiological availability of water is very important factor in coprophilous fungal succession.

Dung contains large quantities of readily available carbohydrates which is major factor favouring the rapidly growing Zygomycetes. Soluble carbohydrates and Zygomycetes are absent initially from most other substrates (Hudson, 1968). The level of soluble carbohydrates varies in different dung. According to an estimte, horse dung contains 4.45 per cent water soluble organic matter (Waksman et al., 1939). The percentage of water soluble organic matter and the time for which they prevail during the period of decomposition is an important factor in determining the colonization of fungi especially mucorales. The other constituents of dung like hemicellulose, cellulose and lignin are all important in bringing about the fungal colonization (Pasricha, 1986).

Dung contains about 4 per cent nitrogen which is higher than any other substrate (Lambourne and Reardon, 1962). This is due to the presence of bacteria and protozoa in the intestine which are subsequently killed and digested. Many microorganisms in the intestine are capable of synthesizing water soluble vitamins and growth factors (Webster, 1970). Coprogen, a factor required by *Pilobolus* and other fungi for growth and reproduction is found only in dung (Hesseltine et al., 1953; Pidacks et al., 1953). It is produced by some species of fungi, actinomycetes and bacteria in the intestine as well as in dung (Hesseltine *et al.*, 1953; Lyr, 1953; Neilands, 1957).

Substrate relartionships

Dung surface carries a complex microflora with much interaction between various colonizers (Harper and Webster, 1964; Webster, 1970). The fungi interact with each other and with other organisms inhabiting dung competing for space and nutrients by stimulating or inhibiting spore germination or by entering into some form of antagonism or even by altering the physical structure of dung (Hudson, 1968). Harper (1962) failed to detact germinated spores in dung of rabbit fed with food heavily loaded with spores of coprophilous fungi. He also showed that the latent period for germination at room temperature following 5 h of incubation at 37°C in alkaline pancreatine varied between 4-12h for the foungi studied. This has been related to the time the food is normally held within the gut; for rabbit 4-6 h (Mangold, 1950; Brandt and Thacker, 1958) for cattle 25-40h (Sineshckekow, 1965) or even 72h (Duke, 1942). spore germination of many coprophilous fungi is generally activated by passing through the alimentary canal of herbivorous animals. Moringa *et al.* (1981) indicated that spore germination of *Pilobolus crystallinus* was stimulated by both culture filtrates of *Streptomyces* spp. and *Bacillus* spp. which were isolated from deer dung. The competition is important in limiting the duration and intensity of fruiting of particular species (Harper and Webster, 1964). Lodha (1971) observed that presence of bacteria in culture increased fruiting of *Podospora decipiens.* Page (1959) demonstrated that *Pilobolus kleinii* fruited better in the presence of *Mucor Plumbeus* because of the production of a stimulating substance, ammonia. *Cladosporium herbarum* and *Penicillium expansum* were also found to produce ammonia which stimulates *P. kleinii* (Singh and Webster, 1972). Harper and Webster (1964) found that bacteria reduced the yield of sporangia of *Pilobolus* sp. on the fifth day after incubation but after 14 days of incubation the bacterial pellets produced higher yields of sporangia than the non-bacterial pellets.

Harper and Webster (1964) found that there are interspecific fungal competition of antagonism resulting in premature cessation of fruiting of certain coprophilous fungal species. They found that *Coprinus hepteremus* growing on rabbit dung was antagonistic to other coprophilous species like *Pilobolus crystallinus, Ascobolus crenulatus, Pilaria anamala.* Its presence caused a drastic reduction in sporangial production in *Pilobolus* and apothecia formation in *Ascobolus.* Ikediugwu and Webster (1970 a, b) also confirmed the findings and they showed that suppression of fruiting in *Pilobolus crystallinus* and *Ascobolus crenulatus* was found to be by hyphal interference. Hyphal interference was also noted in antagonism operated by *Peniophora gigantea* against *Heterobasidium annosum* (Ikediugwu *et al.*, 1970). The effect of the hyphal interference factor was due to a drastic alteration of the permeability of the cell membrane leading to death of the cells. The studies on effect of antagonism by *Coprinus heptemerus* on other coprophilous fungi like *Mucor* sp., *Chaetomium globosum, Podospora* sp.

Sordaria sp., *Stilbella erythrocephala,* and *Trichoderma viride* showed that all these organisms except *Stilbella erythrocephala* were sensitive to its hyphal interference. It was also observed that *S. erythrocephala* produces a diffusible antibiotic which limits the growth of *C. heptemerus* and *A. crenulatus. Stilbella erythrocephala* was the dominant fungus on all pellets of rabbit dung while other fungi grew poorly (Webster, 1970). Singh and Webster (1973) confirmed the findings of Webster (1970) and showed that *S. erythrocephala* which is a common fungus on rabbit pellet caused varying degrees of inhibition of some representatives of each group of coprophilous fungi tested. They also extracted the antibiotics from culture filtrates of *S. erythrocephala* which did not resemble any of the twelve known antibiotics. Ikediugwu and Webster (1970 b) also found that species of Basidiomycetes particularly *Bolbitus vitellinus, Stropharia semiglobate* and *Penaeolus* sp. showed hyphal interference almost to same extent as *C. heptemerus.* Lodha (1971) observed that *Rhizopus nigricans* and *Cunninghamella* sp. took almost double time to sporulate in the presence of *Chaetomium diversum. Strattonia* sp. also did not fruit in the presence of *C. diversum.*

Apart from fungi and bacteria a number of other dung inhabiting microorganisms such as protozoa, nematodes, molluscs, mites and insects are also present. These organisms do inhibit the growth of fungi by competing for nutrition. A large number of protozoa and nematodes are known to act as hosts for a variety of predacious moniliales and Zoopagales (Juniper, 1953; Peach and Juniper, 1955; Duddington, 1957). A large number of mites and other soil inhabiting animals eat different fungal spores (Dowding, 1931). Larvae of sciarid fly feed upon different coprophilous fungi but avoid ascocarps of *Chaetomium bostrychodes* because of the hair on its perithecia (Wicklow, 1979). These sciardic flies, thus have a role in decomposition of rabbit faeces (Helsel and Wicklow, 1979). So the coprophilous fungi-detritiivore interactions are important in generating diversity among dung inhabiting organisms. Competition especially biological antagonism is important in limiting the duration and intensity of fruiting of a particular fungal species, thus, influencing the organization of fungal communities. Few studies, however, have examined the potential importance of these competitive situations as they affect substrate decomposition.

Decomposition of coprous substrate

A central process in the life cycle of all organisms is the decomposition of their remainant. There are numerous studies on the decomposition of dead plant (Witkamp, 1966; Reiners and Reiners, 1970; Witkamp and Frank, 1970; Hughes, 1971; Van Cleve, 1971; Sharma and Mukerji, 1976; Shukla and Singh, 1984) and animal tissues or their faeces (Nicholson *et al.* 1966; Castle and MacDaid, 1972; MacDiarmid and Watkin, 1972; Angel and Dickinson *et al.*, 1981), as they form a potential source of nutrients. Degradation of plant and animal residues are generally included under detritus compartment of ecosystem (Odum, 1962). The decomposition of these materials has been largely attributed to the soil microflora, nevertheless, it is well established that a singificant portion of herbivore faeces consits of coprophilous fungi. There is a greater microbial activity or higher number of decomposer microorganisms in dung pellets of animals or casts of earthworms than in the surrounding soil (Parle, 1963).

Fungi are the most important agents of decomposition amongst the various organisms involved and these organisms begin their activity as soon as the dung is voided by the animal (Webster, 1970). However, very few stuides have analysed the changing composition of substrate during the process of decomposition (Nicholson *et al.*, 1966; Dickinson *et al.*, 1981). Comparing the rate of decomposition of coprous and non-coprous substrates, Angel and Wicklow (1974) and Nagy and Harrower (1980) reported that coprous substrates show a higher rate of weight loss. In contrast to this, Nicholson *et al.* (1966) found no difference in the rate of decomposition of two types of substrates. It was reasoned that similarity in the rate of disappearance of two substrates could be due to an artifact because microbial decomposition cannot be distinguished from removal of material by soil animals and physical processes.

Faeces incubated on soil showed higher weight loss when compared to those incubated in moist chambers (Nicholson *et al.*, 1966; Wicklow and Moore, 1974; Angel and Wicklow, 1975). In contrary to this, Angel and Wicklow (1974, 1983) found no differences in decomposition rates under the two conditions. Rate of decomposition of laboratory incubated rabbit faeces decreased with time (Angel and Wicklow, 1974). Dry weight losses of different aged dung samples did not show significant differences regardless of the time dung was incubated in the field (Angel and Wicklow, 1983).

For studying the rate of decomposition various indices have been used by different workers. These are based upon the product of decomposition, disappearance of specific plant and animal constituents such as, sugar, cellulose, some of the nitrogenous compounds, the formations of resistant products of decomposition such as accumulation of lignins and their transformation into humus. Nicholson *et al.* (1966) used rate of respiration as indicator of microbial acitivity. Loss in dry weight has been considered important in determining the rate and extent of decomposition. Organic matter, nitrogen, C : N ratio and mineral ion contents have also been measured (Nicholson *et al.*, 1968; Dickinson *et al.*, 1981).

Chemical composition of dung samples exposed in the field showed that relative proportion of some chemical constituents changed after six months. An increase was observed in ash content which indicated utilization of organic fraction from the substrate. Lignin and acid digestible fibre did not change significantly with increased interval of filed exposure (Angel and Wicklow, 1983). Lignin degrading Basidiomycetes facilitate the decomposition by freeing the cellulose components of faeces for action by cellulolytic fungi (Wicklow *et al.* 1980). Corpophilous fungi are ecologically specialized to degrade lignocellulosic residues and have enzyme systems adapted to attacking the substructure of native lignins (Angel and Wicklow, 1974). Reduction in C : N ratio is a good indicatior of dung decomposition. Carbon is oxidised to CO_2 and nitrogen is immobilized by coprophilous microflora resulting in reduced C : N ratio. There is an inverse correlation between dry weight loss of different substrates and their respective C : N ratios (Angel and Wicklow, 1974; Nagy and Harrower, 1980).

Composition for substrate resources may place a limit on the number of competitors that can colonize and degrade a faecal substrate (Wicklow, 1981). On the other hand, Wicklow and Yocom (1981) reported that additon of fungal species to dung resulted in increased competition and prevented the efficient utilization of available resources and thus reduced decomposition rate. Very few workers, however, have examined the importance of these competitive situations affecting the substrate decomposition. Angel and Wicklow (1983) have related greater species abundance of fungal communities to slower rate of decomposition of individual faecal substrate. Difference in the climate from place to place and from year to year, the constitution of the dung the nature of substrate on which the dung is voided and the efficiency of the saprobic fauna and flora involved also determine the rate of disappearance of dung pellets (Dickinson *et al.*, 1981). Wicklow and Hirschfield (1979) presented evidence of a competitive hierarchy among coprophilous fungal populations colonizing cattle faeces. Slower growing and later appearing successional species interfered with growth of all other fungal colonists in cultural tests.

Conclusion

From the foregoing discussion it appears that there is no generalized pattern of succession on dung. Succession has always been based upon appearance of fruiting bodies whereas the mycelial spread of fungi in dung has always been neglected. There are no physiological evidences in support of 'Nutritional hypothesis' on dung. Further, a universal hypothesis based upon nutritional preferences of different fungal groups cannot be emplolyed to explain fungal succession on substrates of different nature as succession is an outcome of interaction of number of factors. Hence, it becomes important to study each factor separately and their interactions be evaluated. Succession on dung under laboratory and natural conditions is not comparable. Further, potential capabilities of coprophilous fungi as a decomposer community should be explored to degrade different organic residues and there is still ample scope for further research on this substrate. This can open up many unexplored vistas which will in turn help in solving many ecological problems

REFERENCES

Aas, O. 1978. Two new coprophilous species of *Saccobolus*. *Norw. J. Bot.* **25** : 65-68.

Anson, A.E., Fisher, P.J. and Kuthubutheen, A.J. 1985. Interaction between *Pilobolus crystallinus* and *Pseudomonas paucimobilis* isolated from rabbit dung. *Trans. Br. mycol. Soc.* **85** (1) : 161-164.

Ajrekar, S.L. and Dharmarajulu, K. 1931. A study of mucorineae of the city of Bombay. *J. Ind. Bot. Soc.* **10** : 29-34.

Angel, K. and Wicklow, D.T. 1974. A Decomposition of rabbit faeces : an indication of the significance of the coprophilous microflora energy flow schemes : *J. Ecol.* **62** : 429-437.

Angel, K. and Wicklow, D.T. 1975. Relationships between coprophilous fungi and faecal substrates in a Colorado grass land. *Mycologia* **67** : 63-74.

Angel, K. and Wicklow, D.T. 1983. Coprophilous fungal communities in semiarid to mesic grasslands. *Can. J. Bot.* **61** : 594-602.

Barrasa, J.M. and Moreno, G. 1984. Coprophilous fungi of Spain. *An. Jard. Bot. Mad.* **41** (1) : 3-18.

Bawcutt, R.A. 1983. Biology of *Syncephalis leadbeateri* sp. nov. (Mucorales) and observations on *Syncephalis*. *Trans. Br. mycol. Soc.* **80** : 219-230.

Bell, A. 1975. Fungal succession on dung of the British tailed opossum in New Zealand. *New Zealand. J. Bot.* **13** : 437-462.

Behera, N., Rikhy, M., Sharma, K.R. and Mukerji, K.G. 1973. Fungi of Delhi XXIII. Some interesting records. *Proc. India Nat. Sci. Acad.* **39** : 710-718.

Berkeley, M.J. 1836. British fungi in W.J. Hokeri British Flora 2. The English flora of Sir James Edward Smith. *Cryptogamia* **5** (2) : London.

Blackwell, M. and Malloch, D. 1991. Life history and arthropod dispersal of a coprophilous Stylopage. *Mycologia* **83** (3) : 360-366.

Bonner, R.D. and Fergus, C.L. 1959. The fungal flora of cattle feeds. *Mycologia* **51** : 855-863.

Brandt, C.S. and Thacker, E.J. 1958. A concept of the rate of food passage through the gastero-intestinal tract. *J. Anim. Sci.* **17** : 218-223.

Brown, W.H. and Graff, P.W. 1913. Factors influencing fungus succession on dug cultures. Philipp. *Journ. Sci. C. Bot.* **8** : 21-29.

Brummeler, J. Van. 1967. A world monograph of the genera *Ascobolus* and *Saccobolus* (Ascomycetes, Pezizales). *Persoonia.* Suppl. **1** : 1-260.

Buffin, N and Hennebert. G.L. 1985. *Basifimbria spinosa,* a new species, a new pleoanamorphic coprophilous hyphomycete. *Proc. Indian Acad. Sci. Plant Sci.* **94** : 259-268.

Buller, A.H.R. 1909-1934. *Researches on fungi* vols. 1-6 Longman, Green and Company, London.

Burges, A. 1939. Soil : fungi and root infection. *Proteria* **8** : 35.

Burges, A. 1958. *Microorganisms in the soil.* Butchinson & Co. London.

Cain, R.F. 1934. Studies of coprophilous sphaeriales in Ontario Univ. Toronto Stud. *Biol. Ser.* **38** : 1-126.

Cain, R.F. 1950. Studies of coprophilous Ascomycetes, 1. *Gelasinospora. Can. J. Res. C.* **28** : 566-576.

Cain, R.F. 1956a. Studies of coprophilous Ascomycetes, 2. *Phaeotrichum,* a new cleistocarpous genus in a new family and its relationships. *Can. J. Bot.* **34** : 675-687.

Cain, R.F. 1956b. Studies of coprophilus Ascomycetes, 4. *Tripterospora* a new cleistocarpous genus in a new family. *Can. J. Bot.* **34** : 699-710.

Cain, R.F. 1957. Studies of coprophilous Ascomycetes, 6. species from the Hudson Bay area. *Can. J. Bot.* **35** : 255-268.

Cain, R.F. 1961a. Studies of soil fungi, 3. New species of *Coniochaeta, Chaetomidium* and *Thielavia. Can. J. Bot.* **39** : 1231-1235.

Cain R.F. 1961b. Stuides of corpophilous Ascomycetes 7. *Preussia. Can. J. Bot.* **39** : 1633-1666.

Cain, R.F. 1961c. *Anxiella* and *Diplogelasinospora,* two genera with cleistothecia and pitted spores. *Can. J. Bot.* **39** : 1667-1677.

Cain, R.F. 1962. Studies of coprophilous Ascomycetes B. New Species of *Podospora. Can. J. Bot* **40** : 447-490.

Castle, M.E. and MacDaid, E. 1972. The decomposition of cattle dung and its efffect on pasture. *J. Br. Grassland Soc.* **27** : 133-137.

Dickinson, C.H. and Underhay, V.H.S. 1977. Growth of fungi in cattle dung. *Trans. Br. mycol. Soc.* **69** : 473-477.

Dickinson, C.H., Underhay, V.S.H. and Ross, V. 1981. Effect of season, soil fauna and water control on the deciomposition of cattle dung pats. *New Phytol.* **88** : 129-141.

Dissing, H. 1992. Notes on the coprophilous pyrenomycete *Soprormia fimetaria Persoonia* **14** (4) : 389-394.

Dowding, E.S. 1931. The sexuality of *Ascobolus stercocarius* and transportation of the oidia by mites and flies. *Ann. Bot.* **45** : 621-637.

Duddington, C.L. 1957. The predacious fungi and their place in microbial ecology. *In microbial Ecology.* 7th Symp. Soc. Gen. Microbiol. (Eds. Williams, R.E.O. and Spicer, C.C.) pp. 218-237. Cambridge university Press.

Duke, S.H.H. 1942. *The physiology of domestic animals.* Ithaca, N.Y. Constock Publishing Company.

Ebersohn, C. and Eicker, A. 1992. Coprophilous fungal species composition and species diversity on various dung substrates of African game animals. *Bot. Bull. Acad. Sin* (Taipei) **33** (1) : 85-96.

Eckbald, F.E. 1968. The genera of the operculate Discomycetes. *Nytt. Mag. Bot.* **15** : 1-191.

Fernier, H. 1954. Un Bombardia nouveau sur manioc. Remargues sur. lanatomie de la paron peritheciale de queleques Bombardia *et al* differenciation des ascospores chezles Sordariaceae. *Rev. Mycol.* (N.S.) **19** : Suppl. Colon. **1** : 1-19.

Fries, L. 1956. Studies in the physiology of *Coprinus.* II. Influence of pH, metal factors and temperature. *Svensk. Bot. Jidskr.* **50** : 47-96.

Furuya, K. and Udagawa, S. 1972a. Coprophilous pyrenomycetes from Japan I. *J. Gen. Appl.* Microbiol. **018** : 433-454.

Furuya, K. and Udagawa, S. 1972b. Coprophilous Pyrenomycetes from Japan II. *J. Gen. Appl. Microbiol.* **18** : 455-467.

Furuay, K. and Udagawa, S. 1973. Coprophilous Pyrenomycetes from Japan III. *Trans. mycol. Soc. Japan* **14** : 7-30.

Furuya, K. and Udagawa, S. 1977a. Coprophilous Pyrenomycetes of Japan IV. *Trans. Mycol. Soc. Japan* **17** : 248-261.

Furuya, K. and Udagawa, S. 1977b. New species of *Gelasinospora* and *Anixiella. Trans. mycol. Soc. Japan* **17** : 313-320.

Garcia. Z. and Noemi, 1978. Coprophilous sphaeriales from Urguay. II. *Rev. Biol.* Urug **3** (2) : 153-160.

Garrett, S.D. 1950. Ecology of root inhabiting fungi. *Biol. Rev.* **25** : 220-254.

Garrett, S.D. 1951. Ecological groups of soil fungi : a survey of substrate relationships. *New Phytol* **50** : 149-166.

Garrett, S.D. 1963. *Soil Fungi and soil fertility.* Pergamon Press, Oxford.

Garrett, S.D. 1966. Cellulose-decomposing ability of some cereal foot rot fungi in relation to thier saprophytic survival. *Trans. Br. mycol. Soc.* **49** ; 57-68.

Gibbs, T. 1909. Note on the coprophilous fungi. *J. Linn. Soc.* (Bot.) **38** : 413-417.

Ginai, M.A. 1936. Further Contribution to our knowledge of Indian coprophilous fungi. *J. Ind. Bot. Soc.* **15** : 269-284.

Griffin, D.M. 1960. Fungal colonization of sterile hair in contact with soil. *Trans. Br. Mycol. Soc.* **43** : 583-596.

Griffith, D. 1901. The North American Sordariaceae. *Mem. Torrey. Bot. Club* **11** : 1-134.

Harper, J.E. 1962. *A. comparative ecological study of the fungi on rabbit dung.* Ph.D. Thesis, Sheffield University.

Harper, J.E. and Webster, J. 1964. An experimental analysis of the coprophilous fungal succession. *Trans. Br. Mycol. Soc.* **47** : 511-530.

Harrower, K.M. and Nagy, L.A. 1979. Effects of nutrients and water stress on growth and sporulation of coprophilous fungi. *Trans. Br. Mycol. Soc.* **72** : 459-462.

Helsel, E.D. and Wicklow, D.T. 1979. Decomposition of rabbit faeces. Role of sciarid fly *Lycoriella mali* (Diptera : Sciaridae) in an energy transformation. *Can. Entomol.* **111** : 213-218.

Hesseltine, C.W. 1955. Genera Mucorales with notes on thier synonymy. *Mycologia* **47** : 344-363.

Hering, J.F. 1965. Succession of fungi in the litter of a lake district of Oakwood. *Trans. Br. Mycol. Soc.* **48** (3) : 391-408.

Hesseltine, C.W., Whitehill, A.R., Pidack, C., Tenhagen, M., Bohonos, M., Hutchings, B.L. and William, J.H. 1953. Coprogen, a new growth factor present in dung required by *Pilobolus. Mycologia* **45** : 7-19.

Hogg, B.M. and Hudson, H.J. 1966. Micro-fungi on leaves of *Fagus sylvatica* 1. The micro-fungal succession. *Trans. Br. Mycol Soc.* **49** : 185-192.

Hudson, H.J. 1968. The ecology of fungi on plant remains above the soil. *New Phytol.***67** : 837-874.

Hudson H.J. and Webster, J. 1958. Succession of fungi on decaying stems of *Agropyron repens. Trans. Br. Mycol. Soc.* **41** : 165-177.

Hughes, M.K. 1971. Seasonal caloric values from a deciduous Woodland in England. *Ecology* **52** : 923-926.

Ikediugwu, F.E.O. and Webster, J. 1970a. Angagonism between *Corpinus heptemerus* and other coprophilous fungi. *Trans. Br. Mycol. Soc.* **54** : 181-204.

Ikediugwu, F.E. and Webster, J. 1970b. Hyphal interference in a range of coprophilous fungi. *Trans. Br. Mycol. Soc.* **54** : 205-210.

Ikediugwu, F.E.O., Dennis, C. and Webster, J. 1970. Hyphal interference by *Peniophora gigantea* against *Heterobasidion annosum. Trans. Br. mycol. Soc.* **54** : 307-309.

Ingold, C.T. 1928. Spore discharges in *Podospora curvula Ann. Bot.* **42** : 567-570.

Iyer, Rohini, Ghosh, S.K. and Sarbhoy, A.K. 1973. Studies on coprophilous Fungi. *Proc. Indian Nat. Sci. Acad.* **39** : 199-202.

Jalink, L, 1992. *Poronia erici,* a small species of sphaeriales (Xylariaceae) in the Netherlands. *Coolia* **35** (3) : 95-98.

Jeng, R. S. and Cain, R.F. 1977. *Rhytidospora* a new cleistocaripous genus of the Melanosporaceae. *Mycotaxon* **5** : 278-282.

Jeng, R.S. and Krug, J.G. 1977a. New records and new species of coprophilous Pezizales from Argentina and Venezuela. *Can. J.Bot.* **55** (24) : 2987-3000.

Jeng, R.S. and Krug, J.G. 1977b. A new species of *Oroilia* from Venezuela. *Mycologia* **69** (4) : 832-835.

Juniper, A.J. 1953. Some predacious fungi occurring in dung. *Trans. Br. mycol. Soc.* **36** : 356-361.

Kar, A.K. and Pal, K.P. 1968. Some coprophilous Discomycetes of Eastern Himalayas (India). *Mycologia* **60** : 1086-1092.

Kaur, P. and Chauhan, U.K. 1991. Coprophilous fungal succession on different kinds of dung. *Geobios* **18** (4) : 155-159.

Kelkar, P.V. 1971. Coprophilous fungi of Maharashtra, India I. –The genus *Pilobolus.* Tode Maharashtra-*Vidhyan Mandir, Patrika* **6** : 64-67.

Krug, J. C. and Cain, R.F. 1972. Additions to the genus *Arnium. Can, J. Bot.* **50** : 357-373.

Kuthubudeen, A.J. and Webster, J. 1986. Effects of water availability on germination, growth and sporulation of coprophilous fungi. *Trans. Br. Mycol. Soc.* **86** (1) : 77-91.

Lambourne, L.J. and Reardon, T.F. 1962. Use of seasonal regressions in measureng feed intake of grazing animals. *Nature* **196** : 961-962.

Larsen, K. 1971. Danish endocoprophilous fungi and their sequence of occurrrence. *Bot. Tids* **66** : 1-32.

Liou, S. and Chingchen, Z. 1977. Preliminary studies in coprophilus discomycetes in Taiwan. *Taiwania* **22** : 44-58.

Locquin-Linard, M. 1984. 3 new corpophilous species of Lasibotidium (Ascomycetes, Eoterfeziaceae). *Cryptogam Mycol.* **4** (3) : 283-290.

Locquin-Linard, M. 1990. *Fennellia monodii* new species a new coprophilous Ascomycetes species from arid regions of Africa. *Mycotaxon.* **39** : 9-15.

Lodha, B.C. 1964. Studies on coprophilous fungi I. *Chaetomium. J. Indian Bot. Soc.* **43** : 121-140.

Lodha, B.C. 1971. Studies on coprophilous fungi IV. Some cleistothecial Ascomycetes. *J. Indian Bot. Soc.* **50** : 196-208.

Lodha, B.C. 1974. Decomposition of digested litter. *In* : C.H. Dickinson and G.J.F. Pugh (eds.) *Biology of Plant litter decomposition* Vol. I. pp. 213-241. Acad. Press, London and New York.

Lundqvist, N. 1960. coprophilous Ascomycetes from Northern Spain. *Sv. Bot. Tidsk.***54** : 523-529.
Lundqvist, N. 1964a. *Fimetariella,* a new genus of coprophilous Pyrenomycetes. *Bot. Nat.* **117** : 238-248.
Lundqvist, N. 1964b. *Anopodiu,* a new genus of coprophilous pyrenomycetes with apically pedicellate spores. *Bot. Nat.* **117** : 355-365.
Lundqvist, N. 1973. Studia fungerum fini I. New records of *Podospora* and a new species, *P. papilionaceae. Sv. Bot. Tidsker.***67** : 34-52.
Lundqvist, N. 1974. Studia Fungorum Fimi II. New records of *Arnia* and a new species, *A. bellum. Sv. Bot. Tidskr.* **68** : 289-303.
Lundqvist, N. 1992. *Wawelia effusa,* new species Lundqvist (Xylariaceae). *Persoonia* **14** (4) : 417-423.
Lyr. H. 1953. Zur Konntrrus der Erna Brungsphysiologie der Gattung *Pilobolus. Arch Microbiol.* **19** : 402-434.
MacDiarmid B.N. and Watkin, B.R. 1972. The cattle dung pats. 2-Effect of a dung patch on the chemical status of the soil, and ammonia nitrogen losses from the patch. *J. Grassland Soc.* **27** : 43-48.
Mahju, N.A. 1933. A contribution to our knowledge of Indian coprophilous fungi. *J. Ind. Bot. Soc.* **12** : 153-164.
Malloch, D. and Cain, R.F. 1970. Five new genera in the new family Pseudourotiaceae. *Can. J. Bot.* **48** : 1815-1825.
Malloch, D. and Cain, R.F. 1971. Four New genera of cleistothecial Ascomycetes with hyaline ascospores. *Can. J. Bot.* **49** : 847-854.
Malloch, D. and Cain, R.F. 1972. Trichocomataceae : Ascomycetes with *Aspergillus, Paecilomyces,* and *Penicillium* imperfect states. *Can. J. Bot.* **50** : 2613-2628.
Mangold, E. 1950. Die fulter Durchgangzeitung beim Kaninghen. *Arch. Tierernahr. Tierz* **1** : 136-147.
Mehrotra, Baijal, Kelkar, P.V. 1971. Piptocephalis Indica sp. nov. and Piptocephalis sp. from India. *Sydowia* **17** : 171-173.
Menon, R. and Mukerji, K. G. 1983. Fungi of Delhi XXXVI. Some interesting records from India. *Bibliotheca Mycol.* **91** : 291-300.
Massee, G. and Salmon, E.S. 1901. Researches on coprophilous fungi.I. *Ann. Bot.* **15** : 313-357.
Massee, G. and Salmon, E.S. 1902. Researches on Coprophilous fungi II. *Ann. Bot.* **16** : 57-93.
Minoura, K. and Yamada, K. 1976. Notes on coprophilous Discomycetes in Japan, I. *Trans. Mycol. Soc. Japan* **187** : 321-326.
Mitchell, D.T. 1970. Fungus succession on dung of south African Ostrich and Angora goat. *J.S. African Bot.* **36** : 191-198.
Mirza, J. H. and Cain, R.F. 1969. Revision on the genus *Podospora. Can. J. Bot.* **47** : 1999-2048.
Molliard, M. 1903a. Role des bacteries dans la production des perithecia des *Ascobolus. C.r. hebd. Searc. Acad. Sci.*, Paris **86** : 899-901.
Molliard, M. 1903b. Sur une condition qui favorise la production des peretheces chez les *Ascobolus. Bull. Soc. Mycol.* Fr. **19** : 150-152.
Moreau, C. 1953. Les genres Sordaria et Pleurage. *Encyl. Mycol.* **25** : 1-331.
Morinaga, T., Tanaka, M. and Minoura, K. 1981. Stimulation of spore germination in *Pilobolus crystallinus* by broth filtrate. *Trans. mycol. Soc.* (Japan) **22** : 277-282.
Mukerji, K.G. 1968a. Fungi of Delhi-II. Some interesting records. *Proc. Natl. Inst. Sci.*, India **348** : 71-81.
Mukerji, K.G. 1968b. Fungi of Delhi-IV. A new sps. of *Piptocephalis* from India. *Mycologia* **60** : 326-330.
Mukerji, K.G. 1969. Fungi of Delhi-VI. Two members of Mucorales. *Ceska Mycol.* **23** : 65-67.
Mukerji, K.G. 1970. Fungi of Delhi-III. Some interesting Ascomycetes. *Mycopath. Mycol. Appl.* **42** : 277-280.
Mukerji, K.G. 1978. Taxonomy of chaetomiales in relation to its morphology and citology. C.V. Subramaniam (ed). *Taxonomy of fungi Proc. Inter-natl.* Symp. on Taxonomy of fungi. University of Madras Part I, 1973, 258-261.

Mukerji, K.G. and Singh, N. 1974. Fungi of Delhi XXI, *Chaetomium delhianum* sp. n. *Friesia* **10** : 265-269.

Nagy, L.A. and Harrower, K.M. 1979. Analysis of two southern hemisphere coprophilous fungal succession. *Trans. Br. mycol. Soc.* **72** : 69-74.

Nagy, L.A. and Harrower, K.M. 1980. Coprous and non-coprous decomposition. *Trans. Br. Mycol. Soc.* **74** : 639-641.

Nannenga Bremekamp, N.E. Mukerji, K.G. and Singh, N. 1979. On two coprophilous mycomycetes from India. Proc. K. Ned. Akad. *Wet. Ser. C. Biol. Med. Sci.* **82** (2) : 217-222.

Narendra, D.V. 1973. Studies into coprophilous fungi of Maharashtra (India) III. *Sporormia* and *Sporormiella,* two rare Ascomycetes from sheep dung. *Nova Hedwigia* **24** : 481-486.

Narendra, D.V. 1974. Studies into coprophilous fungi of Maharashtra (India) II. Two rare and interesting Ascomycetes. *J. Unv. Bombay Sci. Sect.* **42** : 115-118.

Neilands, J.B. 1957. Some aspects of microbial iron metabolism. *Bact. Rev.* **21** : 101-111.

Nicholson, P.B., Bocock, K. L. and Heal, O.W. 1966. Studies on the decomposition of the faecal pellets of a millipede (*Glomeris marginata* (Villers). *J. Ecol.* **54** : 755-767.

Odum, E.P. 1962. Relationship between structure and function in the ecosystem. *Jap. J. Ecol.* **12** : 108-118.

Page, R.M. 1959. Stimulation of a sexual reproduction of *Pilobolus* by *Mucor plumbeus. Amer. J. Bot.* **46** : 579-585.

Parle, J. N. 1963. A microbiolgical study of earthworm casts. *J . Gen. Microbiol.* **31** : 13-22.

Pasricha, R. 1986. *Coprophilous fungi : Succession and its role in decomposition.* Ph. D. thesis, Delhi University.

Pathak, R. K. and Agarwal, S.C. 1977a. Studies on coprophilous ascomycetes IV. Production of amylase enzyme. *Trans. Mycol. Soc. Japan.* **18** : 220-223.

Pathak, R.K. and Agarwal, S.C. 1977b. Studies on coprophilous ascomycetes VI. Response to some volatile compounds. *Trans. mycol. Soc. Japan.* **18** : 298-301.

Pasricha, R. and Mukerji, K.G. 1987. Studies on Indian Coprophilous Fungi IV. Two interesting members of Mucorales. *Nova Hedwigia* **44** : 523-526.

Peach, M. and Juniper, A.J. 1955. *Stylopage areae* Drechsler var. magna var. nov. *Trans. Br. mycol. Soc.* **38** : 431-434.

Pidacks. C., Whitehill, A.R., Preuss, L.M., Hesseltine, C.W., Bohonos, N. and William, J.H. 1953. Coprogen, the isolation or a new growth factor required by *Pilobolus* species. *J. Amer. Chem. Soc.* **75** : 6064-6065.

Rajeevalochana, 1980. *Taxonomy and Ecology of coprophilous Fungi.* Ph.D. thesis, Delhi, University.

Rajeevalochana and Mukerji, K.G. 1980. A dwarf strain of *phycomyces blakesleanus* from deer dung. *Phycomyces* **4** : 83-84.

Rangaswami, G. and Chandrasekaran, A. 1963. Cellulolytic properties of four species of soil fungi. *Indian J. Microbiol.* **3** : 45-50.

Rayner, A.D.M. 1978. Definition of succession. *Bull. Br. Mycol. Soc.* **12** : 123-124.

Reese, E.T. and Levinson, H.S. 1952. A comparative Study of the breakdown of cellulose by microorganisms. *Physiol. Plantarum* **5** : 345-366.

Reiners, W.A. and Reiners, N.M. 1970. Energy and nutrient dynamics of forest floors in three Minnesota forests. *J. Ecol.* **58** : 497-519.

Richardson, M.J. 1972. Coprophilous Ascomycetes on different dung types. *Trans. Br. Mycol. Soc.* **58** : 37-48.

Richardson, M.J. and Watling, R. 1968. Keys to fungi on dung. *Bull. Br. Mycol. Soc.* **2** : 18-43.

Richardson, M.J. and Watling, R. 1969. Keys to fungi on dung. *Bull. Br. Mycol. Soc.* **3** : 86-88.

Richardson, M.J. and Leadbeater, G. 1972. *Piptocephalis fimbriata* and observations on the occurrence of *Piptocephalis* and *Syncephalis. Trans. Br. mycol. Soc.* **58** : 205-215.

Saxena, A.S., Mukerji, K.G. and Aggarwal, M.K. 1968. Spread of disease due to fungi. *Aspects of allergen and Appl. Immun.* **2** : 175-180.

Saxena, A.S. and Mukerji, K.G. 1970a. *Kernia bifurcotricha* sp. nov. *Trans. Br. mycol. Soc.* **54** : 146-148.

Saxena, A.S. and Mukerji, K.G. 1970b. Fungi of Delhi XIV. Imperfect state of *Kernia geniculotricha* seth. *Acta. Bot. Neerl.* **19** : 49-52.

Saxena. A.S. and Mukerji, K.G. 1970c. *Didymostilbe ellissi Trans. Br. Mycol. Soc.* **55** : 503-504.

Saxena, A.S. and Mukerji, K.G. 1970d. Fungi of Delhi XV. *Lophotrichus indicus* sp. nov. *Acta Bot. Neerl.* **19** : 722-728.

Saxena, A.S. and Mukerji, K.G. 1972. Fungi of Delhi XVIII. A new species of *Chaetomium. Mycologia* **64** : 1328-1330.

Saxena, A.S. and Mukerji, K.G. 1973a. Fungi of Delhi XIX. *Bahupaathra minuta* sp. n. *Mycopath. Mycol. appl.***49** : 201-204.

Saxena, A.S. and Mukerji, K.G. 1973b. Fungio of Delhi XVI Further additions to Indian species of *Chaetomium. Ceska Mycol.* **27** : 162-164.

Saxena, A.S. and Mukerji, K.G. 1973c. Fungi of Delhi XVIII. Three unrecorded members of Ascomycetes. *Ceska Mycol.* **27** : 165-168.

Schimidt, A. 1912. Die verbreitung der copro Pilze schlesiens Diss. Bot. Garaten Univ. *Breslaci.* pp. 1-82.

Shukhla, AN. and Singh, I.D. 1984. Siodegradiation of *Shorea robusta* Gaertn. Leaf litter and the cycling of minerals in a tropical soil forest. *Plant and Soil* **81** : 403-409.

Sineshchckov, A.D. 1965. Digestgion and assimilation processes in the stomach and intestines of ruminants. *In Physiology of digestion of the ruminant.* (Ed. R.W. Dougherty). pp. 254-261. London, Butterwood.

Singh. N. 1978. *Taxo-ecological studies of Indian coprophilous fungi.* Ph.D. Thesis, Delhi University.

Singh, N. 1978. Taxo-ecological studies of Indian coprophilous fungi I. Some rare records. *J. Indian. Bot. Soc.* **58** : 163-167.

Singh, N. and Mukerji, K.G. 1979. Studies on Indian coprophilous fungi. I. Some rare records. *J. Indian. Bot. Soc.* **58** : 163-167.

Singh, N. and Webster, J. 1972. Effect of coprophilous species of *Mucor* and bacteria on sporangial production of *Pilobolus. Trans. Br. Mycol. Soc.* **59** : 43-49.

Singh, N. and Webster, J. 1973. Antagonism between *Stilbella erythrocephala* and other coprophilous fungi. *Trans. Br. Mycol. Soc.* **61** : 487-495.

Siu, R.G.H. and Reese, E.T. 1953. Decomposition of cellulose by microorganisms. *Bot. Rev.* **19** : 377-416.

Solans, M.J. 1984. Novitates for the spanish corpophilous mycoflora 2. The genus *Podospora* (Sordariaceae). *An. Jard Bot. Madr.* **41** (2) : 247-250.

Spalding, D.H. 1963. Production of Pectinolytic and cellulolytic enzymes by *Rhizopus stolonifer. Phytopath.* **53** : 929-931.

Spegazzini, C. 1899. Fungi argentini nov. *V. criticia. Anal. Mus. Nac. Buenos Aires.* **6** : 81-365.

Subramaniam, C.V. and Lodha, B.C. 1964a. The interesting hyphomycets. *Can. J. Bot.* **42** : 1057-1063.

Subramaniam, C.V. and Lodha, B.C. 1964b. *Antonie Van Leeuwenhoek* **30** : 317-330.

Subramaniam, C.V. and Lodha, B.C. 1968. Two interesting coprophilous fungi from India. *Curr. Sci.* **37** : 245-248.

Subramaniam, C.V. and Chandrasekhara, K. V. 1977. *Lindqvistia,* a new Hyphomycete genus. *Bol. Soc. Argent. Bot.* **18** : 145-151.

Subramaniam, C.V. and Chandrasekhara, K.V. 1979. Two interesting coprophilous hyphomycetes. *Kavaka* **6** : 41-48.

Taj-Al-Deen, S.J., Al-halbeb, E.K. and Abdullah, S.K. 1990. Cellulolytic activity of coprophilous fungi. *Cryptogam Bot.* **2** (1) : 25-29.

Thaysen, A.C. and Bunker, H.J. 1927. The microbiology of cellulose, hemicellulose pectin and gums. Oxford Univ. Press.

Toth, S. 1963. Data to the knowledge on the coprophilous microscopic fungi in Hungary I. *Ann. Hist. Nat. Mus. Nat. Hung.* **55** : 181-185.

Toth, S. 1965. Data to the knowledge of coprophilous microscopic fungi in Hungary II. *Ann. Hist. Nat. Mus. Nat. Hung.* **57** : 147-157.

Toth, S. 1967. Data to the knowledge of coprophilous microscopic fungi. III. *Ann. Hist. Nat. Mus. Nat. Hung.* **59** : 117-123.

Tubaki, K. 1954. Studies on the Japanese Hyphomycetes. I. Coprophilous group. *Nagaoa* **4** : 1-20.

Upadhyay. H.P. 1973. *Helicostylum* and *thamnostylum. Mycologia* **65** : 733-751.

Underhay, V.H.S. and Dickinson, C.H. 1978. Water, Mineral and energy fluctuation in decomposing cattle dung pats. *J. Br. grassland Soc.* **33** : 189-196.

Valldosera, M. Guarro, J. and Figueras, M.J. 1991. Two interesting coprophilous fungi from Spain. *Mycol. Res.* **75** (2) : 243-246.

Van Cleve, K. 1971. Energy and Weight loss functions for decomposing foliage in birch and asper forests, in interior Alaska. *Ecology* **52** : 720-723.

Vander Drift, J. and Witkamp, M. 1959. The significance of the breakdown of oak litter by *Eniocyla pusille. Arch. Neerl Zool.* **13** : 486-492.

Von. Arx, J.A., Mukerji, K.G. and Singh, N. 1978a. *Leucosphaera,* a new genus of the Pseudoeurotiaceae. *Persoonia.* **18** : 144-146.

Von Arx, J. A., Mukerji, K.G. and Singh, N. 1978b. A new coprophilous Ascomycetes from India. *Persoonia.* **18** : 144-146.

Waksman, S.A. 1938. *Humus, Origin and Chemical Composition and importance in nature.* Baltimore Williams and Wilkins.

Waksman, S.A., Cordos, T.C. and Hulpoi, N., 1939. Influence of temperature upon the microbiological population and decomposition processes in composts of stable manure. *Soil Sci.* **47** : 83-114.

Waraitch, K.S. 1976. A contribution to the knowledge of coprophilous Pezizales of India. *Sydowia Ann. Mycol.* **29** (1-6) ; 1-9.

Watling, R. 1963. The fungal succession on hawk pellets. *Trans. Br. mycol. Soc.* **46** : 81-90.

Webster, J. 1956. Succession of fungi on decaying cocks foot culms. *J. Ecol.* **44** : 517-544.

Webster, J. 1957. Succession of fungi on decaying cocks foot culms. II. *J. Ecol.* **45** : 1-30.

Webster, J. 1970. Coprophilous fungi. *Trans. Br. Mycol. Soc.* **54** : 161-180.

Webster, J. and Dix, N.J. 1960. Succession of fungi on decaying cocks foot culms II. *Trans. Br.Mycol. Soc.* **43** : 85-99.

Wicklow, D.T. 1979. Hair ornamentation and predator defense in *Chaetomium. Trans. Br. Mycol. Soc.* **72** : 107-110.

Wicklow, D.T. 1981. The Coprophilous fungal community : a mycological system for examining ecological ideas. *In* : *The fungal community : its organiszation and role in the ecosystem* (Eds. Wicklow, D.T. and Canoll, G.C.) Marcel Dekker, New York. pp. 47-76.

Wicklow, D.T. and Angel, K. 1980. Fungal community expression in Lagomorph versus Ruminant faeces. *Mycologia* **72** ; 1016-1021.

Wicklow, D.T. and Angel, K. 1983. Some reproductive characteristics of coprophilous Ascomycetes in three prairie ecosystems. *Mycologia* **75** : 1070-1073.

Wicklow, D.T. and Hirschfield, B.J. 1979. Evidence of a competitive hierarchy among coprophilous fungal populations. *Can J. Microbiol.***25** : 855-858.

Wicklow, D.T. and Moore, V. 1974. Effect of incubation temperature on the coprophilous fungal succession. *Trans. Br. Mycol. Soc.* **62** : 411-415.

Wicklow, D.T. and Yocom, D.H. 1981. Fungal species numbers and decomposition of rabbit faeces. *Trans. Br. Mycol. Soc.* **76** : 29-32.

Wicklow, D.T. Detroy, R.W. and Adams, S. 1980. Differential modification of the lignin and cellulose components in wheat straw by fungal colonists of ruminant dung : Ecological implications. *Mycologia.* **72** : 1065-1076.

Witkamps. M. 1966. Decomposition of leaf litter, in relation to environment, microflora and mircrobial respiration. *Ecology* **47** : 194-201.

Witkamp, M. and Frank, M.L. 1970. Effects of temperature, rainfall and fauna on transfer of 137 cs, K. Mg and mass in consumer decomposer microcosms. *Ecology* **51** : 465-474.

Wood, S.N. and Cooke, S.N. 1984. Use of semi-natural resource units in experimental studies on coprophilous fungi. *Trans. Br. Mycol. Soc.* **83** : 337-374.

Yocom, D.H. and Wicklow, D.T. 1980. Community differentiation along a dune succession : An experimental approach with coprophilous fungi. *Ecology* **61** : 868-880.

Chapter 4

Microbial Biomass and Activity in Tropical Agroecosystems under Different Cropping Practices

N. Behera AND **Pramila Prasad**

School of Life Sciences Sambalpur University, Burla-768 019, Orissa, India

Abstract. Effect of different crop practices such as single cropping (SC), double cropping (DC) and crop rotation (CR) on soil microbial biomass and activity was investigated in three Indian tropical cropfields. Mean microbial biomass C was estimated to be 270, 458 and 611 $\mu g\ g^{-1}$ soil with microbial C:organic C ratio of 2.33, 2.93 and 4.67 per cent in SC, DC and CR field respectively. Mean basal soil respiration rate was maximum in CR (0.4 μgCO_2-C g^{-1} soil h^{-1}) and minimum in SC (0.28 $\mu g\ CO_2$-C g^{-1} soil h^{-1}) and showed positive relationship with the microbial biomass C (r^2=0.92 $p<0.05$). Analysis of microbial metabolic quotient (q CO_2) indicated that the amount of CO_2-C lost through microbial respiration per unit microbial C was 1.5 to 1.7 times more in SC and DC than in CR.

Key words : *Crop rotation, Organic carbon, Microbial biomass, soil respiration*

Introduction

The soil microbial biomass constitutes a transformation matrix for all the natural organic material in the soil and acts as a labile reservoir of plant available nutrients (Jenkinson and Ladd, 1981). According to Powlson *et al.* (1987) soil microbial biomass responds much more rapidly than the total organic matter to any change in organic inputs. Thus microbial biomass can be considered as an ecological marker for analysing and predicting the long term effects of perturbations in soil subsystem (Smith and Paul, 1990).

Cultivation and types of agricultural practices exert significant influence on microbial carrying capacity of soil (Smith and Paul, 1990). As reported by Ayanaba *et al.* (1976), Van Veen and Paul (1981) and Elliot (1986), cultivation of native soils lead to loss of organic matter content through the alteration of soil structural stability. Different agricultural management practices such as tilling, ploughing, green manuring and chemical fertiliser application can cause a variation in the soil organic matter and constrain native soil biomass level (Doran, 1980; Lynch and Panting 1980; Lynch, 1982; Schnurer *et al.,* 1985; Gupta and Germida 1988; Insam *et al.*, 1991; Hassink *et al.*, 1991; Bloem *et al.*, 1992). Besides cropping practices such as monocropping and multicropping through crop rotation are also reported to exhibit decisive influence on soil microbial biomass and activity (Bolton *et al.*, 1985; Carter, 1986; McGill *et al.*, 1986; Dalal and Mayer, 1987; Granaststein *et al.*, 1987; Anderson and Domsch 1989; Insam *et al.*, 1989). However, there are no reports available on these aspects from Indian tropical soil inspite of the fact that the microbial pool has a significant role in soil nutrient conservation mechanisms in the tropics (Singh *et al.*, 1989). Aim of the present study was to assess the impact of different cropping practices on the level of soil microbial biomass and activity in Indian tropical agroecosystem.

Study Sites

Three types of irrigated crop fields undergoing different crop practices such as single cropping (SC), double cropping (DC) and crop rotation (CR) (Table 4.1) selected for the present study are located around the Sambalpur town of Western Orissa, India (at 21°22'N Lat. and 84°E Long.). The soils of the crop fields were residual ultisol, sandy loam in texture and dark greyish in colour. In CR field, rice is cultivated every alternate year followed by pulse and vegetables. All the three fields are manured with organic green compost (1 t $ha^{-1}yr^{-1}$) during June and chemical (NPK) fertilizer 0.3 t $ha^{-1}yr^{-1}$) during February.

The climate of the area is tropical monsoonal with three distinct seasons : summer (March to June), rainy (July to October) and winter (November to February). Mean monthly maximum

TABLE 4.1. *Nature of crop practices in different crop field.*

Cropping practices	Crop cultivated	Cultivation period
1. Single Cropping (SC)	*Oryza sativa* L.	January-April
2. Double Cropping (DC)	*Oryza sativa* L.	July-November
	Oryza sativa L.	January-April
3. Crop Rotation (CR)	*1st year*	
	Oryza sativa L.	July-October
	2nd year	
	Pulse	October-January
	Cyamopsis tetragonoloba, Taub	
	Vegetables	February-May
	Solanum melongena Linn.,	
	Lycopersicum esculentum, Will.	
	Abelmoschus esculentus, Miller	
	Brassica Oleracea var. capitata, Linn	

temperature ranges from 20°to 39°C and minimum from 10° to 25°C. The average annual rainfall of the area is 1400 mm., out of which 80 per cent falls during rainy season.

Sampling procedure

All the cropfields were sampled at bimonthly intervals starting from June 1991 to October 1992. Five soil samples were collected at random from 0-20 cm. soil depth of each field. After removing large pieces of plant materials by hand, field moist soil samples were sieved through a 2 mm. mesh screen. Each soil sample was divided into two parts. One part in the field moist condition was used for the analysis of microbial biomass and other part was air dried for all other soil analysis.

Soil analysis

Textural analysis of soil samples were carried out using soil sieve of different mesh sizes (Misra, 1968). Soil pH was measured by pH meter (Elico, India) using soil suspension (1:2 soil water ratio). Organic carbon was determined by Walkley and Black's rapid titration method and total nitrogen by Kieldahl technique (Jackson, 1958).

Microbial biomass analysis

Field moist soil samples were stored at room temperature (28 ± 4°C) for a week to settle down the respiration. Soil microbial biomass carbon was determined using the chloroform fumigation incubation technique of Jenkinson and Powlson (1976) except that the liquid $CHCl_3$ was used instead of vapour and CO_2-C evolved from fumigated soil during 10-20 days was taken as control (Merckx *et al.*, 1985; Srivastava and Singh, 1988). Microbial biomass C was calculated by dividing the flush of CO_2-C (CO_2-C evolved from fumigated soil during 0-10 days minus that evolved from same soil during 10-20 days) by a kc (fraction of biomass C evolved as CO_2-C during the first 10 days of incubation) value of 0.45 (Jenkinson and Ladd, 1981).

Basal soil respiration

Basal respiration of collected soil samples were measured following alkali absorption technique (Witkamp, 1966). For this, sieved soil samples of 500 g were placed in glass container (2L capacity) and were remoistened to its 60 per cent water holding capacity by adding sterile water. The samples were kept incubated at room temperature (28 ± 4°C) for 1 week to settle down the respiration after which CO_2-C evolution was measured continuously for 6 hours everyday for a period of 5 days. By relating soil basal respiration rate to the corresponding microbial biomass C, microbial metabolic quotient (qCo_2 : g CO_2-C g^{-1}MB-C h^{-1}) was calculated (Anderson and Domsch, 1990).

All the results were expressed on an oven-dry soil basis (105°C, 24-h. Statistical analysis of the results were carried out following Snedecor and Cochran (1967).

TABLE 4.2. *Soil physico-chemical characteristics of crop fields under different cropping practices.*

	Single crop	Double crop	Crop rotation
Soil texture (%)			
Sand (70.2 mm)	71 ± 5	68 ± 3	64 ± 4
Silt (0.2-0.1 mm)	17 ± 2	19 ± 3	18 ± 2
Clay (>0.1 mm)	12 ± 3	13 ± 2	18 ± 1
Bulk density (g cc^{-1})	1.332 ± 0.121	1.561 ± 0.118	1.601 ± 0.146
Water Holding Capacity (%)	39 ± 3	41 ± 3	43 ± 2
pH	7.01 ± 0.02	6.59 ± 0.12	6.74 ± 0.11
Organic Carbon (%)	1.160 ± 0.461	1.559 ± 0.356	1.307 ± O.226
Total Nitrogen (%)	0.119 ±0.006	0.148 ± 0.011	0.164 ± 0.006

TABLE 4.3. *Soil organic carbon (mg g^{-1} soil) and microbial biomass (μgCg^{-1}soil) in cropfields under different crop practices.*

Sampling	Single Crop		Double Crop		Crop rotation	
month	Org-C	MB-C	Org-C	MB-C	Org-C	MB-C
1991						
June	6.80 ± 0.24	183 ± 21	9.58 ± 0.66	303 ± 17	10.34 ± 0.47	480 ± 41
August	18.22 ± 1.19	314 ± 20	20.92 ± 0.63	485 ± 29	15.99 ± 0.13	627 ± 39
October	14.51 ± 1.39	335 ± 23	15.25 ± 0.20	478 ± 37	14.70 ± 1.03	640 ± 25
December	11.80 ± 1.49	347 ± 21	14.42 ± 0.99	550 ± 17	13.66 ± 0.64	653 ± 12
1992						
February	7.78 ± 0.30	255 ± 18	12.77 ± 1.91	460 ± 24	11.49 ± 1.36	756 ± 36
April	6.83 ± 1.41	200 ± 23	17.78 ± 2.28	532 ± 31	11.34 ± 0.34	547 ± 29
June	6.14 ± 0.22	174 ± 10	13.35 ± 2.95	370 ± 37	9.65 ± 0.57	490 ± 11
August	18.09 ± 1.32	297 ± 10	21.05 ± 0.97	487 ± 17	15.67 ± 0.52	598 ± 65
October	14.19 ± 1.44	322 ± 14	15.16 ± 0.28	460 ± 22	14.76 ± 0.96	710 ± 16
Mean (X)	11.60	270	15.59	458	13.07	611
CV (%)	39.75	23.89	22.8	15.89	17.26	14.41

Results

Some of the soil physicochemical characteristics of the cropfields under different crop practices are mentioned in Table 4.2. Among the three cropfields, in the soil of the CR, there observed an increase in the clay fraction, bulk density, water holding capacity and total nitrogen content. The soil pH of DC and CR were slightly in acidic range, where as that of SC was nearly neutral.

Bimonthly data of soil organic carbon (Org-C) and total microbial biomass (MB-C) are represented in Table 4.3. The amount of Org-C which fluctuated with respect to different sampling months exhibited peak during August in all the sites. The range of Org-C was observed to be more in DC than that of other two sites. Analysis of variance indicated that the differences in the level of Org-C within different sampling months and between three sites were significant ($p < 0.05$). The mean Org-C of SC, DC and CR were 11.60, 15.59 and 13.07 mg g^{-1} soil respectively with maximum coefficient of variation in SC (39.75%) and minimum in CR (17.26%). The range of MB-C was comparatively less in SC than DC and CR and among the three fields, CR exhibited maximum MB-C. The mean value of MB-C in CR and DC were respectively 2.3 and 1.7 times higher than that in SC. In all the three fields, the level of MB-C was found to be at its minimum during June and maximum during December in SC and DC and during February.in CR. The variations of MB-C within different sampling months in each field and between different fields were observed to be significant ($p<0.05$) through the analysis of variance. The bimonthly variation of MB-C exhibited positive relationship with variation of soil Org-C in SC and DC according to the following regression equations:

$$\text{MB-C} = 140 + 11.87 \text{ Org-C}, (r^2 = 0.64\ p < 0.05 \text{ in SC})$$
$$\text{MB-C} = 257 + 12.94 \text{ Org-C}, (r^2 = 0.40\ p < 0.05 \text{ in DC})$$

However in CR, the relationship between the variation of MB-C and Org-C was not significant ($r^2 = 0.25$).

From the mean values of MB-C and Org-C mean ratio of MB-C: Org-C (%) was calculated for the respective cropfields. The ratio for SC, DC and CR were 2.33, 2.93 and 4.67 per cent respectively. Thus per unit Org-C, the quantity of MB-C supported was maximum in the CR and minimum in the soil of SC.

Table 4.4 presented the data of basal soil respiration rate (BSR) and microbial metabolic quotient (qCO_2) in three different cropfields during different months of sampling. the rate of respiration in CR were comparatively higher than that in SC and DC. Analysis of variance indicated that the bimonthly rate respiration was significant ($p < 0.05$) both within and between different cropfields. Bimonthly data of MB-C ($\mu g\ Cg^{-1}$ soil) and respiration rate ($\mu g\ CO_2\text{-}C\ g^{-1}$ soil h^{-1}) of respective sites also exhibited positive relationship with each other through the linear equations :

$$\text{BSR} = 0.122 + 0.588 \times 10^{-3} \text{ MB-C}, (r^2 = 0.98, P < 0.01 \text{ in SC})$$
$$\text{BSR} = 0.050 + 0.655 \times 10^{-3} \text{ MB-C}, (r^2 = 0.94, P < 0.01 \text{ in DC})$$
$$\text{BSR} = 0.030 + 0.605 \times 10^{3} \text{ MB-C}, (r^2 = 0.98, P < 0.01 \text{ in CR})$$

TABLE 4.4. *Basal soil respiration (µg CO_2-C g soil^{-1} h^{-1} × 10^{-3}) and microbial metabolic quotient (g CO_2-C g MB-C^{-1} h^{-1} * 10^{-3}) in cropfields under different crop practices*

Sampling month	Single crop		Double Crop		Crop rotation	
	BSR	qCO_2	BSR	qCO_2	BSR	qCO_2
1991						
June	225 ± 7	1.230	245 ± 6	0.809	323 ± 8	0.673
August	294 ± 6	0.936	373 ± 6	0.769	395 ± 12	0.629
October	317 ± 7	0.946	395 ± 7	0.826	420 ± 5	0.656
December	336 ± 8	0.968	394 ± 15	0.611	433 ± 9	0.663
1992						
February	274 ± 7	1.074	346 ± 11	1.074	487 ± 23	0.644
April	246 ± 7	1.230	396 ± 9	1.229	355 ± 11	0.649
June	224 ± 6	1.287	287 ± 8	1.286	333 ± 6	0.679
August	294 ± 9	0.989	365 ± 8	0.989	391 ± 12	0.654
October	312 ± 5	0.969	346 ± 7	0.969	466 ± 10	0.487
Mean ($\bar{x}$)	280	1.070	350	0.951	400	0.637
CV(%)	13.72	12.41	14.08	21.91	13.42	8.62

Bimonthly data of BSR and MB-C of all the sites, when pooled together, the derived regression equation quantifying the relationship was :

$$BSR = 0.159 + 0.412 \times 10^{-3} MBC \ (r^2 = 0.92 \ p < 0.01)$$

Range of microbial metabolic quotient (qCO_2-C g^{-1} MB-C hr^{-1} × 10^{-3}) was observed to be relatively more in SC and DC than CR. The mean values of qCO_2 in SC, DC and CR were 1.07, 0.951 and 0.637 respectively. Thus the amount of CO_2-C lost through microbial respiration per unit microbial C was 1.7 and 1.5 times more in SC and DC respectively than in the soil of CR. Analysis of variance indicated that the variation of qCO_2 with respect to different sampling months was not significant.

However, the differences of qCO_2 between three sites were significant ($p < 0.05$). Relationship between the bimonthly data of qCO_2 and BSR of all the sites was observed to be inverse through the regression :

$$qCO_2 = 1.389 \times 10^{-3} - 1.13 \ MB\text{-}C \ (r^2 = 0.59 \ p < 0.05).$$

Discussion

The range of MB-C (174-756 µgC g^{-1} soil) estimated in the present study is comparable with that of Srivastava and Singh (1989, 1991) for some Indian cropfield soils. Comparatively higher range of MB-C and higher ratio of MB-C : Org-C in the CR than SC and DC is in agreement with the views of Anderson and Domsch (1986) and Insam *et al.*(1989) that the soil under crop rotation supports more MB-C per unit Org-C. Higher quantities of crop residues and relatively higher chemical diversity of organic matter input from crop plants may be some of the reasons which can explain such quantitative changes in the size of soil microbial pool in the agricultural field under crop rotation (Anderson and Domsch, 1990).

Soil textural differences specifically an increase in the clay fraction can also be another reason for the increase in the size of the soil MB-C in CR. An increase in the clay fraction in soil leads to the formation of more stable aggregates and this provides an environment for closer interaction between microbes and products of their decay (Gregorich *et al.* 1991) supporting a more diverse microbial community and biomass. Van Veen *et al.* (1985) attributed this to the preservation capacity of the soil to preserve and protect microorganisms against microbivore predation.

From the correlation analysis between MB-C and Org-C, it was evident that the variability of Org-C could explain 64 per cent of the variability of MB-C in soil. The corresponding figure in DC was 40 per cent whereas in CR the relationship was not significant. Comparatively less variation of Org-C in DC and CR (C.V. = 23 & 17%) may explain such poor or lack of correlation between MB-C and Org-C.

Microbial activity as evident from the data of the basal soil respiration (BSR) showed significant positive relationship with MB-C in all the fields. The relationship indicated that 94-98 per cent of the variation in BSR in each site would be explained due to the variation in MB-C. Basal soil respiration is considered to be an index of the process of decomposition with respect to carbon cycle (Coleman and Sasson, 1978) and is a measure of microbial turnover rate in soil (Insam *et al.*, 1991). Comparatively higher rate of respiration thus indicated higher decomposition activity and microbial turnover in soil of CR over SC and DC. Significant positive relationship between soil respiration rate and MB-C is in accordance with the earlier observations of Basu and Pati (1991) and Basu *et al.* (1991) for some Indian tropical forest and grassland soils.

Metabolic quotient (qCO_2) is an useful means for comparison of metabolic status of microbial community in quantitative term (Anderson and Domsh, 1990). In the present study, mean metabolic quotient was calculated to be maximum in SC and minimum in CR soil. According to Insam (1990), typically high qCO_2 is found in soil with a recent input of easily degradable substrate. Such substrate would induce microflora composed of mainly r-strategy ecotypes (Lynch, 1984) which respire more CO_2-C per unit degradable carbon than 'K'-strategists. 'K' strategists usually prevail in soil that have evolved more complex detritus food web. Since the CR field having a heterogenous input of organic substrate has such feature of complex detritus food web, it's soil is expected to be predominated by microbial community of 'K'-strategy ecotypes. This explained the high and low qCO_2 in SC and CR soil respectively.

A decline in qCO_2 with increase in the level of MB-C as evident from the negative relationship between the two, is in agreement with the observation of Santruckova and Straskrava

(1991). Relatively lower range of qCO_2 inspite of higher range of MB-C and MB-C: Org-C ratio in CR can be more conveniently explained on the basis of the concept "economic metabolism" of soil microorganisms proposed by Anderson and Domsch (1990) *i.e.* in the more complex CR soil microorganisms could have evolved a higher yield of "coefficient (Y) leading to the establishment of larger MB-C pool. The more efficiently the soil microorganisms function, the less carbon per unit microbial biomass is lost by respiration (Insam 1990). Thus the rise in qCO_2 from CR to SC reflects a decline in the efficiency of substrate/energy utilisation by soil microbial community. This has important implications in the maintenance and management of the soil fertility of tropical agroecosytem with particular reference to agricultural practices for sustainable production.

Acknowledgement

Authors thank Head, School of Life Sciences, Sambalpur University, for providing necessary laboratory facilities to carry out the work. One of the authors (PP) is grateful to the Secretary, Education and Youth Services, Govt. of Orissa for grant of study leave.

REFERENCES

Anderson, T.H. and Domsch, K.H. 1986. Carbon link between microbial biomass and soil organic matter. p. 467-471. In: *Proceedings of the IVth International Symposium on Microbial Ecology* (Eds. F. Megusar & M.Gantar). Slovene Society for Microbiology, Ljubljana.

Anderson, T.H. and Domsch, K.H. 1989. Ratios of microbial biomass carbon to total organic carbon in arable soils. *Soil Biol. Biochem.* **21** : 471-479.

Anderson T.H. and Domsch K.H. 1990. Applications of ecophysiological quotients (qCO_2 and qD) on microbial biomasses from soils of different cropping histories. *Soil Biol. Biochem.* **22** : 251-255.

Ayanaba, A. Tuckwell S.B. and Jenkinson, D.S. 1976. The effect of clearing and cropping on the organic reserves and biomass of tropical soils. *Soil Biol. Biochem.* **8** : 519-525.

Basu, S and Pati D.P. 1991 Soil respiration in relation to microbial population and biomass in a grassland soil from Orissa. *J. Ecobiol.* **3** : 267-273.

Basu, S. Joshi, S.K. Pati, D.P and Behera N. 1991. Soil respiration in relation to microbial biomass in a tropical deciduous forest floor from India. *Rev. Ecol. Biol. SOL.* **28** : 377-386.

Bloem, J. DeRutter, P.C. Koopman, G.J. Lebbink, G. and Brussard, L. 1992. Microbial numbers and activity in dried and rewetted arable soil under integrated and conventional management. *Soil Biol. Biochem.* **24** : 655-665.

Bolton Jr, H. Elliot, L.F. Papendick R.I. and Bezdicek, D.F. 1985. Soil microbial biomass and selected soil enzyme activities : Effect of fertilization and cropping practices. *Soil Biol. Biochem.* **17** : 297-302.

Carter, M.R. 1986. Microbial biomass as an index for tillage-induced changes in soil biological properties. *Soil Till. Res.* **7** : 29-40.

Coleman, D.C. and Sasson, A. 1978. Decomposer subsystem. P. 609-655. In : *Grasslands System Analysis and Man* (Eds. A.J.Breymeyer & G.M.Vandyne), Univ. Cambridge Press, Cambridge.

Dalal, R.C. and Mayer, R.J. 1987. Long term trends in fertility of soils under continuous cultivation and cereal cropping in southern Queensland VII. Dynamics of nitrogen mineralization potentials and microbial biomass. *Aust. J. Soil Res.* **25** : 461-472.

Doran, J.W. 1980. Microbial changes associated with residue management with reduced tillage. *Soil Sci. Soc. Am. J.* **44** : 518-524.

Elliot, E.T. 1986. Aggregate structure and C,N and P in native and cultivated soils. *Soil Sci. Soc. Am. J.* **50** : 627-663.

Granatstein, D.M. Bezdicek, D.F. Cochran, V.L. Elliot, L.F. and Hammel J. 1987. Long-term tillage and rotation effects on soil microbial biomass C and N. *Biol.Fertil. soils.* **5** : 265-270.

Gregorich, E.G. Voroney, R.P. and Kachanoski, J. 1991 Turnover of carbon through microbial biomass in soils with different textures. *Soil Biol. Biochem.* **23** : 799-805.

Gupta, V.S.S.R. and Germida, J.J. 1988. Distribution of microbial biomass and its activity in different soil aggregate size classes as affected by cultivation. *Soil Biol. Biochem.* **20** : 777-786.

Hassink, J. Lebbink, G. and Van Veen, J.A. 1991 Microbial biomass and activity of a reclaimed palder soil under a conventional or a reduced input farming system. *Soil Biol. Biochem.* **23** : 507-513.

Insam, H. 1990. Are the soil microbial biomass and basal soil respiration governed by the climatic regime. *Soil Biol. Biochem.* **22** : 525-532.

Insam, H. Parkinson, D. and Domsch, K.H. 1989. The influence of macroclimate on soil microbial biomass levels. *Soil, Biol Biochem.* **21** : 211-221.

Insam, H, Mitchell, C.C. and Dormaar, J.F. 1991. Relationship of soil microbial biomass and activity with fertilization practice and crop yield of three ultisols. *Soil Biol. Biochem.* **23** : 459-464.

Jackson, M.L. 1958. *Soil Chemical Analysis.* Prentice Hall of India Pvt. Ltd, New Delhi, 498 p.

Jenkison, D.S. and Powllsson, D.W. 1976. The effects of biocidal treatmeents on metabolism in soil-V.A method for measuring soil biomass. *Soil Biol. Biochem.* **8** : 209-213.

Jenkinson, D.S and Ladd, J.N. 1981. Microbial biomass in soil measurement and turnover. pp. 450-471. *In Soil Biochemistry* (Eds. E.A. Paul & J.N. Ladd). Marcel Dekker, New York.

Lynch, J.M. and Panting L.M. 1980. Cultivation and the soil biomass. *Soil Biol. Biochem.* **12** : 29 -33.

Lynch, J.M. 1982. Effects of season, cultivation and N Fertilizer on the size of the soil microbial biomass. *J. Sci. Food Agril.* **33** : 249-252.

Lynch, J.M. 1984. Interaction between biological processes, cultivation and soil structure. *Plant Soil.* **76** : 307-318.

Mc Gill, W.B. Cannon, K.R. Robertson, J.A. and Cook, F. 1986. Dynamics of soil microbial biomass and water soluble organic C in Breton L after 50 years of cropping to two rotations. *Can. J. Soil Sci.* **66** : 1-19.

Merckx, R. Den Hartog, A. and Van Veen, J.A. 1985. Turnover of root derived material and related microbial biomass formation in soil of different texture. *Soil Biol. Biochem.* **17** : 565-569.

Misra. R. 1968. *Ecology work book.* Oxford and I.B.H. Publ. Co., New Delhi, 244 p.

Powlson, D.S. Brookes, P.C. and Christensen, B.T. 1987. Measurements of soil microbial biomass provides an early indication of changes in the total soil matter due to straw incorporation. *Soil Biol. Biochem.* **19** : 159-164.

Santruckova, H and Straskrava, M. 1991. On the relationship between specific respiration activity and microbial biomass in soils. *Soil Biol. Biochem* **23** : 525-532.

Schnurer, J. Clarholm, M. and Rosswall, T. 1985. Microbial biomass and activity in agricultural soils with different organic matter contents. *Soil Biol. Biochem* **17** : 611-618.

Singh, J.S. Raghubanshi, A.S. Singh, R.S. and Srivastava, S.C. 1989. Microbial biomass acts as a source of plant nutrients in dry tropical forest and savanna. *Nature* **338** : 499-500.

Smith, J.L. and Paul, E.A. 1990. The significance of soil microbial biomass estimations. pp. 357-376. *Soil Biochemistry* (Eds. J.M. Bollag & G. Stotzky), Marcel Dekker, New York.

Snedecor, G.W. and Cochran, W.G. 1967. *Statistical Methods.* The Iowa State Univ. Press. Ames.

Srivastava S.C. and Singh, J.S. 1988. Carbon and Phosphorus in the soil biomass of some tropical soils of India. *Soil Biol. Biochem.* **20** : 743-747.

Srivastava, S.C. and Singh, J.S. 1989. Effect of cultivation on microbial biomass C and N of dry tropical forest soil. *Biol. Fertil. Soils.* **8** : 343-348.

Srivastava, S.C. and Singh, J.S. 1991. Microbial C, N and P in dry tropical forest soils : effects of alternate land uses and nutrient flux. *Soil Biol. Biochem.* **23** : 117-124.

Van Veen, J.A. and Paul, E.A. 1981. Organic C dynamics in grassland soils. 1. Background information and computer simulation. *Can. J. Soil Sci.,* **61** : 185-201.

Van Veen, J.A. Ladd, J.N. and Amato, K. 1985. Turnover of carbon, nitrogen through the mircobial biomass in sandy loam and clayey soil incubated with [^{14}C (U)] glucose and [^{14}N] $(NH_4)_2$ SO_4 under different regimes. *Soil Biol. Biochem.* **17** : 747-756.

Witkamp, M. 1966. Rates of CO_2 evolution from the forest floor. *Ecology.* **47** : 492-494.

Chapter 5

Community Structure of Soil Oribatida of a Young Rubber Plantation and an Adjacent Wasteland in Tripura (India)

Tanmay Bhattacharya[1] AND Prabir Chakraborti[2]

1. Department of Zoology, Vidyasagar University, Midnapore-721 102, West Bengal
2. Department of Zoology, Women's College, Agartala-799 001, Tripura, India

Abstract. Oribatida of a four year old rubber plantation and an adjacent wasteland in Tripura (India) was studied for two consecutive years. In total 46 oribatid species have been recorded from the study area with 29 species encountered in both the sites. *Scheloribates praencisus interruptus* was the most common species in both the sites followed by *Haplochthonius simplex* in the wasteland and *Cosmochthonius lanatus diversiseta* in the rubber plantation. In the rubber plantation species diversity (H) was significantly lower and dominance index (DI) was significantly higher as compared to the wasteland. The findings suggest the rubber plantation in its initial stage represents a more stressed habitat than the fallow wasteland for the oribatid community.

Key words : *Community structure, Soil oribatid Mites, Wasteland, Rubbers-plantation, Tripura.*

Introduction

Oribatid mites constitute a very significant component of forest ecosystem where they play important role in the decomposition process and in the maintenance of soil fertility. These can also be utilised as an effective biological tool in assessing the degree of anthropogenic influences like afforestation and silvicultural practices. In spite of their ecological importance soil inhabiting oribatids have not received the desired attention in India. Though some progress has been made towards the understanding of the effects of afforestation and silvicultural practices on the soil fauna of the tropical forest and plantation through the investigation of Beck (1964, 1967), Maladague and Hilger (1963), Di Castri (1963), Imadate and Kira (1964), Ogino *et al.* (1965), Plowman (1979), Holt (1981, 1985) and Holt and Spain (1986) yet in India information on the ecology of soil fauna of forest is very scanty. Singh and Singh (1975), Chaudhuri and Banerjee (1975), Prabhoo (1976), Hazra (1978), Bhattacharya *et al.* (1981, 1985), Joy *et al.* (1986), Sinha *et al.* (1988), Boudjemma *et al.* (1991) have studied different aspects of the ecology of soil microarthropods of tropical forests and plantation in India.

The present communication deals with the community analysis of soil Oribatida of a rubber plantation and an adjacent wasteland in Tripura. The main objective of the present investigation is to assess the impact of anthropogenic interference on the oribatid mites of the soil in an afforested area by way of the introduction of an exotic plant species viz., *Hevea brasiliensis* Muell-Arg.

Material and Methods

The present investigation has been conducted at Ballamukha village (23° 19N and 91° 27E, elevation 50 m ASL) of Belonia sub-division in South Tripura District of Tripura State in North East India. The study site which was lying fallow after repeated slash and burn cultivation comprised of undulating landscape with uplands locally called 'tilla'. In 1981 a portion of this landscape, hence-forth designated as 'rubber plantation', was put under rubber cultivation. Soil samples have been collected from an adjacent 'wasteland' which is still unutilised.

In each site a 5m × 5m plot was selected for the purpose of sampling. Random samples were collected on the same date from both the sites at monthly interval for two consecutive years between May, 1985 and April, 1987. During each sampling occasion from each site 10 samples of 5 cm × 5 cm dimension up to the depth of 10cm were collected using a rectangular sampler for the extraction of oribatid mites. Thus from each site 240 samples were collected during the entire period of investigation.

Soil oribatid mites were extracted from the soil samples with the help of a modified "Tullgren apparatus" fitted with 20 funnels. 40 watt electric bulbs were used as light and heat source and the extraction procedure continued for 24 hours. Oribatid mites along with other soil inhabiting microarthropods were collected in glass vials containing 70 per cent alcohal.

Abundance (number/sample), relative abundance and frequency of all the oribatid species were analysed. Dominance of species was ascertained on the basis of relative abundance (RA) using Engelmann's scale (Engelmann, 1973) Species with RA > 3.2 per cent were considered as 'common' species and were subjected to further analysis such as estimation of standard error and 't' test for comparison of abundance between the sites and index of association between species (Southwood, 1978). Sorensen's quotient of similarity (Sorensen, 1948), species richness index (Menhinick, 1964), species diversity index (Shannon and Weaver, 1963) and dominance index (Mc Naughton and Wolf, 1970) were also analysed for the purpose of comparison of oribatid community of the two sites. Table 5.1 gives a comparative account of the physico-chemical properties of the soil of the sites.

Results

Soil of the two sites under study varied little in their edaphic composition. Only potassium was significantly more in the rubber plantation and water holding capacity was significantly greater in

TABLE 5.1. *Physico-chemical properties of soils of rubber plantation and wasteland.*

Edaphic parameters		Wasteland x ± SE (Range)	Rubber plantation x ± SE (Range)	't'
Colour		Brown 10YR 5/3	Brown 10YR 5/3	
Texture		Sandy loam	Sandy loam	
	Sand %	63.49 (63.78-63.19)	67.75 (65.96-69.54)	
	Silt %	21.54 (21.0-22.08)	15.46 (16.70-14.21)	
	Clay %	16.06 (14.72-17.32)	16.78 (16.24-17.32)	
Temperature (^{o}C)		27.2 ± 0.67 (21-32)	27.58 ± 0.65 (22-34)	-0.340
Water holding capacity (%)		51.55 ± 1.35 (34.22-65)	47.51 ± 0.61 (41.0-52.53)	2.727*
Electrical conductivity (μ mohs/cm)		10.35 ± 0.37 (6.52-14.32)	10.07 ± 0.19 (8.96-12.88)	0.673
Moisture content (%)		14.24 ± 1.11 (7.0 - 23.87)	13.26 ± 0.99 (4.08-23.12)	0.597
pH		6.41 ± 0.1 (5.5-7.74)	6.4 ± 0.09 (5.4-7.5)	0.144
Organic carbon (%)		1.58 ± 0.06 (1.08-2.16)	1.71 ± 0.04 (1.01-2.07)	-1.803
Total nitrogen (%)		0.096 ± 0.02 (0.031-0.155)	0.114 ± 0.008 (0.031-0.190)	-1.765
Available phosphate (ppm)		33.75 ± 3.61 (15-70)	31.79 ± 3.52 (15-70)	0.389
Potassium (ppm)		54.63 ± 3.95 (24-85)	70.56 ± 3.74 (29-100)	-2.928*
C : N ratio		21.12 ± 2.66 (7.18-52.68)	18.39 ± 2.20 (6.73-48.4)	0.791

* $P < .05$

TABLE 5..2. *List of oribatid species encountered in study sites*

Sl. No.	Oribatid species	WL	RP
	Family Mesoplophoridae		
1.	*Mesoplophora pectinata* Mahunka, 1973	+	+
2.	*M. crassisetosa* Subias and Sarker, 1983	+	-
	Family Phthiracaridae		
3.	*Haplophorella scapellata* Aoki, 1965	+	-
	Family Gehypochthoniidae		
4.	*Gehypochthonius* sp.	+	+
	Family Hypochthoniidae		
5.	*Hypochthonius rhadamanthus* Jacot, 1936	+	-
6.	*Malacoangelia similis* Sarkar and Subias, 1982	+	-
7.	*M. remigera* Berlese, 1913	+	-
	Family Cosmochthoniidae		
8.	*Cosmochtonius lanatus diversiseta* Sarkar and Subias, 1982	+	+
9.	*C. plumatus suramerican* Hammer, 1958	+	+
	Family Haplochthoniidae		
10.	*Haplochthonius simplex* Willmann, 1930	+	+
11.	*Haplochthonius* sp.	+	+
	Family Lohmanniidae		
12.	*Cryptacarus dendrisetosus* Bhattacharya *et al.*, 1972	-	+
13.	*Vepracarus cornutus* Sarkar and Subias, 1984	-	+
14.	*Paulianacarus foliatus* Mondal and Chakraborty, 1982	+	+
	Family Trhypochthoniidae		
15.	*Archegozetes magnus longisetosus* Aoki, 1965	+	-
	Family Eremulidae		
16.	*Eremulus nigrisetosus* Hammer, 1958	+	+
	Family Carabodiidae		
17.	*Phyllocarabodes* sp.	+	+
	Family Tectocaphiidae		
18.	*Tectocepheus velatus sarekensis* Tragardh, 1910	+	+
19.	*T. latilamellaris* Karder, 1974	-	+
	Family Octocephiidae		
20.	*Dolicheramaeus lineolatus* Hammer, 1979	+	+
	Family Oppiidae		
*21.	*Multioppia gracilis* Hammer, 1972	+	+
22.	*Oxyoppia polynesia* Hammer, 1972	+	+

Contd. Next page

Sl. No.	Oribatid species	WL	RP
*23.	*Ramusella (Ramusella) chulumeneinsis* Hammer 1968	+	+
*24.	*Brachyoppia brevipectinata* Kok, 1967	+	+
*25.	*Arcoppia mcadami* Balogh and Balogh, 1968	+	+
*26.	*A. fenestralis orientalis* Balogh and Balogh, 1986	-	+
27.	*Stachyoppia granulosa* Subias and Sarker,1983	-	+
	Family Suctobelbidae		
*28.	*Suctobelba amurica* Krivolutasky, 1966	+	-
*29.	*Novosuctobelba dentissima* Hammer, 1977	-	+
	Family Mechadobelbidae		
30.	*Mechadobelba symmetrica* Balogh, 1958	+	+
	Family Chaunoproctidae		
31.	*Chaunoproctus abali* Bhaduri *et al* 1975	-	+
	Family Oribatulidae		
*32.	*Oribatula unispinata* Hammer, 1958	+	-
33.	*Scheloribates praencisus interruptus* Berlese, 1916	+	+
34.	*S. fimbriatoides* Hammer, 1977	+	+
*35.	*S. fimbriatus* Thor, 1930	+	+
	Family Haplozetidae		
36.	*Rostrozetes foveolatus* Sellnick, 1925	+	+
37.	*Xylobates* sp.	+	+
*38.	*Tentaculozets* sp.	+	+
*39.	*Haplozetes minimicoma* Beck, 1964	+	-
	Family Ceratozetidae		
40.	*Hypozetes laysanensis* Aoki, 1964	-	+
	Family Oribatelloidae		
41.	*Lamellobates pallustris* Hammer, 1958	+	+
42.	*Paralamellobates ceylanicus* (Oudemann, 1915)	+	+
	Family Galumnidae		
43.	*Galumna flabellifera* Hammer, 1958	+	+
*44.	*Pergalumna magnipora* capensis Engelbrecht, 1972	+	+
45.	*Trichogalumna* sp.	+	+
*46.	*Disparagalumna* sp.	+	+
	Total number	38	37
	Quotient of similarity (%)	77.33	

+= Present — = Absent * = New record from India
WL= Wasteland RP = Rubber plantation

the wasteland (Table 5.1).

Oribatida is the most common group of microarthropods in the study sites constituting 26% and 23.3 per cent of the total micro-arthropods and 49.6 per cent and 48.5 per cent of Acari in the wasteland and rubber plantation respectively. Oribatid fauna of the two sites consisted of 46 species under 37 genera and 21 families (Table 5.2). Two genera and 13 species are being reported for the first time from Tripura as well as from India (marked by in Table 5.2). 38 species of oribatids were collected from the wasteland and 37 from the rubber plantation. Of these 29 species were common in both the sites and as such Sorensen's quotient of similarity is 77 per cent which indicates that both the sites are strongly similar in respect of oribatid fauna.

In the wasteland there were 4 dominant, 5 subdominant, 9 racedent and 20 subracedent species as compared to 1 eudominant, 1 dominant, 6 subdominant, 5 racedent and 24 subracedent species in the rubber plantation (Table 5.3). *Scheloribates praencisus interruptus* was the most common species in both the sites in terms of abundance and relative abundance. This was followed by *Haplochtonius simplex* in wasteland and *Cosmochthonius lanatus diversiseta* in the rubber plantation. *Arcoppia mcadami* and *Xylobates* sp. were significantly more in number in the wasteland and *S. praencisus interruptus* was the most frequent species and was followed by *H. simplex* and *C. lanatus diversiseta* in the wasteland and *C. lanatus diversiseta* and *H. simplex* in the rubber plantation in that order of frequency. Species diversity index (H) was significantly higher ($P<0.01$) in the wasteland and dominance index (DI) was significantly higher ($P<0.05$) in the rubber plantation (Table 5.4). All the common species in both the sites had negative association between themselves (Table 5.5).

Discussion

Physico-chemical nature of the soil of two sites did not show significant difference. Close proximity of the two sites and very young age of the rubber stand might be the reason for such similarity. Significantly higher potassium content of the soil in the rubber plantation as compared to the wasteland may be attributed to the prescribed burning of the vegetation and application of fertilizer prior to the afforestation. Lower water holding capacity of the rubber plantation might be due to the reduced pore space which might have been caused by the prescribed burning as reported by Metz and Farrier (1973).

Sorensen's quotient of similarity (77%) indicates that both the sites are strongly similar in oribatid faunal composition. This can almost certainly be attributed to the proximity of the sites and is suggestive of the fact that the rubber afforestation could not alter the oribatid community composition at least in the initial stage. Notwithstanding such strong faunal similarity there are 8 species which are present only in the rubber plantation and 9 species which are restricted to the wasteland. Afforestation perhaps has caused disappearance of some species while it has created favourable environment for the existence of others. Bhattacharya *et al.* (1981), Aoki and Harada (1985) and Harada and Aoki (1986) have also found that man made plantation result into the addition and alteration of oribatid species.

TABLE 5.3. *Frequency (F), abundance (x) and relative abundance (RA) of oribatid species in wasteland and rubber plantation.*

Sl. No. of oribatid species vide Table 5.2	Wasteland			Rubber plantation			
	Frequency (F)	Abundance (x ± SE)	Relative abundance (RA)	Frequency (F)	Abundance (x ± SE)	Relative abundance (RA)	't'
1	2	3	4	5	6	7	8
1.	0.8 r	0.008	0.4 SR	0.42 r	0.004	0.9 SR	
2.	0.4 r	0.42	0.004SR				
3.	1.7 r	0.02	0.9 SR				
4.	8.3 o	0.13±0.03	5.5 SD	8.3 o	0.16±0.05	6.5 SD	-0.514
5.	0.8 r	0.008	0.4 SR				
6.	0.8 r	0.008	0.4 SR				
7.	0.4 r	0.004	0.2 SR				
8.	16.7 f	0.27±0.04	11.8 D	18.8 f	0.31±0.04	13.1 D	-0.707
9.	2.6 r	0.03	1.6 R	1.7 r	0.013	0.6 SR	
10.	18.8 f	0.36±0.06	15.9 D	13.8 o	0.24±0.04	9.4 SD	1.664
11.	9.2 o	0.15±0.03	6.9 SD	5.8 o	0.11±0.03	5.7 SD	1.178
12.				0.4 r	0.004	0.2 SR	
13.				0.4 r	0.004	0.2 SR	
14.	6.7 o	0.10±0.02	4.4 SD	7.9 o	0.12±0.02	5.1 SD	-0.707
15.	0.4 r	0.004	0.2 SR				
16.	2.1 r	0.025	1.1 R	2.5 r	0.025	1.1 R	
17.	1.7 r	0.029	1.2 R	0.4 r	0.004	0.2 SR	
18.	0.8 r	0.008	0.4 SR	0.8 r	0.008	0.4 SR	
19.				0.4 r	0.012	0.6 SR	
20.	2.9 r	0.033	1.4 R	0.4 r	0.004	0.2 SR	
21.	1.7 r	0.29	1.2 R	0.8 r	0.008	0.4 SR	
22.	0.8 r	0.012	0.5 SR	2.1 r	0.033	1.5 R	
23.	4.6 r	0.07±0.02	3.2 SD	7.1 o	0.08±0.02	3.3 SD	-0.353

Contd. next page

1	2	3	4	5	6	7	8
24.	0.8 r	0.012	0.5 SR	0.4 r	0.004	0.2 SR	
25.	5.4 o	0.08±0.02	3.7 SD	2.5 r	0.03	1.5 R	2.683*
26.				0.4 r	0.004	0.2 SR	
27.				0.4 r	0.004	0.2 SR	
28.	1.3 r	0.012	0.5 SR				
29.				0.4 r	0.004	0.2 SR	
30.	0.8 r	0.008	0.4 SR	0.4 r	0.004	0.2 SR	
31.				0.4 r	0.004	0.2 SR	
32.	0.4 r	0.004	0.2 SR				
33.	25.4 f	0.39±0.04	16 D	32.1 a	0.72±0.15	32.3 E	-2.125*
34.	0.4 r	0.002	0.2 SR	0.4 r	0.004	0.2 SR	
35.	4.2 r	0.062	2.7 R	2.1 r	0.02	1 R	
36.	2.1 r	0.047	1.8 R	1.7 r	0.037	1.7 R	
37.	13.8 o	0.23±0.04	10.1 D	7.9 o	0.12±0.03	4.6 SR	2.2*
38.	0.4 r	0.004	0.2 SR	0.4 r	0.004	1. R	
39.	0.4 r	0.004	0.2 SR				
40.				0.8 r	0.008	0.4 SR	
41.	4.6 r	0.045	2.1 R	4.2 r	0.05	2.4 R	
42.	0.8 r	0.013	0.5 SR	0.7 r	0.013	0.6 SR	
43.	2.1 r	0.017	0.7 SR	5.4 o	0.091	0.3 SR	
44.	1.7 r	0.03	1.2 SR	0.8 r	0.016	0.7 SR	
45.	0.8 r	0.012	0.5 SR	0.4 r	0.004	0.2 SR	
46.	0.4 r	0.008	0.5 SR	0.4 r	0.004	0.2 SR	

E = Eudominant, D = Dominant, SD = Subdominant,
R = Racedent, SR = Subracedent
a = abundant, f = frequent, o = occasional, r = rare
't' = Estimated for 'common' species (RA > 3.2) only
*P < 0.05

TABLE 5.4. *A comparison between some biocoenological parameters of oribatid community of the two sites*

Parameters	Wasteland	Rubber plantation	't'
Abundance	2.45 ± 0.17	2.35 ± 0.20	0.381
Number of species	38	37	
Dominance index (DI)	47.83 ± 2.67	56.25 ± 2.45	-2.323*
Diversity index (H)	0.82 ± 0.03	0.71 ± 0.02	3.055**
Quotient of similarity	77%		

* P < 0.05   ** P < 0.01

TABLE 5.5. *Index of association (Iai) between nine 'common' oribatid species.*

RUBBER PLANTATION

WASTE LAND	4	8	10	11	14	23	25	33	37	RUBBER PLANTATION
4	x	-0.912	-0.806	-0.903	-0.830	-1	-1	-0.676	-0.76	4
8	-0.916	x	-0.751	-0.897	-0.669	-0.958	-1	-0.764	-0.825	8
10	-0.904	-0.531	x	-1	-0.731	-0.893	-1	-0.715	-0.829	10
11	-0.575	-0.877	-0.906	x	-0.843	-1	-1	-0.958	-0.921	11
14	-1	-0.818	-0.915	-0.1	x	-0.836	-0.941	-0.887	-0.857	14
23	-0.795	-0.925	-0.891	-0.1	-0.906	x	-1	-0.895	-0.836	23
25	-0.846	-0.880	-0.947	-0.925	-0.636	-0.513	x	-0.937	-0.959	25
33	-0.968	-0.540	-0.633	-0.725	-0.752	-0.807	-0.901	x	-0.778	33
37	-0.863	-0.650	-0.586	-0.622	-0.975	-0.817	-1	-0.281	x	37
	4	8	10	11	14	23	25	33	37	

WASTELAND

4, 8, 10, 11, 14, 23, 25, 33, 37, - are serial number of oribatid species vide Tab. 5.2.

Tropical forests are characterised by rich diversity of species (Williums, 1942; Di Castri, 1963). As compared to a true tropical forest the number of oribatid species found in the rubber plantation is quite low and this is most likely because of the monocropped nature and young age of the stand. Edwards and Lofty (1969) has also opined that monoculture can eliminate those animal species which are associated with other plants. However, the number of oribatid species recorded in this study is higher as compared to the similar studies by Singh and Mukherji (1971), Gupta and Mukherji (1976), Stanton (1979), Bhattacharya *et al.* (1982), Sanyal (1982) and Bhattacharya *et al.* (1985) in tropical regions but were lower as compared to the findings of Fujikawa (1970), Schenker (1984) and Kaneko (1985) who worked in temperate regions.

As in the present study Corpuz-raros (1980) also found that *S. praencisus interruptus* was the most abundant species in most of the habitats in Philippines. His findings relating to the relative abundance of this species in mono culture stand and forests are in agreement with the present findings. This species is one of the most common species of soil oribatids in the Eastern and North Eastern India (Bhattacharya *et al.* 1981, 1982; Bhattacharya *et al.* 1985; Bhattacharya and Bhattacharya, 1985), is circum tropical in distribution (Corpuz-raros, 1980) and seems to possess extraordinary adaptability and capability to tolerate wide range of environmental gradients.

Lack of significant positive association between the 'common' species, as revealed by the index of association, is indicative of different environmental requirements or active competition between the species involved. Similar negative association for most of the dominant species in an *Acacia* plantation has previously been shown by Bhattacharya *et al.* (1982).

Importance of a species in a community and the role played by it can be analysed with the help of dominance index (DI). According to Karr (1971) DI for avifauna increases with the increase in harshness of the environment and decreases with vegetational development. Bhattacharya *et al.* (1981) also made similar observations for oribatid fauna. Significantly higher DI in the rubber plantation as compared to the wasteland in the present investigation implies that the rubber plantation is perhaps offering a comparatively harsher environmental condition as compared to the wasteland for oribatid community and as such the functional influence has concentrated on only one or two species. This explains why two dominant species (including one eudominant species) were encountered in the rubber plantation as compared to four dominant species in the wasteland. The relative harshness of the environment of the rubber plantation might have been due to the human interference in the form of intermittent weeding in that site and consequent lack of dense cover of undergrowth as well as due to disturbance of soil profile.

Although the sites under consideration had a strongly similar oribatid faunal composition. species diversity index (H) was significantly lower in the rubber plantation as compared to the wasteland, Species diversity tends to be low in simple less stable ecosystem (Fischer, 1960) and shows marked correlation with vegetational structure and physiognomy (Whittaker, 1970). Stanton (1979) noted that mite diversity in forest is greater as compared to the grassland and Anderson and Hall (1977) and Bhattacharya *et al.* (1981) found that species diversity of Oribatida can also be correlated with the stability and complexity of the habitats. Thus findings of the present investigation suggests that the rubber afforestation is capable of significantly altering the community composition of Oribatida which is reflected in the change in the species diversity and dominance indices. Due to various agricultural manipulations rubber plantation in its initial phase provides a relatively less stable habitat with more harsh environment for soil fauna which in turn results into a structurly less complex community as compared to the nearby wasteland.

Present study reveals that the rubber plantation in its initial phase of development provides a more stressed habitat for the soil inhabiting Oribatida which is also reflected in the species diversity and dominance indices. Therefore it may not be wrong to assume that analysis of community structure of soil inhabiting Oribatida can provide useful clues regarding the impact of human interference on the environment in the from of silviculture management. However, further study in rubber plantation sites with varying age could perhaps distinctly indicate its impact on faunal density and diversity to develop a biological tool for system assessment and management.

Acknowledgements

Authors are grateful to Prof. H.G. Sengbusch, Institute of Arthropodology and Parasitology, Georgia Southern College, Georgia, U.S.A. for his valuable suggestions. Financial assistance from University Grants Commission, New Delhi (Ref. No. F. 8-7(16)/87/SR II) to P.C. is also acknowledged.

REFERENCES

Anderson, J.M. and Hall, H. 1977. Cryptostigmata species diversity and soil habitat structure. *Ecol. Bull.* Stockholm, **25** : 473-475.

Aoki, J. and Harada, H. 1985. Formation of environmental protection forests and change in soil fauna, specially oribatid mites. *Bull. Environ. Sci. Tech.* **12** (1) : 125-135.

Beck, L. 1964. Tropische Boden-fauna im Wechsel Von Regeneund Trockenzeit. *Nature & Museum,* **94** : 63-71.

Beck, L. 1967. Die Boden-fauna des neotropischen Regenwaldes. *Atas do Simposio sobre a Biota Amazonica,* **5** (Zool.) : 97-101.

Bhattacharya, T. and Bhattacharya, J. 1985. Impact assessment of sensitiveness of soil Oribatid species to the waste water from a coal distillation plant. *Entomon.* **10** (4) : 267-270.

Bhattacharya, T., Halder, G. and Saha, R.K. 1985. Soil micro arthropods of a rubber plantation and a natural forest. *Environment and Ecology.* **3** (2) : 143-147.

Bhattacharya, T., Joy, S. and Joy, V.C. 1981. Community structure of soil Cryptostigmata under different vegetational conditions at Santiniketan. *J. Soil Biol. Ecol.* **1** : 27-42.

Bhattacharya, T., Joy, S. and Joy, V.C. 1982. Cryptostigmatid population of some cultivated and uncultivated soils. *Proc. Symp. Ecol. Anim. Popul. Zool. Surv. India,* Pt. **4** : 75-90.

Boudjema, G., Jullen, J.M., Sarkar, S. and Cancela da Fonseca, J.P. 1991. Étude par I'analyse Statistique multi-lineare de I'impact des facteurs physico-chemiques sur I'abondance des Microarthropodes édaphiques d'une forêt de mousson en Inde orientale. *Rev.Écol. Biol. Sol.* **28** (3) : 303-322.

Chaudhuri, D.K. and Banerjee, S. 1975. Qualitative and quantitative composition of Acari and Collembola in relation to soil organic matter microbes complex. *Oriental Insects.* **9** : 313-316.

Corpuz-raros, L.A. 1980. Philippine Oribatei (Acarina) V. Scheloribates Berlese and related genera (Oribatulidae). *J. Biol.*, Philippines, **9** (2-3) : 69-245.

Di Castri, F. 1963. Etat de nos connaissances sur les biocenoses edaphiques due Chili. p 375-385. In : *''Soil Organisms''* (eds. J. Docksen and J. Van der Drift). North Holland, Amsterdam.

Edwards, C.A. and Lofty, J.R. 1969. The influence of agricultural practice on soil microarthropod populations. p 237-247. In : *''The Soil Ecosystem''* (ed. J. G. Sheals). Syst. Assoc., London.

Engelmann, H.D. 1973. Undersuchangen Zur Erfassung Pedozoo-gener Komponent in definicten Okosystem. Forschungsber, Staatl. Mus. Naturkde, Gorlitz. *J. Acta Hydrobiol.* **23** (4) : 349-361.

Fischer, A.G. 1960. Latitudinal variation in organic diversity. *Evolution.* **14** : 64-81.

Fujikawa, T. 1970. Relation between oribatid fauna and some environments of Nopporo National Forest in Hokkaido (Acarina : Cryptostigmata). III Oribatid fauna in soils under four different vegetations. *Appl. Ent. Zool.* **5** (2) : 69-83.

Gupta, G.P. and Mukherji, S.P. 1976. Qualitative composition of the fauna of cultivated and uncultivated soil. *Indian J. Ent.* **38** (4) : 313-314.

Harada, H. and Aoki, J. 1986. Oribatid mite communities in the soil of sandbreak plantation along the shonan beach. *Bulletin of the institute of Environmental Science and Technology.* **13** (1) : 207-215.

Hazra, A.K. 1978. Ecology of Collembola in a deciduous forest floor of Birbhum district, West Bengal in relation to soil moisture. *Oriental Insects.* **12** : 265-274.

Holt, A. 1981. The vertical distribution of cryptostigmatid mites, soil organic matter and macroporosity in three North Queensland rainforest soils. *Pedobiologia.* **22** : 202-209.

Holt, J.A. 1985. Acari and Collembola in the litter and soil of three North Queensland rainforests. *Australian Journal of Ecology.* **10** : 57-65.

Holt, J.A. and Spain, A.V. 1986. Some biological and chemical changes in a North Queensland soil following replacement of rainforest with *Araucaria cunninghamii* (Coniferae : Araucariaceae). *Journal of Applied Entomology.* **23** : 227-237.

Imadate, G. and Kira, T. 1964. Notes on the soil microarthropod collection made by the Thai-Japanese biological expedition, 1961-1962. Description of collecting sites, group contents and some ecological characters and 180 samples from South East Asia. *Nature and Life in South East Asia.* Fauna and flora Research Society, Kyoto, 81-111.

Joy, V.C., Ray, N. and Bhattacharya, J. 1986. Phytogenetic composition of litter and soil inhabiting microarthropods in an artificial forest at Santiniketan. *Proceedings of the III Oriental Entomology Symposium.* Vol. I : Trivandrum, India, 193-202 p.

Kaneko, N. 1985. A comparison of oribatid mite communities in two different soil types in a cool temperate forest in Japan. *Pedobiologia.* **28** : 255-264.

Karr, J.R. 1971. Structure of Avian communities in selected Panama Illinois habitats. *Ecological Monographs.* **41** (3) : 207-233.

Maladague, M.E. and Hilger, F. 1963. Observations faunisteques et micro biologiques dans quelques biotopes forestrier equatoriux. p. 368-374. In : *''Soil Organisms''* (eds. J. Doeksen and J. Van der Drift). North Holland, Amsterdam.

Mc Naughton, S.J. and Wolf, L.L. 1970. Dominance and niche in ecological systems. *Science.* **167** : 131-136.

Menhinick, E.F. 1964. A comparison of some species diversity indices applied to samples of field insects. *Ecology.* **45** : 859-861.

Metz, L.J. and Farrier, M.H. 1973. Prescribed burning and population of soil mesofauna. *Environ. Entomol.* **2** : 433-440.

Ogino, K.P., Saichuae, P. and Imadate, G. 1965. Seasonal changes of soil microarthropod populations in central Thailand. *Nature and life in South East Asia.* **4** : 303-315.

Plowman, K.P. 1979. Litter and soil fauna of two Australian subtropical forests. *Aust. J. Ecol.* 4 : 87-104.

Prabhoo, N.R. 1976. Soil microarthropods of a virgin forest and adjoining tea fields in Western Ghats in Kerala-a brief ecological study. *Oriental Insects.* **10** : 435-442.

Sanyal, A.K. 1982. Soil oribatid mites and their relation with soil factors in West Bengal. *J. Soil. Biol. Ecol.* **2** : 8-17.

Schenker, R. 1984. Spatial and seasonal distribution patterns of oribatid mites (Acari : Oribatei) in a forest soil ecosystem. *Pedobiologia.* **27** : 133-149.

Shannon, C.E. and Weaver, W. 1963. *The mathematical theory of communication.* Urbana, University of Illinois press, 177 p.

Singh, J. and Mukherjee, S.P. 1971. Qualitative composition of soil arthropods in some fields at Varanasi (India). *Oriental Insects.* **5** (4) : 487-494.

Singh, J. and Singh, U.R. 1975. An ecological study of soil microarthropods from soil and litter of tropical deciduous forest of Varanasi (India). *Trop. Ecol.* **16** : 81-85.

Sinha, P.B., Sen, N.S. and Jafri, H.A. 1988. On the seasonal population fluctuations of Collembola and Acarina in a deciduous forest at Ranchi, *J. Environ. Biol.* **9** (1) (Suppl.) : 79-83.

Sorensen, T. 1948. A method of stabilising groups of equivalent amplitude in plant sociology based on the similarity of species content and its application to analysis of vegetation on Danish commons. *Biol. Skr.* **5** : 1-34.

Southwood, T.R.E. 1978. *Ecological Methods.* ELBS and chapman and Hall, London, 19 p.

Stanton, N.L. 1979. Patterns of species diversity in temperate and tropical litter mites. *Ecology.* **70** : 295-304.

Whittaker, R.H. 1970. *Communities and ecosystem.* Macmillan Company, New York, 58 p.

Williums, E.C. 1942. An ecological study of the floor fauna of Panama rainforest. *Bull. Chicago Acad. Sc.* **6** : 63-124.

Chapter 6

Ecological Studies on Soil Mites (ACARI) in India : A Review

A.K. Sanyal

Zoological Survey of India, 'M' Block, New Alipur, Calcutta-700 053, India

Abstract. The paper reviews the research work carried out in India on density, species diversity, seasonal fluctuation, vertical distribution, faunal make-up etc. of soil mites in relation to physico-chemical factors, vegetation and soil microbes, agriculture practices, land pollution, deforestation and fire.

Key words : *Soil mites, Physico-chemical factors, soil microbes, vegetation, agriculture practice, land pollution, seasonal fluctuation, vertical distribution, faunal make-up.*

Introduction

The study of soil fauna has been a long neglected field of zoology as biologists have been slow and hesitant to accept soil as a medium for ecological studies particularly because of the complexity of soil as a biological system. With the exception of a very few groups of soil dwelling animals little was brought to light till the first few decades of this age. The obstacles for study of soil animals were the cryptozoic mode of life and complexity in collecting animals for quantitative and qualitative analysis. Considerable progress has been made in the past two decades in resolving some of the major methodological problems. The occurrence of several millions of animals per square metre of soil poses problems in extraction, preparation and study. Among these different groups of soil organisms the most important biotic component is mite and considering the above fact it is essential of emphasize the need for studying the abundance of these microarthropod in different soil and consequently with them ecological studies of various nature have been made in different parts of the world.

Early work relating to the soil fauna was attempted by Diem (1903) on the soil fauna of Alps, and a preliminary quantitative study of soil invertebrates in a grassland was also made by McAtee

TABLE 6.1. *Ecological Studies on Soil Mites in Different Parts of India*

Author	Soil fauna studied	Sample unit size and collection method employed	Habitat, soil type and locality
1	2	3	4
Trehan 1945	Microarthropods including mites	22.8 × 22.8 × 22.8 cm, separation by washing the soil	Cotton field and fallow land, sandy-loam soil, loyalpur, Punjab.
Mukherjee & Singh 1970	Microarthropods including mites	7.62 × 10.16 × 22.8 cm, Ladell's apparatus	Rose garden, alluvial soil, Varanasi, Uttar Pradesh.
Singh & Mukherjee, 1971, 1973	Microarthropods including Cryptostigmata, Mesostigmata, Prostigmata, Astigmata	7.62 × 10.16 × 22.8 cm, Ladell's apparatus	Fields with rose, sugarcane, falsa and mixed vegetation alluvial soil, Varanasi, Uttar Pradesh.
Banerjee 1974a, 1974b, 1976, 1988	Cryptostigmata Mesostigmata, Prostigmata, Astigmata	10 cm² in surface area, modified Tullgren funnels & high gradient extraction apparatus	Fields with mixed vegetation, alluvial soil, Calcutta and Burdwan, West Bengal.
Choudhuri & Banerjee 1975, 1977	Collembola, Cryptostigmata, Mesostigmata, Prostigmata, Astigmata	10 cm² × 10cm, high gradient extraction apparatus	Fields with mixed vegetation, alluvial and laterite soils, Burdwan, Bankura, Hooghly and Birbhum, West Bengal.
Singh & Pillai 1975, 1981	Microarthropods including Cryptostigmata, Mesostigmata, Prostigmata, Astigmata	10 × 7.5 × 22.5 cm, Ladell's apparatus	Banana field, citrus orchard, sorghum and berseem fields, fallow land, alluvial soil, Varanasi, Uttar Pradesh.
Singh (J) & Singh 1975	Microarthropods including mite in total	10 × 7.5 × 22.5 cm, Ladell's apparatus	Mixed forest, alluvial soil, Varanasi, Uttar Pradesh.

(Contd. next page)

(Contd. Table 6.1)

1	2	3	4
Singh (U.R.) & Singh 1975	Microarthropods including mite in general	10 × 7.5 × 22.5 cm, Ladell's apparatus	Grassland, Himalayan soil, Uttar Pradesh.
Prabhoo 1976	Microarthropods including Cryptostigmata, Mesostigmata	2.5 × 2.5 × 10 cm, Berlese-Tullgren Funnels	Mixed forest and tea plantation, sandy-loam soil, Trivandrum, Kerala.
Joy & Bhattacharya 1977	Cryptostigmata	5 × 5 × 10 cm, Tullgren funnels	Grassland, laterite soil, Birbhum, West Bengal.
Bhattacharya & Joy 1978, 1980	Microarhropods including Cryptostigmata, Mesostigmata, Prostigmata	5 × 5 × 10 cm, Tullgren funnels	Paddy field, fodder field, fallow grassland, wheat field, laterite soil, Birbhum, West Bengal.
Bhattacharya 1979	Microarthropods including Cryptostigmata, Mesostigmata, Prostigmata	Sample unit size not specified, modified Tullgren funnels	Eroded laterite area, unclutivated and wastelands, mixed forest and cultivated area, laterite soil, Birbhum, West Bengal.
Bhattacharya & Raychaudhuri 1979	Microarthropods indluding Cryptostigmata, Prostigmata	5 × 5 × 5 cm Tullgren funnels	Fields with mixed vegetation, laterite soil, Birbhum, West Bengal.
Choudhuri & Pande 1979, 1982	Cryptostigmata, Mesostigmata	Sample unit size not specified, modified Tullgren funnels	Fields with mixed vegetation, sandy-loam soil, Darjeeling (2225 m), West-Bengal.
Bhattacharya, Bhattacharya & Banerjee 1980	Microarthropods including Cryptostigmata, Mesostigmata, Prostigmata, Astigmata	5 × 5 × 5 cm, Tullgren funnels	Smoke affected field with mixed vegetation, laterite soil, Birbhum, West Bengal.
Bhattacharya, Joy & Joy 1980, 1981	Cryptostigmata	5 × 5 × 10 cm, Tullgren funnels	Paddy fields, laterite soil, Birbhum, West Bengal.

(Contd. next page)

(Contd. Table 6.1)

1	2	3	4
Bhattacharya & Bhattacharya 1981, 1984	Cryptostigmata	5 × 5 × 10 cm, Tullgren funnels	Polluted and non-polluted sites, alluvial soil, Durgapur, West Bengal
Ghatak & Roy 1981, 1991	Cryptostigmata, Mesostigmata, Prostigmata	Sample unit size not specified, modified Tullgren funnels	Cultivated and forest lands, alluvial soils, Hooghly, West Bengal
Pai & Prabhoo 1981	Microarthropods including Cryptostigmata, Mesostigmata, Prostigmata	12.57 cm^2 in surface area, Berlese-Tullgren funnels	Paddy field and field with mixed vegetation south Kerala.
Sanyal 1981a, 1981b, 1982, 1991a, 1991b	Cryptostigmata, Mesostigmata, Prostigmata	17.36 cm^2 in surface area, depth 5 cm, modified Macfadyen expedition funnels	Fields with mixed vegetation and cultivated fields, deltaic saline soil, 24-Parganas (South), West Bengal.
Hazra, Mukhopadhya, Guha & Sanyal 1982	Microarthropods including Crypotstigmata, Mesostigmata, Prostigmata	8 cm^2 inner diameter of the corer, modified Tullgren funnels	Non-polluted and polluted fields with mixed vegetation, alluvial soil, Durgapur, Burdwan, West Bengal.
Sanyal & Bhaduri 1982	Cryptostigmata	17.36 cm^2 in surface area, depth 5 cm, modified Macfadyen expedition funnels	Fields with mixed vegetation, daltaic saline soil, 24-Parganas (South), West Bengal.
Mitra, Hazra, Sanyal & Modal 1983	Collembola, mites in total	5 × 5 × 5 cm, modified Macfadyen expedition funnels	Grassland and drainage, alluvial soil, Calcutta.
Sanyal & Sarkar 1983	Cryptostigmata	5 × 5 × 5 cm, modified Macfadyen expedition funnels	Cultivated and uncultivated fields, deltaic saline soils, 24-Paraganas (South), West Bengal.
Hazra & Sanyal 1989	Collembola, Cryptostigmata, Mesostigmata	5 × 5 × 2.5 cm, modified Macfadyen expedition funnels	Drainage site, alluvial soil, Calcutta.

(Contd. next page)

^Contd. Table 6.1)

1	2	3	4
Sengupta & Sanyal 1989	Cryptostigmata, Mesostigmata	Sample unit size not specified, modified Berlese-Tullgren funnels	Fields with mixed vegetation, alluvial soil, different places of south Bihar.
Hazra & Choudhuri 1990	Microarthropods including Cryptostigmata, Mesostigmata, Prostigmata	8 cm² cross sectional diameter, modified Macfadyen expedition funnels	Deforested area fallow land, agricultural fields, reserve forest, irrigtion cannal and sewage canal embankment, polluted lands, highway site, alluvial soil, different places of West Bengal.
Alfred, Darlong, Hattar & Pual 1991	Collemabola, Cryptostigmata, Mesostigmata Prostigmata, Astigmata	5 × 5 cm², depth 10 cm, modified Tullgren funnels	Pine forest, deciduous forest orchard, cultivated fields (Jhum, Terrace) and fallow land, sandy-loam soil, Khasi Hills, Meghalaya.
Banerjee & Sanyal 1991	Cryptostigmata	10 cm² in surface area & 10 cm depth modified Tullgren funnels	Mixed forest, laterite soil, Birbhum, West Bengal.
Haq & Ramani 1991	Microarthropods including Cryptostigmata, Mesostigmata, Prostigmata	2 × 2 × 10 cm, modified Tullgren funnels	Bamboo grove and grassland, sandy-loam soil, Kerala.
Hazra 1991	Microarthropods including Cryptostigmata, Mesostigmata, Prostigmata	8 cm² cross sectional diameter of the corer, modified Macfadyen expedition funnels	Reserve forest and deforested sites alluvial soil, Nadia, West Bengal.
Joy & Chakravorty 1991	Collembola, Cryptostigmata, Prostigmata	Sample unit size not specified, Tullgren funnels	Cultivated fields, laterite soil, Birbhum, West Bengal.

(Contd. next page)

(Contd. Table 6.1)

1	2	3	4
Majumder & Deb 1991, 1991b	Microarthropods including Cryptostigmata, Mesostigmata	150cc, modified Tullgren funnels	Cultivated and uncultivated fields, alluvial soil, West Dinajpur, West Bengal.
Prabhoo, Pai & Namboory 1991	Cryptostigmata, Mesostigmata, Prostigmata	25 $cm^2 \times$ 20 cm, Tullgren funnels	Tropical evergreen forest, sandy-loam soil, Western Ghats, Kerala.
Sarkar 1991	Microarthropods including Cryptostigmata, Mesostigmata, Prostigmata	7 × 7 × 7 cm, modified Tullgren funnels	Field with mixed vegetation, laterite soil, Tripura.
Sengupta & Sanyal 1991	Microarthropods including Crypotstigmata, Mesostigmata, Prostigmata	5 × 5 × 5 cm, modified Macfadyen expedition funnels	Paddy field, grassland, alluvial soil, Hooghly West Bengal.
Sheela & Haq 1991	Cryptostigmata	2 × 2 × 10 cm, modified Tullgren funnels	Twelve different vegetated fields sandy-loam soil, Calicut, Kerala.
Hattar, Alfred & Darlong 1992	Collembola, Cryhptostigmata, Mesostigmata	5 × 5 × 10 cm, modified Tullgren funnels	Field with Jhum cultivation, pin forest, sandy-loam soil, Khasi Hills, Meghalaya.
Sanyal & Sarkar 1993	Cryptostigmata	17.36 cm^2 in surface area, depth 5 cm, modified Macfadyen expedition funnels	Banana, Jamun and bamboo plantation, alluvial soil, Howrah, West Bengal.

(1907) who computed 13.7 million invertebrates per acre by examining an area of 4 sq.ft. of the grassy meadow at Washington. Later paedobiologists from different parts of the world concentrated their attention on the effect of environmental conditions and soil microclimate on faunal make up.

The first study on ecological aspect of soil microarthropods including mites in India was done by Trehan (1945). Then after a long gap of about three decades Mukherjee and Singh (1970) at the Banaras Hindu University, Uttar Pradesh initiated the studies on soil microarthropods in relation to different edaphic factors and soil ecosystems. Later several workers took up this field of studies and published a number of papers (Table 6.1). It may also be mentioned here that the works so far done are restricted to the soil ecosystems of only four Indian states like West Bengal, Uttar Pradesh Tripura and Kerala. More studies on ecology of soil inhabiting mites in India are needed to frame a clear picture of population dynamics and seasonal abundance of these tiny animals of great economic importance.

REVIEW OF ECOLOGICAL STUDIES ON SOIL INHABITING MITES IN INDIA

I. Studies inrelation to mite population and physico-chemical factors of soil

(A) *Physical factors*

The physical factors of soil like temperature, moisture content, relative humidity (R/H), pH, electrical conductivity, rainfall, etc. have profound influence on the life of soil mites. To ascertain the role of these edaphic factors on mite population it might be assumed that the factors exert significant or insignificant and direct or indirect effects either singly or in cumulative way. Several authors in India have attempted to study the effect of these factors on different groups of soil inhabiting mites (Table 6.2).

Among the edaphic factors studied, temperature showed marked variation with change of season. The review of the studies so far done indicates that soil temperature in most of the cases had a positive correlation with the mite population as a whole and also separately with cryptostistimata and mesostigasmata. But this correlation did not always appear to be significant as there were some species having preference for low temperature and as such high temperature decreased the population density while others exhibited a higher range of temperature tolerance and as a result population increased at high temperature.

The moisture content of soil also showed wide range of variation and significant positive correlation with the total population of mites as well as with individual mite groups. So it may be concluded that mite population is always high in soil with sufficient moisture content. But in one study made by Choudhury and Pande (1982) a significant negative correlation with the moisture content of soil was observed.

The pH value of different types of soil sample were perhaps well within the tolerance range of most of the species as reported by the workers. From the data it was noticed that a nearly neutral to slightly alkaline soil was favourable for population increase of mite. The correlation-coefficient analysis revealed that the pH in most cases showed significant negative correlation with the mite population and it might be noted that higher pH inhibited the population increase.

Relative humidity of soil was also an important factor which influenced the growth of microarthropod population including mites. But surprisingly in most of the ecological studies this factor was not taken into consideration. However, the analysis of the studies so far made showed

Table 6.2. *Studies on Correlation Between Mite Population and Physical Factors of Soil in India*

Author	Group of mites studied	Temperature	pH	Moisture	R/H	ECE	Rain fall	Fungi
	1	2	3	4	5	6	7	8
Trehan 1945	Microarthropods including mite	--	*	++	*	*	*	*
Mukherjee & Singh 1970	Microarthropods including mite	*	*	++	*	*	*	*
Choudhuri & Banerjee 1977	Cryptostigmata	+	--	++	*	--	*	*
Bhattacharya & Raychaudhuri 1979	Cryptostigmata	+	--	++	++	--	+	*
Choudhuri & Pande 1979, 1982	Cryptostigmata Prostigmata	++/ —	--	--	*	*	*	*
Joy & Bhattacharya 1981	Cryptostigmata	*	++	++	*	*	*	*
Sanyal 1981a	Cryptostigmata	*	+	*	*	+	*	*
Sanyal 1981b	Cryptostigmata	++	--	++	*		*	*
Sanyal 1982	Cryptostigmata	+	*	+	*	*	*	*
Sanyal & Bhaduri 1982	Cryptostigmata	+	*	+	*	*	*	*
Sanyal & Sarkar 1983	Cryptostigmata	+	+/–	*	+/-	+/-	*	*
Sanyal & Sarkar 1993	Cryptostigmata	--	--	--	*	*	*	*
Banerjee 1988	Cryptostigmata Mesostigmata	+	–	++	*	*	*	*
Hazra & Choudhuri 1990	Mesostigmata, Prostigmata	+/-	-/+	++	*	*	*	++
Ghatak & Roy 1991	Total Mite Fauna	*	*	++	*	*	*	*
Hazra 1991	Cryptostigmata Mesostigmata Prostigmata	*	*	++	*	*	*	*
Haq & Ramani 1991	Cryptostigmata Mesostigmata Prostigmata	*	*	*	*	*	++	*

(Contd. next page)

(Contd. Table 6.2)

	1	2	3	4	5	6	7	8
Sanyal 1991	Cryptostigmata-	+	+	*	*	*	*	*
Sarkar 1991	Cryptostigmata	--	++	--	*	*	**	*
Sengupta & Sanyal 1991	Cryptostigmata Mesostigmata Prostigmata	-	*	-	-	*	*	*
Sheela & Haq 1991	Cryptostigmata	*	*	++	*	*	*	*

+ = Positive correlation but not significant
++ = Postitive singificant correlation
- = Negative correlation but not significant
-- = Negative significant correlation
+/- = Either of two
* = Not studied

that the relationship between R/H and mite population was either positive or negative. The mites subsist upto a particular limit of R/H and in 100 per cent R/H i.e. in case of innundation these organisms could not survive.

The moisture content and R/H of soil are directly dependent or rainfall. The influence of rainfall as a single entity on the oribatid mite was studied by Bhattacharya and Raychaudhuri (1979) and they opined that rainfall had a positive correlation with this group of mites.

The other edaphic factor electrical conductivity was seriously viewed while studying the ecology of oribatid mites of saline soils in West Bengal. The studies clearly indicated that electrical conductivity as a single factor has insignificant positive or negative relationship with oribatid mite population.

(B) *Chemical factors*

The soil contains various chemical substances which play great influences on life system of soil inhabiting mites. Of these elements organic matter, nitrate and phosphate are considered to be most important and their relationships with mite population have been studied by several workers (Table 6.3).

The organic matter content of laterite, saline and alluvial soils were estimated and their relationship with either total mite population or with the population of oribatid and mesostigmatid mites were analysed. The result of various studies clearly showed that population density of total mite or oribatid mesostigmatid mite population was positively correlated with the organic matter content of soil.

TABLE 6.3. *Studies on correlation between Mite Population and Chemical Factors of Soil in India*

Author	Group of mites studied	Organic Carbon	Nitrate	Available phosphate
Mukherjee & Singh 1970	Microarthropods including mites	++	*	*
Banerjee 1974b	Cryptostigmata	++	*	*
Banerjee 1976	Microarthropods including mites	*	++	*
Choudhuri & Banerjee 1975	Total mite	++	*	*
Choudhuri & Banerjee 1977	Cryptostigmata	++	++	++
Singh & Pillai 1975	Cryptostigmata Mesostigmata Prostigmata	++	*	*
Bhattacharya & Raychaudhuri 1977, 1979	Cryptostigmata	+	*	*
Choudhuri & Pande 1979	Cryptostigmata Mesostigmata	++	++	++
Choudhuri & Pande 1982	Cryptostigmata Mesostigmata	++	++	- -
Joy & Bhattacharya 1981	Cryptostigmata	++	*	*
Sanyal 1981a, 1981b, 1982	Cryptostigmata	++	*	-/+
Banerjee 1988	Cryptostigmata Mesostigmata Prostigmata Astigmata	++	*	*
Hazra & Choudhuri 1990	Mesostigmata Prostigmata	++	++	*
Banerjee & Sanyal 1991	Cryptostigmata	++	*	*
Ghatak & Roy 1991	Total mite	++	*	*
Hazra 1991	Microarthropods including mites	++	++	*
Sanyal 1991b	Cryptostigmata	++	*	*
Sarkar 1991	Cryptostigmata	+	+	+

\+ = Positive correlation but not significant
++ = Positive significant correlation
\- = Negative correlation but not significant
— = Negative significant correlation
-/+ = Either of two
* = Not studied

The influence of nitrate content of soil on mite population was studied by Banerjee (1976), Choudhuri and Banerjee (1977) and Choudhuri and Pande (1979, 1982). In all the studies the correlation between nitrate and total microarthropod or mite population dominated by cryphstigmatids or only the cryptostigmatid mites wee analysed. Hazra and Choudhuri (1990) studied the effect of nitrate on menstigmatid and prostigmatid mites. The studies clearly revealed that the population was always strongly and positively correlated with soil nitrate content.

Only a few attempts have so far been made to find out the interrelationship between population of total mite or any single group of mite and available phosphate content of soil. The analysis of the observations indicated that in most of he cases there were significant positive correlation between the variables.

II. Studies on relationship between mite population and biotic components of soil

(i) *Mite population and soil microbes*

The soil meso and microfauna depend on micro-organisms like bacteria, fungi, algae etc. as the source of food. Further, the formation of organic matter in soil is related to these microorganisms. Considering this fact, Banerjee (1974a, 1974b, 1976) and Choudhuri and Banerjee (1975) estimated the soil microbe complex and indicated a direct proportional interdependence between the soil mesofauna and microbe population. Hazra and Choudhuri (1990) concluded that fungi population in nonpolluted uncultivated soils was significantly and positively correlated with mesostigmatid and prostigmatid population, but in polluted soil ecosystem opposite result was found.

(ii) *Mite population and type of vegetation*

Banerjee (1974a) opined that growth of vegetation exerted impacts on population make-up of soil mites. Singh and Mukherjee (1971) reported maximum number of cryptostigmatid mite in uncultivated field in comparison to other study sites like rose garden, sugarcane field and falsa field. But mesostigmatid mites showed almost similar number in all the four study fields. Singh and Pillai (1981) in their study found maximum and minimum number of cryptostigmatid mites in banana and citrus fields respectively. The mesostigmatids were found to occur in greater number in citrus field. The fodder field recorded minimum population of mesostigmatid and astigmatid mites. They also observed that prostigmatids were more in number in citrus filed and less in banana field which also recorded maximum number of astigmatid mites.

Bhattacharya and Joy (1978) studied the habitat preference of three groups of mite *viz.*, cryptostigmata, mesostigmata and prostigmata. It was found that cryptostigmata was predominant in eroded laterite, fodder and grassland sites, whereas prostigmata was found to dominate the paddy field. Bhattacharya, Joy and Joy (1981) found maximum number of cryptostigmatid mites in banana field followed by grassland, acacia forest and paddy field.

Sengupta and Sanyal (1989) studied the distribution of soil mites in six different habitats and reported that Sal forest soil supported largest number of genera in comparison to other vegetational sites.

Sanyal and Sarker (1993) reported that banana plantation harboured maximum cryptostigmatid population followed by jamun and bamboo sites.

III. Studies on effect of agricultural practices on mite population

Several attempts have been made by the soil zoologists to find out the impact of agricultural practices on soil fauna. Trehan (1945), Singh and Pillai (1975, 1981), Prabhoo (1976), Bhattacharya (1979), Bhattacharya and Joy (1978), Ghatak and Roy (1981, 1991), Pai and Prabhoo (1981), Singh and Pillai (1981), Alfred *et al.* (1991) and Hattar *et al.* (1992) studied the qualitative and quantitative aspects of ecological investigations inrelation to either total microarthropods including mite or only mite population as a whole and cumulative effects of agricultural practices like irrigation, ploughing, crop rotation, use of fertilizer and pesticide, etc. The results clearly indicated that different agricultural practices jointly exerted adverse effects on the population. However, Sengupta and Sanyal (1991) observed that ploughing and hand weeding did not reduce the number of mites in paddy field. But they reported drastic reduction in number and species diversity due to inundation of paddy field, resulting from irrigation.

Bhattacharya, Joy and Joy (1981), Majumdar and Deb (1991a, 1991b), Sanyal (1991) and Sanyal and Sarkar (1991a) recorded sharp decline in population of cryptostigmatid and mesostigmatid mites due to different agricultural practices. Of the various factors irrigation was regarded as one which caused innundation and loss of population in all the studies.

Fertilizers and pesticides are regularly used in agricultural fields in India. Sengupta and Sanyal (1991) found that fertilizer favoured the population build up of mesostigmatid mites in paddy field. Bhattacharya and Joy (1980) studied the impact of Bladex-G (Amine 2,4-D) upon the density of soil microarthropod population in a wheat field. They recorded that herbicide exerted ill effects on prostigmatids, whereas cryptostigmatid and mesostigmatid mites maintained a more or less stable population. While studying the effect of some insecticides Joy and Chakravorty (1991) found that the density of total microarthropod and also of acarina suffered a statistically significant and persistent decline in the aldrin 30EC (0.25%) and endosulfan 35EC (0.33%) treated soil of wheat fields. But dimethoate 30EC (0.125%) and phosphamidon 85EC (0.03%) applied in mustard fields produced only a temporary decline. They also reported that *Lancetoppia* sp., a common cryptostigmatid mite species, could completely escape the ill effects of endosulfan.

IV. Studies on the effects of land pollution on mite population

Bhattacharya, Bhattacharya and Banerjee (1980) reported that smoke emited from a husking mill adversely affected the population of soil microarthropods including mites.

Bhattacharya and Bhattacharya (1981, 1984) studied the community structure of soil oribatid mites influenced by industrial effluents and reported a considerable decrease in species richness and diversity in polluted areas. The elimination of many species in polluted soils indicated that there might be some impact sensitive species which could be used as bioindicators of land pollution.

While studying the effects of industrial effluents on population structure of soil microarthropods Hazra *et al.*(1982) observed that prostigmatid mites were completely absent in polluted plots. They also reported that the population of mesostigmatid mites remained almost stable in polluted sites, whereas the number of cryptostigmatids decreased strikingly due to adverse effect of pollutants like heavy metals.

Hazra and Choudhuri (1990) studied the effect of three heavy metals like copper, chromium and lead present in industrial effluents at different polluted sites. The result clearly showed that the pollutants were responsible for diminishing the population of cryptostigmatid, mesostigmatid and prostigmatid mites with an exception in a fallow land site where prostigmata showed positive correlation with copper.

Sanyal (1991) also found that orbatid fauna of polluted site was poor in quality and quantity as compared to non-polluted site.

V. Studies on seasonal fluctuation of soil mite population

The review of the studies clearly indicated that the population of mites as a whole or the cryptostigmatid and mesostigmatid as separate mite group showed an irregular trend of fluctuation depending upon the soil type, climate and different physico-chemical factors of soil. The population fluctuation of prostigmatid and astigmatid mites of Indian soil have not yet been studied.

Choudhuri and Pande (1979), Banerjee (1976, 1988), Hazra and Sanyal (1989), Ghatak and Roy (1991), Hazra (1991) and Hattar *et al.* (1992) observed the peak population of total mites during monsoon months (July-August) and maximum population was recorded in December in Darjeeling and Meghalaya hills and in May in gangetic alluvial soils.

The population fluctuation of oribatid mites varied greatly with the type of soil. In alluvial and sandy-loam soils oribatids were most abundant during July-August and minimum in April-May (Banerjee, 1974a, 1974b, Choudhuri and Banerjee, 1977, Sheela and Haq, 1991) with an exception of peak population in December (Sengupta and Sanyal 1991, Sanyal and Sarkar, 1993) and minimum in August (Sanyal and Sarkar, 1993). Sengupta and Sanyal (1991) also recorded highest population of mesostigmata in September in uncultivated field.

In laterite soil a different picture was observed which indicated post-monsoon (September-October) peak and pre-monsoon (May-June) minima (Bhattacharya and Raychaudhuri, 1979). Sanyal (1981a, 1981b, 1982, 1991a, 1991b) recorded maximum and minimum number of oribatid mites in saline soil during May and December-January respectively.

Sanyal and Sarkar (1983) and Sanyal (1991a) reported pre-monsoon months (April-May) as the period for maximum and July-September for minimum oribatid population in the paddy fields in the area of saline soil. But in the paddy field in laterite soil oribatid population showed post-monsoon peak and winter minima (Bhattacharya, Joy and Joy, 1980). In a study in the paddy field with alluvial soil Sengupta and Sanyal (1991) observed the population maxima of oribatei in October-December and it was in low level in July. The other study by Sanyal (1991b) in alluvial soil polluted with heavy metals showed the highest population of oribatid in February and lowest number in August-September.

Hattar *et al.* (1992) observed maximum and minimum number of mites in February and October respectively in cultivated fields in the hills of Meghalaya.

VI. Studies on faunal make-up of soil mites

The analysis of the studies done in uncultivated and cultivated fields clearly showed that the number of mites always predominated the other microarthropods [Singh and Mukherjee , 1973, Choudhuri and Banerjee, 1975, Singh and Pillai, 1975, Singh (U.R.) and Singh, 1975, Bhattacharya and Joy, 1978, Choudhuri and Pande, 1979, 1982, Bhattacharya, Joy and Joy, 1980, Pai and Prabhoo, 1981 and Sanyal, 1991b].

Among the four suborders of mites cryptostigmata always predominated others followed by mesostimagat, prostimagat and asitgmata (Singh and Mukherjee, 1971, Singh (J) and Singh, 1975, Banerjee, 1988, Choudhuri and Banerjee, 1975, Prabhoo 1976, Bhattacharya and Joy, 1978, Bhattacharya, 1979, Alfred *et al.* 1991, Hattar *et al.*, 1992). Singh and Pillai (1981) reported mesostigmatid mite as dominant over other groups. Sengupta and Sanyal (1989, 1991) observed mesostigmata to predominate over others in paddy field. But Hattar *et al.* (1992) reported cryptostigmata as dominant over mesostigmata in cultivated fields. The qualitative analysis of soil mite population was done extensively by several workers in India. The results showed that the genus *Scheloribates* occupied the highest position both in quantity and quality among cyyptostigmatid genera found to occur in various soil ecosystem. The other dominant genera under this suborder were *Lamellobates, Epilohmannia, Oppia, Peloribates* and *Brachioppia.* The mesostigmatid genera like *Asca, Pachylaelaps* and *Hypoaspis* were recorded as dominant in most of the studies (Singh and Mukherjee, 1971, Banerjee, 1974a, 1988, Choudhuri and Banerjee, 1977, Bhattacharya and Bhattacharya, 1981, 1984, Choudhuri and Pande, 1982, Bhattacharya, Joy and Joy, 1980, 1981, Sanyal, 1981a, 1981b, 1982, 1991a, 1991b, Sengupta and Sanyal, 1989 and Sanyal and Sarkar, 1983, 1993).

VII. Studies on vertical movements of soil mites

The studies on vertical movements of soil mites showed that mites were always present in maximum number in upper most layers (0-5cm) of soil (Choudhuri and Banerjee, 1975, Singh (J) and Singh, 1975, Singh (U.R.) and Singh, 1975, Joy and Bhattacharya, 1977, Bhattacharya and Joy, 1978, Hazra and Choudhuri 1990, Hazra, 1991, Alfred *et al.* 1991 and Hattar *et al.* 1992).

VIII. Studies on effect of deforestation and burning on soil mites

Hazra (1991) studied the effects of deforestation on the soil microarthropod fauna in two forest floor sites. He observed decrease in the population percentage of mites in deforested site as compared to Reserve forest.

Prabhoo, Pai and Namboory (1988) studied the ecology of soil mite fauna in a fire prone tropical forest. They found that burning adversely affected the mite population. It drastically decreased the population in the upper 0-5 cm layer of soil.

Acknowledgements

The author is grateful to the Director, Zoological Survey of India, Calcutta for providing facilities, Thanks are also due to Shri D. Sengupta, Darjeeling Government College for help and suggestion.

REFERENCES

Alfred, J.R.B., Darlong, V.T., Hattar, S.J.S. and Paul, D. 1991 Micro-arthropods and their conservation in some North-east Indian soil. p. 309-319. In : *Advances in Management and Conservation of Soil Fauna* (Eds. G.K. Veeresh, D. Rajagopal and C.A. Viraktamath), Bangalore.

Banerjee, S. 1974a. Qualitative composition and seasonal fluctuation of Oribatei (Acarina) in Burdwan soil, West Bengal (India). *Acta Arachnol.* **25** (2) : 68-72.

Banerjee, S. 1974b. Oribatid mites and organic carbon content of soil. *Acta Arachnol.* **26** (1) : 47-50.

Banerjee, S. 1976. Mocroarthropod and nitrate content of soil. *Bull. Indian Soc. Soil Sci.* **11** : 104-107.

Banerjee, S. 1988. Distribution of acari in relation to soil conditions in 24-Parganas, West Bengal, India. *Progress in Acarology* (Eds. G.P. Channa Basavanna and C.a. Viraktamath), Vol. 1, pp. 451-457.

Banerjee, S. and Sanyal, A.K. 1991. Oribatid mites as bioindicator of soil organic matter. pp. 877-880. In : *Advances in Management and conservation of soil Fauna* (Eds. G.K. Verresh, D. Rajagopal and C.A. Viraktamath), Bangalore.

Bhattacharya, T. 1979. Climate, soil and soil inhabiting arthropods of Shantiniketan and adjacent areas. *Journal of Research,* Visva-Bharati **3** (2) : 12-23.

Bhattacharya, T. and Raychaudhuri, D.N. 1979. Monthly variation in the density of soil microarthropods in relation to some climatic and edapnic factors. *Entomon* **4** (4) : 313-318.

Bhattacharya, T. and Bhattacharya, Jagannath 1981. Role of cryptostigmatid mites in the assessment of industrial pollution. *Journal of the IPHE India* 1981 (3) : 51-56.

Bhattacharya, T., and Bhattacharya, Jagannath 1984. Community structure of soil oribatidea as influenced by industrial waste water. *Entomon* **8** (4) : 337-348.

Bhattacharya, T., Bhattacharya, Jagannath and Banerjee, Runa, 1980. A preliminary survey of the effect of smoke emission from a husking mill on the soil microarthropods. *Newsletter, Soil Biology & Ecology* **1** (2) : 15-17.

Bhattacharya, T. and Joy, Shyamali 1978. A study of the soil microarthropod community of four contrasting sites in Shantiniketan, W. Bengal. *Journal of Research,* Visva-Bharati **2** (2) : 17-28.

Bhattacharya, T. and Joy, V.C. 1980. Impact of Bladex-G (Amine 2,4-D) upon the density of soil microarthropods of a wheat field. *Proc. Symp. Environ. Biol.* Trivandrum, 205-211.

Bhattacharya, T., Joy, S. and Joy, V.C. 1981. Community structure of soil cryptostigmata under different vegetations conditions at Shantiniketan. *J. Soil Biol. Ecol.* **1** : 27-42.

Bhattacharya, T., Joy, V.C. and Joy, Shyamali 1980. Soil inhabiting cryptostigmata Acari) of the rice field ecosystem in relation to agro-technical measures. *Tropical Ecology & Development* : 981-987.

Choudhuri, D.K. and Banerjee, S. 1975. Qualitative and quantitative composition of acari and collembola in relation to soil organic matter-microbes complex. *Oriental Insects* **9** (3) : 313-316.

Choudhuri, D.K. and Banerjee, Somnath 1977. *Soil factors and oribatid mites under conditions of West Bengal.* The University of Burdwan Publication, 88 pp.

Choudhuri, D.K. and Pande, T. 1979. High altitude soil animals and their relationship with soil factors, with special reference to mites. *Rev. Ecol. Biol. Sol.* **16** (2) : 219-226.

Choudhuri, D.K. and Pande, T. 1982. An ecological study of acarines from soil of Himalayan ecosystem. *Geobios new Reports* 1 : 24-26.

Diem, K. 1903. Untersuchungen uber die Bodenfauna in den Alpen. *H.St. Gallen Naturwiss Ges.* 1901; **2** : 234-414.

Ghatak, Tarun Kumar and Roy, Subrata 1981. Acarine fauna of a cultivated field of Hooghly district West Bengal. pp. 24-28. In : *Contributions to Acarology in India* (Ed. G.P. Channa Basavanna), Bangalore.

Ghatak, T.K. and Roy, S. 1991. The role of soil moisture and organic matter on the distribution of acari fauna in a forest floor of Hooghly district, West Bengal. pp. 143-158. In : *Contribution to Acarological Researches in India* (Eds. A.B. Mukherjee, A.K. Somchoudhury and P.K. Sarkar). Kalyani, West Bengal.

Haq, M.A. and Ramani, N. 1991. Population ecology of microarthropods in relation to vegetation and rainfall. pp. 797-803. *In : Advances in Management and Conservation of Soil Fauna* (Eds : G.K. Veeresh, D. Rajagopal and C.A. Viraktamath), Bangalore.

Hazra, A.K. 1991. Effect of deforestation on the soil macro and microarthropod fauna of West Bengal, India. p. 399-411. *In : Advances in Management and Conservation of Soil Fauna* (Eds. G.K. Veeresh, D. Rajagopal and C.A. Viraktamath), Bangalore.

Hazra, A.K. and Choudhuri, D.K. 1990. Ecology of subterranean macro and microarthropod fauna in different degraded and polluted soil environment of West Bengal, *India. Rec. zool. Surv.* India, Occ. Paper No. **120** : 1-295.

Hazra, A.K. Mukhopadhya, P., Guha, D.K. and Sanyal, A.K. 1982. Effects of industrial effluents on population structure of soil microarthropods at Durgapur Steel Plant area : A preliminary report. *Proc. Symp. Ecol. Anim. Popul. zool. Surv.* India pp. **3** : 93-98.

Hazra, A.K. and Sanyal, A.K. Population fluctuation of some predominant species of acarina and collembola on the embankment of a drainage system at Edengardens, Calcutta. *Environment & Ecology* **7** (2) : 366-368.

Hattar, S.J.S., Alfred, J.R.B. and Darlong, V.T. 1992. Soil acarina and collembola in forest and cultivated land of Khasi hills, Meghalaya. *Rec. zool. Surv. India* **92** (1-4) : 89-97.

Joy, S. and Bhattacharya, T. 1977. A qualitative and quantitative survey of soil inhabiting cryptostigmatid mites in four contrasting sites of Shantiniketan, West Bengal. p. 75-76. In : *Proceedings Second Oriental Entomology Symposium* (Ed. T.N. Ananthakrishnan), Madras.

Joy, Vadakepuram Chacko and Chakravorty, Partha Pratim 1991. Impact of insecticides on nontarget microarthropod fauna in agricultural soil. *Ecotoxicology and Environmental Safety* **22** : 8-16.

McAtee, W.L. 1907. Census of four square feet. *Science* **26** : 447-449.

Majumdar, S. and Deb, D.C. 1991a. Composition and relative abundance of soil microarthropods of uncultivated and cultivated fields in the district of West Dinajpur, West Bengal, India. *Journal of Bengal Natural History Society,* New Series **10** (2) : 56-63.

Mazumdar, S. and Deb, D.C. 1991b. Population density of soil microarthropods inrelation to different crop plants. *Journal of Bengal Natural History society, New Series* **10** (2) : 64-68.

Mitra, S.K., Hazra, A.K., Sanyal, A.K. and Mandal, S.B. 1983. Changes in the population structure of collembola and acarina in a grassland and rain-water drainage at Calcutta. pp. 664-667. In : *New Trends in Soil Biology* (Ed. Ph. Lebrun *et al.*)

Mukherjee, S.P. and Singh, J. 1970. Seasonal variations in the densities of a soil arthropod population in a rose garden at Varanasi (India). *Pedobiologia* **10** : 442-446.

Pai, C.G. Anatha and Prabhoo, N.R. 1991. Preliminary observations on the microarthropods fauna of paddy fields and adjoining uncultivated soils in South Kerala UAS Tech. seriess no. 337, 27-32 In : *Progress in soil Biology and Ecology in India* (Ed. G.K. Veeresh), Bangalore.

Prabhoo, N.R. 1976. Soil micro-arthropods of a virgin forest and adjoining tea fields in the Western Ghats in Kerala-a brief ecological study. *Oriental Insects* **10** (3) : 435-442.

Prabhoo, N.R., Pai, C.G.A. and Namboory, K.D. 1988. Ecology of acarofauna of fire prone tropical forests in the Western Ghats in Kerala with special reference to Oribatei (Acari). pp. 475-481. In : *Progress in Acarology* 1 (Eds. G.P. Channa-Basavanna and C.A. Viraktamath), Bangalore.

Sanyal, A.K. 1981a. Ecology of soil oribatid mites in an uncultivated field of gangetic delta of West Bengal in relation to soil pH and salinity. USA Teach. Series No. **37** : 107-112. In : *Progress in Soil Biology and Ecology in India,* (Ed. G.K. Veeresh), Bangalore.

Sanyal, A.K. 1981b. Qualitative and quantitative composition of Orivatei in gangetic delta of West Bengal in relation to edaphic factors. *Bull. zool. Surv. India* **4** (3) : 295-307.

Sanyal, A.K. 1982. Soil oribatid mites and their relation with soil factors in West Bengal. *J. Soil Biol. Ecol.* **2** (1) : 8-17.

Sanyal, A.K. 1991a. Influence of agricultural practices on the population of soil mites in West Bengal, India. pp. 333-340. In : *Advances in Management and conservation of soil fauna* (Eds. G.K. Veeresh, D. Rajagopal. C.A. Viraktamath), Bangalore.

Sanyal, A.K. 1991b. Ecology of soil oribatid fauna in two contrasting environment in Calcutta. pp. 73-86. In : *Contribution to Acarological Researches in India* (Eds ; A.B. Mukherjee, A.K. Som-Choudhury, P.K. Sarkar), Kalyani, West Bengal.

Sanyal, A.K. and Sarkar, B.J. 1983. Qualitative composition and seasonal fluctuation of oribatid mites in saline soil in West Bengal. *Indian J. Acar.* **8** : 31-39.

Sayal, A.K. and Sarkar, B.J. 1993. Ecology of soil oribatid mites in three contrasting sites at Botanical Garden, Howrah, West Bengal. *Environment & Ecology* **11** (2) : 427-434.

Sarkar, Sadhana 1991. Studies on microarthropod community in one undisturbed habitat of Tripura with special reference to oribatid mites. pp. 777-788. In : *Advances in Management and Conservation of Soil Fauna* (Eds. G.K. Veeresh, D. Rajagopal. C.A. Viraktamath), Bangalore.

Sengupta, D. and Sanyal, A.K. 1989. Distribution of soil acari in south Bihar, with special reference to cryptostigmata and mesostigmata-A preliminary report. *J. Acarol.* **11** (1 & 2) : 59-66.

Sengupta, D. and Sanyal, A.K. 1991. Studies on the soil microarthropod fauna of a paddy field in West Bengal, India. pp. 789-796. In : *Advances in Management and Conservation of Soil Fauna* (Eds. G.K. Veeresh, D. Rajagopal and C.A. Viraktamath), Bangalore.

Sheela, K. and Haq, M.A. 1991. Oribatid mites as bioindicators of soil moisture. pp. 871-876. In : *Advances in Management and Conservation of Soil Fauna* (Eds. G.K. Veeresh, D. Rajagopal and C.A. Viraktamath), Bangalore.

Singh, Janardan and Pillai, K.S. 1981. Community structure and biocoenology of soil microarthropods in some tropical fields at Varanasi, India. *Indian J. Soil Biol. Ecol.* **1** : 5-20.

Singh. J. and Singh, U.R. 1975. An ecological Study, of soil microarthropods from soil and litter of tropical deciduous forest of Varanasi (India). *Trop. Ecol.* **16** (2) : 81-85.

Singh, U.R. and Singh, A.K. 1975. A preliminary study of soil microarthropods from a Himalayan grassland biome. *Int. J. Ecol. Environ. Sci.* **1** (13) : 175-177.

Chapter 7

Species Composition, Distribution and Seasonal Variation of Collembola in Two Agroecosystems of Bhubaneswar

B.C. Guru AND A.K. Mohanty

P.G. Deptt. of Zoology, Utkal University Vanivihar, Bhubaneswar, India

Abstract. Species composition, distribution and seasonal variation in the population density of Collembola in two groundnut cultivated fields were studied during September 1989 to February 1991. Altogether six species were collected in this investigation which were common to both the study sites. The population of Collembola decreased significantly from upper soil layers to deeper soil layers. *Hypogastura* sp. was dominant in Field-I and *Isotoma* sp. was dominant in Field-II. Aggregation form of population could not be detected. This seasonal population fluctuation showed one peak in July in Field-I whereas two distinct peaks were observed in July and March in Field-II. In both the study sites minimum population was recorded in January. The maximum number of Collembola per m^2 in Field-I was 3260 ± 83.42 and minimum was 880 ± 23.63. In Field-II during first peak the population per m^2 was 3480 ± 55.5 and it was 6860 ± 61.75 during second peak. The minimum population was 1320 ± 31.72. Field-II contained significantly more number of Collembola. Rainfall, soil temperature and soil moisture in both the study sites showed significant positive correlations with the population number.

Key words : *Collembola, Agroecosystems, Species composition, distribution, seasonal variation.*

Introduction

Various studies on soil arthropods have shown that Collembola represent one of the most abundant groups of soil arthropods. They play a dominant role in soil formation and litter decomposition. There is increasing interest in their functional ecology and in their relation with various human activities. Ecological studies on these organisms in India have long been neglected except some attempts made by Choudhuri and Roy (1967, 1971), Mukharji and Singh

TABLE 7.1. *Details of the time of sowing, germination and harvesting period of groundnut cultivation during the study period in Field-I and II.*

	Field-I			Field-II		
	Sowing	Germination	Harvesting	Sowing	Germination	Harvesting
1st crop	Last week of July. 1989	1st week of Aug. 1989	1st week of Nov. 1989	Last week of July. 1989	1st week of Aug. 1989	1st week of Nov. 1989
2nd crop	Uncultivated			Last week of Dec. 1989	1st week of Jan. 1990	1st week of Apr. 1990
3rd crop	Uncultivated			Last week of Jun. 1990	1st week of Jul. 1990	Mid of Oct. 1990

(1970), Singh and Mukharji (1971, 1973) and some taxonomic work by Prabhoo (1971 a, b). These few informations are indispensable for future works on Collembola. At later stages some extensive works have been done by Pai and Prabhoo (1981), Reddy (1981), Singh and Pillai (1981), Mitra *et al*. (1983), Darlong and Alfred (1984), Hatter and Alfred (1984), Guru *et al.* (1988), Guru and Mohanty (1990), Guru and Panda (1990, 1991), Mitra (1990 a,b) Alfred *et al.* (1991), Guru and Das (1991), Haq and Ramani (1991), Hazra (1991) and Sengupta and Sanyal (1991) indicating the significant role of Collembola in various types of Indian soils. However, there have been paucity of information on the ecological aspects of Collembola in the state of Orissa. Therefore, it was considered worthwhile to investigate the species composition, distribution and seasonal variation of Collembola in two agroecosystems of Bhubaneswar, Orissa.

Study Sites

The study was carried out in two groundnut fields of Bhubaneswar. Field-I was a crop land without irrigation facilities where as Field-II was an irrigated crop land. The source of water to Field-I was only rainfall. So cultivation of groundnut (*Arachis hypogea* L.) was possible once during the study period. But in Field-II groundnut was cultivated for three times during this study period. Table 7.1 gives the details of the time of sowing, time of germination and period of harvesting. The city of Bhubaneswar usually experience a seasonal climate with dry summer followed by heavy rain. Usually four distinct seasons are experienced in a year i.e. March to May (Summer), June to September (Monsoon), October to November (Post monsoon) and December to February (Winter). Table 7.2 gives an account of rainfall, percentage of soil moisture, soil temperature in °C, atmospheric temperature in °C during the study period. The physicochemical characteristics of the two study sites are given in Table 7.3.

Methods

The experiment started in September 1989 and continued till February 1991 (18 months).

TABLE 7.2. *Amount of rainfall, percentage of soil moisture, soil temperature and atmospheric temperature of the two study sites.*

Year & month	Rain fall in mm (No of rainy days)	Percentage of soil moisture		Soil temperature in °C		Atmospheric temperature	
		Field-I	Field-II	Field-I	Field-II	Max.	Min.
1989 September	162.5 (17)	18.5	19.2	32.7	31.4	32.4	25.2
October	41.2 (5)	16.3	16.4	32.6	31.6	33.0	23.1
November	- (-)	15.5	15.0	26.7	27.0	31.9	17.7
December	Trace (1)	14.2	15.2	20.5	21.2	28.5	14.6
1990 January	- (-)	14.0	15.6	18.5	19.3	30.7	15.5
February	59.8 (7)	15.2	14.8	23.0	22.5	31.5	19.5
March	119.5 (7)	16.8	17.0	22.1	21.8	31.6	21.8
April	130.7 (7)	17.2	18.5	26.2	27.1	34.5	24.8
May	124.9 (13)	20.1	20.6	30.9	29.0	34.2	25.7
June	210.6 (16)	18.3	17.6	30.0	29.0	34.2	25.7
July	289.5 (18)	19.2	19.4	28.3	29.5	31.8	25.3
August	468.5 (22)	22.4	22.2	26.1	26.5	32.9	26.0
September	238.1 (17)	23.5	22.5	28.4	27.8	31.5	22.9
October	165.2 (12)	21.2	21.8	31.3	30.9	30.7	20.5
November	246.6 (6)	20.5	20.2	26.5	27.2	29.2	15.6
December	2 (1)	17.2	16.8	19.5	20.1	27.6	16.7
1991 January	60.9 (7)	18.4	17.9	18.7	17.8	27.6	16.7
February	0.6 (1)	15.1	16.2	22.5	21.7	33.8	19.8

TABLE 7.3. *Physical characteristics of the soil of Field-I and Field-II.*

		Feild-I	Field-II
(i)	Composition		
	Sand%	76.5	74.3
	Slit%	9.3	10.0
	Clay%	13.0	14.2
(ii)	Texture	Sandyloam	Sandyloam
(iii)	Colour	Reddish brown	Reddish borwn
(iv)	Bulk density (cm^3/cm^3)	1.7	1.8
(v)	Porosity (cm^3/cm^3)	0.42	0.54
(vi)	pH	5.1-5.6	5.4-5.9
(vii)	Organic carbon (gm%)	0.33	0.55
(viii)	Total nitrogen (gm%)	0.83	0.85
(ix)	Na (gm%)	0.054	0.067
(x)	K (gm%)	0.029	0.033
(xi)	Ca (gm%)	0.423	0.481

Monthly five samples (10 × 10 × 20 cm) were collected at random from both the study sites. Each sample was subdivided into 0-5 cm, 5-10 cm, 10-15 cm, and 15-20 cm, and accordingly twenty sub-samples were collected from each of the study sites every month. The soil samples were kept in polythene bags for transport to the laboratory for processing. Before selecting 10 × 10 cm as the sampling size, some pilot experiments were done by taking 5 × 5 cm and 15 × 15 cm as the sample size. But analysis of variance showed that 10 × 10 cm sample size was the most suitable for efficient extraction of the springtails.

The soil samples were put into a series of modified tullgren funnel and the Collembola were extracted in 70 per cent alcohol. The total numbers of Collembola in each sub sample were calculated and species were identified under a binocular microscope. The moisture content of the soil was determined using an infrared moisture balance and the pH of the soil was estimated with a pH meter.

Results

Species composition

Table 7.4 gives a list of Collembola species collected from the two study sites. Altogether six species were collected in this investigation and all the species were found in both the study sites. The six species collected belong to three families. Three species belong to the family Isotomidae, two species belong to the family Entomobryidae and there is one representative from the family Hypogastruridae.

TABLE 7.4. *List of Collembolan species collected from Field-I and II*

Name of the species collected	Family	Field-I	Field-II
Hypogastrura sp.	Hypogastruridae	+	+
Isotoma sp.	Isotomidae	+	+
Folsomia brevifurca	Isotomidae	+	+
Isotomina sp.	Isotomidae	+	+
Cyphoderus Javanus	Entomobryidae	+	+
Lepidocyrtus sp.	Entomobryidae	+	+

"+" indicates the presence of the species.

TABLE 7.5. *Percentage of total population of Collembola in four different soil layers of the two study sites*

Name of the study site	Percentage of Collembola			
	0-5 cm soil profile	5-10 cm soil profile	10-15 cm soil profile	15-20 cm soil profile
Field-I (Non-irrigated)	38.48	27.21	21.42	12.86
Field-II (Irrigated)	32.48	29.86	22.32	15.31

Distribution

Table 7.5 gives an account of the percentage of total population of Collembola in four different soil layers of the two study sites. It was observed that in both the study sites the maximum number of Collembola were confined to the 0-5 cm soil profile. There was a gradual trend of decrease in the population number towards the deeper layers. Field-I had 38.48 per cent, 27.2 per cent, 21.42 per cent and 12.86 per cen;t of total population in the 0-5 cm, 5-10 cm, 10-15 cm and 15-20 cm soil profile respectively. The corresponding values in Field-II were 32.48 per cent, 29.86 per cent, 22.32 per cent and 15.31per cent. These values were calculated on the averages of the eighteen months study period. Table 7.6 gives the distribution of a particular species of Collembola in four different soil layers of the two study sites. Usually all the species in both the study sites gradually decreased in their number towards the deeper layers with the following exceptions. In Field-II *Isotomina* sp., *Cyphoderus javanus* and *Lepidocyrtus* sp. were

TABLE 7.6. *Vertical distribution of different species of Collembola with their average numbers per m² over 18 months for four depths in Field-I and II.*

	Sp.1 Hypogastrura Sp.	Sp.2 Isotoma Sp.	Sp.3 Folsomia brevifurca	Sp.4 Isotomina Sp.	Sp.5 Cyphoderus javanus	Sp.6 Lepidocyrtus Sp.
Field-I						
0-5 cm per m²	35.93% (3795)	37.09% (3060)	27.19% (1463)	40.74% (1470)	64.34% (1610)	50.55% (1102)
5-10 cm per m²	26.56% (2805)	27.93% (2305)	28.01% (1507)	25.83% (932)	28.37% (710)	26.65% (581)
10-15 cm per m²	23.67% (2500)	21.75% (1795)	28.43% (1530)	20.39% (736)	7.27% (182)	9.95% (217)
15-20 cm per m²	13.82% (1460)	13.21% (1090)	16.35% (880)	13.02% (470)	0.0% (00)	12.84% (280)
Field-II						
0-5 cm per m²	35.98% (2850)	38.46% (4850)	24.86% (1380)	26.19% (1100)	31.48% (3212)	26.13% (986)
5-10 cm per m²	26.13% (2070)	25.8% (3253)	21.98% (1220)	28.80% (1210)	41.22% (4206)	33.35% (1261)
10-15 cm per m²	19.19% (1520)	20.66% (2605)	29.72% (1650)	23.80% (1000)	22.38% (2284)	21.71% (821)
15-20 cm per m²	18.68% (1480)	15.06% (1900)	23.42% (1300)	21.19% (890)	4.9% (500)	18.78% (710)

TABLE 7.7. *Population percentage of different species of collembola in a particular soil layer of Field-I (non-irrigated)*

Sl. No.	Name of the species	% in 0-5 cm	% in 5-10 cm	% in 10-15 cm	% in 15-20 cm	% of the species in the Habitat
1.	Hypogastrura Sp	30.36	31.73	35.91	34.92	32.5
2.	Isotoma Sp.	24.48	26.07	25.79	26.07	25.4
3.	Folsomia brevifurca	11.71	17.04	21.98	21.05	16.6
4.	Isotomina Sp.	11.76	10.54	10.57	11.24	11.1
5.	Cyphoderus javanus	12.88	8.03	2.61	0.0	7.7
6.	Lepidocyrtus Sp.	8.81	6.57	3.11	6.69	6.7

TABLE 7.8. *Population of different species of Collembola in a particular soil layer of Field-II (Irrigated).*

Sl. No.	Name of the species	% in 0-5 cm	% in 5-10cm	% in 10-15 cm	% in 15-20 cm	% of the species in the Habitat
1.	Hypogastrura Sp.	19.81	15.65	15.38	21.82	17.9
2.	Isotoma Sp.	33.72	24.60	26.36	28.02	28.5
3.	Folsomia brevifurca	9.59	9.22	16.70	19.17	12.5
4.	Isotomina Sp.	7.64	9.15	10.12	13.12	9.5
5.	Cyphoderus javanus	22.33	31.81	23.11	7.37	23.0
6.	Lepidocyrtus Sp.	6.81	9.53	8.30	10.47	8.6

more in number in 5-10 cm soil profile, whereas *Folsomia brevifurca* was more in number in 10-15 cm soil profile.

Tables 7.7 and 7.8 give the population percentage of different species of Collembola in a particular soil profile of Field-I and II respectively. In Field-I *Hypogastrura* sp. was recorded as the dominant species which constituted about 32.5 per cent of the total population. When the individual soil profiles were considered separately *Hypogastrura* sp. was found to be dominant in all the soil profiles. It constituted 30.36 per cent, 31.73 per cent, 35.91 per cent and 34.92 per cent in 0-5 cm, 5-10 cm, 10-15 cm and 15-20 cm respectively. In Field-II when the total habitat was considered *Isotoma* sp. was found to be dominant constituting about 28.5 per cent of the total population. But this species was not dominant in 5-10 cm soil profile. In this profile *Cyphoderus javanus* constituted about 31.81 per cent of the population where as *Isotoma* sp. constituted 24.6 per cent of the population. In rest of the soil profiles *Isotoma* sp. was the dominant constituting 33.72 per cent, 26.36 per cent and 28.02 per cent in 0-5 cm, 10-15 cm and 15-20 cm soil profiles respectively.

To detect whether or not aggregation phenomenon exists in this population coefficient of variation proposed by Salt and Hollick (1946) was calculated. CV values were calculated in both the study sites in all the soil profiles in each month. The CV values are given in Tables 7.9 and 7.10 for Field-I and II respectively. The values were always less than one. So aggregated form of population could not be detected in this investigation.

Figure 7.1 shows the population fluctuation of Collembola in the two study sites. The total number of Collembola per m^2 in the two study sites during the eighteen months study period have been given in Table 7.11 (Field-I) and Table 7.12 (Field-II). The population fluctuation in Field-I

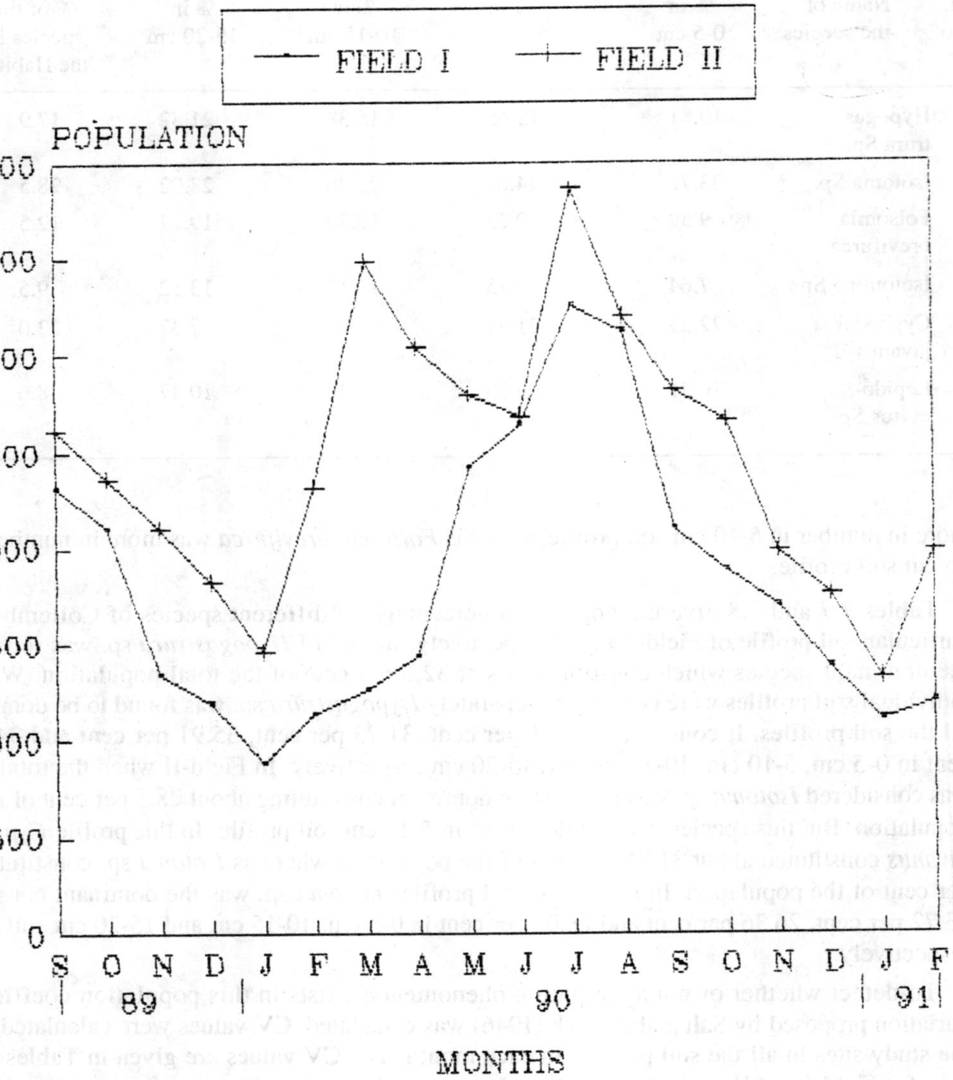

Fig. 7.1. Seasonal variation of total number of Collembola per m² in Field-I and II

TABLE 7.9. *Coefficient of variation values in Field-I for 18 months study period.*

Year & Month		Value in 0.5 cm soil depth	Value in 5-10 cm soil depth	Value in 10-15 cm soil depth	Value in 15-20 cm soil depth	Value in total
1989	Sep.	0.174	0.225	0.283	0.475	0.106
	Oct.	0.225	0.082	0.101	0.353	0.104
	Nov.	0.353	0.176	0.263	0.248	0.097
	Dec.	0.316	0.220	0.380	0.342	0.105
1990	Jan.	0.261	0.475	0.304	0.391	0.120
	Feb.	0.370	0.321	0.248	0.638	0.175
	Mar.	0.124	0.263	0.298	0.353	0.090
	Apr.	0.493	0.261	0.176	0.235	0.088
	May.	0.078	0.166	0.211	0.220	0.087
	Jun.	0.082	0.422	0.203	0.199	0.102
	Jul.	0.211	0.150	0.225	0.244	0.114
	Aug.	0.102	0.197	0.303	0.310	0.103
	Sep.	0.235	0.210	0.220	0.464	0.148
	Oct.	0.125	0.224	0.161	0.342	0.139
	Nov.	0.150	0.200	0.235	0.505	0.147
	Dec.	0.139	0.124	0.500	0.391	0.141
1991	Jan.	0.160	0.159	0.464	0.372	0.152
	Feb.	0.159	0.298	0.612	0.342	0.150

showed one peak in a year where as Field-II showed two peaks in a year. The two peaks observed in the Field-II were in the months of March and July and the peak observed in Field-I was in July. In both the sites the lowest populations were recorded in January. In Field-II after January the population suddenly increased to reach the peak in March. Again it decreased slowly till June. After June the population increased to reach another peak in July after which the population gradually decreased to reach the minimum in the month of January. In Field-I, from the lowest population in January it increased gradually to reach the peak in July and again decreased slowly to come to the minimum in January. The total number of Collembola per m^2 in Field-I was 3260 ± 83.42 during the peak period and it was only 880 ± 23.63 during January 1990. In Field-II during the first peak the population per m^2 was 3480 ± 55.5 and it was 3860 ± 61.75 during the second peak. The lowest population was 1320 ± 31.72 in January 1991. The average value of total Collembola per m^2 of all the eighteen months was calculated for both the study sites. It was 1804.44 ± 164.26 (Field-I) and 2458.88 ± 157.84 (Field-II). After application of 't' test it was observed that there was significant difference between the population of Field-I and Field-II ($P < 0.05$).

The 'r' values were calculated to examine the coefficient of correlation of rainfall, soil temperature and soil moisture with the population number of Collembola in both the study sites.

TABLE 7.10. *Coefficient of variation values in Field-II for 18 months study period.*

Year & Month		Value in 0.5 cm soil depth	Value in 5-10 cm soil depth	Value in 10-15 cm soil depth	Value in 15-20 cm soil depth	Value in total
1989	Sep.	0.135	0.167	0.283	0.316	0.087
	Oct.	0.265	0.211	0.159	0.356	0.091
	Nov.	0.181	0.172	0.316	0.592	0.111
	Dec.	0.224	0.183	0.353	0.570	0.108
1990	Jan.	0.310	0.176	0.421	0.298	0.080
	Feb.	0.199	0.142	0.291	0.176	0.082
	Mar.	0.141	0.100	0.214	0.154	0.071
	Apr.	0.164	0.090	0.172	0.247	0.082
	May.	0.159	0.111	0.236	0.259	0.076
	Jun.	0.197	0.114	0.204	0.316	0.065
	Jul.	0.162	0.107	0.090	0.166	0.071
	Aug.	0.157	0.167	0.174	0.159	0.067
	Sep.	0.175	0.159	0.196	0.250	0.086
	Oct.	0.090	0.154	0.255	0.329	0.082
	Nov.	0.154	0.117	0.395	0.228	0.110
	Dec.	0.085	0.211	0.248	0.592	0.098
1991	Jan.	0.353	0.259	0.333	0.248	0.107
	Feb.	0.101	0.144	0.199	0.298	0.090

The different 'r' values have been shown in Table 7.13. In all cases significant positive correlations were found. The correlation between the rainfall and population number of Collembola were 0.804 (Field-I) and 0.660 (Field-II). These values were significant at 1 per cent level. The soil temperature and soil moisture showed higher correlation in Field-I than in Field-II. In Field-I both the values were significant at 1 per cent level (0.694 and 0.657) where as these values were 0.576 and 0.486 for Field-II (significant at 1% and 5% level respectively).

Discussion

The distribution and abundance of certain species of Collembola, in a particular ecosystem, play an indicator role of the environmental condition. They have a more extensive distribution than the species of any other insect group. The species composition of Collembola in a mixed population in different places depends largely upon the moisture content of the soil. The multivariate analysis, particularly the analysis of correspondences, repeatedly applied on Collembolan populations (Andre, 1983; Arpin *et al.*, 1984; Bonnet *et al.*, 1977, 1979; Gers, 1982; Gers and Izarra, 1983; Ponge, 1980, 1983; Ponge and Prat, 1982; Poursin and Ponge, 1982, 1984; Prat and Massoud, 1980), allows to detect latent factors that affect the distribution and abundance of Collembolan species.

TABLE 7.11. *Total number of Collembola per m² in different soil layers of Field-I during 18 months study period.*

Year & Month	Population ± Standard Error				
	0-5 cm	5-10 cm	10-15 cm	15-20 cm	Total
1989 Sep.	700 ± 54.92	740 ± 75.03	640 ± 81.46	240 ± 51.12	2320 ± 55.07
Oct.	700 ± 70.90	660 ± 24.56	540 ± 24.56	200 ± 31.70	2100 ± 49.16
Nov.	420 ± 66.50	400 ± 31.70	340 ± 40.10	180 ± 20.05	1340 ± 29.28
Dec.	360 ± 51.12	380 ± 37.51	300 ± 51.12	160 ± 24.56	1200 ± 28.36
1990 Jan.	320 ± 37.51	240 ± 51.12	180 ± 24.56	140 ± 24.56	880 ± 23.63
Feb.	560 ± 92.98	260 ± 37.51	180 ± 20.05	140 ± 40.10	1140 ± 44.68
Mar.	440 ± 24.56	340 ± 40.01	280 ± 37.51	200 ± 31.70	1260 ± 25.42
Apr.	420 ± 92.98	320 ± 37.51	400 ± 31.70	300 ± 31.70	1440 ± 28.36
May.	900 ± 31.70	600 ± 44.84	540 ± 51.12	380 ± 37.51	2420 ± 47.30
Jun.	1020 ± 37.51	640 ± 121.15	560 ± 51.12	420 ± 37.51	2640 ± 60.46
Jul.	1300 ± 123.48	760 ± 51.12	700 ± 70.90	500 ± 54.92	3260 ± 83.42
Aug.	1200 ± 54.92	800 ± 70.90	700 ± 95.12	420 ± 58.46	3120 ± 72.07
Sep.	920 ± 97.21	620 ± 58.46	380 ± 37.51	180 ± 37.51	2100 ± 69.57
Oct.	800 ± 44.84	580 ± 58.46	340 ± 24.56	160 ± 24.56	1880 ± 58.56
Nov.	760 ± 51.12	500 ± 44.84	300 ± 31.70	140 ± 31.70	1700 ± 56.16
Dec.	600 ± 37.51	440 ± 24.56	200 ± 44.84	140 ± 24.56	1380 ± 43.79
1991 Jan.	520 ± 37.51	280 ± 20.05	180 ± 37.51	120 ± 20.05	1100 ± 37.62
Feb.	560 ± 40.10	280 ± 37.51	200 ± 54.92	160 ± 24.56	1200 ± 40.41

Hazra and Choudhuri (1983) investigated the Collembolan communities in cultivated and uncultivated sites of West Bengal in relation to three major soil factors. Collembolan fauna encountered in their study belonged to fourteen genera out of which five genera were common to both the fields. Mitra *et al.* (1983) while doing some preliminary observations on the effects of rotation of crops and fertilizers on Collembola in the soils of West Bengal have described nine species from the paddy fields, six species from the wheat fields and five species from the jute fields. Singh and Mukharji (1971) studied the qualitative composition of soil arthropods in some fields of Varanasi and described nine species of Collembola from the uncultivated fields. Pai and Prabhoo (1981) while studying the microarthropods associated with the post-harvest decay of paddy tillers in Kerala reported five species of Collembola. Guru *et al.* (1988) while working on a paddy field reported six species. Guru and Das (1991) working on three different agricultural fields reported not more than six species in any one of their fields. The present findings can well be compared with all the above cited findings except Hazra and Choudhuri (1983).

Distribution of Collembola is regulated due to the seasonal fluctuation in the temperature and moisture content of the soil causing downward migration. Gisin (1943), Nielsen (1949), Kuhnelt (1955), Delamare-Deboutteville (1951) have shown that the higher population densities occur in

TABLE 7.12. *Tatal number of Collembola per m² in different soil layers of Field-II during 18 months study period.*

Year	Month	Population ± Standard Error				
		0-5 cm	5-10 cm	10-15 cm	15-20 cm	Total
1989	Sept.	840 ± 51.12	780 ± 58.46	640 ± 81.46	360 ± 51.12	2620 ± 51.05
	Oct.	780 ± 92.98	700 ± 66.51	560 ± 40.10	320 ± 51.12	2360 ± 48.40
	Nov.	720 ± 58.46	660 ± 51.12	500 ± 70.90	220 ± 58.94	2100 ± 52.27
	Dec.	580 ± 58.46	620 ± 51.12	420 ± 66.51	200 ± 51.12	1820 ± 44.20
1990	Jan.	420 ± 58.46	400 ± 31.70	360 ± 68.00	280 ± 37.51	1460 ± 26.44
	Feb.	760 ± 68.00	700 ± 44.84	460 ± 60.16	400 ± 31.70	2320 ± 42.69
	Mar.	1120± 70.90	1000 ± 44.84	780 ± 75.03	580 ± 40.10	3480 ± 55.50
	Apr.	1000 ± 73.68	920 ± 37.51	660 ± 51.12	460 ± 51.12	3040 ± 56.35
	May.	820 ± 58.46	900 ± 44.84	640 ± 68.00	440 ± 51.12	2800 ± 48.14
	Jun.	800 ± 70.90	780 ± 40.10	600 ± 54.92	500 ± 70.90	2680 ± 39.27
	Jul.	1280 ± 92.98	1060 ± 51.12	920 ± 37.51	600 ± 44.84	3860 ± 61.75
	Aug.	960 ± 68.00	980 ± 73.68	700 ± 54.92	560 ± 40.10	3200 ± 48.40
	Sep.	900 ± 70.90	820 ± 58.46	580 ± 51.12	520 ± 58.46	2820 ± 54.75
	Oct.	920 ± 37.51	740 ± 51.12	580 ± 66.51	420 ± 58.46	2660 ± 49.37
	Nov.	740 ± 51.12	600 ± 31.70	400 ± 70.90	240 ± 24.56	1980 ± 48.94
	Dec.	640 ± 24.56	540 ± 51.12	360 ± 40.10	220 ± 58.46	1760 ± 38.73
1991	Jan.	400 ± 63.41	440 ± 51.12	300 ± 44.84	180 ± 20.05	1320 ± 31.72
	Feb.	700 ± 31.70	580 ± 37.51	420 ± 37.51	280 ± 37.51	1980 ± 40.06

TABLE 7.13. *Coefficient of correlation values between different physical parameters and Collembola in population in Field-I and II.*

	Population number	
	Field-I	Field-II
Rain fall	0.804*	0.660*
Soil temperature	0.694*	0.576*
Soil moisture	0.657*	0.486**

* Significant (P<0.01).
** Significant (P<0.05).

the upper layers of the soil. Exception occurs only in places where sufficient food is present at greater depths. Sheals (1957) studied the vertical, horizontal and seasonal distributions of Acarines and Collembolans in some uncultivated soils of Glasgow. In most cases the populations densities were significantly greater in 0-3 inch layer than in the 3-9 inch layer. Huther (1983) also found that the Collembola were dominant in the soil surface while in the soils it was the mites. Peterson (1980) and Hagvar (1983) had also shown that the higher densities of population occurred in the upper layers of the soil. Takeda (1978) also showed that majority of Collembola were concentrated in the upper 0-4 cm soil layer. The concentration of individuals in this layer was mainly due to the soil structure and the springtails were limited to the soil organic layer. He calculated as many as 40,000 individuals per m^2 in the 0-5 cm layer where as only 920 individuals per m^2 in the 4-8 cm soil profile. Only 4 per cent of the total population were found in the deeper layers which is very close to the present investigation. Hazra and Choudhuri (1983) reached in a similar conclusion regarding the vertical distribution of Collembola and observed the Collembola fauna inhabiting the uppermost 0-3 cm layer in an uncultivated site constituted about 59.15 per cent of the total population while the minimum percentage (3.7%) was recorded from the deepest layer of the cultivated site. Holt (1985) while working on the Acari and Collembola in the litter and soil of three north Queensland rain forests observed that 53-75% of total population of Collembola were found in the upper soil layer at each site. Greenslade and Greenslade (1968) also had similar findings. Wood (1967a, b) found the greatest densities of Acari and Collembola in the upper 4 cm. They recorded 87-91 per cent of the total population as surface concentrations. Hutson and Veitch (1983) worked on the Collembola and Acari in the soil and litter of three south Australian forests and recorded a mean of 67-75 per cent of the population inhabiting the upper 4 cm of the soil and only 10-15% of the population inhabited the 4-8 cm soil layer. Guru *et al.* (1988), Guru and Das (1991) and Mallick (1989), working on the distribution of Collembola in different agroecosystem recorded maximum percentage (54-63%) of Collembola population in 0-5 cm soil profile. They also observed gradual trend of decrease in the population towards deeper layers. In this investigation also more number of Collembola were confined to the uppermost layer (0-5 cm) with a gradual decrease towards the deeper layers. However the 0-5 cm soil profile had only 38.4 per cent8% and 32.48 per cent of total population in Field-I and II respectively.

Hazra and Choudhuri (1983) while investigating the Collembolan communities in cultivated and uncultivated sites of West Bengal found the predominant Species being *Cyphoderus javanus* _in unclutivated and *Isotomina thermophila* in the cultivated site. Mitra *et al.* (1977) while working on the ecology of Collembola in the Eden gardens of Calcutta found the following species to dominate the population in various plots (*Pseudosinella* sp. in grass; *Megalothorax minimum* in bamboo, banana and sansevieria sites; *Cyphoderms albinus* in the Ixora and *Isotomina thermophila*in drainage site). Mitra *et al.* (1983) while observing the effects of rotation of crops and fertilizers on collembola showed that *Cryptopygus thermophilus* remained associated with rotation of all the three crops he had studied. Singh and Mukharji (1971) studied the qualitative composition of soil arthropods in some fields at Varanasi and recorded the most common species being *Folsomides parvulus, Isotomides* sp. and *Oncychiurus armatus.* Takeda (1981) while investigating the effects of shifting cultivation on the soil mesofauna with special reference to Collembola populations in the north-east Thailand had reported *Isotomiella minor* and *Isotomina thermophila* to be the dominant forms in the cultivated site. Guru *el at.* (1984), Guru and Das (1991) and Mallick (1989) all recorded *C. thermophiluus* as the dominant species in various agricultural fields of Bhubaneswar. In this investigation *Hypogastrura* sp. and *Isotoma* sp. were found to be dominant in Field-I and II respectively. The species *Cryptopygus thermophilus* which was the dominant species of various earlier works in different agricultural fields of Bhubaneswar was never found during the course of this investigation. When the different species were considered separately it was found that more percentage of a particular species was confined to the uppermost layer, decreasing in its percentage towards the deeper later, with slight exceptions. Interestingly all the species were recorded from all the soil profiles in both the study sites with only one exception when *Cyphoderus javanus* was not recorded from 15-20 cm soil profile of Field-I.

Glasgow (1939), Macfadyen (1952, 1957), Raw (1956), Hughes (1962), Haarlov (1960) and Poole (1961) have found that microarthropods tend to be aggregated in the soil. Glasgow (1939) also estimated the sizes of aggregation of Collembola and found that *Onychiurus armatus* aggregated Between 7.6 cm and 30.5 cm in diameter while *O. ambulans, Tullbergia quadrispina* and *T. Krausbueri,* the diameter was greater than 30.5 cm. Joosse (1970) found aggregated distribution of Collembola. Verhoef and Selm (1983) also found aggregated distribution for different Collembolan species. However, Usher (1969) found the distribution of Collembola within a block of samples to be of three types : (a) uniform; (b) random; (c) aggregated. Panda (1987) did not find aggregated form of population in a crop field. Aggregated form of collembola population also could not be detected in the present investigation.

There are several works dealing with the seasonal variation of soil arthropods. Sheals (1957) observed higher population of Collembola in autumn and winter. Some workers like Agrell (1941), Hammer (1944) and Stockli (1957) had found the peak pupulation of springtails to occur during summer months. Mukharji and Singh (1970) working on the seasonal variation in the popultion densities of soil arthropod in a rose garden at Varanasi recorded the maximum number of Collembola in the months of March, August and January showing three distinct peaks in a year. Mitra *et al.* (1977) also observed two peaks in a grass land. Choudhury and Roy (1972) observed only one peak during June-August in some soils of West Bengal. Hazra and Choudhuri (1983) reported that the predominant forms of Collembola obtained from the cultivated and uncultivated fields of West Bengal exhibited an irregular trend of fluctuation usually showing maximum in July, minimum in May and slightly higher in December-January. Alfred *el al.* (1991) showed an unimodal peak with maximum number of Collembola during monsoon to post monsoon in different agroecosystems. Although the magnitude of population fluctuation in Collembola varied in different systems in most of the cases, there was a sharp rise in abundance of collembola in the post-monsoonic period (Mukharji and Singh, 1970; Choudhuri and Roy, 1971, 1972; Prabhoo, 1976). Further works from other areas had also shown that soil fauna attained its maximum number at a period when soil moisture was at its optimum, coinciding with rainfall (Rapoport and Najt, 1966). Keeping this in mind coefficient of correlations were calculated between rain fall, soil moisture, and population.

Rainfall and soil moisture both showed significant positive correlations with population of Collembola in both the study sites. Guru *et al.* (1988) recorded two distinct peaks in November and February in a paddy field and recorded the minimum population in summer. Guru and Das (1991) working in different agricultural fields observed two peaks. They observed the first peak during February-March and the second one during October-November. In this investigation in Field-I (Non-irrigated) only one peak was observed in the month of July and in Field-II (Irrigated) there were two peaks, one in March and one in July, coinciding with some of the earlier works. The minimum population in this investigation were recorded in the month of January in both the study sites, which did not agree with most of the earlier findings where they observed the minimum population in summer months of May. This deviation can be explained when one looks to the amount of rainfall in different months of the study period. During the course of this investigation the months of March, April and May (1990) had considerable amount of rainfall which might be the cause of getting higher number of Collembola duirng these months.

Takeda (1981) while studying the effects of shifting cultivation on Collembolan population in Thailand soils observed 960 individuals per m^2 in a crop field. Pai and Prabhoo (1981) recorded the maximum density of Collembola as 340 per m^2 in the paddy field and 8949 per m^2 in the uncultivated sites. Similar observations were also made by Hazra and Choudhuri (1983) while they compared the Collembolan densities in the cultivated and uncultivated sites. In contrast, the observations of Singh and Mukharji (1971) has shown that the denisty of collembola was lower in uncultivated fields than that of the cultivated fields. Loring *et al.* (1981) while studying extensively the effects of defferent tillage practices on Collembola and Acarina populations in the crop fields of USA had concluded that the conventional tillage systems stimulated an increase of Collembola population and this was in response to favourable changes in nutrient availability, pore space and other physical parameters of the soil. Guru *et al.* (1988) working on a paddy field located at Bhubaneswar recorded maximum population of 5100/m^2 and minimum population of 1000/m^2. Guru and Das (1991) worked on the population of Collembola in three different agricultural fields again from the same city of Bhubaneswar. They recorded as high as 11300/m^2 from one field and in contrast in the other two fields the maximum population were as low as 1800/m^2 and 1220/m^2. In this investigation the maximum population recorded was 3260/m^2 and 3860/m^2 in Field-I and II respectively. From the above findings and discussion it seems evident that population density of Collembola in different cultivated fields does not follow any uniform pattern. In some places the density is very low while in others it is very high. This may be due to different types of agricultural practices at different places. In some fields fertilizers and pesticides are sometime not used where as in some fields these are used in high doses. These factors are definitely responsible in upsetting the total population of Collembola in a particular agroecosystem.

Mitra *et al.* (1977), Bhattacharya and Raychaudhuri (1979) found positive correlation between soil temperature and Collembola population. Takeda (1978) observed insignificant posititive correlation for some species and negative correlation for some other species. Hazra and Choudhuri (1983) commented upon the direct influence of the temperature on the distribution pattern of Collembola and concluded that temperature alone did not show any significant correlation with Collembola population. In this investigation, all the parameters like rainfall, soil moisture and soil temperature showed significant positive correlations in both the fields. Which is similar to many of the earlier findings. However it might be concluded that the cumulative effects of various edaphic and climatic factors other than any individual factor is responsible for the variation in Collembola population in different agroecosystems.

Acknowledgements

The authors wish to thank Head, P.G. Deptt. of Zoology for providing necessary laboratory facilities. A.K. Mohanty wishes to thank Utkal Univeristy for providing UGC research fellowship during the course of investigation.

REFERENCES

Agrell, I., 1941. Zur Okologie der Collembolen. Untersuchungen im schwedischen Lapland. *Opuscula Entomol.*, supp. III : p. 236.

Alfred, J. R.B. Darlong, V. T. Hattar S.J.S. and Paul, D. 1991. Micro arthropods and their conservation in some North-East Indian soil. pp. 309-319 In : *Advances in management and conservation of soil fauna.* (Eds. G. K. Veeresh *et al.*) Oxford & IBH Publishing Co. Pvt. Ltd. New Delhi.

Andre, H.M., 1983. Notes on the ecology of corticolous. epiphyte dwellers. 2. Collembola. *Pedobiologia.* **25** : 271-278.

Arpin, P., Ponge, J.F. Dabin, B. and Mori, A. 1984. Utilisation des nembatodes Mononchida et des Collemboles pur characteriser des phenomenes pedobiologiques. *Rev. Ecol. Biol. Sol.* **21** : 242-268.

Bhattacharya, T., and Raychaudhuri, D.N. 1979. Monthly variation in the density of soil micro-arthropods in relation to some climatic and edaphic factors. *Entomon.* **4** (4) : 313-318.

Bonnet, L., Cassagnau, P. and Deharveng, L. 1977. Influence du deboisement et du rebiosement sur les biocenoseș de collemboles dans quelgues sols pyreneens. *Bull. Ecol.* **8** : 321-332.

Bonnet, L., Cassagnau, P. and Deharven, L. 1979. Recherche d'une methodologie dans I'analyse de la rupture des equilibres biocenotiques : applications aux collemboles edaphiques des pyrenees. *Rev. Ecol. Biol. Sol.* **16** : 373-401.

Choudhuri, D.K., and Roy, S. 1967. Quantitative composition of the Collembolan fauna of some uncultivated fields is Nadia District (West Bengal) with corrrelation between monthly population and individual soil factor. *Rev. Ecol. Biol. Sol.* **4** (3) : 507-515.

Choudhuri, D.K., and Roy, S. 1971. Distribution of Collembola in relation to soil conditions in the district of Jalpaiguri, West Bengal (India). pp. 289-296 In : *Organisms du sol et production primaire.* Proc, IV Coll. Pedobiologiae. INRA, Paris.

Choudhuri, D.K., and Roy, S. 1972. An ecological study on Collembola of West Bengal (India). *Rec. Zool. Surv. India.* **66** : 81-101.

Darlong, V.T., and Alfred, J.R.B. 1984. A comparative seasonal population dynamics of a dominant Collembolan species in a pine forest and jhum soils of Meghalaya. Proc. III Oriental Entomol. Symp., p. 105.

Delamare Deboutteville, C., 1951. Microfauna du sol des pays tempers et tropicaux. *Vie at Milieu, Suppl. 1,* (Quoted from Bullock, J.A., *Trop. Ecol.,* 1967, 74-87). p. 360.

Gers, C., 1982. Icidence de la simplifcation du travail du sol sur la microfaune edaphique hivernale : donnees preliminaries. *Rev. Ecol. Biol. Sol.* **19** : 593-604.

Gers, C., and Izarra, D.C. 1983. Rupture de I'quilibre biocenotique des populations de Collemboles a la station de ski ke Calmazeille-Formigueres (Pyrenees-/Orientales) *Bull. Soc. Hist. Nat.*, Toulouse. **119** : 63-69.

Gisin, H., 1943. Okologie and Lebensgemeinschaften der Collembolan in schweizerischen Exkursionsgebiet basel. *Rev. Suiss. Zool.* **50** : 131-224.

Glasgow, J.P., 1939. A population study of substerrenean soil Collembola. *J. Anim. Ecol.* **8** : 323-353.

Greenslade, P.J.M., and Greenslade, P. 1968. Soil and litter fauna densiteis in the solomon islands. *Pedobiologia.* **7** : 362-370.

Guru, B.C., and Das, B. 1991. Species diversity and population size of Collembola in some cultivated fields. pp. 75-89 In : *Advances in management and conservation of soil fauna.* (Eds. G.K. Veeresh *et al.*), Oxford & IBH publishing Co. Pvt. Ltd., New Delhi.

Guru, B.C., and Mohanty, M.C. 1990. Growth of two species of Collembola (Apterygota : Insecta) in three different culture methods. *Agrokemia Es Talajtan, Tom,* 39 No. **3-4** : 576-582.

Guru, B.C., and Panda, S. 1990. Species diversity and pattern of distribution of Collembola (Apterygota : insecta) in cultivated and uncultivated sites of Orissa, India. *Agrokemia Es Talajtan, Tom,* 39, No **3-4** : 569-575.

Guru, B.C., and Panda, S. 1991. The role of cryptopygus theremophilus (Collembola) in regulating C/N ratio. pp. 99-107 In : *Advances in management and conservation of soil Fauna.* (Eds. G.K. Veeresh *et al.*) Oxford & IBH Pubhishing Co. Pvt. Ltd., New Delhi.

Guru, B.C., Panda, S. and Mohapatra, P. 1988. Species composition, vertical distibution and seasonal variation of Collembola in cultivated and uncultivated sites of Eastern Orissa. *J. Soi. Biol. Ecol.* **8** (2) : 104-116.

Haarlo, N., 1960. Microarthropods from Danish soils. Ecology, Phenology. *Oikos,* Supplementum. **3** : 1-76.

Hagvar, S., 1983. Collembola in Norwegian condiferous forest soils. II. Vertical distribution. *Pedobiologia,* **25** : 383-401.

Hammer, M. 1944. Studies on the Oribatids and Collemboles of Greenland. *Medd. Greenland.* **141** : 1-210.

Haq, M.A., and Ramani, N. 1991. Population ecology of microarthropods in relation to vegetation and rainfall. pp. 797-803 In : *Advances in management and conservation of soil fauna* (Eds. G.K. Veeresh *et al.*), Oxford & IBH Publishing Co. Pvt. Ltd., New Delhi.

Hattar, S.J.S., and Alfred, J.R.B 1984. A population dynamic study and community analysis of Collembola in pine forest soils of Meghalays, N.E. India. Proc. III Oriental Entomol. Symp.,104 p.

Haazra, A.K., 1991. Effects of deforestation on the soil macro and microarthropods fauna of West Bengal, India.

Hazra, A.K., and Choudhuri, D.K. 1983. A study of Collembola communities in cultivated and uncultivated sites of West Bengal in relation of three major soil factors. *Rev. Ecol. Biol. Sol.* **20** (3) : 385-401.

Holt, J.A., 1985. Acari and Collembola in the litter and soil of three north Queensland rain forests. *Aust. J. Ecol.* **10** : 57-65.

Hughes, R.D., 1962. The study of aggregated populations. pp. 51-55 In : *Progress in Soil Zoology* (Ed. P.W. Murphy), 51-55.

Huther, W., 1983. Collembolan population brasilianische Regenwalder and hire Beeinflussung durch den Menschen. *Pedobiologia.* **25** : 317-322.

Hutson, B.R., and Veitch, L.G. 1983. Mean annual population density of Collembola and Acari on the soil and litter of three indigenous South Australian forest. *Aust. J. Ecol.* 8, 113-126.

Joosse, E.N.G., 1970. The formation and biological significance of aggregations in the distribution of Collembola. *Neth. J. Zool.* **20** : 299-314.

Kuhnelt, W., 1955. A brief introduction to major groups of soil animals and their biology. pp. 29-43 *In Soil Zoology* : (Ed. Kevan, D. Keith Mc E.) London.

Loring, S.J., Snider, R.J. and Robertson, L.S. 1981. The effects of three tillage practices on Collembola and acarina populations. *Pedobiologia.* **22** : 172-184.

Macfadyen, A., 1952. The small arthropods of a Molinia fen at Cothill. *J. Anim. Ecol.* **21** : 87-117.

Macfadyen, A., 1957. Animal Ecology, ''Aims and Methods'', Pitman, London.

Mallick, S.C., 1989. Ecology of microarthropods with special reference to Collembola (Apterygota : Insecta) in some cultivated fields of Bhubaneswar (Orissa). M. Phil. Thesis, Utkal University.

Mitra, S.K. 1990 a. On *Idiomerus* Imms (1912) from India (Collembola : Entomobryidae : Paronellinae). *Rec. Zool. Surv. India.* **86** (1) : 69-82.

Mitra, S.K., 1990 b. On the concepts of *Paronana* womrsley (1939) and *Parasalina* salmon (1944) (Collembola : Entomobryidae : Paronelliane) from New Zealand. *Rec. Zool. Surv. India.* **86** (1) : 119-144.

Mitra, S.K., Hazra, A.K. and Sanyal, A.K, 1977. Ecology of Collembola at the Eden Gardens, Calcutta. *Ecol. Bull.* (Stockholm) **25** : 539-544.

Mitra, S.K., Hazra, A.K. Sanyal, A.K. and Mandal, S.B. 1983. Changes in the population structure of Collembola and acarina in a grassland and rainwater drainage at Calcutta. pp. 664-667 In : *New Trends in Soil Biology.* (Ed. Ph. Lerbrun *et al.*)

Mukharji, S.P., and Singh J. 1970. Seasonal variation in the densities of a soil arthropod population in a rose garden at Varanasi (India). *Pedobiologia.* **10** : 442-446.

Neilsen, C.O., 1949. Nat Jutland. 2, 1-131. Quoted from ''Collembola'' pp. 397-411 In : *Soil Biology* (Eds A. Burges and F. Raw), Academis Press, London.

Pai, C.G.A., and Prabhoo, N.R. 1981. General Soil Fauna. Preliminary observation on the microarthropod fauna on paddy fields and adjoining uncultivated soils of South Kerala. pp. 27-32 In : *Progress in Soil Biology and Ecology in India,* UAS Tech.

Panda, S., 1987. Ecology of Collembola (Apterygota : insecta) with special reference to their role in decomposition process. Ph. D. Thesis submitted to Utkal Univ., Bhubaneswar.

Peterson, H., 1980. Population dynamics and metabolic characterisation of Collembola species in a beach forest ecosystem. pp. 806-833 In : *Soil Biology as related to Land use practices* (Ed. D.L. Dindal), Proc. VII Int. Soil Zool. Colloq.

Ponge, J.F., 1980. Les biocenoses des collemboles de la foret de senart. pp. 151-176 In : *Actualites d'ecologie forestiere,* (Ed. P. Pesson), Gauthier-Villars, Paris.

Ponge, J.F., 1983. Les collemboles, indicateurs du type d' humus en milieu forestier. Resultats obtenus au sud de Paris. *Acta Ecologica, Ecol. Gener.* **4** : 359-374.

Ponge, J.F., and Prat, B. 1982. Les collemboles indicateurs du mode d'humification dans les peuplements resineux, feuillus et melanges : resultats obtenus en foret d' Orleans. *Rev. Ecol. Biol. Sol.* **19** ; 237-250.

Poole, T.B., 1961. An Ecological study of Collembola in a coniferous forest soil. *Pedobiologia.* **1** : 113-137.

Poursin, J.M. and Ponge, J.F. 1982. Comparison de trois types de peuplements forestiers (feuillu, mixte, resineux) a l' aide de la mesofaune du sol (Collemboles et Oribates). *CR. Acad. Sci. Paris, Serie* III, **294** : 1021-1024.

Poursin, J.M, and Ponge, J.F. 1984. Etude des peuplements de microarthropodes (Insectes Collemboles et Acariens Oribates) dans trois hums forestiers acides de la Foret d' Orleans (Loiret, France), *Pedobiologia.* **26** : 403-414.

Prabhoo, N.R., 1971 a. Soil and litter Collembola of South India. 1. *Arthropleona. Oriental Insects.* **5** (1) : 1-46.

Prabhoo, N.R., 1971 b. Soil and litter Collembola of South India. Symphypleona. *Oriental Insects.* **5** (2) : 243-262.

Prabhoo, N.R., 1976. Microarthropods virgin forest and adjoining tea fields in the Western Ghats in Kerala. *Oriental Insects.* **10** : 435-442.

Part, B., and Massoud, Z. 1980. Etude de la communaute des collemboles dans un sol forestier. I. Structure du peuplement. *Rev. Ecol. Biol. Sol.* **7** : 199-216.

Rapoport, E., and Najt, J. 1966. Ecologia de los microarthropods en suelos gley and solonchak de bahia blanca, Argentina Monografis, progress en Biologia del Suelo, Actas Del Primer Coloqio Latin Americano de Biologia del Suelo. pp. 521-614.

Raw. F., 1956. The abundance and distribution of protura in grassland. *J. Anim. Ecol.* **25** : 15-21.

Reddy, M.V., 1981. Faunal composition of summer population of soil microarthropods of tea nursery beds at Nayabunglow, N.E. India. *Ind. J. Ecol.* **8** (1) : 108-111.

Salt, G., and Hollick, F.S.J. 1946. Studies of wireworm population, spatial distribution. *J. Expt. Biol.* **23** : 1-46.

Sengupta, D., and Sanyal, A.K. 1991. Studies on the soil microarthropod fauna of a paddy field in West Bengal, India pp. 789-796. In : *Advances in management and conservation of soil fauna* (Eds. G.K. Veeresh *et al.*), Oxford & IBH Publishing Co. Pvt. Ltd., New Delhi.

Sheals, J.G., 1957. The Collembola and Acarina of uncultivated soil. *J. Anim. Ecol.* **26** : 125-134.

Singh, J., and Pillai, K.S. 1981. Community structure and biocoenology of soil microarthropods in some tropical fields at Varanasi, Ind. *India. J. Soil. Biol. Ecol.* 115-120.

singh, J., and Mukharji, S.P. 1971. Qualitative compostion of soil arthropods in some fields at Varanasi, India. *Oriental Insects.* **5** : 487-494.

Singh, J., and Mukharj , S.P. 1973.Population size arthropods in some fields at Varanasi, India. *Sci and Cult.* **39** : 498-500.

Stockli, A., 1957. Die Metazoenfauna non Wiesen and Ackerboden aus der Umgebung von Zurich. Landw. Jb. Scheweiz. b (n.s.), 571-595.

Takeda, H., 1978. Ecological studied of Collembolan populations in a pine forest soil. II. Vertical distribution of Collembola. *Pedobiologia.* **18** : 22-30.

Takeda, H., 1981. Effects of shifting cultivation on the soil meso-fauna with special reference to Collembolan populations in the North East Thailand. Mem. Coll. of Agric. Kyoto Univ., No. **118** : 45-60.

Usher, M.B., 1969. some properties of the aggregations of soil Arthropods : Collembola. *J. Anim. Ecol.* **38** : 606-622.

Verhoef, H.A., and Selm, A.J. Van 1983. Distribution and population dynamics of Collembola in relation to soil moisture. *Holarctic Ecology.* **6** : 387-394.

Wood, T.G., 1967 (a). Acari and collembola of Moorland soils from Yorkshire, England. I. Description of the sites and their population. *Oikos.* **18** : 102-117.

Wood, T.G., 1967 (b). Acari and Collembola of Moorland soils from Yorkshire, England. II. Vertical distribution in 4 grassland soils. *Oikos.* **18** : 137-140.

Chapter 8

Enchytraeid Fauna of the Shore of Lake Balaton

Klára Dózsa-Farkas

Department of Systemetic Zoology and Ecology of the Eötvös Loránd University, 1088 Budapest, Puskin utca 3. Hungary

Abstract. Enchytraeid fauna of the shoreline of the largest shallow lake of Europe, Lake Balaton was investigated during 3 years. A total number of 136 samples from 35 localities belonging to different biotopes were analysed. A total number of 21 enchytraeid species belonging to 9 genera were found; three of them are new for the Hungarian fauna /*Achaeta pannonica* Graefe, 1989; Henlea heleotropha, Niels. et Christ., 1959; *Marionina spicula* (Leukart, 1847)/. All the other species-except for Henlea ventriculosa (Udekem, 1854)-are new for the fauna of Balaton. There were no enchytraeid species found in the mud and on the waterplants. Only floating, decaying reed roots and the reed stems standing in the wwater were found to be suitable habitats for enchytraeids. Bryum, covering the reed stems seemed to be an interesting microhabitat, with a dominance of *Cognettia glandulosa* (Mich., 1888). The further the samples were taken away from the permanent water covered area the more enchytraeid species were found in them. Species with a widest distribution were found to be : *Enchytraeus minutus* Niels. et Christ., 1961; *H. ventriculosa* and Marionina argentea (Mich., 1889).

Key words : Lake Balaton, Oligochaeta, Enchytraeid fauna of different biotopes

Introduction

Lake Balaton is the largest shallow lake of Europe situated in Transdanubia (in the western part of Hungary), at an altitude of 104.8 m above the Adriatic sea level (Fig. 8.1). The lake has a length of 78 km, a mean width of 7.6 km and a mean depth of 3.25 m. The minimum width (1.5 km) and maximum depth (10.2 m) are at Tihany straits (Herodek *et al.,* 1988).

Studies on the fauna of the lake identified 1200 invertebrate species existing in it. Mots of

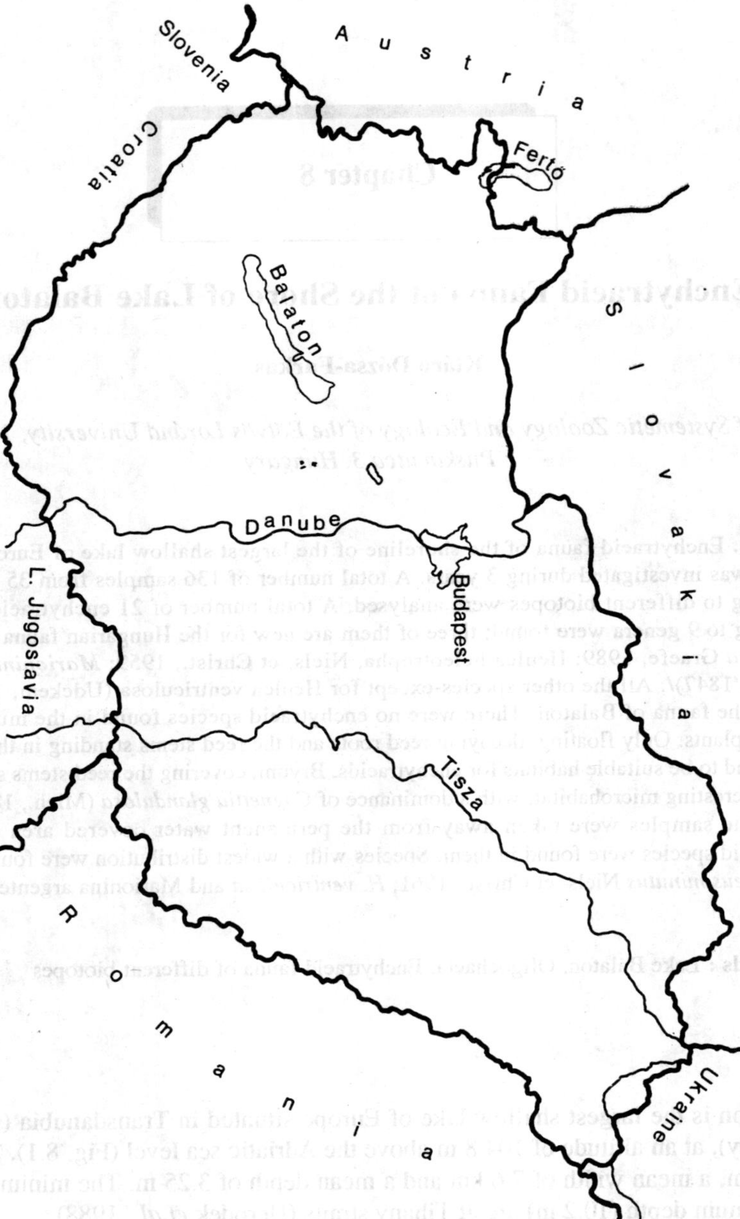

Fig. 8.1. Localization of Lake Balaton in Hungary

them are inhabitants of the littoral zone of this eutrophic lake (Herodek *et al.*,1988). Conserning the echytraeid fauna of the shoreline, there are only two data published : Vangel (1897) detected *Henlea ventriculosa* at 3 localities; and Apathy (1918) described *Enchytraeus albidus* in ''Fauna Regni Hungariea'' as a Ccharasteristic oligochaetic species for the shore fauna. Detailed investigations on enchytraeid fauna either of lake Balaton, or of other shallow lakes having a similarly big surface have not been carried out yet.

Our aim was to described the enchytraeid fauna of the lake and the differences between the special assemblage occurring in certain biotopes of the shoreline.

Meterial and Methods

Our investigations were carried out during 3 years (1990-1992). The lake and its surrounding is a favoured recreation area, with artificial beaches and holiday houses. Our samples were taken from natural like sites such as reed-grown zones and untouched beaches if it was possible, but enchytraeid fauna of the offshore bars in beaches was analysed as well. A total number of 136 samples from 35 localities were analysed around the Balaton (Fig. 8.2), taken from different biotopes and in different seasons.

The different samples can be devided into 9 biotope-types :

I. Mud from deeper than 50 cm depth;
II. Mud from 20-25 cm layer;
III. (a) Water plants, reed-grass, Potamogeton, wrack-coats;
(b) Floating, decaying bundles of reed-roots;
(c) Stem of reed;
IV. Wet mud from the littoral zone and reed-root debris form shoals;
V. Bryum form the stem of reeds;
VI. Offshore bars, wet
VII. Offshore bars, dry
VIII. Reed-bed, temporarily flodded reed-bed and sedge vegetation
IX. Soil from fields, and from the surrounding grasslands.

Qualitative samples were taken only. Muddy samples were taken by Hangrove and Ekmann samplers. The worms were extracted by O'Connor method and identified alive. The identified enchytraeids were preserved in 75 per cent alcohol.

List of localities (in parentheses the biotope-types)

1. *Balatonkenese I.* 25.10.1991.
   1. Soil from grassland at the skirts of reed-bed (IX).
   2. Offshore bar of reed-grass (VI).
   3. Offshore bar of reed (VII).

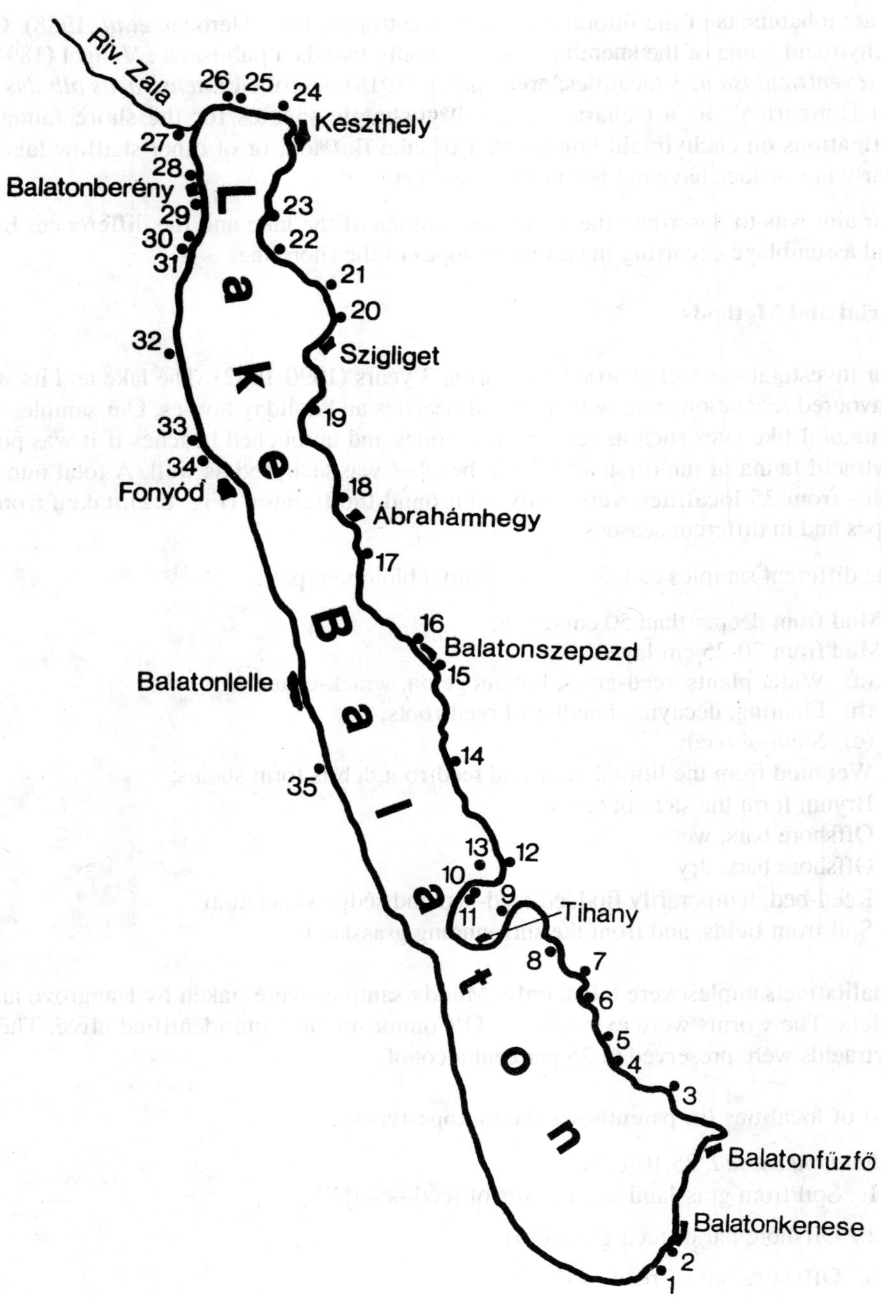

Fig. 8.2. Sample plots around Lake Balaton

2. *Balatonkenese II.* 25.10.1991.
    4. Offshore bar of reed-grass (VI).
    5. Offshore bar of reed (VII).
    6. Mud from among littoral stones (IV).

3. *Between Füzfö and Alsóörs* 22.9.1992.
    7. Offshore bar (VII).
    8. Mud from the littoral zone (IV).

4. *Between Káptalanfüred and Alsóörs: around the board walk to the reed bed* 13.10.1992.
    9. Bryum from the stem of reeds (V).
    10. From the surface of waterplants Hydrochar is morsus-ranae and Ricchia fluitans (IIIa).
    11. Soil of reed-bed (VIII).
    12. Soil from reed-bed with Carex (VIII).

5. *Between Káptalanfüred and Alsóörs, nearer to Alsóörs, around the board walk to the reed-bed*
    13. Stem of reeds and Myriophyllum. (III c) 4.9.1990.
    14. Bryum from the stem of reeds (Bryales) (V) 4.9.1990.
    15. The same as Nr. 14 (V) 28.11.1991.
    16. The same as Nr. 14 (V) 28.10.1992.
    17. Stem of reeds (III c) 28.11.1991.
    18. Soil from reed-bed with Carex (VIII) 28.11.1991.
    19. Soil from grassland (IX) 28.11.1991.
    20. Mud and reed-grass from low waer at the end of the board walk (III a) 28.10.1992.
    21. The same as Nr. 18 (VIII) 28.10.1992.

6. *Alsóörs, camping* 22.9.1992.
    22. Moist soil from the littoral zone (IV).

7. *Paloznaki Bay* 29. 4. 1991.
    23. Soil and reed debris decaying in reed bed (VIII).
    24. Leaf litter and soil from the stem of a willow (IX).
    25. Soil from grassland (IX).

8. *Balatonfüred, at the statue of Darányi, from ship,* 16.5.1990.
    26. 510 m from the shoreline, from a depth of 3.1 m, the upper 3-5 cm thick layer of mud (I).
    27. The same as Nr. 26., mud from 20-25 cm (I).
    28. 250 m from the shoreline, the upper 2-3 cm thick layer of mud (II).
    29. Soil from a little (1.5 m^2) reed island near the sore (VI).
    30. Underwater part of the reed-island (IV).
    31. Floating decaying reed roots (III b).
    32. Offshore bars of reed (VI).

33. Myriophyllum from the water (III a).

9. *Tihany, nearby the beach* 4.9.1990.
   34. Offshore bar of reed-grass (VI).
   35. Bulky offshore bar of reed (VII).

10. *Bozsai Bay, beach* 4.9.1990.
   36. Stem of reeds (III c).
   37. Offshore bar of reed-grass (VI).
   38. Bulky offshore bar of reed (VII).

11. *Bozsi Bay, not far of the beach* 13.10.1992.
   39. Mud from the littoral zone (IV).
   40. Plant debris from reed bed among stones (IV).
   41. Offsore bar of reed (VII).
   42. Offshore bar of reed-grass (VI).
   43. Offshore bar of reed (VII).

12. *Bozsai Bay, at the railway station of Aszófő, reed-bed*
   44. Wet bebris of reed (VI) 16.5.1990.
   45. The same as Nr. 44. (VI). 29.4.1991.
   46. Wet mud and reed debris (IV) 16.5.1990.
   47. Soil from field with Iris pseudicorus (IX) 16.5.1990.
   48. The same as Nr. 47. (IX) 29.4.1991.
   49. The same as Nr. 47. (IX) 29.4.1991.
   50. Soil from reed-bed with Carex (VIII) 16.5.1990.

13. *Bozsai Bay, from ship* 16.5.1990.
   51. 500 m from the shoreline, the upper 2-3 cm thick layer of mud (I).
   52. Mud sample from the depth of 20-25 cm of it (I).
   53. At the same place, the upper 2-3 cm thick layer of an other mud-sample (I).
   54. 250 m from the shoreline, mud from the depth of 2-5 cm (I).
   55. Mud near the reed-bed, the upper 2-5 cm layer (II).
   56. Coating on the stem of reeds (III a).
   57. Floating reed roots (III b).
   58. Grass-roots and debris of reed from the littoral zone (VII).
   59. Debris of reed in low water (IV).

14. *Balatonakali, Dövicse landing place* 13.10.1992.
   60. Offshore bar of alga (VI).
   61. Wet mud from lottoral zone (IV).
   62. Offshore bar and mud (IV).
   63. Soil from under a willow (IX).

15. *Balatonszepezd* I. 25.10.1991.
   64. Wet offshore bar (VI).

65. Mud from the littoral zone (IV).

16. *Balatonszepezd* II. 28.10.1992.
   66. Wrack and reed-grass debris from the stony shore (VI).
   67. Stony soil from under grass (IX)
   68. Bundle of reed-roots in the lottoral zone (VII).

17. *Balatonrendes* 9.7.1990.
   69. Soil from reed-bed with Typha angustipholi (VIII).
   70. Soil from red-bed with Carex (VIII).

18. *Abrahámhegy*, 28.10.1992.
   71. Bryum from stem of reeds, near the board walk to the reed-bed (Bryales) (V).
   72. Reed debris near the shore (VII).
   73. Soil from between reed-bed and potato field (IX).

19. *Badacsonylábdihegy* 9.7.1990.
   74. From the surface of Hydrochar is morsus-ranae (III a).
   75. Stems of reeds standing in water (III c).

20. *Sziqliqeti Bay, Anglers-camp,* 29.4.1991.
   76. Floating bundle of reed (III b).
   77. Offshore bar of reed (VII).
   78. Soil from under willows and bushes (IX).

21. *Sziqligeti Bay,* 29.4.1991.
   79. Soil of a little reed island, at the end of the board walk (VIII).
   80. Reed-root bundle from water in the littoral zone (III b).
   81. Reed debris from low water (IV).
   82 Soil of a clump standing out the water, in a reed-bed destroyed by fire (VIII).
   83. Soil from under willow-bushes (IX).
   84. Sand from under willow-bushes (IX).
   85. Soil from shore with tall weeds (IX).
   86. Soil from under grassy bundles (IX).
   87. Soil from under a willow (IX).
   88. Soil of a reed-bed on the shore (VIII).

22. *Balatongyörök* 9.7.1990.
   89. Mud with alga (IV).
   90. Offshore bar of reed and reed-grass (VI).
   91. Soil from reed-bed with Carex (VIII).

23. *Under the Szent-Mihály Hill* 28.10.1992.
   92. Offshore bar of reed-grass (VI).
   93. Offshore bar of reed (VII).
   94. Soil of a reed-bed with Typholia angustipholia (VIII).

95. Soil from under willows (IX).
96. Loamy soil from under willow-bushes (IX).

24. *Keszthely után* 25.10.1991.
97. Offshore bars of reed (VII).
98. Very thick offshore bar of reed (VII).

25. *Fenékpuszta* I. 15.7.1991.
99. Stem of Typha angustipholia (VIII).
100. Soil from reed-bed (VIII).

26. *Fenékpuszta* II. 26.5.1992.
101. Very thick offshore bar or reed (VII).
102. Mud and sand from the littoral zone (IV).
103. Decaying wood and soil from the stem of an old willow (IX).

27. *Estuary of th river Zala, Balatonberény, at the bridge* 25.10.1991.
104. Stem of Glicer is maxima standing in water (III c).
105. Mud from littoral zone (IV).
106. Soil from reed-bed (VIII).
107. Moist soil from the riverside (VIII).

28. *Balatonberény* I. 26.5.1992.
108. Bryum from stem of Carex in reed-bed (V).
109. Soil from reed-bed with Carex (VIII).
110. Mud from the littoral zone (IV).
111. Offshore bar with shells of Dreissena polymorpha (VI).

29. *Balatonberény II.*
112. Offshore bar or reed (VII) 25.10.1991.
113. Offshore bar of reed-grass (VI) 15.7.1991.
114. Offshore bar of reed (VII) 15.7.1991.
115. Soil from under willows (VIII) 15.7.1991.
116. Shells of Dreissena and reed-grass (VI) 15.7.1991.
117. Soil from reed with Carex (VIII) 15.7.1991.
118. Offshore bar of reed (VII) 17.11.1992.
119. Offshore bar on the edge of reed-bed (VII) 17.11.1992.
120. Offshore bar of reed (VII) 23.7.1992.

30. *Balatonmáriafürdö I.* 26.5.1992.
121. Soil from reed-bed (VIII).
122. Stem of Gliceria maxima (VII).

31. *Balatonmáriafürdö II.* 17.11.1992.
123. Offshore bar (VII).
124. Leaf litter under a willow on the shore (VII).

32. *Balatonfenyves, terminal of narrow-gauge railway* 26.5.1992.
    125. Soil from reed-bed (VIII).
    126. Offshore bar of reed and reed-grass (VI).
    127. Mud from low water (II).
    128. Floating bundle of reed-roots (III b).
    129. Offshore bar of reed (VII).

33. *Bélatelep, beach* 24.7.1990.
    130. Bulky offshore of reed (VII).
    131. Offshore of reed-grass (VI).
    132. Bundle of reed-roots on the shore (VII).
    133. Offshore bar of reed (VII) 17.11.1992.
    134. Mud from the littoral zone (IV) 17.11.1992.

34. *Bélatelep, railway station, pier* 24.7.1990.
    135. Bundle of reed-roots on the shore (VIII).

35. *Between Balatonlelle and Balatonszemes* 17.11.1992.
    136. Offshore bar of reed (VII).

Results and discussion

As a result of our investigations a total number of 21 enchytraeid species belonging to 9 genera were found in the shore fauna of Balaton. Three of them are new for the Hungarian fauna (Achaeta pannonica, Henlea heleotropha and *Marionina spicula*). All the other species-except for *Henlea Ventriculosa*-are new for the funa of Balaton. Although Apáthy (1918) designated that *Enchytraeus albidus* was characteristic for the lake, during this study this species was not found. This fact could be explained in two ways : either by misidentification or by the fact that *E. albidus* was found to inhabit only in dunghills, composts in Hungary so those specimens that were recorded by Apáthy might originated from such biotopes.

There were no enchytraeid species found in the mud (biotopetype I and II) though there are some aquatic species recorded in the literature. Biotopes such as waterplants, reed-grass, coats on reeds are also unfavoured by enchytraeids (Table 8.1, columns I, II and IIIa). Among those sample sites from the biotopetype III only floating, decaying reed roots were found to be suitable habitats for enchytraeids (column IIIa), there were 6 species found (*Cernosvitoviella minor, Enchytraeus buchholzi, E. minutus, Henlea ventriculosa, Marionina argentea and M. riparia*). In the stems of reeds two more species occurred : Cognettia glandulosa and *Fridericia bulboides*. The number of the specimens in this samples was extremely low.

Data from the samples belonging to group IV differed from each other. Altogether 1-7 species occurred there in the reed debris, enchytraeids were missing from 38 per cent of the wet mud samples while other sites showed an extremely high diversity. For example we found 11 species in one sample at Balatonberèny. In this case the shoreline is natural like without an artificial

Table 8.1. The sum and occurrence of species in the different biotope types (in per cent to the sample-number)

	Occurrence of the species in the diffrent biotope types (in % of the samples)									
	I.-II.	III.a	III.b	III.c	IV.	V.	VI.	VII.	VIII.	IX.
Achaeta pannonica			40	20	5				4	5
Cernosvitoviella minor					5	17	18	7	15	
Cognettia glandulosa				40	5	66	6	7	37	17
Buchholzia fallax									4	11
B. appendiculata								4	4	5
B. juv.					22		12	30	21	56
Enchytraeus minutus			60	100	61	83	71	89	71	56
E. bulbosus										11
E buchholzi			20		22		29	30	42	78
Fridericia bulbosa					5		6		8	11
F.bulboides				40	17		18	11	22	33
F.galba					5		6		8	22
F. paroniana								4	8	28
F. perrieri									8	11
Henlea heleotropha					5			7	21	17
H. nasuta							12	19	4	56
H. perpusilla				20	17	17	24	41	50	78
H. ventriculosa			60	60	67	17		89	58	30
Hemifridericia parva					5			4		10
Marionina agentea			40	60	28	33	53.	33	75	45
M. riparia			60	80	50	67	53	41	25	5
M. spicula					5		6	7		
Number of species			6	8	16	7	13	16	19	20
Mean number of spec.			2.6 1.4	4.2 0.5	3.3 7.6	3.0 2.3	3.7 4.5	4.2 4.6	4.9 4.0	5.8 6.7

shorewall-perhaps this is one of the last natural shore-sections of Balaton-and the wet mud from the littoral zone is rich in organic material. The absence of enchytraeid species in some samples can be explained with the strong stink of hydrogen sulphide. Relative dominance level estimations showed species of genus *Marionina* to be dominant in the very wet habitats.

An interesting difference was found between the fauna of samples taken from the underwater and the above water parts of a small reed island. In the latter case there were 3 times more species found than in the former habitat. A sort of isolation of two sometimes coexisting *Enchytraeus* species was observed. since *E. minutus* inhabiting mostly in wet biotopes was dominant in the underwater part, cosmopolitan *E. buchholzi* with a wide tolerance was mostly abundant in the upper sample.

Bryum, seemed to be an interesting microhabitat (biotopetype V). 7 species were found there, among them *Cognettia glandulosa* was dominant, with an outstandingly high number in those samples taken in autumn. *C. glandulosa* is one of the few enchytraeid species, which can reproduce by fragmentation. *C. glandolosa* reproduces itself mostly by this way and becomes maturated for a short time once in a year only. This fact can explain, that maturated specimens were found only once in spring and never before, but and even those specimens had been reducing genital organs. Some species of genus *Buchholzia* (also occurring at lake Balaton) have the same way of reproduction, but they become maturated twice a year (therefore this 2 species were signed by B.juv.).

The further we go away from the open water the more species occurred in the samples. For example whereas in the wet offshore bars (biotope type VI) the number of the species was 13, in the dryer one (biotope type VII) it reached 16. At the same time the number of marionina specimens-first of all of M. riparia-decreased.

The shorter the flooding period of the reed bed is, the better environmental conditions there are for enchytraeids since the soil becomes richer in organic matter. So that on the shore-sections covered with reed (biotope types VIII and IX) *Fridericia* species could be found as well. The highest species number was found in two samples from these biotopes. One of them consisting of 13 species was taken from the edge of the reed-bed with Iris pseudocorus in Bozsai Bay ; the other one was sampled from a grassland in Szaloki Bay : 12 species were found. Not only the number of species was high in these sites but also the individual numbers per sites.

Summarising the data set in terms of distribution of the species we can conclude, that *Enchytraeus minutus* (Table 8.2), was found to have the widest distribution occurring in 76.4 per cent of the samples containing *Enchytraeus,* (89 per cent of those samples from dryer offshore bars contained this species) and being dominant species quite often. The second most abundant enchytraeid species is *Henlea ventriculosa* occurring in 65 per cent of the samples. Interestingly this cosmopolitan species was found to live in very different biotopes in Hungary, such as in floating moors, in forest soils and in dolomite steppes. Four *Henlea*-species occurred together though *H.ventriculosa* and *H. perpusilla* can not be abundant in the same habitat together.

TABLE 8.2. *Occurance of species in per cent of the samples containing enchytraeids*

	%
Achaeta pannonica-Graefe, 1989	2.7
Cernosvitoviella minor-Dózsa-Farkas K., 1990	12.7
Cognettia glandulosa	20.0
Buchholzia fallax-Niels. & Christ., 1959	1.8
B. appendiculata-(Buchholz, 1862)	2.7
B. juv.	26.4
Enchytraeus minutus-Niels. & Christ, 1961	76.4
E. bulbosus-Niels. & Christ, 1963	1.8
E.buchholzi-Vejd., 1879	37.0
Fridericia bulbosa-(Rosa, 1887)	5.4
F. bulboides-Niels. & Christ., 1959	20.9
F. galba-(Hoffmeister, 1843)	7.2
F. paroniana-Issel, 1904	7.2
F. perrieri (Vejd., 1877)	3.6
Henlea heleotropha-steph., 1922	10.0
H. nasuta-(Eisen, 1878)	16.4
H. perpusilla-Friend, 1911	41.0
H. ventriculosa-(Udekem, 1954)	65.4
Hemifridericia parva-Niels. & Christ., 1959	3.6
Marionina argentea-(Mich., 1889)	50.0
M. riparia-Bretscher, 1899	42.0
M. spicula-(Leuckart, 1847)	3.6

Either one or the other of them was dominant. On the third place the Marionina argentea occurred most frequently being found in 50 per cent of the samples. This is also a cosmopolitan species, occurring in the wet soils.

Species being new for the Hungarian fauna (*Achaeta pannonica, H. heleotropha* and M. spicula) were found in many samples, but they never were dominant ones.

Cernosvitoviella minutus was described as a new species in 1990 from the Sphagnum-moor of Bábtava (Csaroda) and from the floating moor of Szigetcsép (Dózsa-Farkas, 1990). This species was found in the 12.7 per cent of the samples studied and was missing from the dryest samples (biotope-type IX) only. Since species of this genus are mostly hidrofil ones this sort of distributional pattern can easily be explained. Considering our data, *C. minutus* is expected to have a wide distribution in Hungary.

Unfortunately because of the shoreline management policy the tourist industry and especially because of the building of concrete shorewalls the autoregulation of the lake can not work any more. The deposit can not form bars on the shore any more and can not be consumed by saprophagous animals such as enchytraeids. This leads to further eutrophication.

Occurrence of species

Achaeta pannonica Graefe, 1989
102, 115, 121.

Cernosvitoviella minor Dózsa-Farkas, 1990
14, 29, 31, 32, 35, 59, 66, 70, 75, 77, 80, 82, 88, 131.

Cognettia glandulosa (Michaelsen, 1888)
11, 12, 13, 14, 16, 17, 23, 24, 25, 32, 69, 70, 71, 72, 77, 79, 81, 83, 88, 91, 99, 107, 108.

Buchholzia fallax (Michaelsen, 1887)
25, 47, 117.

Buchholzia appendiculata (Buchholz, 1862)
25, 106, 123.

Buchholzia Juvenil
1, 5, 19, 25, 29, 41, 44, 46, 47, 48, 49, 58, 65, 67, 73, 78, 85, 87, 101, 102, 103, 106, 109, 110, 111, 112, 120, 121, 123, 129, 130.

Enchytraeus minutus Nielsen & Christensen, 1961
2, 4, 5, 6, 7, 8, 11, 12, 13, 14, 15, 16, 17, 19, 21, 22, 23, 29, 34, 35, 36, 37, 38, 40, 41, 42, 43, 44, 45, 46, 47, 48, 49, 50, 58, 64, 65, 66, 67, 68, 70, 71, 72, 75, 76, 77, 80, 81, 82, 83, 84, 88, 89, 90, 91, 93, 94, 95, 96, 97, 98, 99, 100, 101, 102, 104, 105, 107, 108, 110, 112, 114, 115, 116, 117, 118, 119, 120, 124, 128, 129, 130, 131, 132, 135, 136.

Enchytraeus buchholzi vejdovsky, 1879
1, 5, 12, 19, 21, 24, 25, 29, 30, 31, 34, 44, 45, 46, 47, 49, 58, 63, 65, 67, 70, 73, 77, 82, 83, 86, 87, 96, 99, 100, 101, 102, 103, 109, 111, 114, 115, 116, 120, 121, 123, 129, 130.

Enchytraeus bulbosus Nielsen & Christensen, 1963
1, 73.

Fridericia bulbosa (Rosa, 1887)
25, 44, 47, 48, 49, 102, 106, 121.

Fridericia Bulboides nielsen & Christensen, 1959
15, 17, 25, 44, 47, 50, 64, 65, 70, 72, 82, 83, 85, 86, 91, 101, 102, 104, 105, 106, 109, 111.

Fridericia galba (Hoffmeister, 1843)
64, 65, 67, 73, 87, 106, 115, 121.

Fridericia paroniana Issel, 1904 (non Nielsen & Christensen, 1959)
47, 48, 63, 67, 78, 87, 106, 120, 121.

Fridericia perrieri (Vejdovsky, 1877)
83, 84, 106, 107, 109, 121, 125.

Henlea heleotropha Stephenson, 1922, Nielsen & Christensen, 1959
5, 25, 48, 86, 101, 102, 106, 107, 109, 121, 125.

Henlea nasuta (Eisen, 1878)
1, 5, 19, 34, 47, 48, 67, 73, 78, 83, 84, 95, 101, 103, 106, 111, 123, 129, 135.

Henlea perpusilla Friend, 1911
3, 5, 12, 17, 18, 21, 24, 25, 34, 41, 44, 45, 46, 47, 48, 50, 51, 58, 63, 67, 77, 79, 82, 83, 84, 85, 86, 87, 90, 95, 96, 98, 100, 101, 102, 103, 106, 107, 108, 109, 110, 112, 115, 117, 123, 129, 130, 135.

Henlea ventriculosa (Udekem, 1854)
2, 3, 4, 5, 6, 7, 8, 12, 18, 21, 22, 23, 25, 29, 30, 31, 34, 35, 36, 37, 38, 41, 43, 44, 46, 47, 48, 50, 57, 58, 64, 65, 66, 71, 75, 76, 77, 79, 81, 82, 83, 86, 88, 89, 90, 91, 93, 94, 96, 97, 98, 101, 102, 104, 105, 107, 110, 111, 112, 114, 116, 118, 119, 120, 124, 125, 128, 129, 130, 132, 133, 134, 135, 136.

Hemifridericia parva Nielsen & Christensen, 1959
25, 46, 47, 101.

Marionina argentea (Michaelsen, 1889)
2, 6, 11, 12, 16, 17, 18, 21, 22, 23, 24, 25, 29, 32, 34, 36, 43, 45, 46, 47, 48, 49, 58, 59, 64, 70, 72, 75, 76, 77, 79, 82, 83, 85, 88, 90, 91, 93, 95, 99, 100, 101, 102, 103, 107, 108, 111, 117, 120, 121, 123, 124, 125, 126, 128, 131, 135.

Marionina riparia Bretscher, 1899
2, 6, 7, 8, 11, 13, 14, 16, 22, 24, 34, 35, 36, 42,43, 50, 51, 57, 59, 64, 66, 70, 71, 72, 75, 76, 77, 81, 89, 90, 92, 93, 94, 98, 102, 104, 106, 107, 108, 110, 111, 119, 128, 131, 132, 133, 136.

Marionina spicula (Leuckarti, 1847)
8, 115, 120, 135.

Acknowledgement

I am grateful to Dr. Jenó Ponyi for the organisation of the investigations. This research was supported by the Hungarian Academy of Sciences (AKA 89-695 and OTKA 1878).

REFERENCES

Apáthy. I. 1818. Vermes. Annelides. *Fauna Regni Hungariae* 8-13.

Dósza-Farkas, K. 1990. New Enchytraeid species from Sphagnum-bogs in Hungary (Oligochaeta : Enchytraeidae). *Acta Zool. Hung.* **36** (3-4) : 265-274.

Graefe, U. 1989. Systematische Untersuchungen an der Gattung Achaeta (Enchytraeidae, Clitellata) 2. Beschreibung von vier neuen Arten). *Mitt. Hamb. Zool. Mus. Inst.* **86** : 127-131.

Herodek, S. Lackó, L. and Virág, A. 1988. *Lake Balaton, research and management.* (Ed. K. Misley). Budapest, 110 p.

Nielsen, C.O. and Christensen, B. 1959. The Enchytraeidae. Critical revision and taxonomy of European species. *Natura Jutland.* **8-9** : 1-160.

Vángel, E. 1897. Gliedewürmer. (Annelides). *II. Band, I. Theil : Resultate der wissenschaftlichen Erforschung des Balatonsees.* 1-7.

Chapter 9

The Distribution of Earthworms (*Annelida, Oligochaeta*) in North America

John W. Reynolds

School of Natural Resources, Sir Sandford Fleming College Lindsay, Ontario, Canada K9V 5E6

Abstract. A discussion of biogeography is presented, together with its relationship to ecology. Twelve megadrile families are presently known in North America. The major agent for introduction and migration inter- and intracontinentally is man. All twelve families are found south of the southern limit of Pleistocene glaciation in North America. The exotic species in this unglaciated region may result from a coterminous period. A discussion of previous explanations for earthworm migration to North America is presented. Arguments are made for the Post Quaternary Introduction Theory, to try to explain the sole presence of exotic European Lumbricidae in the glaciated portions of North America.

Key words *: Oligochaeta, Earthworm, Biogeography, Megadrile, Lumbricidae.*

Introduction

Data for this discussion were obtained by the author from his personal published and unpublished collections and surveys in North America, including over 150,000 specimens from 35 states (U.S.A.) and 10 provinces and territories (Canada). A summary of the published surveys of North American megadriles can be found in Reynolds and Wetzel (1993).

There are three steps of biogeographical theory (Udvardy, 1969). The first is *descriptive* or *faunistics,* the gathering of facts and enumeration of animals in an area, including research and literature surveys. This seldom achieved stage seeks to present a clear distributional picture of all animals in all areas. The second stage is the *classification of data,* or grouping of the distributional data according to as many different ploints of view as a particular investigation may find necessary, including a comparison of distributional ranges of phylogenetically related groups. This stage also analyzes the fauna of a geographically, ecologically or historically uniform area. The third and final stage, *causal analysis,* tries to explain the reason for the present distribution as it has been presented. It is from this stage that the principles of biogeography emerge. At present, in North America, we are probably somewhere between the first two stages with megadriles (Reynolds, 1975 a, 1976 a).

The study of the status quo of animal distributions and of the grouping of these distributions acacording to different principles is a study of momentary phenomena. Faunistic and regional zoogeography is basically static zoogeography ; present distribution is the result of processes which move the animals in space and as a consequence of which the distributional picture changes with the passage of time. Thus, causal zoogeography is identical to dynamic zoogeography. Causations of distributions are dynamic processes. As long as our inquiry probes into the reasons for the arrival and settling of a species in a certain area, we are within the field of dynamic zoogeography. As soon as we ask ourselves why and how an animal is able to live in a particular area, we leave dynamic zoogeography; as a matter of fact, we leave zoogeography entirely, and our question receives an answer along ecological lines. For example, the lack of an animal species in a certain area may be caused by ecological (it cannot exist there) or zoogeographical (it has not arrived there) reasons. The answers pertaining to dynamic zoogeography are often ecological, but here ecological cause is followed by movement, individual or population, of the subject. When the ecological cause is followed by a positive or negative response of the organism or population and when a spatial shift is NOT involved, the whole phenomenon is within the sphere of ecology. When distributional phenomena change over a longer time span, and when the functional relationships of the organism and the environment result in adaptive changes, we come to the meeting ground of dynamic zoogeography and evolutionary studies. The latter, however, are usually assigned to the taxonomist, for evolutionary changes are most easily discerned by noting morphological changes. Lindroth (1962) expressed the relationship of these two disciplines as follows : zoogeography always depends on taxonomy to know WHAT to study; but taxonomy also depends on zoogeography, since geographic speciation is the accepted norm of the formation of its basic working unit.

Every animal species originated from a few ancestors in a limited area; if a particular species is now found to be widespread, it must of necessity have reached parts of its present range at an earllier period. The first aspect of dynamic zoogeography pertains to *dispersal.* If we know the details of dispersal processes, we can explain much about the presence or absence of animals. Dispersal may result as a by-product of other important phenomena belonging to the biological habitus of the animal, or it can result from distinct, adaptive characteristics of the species that directly assist dissemination into wider areas. In dispersal study, both the intrinsic abilities of an animal to spread and the opportunities the environment offers for spreading are equally important. Though every animal species has the capability to migrate, dispersing individuals must find a suitable area in which to settle and reproduce through many generations. When one studies the process of settling, or colonization, one must scrutinize the ecological factors that make existence possible in a given area, as well as the adaptations and limitations of a species - structural, physiological, behavioural, population dynamical, *etc.* - which enable it to start a new population and survive (successfully) in the newly colonized area. Factors of dispersal, as well as factors of existence in an area influence the size, extent and dynamism of the distributional range of the animal.

Discussion

North America, currently with members in twelve megadrile families (excluding the Enchytraeidae), has probably the greatest variety of terrestrial earthworm fauna of any continent. On closer examination, we can see the diversity of origins from which these groups come.

The ACANTHODRILIDAE were traditionally represented by members of five genera in North America south of the glacial boundary : *Argilophilus, Diplocardia* and *Megascolecides* (Nearctic), *Microscolex* (Neotropical) and *Pontodrilus* (Circumtropical). *Argilophilus* and *Megascolecides* seem to be more closely related to Oriental and Australian worms, often called "megascolecid". Recent accounts of the Megascolecidae in western North America (McKey-Fender, 1982; McKey-Fender and Fender, 1982, 1990) removed many of these species which had been traditionally placed in the genus *Argilophilus* by Gates (1977) and others. The first two genera to be erected were *Arctiostrotus* (McKey-Fender and Fender, 1982) and *Kincaidodrilus* (McKey, 1982) from the classical *Plutellus.* More recently, these authors (1990) erected *Driloleirus* (from *Megascolecides*) and *Chetcodrilus, Drilocheria, Macnabodrilus, Nephrallaxis* and *Toutellus* (from *Plutellus*). The species of the western Argilophilini are found along a narrow strip of the Pacific coast northward from San Francisco to central Washington, and seem to be eliminated from the natural populations when exotic species arae introduced,*e.g.* Lumbricidae (Fig. 9.1). Recent collections of *Arctiostrotus* have been obtained from central Idaho. The species of *Diplocardia* are restricted to continental U.S.A. and northern Mexico (Fig. 9.1). Both *Microscolex* and *Pontodrilus* are limited to a few field records in southern U.S.A. Climatic conditions in North America will keep them from spreading out of this area. There have been a couple of greenhouse records of *Microscolex* north of the southern limit of Pleistocene glaciation (Fig 9.2).

Only a few collection of the Palaearctic *Criodrilus* (CRIOBRILIPAE) have been recorded in Northern America and they were in the mid-Atlantic region near the glacial boundary.

The Ephiopian species *Eubrilus eugeniae* (EUDRILIDAE) (Fig. 9.2) was introduced into southern U.S.A. during the last couple of decades, as an additional source of fish-bait. As this species becomes more popular as bait, its distribution in the southeastern U.S.A. should spread.

Six additional families are also restricted to the southeastern and unglaciated portions of the U.S.A., and again probably due to climatic conditions in North America (Fig. 9.3) :

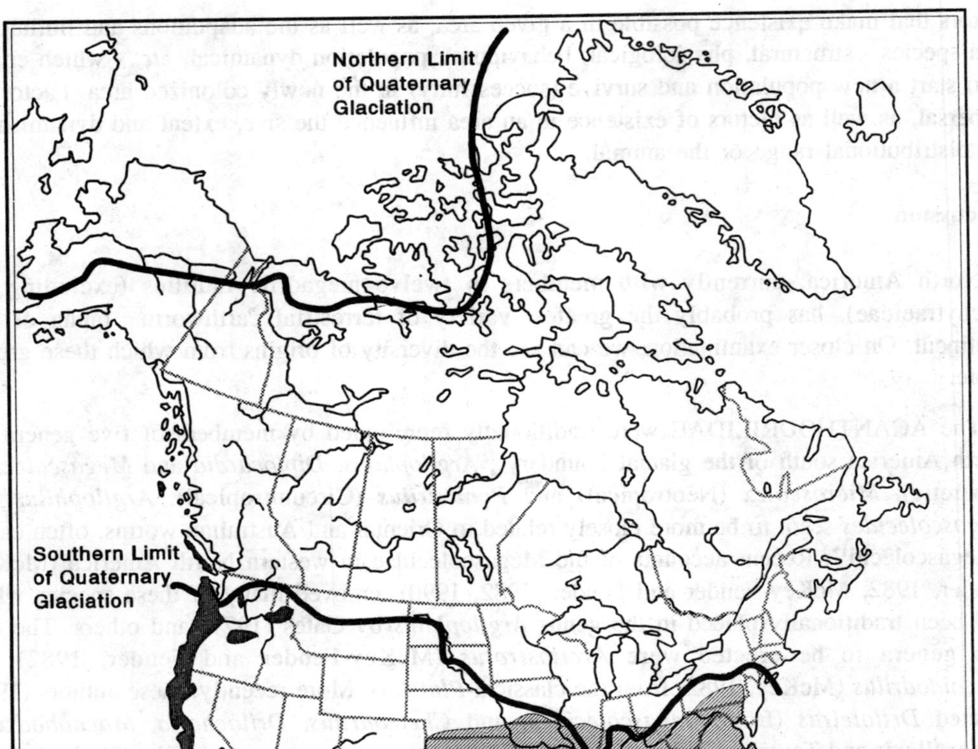

Fig. 9. 1. The North American distribution of the earthworm tribe Argilophini and the genus *Diplocardia.*

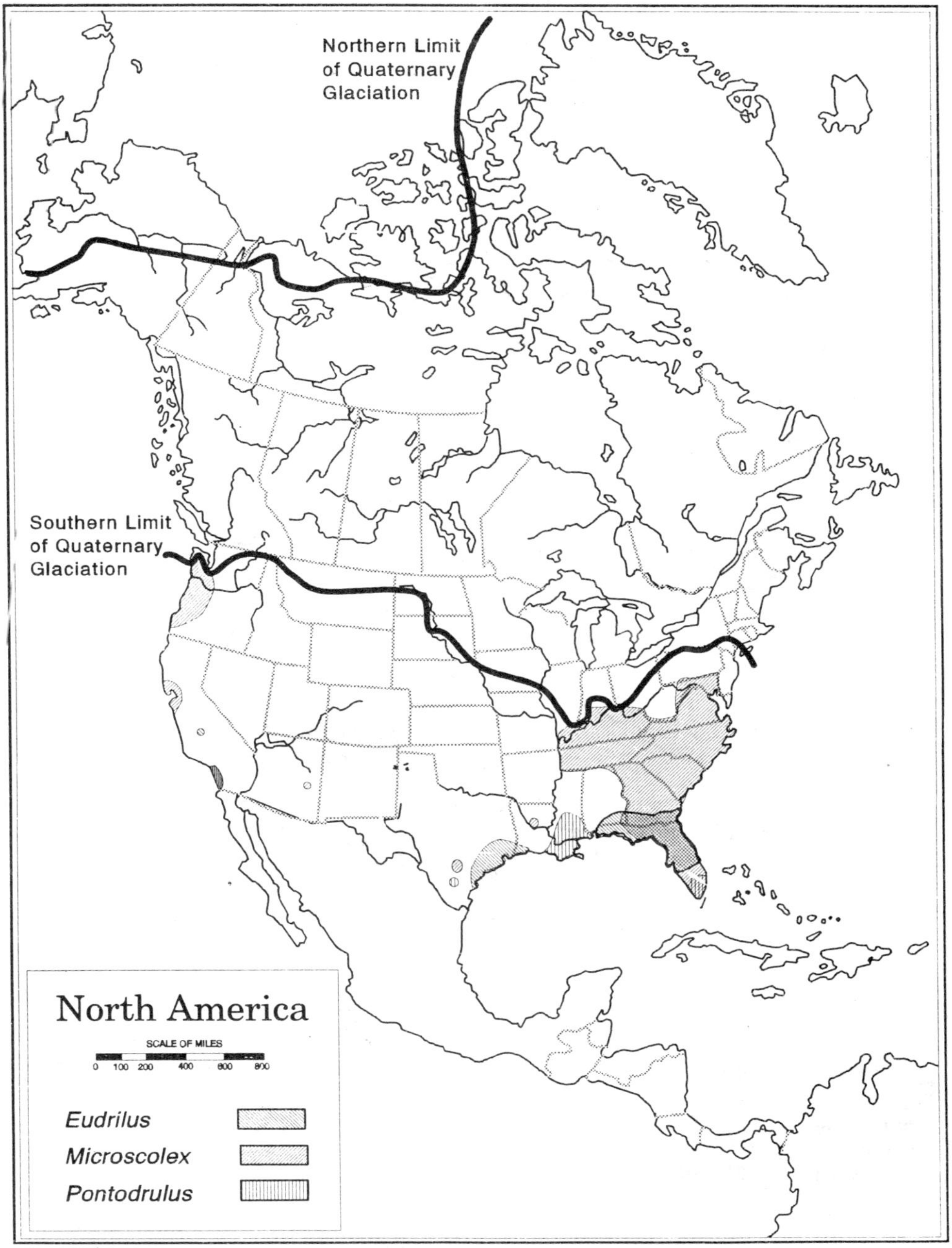

Fig. 9.2. The North American distribution of the genera *Eudrilus, Microscolex* and *Pontodrilus.*

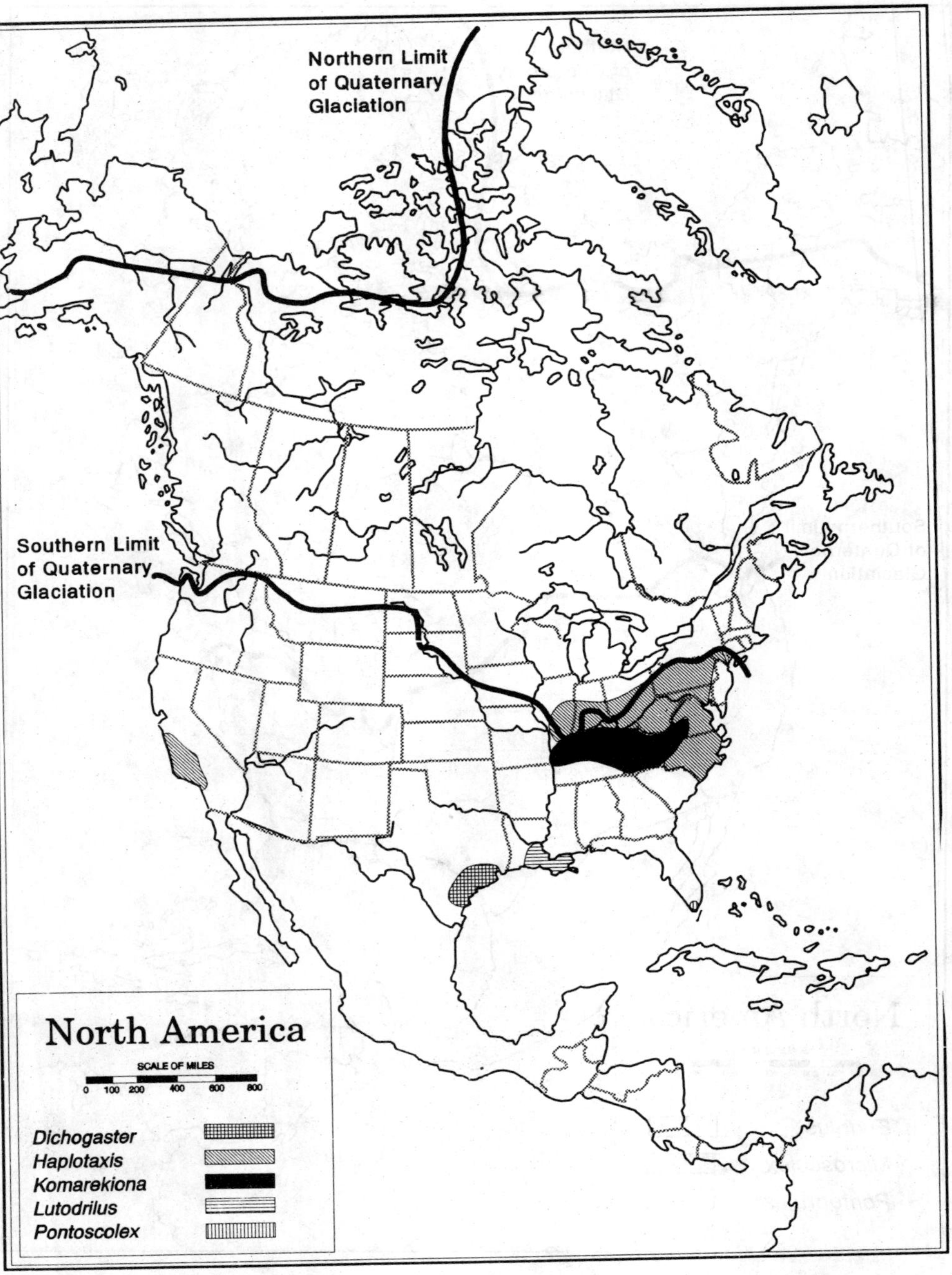

Fig. 9.3. The North American distribution of the genera *Dichogaster, Haplotaxis, Komarekiona, Lutodrilus* and *Pontoscolex.*

GLOSSOSCOLECIDAE - *Pontoscolex* (Neotropical), HAPLOTAXIDAE - *Haplotaxis* (Nearctic and Palaearctic), LUTODRILIDAE - *Lutodrilus* (Nearctic), KOMAREKIONIDAE - *Komarekiona* (Nearctic), OCNERODRILIDAE - *Eukerria* and *Ocnerodrilus* (Neotropical) and OCTOCHAETIDAE - *Dichogaster* - (Oriental).

The Oriental genus *Pheretima s.l.* (now split into *Amynthas, Metapheretima, Metaphire* and *Pithmera* by Sims and Easton, 1972) (MEGASCOLECIDAE) is now obtained frequently "in nature" in the southern U.S.A. (Fig. 9.4). Many greenhouse escapees over the years have contributed to populations a little north of the glacial boundary. Here again, climatic conditions should be a barrier to further northern migration.

The SPARGANOPHILIDAE (Nearctic) (Fig. 9.5) are closely related to the Komarekionidae and Lumbricidae (Reynolds and Cook, 1993). *Sparganophilus* is represented by one species in other continents, as a result of human transportation since 1500 A.D. (Reynolds *et al.*, 1974). Except for the anthropochorous *S. eiseni,* the other ten species in this monogeneric family are restricted to the southern U.S.A. and/or northern Central America (Reynolds, 1980).

The distribution of the families discussed so far, whether endemic or exotic, has been straight forward, with their distribution almost exclusively south of the southern limit of Pleistocene glaciation except for recent human transportation.

The question of megadrile distribution, particularly in North America, has long been one of controversy, even among oligochaetologists around the world. The LUMBRICIDAE is a good example of migratory controversy, since members of many of its genera have been transported all over the globe. Except for the ENCHYTRAEIDAE (which are not dealt with in this discussion), the natural populations of European megadrile fauna are composed almost entiely of LUMBRICIDAE, while the other continents have many other indigenous groups. Until the mid 1970's, there were 15 genera in the LUMBRICIDAE : *Allolobophora, Aporrectodea, Bimastos, Dendrobaena, Dendrodrilus, Eisenia, Eiseniella, Eiseniona, Eisenoides, Eophila, Helodrilus, Lumbricus, Octodrilus, Octolasion* and *Satchellius* (Reynolds and Cook, 1977, 1981). For years, several authors (Bouché, 1972; Gates,1970, 1979; Reynolds *et al.*, 1974 and Mrsić, 1991) have argued that the classical belief of Stephenson (1930, p. 905) that "The Lumbricidae are a recently evolved ... group" based on the relatively few genera, of which most were so widely distributed around the world, was incorrect. These authors stated that, in time, careful collecting would yield many additional genera in this family. During the past two decades, 18 new genera have been described : *Ethnodrilus, Kritodrilus, Nicodrilus, Orodrilus, Prosellodrilus* and *Scherotheca* (Reynolds and Cook, 1977); *Fitzingeria* and *Murchieona* (Reynolds and Cook, 1981); *Allolobophoridella, Alpodinaridella, Meroandriella, Octodriloides, Perelia (=Svetlovia)* and *Proctodrilus* (Reynolds and Cook, 1989); *Creinella, Healyella, Italobalkaniona* and *Karpatodinariona* (Reynolds and Cook, 1993). Of these 18 most recently

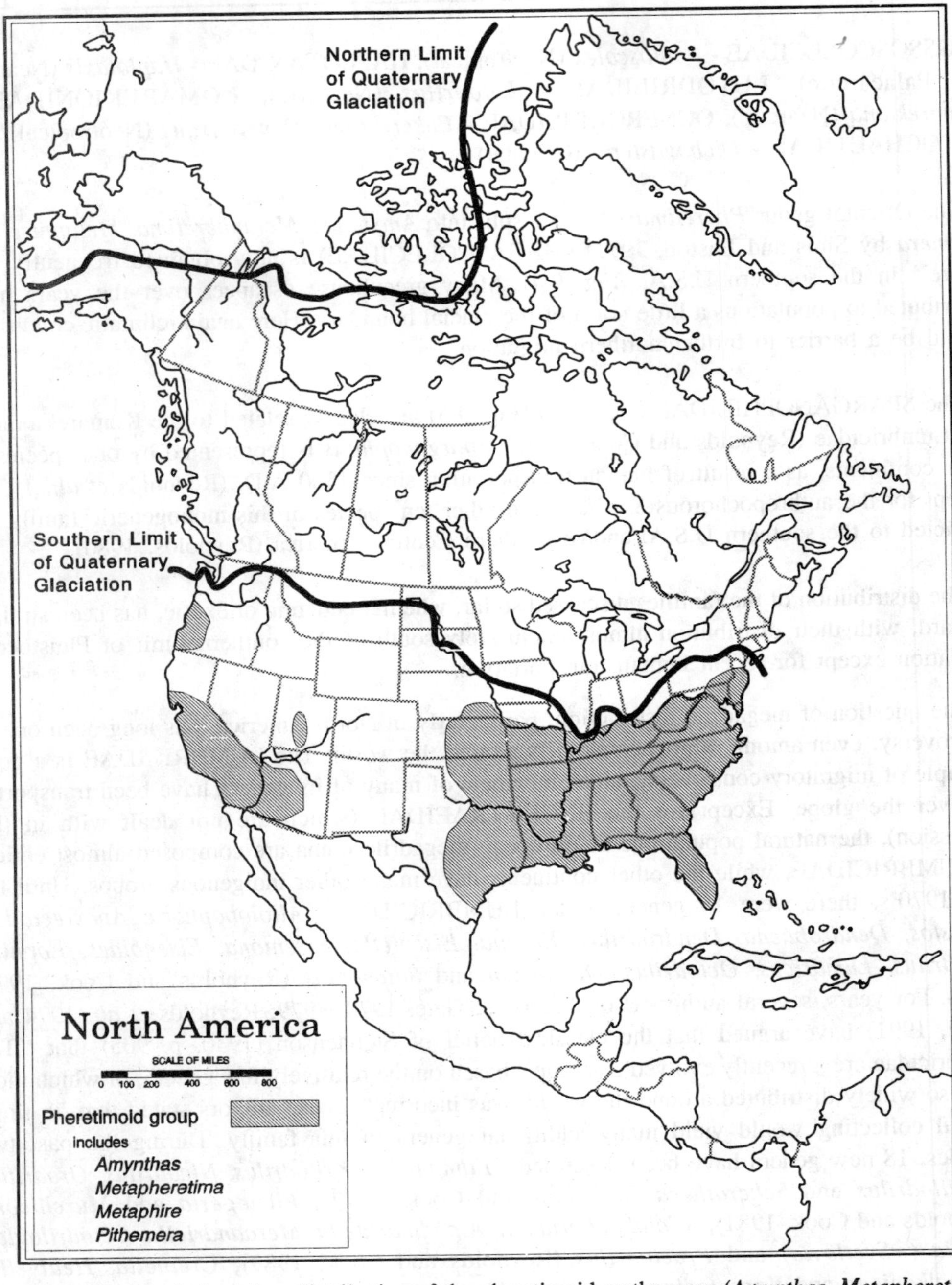

Fig. 9.4. The North American distribution of the pheretimoid earthworms (*Amynthas, Metapheretima, Metaphire* and *Pithemera*).

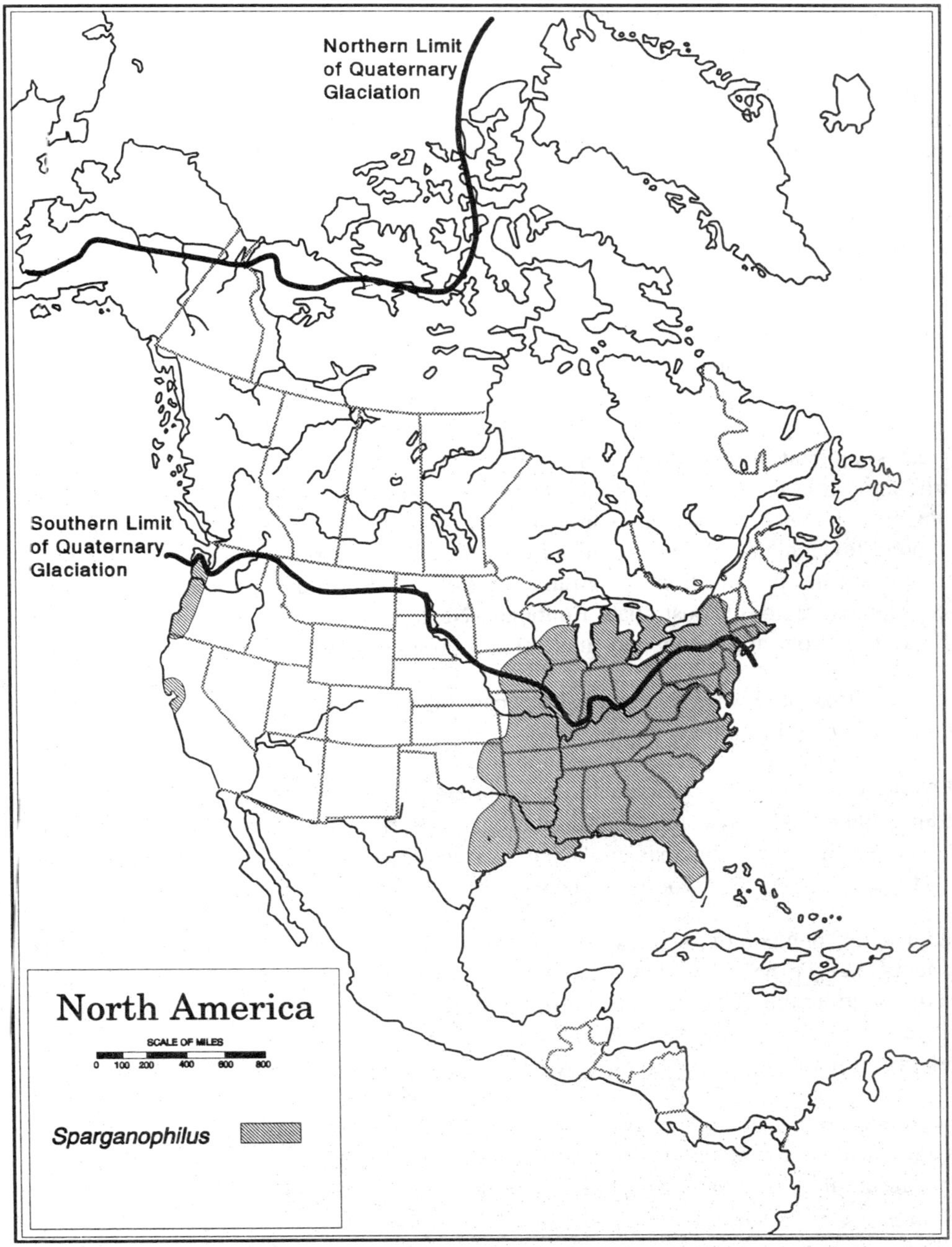

Fig. 9.5. The North American distribution of the genus *Sparganophilus*.

described genera, only the monospecific *Allolobophoridella eiseni* (as *'Lumbricus' eiseni*) and *Murchieona muldali* (as *Bimastos muldali*) have been recorded from isolated collection in North America, Michigan and Tennessee, (Reynolds *et al.*, 1974); the rest are restricted to Europe.

In the original 15 genera, the genera *Eophila* and *Helodrilus* are restricted to Europe and a few adjacent areas, such as North Africa. The taxonomic position of *Eiseniona* is in doubt, but its species have also been reported only from Europe. Only one collection of the monospecific *Satchellius* has been recorded outside Europe, that of Davies (1954) from New Jersey. Two genera, *Bimastos* and *Eisenoides,* are Nearctic and, with the exception of *B.parvus,* which has been carried around the world by man, are restricted to North America, and prinicipally to the southeastern U.S.A.

The LUMBRICIDAE of the Palaearctic region, primarily Europe, are a diverse group. With th exception of the Nearctic genera, the number and variety of species in the remaining 30 genera are greater than anywhere else in the world. This fact would support the European claim to the centre of origin for the Lumbricidae (Mrsic 1991). As with most plant and animal groups, the greatest amount of diversity is usually recorded in the centre of origin (Allard, 1960). This is particularly true for sexual forms, where parthenogenesis and rapidly mutating forms are not prominent. The only exclusively parthenogenetic lumbricid genus is the nearctic *Bimastos.* There are a few male sterile morphs in some of the other lumbricid genera (*cf* Reynolds, 1974). No one seems to question the transportation of European lumbricids to the other biogeographical regions : Australian, Ethiopian, Neotropical or Oriental. But when it comes to the Nearctic region, arguments and interesting theories abound.

There are two main ways that earthworms can migrate : (1) they move by themselves, or (2) someone or something moves them.

Obviously, earthworms' native ability to migrate is somewhat limited. A discussion of geographic and topo-environmental limitations to earthworm migration was presented by Reynolds (1973a). The principal method of earthworm migration has been transportation by man. This has helped earthworms to overcome the many barriers in their path.

There are a number of theories to explain geographical distribution. The Permanence of Continents Theory has provided a satisfactory explanation for the interpretation of most Cenozoic distributions (George, 1962), but this theory will not help to explain the presence of European Lumbricidae in North America today. Also, this theory has been reduced in importance, if not dismissed, by relatively recent factors favouring the Continental Drift Theory.

If one accepts the hypothesis that Quaternary glaciation (*cf* Fig. 9.6), ending some 11,000 years ago, destroyed earthworms in Iceland, Greenland, Canada and many of the United States, one can refute the other theories as well. The two main theories used to explain geographical distribution are : (1) land bridges, and (2) continental drift.

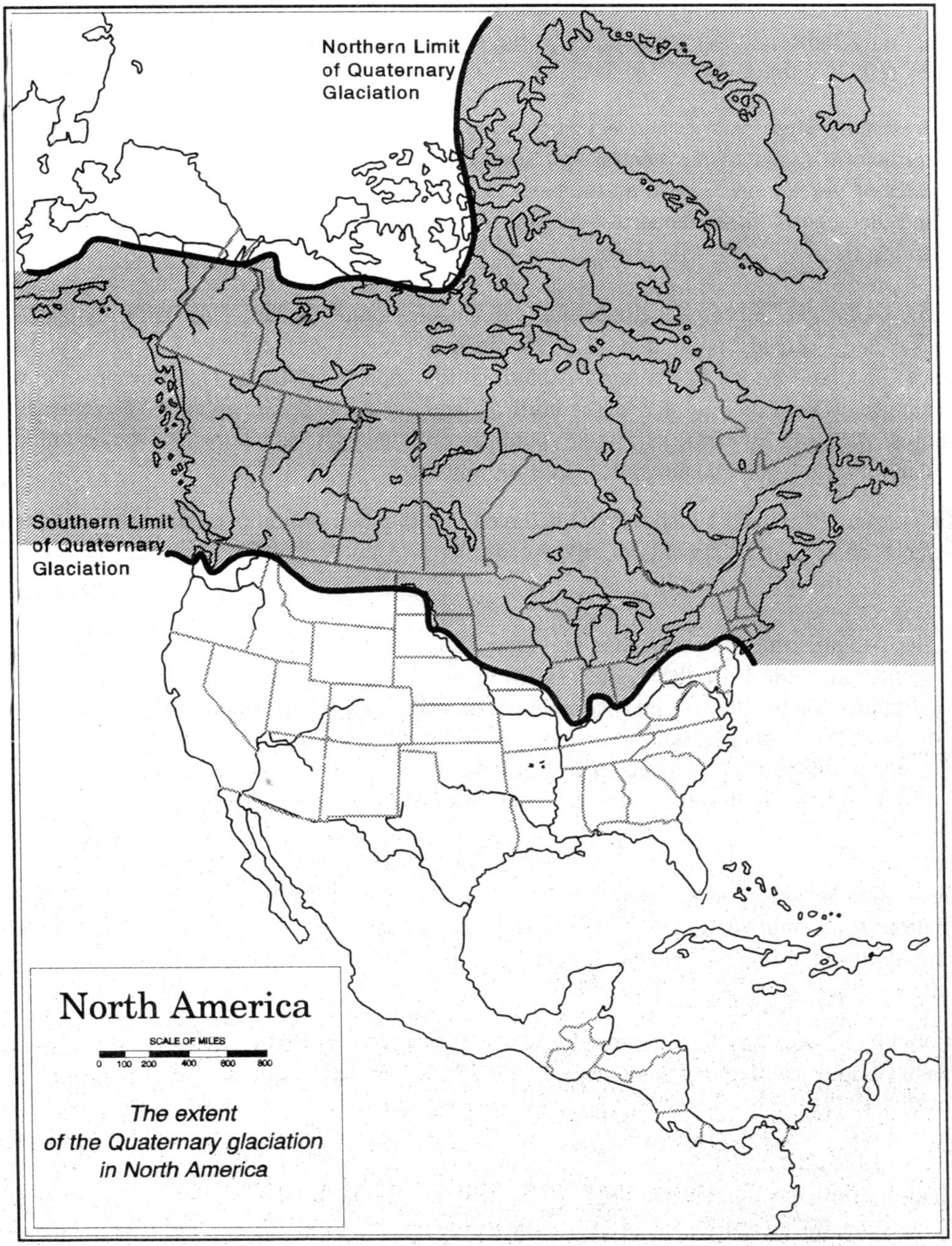

Fig. 9.6. The extent of the Quaternary glaciation in North America.

Explaining lumbricid occurrences in North America by continental drift and land bridges, Omodeo (1963) wrote :

> Personally I prefer a certain eclecticism and would tend to explain the genesis of earthworm fauna of the central and southern countries facing each other on both sides of the Atlantic Ocean on the basis of Wegener's Theory [of Continental Drift] and the genesis of the fauna of the North Atlantic countries on the basis of the land-bridge theory.

In light of more recent developments, we can no longer accept Omodeo's explanation. Geologists have already pointed out (*cf* Wright and Frey, 1965) that there is no supporting evidence for a land bridge in the north Atlantic at the appropriate time. This European view of their lumbricid fauna moving across the north Atlantic to Iceland, Greenland and eventually to the shores of North America, surviving countless millennia on bare nunataks seems unlikely, particularly for soft-bodied, moisture-requiring animals.

Michaelsen (1921) and Voisin (1961) believed that the distribution of earthworms confirmed Wegener's Theory of Continental Drift. According to George (1962), evidence for the drift theory was confusing :

> The Atlantic and Indian Oceans may date from the Jurassic days at least, and this would make the drift theory necessarily a much earlier happening than Wegener estimated. Doubts have been cast on the SIAL-SIMA theory, since parts of the floors of the oceans are now known to be SIAL like the continents. On the other hand recent work on rock magnetism has brought up the whole problem of continental drift in a new light and further information must be awaited.

There was much circular reasoning relating to the drift theory. The fitting of continents by geologists was seized upon by biologists to explain animal distribution, while others used the distribution of the animals to support the movement of continents. It would seem easier to have the animals move around the world, rather than the continents moving around collecting the animals.

As obvious as this last statement might seem, it appears to be incorrect. The new information to which George referred has been called the Theory of Plate Tectonics (Dewey, 1972 and McKenzie, 1972). This more recent theory is a unification of sea-floor spreading and the continental drift theory. Only in recent decades have the data been available to place continental drift on a firm foundation (Dewey, 1972). The two instrumental developments which made the formulation of this theory possible were : (1) global installation of standardized seismic stations, and (2) systematic measurement of the earth's magnetic field at the sea surface (McKenzie, 1972).

In any case, most proponents of the drift theory are talking about the Carboniferous (late Palaeozoic) through the late Cenozoic Eras, which are too far in the past, if one accepts the premise that Quaternary glaciation destroyed North American earthworm fauna in the areas it covered (Fig. 9.6), including possibly areas a little further to the south.

Stephenson (1930) recorded some 123 species in the large family Lumbricidae. Since then, many have been placed into synonymy, while many new species have been described. According to Easton (1983), his global check-list included 386 species and subspecies. One of the most recent accounts (Mrsic, 1991) brings the world list to approximately 450 species and subspecies. Yet, in this large family, there are only about 12-13 species which exhibit a widespread distribution (peregrine). All of these species are now recorded in the earthworm fauna of North America. If these species were eliminated during the Quaternary glaciation, how did they become established in North America today?

Let us examine the North American distribution of some of the peregrine European lumbricids.

Five species of *aporrectodea* have been obtained in North America (Fig. 9.7). The distribution of *Ap. longa* seems to be centred in the Canadian maritimes. Very few specimens have been obtained south of the glacial boundary. *Ap. trapezoides* is found primarily south of the glacial boundary in the unglaciated portions of the U.S.A., where it is > 95% of the *Aporrectodea* obtained. When obtained north of the glacial boundary, it ranks third or fourth in frequency and abundance for the genus. *Ap. tuberculata* is the dominant worm in Canada and decreases greatly in frequency and abundance as one proceeds southward, where it is rarely collected in the southeastern U.S.A. A recent study (Reynolds and Reynolds, 1992) suggests that the dominance of *Ap. tuberculata* in Quebec may be superseded by *Ap. turgida*. *Ap. turgida* is found north and south of the glacial boundary, rarely in a dominant position, and infrequently at the extreme (north and south) of North American megadrile distributions. A great deal of confusion has centred around this group in the past, since all four species have at one time or another been lumped under the ''*species" Allolobophora caliginosa*. However, detailed mapping of these four species in North America illustrates definite distributional patterns. *Aporrectodea rosea* is a highly adaptable species and is found throughout North America. Its frequency and abundance greatly diminishes as one proceeds southward into the unglaciated portions of North America.

The species of the nearctic genus *Bimastos* are restricted to the southeastern U.S.A. (Fig. 9.8). A few collections have been obtained just north of the glacial boundary, *e.g.* southern Illinois. Collections outside the southern refugium are primarily the anthropochorous *B. parvus* (Reynolds, 1977a). Niche utilization illustrates an obvious pairing of *Bimastos* spp. with *Dendrodrilus rubidus,* and to a lesser extent with *Dendrobaena octaedra*. The latter (*cf* Fig. 9.8) are obtained principally north of the glacial boundary, but some limited collections have been secured in southern North America. Many species of *Bimastos* live beneath the bark, where they

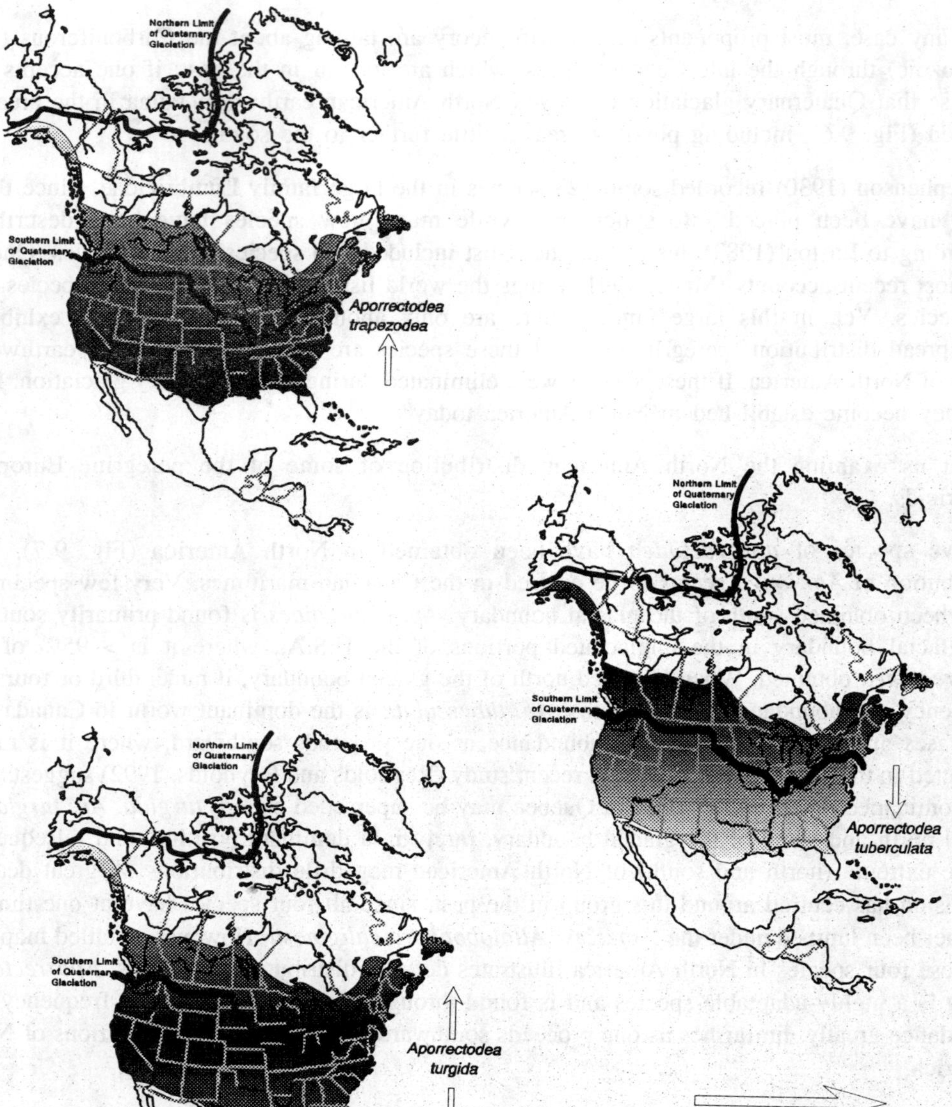

Fig. 9.7. The North Amearican distribution of genus *Aporrectodea.*

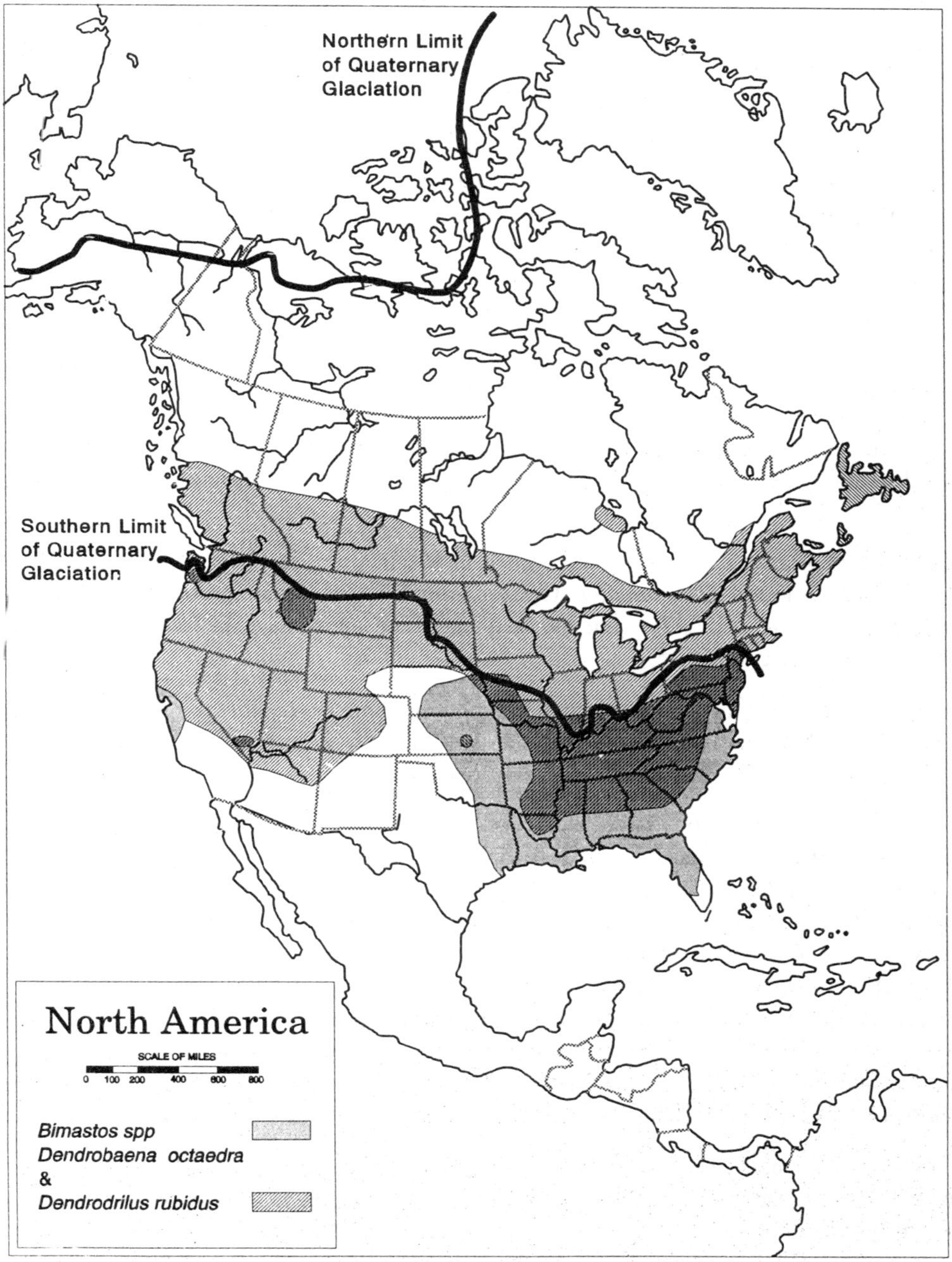

Fig. 9.8. The North American distribution of the genera *Bimastos, Dendrobaena* and *Dendrodrilus.*

are involved with decomposition of the fallen logs, *etc.* In the niche north of the glacial boundary, or just south of this line, where *Bimastos* is absent or a minor part of the earthworm fauna, this decomposition role is carried out by primarily by *Dd. rubidus.* Where *Dd. rubidus* is found further south, it is frequently obtained in other habitats (as it also utilizes in the north), such as in grass clippings and compost. Since there is a much greater diversity of earthworm species south of the glacial boundary, *i.e.* 4-5 times, the habitat utilization in the southeastern U.S.A. is narrower.

Dendrobaena octaedra (Fig. 9.8) is found throughout North America, but in fewer locations and in lower numbers than species of *Bimastos* and *Dendrodrilus.* The main centre of this surface dwelling species is northeastern North America, with very few collections much below the glacial boundary.

The nearctic species *Eisenoides carolinensis* and *Es. lonnbergi* are almost exclusively found in the unglaciated portion of eastern North America. Three small collections have been reported in arboreta north of the glacial line.

Two species of *Octolasion* have been recorded in North America, but very few locations and specimens have included *O. cyaneum.* O. *tyrtaeum* is very abundant in the area around Tennessee (Reynolds *et al.,* 1974), but the frequency and abundance declines rapidly as one radiates out from this centre.

The last major genus of lumbricid to be dealt with here is *Lumbricus.* Four species are present in North America, with quite distinct distributions. *L. castaneus* has a distribution centred in and almost limited to the Canadian maritimes and New England (Reynolds, 1977a). A few small isolated collections have been found outside this area. Until 1975, only a single specimen of *L. festivus* was recorded in North America (Langmaid, 1964). Collections during the early 1970's extended this species' range to the Gaspé (Reynolds, 1975b), southern Quebec (Reynolds, 1976), northern Quebec (Reynolds & Reynolds, 1992) and southern British Columbia (Reynolds, 1977b). *L. rubellus* is one of the most widely distributed and abundant earthworms of northeastern North America. Very few specimens have been found south or west of Tennessee. *L. terrestris* is also widely distributed in northeastern North America, but does not exhibit as great an abundance as *L. rubellus.* The use of *L. terrestris* as fish-bait has helped disseminate this species. The close association between *L. terrestris* and human habitation is quite noticeable in southern North America.

The answer to the earlier question seems to be that the Palaearctic Lumbricidae have been introduced to North America by European settlers since 1500 A.D. There are essentially no endemic earthworms in Canada, although a very few explainable exceptions have been recorded : one specimen of *Arctiostrotus perrieri* in the Queen Charlotte Islands of British Columbia (McKey-Fender & Fender, 1982), four specimens of *Bimastos parvus* from Ontario (Reynolds,

1972), one specimens of *B. beddardi* from Quebec (Reynolds & Reynolds, 1992) and about seven limited collections of *Sparganophilus eiseni* in Ontario (Reynolds, 1977a) and Quebec (Reynolds & Reynolds, 1992). In the United States, there are a few endemic species north of the southern limit of the Quaternary glaciation. The exceptions are trade routes along natural waterways, such as the Mississippi River waterway and its tributaries. This helps explain the invasion of endemic species of *Bimastos, Diplocardia* and *Sparganophilus* just north of this line. The colonization of North America, north and south, was primarily by Europeans who brought potted plants, *etc.* with them (Gates, 1966). The earthworm fauna of North America, north of the Quaternary glacial boundary, is composed almost solely of European anthropochores. Some of these species are present south of the glacial boundary, but in much fewer numbers. These species, now established in North America and Europe south of the glacial boundary, may result from a coterminous period in preglacial time.

Evidence from collections during the past three decades seems to refute the ideas that endemic nearctic megadriles and some exotic European lumbricids followed the retreating glaciers, as is widely assumed. Results from published natural collections (*i.e.* not from arboreta, greenhouses,*etc.*) from northern regions, excepting those immediately north of the glacial boundary, do not indicate the establishment of nearctic endemic species [Reynolds, 1975 (b,c,d,e); 1976 (b,c,d); 1977 (a,c); 1978; Hilton & Reynolds, 1983 and Reynolds & Reynolds, 1992)].

Two published surveys which illustrate the Post Quaternary Introduction Theory for Palaearctic lumbricid distribution in North America were presented by this author. On the mid-Atlantic coast of North America, there are two small states (Delaware and Rhode Island). The earthworm fauna of Delaware (Reynolds, 1973b) is > 90% exotic, but at least four nearctic species in three genera appear in the state's list. In the case of Rhode Island (Reynolds, 1973c), 100% of the earthworm fauna is exotic Palaearctic Lumbricidae. The percentage of nearctic earthworm species, abundance, density and frequency, increases as one proceeds southward from the glacial boundary. These two states are approximately 325 kilometres apart, with no major topo-environmental barriers between them. The climate and vegetation, *etc.*are similar in both states. Delaware is just south of the glacial boundary, while Rhode Island is just north of the line. Obviously, earthworm migration or introduction to new areas within North America requires some vector. Independent migration of some 300 kilometres in about 11,000 years should not be too difficult a task for an animal. Apparently, this has not occurred in eastern North America.

If the megadrile fauna did not move far north from below the region of Quaternary glaciation, and the present megadrile fauna in the region north of the Quaternary glacial boundary is solely or almost exclusively European, then the earthworms presently in that region must have been introduced by European settlers since 1500 A.D.

Acknowledgements

I wish to express my appreciation to W.M. Reynolds of the Oligochaetology Laboratory, Lindsay, for reviewing the manuscript, and her comments, criticisms, and suggestions. I am very grateful to Susan Mantle of the CARMA Centre of Sir Sandford Fleming College for producing the figures in this text. After the last draft of the manuscript was prepared, I had the opportunity to consult with William M. Fender of Soil Biology Associates of McMinnville, Oregon and Samuel W. James of Maharishi International University of Fairfield, Iowa. I am pleased to have been able to add their new unpublished range data in order to present a most complete picture of the North American distribution to date.

REFERENCES

Allard, R.W. 1960. *Principles of plant breeding.* John Wiley & Sons, Inc., New York, xi + 485 p.

Bouché, M.B. 1972. *Lombriciens de France écologie et systematics.* Inst. Nat Rech. Argon., Paris, 671 p.

Davies, H. 1954. A preliminary list of the earthworms of northern New Jersey with notes. *Breviora, Mus. Comp. Zool. no.* 26, 1-13.

Dewey, J.F. 1972. Plate tectonics. *Sci. Amer.* 226 (5) : **56**-66,68.

Easton, E.G. 1983. A guide to the valid names of Lumbricidae (Oligochaeta). pp. 475-485. In *Earthworm ecology from Darwin to vermiculture* (Ed. Satchell, J.E.) Chapman and Hall, London, xvii + 495 p.

Gates, G.E. 1966. Requiem - for megadrile utopias. A contribution Howard the understanding of the earthworm fauna of North America. *Proc. Biol. Soc. Wash.* **79** : 239-254.

Gates, G.E. 1970. Miscellanea Megadrilogica VII. *Megadrilogica* **1** (2) : 1-14.

Gates, G.E. 1979. South Dakota does have earthworms! *Megadrilogica* **3**(9) : 165-166.

George, W. 1962. *Animal geography.* Heinemann Educational Books Ltd., London, x + 142 p.

Hilton, D.F.J. and Reynolds, J.W. 1983. The earthworm (Oligochaeta : Lumbricidae) fauna of the cypress Hills Provincial Park, Alberta. *Megadrilogica* **4** (4) : 91-92.

Langmaid, K.K. 1964. Some effects of earthworm invasion in virgin podzols. *Can. J. Soil Sci.* **44** : 34-37.

Lindroth, C.H. 1962. Forward. pp. 3-5. In : *Taxonomy and geography* (Ed. D. Nichols), Syst. Assoc. Publ., London, no. 4.

McKenzie, D.P. 1972. Plate tectonics and sea-floor spreading. *Amer. Sci.* **60**(4) : 425-435.

McKey-Fender, D. 1982. Doctor Kincaid's other worm. *Megadrilogica* **4** (3) : 86-88.

McKey-Fender, D. and Fender, W.M. 1982. *Arctiostrotus* (gen. nov.). I. The identity of *Plutellus perrieri* Benham, 1892 and its distribution in relation to glacial refugia. *Megadrilogica* **4** (3) : 81-85.

Mckey-Fender, D. and Fender, W.M. 1990. Oligochaeta : Megascolecidae and other earthworms from Western North America. pp. 357-386. In : *Soil Biology Guide* (Ed. D.L. Dindal). John Wiley & Sons, New York.

Michaelsen, M. 1921. Die Verbreitung der Oligochäten im Lichte der Wegener' schen Theorie der Kontinentalverschiebung. *Verhandl. Naturwiss. Ver. Hamburg,* 37 p.

Mrsic, N. 1991. *Monograph on earthworms (Lumbricidae) of the Balkans.* Slovenska Akademija Znanosti Umetnosti, Ljubljana, 757 p.

Omodeo, P. 1963. Distribution of the terricolous oligochaetes on the two shores of the Atlantic. pp. 127-141. In : *North Atlantic biota and their history* (Eds. Löve and Löve), Pergamon press, New York.

Reynolds, J.W. 1972. Earthworms (Lumbricidae) of the Haliburton Highlands, Ontario, Canada. *Megadrilogica* **1** (3) : 1-11.

Reynolds, J.W. 1973a. Earthworm (Annelida, Oligochaeta) ecology and systematics. pp. 95-120. In : *Proc. Ist Soil Microcommunities Conf.* (Ed. D.L. Dindal). Nat. Inform. Tech. Serv., Springfield, VA.

Reynolds, J.W. 1973b. The earthworms of Delaware (Oligochaeta : Acanthodrilidae and Lumbricidae). *Megadrilogica* **1**(5) : 1-4.

Reynolds, J.W. 1973c. The earthworms of Rhode Island (Oligochaeta: Lumbricidae). *Megadrilogica* **1** (6) : 1-4.

Reynolds, J.W. 1974. Are oligochaetes really hermaphroditic amphimictic organism? *Biologist* **56** (2) : 90-99.

Reynolds, J.W. 1975a. Die biogeografie van Noorde-Amerikaanse (Oligochaeta) noorde van Meksiko - I. *Indikator* **7** (4) : 11-20.

Reynolds, J.W. 1975b. Les lombricidés (Oligochaeta) de la Gaspésie, Québec. *Megadrilogica* **2** (4) : 4-9

Reynolds, J.W. 1975c. Les lombricidés (Oligochaeta) d'île d'Orléans, Québec. *Megadrilogica* **2** (5) : 8-11.

Reynolds, J.W. 1975d. Boiteagan (Oligochaeta : Lumbricidae) Cheap breatunn. *Megadrilogica* **2** (6) : 4-10.

Reynolds, J.W. 1975e. The earthworms (Oligochaeta : Lumbricidae) of Prince Edward Island. *Megadrilogica* **2** (7) : 4-10.

Reynolds, J.W. 1976a. Die biogeografie van Noorde-Amerikaanse (Oligochaeta) noorde van Meksiko - II. *Indikator* **8** (1) : 6-20.

Reynolds, J.W. 1976b. The distribution and ecology of the earthworms of Nova Scotia. *Megadrilogica* **2** (8) : 1-8.

Reynolds, J.W. 1976c. Catalogue et clé d' identification des lombricidés du Québec. *Nat. can.* **103** (1) : 21-27.

Reynolds, J.W. 1976d. A preliminary checklist and distribution of the earthworms of New Brunswick.*N.B. Nat.* **7** (2) : 16-17.

Reynolds, J.W. 1977a. The earthworms (Lumbricidae and Sparganophilidae) of Ontario. *Life Sci. Misc. Publ., Roy. Ont. Mus.*, x + 141 p.

Reynolds, J.W. 1977b. Le ver québécois (*Lumbricus festivus)* envahit la Colombie-Britannique. *Can. Field-Nat.* **91** (4) : 395-396.

Reynolds, J.W. 1977c. The earthworms of Massachusetts (Oligochaeta : Lumbricidae, Megascolecidae and Sparganophilidae). *Megadrilogica* **3** (2) : 49-54.

Reynolds, J.W. 1978. A contribution to our knowledge of the earthworm fauna of North Dakota. *Megadrilogica* **3** (8) : 148-149.

Reynolds, J.W. 1980. The earthworm family Sparganophilidae (Annelida, Oligochaeta) in North America. *Megadrilogica* **3** (12) : 189-204.

Reynolds, J.W. and Cook, D.G. 1977. *Nomenclatura Oligochaetologica : a catalogue of the names, descriptions and type specimens of the Oligochaeta.* University of New Brunswick, Fredericton, x + 217 p.

Reynolds, J.W. and Cook, D.G. 1981. *Nomenclatura Oligochaetologica Supplementrum Primum : a catalogue of the names, descriptions and type specimens of the Oligochaeta.* University of New Brunswick, Fredericton, v + 39 p.

Reynolds, J.W. and Cook, D.G. 1989. *Nomenclatura Oligochaetologica Supplementrum Secundum: a catalogue of the names, descriptions and type specimens of the Oligochaeta.* New Brunswick Museum Monographic Series (Natural History) No. 8, Saint John, v + 37 p.

Reynolds, J.W. and Cook, D.G. 1993. *Nomenclatura Oligochaetologica Supplementrum Tertium : a catalogue of the names, descriptions and type specimens of the Oligochaeta.* New Brunswick Museum Monographic Series (Natural History) No. 9, Saint John, vii + 33 p.

Reynolds, J.W. and Reynolds, K.W. 1992. Les vers de terre (Oligochaeta : Lumbricidae et Sparganophilidae) sur la rive nord du Saint-Laurant (Québec). *Megadrilogica* **4** (9) : 145-161.

Reynolds, J.W. and Wetzel, M.J. 1993. North American megadriles (continental America north of Mexico). p. xx-xx. In : *Common and scientific names of invertebrates from the United States and Canada: Annelida* (Eds. K.A. Coates, Wetzel, M.J., Madill, J., Gelder, S.R., Reynolds, J.W. and Brinkhurst, R.O.) Amer. Fish. Soc. Spec. Publ. 17 (in press).

Reynolds, J.W., Clebsch, E.E.C. and Reynolds, W.M. 1974. The earthworms of Tennessee (Oligochaeta). I. Lumbricidae. *Contr. N. Amer. Earthworms, no. 3, Bull. Tall Timbers Res. Stn.*, no 17, viii + 133 p.

Sims, R.W. and Easton, E.G. 1972. A numerical revision of the earthworm genus *Pheretima* auct.

(Megascolecidae : Oligochaeta) with the recognition of new genera and an appendix on the earthworms collected by the Royal Society North Borneo Expedition. *Biol. J. Linn. Soc.* **4** (3) : 169-268.

Stephenson, J. 1930. *The Oligochaeta.* Clarendon Press, London, xvi + 978 p.

Udvardy, M.D.F. 1969. *Dynamic zoogeography with special references to land animals.* Von Nostrand Reinhold Co., New York, xviii + 445 p.

Voisin, A. 1961. Better grassland sward. pp. 32-37. In : *The challenge of earthworm research* (Ed. R. Rodale.). The Soil and Health Foundation, Emmaus, PA.

Wright, H.E. and Frey, D.G. 1965. *The Quaternary of the United States.* Princton University Press, Princton, x + 922 p.

Chapter 10

Development of Vermiculture Technology in Karnataka State

Radha K. Kale

Department of Zoology, University of Agricultural Sciences (GKVK), Bangalore

Abstract. An overview of earthworm research undertaken in India and development of vermitechnology research and its transfer from laboratory to land with particular reference to the state of Karnataka has been highlighted.

Key words : *Vermiculture, Vermicomposting, Earthworm, Eudrilus euginae*

Preamble :

An ecologist's View as ''All flesh is grass and all fish diatoms'' upholds the dependency of animal life either directly or indirectly on the plant life. It is the lush growth of the plants and their yield fulfill the dream of providing bread for all mankind. The thought of the day for providing for all mankind should not be from the alms that fall from rich people but through worldwide hums production.

The great biologist Roul France was one of the pioneer scientists to attribute the importance of humus in bringing about the healthy growth of plants. He was the one to coin the term ''Edaphon'' to the soil organisms and to show that they thrive and develop as long as there is humus. But with the intensive use of chemicals as fertilizers all-over the world, the need of humus in the maintenance of life activities in the soil strata was totally neglected. The drastic hampering of life functions in the soil has led to the decline in the soil productivity. Dash (1978) has highlighted the role of earthworms in the degradation of organic matter and stimulation of microbial activity in the soil. He estimated about 247 tonnes of cast production by worms per hectare of land per annum. In his reprot he has suggested the possibilities of using earthworms in pits to hasten up the process of decomposition.

It was Darwin (1881) who in his book explained the process of formation of vegetable mould by the worm activity. Later on, for several decades the earthworm research revolved around the taxonomy, physiology and population ecology. In the Indian sub-continent earthworm research picked up momentum only in the later part of the 20th century. The pioneer contribution from the microbiologist J.V. Bhat to the earthworm microbe association in the early 1950 was the major contribution of the times towards the applied biology of earthworms. Rao, K.P was responsible to indoctrinate the study of environmental physiology of earthworms, especially the effect of temperature acclimation on the metabolic pathways and neurohumoral secretion. Dash and his associates concentrated their studies on earthworm ecology, mainly on the aspects of bioenergetics. Similarly Nanda and his group have contributed to the histochemistry of different systems of earthworms. Thus these four senior scientists were responsible for the development of earthworm research and its application in this country.

Earlier in the preparation of Farmyard manure, the farmers were using more of cattle dung along with the other farm wastes in the compost pit. The time required for the composting of the materials in the pits normally is 3-6 months or more depending on the weather conditions. Due to the duration of time required for the pit composting and lure for higher yield led the farmers to depend more and more on chemical fertilizers. But the realisation on the adverse effects on continuous application of only chemical fertilizers has made them look for an alternative method for it composting to produce the required bulk quantity of organic manure to their fields. At this juncture it was made known to them that any farm waste available in their own land can be converted into compost in the presence of worms.

The repeated studies have shown that it is possible to bring down the amount of cowdung in the production of compost as the cowdung finds use elsewhere in the life of farmers. Table 10.1. provides the list of organic wastes that have been used for the production of vermicompost.

Since the use of selected species of earthworms for organic waste degradation is tested and proved, both the small farmers with a few acres of cultivable land and the plantation owners are experimenting in their lands with the vermicompost. Other groups which are actively involved are the voluntary organizations like Mythri Seva Samiti and Centre for Environment Education who are experimenting by collecting the segregated waste from the houses for vermicomposting of organic wastes at the community centres. This programme is also slowly gaining popularity and has taken its first fruitful steps in Bangalore city. The industrial units like Hindustan Aeronautics, Bangalore, MICO, Bangalore, HMT, Bangalore have shown keen interest in composting of the organic waste produced in their premises. The Military Dairy Farm, Bangalore has started this programme for their dairy waste. The Spices Board a Central Govt. organization in its Sakaleshpur Research Centre has successfully cultured worms for composting of the cardamom waste. They got very encouraging results in their nurseries on vermicompost application. The gradual spread of technology in a decade's time in Karnataka is provided in Table 10.2.

Among the agro-based industries who have given a serious thought to make use of all the organic wastes in their units into vermicampost in Karnataka are M/S Khoday Group of companies. The pressmud, fine bagasse and the ash from the sugar factory has been tested for vermicomposting. A good population increase of worms was observed when these materials rich

TABLE 10.1. *The list of Organic Wastes tested as feed mix* for Eudrilus eugeniae *for Vermicompost Production*

Sl.No.	Organic Waste Mix	Ratio of the ingredients in mix (vol)	Protreatment for the mix before introducing worms
1.	Cowdung + kitchen waste	4 : 1	Mix with 40 per cent moisture left for one week before introducing worms.
2.	Cowdung + green leaves (Hedge cuttings and lawn trimmings)	5 : 3	Finely chopped material mixed with dung and left for 15 days with 40-50 per cent moisture for 15 days.
3.	Cowdung + green leaves + Vegetable waste groundunt cake	10 : 5 : 2 : 2.5	Same as above (2)
4.	Cowdung + coirwaste + Fresh pongamia leaves	4 : 2 : 1	Coir waste soaked in water for two weeks and after draining the water mixed with the other materials and entire mix left for two more weeks before introducing worms.
5.	Cowdung + coir waste + Fresh neem leaves	4 : 2 : 1	- Same as in 4 -
6.	Cowdung + coir waste + Fresh glycerdia	4 : 2 : 1	- Same as in 4 -
7.	Cowdung + coir waste + dry litter	4 : 2 : 1	- Same as in 4 -
8.	Cowdung + partnenium weed	2 : 3	The weed was uprooted before the on set of flowers and the mix was left for two weeks before introducing worms with 40 per cent moisture
9.	Litter (leaf fall of different trees) + Eucalyptus bark	4 : 1	The Eucalyptus bark and leaves were soaked separately in water for 3 weeks. The water was changed once in three days and later mixed with other litter before introducing worms
10.	Litter (leaf fall of different trees) + Eucalyptus leaves dry)	6.5 : 1	
11.	Press mud from sugar factory + fine bagasse + litter + fly ash from factory	8 : 3 : 2 : 1	The mix was maintained with 40 per cent moisture for two weeks before introducing worms

Table 10.2. *The spread of Vermicomposting Technology from UAS, Bangalore to different places from 1983-1993 (June)*

Year	No. within the State	No. outside the State
1983	2	MAH/1
1984	1	TN/1 AP/1 MAH/1
1985	1	TN/1 KER/1 MAH/1
1986	0	0
1987	1	AP/1
1988	0	MAH/1
1889	0	TN/1 KER/1 PUN/1
1990	4	TN/3 MAH/1
1991	51	KER/4 AP/4 MAH/3 TN/6
1992	290	KER/10 MAH/5 GUJ/2 AP/10 TN/10 PUN/1 MAN/1 DEL/1 HP/1 MP/1 RAJ/1 ORI/3 BEN/1 UP/1 HAR/1
1993 (June)	230	TN/14 KER/15 AP/8 ORI/4 GUJ/1 PUN/1 MAH/8 DEL/1 BEN/1 SIK/1

AP-Andhra Pradesh, BEN-Bengal, DEL-Delhi GUJ-Gujarat, HAR-Haryana, HP-Himachal Pradesh KER-Kerala, MAH-Maharashtra, MAN-Manipur, MP-Madhya Pradesh, PUN-Punjab, RAJ-Rajasthan, SIK-Sikkim, TN-Tamil Nadu, UP-Uttar Pradesh.

in nitrogen and other neutrients were used for vermicomposting. Similarly feeler studies were carried out on their distillery effluent and it was found that this effluent can become the required moisture source for other organic materials used for composting process. Based on the findings, they have started a vermicomposting unit to convert the organic wastes into vermicompost and worm biomass. In future the work will be extended to make it an integrated, multi directional unit for vermicompost production, worm meal production for animal protein, in land fisheries and mushroom cultivation.

UAS, Bangalore as the centre for earthworm research in Karnataka

The work at UAS, Bangalore on earthworms started in 1974 on the ecological aspects. The reseach at Agricultural University is meaningful only when the research findings have the direct application to improve the farm practices. Keeping this view in mind, the relationship between the distribution of earthworms and their species diversity to soil physicochemical characters was studied (Kale and Krishnamoorthy, 1977; 1978; 1981; Bano and Kale, 1991). Along with this, the laboratory studies carried out on their vertical distribution, food preferences provided the information on their niches (Kale and Krishnamoorthy, 1981; Kale *et al*, 1977; 1981; 1982) Similarly the selective feeding of worms and thier differential effect on the chemical parameters in recovered cast was also studied under laboratory conditions (Kale and Krishnamoorthy, 1981). The laboratory studies on the growth and population increase in different species of earthworms exhibited a set trend (Kale and Bano, 1985). These studies were carried out for almost a decade to exploit their ability for selective feeding and cast production to the field of applied biology and biotechnology.

While carrying out year round research studies on the population structure of three species of earthworms at a garden in Bangalore, *Perionyx excavatus* was found exclusively in a biogas slurry pit of the same garden. Though the slurry was applied to the standing crops quite often, the population of the earthworms was never sighted in the open field except in the pit. This evoked the interest to develop this species on animal waste and plant residues (Kale *et al*, 1982).

The cocoons of *Perionyx excavatus* were exchanged to the cocoons of a Nigerian earthworm *Eudrilus eugeniae* in 1982. Since then the work on suitability of this species in organic waste degradation is simultaneously going on in East Africa (Mba 1983) Syracuse University, USA (Neuhauser *et al*, 1979; Hartenstein and Bisesi, 1989) and South Africa (Vijoen and Reinecke, 1989) along with UAS, Bangalore (Kale and Bano 1988). After studying the probability of *Eudrilus eugeniae* to hasten up the process of break down of farm waste to provide a good quality of organic fertilizer, the technology of vermicompost production was released by the University of Agricultural Sciences, Bangalore in the year 1984.

The use of organic wastes for vermicompost production

The release of technology brought an awareness among the farming community, the agro-based industrial units of the states as well as of other states. After the release, the response was low in the first 5-6 years and from 1989-90 the trend for the know-how of the production of vermicompost has taken a big leap. The direct dialogue between the farmers and others led to try out the different degradable materials and also the various methods to find suitable applicability of the technology with minimum expense.

The Financial Support for the Development of Vermiculture Technology

The University of Agricultural Sciences, Bangalore supported the research work on earthworm biomass production through staff research grants from the year 1982 to 1986. During this period the mass culturing of *Eudrilus eugeniae* was established. Organic waste that was used at that time was mostly animal waste like cattle dung, sheep droppings, horse dung

TABLE 10.3. *List of some of the cash crops tested for the Performance of Vermicompost over Recommended Doses of Chemical Fertilisers.*

Sl.No	Cash Crops	Inferences
I. Vegetables		
(1)	Brinjal Varieties	Yield Improved with reduction of chemicals to half of the quantity with just half the quantity of V. Comp. Suggesting the possibility to reduce the chemical application.
(a)	IIHR Variety	
(b)	V1-ARKA SHIRISH	
(c)	V2-ARKA KUSUMAKAR	
(2)	Tomato Varieties	In both the varieties no significant change seen with the teratment suggesting the possibilities for reducing the chemcial application or aplication of only vermicompost.
(a)	IIHR Variety	
(b)	V1-ARKA VIKAS	
(c)	V2-ARKA SURABH	
(3)	Radish Variety II HR Radish.	Using 1/2 the concentration of chemicals with vermicompost showed a significant increased in yield over other treatments. The control and only vermicompost gave the same yield, Suggesting the possibility of limiting the use of chemical fertilisers.
(4)	Carrot Variety II HR Variety-14R205	None of the treatments showed any significant variation in the yield, suggesting the possibility of either reducing or eliminating the chemicals on large scale production of vermicompost.
II Oil Seed Production		
(1)	Ground nut JL 24 Variety	No significant change seen in various parameters, suggesting the possibility for reducing the chemical fertiliser application or using only vermicompost for better results.
(2)	Soyabean Hardy Variety	No significant change observed except for the number of root branches. The worm cast showed gradual increase in the no. of root branches over control and other combinations with chemicals.

Contd. next page

Contd. Table 10.3

Sl.No	Cash Crops	Inferences
(3)	Sunflower	Irrespective of the chosen
(a)	V1-EC68415	varieties the vegetative characters
(b)	V2-Morden	remained unaltered in the different combinations of applied fertilisers for both the varieties. The worm cast suplemented with FYM to the level of N of the reocmmended dose of chemicals and half the dose of worm cast and half of chemicals increased the yield of seeds significantly.
III Mulberry		
(1)	M5 Variety	Vermicompost applicaiton showed increased stem girth with significant results over control and other treatments. No other significant changes observed among other parameters suggesting use of vermicompost alone or reducing chemicals.
IV Ornamental Plants		
(1)	Balsam	The worm cast application which had no effect on vegetative growth, improved the flowering in Balsam plants.
(2)	Zenia	The worm cast application alone cannot bring about any remarkable change but on mixing with small quantity of chemicals induced profuse branching and flowering.
(3)	Celotia	The worm cast with small quantity of chemicals brought out increased height of plants but the other parameters remianed unaltered.
(4)	Marigold	The performance of fertilisers remained the same.
(5)	Lady's lace	No significant effect of fertilisers on the referred parameters.

along with vegetable waste (Kale and Bano 1988). It was also found that poultry waste was toxic to worms when used directly whereas the biogas slurry of the poultry waste served as good source for biomass production (Kale 1986). Worm meal was prepared by drying at 60°C after clearing their gut. The dried worms were powdered and used as protein source in the feed of broiler chicks and in fish feed (Kale, 1986; Nandeesh *et al* 1988). Some preliminary field trials were also carried out using the vermicompost for paddy crop.

The financial support received from the University Grant Commission, New Delhii from 1984 to 1988, helped to take up a survey of earthworm faunal distribution in Karnataka (Bano and Kale 1991; Kale and Bano, 1992). Some of the indigenous worms were also maintained as cultures to study their life cycles (Kale and Bano 1988).

The Karnataka State Council for Science and Technology provided the financial support to take up the work at field level and test its application on the yield of some vegetable crops and ornamental plants. The studies carried out showed the possibility of controlling the use of chemical fertilizers to these crops. Various crops to which the vermicompost was used as the source of organic fertilizer and the inferences drawn from the results is summarized in Table 10.3. The report submitted to KSCST, provided the opportunity for other organizations to get the financial assistance from KSCST to adopt the technology of vermicomposting for urban solid waste management.

The recent grants received from the Indian Council of Agricultural Research has strengthened the unit at the Agricultural University for mass production of earthworms and the compost. Earthworms and compost have reached the nook and corner of the Karnataka state and also other parts of the country. It is on the cards to take up the dryland improvement programme in near future.

The latest development is of the collaborative programme planned with the Karnataka Pollution Control Borad. The feeler studies are being carried out in the laboratory using the effluent samples supplied by the Board to check for the ability of earthworms to withstand certain level of toxicity in the effluents and detoxify them in the process.

Thus, earthworm research from the laboratory of Univeristy of Agricultural sciences, Bangalore has reached its destinations such as other Agricultural universities, other industrial organizations and fields of farmers. Simultaneously it is providing the oragnic manure to fulfill the need of the day for sustenance of life in the soil.

Prof. M.C. Dash in the year 1984 was the centre figure for bringing together all the like minded scientists working on the organic waste degradation and one part of the convention was exclusively for verms and vermicomposting. Today, I am presenting this step-wise development of vermicomposting in Karnataka as a tribute to his contribution to the field of earthworm ecology for this commemorative volume.

REFERENCES

Bano, K. and Kale, R.D. 1991. Earthworm fauna of south Karnataka. ''Advances in management and conservation of soil fauna (Eds. G.K. Veeresh; D. Rajagopal and C.A. Viraktamath) Oxford and IBH Pvt. Ltd. pp 627-634.

Darwin, C.R. 1881. ''*The formation of vegetable mould through the action of worms with observations on their habits.*'' John Murray, London. 298p.

Dash, M.C. 1978. Role of earthworms in the decomposer system in ''*Glimpses of ecology*'' (Eds. J.S. Singh and B. Gopal) India International, Scientific Publication pp 399-406.

Hartenstein, R. and Bisesi, M.S. 1989. Use of earthworm biotechnology for the management of effluents from intensively housed livestock. In ''*Out look on Agriculture*'' Pergamon Press, Great Britain pp. 72-75.

Kale, R.D. 1986. Earthowrm feed for poultry and aquaculture. In proc. Nat. Sem, Org. Waste Utilitz. Vermicomp. Part B : *Verms and Vermicomposting.* (Eds. M. C. Dash; B. K. Senapati and P.C. Misra) Sri Artatrana Ront for Five star Printing Press. Burla, India pp 137-145.

Kale, R.D. and Babo, K. 1985. Laboratory propagation of some indigenous species of earthworms. *J.Soil Biol. Eco.* **5** : 20-26.

Kale, R.D. and Bano, K. 1985. Earthworm cultivation and culturing technique for production of ''Vee Comp. 83 E UAS and Vee meal 83 P USA. *Mys. J. Agri. Sci.* **22** : 339-344.

Kale. R.D. and Bano, K. 1992., Niche divergence-A limiting factor for recommendation of earthworm for Biotechnology. *Proc. Natl. Sem. Org. farming.* Apl. 18-19. pp 42-44.

Kale, R.D. and Krishnamoorthy, R.V.. 1977. Distribution and abundance of earthworms in Bangalore. *Proc. Indian Acad. Sci.* **87B** : 23-25.

Kale, R.D. and Krishnamoorthy, R.V. 1978. Distribution of earthworms in relation to soil condition in Bangalore. In ''*Soil Biology and Ecology in India.* (Eds. C.A. Edwards and G. K. Veeresh) pp : 63-69.

Kale, R.D. and Krishnamoorthy, R.V. 1981. What affects the abundance and diversity of earthworms in soil? *Proc. Indian Acad. Sci.* **90** : 117-121.

Kale, R.D.; Bano, K. and Krishnamoorthy, R.V. 1977. Feeding zones and interspecific zonation in earthworms. *Curr. Sci.***46** : 79.

Kale, R.D. Bano, K and Krishnamoorthy, R.V. 1981. Anachorisis of earthworms. *Bambay. Nat. Hist. Soc.* **78** : 400-402.

Kale, R.D.; Bano, K and Krishnamoorthy, R.V. 1982. Potential of *Perionyx excavatus* for utilisation of organic wastes. *Pedobiologia.* **23** : 419-425.

Mba, C.C; 1983. Utilization of *Eudrilus eugeniae* for disposal of Cassava peel. In ''Earthworm ecology. From Darwin to vermiculture'' (Ed. J. E. Satchell) Chapman and Hall Lond pp 315-321.

Nandeesha, M.C; Srikant, G.K; Basavaraja, N; Keshavanath, P; Verghese, T. J; Bano K; Roy, A.K. and Kale R.D. 1988. Influence of earthworm meal on Growth and fresh quality of common corp. *Biological water* **26** : 189-198.

Neuhauser, F. Kaplan, D.L. and Hartenstein. R. 1979. Life history of earthworm *Eudrilus eugeniae. Rev. Ecol. Biol. sol* **16** : 515-534.

Vijoen S.A. and Reinecke, A.J. 1989. The number, size and growth of hatchings of the African night crawler *Eudrilus eugeniae. Rev. Ecol. Biol. sol* **26** : 1-12.

Chapter 11

Phytosociology and Earthworm Production in the Ground Floor of Casuarina and Cocoanut Plantations in Coastal Orissa

Malaya K. Misra AND Manasi Ray

Ecology & Floristic Laboratory Department of Botany, Berhampur University, Berhampur-760 007, Orissa, India

Abstract. Ecological studies of the ground cover vegetation of Casuarina (site I) and Cocoanut (site II) plantations at Berhampur, Orissa, India were made during the monsoon period of 1988. Maximum 18 species in site I and 34 species in site II were recorded. The similarity index for both the sites was 0.576. Above ground green biomass peaks were 68 g m^{-2} for site I and 148 g m^{-2} for site II while the standing dead peaks were 28 and 41 g m^{-2} for site I and site II respectively. Maximum belowground biomass was 285 g m^{-2} in site I and 192 g m^{-2} in site II. Net aboveground primary productivity for site I and site II were 81 and 154 g m^{-2} yr^{-1}. The dry weight of earthworms was 21.7 g m^{-2} and 24.3 g m^{-2} for site I and site II respectively. The secondary productivity of earthworms in site I and site II were 21.7 and 24.3 g dry wt m^{-2} yr^{-1} respectively.

Key words : *Ground vegetation, Casuarina, Cocoanut, biomass, population, earthworms, primary production, secondary production.*

Introduction

For the management and maintenance of any plantation system, a thorough knowledge on the ground cover vegetation, soil organisms, microclimate and their interaction with the environment, is necessary. The structure and functioning of the ground cover vegetation under different plants varies from place to place and depends on many factors. Very little work is available on the secondary producer of earthworms in the tropical plantations, which are mostly confined to grasslands or woodlands (Dash and Patra 1977, Senapati 1980, Dash and Senapati 1983, Lavelle 1983, Senapati, and Dash 1984, Mishra and Dash 1984, Misra and Tripathy 1988).

Earthworms are present everywhere except in arid and frozen regions. In this paper an attempt has been made to evaluate the plant structure, net primary productivity and density, biomass and secondary productivity of earthworms in the ground vegetation of Casuarina and Cocoanut plantations at Berhampur.

Study sites

Two experimental sites, Casuarina (site I) and Cocoanut (site II) plantations were selected for the study druing the wet period of 1988. These sites were situated at Bhanja Bihar (19° 16'N, 84° 53'E) at a distance of about 12 km on the east side of Berhampur city and 5 km away from the Bay of Bengal. Bhanja Bihar is situated on an upland about 30 m above the mean sea level. These two sites were open grazing lands (grassland/rangeland) before the plantations were raised. The age of the Casuarina plantation was about 25 years and that of Cocoanut was about 12 years. In Casuarina plantation less ground space was available compared to Cocoanut plantation.

The climate of the area is monsoonic. On the basis of temperature and rainfall the year can broadly be divided into 3 season, viz. the summer (March-June), the rainy (July-October) and the winter (November-February). The total rainfall during the monsoon period (May-October) of 1988 was 1313 mm and maximum rainfall (358 mm) was received during October. The minimum and maximum temperatures were 23.53 and 33.45°C respectively during October 1988.

The physical and chemical characteristics of the soils of the experimental sites are shown in Table 11.1. The soils were acidic. There was little variation in the property of soils at different horizons (depths). Soil water content fluctuated between 3.7 and 18.3 per cent in site I and between 4.3 and 17.6 per cent in site II during the study period. The fluctuation was in accordance with the precipitation pattern during that period.

TABLE 11.1. *Physical and Chemical Characteristics of the soil in the sampling sites.*

Sites	Depths (cm)	pH	Organic matter (g%)	Total organic carbon (g%)	Total nitrogen (g%)	C/N ratio	K (g%)	Na (g%)
	0-10	6.29	2.0	1.16	0.031	37.42	2.39	2.98
Sites I	10-20	5.80	2.0	1.16	0.031	27.42	2.01	2.46
(Casuarina)	20-30	5.85	1.0	0.58	0.032	18.13	2.39	3.14
	Average	5.98	1.67	0.97	0.031	30.99	2.26	2.86
	0-10	6.19	7.0	4.06	0.028	145.0	2.22	2.35
Site II	10-20	6.22	6.0	3.48	0.027	128.88	1.24	2.12
(Cocoanut)	20-30	6.29	4.0	2.32	0.027	85.92	1.03	1.86
	Average	6.23	5.67	3.29	0.027	119.93	1.50	2.11

Materials and Methods

The study was conducted during the growth period (July to November) of 1988. Sampling of vegetation and collection of earthworms were done on the last date of each month. Quadrats of 50 × 50 cm size, as determined for grassland ecosystem (Misra 1978), were used for the study of aboveground parts. Monoliths of 25 × 25 × 30 cm were excavated and were cut into 3 parts (0-10 cm, 10-20 cm and 20-30 cm deep) for the collection of earthworms and belowground parts.

Primary producers

Plants from both the sites were collected, pressed, dried and identified. Basing on these specimens the floristic list was prepared. For the phytosociological analysis 25 quadrats were laid down, on every sampling date, on the ground vegetation of the two sites and the frequency, density, basal area and importance value index (IVI) were determined following Misra (1968). The average IVI for all the sampling periods were taken into consideration. Plant species diversity (H) was calculated using Shannon and Wiener index (Pielou 1975). Dominance index (C) was calculated by means of Simpson's (1949) formula. Similarity index for the two sites was calculated following Sorensen (1948).

The vegetation was clipped from both the sites very close to the ground from 10 randomly located plots: 5 from site I and 5 from site II; on each sampling date. Materials clipped were brought to the laboratory and separated into green and standing dead. Belowground parts were collected from the centre of the clipped plots by excavating the soil and they were collected from different depths. Water was poured into the soil and passed through a 2 mm sieve to collect the roots. The roots of Casuarina in site I and Cocoanut plants in site II were separated from the herbaceous roots. All the plant samples were dried in a hot air oven for a minimum period of 24 hours at 80°C.

Aboveground net primary productivity (ANPP) was estimated by summation of increments in the aboveground green biomass during the sampling period plus summation of increments in standing dead biomass for only those periods when the increment in standing dead could not be accounted for by a decrease in live biomass (Singh *et al.* 1975).

Secondary producers

Earthworms were hand sorted from the soil excavated for the belowground parts on every month as per Ali *et al.* (1973) and Dash & Patra (1977), during morning hours. They were separated as per the depth (0-10, 10-20 and 20-30 cm). The worms were predominated by *Lampito mauritii* Kinberg in the experimental sites. Earthworms thus collected were brought to the laboratory, washed thoroughly and counted. Mature and immature worms were taken together for the study. Earthworms were kept in ice buckets and after they became immobile, the measurement (length) was teken, at resting stage. Worms were dried for 24 h at about 70°C after gut clearence, then the dry weight of the tissue was taken.

TABLE 11.2. *The average Importance Value Index (IVI) for different species of the ground vegetation of Casuarina (Site I) and Cocoanut (Site II) plantations during the wet season of 1988.*

Name of the species	Average IVI	
	Site I	Site II
Grasses		
1. *Alloteropsis cimicina*	-	3.6
2. *Arstida setacea*	11.3	14.4
3. *Bothriochloa odorata*	11.2	5.8
4. *Bothriochloa pertusa*	-	4.3
5. *Chrysopogon aciculatus*	35.2	9.1
6. *Cynodon dactylon*	20.6	-
7. *Dactyloctenium aegyptium*	-	7.2
8. *Eragrostis coarctata*	4.2	6.4
9. *Heteropogon contortus*	13.3	20.7
10. *Sporobolus diander*	20.6	13.8
Forbs		
11. *Amaranthus gangeticus*	-	7.6
12. *Andrographis echioides*	-	5.1
13. *Celosia argentia*	-	3.8
14. *Chromolaena odorata*	-	3.0
15. *Commelina benghalensis*	4.2	9.04
16. *Croton bonoplandianus*	-	4.2
17. *Cyperus castaneus*	-	10.2
18. *Cyperus rotundus*	-	5.5
19. *Cyperus triceps*	-	8.2
20. *Desmodium triflorum*	33.7	36.7
21. *Eleocharis atropurpurea*	-	19.4
22. *Evolvulus alsinoides*	-	10.8
23. *Evolvulus numularius*	51.5	-
24. *Fimbristylis acuminata*	17.4	10.8
25. *Hydrocotyl asiatica*	3.0	-
26. *Ionidium suffruticosum*	-	3.2
27. *Justicia simplex*	17.8	12.4
28. *Mariscus squarrosus*	-	3.8
29. *Mimosa pudica*	10.9	6.5
30. *Ocimum basilicum*	7.1	4.5
31. *Oldenlandia corymbosa*	-	3.4
32. *Phyllanthus simplex*	-	3.6
33. *Rungia repens*	-	3.1
34. *Sida cordifolia*	10.6	5.7
35. *Spermacoce hispida*	1.0.9	13.5
36. *Tridax procumbens*	-	4.8
37. *Triumfetta rhomboides*	6.3	-
38. *Vandella cordifolia*	-	7.2

Secondary productivity can be calculated as $P = \Delta b + \Delta g$, where Δb represents change in body weight and Δg reproduction over a period of time (say one year), irrespective of whether it has survived to the end of that period or not (Macfadyen 1967). In the present study the loss of tissue due to mortality or reproduction has not been taken into consideration and the secondary productivity was estimated taking the growth (increment in dry tissue weight in the successive sampling period) only.

Results and Discussion

Vegetation structure

The number of plant species observed in the ground vegetation of Casuarina plantation (site I) was 18 and that of Cocoanut (site II) was 34. Most species observed were common to both the sites. The similarity index calculated taking these species into account was 0.576. In both the sites, density (ind. m^{-2}) of species was very low in July and thereafter the number of individuals increased attaining the peak of 2556 and 2693 ind. m^{-2} in site I and site II respectively, in September. More number of species and higher density in site II than site I may be attributed to availability of more space, light and moisture in site II than in site I. The importance value index (IVI) of the species (Table 11.2) revealed that site I was predominated by *Evolvulus numularius, Cynodon dactylon* and *Chrysopogon aciculatus* while site II by *Desmodium triflorum, Heteropogon contortus* and *Aristida setacea.*

The life-forms (biological spectrum) analysis revealed that site I and II included therophytes 50 per cent and 55.9 per cent, geophytes 5.6 per cent and 14.7 per cent, hemicryptophytes 33.3 per cent and 17.6 per cent, and chamaephytes 11.1 per cent and 11.8 per cent, respectively. Higher percentage of therophytes in these two sites may be attributed to the periodicity of climate, and biotic interference as the two sites were under intensive grazing pressure.

Density based species diversity (H') fluctuated throughout the sampling period (July to November) in both the plots. In site I it fluctuated between 1.75 and 2.096 and site II between 2.402 and 2.730. The dominance index (C) also fluctuated during the study and the minimum was 1.167 and 0.046 while the maximum was 0.243 and 0.210 in site I and II respectively. Higher H' in September and August in both the sites was due to the emergence of maximum number of seedlings, and tillers in some perennial grasses. Diversity index (X) and dominance index (Y) of site I showed a significant inverse relationship ($r = -0.941$, $p \leq 0.01$) and both the parameters were related as per the equation $Y = 0.63-0.22 X$. However, in site II, no significant relationship ($r = 0.536$) between these two parameters was observed.

Plant Biomass

Live green biomass was low in July, then it gradually increased reaching the peak in September and there after it decreased, in both the sites (Table 11.3). However, the green biomass was more in site II in all the sampling periods. Forbs (non-grass) contributed higher percentage to the totl green biomass in site I whereas grass biomass contribution was more in site II (Table 11.3). Higher green biomass during July to September was mainly due to the growth

TABLE 11.3. *Biomass structure of the ground vegetation of Casuarina (Site I) and Cocoanut (Site II) plantations during the wet season of 1988. ± SEM.*

Sampling Period	Aboveground live biomass g m^{-2}					
	Site I			Site II		
	Grass	Forb	Total	Grass	Forb	Total
July	18.4 ± 1.6	25.0 ± 2.3	43.4 ± 2.8	39.2 ± 4.7	55.6 ± 3.9	94.9 ± 4.1
Aug.	27.9 ± 6.0	35.1 ± 8.6	63.0 ± 10.6	61.4 ± 10.6	53.3 ± 4.7	114.7 ± 12.6
Sept.	29.4 ± 5.3	39.1 ± 6.5	68.6 ± 11.8	82.2 ± 11.1	65.5 ± 5.1	147.8 ± 16.1
Oct.	17.4 ± 2.9	30.1 ± 5.5	47.4 ± 6.0	72.2 ± 10.5	26.3 ± 8.4	98.5 ± 7.7
Nov.	10.8 ± 3.0	28.9 ± 4.0	39.8 ± 4.8	45.2 ± 4.2	48.7 ± 3.8	93.9 ± 3.7

	Standing dead biomass g m^{-2}	
	Site I	Site II
July	10.3 ± 0.9	27.0 ± 2.8
Aug.	11.5 ± 3.7	21.0 ± 1.9
Sept.	22.8 ± 7.3	29.0 ± 4.2
Oct.	25.7 ± 3.9	30.9 ± 5.7
Nov.	27.8 ± 4.0	40.9 ± 4.9

	Underground biomass g m^{-2}	
July	177.6	144.7
Aug.	285.3	192.6
Sept.	179.7	142.9
Oct.	137.1	116.5
Nov.	89.6	78.9

pattern of the species which was regulated by rainfall pattern during that period. Peak green biomass values of the two sites showed a significant difference ($t = 4.013$, $p \leq 0.01$). Reasons for more green biomass in site II than site I may be attributed to high soil moisture and organic matter in that site. Moreover, the ground vegetation under Cocoanut plantation (site II) received more light energy than the ground floor of Casuarina plantation.

It was observed that the standing dead (SD) biomass was low in July but gradully increased upto the last sampling period. In site II SD biomass was more compared to site I (Table 11.3). The increasing trend in SD in both the sites was caused due to the continuous mortality of tillers and senescence of lower leaves of perennial grasses. Moreover, most of the annuals completed their phenological (life) cycle during this time. Statistical comparison of the peak SD biomass values of both the sites showed significant difference ($t = 17.75$, $p \leq 0.001$). More SD biomass in site II was due to the presence of more grasses (biomass) in that site. Moreover, contribution of forbs to SD was less significant, indicates that the dead parts of forbs direclty enter into the litter compartment.

Underground biomass of the two sites are shown in Table 11.3. Peak underground biomass in August shows that the peak growth period favoured the growth of roots. Low root biomass in the latter months was probably due to translocation of more food materials from roots to aboverground parts. Moreover, higher rate of death and decay of root parts during this period was probably another reason for the decreased root biomass. Peak underground biomass values of the two sites showed not significant difference ($t = 0.565$). Maximum root biomass in the top soil was recorded (Ray 1989) because almost all the species of the gorund vegetation were annual herbs, their roots spread mostly on the top horizon (0-10 cm soil depth).

Earthworm population

Earthworm population (individuals m^{-2}) in both the sites first increased from July upto September then decreased in subsequent sampling periods (Table 11.4). Minimum population was recorded in November in both the sites. This was due to non-availability of optimum soil moisture (Ray 1989). Because of cutaneous respiration earthworms are extremely sensitive to drought and as soon as soil moisture reaches a given threshold, usually a little above the wilting point for plants earthworms becomes inactive (Lavelle 1983). It was observed that the top soil (0-10 cm) in both the sites, carried maximum number of earthworms. The density decreased from top to deeper soil (Table 11.4).

The size of the earthworm population, growth rate and fecundity is being mostly influenced by the kind and amount of food materials available to them (Watanabe & Tsukamoto 1976, Edward & Lofty 1977, Senapati & Dash 1984). Earthworms were omnivorous but often selective in their food habits but surface dwelling earthworms take the food materials more selectively and deep soil dwelling worms ingest as such (Senapati and Dash 1984). Earthworms derive their nutrition from various substances such as organic matter, living bacteria, fungi, etc. Fungal feeding by earthworms was reported by Piearce (1978) and Dash *et al.* (1979). Thus the increase in

TABLE 11.4. *Variation in earthworm density and biomass in the ground vegetation of Casuarina (Site I) and Cocoanut (Site II) vegetation during the monsoon period of 1988.*

Sampling period	Site-I			Soil depth (cm)	Site II			
	0-10	10-20	20-30	Total	0-10	10-20	20-30	Total
Earthworm population (Ind. m^{-2})								
July	64.0 ± 7.6	21.3 ± 4.4	10.7 ± 4.4	96.0	32.0 ± 7.6	26.7 ± 4.4	10.7 ± 4.4	69.3
Aug.	64.0 ± 7.6	37.3 ± 11.5	26.7 ± 4.3	128.0	64.0 ± 7.6	42.7 ± 15.7	32.0 ± 7.6	138.7
Sept.	90.7 ± 11.5	48.0 ± 15.1	16.0 ± 7.6	154.7	101.3 ± 11.5	37.3 ± 8.7	26.7 ± 4.4	165.3
Oct.	58.7 ± 8.7	32.0 ± 7.5	21.3 ± 4.4	112.0	37.3 ± 4.4	21.3 ± 4.3	16.0 ± 7.6	74.7
Nov.	64.0 ± 7.6	16.0 ± 7.6	16.7 ± 4.3	70.6	26.7 ± 4.3	10.6 ± 7.4	10.7 ± 4.4	53.3
Earthworm biomass (g dry wt. m^{-2})								
July	4.8 ± 1.9	1.6 ± 0.7	1.7 ± 0.6	8.1	3.2 ± 1.5	1.4 ± 0.6	0.9 ± 0.4	5.5
Aug.	9.6 ± 1.5	2.8 ± 1.3	0.1 ± 0.04	12.5	12.3 ± 1.2	4.6 ± 1.7	1.8 ± 0.6	18.7
Sept.	10.7 ±1.2	5.9 ± 1.7	5.1 ±1.8	21.7	16.0 ± 2.0	5.0 ± 1.9	3.3 ±1.5	24.3
Oct.	10.1 ± 0.4	0.4 ± 0.1	0.1 ± 0.08	10.6	5.9 ± 0.4	0.8 ± 0.3	0.3 ± 0.04	6.9
Nov.	5.9 ± 1.7	1.7 ± 0.6	0.2 ± 0.07	7.8	4.3 ± 1.6	0.2 ± 0.06	0.1 ± 0.04	4.6

earthworm population during the rainy season was due to the favourable environmental conditions such as increase in plant density and biomass, and soil temperature and moisture during this time. Many workers (Dowdy 1944, Gerard 1967, Nakamura 1968, Dash & Patra 1977, Lavelle 1983) have stressed the importance of soil moisture and temperature on earthworm activity. In the present study number of earthworms (Y) were significantly correlated with soil moisture content (X) ($r = 0.913$, $p < 0.05$, site I; $r = 0.878$, $p < 0.05$, site II) and these two factors were related as per the equation : $Y = 67.9 + 4.95 \times$ for site I and $Y = 20.8 + 7.2 \times$ for site II. This confirms the findings of Dash and Patro (1977) and Mishra and Dash (1984), which stress the importance of soil moisture for earthworm growth.

The present earthworm population figure both in Casuarina (71-155 ind. m^{-2}) and Cocoanut (53-165 ind. m^{-2}) plantations was although much poorer but still higher than the values of 18-31 earthworms m^{-2} reported for *Picea* raw humus and 23-81 ind. m^{-2} for *Fagus* raw humus in Denmark Bornebusch 1930). The figures (12-15 ind. m^{-2}) reported for a humid Panama grassland are much lower (quoted in Lavelle 1983). Lavelle (1983) quoted much higher figures (188-400 ind. m^{-2}) for Lamto (Ivory coast) grass and shrub savannas and 400-582 ind. m^{-2} for Foro Foro (Ivory coast) tree savanna. The density (64-800 ind. m^{-2}) obtained by Dash & Patra (1977) for a tropical grassland of Orissa are much higher than the present finding. In western Orissa, the mean annual earthworm density ranged from 174 ind.m^{-2} in grazed pasture to 247 ind. m^{-2} in the ungrazed site (Senapti 1980). At Berhampur (sougth Orissa) still lower figures (16-112 ind. m^{-2}) were reproted by Misra & Tripathy (1988) for the unfertilized and fertilized grasslands. Misra & Dash (1984) reported 24-131 earthworms m^{-2} for a subtropical dry woodland of western Orissa.

Earthworm Biomass :

There was an increase in earthworm biomass (dry weight) from July to September and thereafter it decreased (Table 11.4). Peak biomass during september was due to maximum number and length of earthworm during that period. Moreover, death and decay of plant parts including roots during this period may also be attributed to the growth of earthworms. Dry weight of earthworms (Y) showed positive correlation with the number of earthworms ($r = 0.911$, $p \leq 0.05$ site I; $r = 0.996$, $p \leq 0.001$ site II) in both the sites. They were related according to the equation $Y = -5.9 + 0.16\ x$ in site I and $Y = -6.1 + 0.18\ x$ in site II. Biomass and length of earthworms were not significantly correlated ($r = 0.619$, site I, $r = 0.431$ site II). Soil moisture content (x) had significant correlation with earthworm dry biomass in both the sites ($r = 0.987$, $p \leq 0.001$; $Y = 3.54 + 0.96\ x$, site I; $r = 0.875$, $p \leq 0.05$; $Y = -2.54 + 1.31\ x$, site II). It shows that soil moisture had important role on earthworm growth. The decrease of earthworm biomass after September was due to the unfavourable environmental conditions especially the soil moisture and temperature. The earthworm biomass reported earlier for some habitats are much lower than the present values. Earthworm biomass of 0.9 - 1.5 g m^{-2} in *Picea* raw humus sites of Denmark (Bornebusch 1930) 3 - 8 g m^{-2} (wet wt.) for the *Andropogon* habitat of Tennessee, U.S.A. (Reynold 1973), and 1.5 - 13.9 g m^{-2} in grassland sites of Berhampur (Misra & Tripathy 1988) have been reported. Dash & Patra (1977) reported 6 - 61 g m^{-2} (wet wt.) for a tropical grassland while Mishra & Dash (1984) reported 7 - 28 g m^{-2} (wet wt.) for a deciduous woodland of Orissa. Higher biomass values (41 - 56 g m^{-2}, wet wt.) were obtained for ungrazed and grazed pastures of Sambalpur (Senapati 1980). At Lamto (Ivory coast) earthworm biomass of 35.9 - 48.6 g m^{-2} wet wt was reported for grass and shrup savannas while very low figures (0.15 - 0.36 g m^{-2}) were reported for humid Panama grassland (Quoted in Lavelle 1983). Plant productivity :

The aboveground net primary productivity (ANPP) was 81.1 g m^{-2} yr^{-1} in site I and 153.8 g m^{-2} yr^{-1} for site II (Table 11.5). The growth period for the ground cover vegetation under casuarina and cocoanut plantations are mostly confined to the rainy season (wet period), thus seasonal in nature (Table 11.4). Therefore, the green biomass for July was considered as productivity for that

TABLE 11.5. *Aboveground net primary productivity and secondary productivity of earthworms for the two sites during wet season of 1988 (in g dry wt. m^{-2})*

	Aboveground plant productivity		Earthworm productivity	
	Site I	Site II	Site I	Site II
July	43.4	94.9	8.1	5.5
Aug.	20.8	19.8	4.4	13.2
Sept.	16.9	39.1	9.2	5.6
Total	81.1	153.8	21.7	24.3

period and the productivity for the growth period was considered as annual productivity. The herbage production during the rainy season was regulated by the combined effect of precipitation pattern, the soil water content, soil pH, soil organic matter, biotic interference, space for light penetration into the ground, etc. More herbage production in sites II may be attributed to the availability of more soil organic matter, space and light. The present ANPP values (81 - 154 g m^{-2}) are less than the grassland productivity values reported by Pandey (1978) for Varanasi (1921 g m^{-2}), Misra & Misra (1984) for Berhampur (956 g m^{-2}), Billore and Mall (1977) for Ratlam (429 g m^{-2}) and Misra & Tripathy (1988) for Berhampur (307 -460 g m^{-2}), but higher than the value (22 - 164 g m^{-2}) reported for a Jodhpur grassland.

Earthworm productivity

Secondary productivity of earthworms was recorded during July to September in both the experimental sites (Table 11.5). Maximum productivity was recorded in September in site I and in August in site II. Thereafter the productivity was not observed probably because of the dry condition of the sites. Total annual productivity was 21.7 g m^{-2} in Casuarina plantation site and 24.3 g m^{-2} in Cocoanut plantation site. As the mortality figure has not been considered this value may be an underestimate. There was not much difference in the productivity of the two sites. Restriction of earthworm productivity during July to september was due to the availability of soil moisture which is very important for the growth of worms (Lavelle 1983). The present findings (22 - 24 g dry wt m^{-2}) are well within the range of the data reported for many habitats. Dash & Patra (1977) reported 126 g m^{-2} yr^{-1} for a tropical grassland of Berhampur, Orissa while Misra & Tripathy (1988) reported 9 gm^{-2} for a unfertilized grassland and 14 g m^{-2} for the fertilized plot at Berhampur. Secondary annual productivity of earthworms in a woodland of western Orissa was reported to be about 22 g m^{-2} and the mortality values was about 52.5 g m^{-2}, thus the total annual production was 77.5 g m^{-2} (13 g dry wt.).

The higher productivity in these sites may be attributed to the environmental conditions. This may be the result of the combined affect of the ground vegetation and the plantation. Ground cover mainly provides the organic matter while the tree vegetation is responsible for shade which helps in maintaining the soil moisture level. Moreover, the dead and decay roots of Casuarina and Cocoanut plants were also a source of food for the earthworms. The secondary productivity of earthworms was about 26.7 per cent and 15.6 per cent of the total aboveground net primary production in site I and site II respectively.

REFERENCES

Ali, M.K., Dash, M.C. and Patrar, U.C. 1973. Estimation of *lampito mauritii* (Oligocheata) population by chemical methods. *Sci. & Cult.* **39** : 558-560.

Billore, S.K. and Mall, L.P. 1977. Dry matter structure and its dynamics in *Sehima* community. II. Dry matter dynamics. *Trop. Ecol.* **18** : 29-35.

Bornebusch, C.H. 1930. *The Fauna of Forest Soils.* Neilsen and Lydicke, Copenhagen.

Dash, M.C. and Patra, U.C. 1977. Density, biomass and energy budget of a tropical earthworm population from a grassland site in Orissa, India. *Rev. Ecol. Biol. Sol.* **14** : 461-471.

Dash, M.C. and Senapati, B.K. 1983. Energetics of earthworm populations in tropical pastures from India. *Ind. Acad. Sci. B (Animal Science)* **92** : 315-321.

Dash, M.C. Mishra P.C. and Behera, N. 1979. Fungal feeding by a tropical earthworm. *Trop. Ecol.* **20** : 9-12.

Dowdy, W.W. 1944. Influence of temperature on vertical migration. *Ecology* **25** : 449-460.

Edward, C.A. and Lofty, J.R. 1972. *Biology of Earthworms.* Chapman and Hall, London, 283 p.

Gerard, B.M. 1967. Factors affecting earthworms in pasture. *J. Ani. Ecol.* **36** : 235-252.

Levelle, P. 1983. The Soil Fauna of Tropical Savannas. II. The Earthworms. pp. 485-504. *In : Tropical Savannas* (Ed. F. Bourliere), Elsevier, Amsterdam.

Macfadyen, A. 1967. Methods of investigation of productivity of invertebrates in terrestrial ecosystems. pp. 383-412. *In : Secondary Productivity of Terrestrial Ecosystem* (Ed. K. Petrusewics) Warszawa.

Misra, R. 1968. *Ecology Workbook,* Oxford and IBH, New Delhi, 244 p.

Misra, M.K. 1978. Phytosociology and Primary Production of a Grassland community at Berhampur, Orissa (India). *Ph. D. Thesis,* Berhampur University, Berhampur.

Misra, M.K. and Tripathy, P.C. 1988. Effect of nitrogen fertilizer on plant and earthworm production in a tropical grassland in Orissa. *Trop. Ecol.* **29** : 61-72.

Misra, M.K. and Misra, B.N. 1984. Biomass and primary production in an Indian grassland. *Trop. Ecol.* **25** : 239-247.

Mishra, P.C. and Dash, M.C. 1984. Population dynamics and respiratory metabolism of earthworm population in a sub tropical dry woodland of western Orissa, India. *Trop. Ecol.* **25** : 103-116.

Nakamura, Y. 1968. Studies on the ecology of terrestrial Oligochaetae. 1. Seasonal variation in population density of earthworms in alluvial soil grassland in Sapporo. *Jap. J. App. Ento. Zool.* **3** : 89-95.

Pandey, T.N. 1978. Seasonal variation in biomass and productivity in a protected grassland of the Chakia forest, Varanasi. *Indian Jour. Ecol.* **5** : 37-42.

Piearce, T.G. 1978. Gut content of some Lumbricid earthworms. *Pedobiologia* **18** : 153-157.

Pielou, E.C. 1975. *Ecological Diversity,* John Wiley & Sons. New York, 165p.

Ray, M. 1989. Ecological study of ground vegetation of Casuarina and Cocoanut plantations in the coastal belt of Orissa. *M. Phil. dissertation.* Berhampur University, Berhampur (unpublished).

Reynolds. J.W. 1973. Earthworm (Annelida : Oligochaetae) : Ecology and Systematics. pp. 95-120. *In : Proceedings of the First Soil Micro-communities Conference* (Ed. D.L. Dindall) N.T.I.S., U.S. Dept. Commerce, Springfield, Virginia.

Senapati, B.K. 1980. Aspects of ecophysiological studies on tropical earthworms (Distribution, population dynamics, production energetics and their role in the decomposition process). *Ph.D. Thesis,* Sambalpur University, Orissa.

Senapati, B.K. and Dash, M.C. 1984. Functional role of earthworms in the decomposer sub-systems. *Trop. Ecol.* **25** : 52-72.

Simpson, E.H. 1949. Measurement of diversity. *Nature,* (England), 163 : 688.

Singh, J.S., Lauenroth, W.K. and Steinhorst, R.K. 1975. Review and assessment of various techniques for estimating net aerial primary production in grasslands from harvest data. *Bot. Rev.* **41** : 181-232.

Sorensen, T. 1948. A method of estimating groups of equal emplitude in plant sociology based on similarity of species content. *Det* Kong. *Danske Vidensk. Selsk. Biol. Skr.* **5** : 1-34 (original not seen).

Watanabe, H. and Tsukamoto, J. 1976. Seasonal change in size, class and age structure of the Lumbricid *Eisenia* foetida population in a field compost and its practical application as the decomposer of organic waste water. *Rev. Ecol. Biol. Sol.* **13** : 141-146.

Chapter 12

Plant Litter Decomposition and Soil Respiration in the Natural and Cultivated Grassland Ecosystems at Kangpokpi, Manipur

P.S. Yadava, N.S. Thoudam AND T. Paukhanmang

Ecology Laboratory, Life Sciences Department, Manipur University, Imphal-795 003, India

Abstract. Soil respiration and rate of litter decomposition were measured by alkali-absorption and litter bag technique respectively during growing season of 1990 in the natural and cultivated grassland ecosystems at Kangpokpi, Manipur (25° 41' N Latitude, 94° 47' E Longitude and 1, 105 m elevation). The cultivated grassland exhibited higher rates of soil respiration than that of natural grassland and varied from 294.60 ± 11.12 to 431.20 ± 11.65 in the natural and 379.35 ± 9.54 to 511.57 ± 10.00 $mgCO_2m^{-2}hr^{-1}$ in cultivated grassland ecosystems. The root and microorganisms contributed 36.37 to 38.42 per cent and 61.57 to 63.24 per cent of the total soil respiration respectively. The time taken for decomposition of 50 and 95 per cent of the plant materials were determined from the decomposition constant. The rate of decomposition is found to be highest in the shoot parts as compared to the underground parts and the nutrient release during the decomposition has also been estimated. The rate of soil respiration exhibited positive correlation with soil temperature, root biomass and rainfall.

Key words : Soil respiration, Litter decomposition, Grassland ecosystem, Kangpokpi, Natural, Cultivated.

Introduction

The majority of the energy for soil, root and microbial respiration is derived from root and litter decomposition. Soil respiration may give an idea of productive efficiency of a particular ecosystem which is reflected by the amount of CO_2 evolved from its substrate in soil (Weber, 1985). The evolution of CO_2 from a decomposer subsystem is directly proportional to the activity of the decomposer biota, mass loss of the substrate and usually, the mineralization of organically bound elements (Carlyle and Than, 1988). The inclusion of experiments into the time-course of decomposition by use of the litter-bag technique has been regarded as a more suitable basis index of decomposition than the turnover coefficient or soil respiration (Anderson and Swift, 1983). Field methods of soil respiration measurement are used to characterize soil ecosystems (Buyanovski *et al.,*1986) and it is only possible to trap CO_2 from rhizophere organisms under actual growth conditions (Smucker and Safir, 1986). Singh and Gupta (1977) have reviewed litter decomposition and soil respiration of terrestrial ecosystems. Both these aspects havse received little attention except for a few studies conducted in India at Kurukshetra (Gupta and Singh 1981a, 1981b); Cherrapunji (Subhash and Ramakrishnan, 1988) and Shiroy (Laishram I. Devi, 1989). Several reports are available from the different parts of the world in the grassland ecosystems on soil respiration (Witkamp and Frank 1969, Howard and Howard, 1979; Chapman, 1979; Valerie, Orchard and Cork, 1983; Behera and Pati, 1986; Weber, 1990; Yadava and Kakati, 1990; Basu and Pati, 1991; Rochette *et al.*, 1991; Hingane, 1991 and Edwards 1991) and on litter decomposition (Herman *et al.*, 1977; Nicholas *et al.*, 1985; Mitchell *et al.*, 1986; Fisher, 1990; Steinburger *et al.*, 1990 and Fahey *et al.*, 1991.

The Paper reports on the microbial respiration, root respiration and total soil respiration, physico-chemical characteristics of soil, rate of litter decomposition, and nutrients (specially N, P and Lignin) releases through litter decomposition in natural and grassland ecosystem at Kangpokpi, Manipur. An attempt has also been made to correlate soil respiration, root and microbial respiration with abiotic variables.

Material and Methods

The study site is situated within the Regional Cross-Breed Cattle Breeding Farm, Turibari, Kangpokpi (25° 41' N lat., 94°47' E long.) which comprises of both cultivated and natural grasses area at an altitude of 1,105 m. The climate is typically monsoonic with an annual average rainfall of 1910. 75 m. The cultivated grassland comprised of *Panicum maximum* and natural grassland was dominated by *Imperata cylindrica, Axonopus compressus* and *Chrysopogon aciculatum.*

Soil respiration rate were measured by alkali-absorption method (Singh and Gupta, 1977; Grahammer *et al.*, 1991). The litter decomposition was studied using a litter bag-technique (Shanks and Olson, 1961; Anderson and Ingram, 1989). The physico-chemical characteristics of soils, plant meterial were analysed following the method of Anderson and Ingram (1989) lignin per cent was calculated by the method of Miller (1959). The decomposition constant was computed as per the formula given by Olson (1963).

Results And Discussion

The physico-chemical characteristics of soil were given in table 12.1. The soils of both grasslands were acidic in nature (pH 4.6 to 6.0) with organic carbon (%) varied from 0.50 to 2.26; total nitrogen varied from 0.08 to 0.258 per cent C/N ratio range from 5.39 to 11.78 per cent and available phosphorus ranged from 0.0011 to 0.00233 per cent. The higher nutrient contents in the cultivated grassland soil was due to the direct regular addition of cattle dung with decomposed

TABLE 12.1. *Physico-chemical characteristics of soils under natural and cultivated grasslands*

Physico-chemical characteristics of soils	Study area	
	Natural grassland ecosystem	Cultivated grassland ecosystem
1. Soil pH	4.9 ± 0.1 - 6.0 ± 0.1	4.6 ± 0.05 - 5.7 ± 0.05
2. Moisture content (per cent)	19.6 ± 1.4 - 39.4 ± 5.7	21.51 ± 1.59 - 42.32 ± 5.2
3. Bulk density (g/c.c.)	1.28 ± 0.01–1.43 ± 0.34	1.01 ± 0.05 - 1.28 ± 0.05
4. Soil porosity (per cent)	44.3 ± 6.9 - 51.0 ± 4.1	50.69 ± 2.04 - ± 57.85 ± 5.57
5. Organic carbon (per cent) (C)	0.51 ± 0.05 - 1.62 ± 0.14	0.50 ± 0.21 - 2.26 ± 0.12
6. Total Nitrogen (per cent) (N)	0.09 ± 0.01 - 0.18 ± 0.02	0.08 ± 0.01 - 0.25 ± 0.01
7. C/N ratio (per cent)	5.39 ± 1.02 - 9.37 ± 0.68	6.06 ± 1.45 - 11.78 ± 1.31
8. Available phosphorus (per cent)	0.00113 ± 0.0001 -0.0033 ± 0.0002	0.0011 ± 0.0002 - 0.00233 ± 0.0008

organic matter of straw; whereas, in the natural grassland soil, no such addition takes place.

Microbial respiration

The microbial respiration varied from 195.72 ± 8.40 to 273.06 ± 13.06 mgCO_2 m^{-2} hr^{-1} in the natural and 222.94 ± 9.52 to 340.46 ± 11.22 in the cultivated grassland soil (Figure 12.1). The rate of respiration by microorganisms contributed 61.57 to 63.24 per cent ot the total soil respiration in the cultivated and natural grassland respectively. The rate of microbial respiration in both grassland were positively correlated with soil temperature and explained 32.83 per cent variability in microbial respiration due to variability in the soil temperature ($r = 0.573$; $p < 0.05$). The microorganisms present in the soil play a major role in releasing CO_2 by metabolizing organic debris. Of the total soil respiration 60 per cent was due to microbial metabolism in the tall-grass praire (Gupta and Singh 1981b). Weber (1990), has also reported the strong relationship between soil surface temperature and soil respiration in the Aspen ecosystem at Eastern Ontario.

Root respiration

The rate of root respiration ranged from 72.59 ± 11.77 to 174.56 ± 6.02 in the natural and 121.41 ± 10.07 to 217.33 ± 4.28 mg$CO_2$$m^{-2}$$hr^{-1}$ in the cultivated grassland (Figure 12.1). The root respiration contributed 36.37 and 38.42 percent to the total soil respiration in the natural and cultivated grassland respectively. It is positively correlated with root biomass ($r = 0.985$; $p < 0.05$) and monthly rainfall ($r = 0.608$; $p < 0.01$) in these two grasslands. With the increase of root biomass, the root respiration also increased.

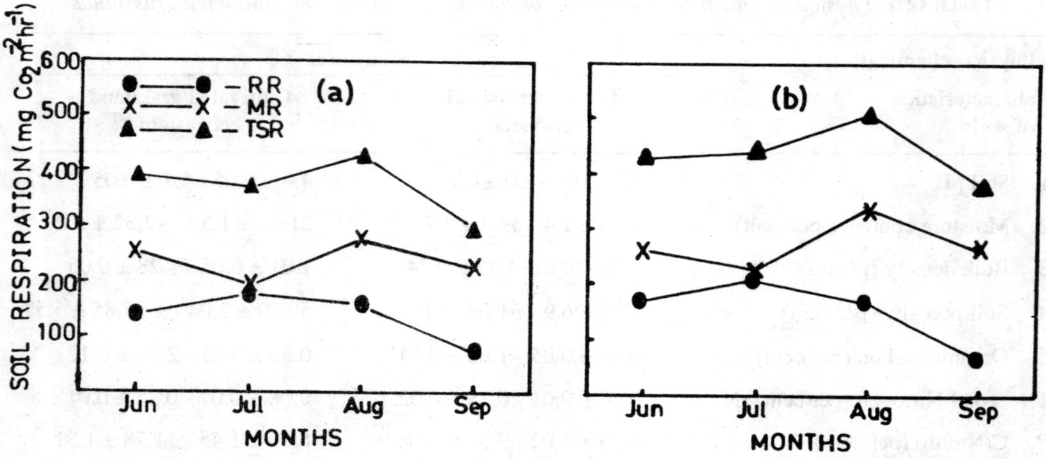

Fig. 12.1. Monthly variations of soil respiration in different site,
(a) Natural grassland,
(b) Cultivated grassland;
RR = Root respiration
MR = Microbial respiration and
TSR = Total soil respiration.

The contribution of root and associated microorganisms to the total soil respiration on tropical grassland and tall grass prairie have been reported to be 42 per cent and 60 percent respectively by Gupta and Singh (1981b), Edwards (1991) and Beyer (1991) have also reported similar values of root respiration rates and root biomass in the *Pinus taeda* stand and arable and forest soils.

Total soil respiration

Total Soil respiration rates varied from 294.60 ± 11.12 to 431.20 ± 11.65 mgCO_2 $m^{-2}hr^{-1}$ in the natural and 379.35 ± 9.54 to 511.57 ± 10.00 mgCO_2 m^{-2} hr^{-1} in the cultivated grassland (Figure 12.1). The higher rate of soil respiration in the present study in comparison to the rate recorded else where (table 12.2) might be due to the favourable effect of soil moisture, moderate temperature or microbial activity and activity of roots in the growing season. Gupta and Singh (1981b) and Behera and Pati (1986) have also reported higher soil respiration during growing season . The higher rate in the cultivated grassland ecosystem may be due to higher root biomass and higher nutrient contents in the soil. In both the grasslands, maximum CO_2 evolution were recorded in the month of August, this may be due to higher rate of microbial respiration and root biomass. Higher rate of soil respiration in the month of August and September was also been reported by several workers (Howard and Howard, 1979; Subhash and Ramakrishnan, 1988; Rochette *et al.,* 1991 and Hingane, 1991).

TABLE 12.2. *Soil respiration rates (mgCO_2 m^{-2} hr^{-1}) from different grassland ecosystems of the world.*

Sl. No. (1)	Grassland ecosystems (2)	Place (3)	Rate (4)	Measurement period (5)	Measurement technique (6)	Author (s) (7)
Temperate						
1.	*Gratiola officianalis cares praecoxsuzae*	Czechoslovakia	300-600	May-October 18 periods	Alkali absorption	Tesarova and Glosen (1972)
2.	Mixed Virgin prairie	Saskatchewan, Canada	26-28	Mid June 1 hr mean	Gas Chromatography	des Jong and Schappert (1972)
3.	*Agropyron koeleria*	Saskatchewan,	135-150	May-Sept. 3-6 hr mean	Infrared gas analysis	Redmann (1974)
4.	Tall grass prairie (Andropogon)	Missouri, USA	0-450	Jan-Sept. weekly mean	Infrared gas analysis	Kucera and Kirkham (1971)
5.	Pacific N.W. bunch grass (Agropyron)	Washington, USA	30-200	Apr.-Dec. 24 hr mean	Alkali absorption	Wildung *et al.* (1975)
6.	Old field (Andropogon)	South Carolina USA	47-332	Oct.-August 24 hr mean	Alkali absorption	Coleman (1973)
Tropical						
7.	Savanna	Costa Rica	329	Feb.-March	Alkali absorption	Schulze (1967)
8.	Tracypogon Savanna	Venezuela	257-532	August	Alkali absorption	Medina and Zelwer (1972)
9.	Mixed grass stand	Kurukshetra, India	55-78	July-May	Alkali absorption	Gupta and Singh (1981)
10.	*Sesbania bispinosa* stand	Kurukshetra, India	49-358	July-May	Alkali absorption	Gupta and Singh (1981)
11.	*Desmostachya bipinnota* stand	Kurukshetra, India	64-389	July-May	Alkali absorption	Gupta and Singh (1981)
12.	*Dichanthium annulatum* Stand	Imphal, India	71-194	Oct.-Dec.	Alkali absorption	Yadava and Kakati (1990)
13.	Tropical grassland	Sambalpur, India	162-348	Feb.-March	Alkali absorption	Behera and Pati (1986)
14.	Cultivated grassland *(Panicum maximum)* Stand	Kangpokpi, India	379-511	June-Sept.	Alkali absorption	Present study
15.	Natural grassland *(Imperata cylindrica)* stand	Kangpokpi, India	294-431	June-Sept.	Alkali absorption	Present study

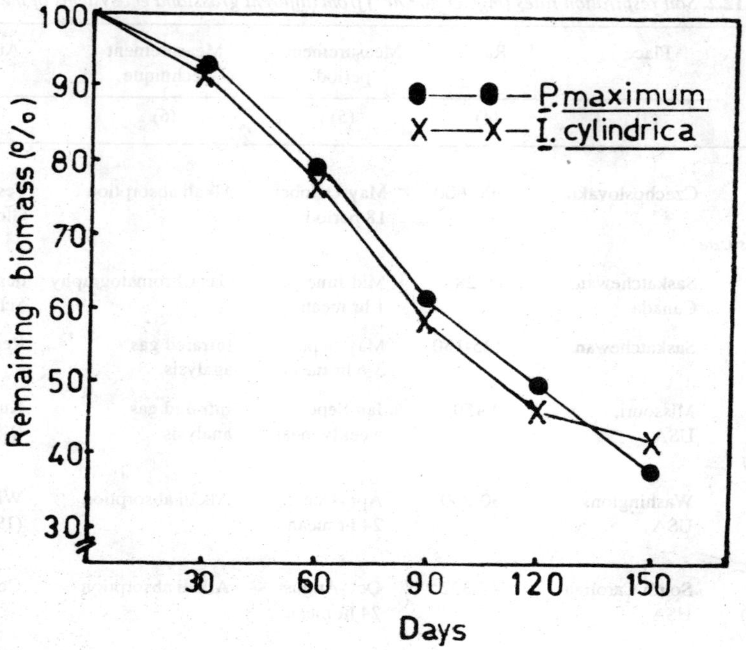

Fig. 12.2. Percentage of original biomass remaining after various months of decomposition in *P. maximum* and I. *cylindrica.*

Litter decomposition

The changes in dry weight loss, and total content of the nutrients of shoot, rhizome and root combined together are expressed as percentages of the initial values remaining at the sampling time in all the replicates. In *P. maximum,* 92.8 per cent of plant material was remained in the first 30 days and subsequently decreased upto 37.7 per cent on the termination of the experiment, whereas, *I. Cylindrica* 92.6 per cent remained in the first 30 days and 42.6 per cent on the termination of 150 days (Fig. 12.2) Decomposition of *P. maximum* and *I. cylindrica* were 62.28 per cent and 57.4 per cent at the end of 150 days respectively. However, the decomposition was slow in the initial months and maximum in the month of August in both the species.

The initial nitrogen content in *P. maximum* and *I. cylindrica* was recorded to be 0.029 and 0.013 g per cent respectively and increased consistently upto August in both the species with exception of *P. maximum* in which it decreased slightly in July at the end of 120 days. Phosphorus content decreased initially in the both species but increased subsequently to maximum value after 60 days and thereafter declined to the lowest value on termination of experiment after 120 days. In both the species the lignin content decreased consistently in the plant litter in subsequent months. At the end of 120 days percentage of lignin was lowest in both the specie (Fig. 12.3).

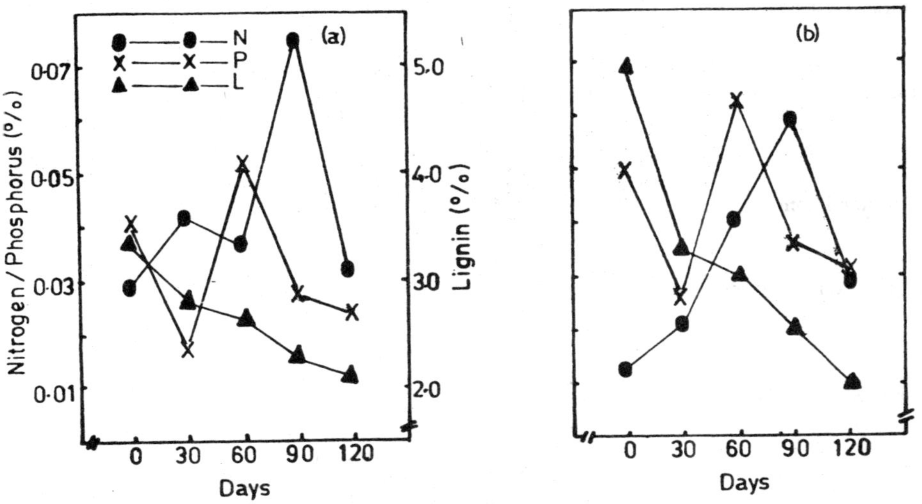

Fig. 12.3. Various percentage of nitrogen, phosphorus and lignin remaining after different days of decomposition.
(a) *P. maximum*
(b) *I. cylindrica;*
N = Nitrogen,
P = Phosphorus and
L = Lignin

The plant litter having high lignin contents shows low rate of decomposition and high initial nitrogen content showed a faster rate of decomposition (Laishram and Yadava, 1988; Teuban, 1991). However, Schaefer *et al.*, (1985) reported no relationship between rate of mass loss and initial lignin and nitrogen content in various species from a North American desert. Besides, the rate of decomposition of litter was influenced by various factors like chemical composition of litter, environmental conditions (Howard and Howard, 1979; Steinberger, Shmida and Whitford, 1990; Conteaux *et al.*, 1991). In this study, the rate of decomposition was recorded to be slow in the *I. cylindrica* as compared to *P. maximum*. This may be due to the higher lignin content (4.95 per cent) and there is evidence that the initial lignin content as a more influential factor than the initial nitrogen content, in determining the rate of litter decomposition. In the present investigation, the value of the decomposition rate constant (K) was recorded to be 0.903 in *P. maximum* and 0.875 in *I. cylindrica* (Table 12.3). The decomposition constant (K) varied from 0.36 to 2.87 in different grasslands of the world and the data reported in this present study are in the range of the other world grasslands.

TABLE 12.3. *Decomposition constant (K) for different grasslands of the world*

SI. No.	Grassland types	Place	Decay constant (K)	Author(s)
1.	Tall grass prairie	Missouri, USA	0.72	Koelling and Kucera (1965)
2.	Grassland communities	WN Dakota, USA	0.36-0.51	Redmann (1975)
3.	Agropyron trachycaulum	Central Utah, USA	0.36	Bleak (1970)
4.	Lupinus alpestris	Central Utah, USA	0.64	Bleak (1970)
5.	Bromus inermis	Central Utah, USA	0.51	Bleak (1970)
6.	Garss fall	USSA	0.36	Zlotin (1970)
Indian Grasslands				
7.	Mixed grass	Kurukshetra, India	2.87	Gupta and Singh (1981)
8.	Mixed grass	Cherrapunji, India	1.98	Subhash and Ramakrishnan (1988)
9.	Cultivated grassland *(Panicum maximum)* stand	Kangpokpi, India	0.903	Present study
10.	Natural grassland *(Imperata cylindrica)* stand	Kangpokpi, India	0.875	Present study

REFERENCES

Anderson, J.M. and Ingram, J.S.I. 1989. *Tropical Soil biology and fertility : a handbook of methods.* Published by CAB International, Wallingford Oxon Ox 10 8DE, UK.

Anderson, J.M. and Swift, M.J. 1983. *Decomposition in Tropical forest.* pp. 287-299. In : S.L. Sutton, T.C. Whitmore and A.C. Chadwick (Editors), Tropical Rain forest; Ecology and Management. Blackwell, Oxford.

Basu, S. and Pati, D.P. 1991. Soil respiration in relation to microbial population and biomass in a grassland soil from Orissa. *J.Ecobiol.* **3**(4) : 267-273.

Behera, N. and Pati, D.P. 1986. Carbon budget of a protected tropical grassland with reference to primary production and total soil respiration. *Rev. Ecol. Biol. Sol.* **23** (2) : 167-181.

Beyer, L. 1991. Intersite Characterization and Variability of soil respiration in different arable and forest soils. *Biol. Fertil. Soils.* **12** : 122-126.

Bleak, A.T. 1970. Disappearance of plant material under a winter snow cover. *Ecology* **49** :142-145.

Buyanovski, G.A., Wanger, G.H. and Gantzer, C.G. 1986. Soil respiration in a winter wheat ecosystem. *Soil Sci. Soc. Am. J.* **50** : 338-344.

Carlyle, J.C. and Than, U.B.A. 1988. Abiotic controls of soil respiration beneath on eighteen-year old *Pinus radiata* stand in South-Eastern Australia. *J. Ecology* **76** : 652-662.

Chapman, S.B. 1979. Some interelationship between soil and root respiration in lowland Calluna heathland in Southern England. *J. Ecology* **67** : 1-20.

Coleman, D.C. 1973. Compartmental analysis of total soil respiration : an explanatory study. *Oikos* **24** : 361-366.

Couteaux Marie - Madeleine, Marianne Mousseau, Marie-Louise Celerier and Pierre Bottner 1991. Increased atmospheric CO_2 and litter quality : decomposition of sweet chestnut leaf litter with animal food webs of different complexities. *Oikos.* **61** : 54-64.

de Jong, E. and Schappert, H.J.V. 1972. Calculation of soil respiration and activity from CO_2 profiles in the soil. *Soil Sci.* **113** : 328-338.

Edwards, N.T. 1991. Root and soil respiration responses to ozone in *Pinus taeda* L. seedlings, *New Phytol.* **118** : 315-321.

Fahey, T.J., Stevens, P.A. Hornung, M. and Rowland, P. 1991. Decomposition and nutrient release from logging residue following conventional harvest of Sitka spruce in North Wales. *Forestry* **64** (3) : 289-301.

Fisher, F.M., Freckman, D.W. and Whitford, W.G. 1990. Decomposition and soil nitrogen availability in Chihuahuan desert field microcosms. *Soil Biol. Biochem.* **22** (2) : 241-249.

Grahammer, K., Jawson, M.D. and Skopp, J. 1991. Day and night soil respiration from grassland. *Soil Biol. Biochem.* **23** : 77-82.

Gupta, S.R. and Singh, J.S. 1981a. The effect of plant species, weather variables and chemical composition of plant material on decompostion in a tropical grassland. *Plant and soil.* **59** : 99-117.

Gupta, S.R. and Singh, J.S. 1981b. Soil respiration in a tropical grassland. *Soil Biol. Biochem.* **13** : 261-268.

Herman, W.A., Mc Gill, W.B. and Dormaar, J.F. 1977. Effects of initial chemical composition on decomposition of roots of three grass species, Can. *J. Soil Sci.* **57** : 205-215.

Hingane, L.S. 1991. Some aspects of carbon dioxide exchange between atmosphere and Indian plant biota, *Climatic change.* **18** : 425-435.

Howard, P.J.A. and Howard, D.M. 1979. Respiration of decomposing litter in relation to temperature and moisture. *Oikos.* **33** : 457-465.

Kucera, C.L. and Kirkham, D.L. 1971. Soil respiration studies in tall grass prairie. *Ecology* **48** : 536-541.

Koelling, M.R. and Kucera, C.L. 1965. Production and turnover relationship in native tall grass prairie. *Iowa State J. Sci.* **39** : 387-392.

Laishram, I. Davi. 1989. Litter decomposition and soil respiration in a forest ecosystem at Shiroy hills, Manipur, *Ph.D. Thesis* Manipur University, Imphal, India.

Laishram, I.D. and Yadava, P.S. 1988. Lignin and nitrogen in the decomposition of leaf litter in a sub-tropical forest ecosystem at Shiroy hills in North-Eastern India. *Plant and Soil.* **106** : 59-63.

Medina, E. and Zelwer, M. 1972. *Soil respiration in tropical plant communities,* pp. 245-269. In : GOLLEY P.M. and F.B. GOLLEY (Eds.) : Tropical ecology with an emphasis on organic production. University of Georgia, Athens.

Miller, F.C. 1959. *Plant Physiology,* Mc Graw Hill Book Company. Inc. London.

Mitchell, D.T., Coley, P.G.F., Webb, S. and Allsopp, N. 1986. Litterfall and decomposition processes in the coastal Fynbos vegetation, South-Western Cape, South Africa, *J.Ecol.* **74** : 977-993.

Nicholas G. Aumen, Peter J. Bottomley and Stan V. Gregory 1985. Impact of nitrogen and phosphorus on 14_c Lignocellulose decomposition processes in the Coastal Fynbos vegetation. South-Western Cape, South Africa. *J.Ecol.* **74** : 977-993.

Olson, J.E. 1963. Energy storage and balance of producers and decomposition in ecological systems. *Ecology* **44** : 322-331.

Redmann, R.E. 1974. Photosynthesis, plant respiration, and soil respiration measured with control environment chambers in the field : III. *Soil respiration.* pp. 37. *Can. Comm. for IBP, Matador Project.* Tech. Rep. No. 60.

Redmann, R.E. 1975. Production ecology of grassland plant communities in Western North Dakota. *Ecol. Manage.* **45** : 83-106.

Rochette, P., Desjardins, R.L. and Pattey, E. 1991. Spatial and temporal variability of soil respiration in agricultural fields. *Can. J. Soil Sci.* **71** : 189-196.

Schaefer, D., Steinberger, Y. and Whiteford, W.G. 1985. The failure of nitrogen and lignin control of decomposition in a North American desert. *Decologia* **65** : 382-386.

Schulze, E. 1967. Soil respiration of tropical vegetation types. *Ecology* **46** : 652-653.

Shanks, R.E. and Olson, J.S. 1961. First year breakdown of leaf litter in Southern Applachian forests. *Science* **134** : 194-195.

Singh, J.S. and Gupta, S.R. 1977. Plant decomposition and soil respiration in terrestrial ecosystems. *Bot. Rev.* **43** (3) : 450-528.

Smucker, A.J.M., Safir, G.R. 1986. *Root and soil microbial interactions which influence the availability of photoassimilate carbon in the rhizosphere.* pp. 203-244. In : mittchell M.J. Nakas JP (eds) Microfloral and faunal interactions in natural and agro-ecosystems. Nijhoff and Junk Publishers, Dordrecht.

Steinberger, Y., Shmida, A. and Whiteford, W.G. 1990. Decomposition along a rainfall gradient in the Judean desert, Israel. *Oecologia* **82** : 322-324.

Subhash C. Ram and Ramakrishnan, P.S. 1988. Litter production and decomposition patterns in seral grasslands at Cherrapunji in North-eastern India. Pedobiologia **32** : 65-76.

Tesarova, M. and Glosev, J. 1972. *Soil respiration in a moist meadow plant community.* In : Ecosystem study on grassland Biome in Czekoslovakia, ed. by M. Rychnovska. IBP/PT-pp. Rep. No. 2 Brno, Czekoslovakia.

Teuben, A. 1991. Nutrient availability and interactions between soil arthropods and microorganisms during decomposition of Coniferous litter : a mesocosm study. *Biol. Fertil. Soils.* **10** : 256-266.

Valerie, A., Orchard and Cork, F.J. 1983. Relationship betwen soil respiration and soil moisture. *Soil Biol. Biochem.* **15** (4) 447-453.

Weber, M.G. 1985. Forest soil respiration in eastern Ortarid jack pine ecosystems. *Can. J. For. Res.* **15** : 1069-1073.

Webber, M.G. 1990. Forest soil respiration after cutting and burning in Immature Aspen ecosystems. *For. Ecol. Manage.* **31** : 1-14.

Wildung, R.E., Garland, T.R. and Buschborn, R.L. 1975. The interdependence effects of soil temperature and water content on soil respiration rate and plant root decomposition in arid grassland soils. *Soil Bios. Biochem.* **7** : 611-632.

Witkamp, M. and Frank, M.I. 1969. Evolution of CO_2 from litter, humur, and subsoil of a pine stand. *Pedobiologia* **9** : 358-365.

Yadava, P.S. and Kakati, L.N. 1990. Soil respiration in a tropical grassland ecosystem during post monsoon period. *J. Curr. Biosci.* **7** (3) : 98-101.

Zlotin, R.T. 1970. In vertebrate animals as a factor of the biological turnover. IV. Collogu. Int. de la Faune dae Sol Digon Paris **194** : 455-462.

Chapter 13

Ecobiology of Marine Cyanobacteria

G. Subramanian AND N. Thajuddin

National Facility of Marine Cyanobacteria, Bharathidasan University, Tiruchirapalli-620 024, (India)

Abstract. Cyanobacteria are considered to be an ideal biotechnological tool because of its importance as feed, medicine and fertilizer and its role in pollution abatement. This necessitates the study of ecobiological properties of cyanobacteria with respect to different hydrobiological parameters. The exploratory study conducted in the Gulf of Manner region of Indian coasts reports 130 cyanobacterial species, of which 26 species are categorized as constant species, 37 as seasonal species and the rest are rare or ephemeral species. The study indicates occurrence of *Trichodesmium* blooms affecting cyanobacterial planktonic diversity. Heterocystous forms are observed to be less in number in comparison to nonheterocystous forms.

Key words : *Cyanobacteria, Distribution, Seasonal variation, Marine Ecosystem.*

Introduction

Cyanobacteria also referred as Blue-Green algae, comprise a unique group of organisms whose distribution around the world is rivailed only by bacteria. These were considered algae because of their microscopic morphology, pigmentation and oxygen evolving photosynthesis. This group along with several other microalgae were treated with contempt as 'scum', since it was believed that these organisms were only of nuisance value-fouling water bodies and causing diseases and toxemia. Ecologists however, started giving them respect as primary producers, in all aquatic ecosystems, and realized that without them no animal population including fishes and prawn could exist. With the advent of electron microscopy and several other sophisticated instruments and techniques, it was realized that the so called 'blue-green algae' are really gram-negative bacteria in their cell structure and chemistry but form a connecting link with algae by their pigment composition and photosynthetic mechanism. These along with bacteria and actinomycetes constitute a group called 'Prokaryotes' where as the rest of the living organisms, irrespective of their plant or animal status, belong to another major group termed 'Eukaryotes'. The discovery about the prokaryotic nature of cyanobacteria coupled with the discovery that a number of these organisms are unique in not only having a trophic independence for carbon but also nitrogen, evoked considerable interest on these organisms around the world.

Of the total estimated area of 510 million sq.km of the earth about 70.68 per cent is occupied by oceans. However, of the total photosynthetic productivity of 555.2 billion tons dry weight/ year of the earth only 35.4 per cent is contributed by the oceans (Bassham 1975). Kesava Das and Desai (1988) while reviewing the recent progress in ocean resource research have quoted an eminent oceanographer as saying "the most competent forecasts available today indicate that the remainder of the 20th century will find people everywhere in the world forced to concentrate upon seven great quests and these are the search for living room, for water, for food, for climate, for minerals, for energy and for knowledge itself. In every one of these areas, the sea is providing a crucial element, so crucial that it is difficult to escape the conclusion that the new age into which we are moving is not only the age of the atom, the electron and space, it is also the new age of the sea".

Cyanobacteria, till recently in the oblivion uncared and unrecognized, have shot into fame and popularity owing to a host of their innate properties that make them ideal organisms for use in a variety of ways to meet our needs and to promise us a bright future. Among these potential biotechnological tools, it is the fresh water forms which had carried the attention of interested workers all along. Now, people have started realizing the scope available to harness the marine forms and to make them perform exacting tasks in varied areas such as matriculate (Mitsui 1975 and Mitsui *et al.* 1981), food, feed (Mitsui 1979, 1980a and 1980b, Venkataraman 1983, Venkataraman and Becker 1985, Srinivasan 1992 and Sundararaman 1992) fuel (Mitsui 1978; 1978a 1980, 1980a and 1980b, Prabaharan and Subramanian 1991 and Prabaharan 1992), fertilizer (Venkataraman and Kaushik 1980), medicine (Mynderse *et al.* 1977, Mitsui 1980a and 1980b, Gerwick *et al.* 1989, Gustafson 1989 and Sivonen *et al.* 1989), industry (Mitsui 1978, Venkataraman 1983, Gaffray and Tel-or 1985 and Tredici *et al.* 1986) and combating pollution (Gale and Wixon 1979, Uma and Subramanian 1990, Subramanian 1991 and Thangeswaran 1991), since the usefulness of cyanobacteria for these purposes has been established.

The basic and fundamental requirement for initiating marine cyanobacterial biotechnology is to first numerate the available natural cyanobacterial wealth and to understand their ecobiological properties. For instance, a strain which is very highly seasonal and restricted in this distribution may pose problems, if found the year mass cultivation is attempted, while a strain that is both versatile as well as constant in its distribution may be more easily amenable to round the year cultivation. Another very useful study would be the understanding of the causative factors of the phenomenon of blooming by certain strains. This would facilitate artificial induction of blooms and rapid mass cultivation.

Desikachary (1959) suggested that probably 20 per cent of all known cyanobacteria occur in saline situations and a majority of them are truly marine. Van Baalen (1961) reported that the cyanobacteria form an important class of marine vegetation. Very little work appears to have been done to understand the cyanobacterial flora of marine environments of our country (Iyengar and Desikachary 1944 and Anand *et al.* 1986). This would become amply clear if one surveys the available literature on Indian marine cyanobacterial flora. There have been a number of sporadic and casual reports on the occurrence of cyanobacteria in marine environments from India such as :

Cleve (1901)	Report on planktons of Indian Ocean
Biswas (1926)	Reporting 85 Taxa from saline environments
Iyengar (1927)	Reporting *Trichodesmium* from Kurusadai Island
Biswas (1927)	Cyanobacterial flora from the salt lakes of Calcutta
Boergesson (1935)	Record of 5 species from the coast of Bombay
Dixit (1936)	Reporting 10 species from sea water near Bombay
Gopala Iyer and Sankara Menon (1936)	32 marine species from Gulf of Mannar
Venkataraman (1957)	Reporting 19 species from Cape Comorin
Anand (1982)	Reporting 9 taxa from Madras coast
Ramachandran (1982)	10 taxa from Porto Novo
Anand and Venkatesan (1985)	17 species of 7 genera from Madras coast
Anand *et al.* (1986)	25 taxa from Madras and south Arcot District
Santra *et al.* (1988)	Reporting 67 taxa from various saline habitats of West Bengal
Subramanian *et al.* (1989) Thajuddin & Subramanian (1991 & 1992)	Reporting a total of 163 species from 48 genera and belonging to 14 families of cyanobacteria from the southern east cast of India

Apart from the paucity of detailed surveys of Marine cyanobacteria from India, even globally most of the reports are restricted not only in area but also in the number of species recorded. There are certain reports such as:

Zaneveld (1972)	Recorded 27 species from American coast
Oah (1973)	Found 25 per cent of the total flora of the saline waters in Chad area to be comprised of cyanobacteria
Moore (1974)	Reported on the epilithic and epiphytic forms from Southern Baffin Island
Pulz (1975)	Studied the marine littoral algae in the Gulf of Baabo

Gonzalez and Para (1975) & Para and Gonzalez (1975)	Reported 161 species of cyanobacteria belonging to 41 genera from the coast of Chille
Potts (1980)	Studied the distribution and taxonomic diversity of cyanobacteria from marine coastal environments of Sinai Peninsula
Marshal (1981)	Reported 16 species of cyanobacteria belonging to 10 genera off the south eastern coast of United States
Khoja (1987)	Reported 19 species of cyanobacteria from Red sea coast of Saudi Arabia

In addition there are several similar reports with smaller number of species from different parts of the world including brines (Hoof & Fremy 1933, Umazaki 1961, Stewart and Pugh 1963, Weffer 1967, Whitton 1968, Petrov 1974, Helferin 1974, John and Lawson 1977, Ralph 1977, Sage and Sullivan 1978, Helferin 1974, John and Lawson 1977, Ralph 1977, Sage and Sullivan 1978, Bosson 1979, Mc Carthy and Carpenter 1979 and Aleem 1978 and 1980).

Hydrography

Fogg (1973) suggested that the ability of cyanobacteria to grow in sea water was presumably related to a preference for alkaline conditions. The tolerance to saline situations shown generally by these algae and the resistance which many species show towards osmotic shock, variation in temperatures and reducing conditions of the environment were particularly suitable for their existence in a variety of intertidal habitats (Stewart 1977, Megue 1977 and Pillai 1954). Desikachary (1959) attributed the capability of cyanobacteria to thrive in halotolerant conditions to the peculiar characteristics of their protoplasm especially plasmolysis and deplamolysis characteristics which are quite distinct from what is seen in other plant cells. Petrov (1974) reported that the most significant factors which affect the vertical and horizontal distribution of benthic forms were transparency, movement of water, temperature, chemical composition and grazing. The substratum ecology as well as specific process of water movement (current and the aperiodical fluctuations of water level) and the characteristic factors which influence the manifold ecological background of the marine littoral vegetation. Dunston and Hosford (1977) also quoted that the factors such as water column stability, temperature and nutrient levels favour bloom formation. Ramachandran (1982) reported that the difference in the composition of algal mats observed in the mangrove area were probably due to the effect of salinity. Sanders (1979) reported that the salinity differences play a major role in deciding the algal assemblages.

The water samples from the Bay of Bengal, Palk Strait, Palk Bay and Gulf of Mannar (Thajuddin and Subramanian 1992) showed variations atleast in some hydrographical parameters (Table 13.1). In general, the water from the Gulf of Mannar was richer in nutrients compared with the other regions. The levels of macroelements such as carbon, nitrogen and

TABLE 13.1. *Analysis of Sea Water from Different Regions*

No.	Parameters	Bay of Bengal	Palk Strait	Palk Bay	Gulf of Mannar
1.	pH	8.4	7.83	7.91	8.21
2.	Salinity ppt.	33.53	33.35	35.66	36.43
3.	Conductivity (at 25°C in millimhos)	33	31	32	32
4.	Sodium Meg/l	417.39	369	452.17	488.15
5.	Potassium Meg/l	9.23	7.17	5.12	10.3
6.	Carbonate mg/l	14	16	12	12
7.	Bicarbonate mg/l	32	30	44	48
8.	Chloride g/l	18.5	18.4	19.69	20.17
9.	Calcium g/l	0.380	0.360	0.360	0.400
10.	Magnesium g/l	1.31	1.366	1.322	1.56
11.	Iron (ppm)	0.37	0.29	0.52	1.67
12.	Zinc (ppm)	0.20	0.09	0.16	0.16
13.	Manganese (ppm)	0.01	0.27	0.24	0.32
14.	Copper (ppm)	0.30	0.26	0.29	0.55
15.	Nitrate µg/l	3500	3000	2200	9600
16.	Nitrite µg/l	0.55	1.11	12.99	14.44
17.	Ammonia µg/l	12	34	23	44
18.	Inorganic phosphate µg/l	4	3	6	4.4
19.	Total phosphorus µg/l	6	8	15	11.1
20.	Sulphate µg/l	1620	1460	1628	2030
21.	Sulphide µg/l	40	65	10	130

TABLE 13.2. *pH and Salinity of Water Samples from Different Areas in each Region of the Southern East Coast*

		Actual sea		Stagnated sea waters		Back waters		Salt pans	
No.	Region	pH	Salinity ppt.	pH	Salinity ppt.	pH	Salinity ppt.	pH	Salinity ppt.
1.	Bay of Bengal	8.4	33.53	8.41	35.5	8.14	32.5	8.25	>40
2.	Palk Strait	7.83	33.35	8.6	38	8.12	32.5	8.2	>50
3.	Palk Bay	7.91	35.66	8.1	36.5	7.89	34.0	8.01	>40
4.	Gulf of Mannar upto Indian Ocean	8.21	36.43	8.5	37	8.0	33.5	8.2	>50

TABLE 13.3. *Region and Area-wise Distribution of Cyanobacterial Flora (Number of species and genera)*

No.	Region	Actual sea	Stagnated sea waters	Back waters	Salt pans
1.	Bay of Bengal	11/7	32/14	21/11	9/8
2.	Palk Strait	13/6	27/14	21/7	20/9
3.	Palk Bay	59/26	47/14	35/13	32/12
4.	Gulf of Mannar upto Indian Ocean	145/45	47/21	42/18	37/15
5.	In the entire survey area	154/46	77/29	58/22	50/19

sulphur as well as microelements such as iron, copper and manganese were significantly higher in this region. This was probably due to a much higher upwelling activity in adjoining Indian Ocean areas (Rochford 1962, El-Sayed and Jitts 1973) compared to the poor upwelling reported in the Bay of Bengal region especially off Madras and Karaikal Coast (Murthy and Varadachary, 1968, Murthy and Reddy 1968). In all regions the pH and salinity showed identical trends. The salinity in back waters was relatively lower than in the open sea, while it was higher in stagnant seawater (Table 13.2). In most areas, the pH was also higher than the open sea. In general the Gulf of Mannar showed higher salinity and pH than the other regions. The higher nutrient content of the water in the Gulf of Mannar region correlated well with the maximum diversity in the cyanobacterial flora (Tables 13.1 and 13.3)

Thajuddin (1991) and Thajuddin and Subramanian (1992) reported different types of shores in the southern east coast of India. The shores of the Bay of Bengal region were mostly sandy, while those of the Palk Strait region were of different types. Along the great lagoons the shores were clayey and particularly in Adhirampattinam area, they were completely muddy and almost inaccessible. The shores towards the Palk Bay from Palk Strait were of coarse sand and rich in organic matter due to the deposition of dense layers of sea grass over vast stretches. However, most part of the Palk Bay area revealed a sandy shore with scattered boulders and masses of coral rocks. In the Gulf of Mannar region the shores varied from place to place. In the north and south coast of Mandapam the shores were rocky due to extensive sand stone. In the Kurusadai island, a portion of the sea shore was essentially of coral reefs. The shores towards the Cape Comorin were essentially sandy with intermittent rocky spots while in Cape Comorin it was partly rocky and partly sandy.

Different types of substrata which occur in the intertidal areas influenced greatly the availability on benthic cyanobacterial flora. Thajuddin and Subramanian (1992) observed the shore in the Bay of Bengal region was essentially sandy and therefore there were only 11 species of cyanobacteria whereas in the Gulf of Mannar region where at several places the shores were rocky (sand stones) or coral reefs, a strikingly large number of 145 species, of course including both benthic as well as planktonic forms were recorded (Table 13.3). Khan (1969) observed the supralittoral and intertidal rock pools along the coast of Oahu, Hawaii, there are two main substrata present, lava and limestone, the former supports the greater number of species of cyanobacteria. Certain species are restricted to one or the other type of the rock types, for instance *Brachytrichia quoyi, Hydrocoleum cantharidosum, H. lyngbyaceum* are found only on lava. Little (1973) and Renaut *et al.* (1975) reported that the cyanobacterial community is most abundant on soft, porous rocks such as sand stones. A hard substratum in not essential for the growth of cyanobacterial mats (Black 1933). Cyanobacteria are apparently always present as epiliths, chasmoliths and endoliths including sometimes as a discrete well-developed cryptoendolith layer (Golubic *et al.* 1981), such as that described by Moul (1957), with Oscillatoria *nigro-viridis, Lyngbya confervoides* and *Phormidium valderianum* at a depth of 6 mm inside the rock.

Thajuddin and Subramanian (1992) reported that in the case of sandy shores, the cyanobacterial population was very poor due to the rough tides, absence of substratum nutrient content of water etc. In the same areas, the stagnated sea water ponds and puddles showed rich populations of cyanobacteria growing in the form of thick mats, because these habitat remain undisturbed for relatively long periods, although once established they may play an important role in binding the sand. *Lyngbya confervoides, L. martensiana, Microcoleus chthonoplasts, M. acutissimus, Oscillatoria salina, O. tenuis, Spirulina subsalsa, S. labyrinthiformis, Pseudanabaena schemidlei* are predominant in these mats. The type and flourishing composition of these communities appear to be determined by a variety of factors. Among these are the chemical composition of the sand, especially whether quartzite or carbonate minerals, nature of the underlying sediments, influence of any water movement though the sand and perhaps also atmospheric humidity and rainfall (Whitton and Potts 1982).

Backwaters and mangrove forests are common along many shores in the tropics and subtropics particularly where there is protection from severe wave action and where major rivers enter the sea (Chapman 1977). Ramachandran (1982) pointed out the mangrove environment supported abundant growth of benthic cyanobacteria and showed the richness and diversity of the species. The probable reasons that could be attributed to this are the organic rich muddy substratum the relatively stagnant shallow water conditions, sheltered nature (hence reduced water movement); and optimum saline conditions (15-30 ppt) which all ensured a good growth of these forms. Thajuddin and Subramanian (1992) reported as many as 58 species of cyanobacterial forms belonging to 22 genera (Table 13.3) in the backwaters and mangrove habitats of the southern east coast of India. The members of family Oscillatoriacea viz. *Spirulina subsalsa, Oscillatoria salina, O.tenuis, O.chalybea, O. nigrobiridis, O. laete-virens, O. cortiana, Phormidium tenue, P. valderianum, P. angustissimum, Lyngbya confervoides, L. martensiana, Microcoleus chthonoplasts* and *M. acutissimus* were the dominant flora in these habitats.

Wood (1965) suggested that the abundance of cyanobacteria in mangrove situations were noticeably associated with the presence of anaerobic mud. He also showed rapid growth of marine cyanobacteria in media with low redox potentials. The decomposition of organic matter in the mangrove environment promotes bacterial production of hydrogen sulfide resulting in a reducing environment. Stewart and Pearson (1970) showed that trace amount of sulfide and the associated reducing conditions were favourable for growth and nitrogen fixing activity of cyanobacteria.

Thajuddin and Subramanian (1992) reported 50 species from 19 genera (Table 13.3) in salt pans with salinity of over 50ppt. Although several workers had earlier reported the occurrence of cyanobacteria in salt pans and salt lakes with very high salinity (Biswas 1926, Woronichin 1929, Aggarwal 1951, Subbaramaiah 1972, Anand and Venkatesan 1985 and Anand *et al.* 1986) none of them have recorded such a large number from hypersaline conditions. Thajuddin and Subramanian (1992a) further identified 28 species of 11 genera belonging to 4 families in the hypersaline environments (salt pans) of Cuddalore (South Arcot district) and Marakkanam (Chengalpattu district) of Tamil Nadu. It was interesting to note that non-heterocystous forms in general, the species belonging to the family Oscillatoriaceae in a particular were dominant in these environments and found growing in 340 ppt. salinity. A striking feature of cyanobacteria in hypersaline environments was that all of them were non-heterocystous. Although explanations for the absence of heterocystous forms in hypersaline environments have been offered in terms of high sulfide content and its toxicity to heterocystous forms (Hoyslew and Pearson 1979), this was subsequently disputed and the anaerobic conditions prevailing in the dark in these environments were believed to exclude the heterocystous forms (Oren and Shilo, 1979, Stal *et al.* 1985). No satisfactory explanation supported with experimental evidence has so far been made. Hof and Fremy (1933) who studied the flora of salt works divided the algae into two physiological groups which they termed halotolerant and halophilic. Halotolerant species are those which cannot grow at sodium chloride concentrations above 3M (175.5 ppt.) and *Calothrix scopularum* is a good example of this type. In laboratory culture, it can withstand but cannot grow at a salinity of 106 ppt. (Stewart 1964a). Other examples are some salt works forms such as *Microcoleus chthonoplasts* and *Lyngbya estuearii.* Halophilic species such as *Spirulina subsalsa,* which can grow at salt concentrations above 3 M occurs commonly in salt works and this ability is probably helped by their prokaryotic cellular organization and absence of large sap vacuoles. Feldmann (1951) divides the algae into euryhaline and stenohaline,-those living in brines as hyperhaline and those of brackish water as hypohaline and he considers that *Microcoleus chthonoplasts* as euryhaline.

The remarkable adaptability of cyanobacteria is, however, well known (Hof and Fremy 1933, Desikachary 1959, Van Baalen 1962, Fogg *et al.* 1973, and Carr and Whitton 1982). *Phormidium valderianum* BDU 30501 was shown to grow in salinities ranging from 0-99ppt. (Prabaharan 1988 and Prabaharan and Subramanian 1990). Thajuddin and Subramanian (1992) reported that as many as 75 of the species belonging to 30 genera have originally been reported from freshwater sources by Geitler (1932), Biswas (1949), Smith (1950), Prescott (1951), Desikachary (1959), Tilden (1968), Hum and Wicks (1980) and Anand (1989). It therefore becomes clear that it is difficult to strictly segregate most of the cyanobacteria into marine and freshwater species as can be done with other algal forms.

Seasonal Variation

Most of the cyanobacteria posses broad tolerance to different environments and are able to survive through most of the seasons. However there may be only one or two occasions during the year when they might be really dominant. This phenomenon known as seasonal variation or succession had to be studied in considerable detail to answer the question of interaction of these species with environment, i.e. whether the occurrence was related to physical changes produced by the organisms themselves (Braarud, 1966). Further since the quality of the water (or any environment) changes anytime in a day and every day in the season and every season in a year, it is necessary to study the biological quality and the potential productivity of a particular environment both seasonally and spatially. Thus the study of seasonal and regional variations of planktonic cyanobacteria forms a fundamental aspect in their ecology. According to Kobletz-Mishke *et al.* (1970) the total annual net primary production by phytoplankton of the world ocean amounts to 15-18 × 10^9 t of carbon.

Most of the earlier work on ecobiology had been at the total planktonic level, primarily aimed at productivity and the other conventional ecobiological aspects (Srinivasan 1946, Dutt *et al.* 1954, Subramanian and Sarma 1960, Ryther *et al.* 1966, Hulbert 1967, Jitts 1969, Devassy and Gopinathan 1970, Marshal 1971, Mefferet and Krambeck 1977, Tilman *et al.* 1982, Smith and Atkinson 1984, Shim *et al.* 1985, Shim and Lee 1987, Maddock *et al.* 1989, Subramaniyam and Bhavanarayana 1989). Studies concentrating essentially on cyanobacteria had been very limited (Borongersma-Sandars 1957, Prabhu *et al.* 1965, Ramamurthy, 1968, Qasim 1970, Kristiansen 1972, Jayaraman 1972, Fogg 1973 and 1982, Sakthivel and Haridas 1974, Dunston and Hosford 1977, Devassy *et al.* 1978 and 1979, Nair *et al.* 1980, Krempin and Sullivan 1981, Marshall 1981, Ramachandran 1982, El-Hag and Fogg 1986, Baker 1987, Devassy 1987, Thajuddin and Subramanian 1990 and Thajuddin *et al.* 1991.

Thajuddin and Subramanian (1990) and Thajuddin (1991) had made a modest attempt to understand certain aspects of ecobiology of natural marine cyanobacterial population in the Gulf of Mannar region of the Indian coast specifically in Mandapam-Rameswaram-Kurusadai island region. This region was chosen for the study based on the results of the survey of the southern east coast of India which showed this area to be (1) an ideal location with a very suitable substratum for benthic cyanobacteria, (2) high in nutrient content of favouring both benthic and planktonic forms and (3) with maximum diversity of cyanobacterial species in the entire survey area.

Six meteorological and 21 physicochemical parameters were considered throughout the two year (1980-1989) period at monthly intervals. All the meteorological parameters probably with the exception of rainfall showed fairly identical patterns during both 1988 and 1989 (Thajuddin, 1991). The rainfall pattern also was fairly similar in both years except that in 1989, a much higher rainfall could be recorded during the monsoon. The physicochemical parameters on the other hand, varied widely during the two years of study. However, among them pH, salinity and chloride level patterns were fairly similar during both years. Magnesium level also showed a fairly similar pattern in both years except in April and May 1989 when it was drastically low. A number of parameters such as carbonate, bicarbonate, sodium, potassium, ammonia, nitrate, nitrite, total phosphorus, sulphate, suphide and manganese exhibited highly variable patterns of availability during the two years. Strikingly, contrasting patterns in the levels of dissolved oxygen, conductivity, calcium, inorganic phosphate, iron, zinc and copper were observed during the two years (Thajuddin, 1991).

In view of the enormity of meteorological and physicochemical factors which influence the biological parameters, it has become extremely difficult to establish strict correlations between them. When such an attempt, was made the problem became more confounding due to differences between their levels in the two successive years. However, certain meaningful conclusions could still be drawn based on the observations. As far as the total phytoplanktonic population was concerned, atmospheric temperature, salinity, bicarbonate, sodium, magnesium, ammonia, inorganic phosphate, total phosphorus and iron had a distinct positive correlation in both years since high levels of these factors coincided with high phytoplanktonic population levels (Thajuddin, 1991). Similarly, wind speed, relative humidity, sea level pressure, dissolved oxygen content and sulphide levels had a distinct negative correlation in both the years. All the remaining factors did not reveal any correlation or at best had some correlation only in one of the two years on the total phytoplankton level. It is not difficult to explain the positive correlation between the factors mentioned above and the phytoplankton levels, in view of available literature. The negative correlation of dissolved oxygen level is rather unexpected, but similar observations of lowering of O_2 concentration in North Sea with large algal blooms have been made by Rosenberg (1985) and Howarth (1988). The negative correlation with sulphide levels and phytoplanktons could well be due to the toxic nature of sulphides to these organisms (Stal *et al.* 1985). The different biological parameters such as total plankton, phytoplankton, zooplankton, cyanobacteria plankton, productivity, etc. exhibited clear cut interrelationships (Thajuddin 1991).

Many of the earlier workers have concentrated their attention only on a limited number of factors in a particular season or area and have indicated the importance of those factors on phytoplankton levels. For example Jayaraman and Shesappa (1957) positively correlated phosphate level with phytoplankton content : Confield (1983), Confield *et al.* (1985) and Sakamoto *et al.* (1989), found both nitrogen and phosphorus influencing phytoplankton abundance; Martin *et al.* (1990) correlated the low phytoplankton level and lack of blooming in Antarctic waters to iron deficiency; Braarud (1951), Pearse and Gunter (1957), Provasoli *et al.* (1958) Smayda (1958) and Munda (1978) have all emphasized the role of salinity in phytoplankton maintenance; Shim and Lee (1987) found water temperature as the most significant factor for the distribution of phytoplanktons.

Several parameters such as temperature, salinity, levels of bicarbonate, sodium, chloride, ammonia, inorganic phosphate, iron and copper had a positive correlation and relative humidity, rainfall, sea level pressure, dissolved oxygen, nitrate and nitrite had a negative correlation to the zooplankton populations (Thajuddin 1991). It appears that the influence of all these factors could be essentially indirect since almost all these factors have identical effects on phytoplankton population and there is an excellent positive correlation between the phytoplankton and zooplankton levels. Studies on zooplankton in marine environments especially in Indian Ocean with reference to temperature, salinity and dissolved oxygen have been made earlier (Bogorov and Vinogradov 1961, Tranter and Kerr 1969, Rao 1973, Vannucci and Navas 1973). The scope of Zooplankton study in our investigation was only to find its influence on productivity, which is linked not only to total phytoplankton levels but also to the percentage and composition of cyanobacterial plankton.

Out of 130 species of cyanobacteria recorded in the ecobiological study area both as benthic and planktonic forms, only 26 species were found growing throughout year during both 1988 and 1989 (Thajuddin 1991). These have been termed "constant species". Thirty seven species were found to grow only between 3-11 months in an year during both years. These were termed "seasonal" species. Eight species were recorded during one or two months in both years and were termed rare species. In addition some species were not only "rare" but also "ephimeral" in the sense that they could be recorded only in one year and not in the other (Thajuddin 1991).

Cyanobacteria based on the density of growth and distribution could be categorized as "most abundant", "abundant" and "sparse". Only 11 out of 130 species were "most abundant" in both years while 2 species namely *Aphanocapsa litteralis and Hydrocoleum cantharidosam* were "most abundant" in 1988 but only "abundent" in 1989 and one species namely *Oscillatoria brevis* was "Most abundant" in 1989 but "abundant in 1988 (Thajuddin 1991). As many as 36 species were "abundant" in both years while 33 species were "sparse" in both years. It was exceedingly interesting to find 9 species were not only "constant" (present throughout year in both years) but also "most abundant" during both years (Thajuddin 1991). Although there are reports on the seasonality and distribution of cyanobacteria in marine habitats (Chapman 1956 and 1971, Ramamurthy 1968, Kristiansen 1972, Dunstan and Hosford 1977, Krempin and Sullivan 1981, Ramachandran 1982, Baker 1987) as study of this kind looking into these aspects of marine cyanobacteria has so far not been reported.

Another interesting observation was that cyanobacteria constituted an important component of total plankton was revealed by an increasing percentage of cyanobacteria corresponding to the increase in total phytoplankton in both years. The cyanobacterial planktonic diversity was adversely influenced by the *Trichodesmium* bloom which occurred during both the years of study. Among the cyanobacterial blooms the marine *Trichodesmium* bloom has attracted the attention of several workers around the world (Sournia 1968, Van Baalen and Brown 1969, Taylor *et al.* 1973, Maruma and Asasoka 1974, Carpenter and Mc Carthy 1975, Saino and Hattori 1978, Mc Carthy and Carpenter 1979, Rueter *et al.* 1979, Bryceson and Fay 1981, Ohki and Fujita 1982, Carpenter 1983, Ohki *et al.* 1986, Rueter 1988). It was generally believed that the *Trichodesmium* bloom appeared and disappeared totally quite infrequently. The occurrence of *Trichodesmium* bloom around Indian coast in general and the study area in particular was reported even much earlier (Chacko 1942 and 1950 Chidambaram and Unny 1944, Chacko and Mahadeven 1956, Qasim, 1970 and 1972, Sakthivel and Haridas 1974, Prabhu *et al.* 1965; Ramamurthy and Krishnamurthy 1968, Jayaraman 1972, Devassy *et al.* 1979, Nair *et al.* 1980, Devassy 1987, Thajuddin and Subramanian 1990 and Thajuddin, 1991).

Trichodesmium was found as a component of the phytoplankton almost throughout the year especially in 1989 with its number reaching the peak during certain months constituting the bloom. It appeared as though there were minor and major blooms routinely every year during certain conditions (Thajuddin and Subramanian 1990). Generally the major blooms which occurred once (1988) or twice (1989) prevailed for longer durations while the minor blooms were short lived. The morphological and some of the physiological and biochemical features of *Trichodesmium erythraeum* during the bloom where characterised with a view to put the bloom into some practical views (Thajuddin *et al.* 1991). Certain physicochemical parameters such as phosphorus, ammonia and iron were correlated with *Trichodesmium* blooms. (Qasim 1970 and 1972, Ramachandran 1982, McCarthy and Carpenter 1983, Devassy 1987, Shim and Lee 1987, Rueter 1988 and Martin *et al.*, 1990) In our study, there was a good positive correlation between the temperature, salinity and the levels of chloride, ammonia, inorganic phosphate, total phosphorus and iron during both the years, whereas a typical negative correlation between the wind speed, rainfall, see level pressure and dissolved oxygen as could be expected were observed (Thajuddin, 1991). Among the various factors which seemed to influence the blooming of *Trichodesmium,* ammonia appears to be more a consequence than a cause of the bloom, because of the very high rates of nitrogen fixation observed in this organism (Thajuddin *et al.* 1991). Devassy *et al.* (1978) found a rapid release of ammonium-nitrogen into sea water towards the end *Trichodesmium* blooms. Gunderson *et al.* (1976) concluded that fixation by *Trichodesmium* and *Richelia* might contribute about 3.8 per cent of the total annual nitrogen requirement of phytoplankton in the vicinity of the Hawaiian islands. With phytoplankton in which *Richelia* was abundant Mahe *et al.* (1974) found rates suggesting that between 2 to 5 per cent or perhaps even upto 30 per cent of the daily nitrogen requirement might be provided by fixation.

When compared to the unicellular and non-heterocystous filamentous forms, the heterocystous forms were very poor in numbers and this may be due to the high levels of combined nitrogen (6-19 mg/l NO_{-3}, 0.4-42 μg/l NO_{-2} and 1-50 μg/l NH_4+) in the sea (Thajuddin and Subramanian 1990) the heterocystous endophytic cyanobacterial form *Richelia intracellularis Encountered inside various species of the diaiom, Rhizosolinia* was found eight months in the year 1988 and 6 months in the year 1989. The absence in the months was probably due to the population of *Trichodesmium* Bloom which brought down the population of other phytoplankton (Jayaraman 1972). In this period, the combined nitrogen phosphorus and the levels of micronutrients like iron and manganese were very low in the sea water. Probably these organisms are efficient scavengers of nutrients and also have the capacity to fix atmospheric nitrogen (Prabaharan and Subramanian, 1991)

Whenever there was an increased population of *Trichodesmium,* it prevented the growth of other phytoplankton including cyanobacterial forms. Chidambaram and Unny (1944) have also referred to the swarming of the Planktonic cyanobacterium *Trichodesmium erythraeum* in the same area and its effects on the local fauna. Qasim (1972) pointed out the *Trichodesmium* generally constitutes a minor component in the marine plankton, but in certain seasons, it becomes extremely abundant and far exceeds any other plant life in the sea. Jayaraman (1972)

TABLE 13.4. *Diurnal Variations in Plankton Population and Primary Production*

Time of Collection	Dissolved Oxygen mg/l (Winkler's Method)	Primary productivity (mgC/m³/hr)			Primary Productivity		Plankton (nos/l)		
		Gross Production	Net Production	Respiration	(^{14}C method) Mg C/m³/hr	Total Plankton	Zoo-Plankton	Phyto-Plankton	Cyano-bacterial forms
12.00 p.m.	5.5967	45.72	0	164.63	4.9111	8283	1233	7000	1533
3.00 a.m.	6.4197	54.88	0	109.73	3.1136	8667	750	7917	2000
6.00 a.m.	6.2551	160.02	22.86	164.63	8.7702	7067	634	6433	2250
9.00 a.m.	7.4074	91.44	68.58	27.43	38.8633	4184	400	3784	1150
12.00 noon	7.4897	228.63	91.44	164.63	28.0485	6361	3344	3017	1200
3.00 p.m.	6.9958	388.66	228.63	192.03	119.5332	17467	1033	16433	3200
6.00 p.m.	6.2551	144.30	91.44	27.43	0.9668	10533	733	9800	1133
9.00 p.m.	5.7613	45.75	0	109.76	8.6637	9067	1100	7967	800

stated that when the *Trichodesmium* bloom occurs, the other phytoplanktonic organisms are totally absent or present in very low levels. This may be due to two reasons : It may be a case of the superiority of the organism competing with other organisms for the nutrient supply from the environment or it may be an instance where the metabolic products of the *Trichodesmium* bloom are creating unfavourable conditions in the environment for the growth of the other organisms. In the coastal waters of Palk Bay and Bay of Bengal, a large number of dead animals were seen following blooms of *Trichodesmium* (Chacko 1942, John and Menon 1942, and Chidambram and Unny 1944) this was probably caused by the decay and putrefaction of the algae in rock pools and other sheltered areas. But according to Devassy (1987), The *Trichodesmium* Red tide had no harmful effect of fish and fisheries and no fish or other animal mortality was observed during his entire study period. *Trichodesmium* can of course, cause death of fish by clogging their gills and the death and decay of other animals lead to depletion of oxygen. In spite of sporadic reports of culturing this organism successfully in the laboratory (Ramamurthy 1972, Ohki and Fujita 1982, Ohki *et al.* 1986) several attempts to culture this organism manipulating the medium and growing conditions did not succeed in our laboratory.

Thajuddin and Subramanian (1992b) studied the diurnal variation of plankton and primary production in Mandapam coast, Gulf of Mannar were studied in an interval of three hours for 24 hrs (Table 13.4). Such studies on diurnal variations in marine environments have recently been carried out in different areas (Bhargava and Dwivedi 1974, Tijssen 1979 and Gouda and Panigrahy 1989). The primary production as measured by light and dark bottle method showed a fair degree of correlation with 14C method. The total planktonic population was maximal at 12 noon and the cyanobacterial as well as other phytoplankton were maximum at 3 pm. coinciding with maximum productivity by 14C method (Thajuddin, 1991 : Thajuddin and Subramanian, 1992b).

The ecobiological studies on the marine cyanobacteria of the Gulf of Mannar region has provided a wealth of information that is not only academically rewarding but also very useful in planning mass cultivation strategies oriented towards the biotechnological exploitation of these least used organisms.

Acknowledgement

The authors are grateful to the Dept. of Ocean Development and Dept. of Biotechnology, Govt. of India of their financial support and to the colleagues in National Facility for Marine Cyanobacteria for physical help and moral support during the entire period of study.

REFERENCES

Aggarwal, S.C. 1951. *The Sambhar lake salt sources.* Govt. of India Press, New Delhi p. 365.

Aleem, A.A. 1978. Contribution to the study of the marine algae of the Red Sea, I-The algae in the neighbourhood of al-Ghardaqa, Egypt (Cyanophyceae, Chlorophyta and Phaeophyta). *Bull. Fac. Sci.* King Abudul Aziz University **2** : 73-88.

Aleem, A.A. 1980. Cyanophyta from Sierra Leone (West Africa). *Bot Mar.* **23** : 59-61.

Anand, N. 1982. Studies on Marine blue-green algae. Abstr. Indian Science Congress Association 69th Session, Mysore.

Anand, N and Venkatesan N. 1985. Note on Blue-green algae from Salt pans. *Seaweed Res. Utlin.* **7** (2) : 101-103.

Anand N., Mohan E., Hopper R.S.S. and Subramanian T.D. 1986. Taxonomic studies on Blue-green algae from certain marine environment. *Sea Weed Res. Utiln.* **9** (1 & 2) : 49-56.

Anand, N. 1989. *Hand book of Blue-green algae of Rice fields of South India.* Bishen Singh Mahendra Pal Singh, Dehara Dun p. 80.

Baker, K.K. 1987. Systematics and Ecology of *Lyngbya* spp. and associated species (Cyanophyta) in a New England Salt Marsh. *J. Phycol.* **23** : 201-208.

Bardach, J.E., Rhyther J.H. and McLaurey W.O. 1972. The farming and husbandry of freshwater and marine organisms. In : *Aqua culture,* Wiley Interscience, New York, p. 868.

Basson, P.W. 1979. Marine algae of the Arabian Gulf coast of Saudi Arabia (first half) *Bot. Mar.* **22** : 47-64.

Bhargava, R.M.S. and Dwivedi, S.N. 1974. Durnal variation in phytoplankton pigments in Zuari estuary, Goa, *Indian J. Marine Sciences* **3** : 142-145.

Biswas, K. 1926. Flora of the salt lake, Calcutta. *J. Dept. Sci.* Calcutta Univ. **8** : 1-57.

Biswas, K. 1949. *Common fresh and brackish water algal flora of India & Burma (Part I & II).* Records of the Botanical Survey of India, Manager of Publications, New Delhi.

Black, M. 1933. The algal sediments of Andros Island, Bahamas. *Phil. Trans. R. Soc. Lond.* B. **222** : 165-192.

Boergesen, F. 1935. A list of Marine Algae from Bombay.. Kg. Danske Videnskabernes Selskabs., *Biol. Medd.* XII, 2.

Bogorov, V.G. and Vinogradov M.E. 1961. Some features plankton biomass distribution in the surface water in the Indian Ocean during the winter of 1959/60. *Okeanol. Issled.* **4** : 72-75.

Braarud, T. 1951. Salinity as on ecological factor in marine phytoplankton. *Physiol. Plant.* **4** : 28-34.

Brasurd, 1966. Salinity as an ecological factor. In : *Marine Biology* (Ed) C.H. Oppenheiman, New York Academy of Sciences. p. 146.

Brongersma-Sanders M. 1957. Mass mortality in the sea. In : *Treatise on marine ecology and paleoecolgy. Vol. I. Ecology* (Ed.) J.W. Hedgepeth, Geol. Soc. Am. Mem. **67** : 941-1010.

Bryceson, I. and Fay, P. 1981. Nitrogen fixation in *Oscillatoria (Trichodesmium) erythreaum* in relation to bundle formation and trichome differentiation. *Marine Biology.* **61** : 159-166.

Carpenter E.J. 1983. Nitrogen fixation by marine *Oscillatoria (Trichodesmium)* in the World's Oceans. In : *Nitrogen in the Marine Environment* (Ed.) Carpenter E.J. & Capone D.G., Academic Press, New York. pp. 65-103.

Carpenter E.J. and Mc Carthy J.J. 1975. Nitrogen fixation and uptake of combined nitrogenous nutrients by *Oscillatoria (Trichodesmium) thiebautii* in Western Sargasso Sea. *Limnol. Ocenogr.* **20** : 389-401.

Carr, N.G. and Whitton, B.A. 1982. *The biology of Cyanobacteria.* Blackwell Scientific Publications. p. 688.

Chacko, P.I. 1942. An unusual incidence of mortality of marine fauna. *Curr. Sci.* **11** : 404.

Chacko, P.I. 1950. Marine Plankton from waters around the Kurusadai island, *Proc. Indian Acad. Sci. B.* **31** : 162-174.

Chacko, P.I. and Mahadeven, S. 1956. *Swarming of Trichodesmium erythaeum in waters around kurusadai island. Gulf of Mannar.* Fish station reports and Year Book. April 1954-March 1955. Government of Madras.

Chapman, V.J. 1956. The Marine Algae of New Zealand. Part. I Myxophyceae and Chlorophyceae. *J. Lim. Soc. Bot.* **55** : 333-501.

Chapman, V.J. 1971. The Marine Algae of Fuji. *Rev. Algol.* **2** : 164-171.

Chapman, V.J. 1977. Wet coastal Eco Systems In : *Ecosystems of the world.* Chapman V.J. (Ed.) Elsevier, Amsterdam. pp. 1-29.

Chidambaram K. and Unny M. 1944. Note on the swarming of the planktonic alga *Trichodesmium erythreaeum* in the Pamban area and its effects on the fauna. *Curr. Sci.* **13** : 263.

Cleve, P.T. 1901. Plankton from the Indian Ocean and the Malay Archipelago. *K. Sv. Vet. Akad. Handl.* XXXV, **5** : pp. 1-58.

Confield, D.E. JR. 1983. Prediction of chlorophyll a concentrations in Florida lakes. The importance of phosphorus and nitrogen. *Water Resour. Bull.* **19** : 255-262.

Confield, D.E., Linda S.B. and Hodgson, L.M. 1985. Chlorophyll-biomass-nutrient relationships for natural assemblages of Florida phytoplankton. *Water Resour. Bull.* pp. 381-391.

Desikachary, T.V. 1959. *Cyanophyta.* Indian Council of Agricultural Research. New Delhi p. 686.

Devassy, V.P. 1987. *Trichodesmium* red tides in the Arabian Sea-contributions in marine sciences. Dr. S.Z. Qasim Sastyabdapurthi felicitation volume, pp. 61-66.

Devassy, V.P. Bhattathiri P.M.A. and Qasim S.Z. 1978. Trichodesmium phenomenon. *Indian J. Mar. Sci.* **7** : 168-186.

Devassy, V.P., Bhattathiri P.M.A. and Qasim S.Z. 1979. Succession of Organisms following *Trichodesmium* phenomenon. *Indian J. Mar. Sci.* **8** : 89-98.

Devassy, V.P. and Gopinathan, C.K. 1970. Hydrobiological features of the Kerala Backwaters during premonsoon and monsoon months. *Fish Tech.* **7** : 190-194.

Dixit, S.C. 1936. The myxophyceae of the Bombay Presidency, India-I. *Proc. Indian Acad. Sci., B, III* pp. 93-106.

Dunston, N.M. and Hosford, J. 1977. The distribution of planktonic blue green algae related to the hydrography of the Georgia bight. *Bull. Mar. Sci.* **27** : 824-829.

Dutt, M., Malhotra, J.C. and Bose, B.B. 1954. *Hydrology and seasonal fluctuations of the planktons in the Hoogly Estuary.* In : Proc. Indo-Pacific Fisheries Council Symp. Marine and Freshwater Plankton. pp. 35-47.

El-Hag, A.G.D. and Fogg, G.E.. 1986. The distribution of coccoid blue-green algae (Cyanobacteria) in the Menai Straits and the Irish Sea. *Br. Phycol. J.* **21** : 45-54.

El-Sayed, S.Z. and Jitts, H.R. 1973. Phytoplankton Production in the southern eastern Indian ocean. In : *The Biology of the Indian Ocean* (Ed.) B. Zeitzschel and S.A. Gerlach S.A. Chapman & Hall Limited London. pp. 131-142.

Feldmann,J. 1951. Ecology of marine algae in Smith, Manual of Physiology Chronica Botanica, Waltham. Mess. 313-334.

Fogg, G.E. 1973. Physiology and ecology of Marine blue-green algae. In : *The blue-green algae.* N.G. Carr and Whitton, B.A. (Ed.) Academic Press, London.

Fogg, G.E. 1982. Marine Plankton. In : *The biology of cyanobacteria.* N.G. Carr and B.A. Witton (Eds.) Blackwell Scientific Publications, London. p. 688.

Fogg, G.E., Stewart, W.D.P., Fay, P. and Walsby, A.E. 1973. *The Blue Green Algae.* Academic Press. p. 459.

Gaarder. T. and Gran, N.H. 1927. Investigation of the production of plankton in the Oslo fjord. repp p-v. Cons Int. Explor. Mer. **42** : 1-48.

Gabbay, R. and Tel-Or E. 1985. Cyanobacterial biomass production in saline media. *Plant and Soil.* **89** : 107-116.

Gale, N.L. and Wixson, B.G. 1979. Removal of Heavy Metals from Industrial effluents by Algae. In : *Development in Industrial Microbiology.* A Publication of the Society for Industrial Microbiology, **20** : 259-273.

Geitler, L. 1932. Cyanophyceae. In : *L Rabenhorst's Kryptogamen flora.* Akademische Verlagsgesellschaft Leipzig. p. 1196.

Gerwick, W.H., Mrozek, C., Moghaddam, M.F. and Agarwal, S.K. 1989. Novel cytotoxic peptides from the tropical marine cyanobacterium *Hormothomnion enteronorphoides* l. Discovery, isolation and initial chemical and biological characterization of the hormothomnins from wild and cultured material. *Experientia,* **45** : 115-121.

Golubic, S., Freidman, E.I. and Schenider, J. 1981. The lithobiontic ecological niche with special reference to microorganisms. *J. Sed. Petrol.* **51** : 475-472.

Gonzalez, M. and Parra, O.O. 1975. Marine blue green algae of Chile.1. Blue green algae of the intertidal zone of Bay of Conception. *Gayana Bot.* **31** : 3-64.

Gouda, R. and Panigrahy, R.C. 1989. Diurnal variation of phytoplankton in Rushikulya estuary, east coast of India. *Indian J. Marine Sciences.* **18** : 246-250.

Gopala Iyer, R. and Sankara Menon, K.S. 1936. Plankton records for the year 1929-30. *Journ. Madras Univ., VIII* (1), pp. 97-139.

Gundersen, K.R., Corbin, J.S., Hanson, C.L., Hanson M.L., Hanson R.B., Russel D.J., Stollar A. and Yamada O. 1976. Structure and biological dynamics of the oligotrophic ocean photic zone of the Hawaiian Islands. *Pacif. Sci.* **30** : 45-68.

Gustafson, K.R., Cardellina II, J.H., Fuller, R.W., Weislow, O.S., Kiser, R.F. and Snader, K.M. 1989. AIDS-Antiviral sulfolipids from cyanobacteria (blue-green algae). *J. of the National Cancer Institute.* **81** : 1254.

Halferin, D.R. 1974. Marine cyanophyta from Puerto Deseado (Bruince of Santacruz Argentina). *De Physis.* **33** : 465-482.

Hof, T. and Fremy, P. 1933. On Myxophyceae Living in strong brines. *Rev. Trav. Bot. Neerland.* **30** : 140-162.

Howarth, R.W. 1988. Nutrient limitation of net primary production in Marine ecosystems. *Ann. Rev.* **19** : 89-110.

Howsley, R. and Pearson, H.W. 1979. pH dependent sulphide toxicity to oxygenic photosynthesis in cyanobacteria, *FEMS Microbiol. Lett.* **6** : 287-292.

Hulbert, E.M. 1967. Some notes on the phytoplankton of the southeastern coast of the United States. *Bull. Mar. Sci.* **17** : 330-337.

Hum, H.G. and Wicks, S.R. 1980. *Introduction and Guide to the Marine Blue-Green Algae.* John Wiley & Sons, New York, p. 194.

Iyengar, M.O.P. 1927. Kurusadai Island Flora. Bulletin of Madras Govt. Museum, New series, Nat. Hist. Section. I, (1), pp. 1-4.

Iyengar, M.O.P. and Desikachary, T.V. 1944. A systematic account of some marine Myxophyceae of the South Indian Coast. *J. Madras Univ.* **16** : 37-68.

Jayarman, R. 1972. *On the occurrence of blooms of blue-green algae and the associated oceanographic conditions in the Northern Indian ocean.* In : Proceedings of the symposium on taxonomy and biology of blue-green algae (Ed.) T.V. Desikachary, University of Madras, pp. 428-432.

Jayaraman, R. and Seshappa, G. 1957. Phosphorus cycle in the sea with particular reference to tropical inshore waters. *Proc. Indian Acad. Sci.* **46** : 110-125.

Jitts, H.R. 1969. Seasonal variations in the Indian Ocean along 110°E. IV. Primary Production. *Aust. J. Mar. Freshwat. Res.* **20** : 65-76.

John, C.C. and Menon, M.A.S. 1942. On the alleged inhibitory effect of *Trichodesmium. Curr. Sci.* **11** : 407.

John D.M. and Lawson, G.W. 1977. The Marine algal flora of the Sierra Leone Peninsula. *Bot. Mar.* **20** : 127-135.

Kesava Das, V. and Desai, B.N. 1988. In search of Ocean Resources-Recent progress. In : "*The service of the nation CSIR's contribution to National Development*". Council of Scientific and Industrial Research, New Delhi.

Khoaja, T.M. 1987. New records of marine algae for the Red Sea coast of Saudi Arabia. *Botanica Marina* **30** : 167-176.

Koblentz-Mishke, O.J., Volkovinsky, V.V. and K anova, J.G. 1970. Plankton primary production of the World Ocean. In : *Scientific exploration of the south Pacific.* W.S. Wooster (Ed.)., Washington, Nat, Acad. Sci. pp. 183-193.

Krempin D.W. and Sullivan, C.W. 1981. The seasonal abundance, vertical distribution and relative microbial biomass of chroococcoid cyanobacteria at a station in southern California coastal wasters. *Can. J. Microbiol.* **27** : 1341-1344.

Kristiansen, A. 1972. A seasonal study of the marine algal vegetation in Tuborg Harbour, the south Denmark. *Bot. Tidsskr.* **67** : 201-204.

Little, M.G. 1973. The zonation of marine supralittoral blue-green algae. *Br. Phycol. J.* **8** : 47-50.

Mackareth, F.J.H. 1963. *Some methods of water analysis for limnologist.* Scient. Publ. Fresh water Biol. Ass. 21.

Mague, T.H. 1977. Ecological aspects of dinitrogen fixation by blue-green algae. In : *A treatise on Dinitrogen Fixation, Section IV.* Hardy R.W.F. Gibson, A.H. (Eds.). Wiley, New York. pp. 85-140.

Mague, T.H. Weare N.M. and Holm-Hansen, O. 1974. Nitrogen fixation in the North Pacific ocean. *Mar. Biol.*, **24** : 109-119.

Marshal H.G. 1971. Composition of the phytoplankton off the south eastern coast of the United States. *Bull. Mar. Sci.* **21** : 806-825.

Marshal, H.G. 1981. Occurrence of blue green algae (cyanophyta) in the phytoplankton of the south eastern coast of the United States. *J. Plankton. Res.* **3** : 163-166.

Martin, J.H., Michael Gordon, R. and Fitzwater, S.E. 1990. Iron in Antarctic waters. *Nature.* **345** : 156-158.

Maruma, R. and Asaoka, O. 1974. *Trichodesmium* in the East China Sea. 1. Distribution of *Trichodesmium thiebautii* Gomont during 1961-1967. *J. Oceanogr. Soc. Japan.* **30** : 298-303.

Mc Carthy, J.J. and Carpenter, E.J. 1979. *Oscillatoria (Trichodesmium) thiebautii* (cyanophyta) in the Central North Atlantic Ocean. *J. Phycol.* **15** : 75-82.

Mc Carthy, J.J. and Carpenter, E.J. 1983. Nitrogen cycling in near surface waters of open ocean. In : *Nitrogen in the marine environment.* (Eds) E.J. Carpenter and D.G. Capone, Academic Press, New York. pp. 487-512.

Mefferet M.E. and Krambeck, H.J. 1977. Planktonic blue green algae of the *Oscilltoria redeki* group. *Arch. Hydrobiol.* **79** : 149-171.

Mitsui, A. 1975. Multiple utilization of tropical and subtropical marine photosynthetic organisms. In : *Proceedings of the III International Ocean Development Conference.* Vol. III. Seino Printing Co. Tokyo. pp. 13-29.

Mitsui, A 1978. Bio-solar hydrogen production. In : *"Hydrogen Energy System"*. Vol. 3 (Eds). Veziroglu, T.N. and Seifritz, W. Pergamon Press, Oxford, pp. 1267-1291.

Mitsui, A. 1978 a. *Marine photosynthetic micro-organisms as potential energy sources : Research on nitrogen fixation and hydrogen production.* Proceedings of the 5th International Ocean Development Conference, Vol. I. IODC organising Committee, Tokyo. pp. 29-52.

Mitsui, A. 1979. Biosaline research : The use of photosynthetic marine organisms in food and feed production. In : *The Biosaline concept. An approach to utilization of unexploited resources.* (Eds). Hollaender A., Allen J.C., Epstein E., San Pietro, A. and Zaborsky, O. Plenum Press, New York. pp. 177-215.

Mitsui, A. 1980. Salt water based biosolar hydrogen production. In : *Abstracts of the symposium on "The production and storage of hydrogen"*. (Ed.) Yonezawa T., Kyoto University, Japan pp. 4-5.

Mitsui, A. 1980a. *Salt water based biological solar energy conversion for fuel, chemicals, fertilizers, food and feed.* Proceedings of Bio-energy council, Washington, D.C. pp. 486-491.

Mitsui, A. 1980b. *Nitrogen fixation and hydrogen production by marine blue-green algae used as a source of food, feed and chemicals.* (Eds.) Gnanam A. and Kahn J. Proceedings of the International symposium on Biological Application of Solar Energy, Madurai, India. pp. 109-115.

Mitsui, A., Murray R., Entenmann, B., Miyazawa, K. and Polk, E. 1981. Utilization of marine blue-green algae and macroalgae in warm water mariculture. In : *"Biosaline Research. A look to the future"*, (Ed.) San Pietro A. Plenum Press, New York. pp. 215-225.

Moore, J.W. 1974. Benthic algae of Southern Baffin Island III Epilithic and epiphytic communities. *J. Phycol.* **10** : 456-462.

Moul, E.T. 1975. Preliminary report on the flora of Onotoa Atoll, Gilbert Islands. *Atoll. Res. Bull.* **57** : 1-48.

Munda, I.M. 1978. Salinity dependent distribution of benthic algae in estuarine areas of Iceland Fjords. *Bot. Mar.* **21** : 451-468.

Murty, C.S. and Varadachari, V.V.R. 1968. Upwelling along the east coast of India. *Bull. Natl. Inst. Sci. India,* **38** : 80-86.

Murty, P.S.N. and Reddy C.V.G. 1968. Distribution of phosphorus in the marine sediments off the east coast of India. *Bull.Natl. Inst. Sci.* India. **38** : 405-410.

Mynderse, J.S. and Moore, R.E. 1977. Antileukemia activity in the Oscillatoriaceae : Isolation of debromoaplysiatoxin from *Lyngbya. Science.* pp. 538-539.

Nair, V.S.,Devassy, V.P. and Qasim, S.Z. 1980. Zooplankton and *Trichodesmium* Phenomenon. *Indian Journal of Marine Sciences.* **9** : 1-6.

Oah, I.A. 973. Algal flora of saline wasters in Kanen area (Chad) 2nd part. *Bondy.* **7** : 25-54.

Ohki, K. and Fujita, Y. 1982. Laboratory culture of the pelagic blue-green alga *Trichodesmium thiebautii* : conditions for unialgal culture. *Mar. Ecol. prog. Sci.* **7** : 185-190.

Ohki, K., Rueter, J.G. and Fujita Y. 1986. Cultures of the pelagic cyanophytes *Trichodesmium erythraeum* and *T. thiebautii* in synthetic medium. *Mar. Biol. (Berlin).* **91** : 9-13.

Oren, A. and Shilo, M 1979. Anaerobic heterotrophic dark metabolism in the cyanobacterium *Oscillatoria limnetica* : sulfur respiration and lactate fermentation. *Arch. Microbiol.* **122** : 77-84.

Parra, O.O. and Gonzalez 1976. Bibliographic review and distribution of the blue green algae of Chile (excluding the Antarctic continent). *Gayana Bot.* **32** : 3-55.

Patrov, Y.E. 1974. The distribution of sea benthic algae as a result of a system of factors. *Bot.Zh.* **59** ; 955-966.

Pearse, A.S. and Gunter, G. 1957. Salinity In : *Treatise on marine ecology and Paleoecology* (Ed.) J.W. Hedgpth Geol. Soc. Am. Mem. **67** : 129-159.

Petrov Y.E. 1974. The distribution of sea benthic algae as a result of a system of fctors. *Bot. Zh.* **59** : 955-996.

Pillai V.K. 1954. Growth requirements of a halophilic blue-green algae *Phormidium tenue* (Menegh.) *Indian J. Fish.* **1** : 130-144.

Potts, M. 1980. blue-green algae (Cyanophyta) in marine coastal environments of the Sinai Peninsula; distribution, zonation, stratification and taxonomic diversity. *Phycologia.* **19** : 60-73.

Prabaharan, D. 1988. Partial characterization of the marine cyanobacterium *Phormidium valderianum* BDU 30501 for biotechnological purposes. *M. Phil Thesis.* Bharathidasan University, Tiruchirapalli.

Prabaharan, D. 1992. Hydrogen production by marine cyanobacteria. *Ph. D. Thesis.* Bharathidasan Univeristy, Tiruchirapalli.

Prabaharan, D. and Subramainan, G. 1990. *Partial characterization of marine cyanobacterium Phormidum valderianum BDU 30501.* Proc. Natl. Symp. Cyanobacterial Nitrogen Fixation, IARI, New Delhi. pp. 477-483.

Prabaharan, D. and Subramanian, G. 1991. *Hydrogen production by marine cyanobacteria.* Abstracts 31st Annual conference of the Association of Microbiologists of India. (Eds.). Kannaiyan S. and Ramasamy K., Tamil Nadu Agricultural University, Coimbatore. p. 137.

Prabhu, M.S, Ramamurthy, S., Kuthalingam, M.D.K. and Dhulkhed M.H. 1965. On an unusual swarming of the planktonic blue-green algae *Trichodesmium* spp. of Mangalore. *Curr. Sci.* **34** : 95.

Prescott, G.W. 1951. Algae of the western great lakes area. *Brown Dubuque lowa.* p. 977.

Provasoli, L. 1958. Nutrition ecology of protozoa and algae. *Ann Rev. Microbiol.* **12** : 279-308.

Pulz, O. 1975. Marine littoral algae in the Gulf of Baabo (Tang King in the DR of Vietnam). *Naturwissenschaft. Miche. Rahie.* **22** : 715-719.

Qasim, S.Z. 1970. Some characteristics of a *Trichodesmium* bloom in the Laccadives. *Deep-sea Research.* **17** : 655-660.

Qasim, S.Z. 1972. *Some observations on Trichodesmium blooms.* Proceedings of the symposium on taxonomy and biology of blue-green algae. (Ed.s) Desikachary, T.V., University of Madras, pp. 433-438.

Ralph, R.D. 1977. The Myxophyceae of the marshes of southern Delaware. *Chesapeake. Sci.* **18** : 208-221.

Ramachandran, S. 1982. Studies on blue green algae from Porto Novo (Lat. 11° 30' N Long. 79° 50'E) marine environments. *Ph. D. Thesis,* Annamalai University.

Ramamurthy, V.D. 1968. Studies on phytoplankton in Porto Novo waters with reference to *Trichodesmium erythraeum* (Ehr.) *Ph. D. Thesis.* Annamalai University.

Ramamurthy, V.D. 1972. *On the culture of the marine blue-green algae Trichodesmium erythraeum.* Proceedings of the symposium on taxonomy and biology of blue-green algae. (Ed.) T.V. Desikachary. University of Madras. pp. 425-427.

Ramamurthy, V.D. and Krishnamurthy, S. 1968. Nitrogen fixation by the blue-green algae, *Trichodesmium erythraeum* (Ehr.) *Curr. Sci.* **37** : 21-22.

Rao, T.S.S. 1973. Zooplankton studies in the Indian Ocean. In : *Biology of Indian Ocean* (Ed.) Zeitzschel B. Berlin-Heidelberg-New York : Springer pp. 243-255.

Rochford, D.J. 1962. Hydrology of the Indian Ocean. II. The surface waters of the southeast Indian Ocean and Arafura Sea in the spring and summer. *Aust. J. Mar. Fresh Wat. Res.* **13** : 226-251.

Rosenberg, R. 1985. Eutrophication-the future marine coastal nuisance? *Mar. Poll. Bull.* **16** : 227-231.

Rueter, J.G. 1988. Iron stimulation of photosynthesis and nitrogen fixation in *Anabaena* 7120 and *Trichodesmium* (Cyanophyceae). *J. Phycol.* **24** : 249-254.

Rueter, J.G., Mc Carthy, J.J. and Carpenter, F.J. 1979. The toxic effect of copper on *Oscillatoria (Trichodesmium) theibautii. Limnol. Oceanogr.* **24** : 258-562.

Ryther, J.H., Hall J.R., Pease A.K., Bakun A and Jones M.M. 1966. Primary production in relation to the chemistry and hydrography for western Indian Ocean. *Limnol. Oceanogr.* **11** : 371-380.

Sage, W.W. and Sullivan, M.J. 1978. Distribution of blue green algae in Mississipi Gulf coast salf marsh. *J. Phycol.* **14** : 333-337.

Saino, T. and Hattori, A. 1978. Diel variation in nitrogen fixation by a marine blue-green algae, *Trichodesmium theibauti. Deep Sea Res.* **25** : 1259-1263.

Sakamoto, M., Hayashi H., Otsuki A., Aoyama K., Watanabe Y., Hanazato T., Iwakuma T. and Yasuno M. 1989. Role of Bottom sediments in sustaining plankton production in a Lake ecosystem. *Ecol. Res.* **4** : 1-16.

Sakthivel, M. and Haridas, P. 1974. Synchronization on the occurrence of *Trichodesmium* bloom and swarming of *Creseis acicula* Rang (Petropoda) *Penilia avirostris* Dana (Cladocera) in the area off Cochin Mahasagar-*Bulletin of the National Institute of Oceanography.* **7** : 61-67.

Sanders, J.G. (1979). The importance of salinity in determining the morphology and composition of algal mats. *Bot. Mar.* **22** : 159-162.

Santra, S.C., Pal, U.C. Maity, H. and Bandyopadhyaya, G. 1988. Blue-green algae in saline habitats of West Bengal. A systematic account. *Biol. Mem.* **14** : 81-108.

Shim, J.H. and Lee W.H. 1987. Distribution of phytoplankton species and associated environmental factors in the southwestern waters of the East Sea (Sea of Japan), Korea : Aeronical correlation analysis. *J. Oceanol. Soc. Kor.* **22** : 34-42.

Shim J.H., Lee, W.H. and Bak S.Y. 1985. Studies on the plankton in the South western waters of the East sea (Sea of Japan) (II) phytoplankton standing crop, nanofraction and primary production. *J. Oceanol. Soc. Kor.* **20** : 37-54.

Sivonen, K., Kononen, K., Carmichael, W.W. Dahlem, A.M. Rinehart, K.L., Kiriranta, J. and Niemela, S.I. 1989. Occurrence of the hepatotoxic cyanobacterium *Nodularia spumigena* in the Baltic sea and structure of the toxin. *Appl. and Envir. Micro.* **55** : 1990-1995.

Smayda, T.J. 1958. Biogeographic studies of marine Phytoplankton. *Oikos* **9** : 158-191.

Smith, G.M. 1950. *The freshwater algae of United States.* Cyanophyceae. Mc-Graw Hill book company. New York pp. 539-604.

Smith, S.V. and Attinson, M.J. 1984. Phosphorus limitation of net production in a confined aquatic net production in a confined aquatic ecosystem. *nature.* **307** : 626-27.

Sournia, A. 1968. L. Cyanophyceae *Oscillatoria (Trichodesmium)*, dans le Mozambique. *Nova Hedwigia.* **15** : 1-12.

Srinivasan K.G. 1992. Feasibility studies of phycoflour and colourant production from marine cyanobacteria. *M. Sc. Disseratation,* Bharathidasan University, Tiruchirapalli,

Srinivasan, K.S. 1946. Ecology and seasonal succession of the marine algae at Mahabalipuram (Seven Pagodas) near Madras. *J. Indian. Bot. Soc. M.O.P. Iyengar Comm.* pp. 267-278.

Stal, L.J., Van Gemerden H. and Krumbeine, W.E. 1985. Structure and development of abenthic marine microbial mat. *FEMS Microbiology Ecology.* **31** : 111-125.

Starmach, K. 1966. Cyanophyta-Sinice, Glaucophyta-Glaukofity Florra Sodkowodna Polski, tom 2, Warzawa. p. 800.

Stewart, W.D.P. 1964a. Nitrogen fixation by myxophyceae from marine environments. *J. Gen. Microbiol.* **36** : 415-422.

Stewart,. W.D.P. 1977. A botanical ramble among the blue-green algae. *Br. Phycol. J.* **12** : 89-115.

Stewart, W.D.P., And Pearson, H.W. 1970. Effects of aerobic and anaerobic conditions on growth and metabolism of blue-green algae. *Proc. R. Soc. Lond.* B. **175** : 293-311.

Stewart, W.D.P. and Pugh, G.F.J. 1963.blue-green algae of a developing salt marsh. *J. Mar. Biol. Ass. U.K.* **43** : 309-317.

Subbaramaiah K. 1972. The biology of the blue-green algae of the sambhar lake salt works. In : *Taxonomy and biology of Blue-green algae* (Ed.) T.V. Desikachary, Madras University. pp. 439-441.

Subramanian, G. 1991. Cyanobacterial Immobilization for distillery effluent treatment. *M. Phill. dissertaiton,* Bharathidasan University. Tiruchirapalli.

Subramanian, G., Sundararaman, M,., Thajuddin, N., Prabaharan, D. and Sekar, S. 1989. Survey of cyanobacterial flora in the southern east coast of India. In : *Advances in Applied Phycology-II* (Eds.) A.C. Shukla. Today & Tomorrow's Printers & Publishers, New Delhi. 307-321.

Subramanian, R. and Sarma, A.H.V. 1960. Studies on the phytoplankton of the west coast of India III : seasonal variation of the phytoplankters and environmental factors. *Indian J. Fish.* **7** : 307-336.

Subramaniyam, M.N.V. and Bhavanarayana, P.V. 1989. Distribution and abundance of phytoplankton in Visakhapatnam harbour. *Indian J. of Marine Sciences.* **18** : 251-258.

Sundararaman, M. 1992. Intra and extracellular substances in marine cyanobacteria. *Ph.D. Thesis,* Bharathidasan University, Tiruchirapalli.

Taylor,B.F., Lee, C.C. and Bunt, J.S. 1973. Nitrogen-fixation associated with marine blue-green algae, *Trichodesmium* as measured by the acetylenereduction technique. *Arch. Microbiol.* **88** : 205-212.

Thajuddin, N. and Subramanian, G. 1990. *Cyanobacterial phytoplanktons of the Gulf of Mannar region.* Proc. Natl. Symposium on Cyanobacterial Nitrogen Fixation. NFBGA, IARI, New Delhi. pp. 457-463.

Thajuddin, N. and Subramanian, G. 1991. New report of marine cyanobacteria from the southern east coast of India. *Phykos.* **30** : 19-23.

Thajuddin, N. and Subramanian, G. 1992. Survey of cyanobacterial flora of the southern east coast of India. *Bot. Mar.* **35** : 305-314.

Thajuddin N. and Subramanian, G. 1992a. *Cyanobacterial flora from the the hypersaline environments.* Abstracts. Association of Microbiologists of India, 33rd Annual conference, Goa University, Goa.

Thajuddin, N. And Subramanian, G. 1992b. *Diurnal variation of plankton and primary production in Mandapam coast, Fulf of Mannar region.* Abstracts. Association of Microbiologists of India, 32nd Annual conference. Madurai Kamaraj University. Madurai.

Thajuddin, N. 1991. Marine cyanobacteria of the southern east coast of India. Survey and ecobiological studies. *Ph. D. Thesis.* Bharathidasan University, Tiruchirapalli.

Thajuddin, N., Sundararaman, M., Prabaharan, D., Sekar S., and Subramanian, G. 1991. *Characterization of Trichodesmium erythraeum from the southern east coast of India.* Abstracts 31st Annual conference of the Association of Microbiologists of India. (Eds.) S. Kannaiyan and Ramasamy, K, Tamil Nadu Agricultural University. Coimbatore.

Tilden, J.E. 1968. *The algae and their life relations-cyanophyceae.* Hafner publishing Co. New York pp. 48-99.

Tijessen. S.B. 1979. Diurnal oxygen rhythm and primary production in the mixed layer of the Atlantic Ocean at 20°N. *Netherlands J. sea Research* **13** (1) : 79-84.

Tilman, D., Kilhman, S.S. and Kilham, P. 1982. Phytoplankton community ecology. The role of limiting nutrients. *Ann. Rev. Ecol. Syst.* **13** : 349-372.

Tranter, D.J. and Kerr, J.D. 1969. Seasonal variations in the Indian Ocean along 11° E.W. zooplankton biomass. *Aust. J. Mar. Freshwat. Res.* **20** : 77-84.

Tredici, M.R, Papuzzo T. and Tomaselli, L. 1986. Out door mass culture of *Spirulina maxima* in sea water. *Appl. Microbiol. Biotechnol.* **24** : 47-50.

Uma L. and Subramanian, G. 1990. *Effective use of cyanobacteia in effluent treatment.* Proc. Natl. Symposium on cyanobacterial Nitrogen Fixation. NFBGA, IARI, New Delhi. pp. 437-444.

Umezaki, H.C. 1961. The marine Blue green algae of Japan. *Mem. College Agriculture,* Kyoto University **83** : 1-149.

Van Baalen, C. 1961. Vitamin B_{12} requirement of marine blue-green algae. *Science,* N.Y. **133** : 1922.

Van Baalen, C. 1962. Studies on marine blue green algae. *Bot. Mar.* **4** : 129-139.

Van Baalen, C. and Brown, R.M. 1969. The ultrastructure of the marine blue-green algae, *Trichodesmium erythreaeum* with special reference to cell wall, gas vacuoles and cylindrical bodies. *Arch. Microbiol.* **69** : 79-91.

Vannucci, M. and Navas, D. 1973. Distribution of Hydromedusae in the Indian Ocean. In : *The biology of the Indian Ocean,* (Ed.) Zeitzschel B. Berlin-Heidelberg-New York : Springer. pp. 273-281.

Venkataraman, G.S. 1957. A list of Marine myxophyceae from Cape Comorin (Kanyakumari) India. *J. Indian Bot. Soc.* **36** : 472-474.

Venkataraman, G.S. 1964. Algalization. *Phykos.* **5** : 164-174.

Venkataraman, G.S. 1977. *Blue-green algae* (A bioferitilizer for rice). Information manual from the Indian Agricultural Research Institute, New Delhi, India.

Venkataraman, G.S. and Kaushik, B.D. 1980. Save on nitrogen fertilizers by the use of algae on rice fields. *Indian farming* **30** (7) : spl. No. 1.

Venkataraman, L.V. 1983. *A monograph on Spirulina platensis-biotechnology and application.* The All India Co-ordinated project on Algae Dept. of Science and Technology, India and the Indo-German Algal Project.

Venkataraman, L.V. and Becker E.W. 1985. *Biotechnology and utilization of algae. The Indian experience.* Dept. of Science and Technology, India. Central Food Technological Research Institute, Mysore, India.

Webber, E.E. 1967. Blue green algae form Massachusetts salt marsh. *Bull.Torrey Bot. Club.* **94** : 98-106.

Wood, E.J.F. 1965. *Marine Microbial Ecology.* Chapman and Hall, London. p. 243.

Whitton, B. A. and Potts, M. 1982. Marine littoral. In : *Biology of Cyanobacterial* (Eds.) Carr, N.G. and Whitton, B.A., Blackwell Scientific Publications. pp. 515-542.

Woronichin, N.N. 1929. Survey of the vegetation of the Bitter Salt Lakes near Piatyorsk. *Comm. Princi. Bot. Garden Toxsk.* 1929. G.: 12.

Whitton, B.A. 1968. Blue-green algae from Sierra Leone. *Nova Hedwigia.* **15** : 203-204.

Zaneveld, J.S. 1972. The marine algae of the American coast between Cape May, N.J. and Cape Hatteras, N.C.I. The Cyanophyta. *Bot. Mar.* **9** : 101-128.

Chapter 14

Phosphorus in Aquatic Systems : An Overview

B.B. Jana AND **S.K. Das**

Fisheries and Limnology Laboratory, Department of Zoology, University of Kalyani, Kalyani 741235, West Bengal, India

Abstract. Phosphorus is essential for all living organisms. Geologic in origin, phosphorus is a macro-nutrient but its availability is extremely low. Phosphorus often acts as limiting element in oligotrophic system but may cause hazards in eutrophic lakes. The subject of phosphorus in aquatic system has been reviewed in terms of P-cycle, phosphorus in water and sediments, mud-water exchange, phosphorus dynamics, role of enzymes in P-cycle, phosphorus in eutrophication, nutrient loading, restoration, quantitative aspect of P-cycle, phosphorus budget and global P-cycle. As a part of restoration programme various approaches have been made to reduce P-input either from point source or from various non-point sources. A number of eutrophication models deal with P Loading and management of eutrophic water bodies.

Key word : *Aquatic System, Phosphorus, Dynamics, Budget*

Introduction

Phosphorus is absolutely necessary to all living organisms; it functions in the building of nucleic acids and in the storage and transfer of energy in cells. It is not exchangeable with other elements in biological systems (Pierrou 1976; Arson et al. 1984). Found in meteorites, rocks, soils and even in the sun's atmosphere, phosphorus is a macronutrient, but its availability is often extremely low (ng g^{-1}). The gross abundance of phosphorus on the earth surface is 0.10 to 0.12 per cent by weight. It is taken up rapidly and concentrated by living organisms; the ratio of phosphorus to other elements is considerably greater in living organisms compared to aquatic environments. Consequently, phosphorus becomes ecologically more important and often act as limiting element in the ecosystem (Hecky & Kilham 1988).

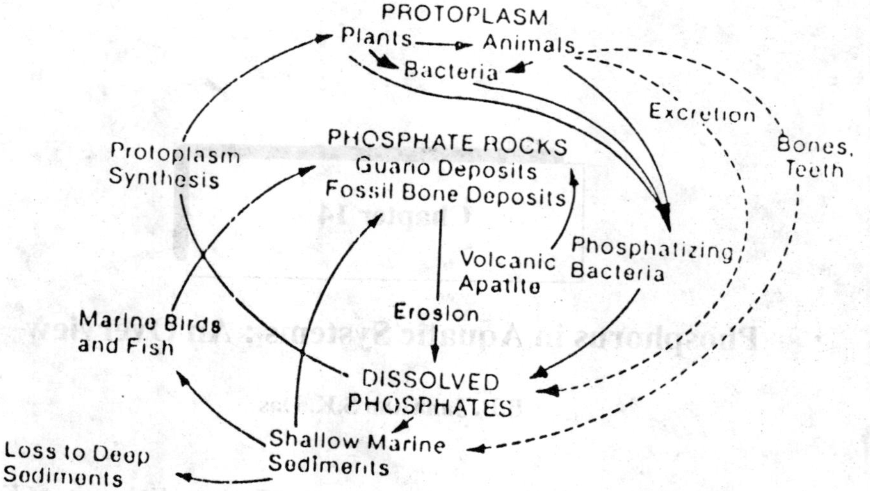

Fig. 14.1. The basic phosphorus cycle

Phosphorus cycle

In typical biogeochemical cycle, phosphorus is transferred to consumers and decomposers as organic phosphate and subsequently is made available as inorganic phosphate for recycling (Fig. 14.1); the driving force being the life process of biosphere which derive energy from the sun (Fuller 1972). Investigations have shown that biological cycling of phosphorus is rapid with extremely short turnover time (Rigler 1956). Using a radio tracer technique, Lean (1973) observed that the biological phosphorus turnover time was longer in great and deep lakes than in small and shallow lakes.

Igneous rock is regarded as the original source of phosphorus on earth. It is in the form of apatite, $Ca_5 (PO_4)^+_3$ united with either OH^-, CL^- or fluoride. The fluroapatite is the main mineral source in igneous material, but there are at least 205 minerals that contain phosphorus (Cole 1988). Subsequent weathering, probably influenced by carbonic acid, liberate much of the phosphorus into the ecosystem. Erosion of phosphorus nodule on the ocean floor is another important source of phosphorus. Biotic sources of phosphorus accumulations are : human wastes, laundry detergents, marine fishes and birds, skeletons of vertebrates, etc.

Phosphorus on liberation is redeposited and becomes in accessible to the cycle, or may be carried to the ocean. It is stated that the flow from continents to seas is the major movement of phosphorus (Cole 1988). Most phosphates become lost to the phosphorus cycle by sedimentation because it is out of reach of major water circulation. Mud may also act as sink to phosphorus in marine and majority of the eutrophic systems (Boyd 1982). In the marine environment, part of the phosphate is deposited in the shallow sediments and then lost to the deep sediments. Biological processes such as formation of teeth and bone are more or less resistant to weathering and thus account for considerable loss from the biotic portion of the cycle. Also, part of animal excretion is responsible for substantial loss of phosphorus from the cycle (Kormondy 1978).

There are several ways by which substantial amount of phosphorus is returned to the cycle, despite the means of returning phosphorus to the cycle are inadequate to compensate for the loss (Odum 1983). For example, there is no extensive uplifting of sediments; transport of fish from sea to land is not adequate to replace phosphorus flowing from land to sea. Thus, there is a disbalance in favour of phosphorus loss in the sea. However, sea birds play an important role in returning phosphorus to the cycle. The tremendous guano-deposits along the western coast of Peru (South America) is a well known example of such return. According to Hutchinson (1952), the contribution of guano deposit is 0.01 Tg P Y^{-1} to terrestrial phosphorus. With industrial and agricultural use of guano beds there has been a reverse flow of phosphorus from the sea to the continents (Cole 1988).

Phosphorus in waters

Phosphorus occurs in natural waters and in wastewater almost solely in the form of various types of phosphates. These forms are commonly classified into orthophosphates, condensed phosphates (pyro-, meta-, and polyphosphates) and organically bound phosphates. These may occur in the soluble form, in particles of detritus, or in the bodies of aquatic organisms. Phosphorus in aquatic system cycles as :

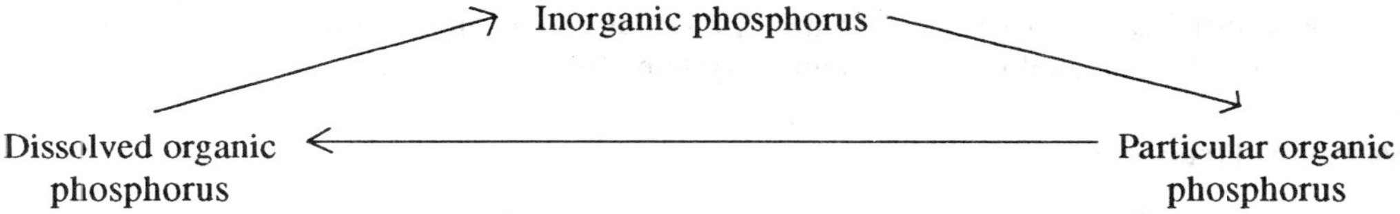

A common concept of the phosphorus cycle assumes that particulate organic phosphate occurs in suspension in living and dead protoplasm and is insoluble, whereas, dissolved organic phosphorus is derived from the particulate matter by excretion and decomposition. The changes are mediated by bacteria. Organic phosphorus eventually becomes soluble inorganic phosphorus through hydrolysis of organic phosphorus by chemical and or biological reactions cycle (Lean 1973). Inorganic phosphorus is eventually incorporated into the biological systems. Both organic and inorganic forms are involved in transformations (Holtan *et al.* 1988). Living plankton organisms and littoral plants excrete soluble organic phosphorus mainly as dissolved organic matter, much of which aggregates to become colloidal and hence particulate organic matter. Another fraction becomes particulate by bacterial incorporation (Cole 1988).

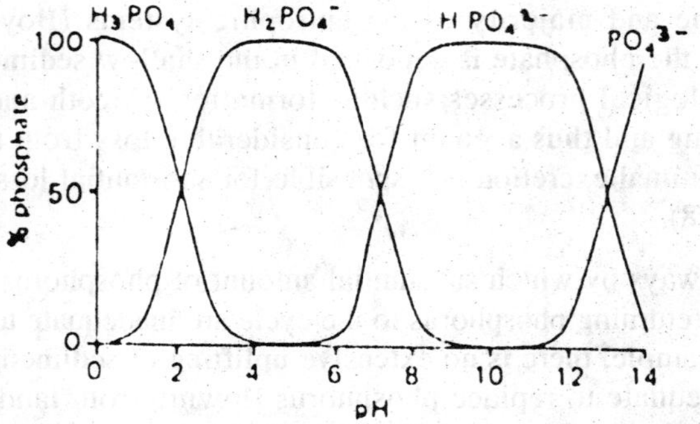

Fig. 14.2. Distribution of different species of phosphate with pH

The inorganic forms of phosphorus in natural waters are ionization products of orthophosphoric acid :

$$H_3PO_4 = H^+ + H_2PO_4^- \; ; \quad K_1 = 10^{-2.13}$$

$$H_2PO_4^- = H^+ + HPO_4^{2-} \; ; \quad K_2 = 10^{-7.21}$$

$$HPO_4^{-2} = H^+ + PO_4^{3-} \; ; \quad K_3 = 10^{-12.36}$$

In natural waters, $H_2PO_4^-$ and HPO_4^{2-} are usually predominant. The dissociation of orthophosphoric acid in aqueous system is a function of pH (Fig. 14.2).

Fractions of phosphorus in water

Phosphorus can be classified (Table 14.1) depending upon the particle size (analytical fractions) or their availability to biological systems (bio-available-P).

Analytical fractions

Because of the fundamental importance of phosphorus as a nutrient and major cellular constituent, much emphasis has been paid on analytical evaluation of its changes in concentration over time. From the analytical point of view, the natural fraction of phosphorus are : total phosphorus (0.45 μm), particulate phosphorus (> 0.45 μm) and soluble phosphorus (<0.45 μm) consisting of soluble reactive phosphorus and soluble unreactive phosphorus (Table 14.1). Made up of many minerals, amorphos precipitates, and sorbed reaction products, particulate-P is in various forms : adsorbed, exchangeable-P, organic-P, precipitates (fertilizer, reaction products with Ca, Fe, Al and other cations), crystalline minerals and amorphous P. The condensed phosphates (pyrophosphate, polyphosphates, metaphosphates) are mainly manmade, discharged with domestic and industrial wastes and also generated by all living organisms are unstable in water, and are slowly hydrolysed to the orthophosphate form (Alexander 1978).

TABLE 14.1. *Fractions of Phosphorus in water*

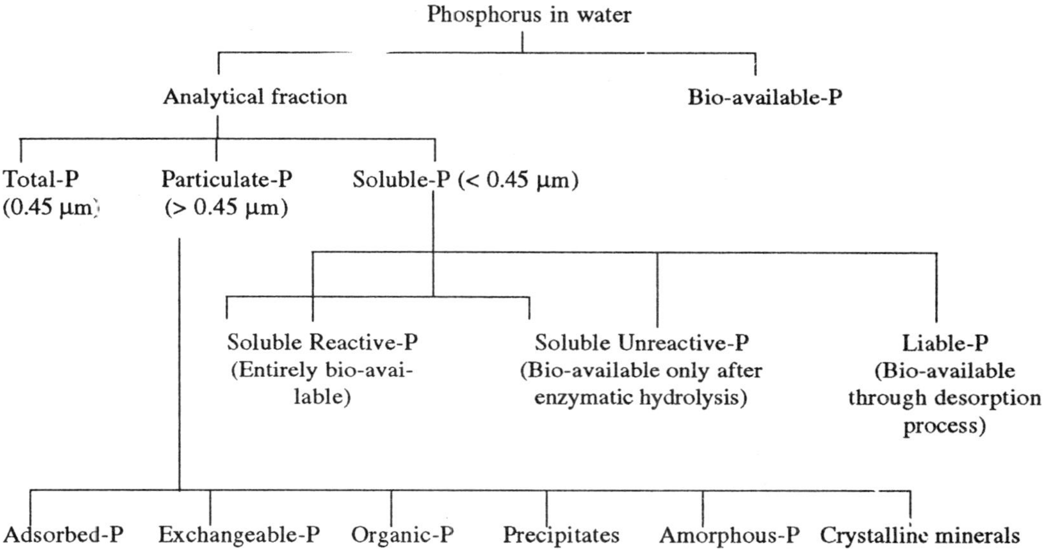

Bio-available fractions

Schaffner and Oglesoy (1978) defined biologically available phosphorus as soluble reactive phosphorus, soluble unreactive and labile phosphorus. Soluble reactive phosphorus is considered to be entirely biologically available, whereas, soluble unreactive phosphorus becomes available only after enzymatic hydrolysis. Labile phosphorus, associated with soil particles may dissolve into an aqueous solution through desorption process. Potentially bio-available phosphorus was found to range from 60 per cent of the total phosphorus in tributaries (Sonzogni *et al.* 1982) to about 70 per cent in municipal waste water (De Pinto *et al.* 1980) ; the type of treatment had no influence on the bioavailable fraction. The measurements of biologically active phosphorus in water are interferred by seasonal variations of input, chemical and biological uptake, especially the algal uptake. The algal uptake is again a function of several environmental factors such as the levels of other essential nutrients, algal species composition and abundance, light, temperature and pH (Holtan *et al.* 1988).

Phosphorus movement in the open water

It has been argued that phosphorus that is added to the open water is rapidly incorporated by the phytoplankton and bacteria. The specific rates of movements of phosphorus have been demonstrated by Rigler (1973) and Lean and Rigler (1974).

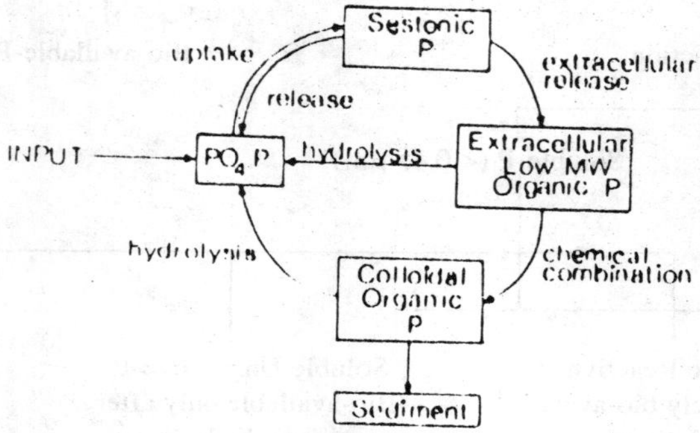

Fig. 14.3. Principal phosphorus pathways in the epilimnion

In the epilimnion phosphorus occurs in following fractions : (a) particulate forms containing the bulk of the phosphorus (>95 per cent), (b) reactive inorganic soluble orthophosphate is very small with an extremely short turnover time (often minutes), (c) from low molecular weight organic phosphorus compound, orthophosphate is released continuously mainly by the activity of free phosphatases and then become available to phytoplankton, (d) soluble macromolecular colloidal phosphorus of a molecular weight > 5,000,000.

There is an exchange mechanism between the inorganic phosphorus and particulate fraction. Some phosphorus is excreted by the microorganisms in the form of low molecular weight compound. Polycondensation of the low molecular weight compound produces the high molecular weight colloidal compound. Both fractions, release phosphate in soluble inorganic fraction which then becomes available for rapid uptake by the plankton (Fig. 14.3). Because of high metabolic activity, bacteria are of major importance for the dynamics of phosphorus cycling of waters (Wetzel 1975).

Role of Enzymes

The role of phosphatases in the phosphorus dynamics of water is well established. Phosphatases are the enzymes which promote the degradation of complex phosphorus compounds into orthophosphate. The occurrence of enzymatic breakdown of organic phosphorus was first suggested by Steiner (1938) but most intensive studies on phosphatases and their function in aquatic systems did not begin until around 1960. During recent decades the development of improved analytical techniques and growing interest in phosphorus turnover in a quatic systems have led a large number of researchers to investigate phosphatase in aquatic systems. An excellent review on the subject has been compiled by Jansson *et al.* (1988).

The total phosphatase activity in lake water results from a mixture of phosphatases localised on the cell surface of algae and bacteria and from dissolved enzymes supplied by autolysis or excretion from algae, bacteria and zooplankton (Jansson *et al.* 1988). A survey on the occurrence or dissolved phosphatase activity in filtered or cell free samples for lakes, algal cultures and zooplankton (Table 14.2) showed that dissolved enzymes are supplied by excretion from grazing algae and zooplankton, but perhaps less so by bacteria. Another important source is the autolysis of dying and disintegrating cells (Overbeck and Babenzien 1964 ; Berman, 1969, 1970). It has been observed that there was no essential difference in the characteristics of dissolved phosphatases from those of seston bound phosphatases (Glew and Heath 1971, Jansson *et al.* 1981).

External lake water phosphatase usually have pH optima in the alkaline region. Acid phosphatases generally active in the internal cell metabolism. Phosphatase activity has been used as a phosphorus deficiency indicator in algae and in natural plankton populations (Olsson 1990). The synthesis of external alkaline phosphatases is often repressed at high phosphate level and depressed at low phosphate concentration. Jones (1972b) examined 16 lakes in the English Lake district in search of indices those are best correlated with summer standing crop and degree of eutrophication. As others have found total-P was on the top of the list, alkaline phosphatase activity of the epilimnetic samples was also an excellent indicator of biomass in the English lakes. In Lake Kinneret (Israel) Berman (1970) observed an annual fluctuations of this enzyme and showed its importance to the biomass of algae and dianoflagellets.

Phosphorus in Sediments

Sediment generally act as sink for phosphorus and play an important role in overall phosphorus metabolism of water bodies. The major source of sediment phosphorus is settled particulate inorganic phosphorus of allochthonous and autocthonous origin consisting of settling materials of dead or live planktonic organisms, excretion products, organic dotritus, precipitated humus substances and calcium carbonates acting as vehicle for phosphate. The sedimentation of algae and the faecal pellets of grazing animals together with microbial uptake tend to accelerate the flux of phosphorus to the sediments (Jackson and Schindler 1975). The second source is dissolved phosphate sorbed to the surface sediment and accumulated in the interstitial water. In rare cases dissolved phosphate derived from seepage is accumulated in the sediments (Holtan *et al.* 1988).

Fractions of Phosphorus in Sediments

Analytical fractions

The total sediment phosphorus can be divided into dissolved and particular forms which can be separated into organic and inorganic forms. Organic-P accounts for a large part of the sediment-phosphorus (Groot & Golterman 1991). Total organic phosphorus in sediments is

TABLE 14.2. *Free dissolved phosphatase activities examined in samples from different sources*

Type of sample	Type of phosphatases assayed (Ac=acid N = neutral Al = alkaline)	Pore-size if filter was used (μm) (- = not specified) (centr. = centrifugation)	% of total phosphatase activity	References
Lake water	Ac	0.20	ca.50	Jansson *et al.*, 1981
Lake water/algal cult.	Ac.Al	0.22	--	Reichardt, 1971
Lake water	Ac.Al	0.45	ca.10-90	Halemejko & Chrost, 1984
Lake water/zooplankton	N	0.20	25	Jansson, 1976
Lake water	N	0.22	--	Jones 1972a,b
Lake water	N	0.45	ca.30	Chrost *et al.*, 1984
Lake water	Al	0.20	15-70	Pettersson, 1980
Lake water	Al	0.22	< 6	Francko, 1983
Lake water	Al	0.22	--	Tiwari & Mishra, 1982
Lake water	Al	0.45	0-57 $\bar{X}$ = 16	Berman, 1970
Lake water	Al	0.45	20-98 $\bar{X}$ = 48	Francko, 1984
Lake water	Al	0.45	4-82 $\bar{X}$ = 16	Gunatilaka, 1984
Lake water	Al	0.45	5-50 $\bar{X}$ = 26	Healey & Hendzel, 1979
Lake water	Al	0.50	often > 50	Wetzel, 1981
Lake water	Al	0.60	7-25	Reichardt *et al.*, 1967
Lake water	Al	centr.	14-61	Stewart & Wetzel, 1982
Pond	Ac.Al	--	--	Overbeck & Babenzien,
Pond	Al	0.45	11-28	Berman & Moses, 1964 1972
Algal cult.	Ac	centr.	29-65	Aaronson & Patni, 1976
Algal cult.	Ac	centr.	--	Antia & Watt, 1964
Algal cult.	Ac.Al	0.45	--	Wynne, 1981
Algal cult.	Ac.Al	centr.	(> cellular)	Aaronson, 1971
Algal cult.	Ac.Al	centr.	--	Patni *et al.*, 1977
Algal cult.	N	0.45	often < 50	Huber *et al.*, 1983
Algal cult.	Al	0.45	10-50	Healey, 1973
Algal cult.	Al	0.45	4-20	Healey, 1985
Algal cult.	Al	0.45	5-80	Healey & Hendzel, 1979
Algal cult.	Al	0.45	--	Lin. 1977
Algal cult.	Al	--	0-ca.10	Kuenzler & Perras, 1965
Algal cult.	Al	--	0-63	Walther & Fries, 1976
Zooplankton	Ac.Al	0.45	--	Boavida & Heath, 1984
Zooplankton	N	--	--	Rigler, 1961.

normally determined as the difference between total sediment phosphorus analysed by acid digestion and the total amount of inorganic phosphorus. Phosphorus bound to inorganic species often constitute the dominant portion of sediment phosphorus in water bodies. According to Hieltjes & Lijklema (1980), the particulte inorganic phosphorus is separated into :

- Loosely bound phosphorus extracted with ammonium chloride,
- Iron and aluminium bound phosphorus extracted with sodium hydroxide.
- Calcium-bound phosphorus extracted with hydrochloric acid.
- The 'rest phosphorus' consisting of organic and inert phosphorus found by difference.

Both inorganic phosphorus (mainly iron-and aluminium phosphate, often referred to as non-apatite) and organic phosphorus decrease strongly with sediment depth because redox potential, pH, temperature, oxygen condition and biological activity decrease with depth favouring dissolution of iron, manganese and aluminium in the deeper sediment layers (Holtan *et al.* 1988).

Since, the metabolic activity is highest at the sediment surface due to high degradability and favourable oxygen conditions, dissolved and sorbed phosphates have their peaks at or near the sediment surface (0-10 cm), whereas, condensed phosphates dominate in the lower parts (>20 cm) (Enell & Lofgren 1988). However, permanent anaerobic sediment surfaces and inverse pH profiles in acidic lakes may change the profile of loosely bound phosphate. As the diagenetic processes transform the sedimented phosphorus according to biological activity, redox conditions and pH into the final sediment composition, the quantitative and qualitative composition of aquatic sediments is influenced by the source of the sediments and the biological, physicochemical environment of the sediments.

Bioavailable fractions

Bioavailable phosphorus is defined as the sum of immediately available phosphorus and the phosphorus which can be transformed into an available form by naturally occurring processes (Bostrom *et al.* 1988). In recent years there has been much interest about qualitative and quantitative utilization of sediment phosphorus by bacteria and algae. Jonge *et al.* (1991) found that biologically available form of phosphorus in the sediment is an order of magnitude lesser than the total phosphorous. The most part of the available phosphorus in the soil is immobilized in the form of complexes of iron, aluminium and calcium. Phillips & Webb (1971) found soluble phosphorus rarely migrates more than 2-3 cm from a fertilizer granuli before being immobilized. Sediments ineffective in binding phosphorus, have moderate pool of potentially bioavailable phosphorus (Stauffer 1985). In lakes with a high phosphorus retention capacity, actual bioavailability of phosphorus differ from the potential bioavailability of phosphorus (Bostrom *et al.* 1988).

Phosphorus in sediment pore-water

The amount of phosphorus in pore waters generally constitute a very small portion (<1 per cent) of the total sediment phosphorus. It is directly exchangeable with overlying water and with the highest chemical mobility (Syers *et al.* 1973). The interstitial P concentration generally exceeds the P concentration in overlying water. Bostrom *et al.* (1982) estimated that pore water concentrations are approximately 5-20 times higher than that of overlying water.

Phosphorus in sediment pore water is highly variable depending upon the space, time and nutrient status of the water bodies (Enell & Lofgren 1988). The phosphorus concentration in sediment water was found to be 0.02 mg l^{-1} for oligotrophic lakes (Brunskill *et al.* 1971), whereas, 0.12 - 3.16 mg l^{-1} for eutrophic lakes (Serruya *et al.* 1974 ; Ripl & Lindmark 1979).

Mud-water exchange

Sediment phosphorus has been a subject of research since early thirty's (Einsele 1936; Ohle 1937 and Mortimer 1941, 1942). Generally, aerobic sediments in oligotrophic lakes have been considered to act as sink, while anaerobic sediments in highly eutrophic lakes have been regarded as source of phosphorus (Bostrom *et al.* 1982). In eutrophic lakes, significant fluxes of P across the sediment-water interface often occur in both directions simultaneously. Over extended periods, such as an annual cycle, there is a net deposit of P in most of the eutrophic lakes (Bostrom *et al.* 1982). More than 95 per cent of the externally loaded P was retained in the eutrophic lake Balaton (Herodek 1984).

Sedimentation of inert allochthonous matter which account for 50 per cent of the P accumulation in the sediments and, setting of autochthonous organic matter and coprecipitation of P with biogenically formed carbonate precipitates as major mechanisms for the transfer of P from lake water to the sediments (Lijklema *et al.* 1983). Apart from this absorption of P account for substantial flux of P from water to sediments.

Release of P from sediments can be viewed as the result of : (1) mechanisms by which particulate P is transformed to dissolved form as phosphate; and (2) mechanisms by which dissolved compounds are transported upwards (Petterssons & Bostrom, 1986).

Phosphorus release into an oxygen rich water column is a result of a complex mixture of processes determined by the physical, chemical and biological characteristics of the sediment along with various forms of turbulence.

Sediment water exchange of phosphorus has variously been described as a simple diffusion from an anaerobic sediment (Jorgensen *et al.* 1973; Kamp-Nielsen 1974), a sorption equilibrium with an oxidized sediment (Kamp-Nielsen 1974), a temperature-dependent mineralization and diffusion (Jorgensen *et al.* 1975), a function of 'active' surface layer (Jorgensen 1976) etc.

High rates of aerobic release are dependent on water turbulence (Ryding 1981, Bostrom *et al* 1982), gas bubblings (Bostrom *et al.* 1982), ground water upwellings (Peters & Vanliere 1985), water currents (Knoblauche 1982), sediment convection currents (Lappalainen 1982) and bioturbation (Neame 1977; Jana & Das 1992; Jana *et al.* 1992).

An overall review of factors and processes that govern P release from lake sediments was done by Bostrom *et al.* (1982). Phosphorus release from aquatic sediments is mediated through desorption and mineralization from sediment particles and microbial degradation of particulate organic phosphorus, which in turn are greatly influenced by temperature (Holdren & Armstrong 1980; Matisoff *et al.* 1981) and groundwater seepage (Van Liere & Mur 1982), pH (Anderson 1974 ; Lijklema 1980).

According to Wetzel (1975) phosphorus is most available and easily released from sediments with a pH of 6 to 7. At lower pH, phosphorus readily combines with iron, aluminium and manganese, while, at pH 6 and above progressively greater amounts of phosphates get associated with calcium as apatites and calcium phosphates. The rate of release of phosphorus from intact sediment cores was measured as a function of the pH of overlying water, yielding the relationship $\log K = 0.54\ pH - 3.94$, $K = mg\ P\ m^{-2}\ day^{-1}$ (Drake & Heaney 1987).

Desorption, especially from iron and aluminium hydroxides is very sensitive to changes in pH. Desorption takes place at a layer of about 1 mm at the sediment-water interface (Van Raaphorst & Brinkman 1984); but the layer involved in mineralization may be much thicker (Lijklema 1985). Release due to mineralization is much more sensitive to downward seepage than release due to desorption (Lijklema 1985). Also, there is an interaction between mineralization and desorption processes, because mineralization influences desorption, and with increasing mineralization activity the thickness of the aerobic zone in the sediment decreases. This promotes desorption of phosphate from iron hydroxides, reducing from Fe (III) to Fe (II). Desorption of inorganically bound P may occur when the sediment surface is in contact with water with a low orthophosphate concentration (Rippey 1977) and phosphorus desorption from Fe (III) increases as pH rises above 8, due to competitive ligand substitution by OH^- (Stauffer & Armstrong 1986).

Bacteria have a direct role in the release of P from the organic and inorganic constituents of sediment. Microbial degradation of organic matter lowers the interstitial pH by release of acidic fermentation products. The consequent mineral dissolution of carbonate minerals allows the mobilization of calcium, magnesium and associated P which is important in controlling P concentrations in hard water lakes (Golterman 1982 ; Lijklema *et al.* 1983; Lofgren & Ryding 1985).

High rates of sediment P release from Lake Vallentunasjon, Sweden, were associated with the onset of anaerobic conditions following temporary stratification. Concurrent reductions in the P associated with organic matter in the sediment and high bacterial respiration, suggested that the P was released by bacterial action (Bostrom *et al.* 1985). The release of P during aerobic decomposition is less significant if bacteria are P limited (Doremus & Clesceri 1982). However, in shallow eutrophic lakes the potential for release of mineralised P is greater due to the microbial decomposition of P rich algae occurs at the sediment surface (Marsden 1989).

Phosphorus fluxes from sediment to water are usually measured as changes in total-P concentrations and may include the resuspension of particulate P unavailable for algal growth (Ryding 1985). Consequently rates of aerobic P release vary considerably from 1-2 mg T P m^{-2} day^{-1} in Esthwaite water, England (Drake & Heaney 1987) and 40 mg T P m^{-2} day^{-1} in Barton Broad, England (Osborne 1981); to 550 mg T P m^{-2} day^{-1} in Wahnback reservoir, West Germany (Knoblauch 1982).

Phosphorus Dynamics

Role of Pelagic invertebrates

Much research have been done to investigate the role of pelagic animals on phosphorus dynamics of water. Pelagic animals not only transform phosphorus, they can also translocate or distribute the recycled phosphorus within the system. According to Wright & Shapiro (1984) *Daphnia pulex* can translocate as much as 1 μg of dissolved phosphorus/15 days by a combination of vertical migration and excretion. The uptake and subsequent release of phosphorus by zooplankton microcrustacea have been suggested as significant regenerative pathways for phosphorus cycling in pelagic system (Rigler 1964, Hargrave & Geen 1968). The study of phosphorus regeneration by zooplankton was studied firstly as chemical analyses of excreted phosphorus by zooplankton as increase in concentration (Barlow & Bishop 1965, Ferranate 1976) and secondly by measuring excretion of ^{32}p (Peters & Rigler 1973 ; Taylor and Lean 1981). The rates of phosphorus excretion are normally inversely related to mean body size of the grazers. It is estimated that the rate of excretion of phosphorus by crustaceans and rotifers is equal to the daily phosphorus requirements of the phytoplankton (Hargrave and Geen 1968). *Daphnia pulex* is known to excrete soluble phosphorus equivalent to 1.5 per cent to 2.5 per cent of its total phosphorus each hour; about 35 per cent to 60 per cent of the total phosphorus in this herbivore was turned over every day as egested and excreted matter (Lehman 1980). In lake Washington, epilimnetic zooplankton supplied 10 times more phosphorus to the primary producers during summer than all external sources of phosphorus (Lehman 1980).

Role of benthic invertebrates

Researches have shown that benthic invertebrates have been found to play an important role in the phosphorus cycle of water. The role of benthic invertebrates on nutrient cycling has extensively been studied by Peter (1977), Kitchell *et al.* (1979), Hakansson & Jansson (1983), Merritt *et al.* (1984) and Matisoff *et al.* (1985). The effect of benthic organisms on the P flux to the water mass are mainly due to active transport of sediment particles and water across the sediment water interface and by alteration of the physical environment like pH, redox properties, etc. (Petr 1977 ; Berner 1980 ; Matisoff *et al.* 1985). It is observed that the influence of tubificids and chironomids on phosphate release was greater than the physical factors like redox potential, mixing intensity, temperature and sediment type (Holdren & Armstrong 1980). A positive correlation has been demonstrated between the amount of benthic macrofauna and liberation of phosphate from sediment cores (Wisniewski & Planter 1985; Starkel 1985).

There are some controversy regarding the role of benthic invertebrates on phosphorus release. Graneli (1979) reported that benthic invertebrates substantially increase the downward oxygen flux. According to Rebsbech *et al.* (1980) molecular oxygen as terminal electron acceptors penetrate to a depth of a few mm in the sediment resulting in more efficient sediment mineralization and liberation of phosphate from particulate organic matter. On the other side, Bostrom (1984) stated that oligochaete worms increase the redox potential of surface sediments containing appreciable amounts of redox-sensitive phosphorus, thereby causing a stronger binding of phosphate (Davis 1974). According to Davis *et al.* (1975) and, Jorgensen & Rebsbech (1985), tubificids could decrease the pore water P concentration in the upper sediment layer as a result of inflow of oxygenated water into the worm burrows and the mean relative abundance of oligochate worms was found to be negatively correlated with total phosphorus concentration (Lang 1990).

Defeacation and excretion of dissolved phosphorus by benthic invertebrates play significant role in the exchange process at the sediment water interface (Andersson *et al.* 1988). For example, tubificid worms have stratigraphic effects in lake sediments (Davis 1974). According to Krezoski *et al.* (1978) tubificid defeacation rate may be similar to or exceed net sedimentation rate. As tubificids excrete phosphorus, sediments with a limited phosphorus binding capacity would loose phosphate in proportion to tubificid biomass (Gardner *et al.* 1981).

Role of Fish

Generally fish is regarded as a component where phosphorus is stored and thereby a substantial amount is lost from the cycling processes. About 70 per cent of the phosphorus in the pelagic may be found in fish tissues of Lake Wingra, USA (Kitchell *et al.* 1975) and 40-50 per cent of the total phosphorus in the open water of Lake Memphremagog, Canada (Nakashima and Leggett 1980). Part of the phosphorus bound in fish tissues is recycled continuously through excretion or discontinuously through decomposition of dead fishes (Andersson *et al.* 1988). In some lakes, phosphorus excretion by fish was estimated to be less important (Andersson *et al.* 1988) and considered to be unimportant compared to the phosphorus demand of the seston (Nakashima and Leggett 1980). However, the situation may be changed in fish ponds with intensive stocking density.

Bottom feeding fish is also important in the phosphorus economy of water bodies (Lamarra 1975), but there is a basic difference between planktivorous and benthivorous fish in their influence in phosphorus cycling (Shapiro and Carlson 1982). Cyprinid benthivorous species like roach and bream may contribute much more than planktivorous perch to the internal release of phosphorus. Lamarra (1975) and Andersson *et al.* (1978) found that the sediment surface is disturbed by benthivorous fish and played a significant role in the phosphorus release in lake enclosure experiments.

According to HR Bacek *et al.* (1961) and Hurlbert *et al.* (1972) fish can also affect phosphorus transformation and translocation through their feeding habits. This is brought about either directly through egestion (Lamarra 1975) or indirectly via food chain triggered by selective predaton (Lynch and Shapiro 1981 ; Perseon 1983).

The direct role of fish in transforming and translocating phosphorus is most pronounced in sallow lakes with dense populations of benthivorous, cyprinid fish. In eutrophicated lakes, the foraging of common carp for animals in the sediment mobilize large amounts of phosphorus (Lamarra 1975). Thus sediment act as a source or a sink depending on the feeding habits of the different fish species (Andersson *et al.* 1988).

Jana *et al.* (1992) have demonstrated the bioturbation effect of bottom grazing common carp (*Cyprinus carpio*) in increasing the fertilizer value of phosphate rock. Bioturbation caused by common carp fry resulted in as high as 72-100 per cent influx of phosphate from bottom soil in presence of phosphate rock but only about 7-8 per cent in the absence of it (Fig. 14.4).

That the bioturbation effect was dependent upon the number of common carp fry introduced was clearly demonstrated in an experimental study of Jana and Das (1992) (Fig. 14.5). They have shown that the level of phosphate coupled with alkaline phosphatase activity tended to rise in a logistic manner with an increase in the number of common carp fry upto 8 but showed no further rise even with rise in fry numbers (Fig. 14.6). In any treatment, alkaline phosphatase activity of water was directly correlated with phosphate level to a certain extent beyond which an inverse relationship was established (Fig. 14.7). Comparing the bioturbation effect of three bottom grazing fishes, in the release of orthophosphate from bottom sediments rich in phosphate rock, Jana & Sahu (1993) have shown that bioturbation activity of the common carp (*Cyprinus carpio*) was highest, even though its body mass was three times less than that of marginal (*Cirrhinus mrigala*) and about the same as that of *Heteropneustes fossilis* (Fig. 14.8). As a result of bioturbation, the amount of citrate soluble phosphate and available phosphate of bottom soil have been observed to decrease by 33 and 15 per cent respectively, whereas, the level of orthophosphate of water increased by 27 per cent. As a result both of them maintained an inverse relationship between them (Fig. 14.9).

From this study it is clear that the bottom grazing fishes have a profound influence on the release of phosphate from otherwise insignificantly soluble phosphate rock by bioturbation. The effect of bioturbation was brought about by the physical disturbance and agitation of phosphate rock enriched sediment surface by the free movement of the carp.

Phosphorus budget

The phosphorus budget has been largely used as an index to environmental protection strategies. In most water bodies input of phosphorus exceeds output because of the accumulation in the sediments. Mass balance calculations in a number of lakes in Sweden have shown that the internal phosphorus load is upto 4 times larger than external loading (Ryding & Forsberg 1977). Further, mass balance studies have demonstrated that the net release of phosphorus can be substantial also from the sediments superimposed by well aerated water columns (Lee *et al.* 1977 ; Ryding and Forsberg 1977 ; Bostrom *et al.* 1988).

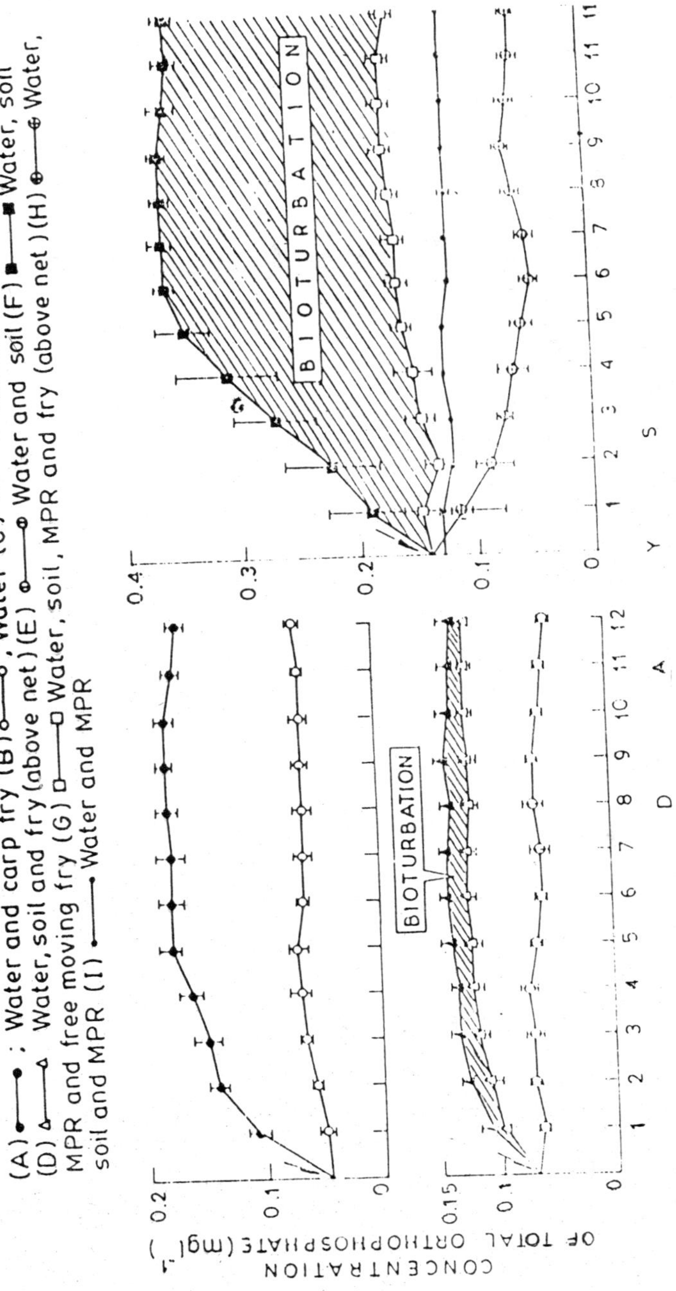

Fig. 14.4. Responses of orthophosphate of water to various treatments.

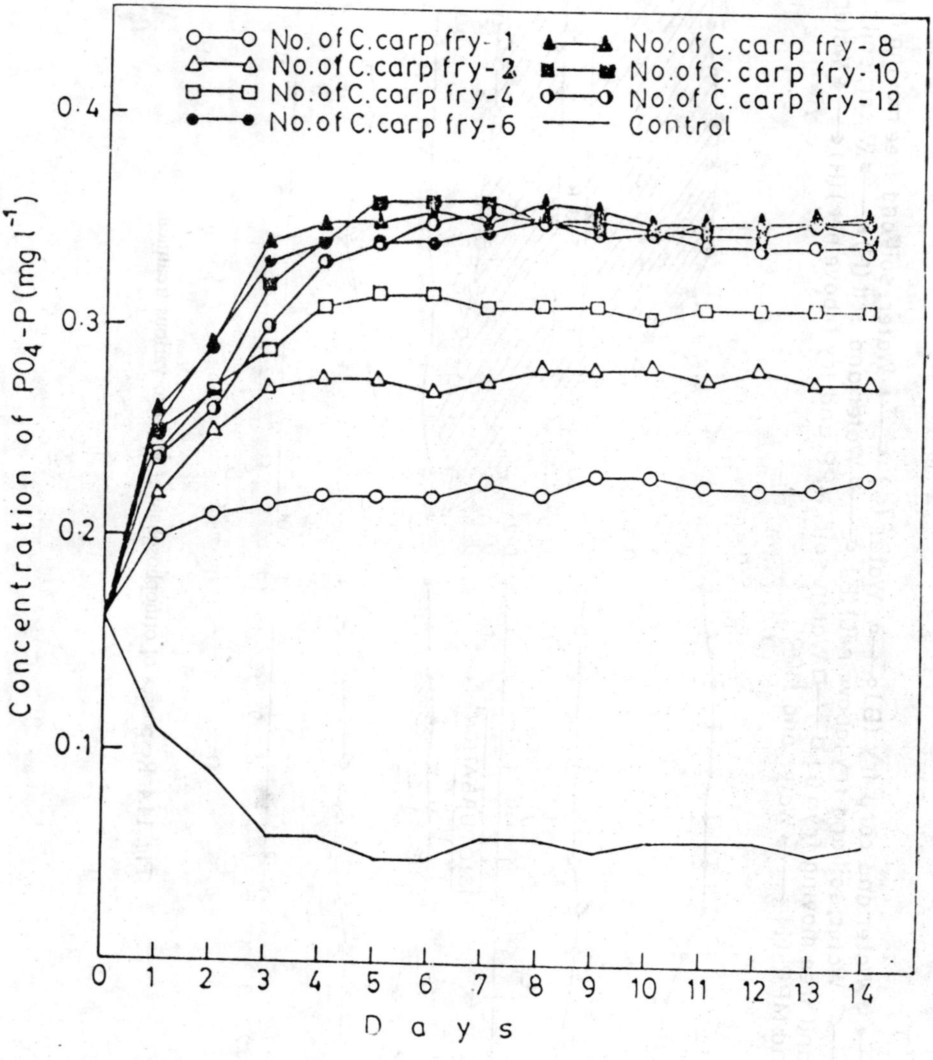

Fig. 14.5. Changes in phosphate level of water in different treatments.

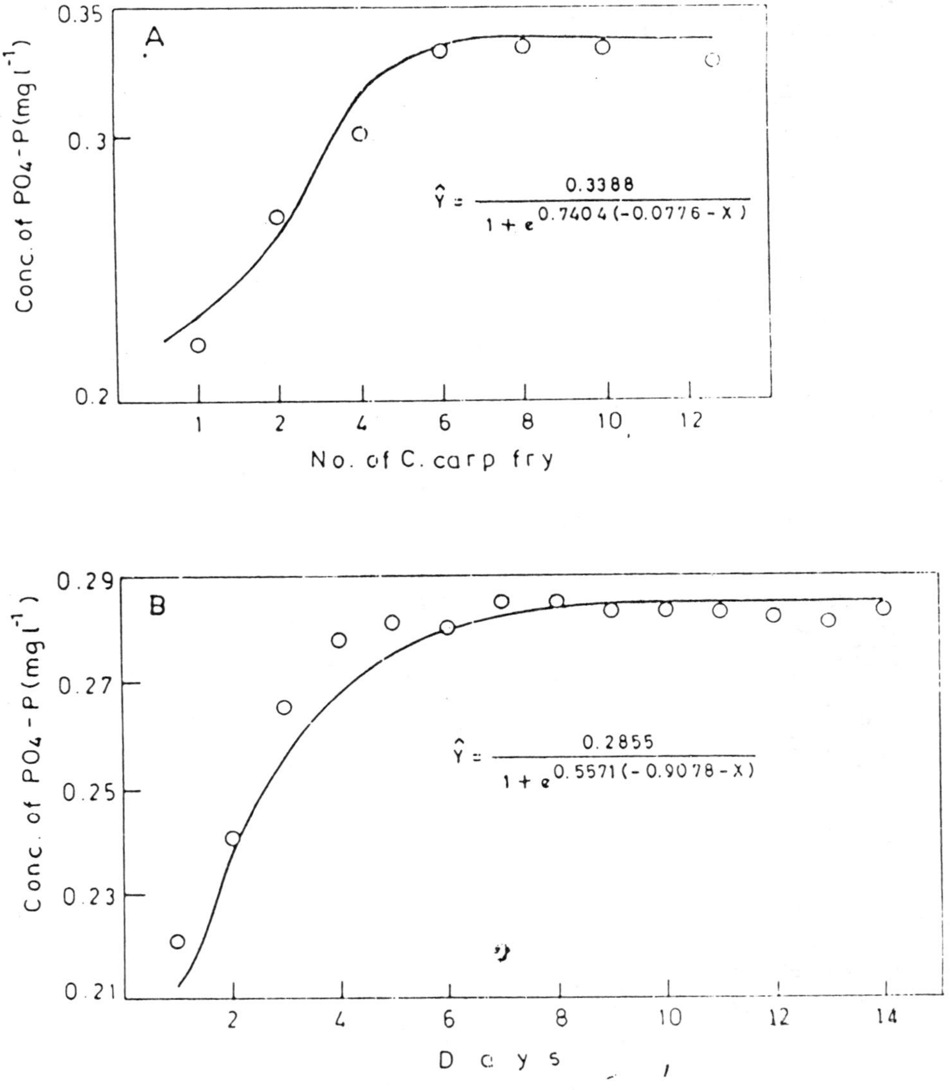

Fig. 14.6. (A) Relationship between the orthophosphate level of water and the number of carp fry stocked
(B) Relationship between the overall mean values of phosphate in the water in various treatments against days of observation.

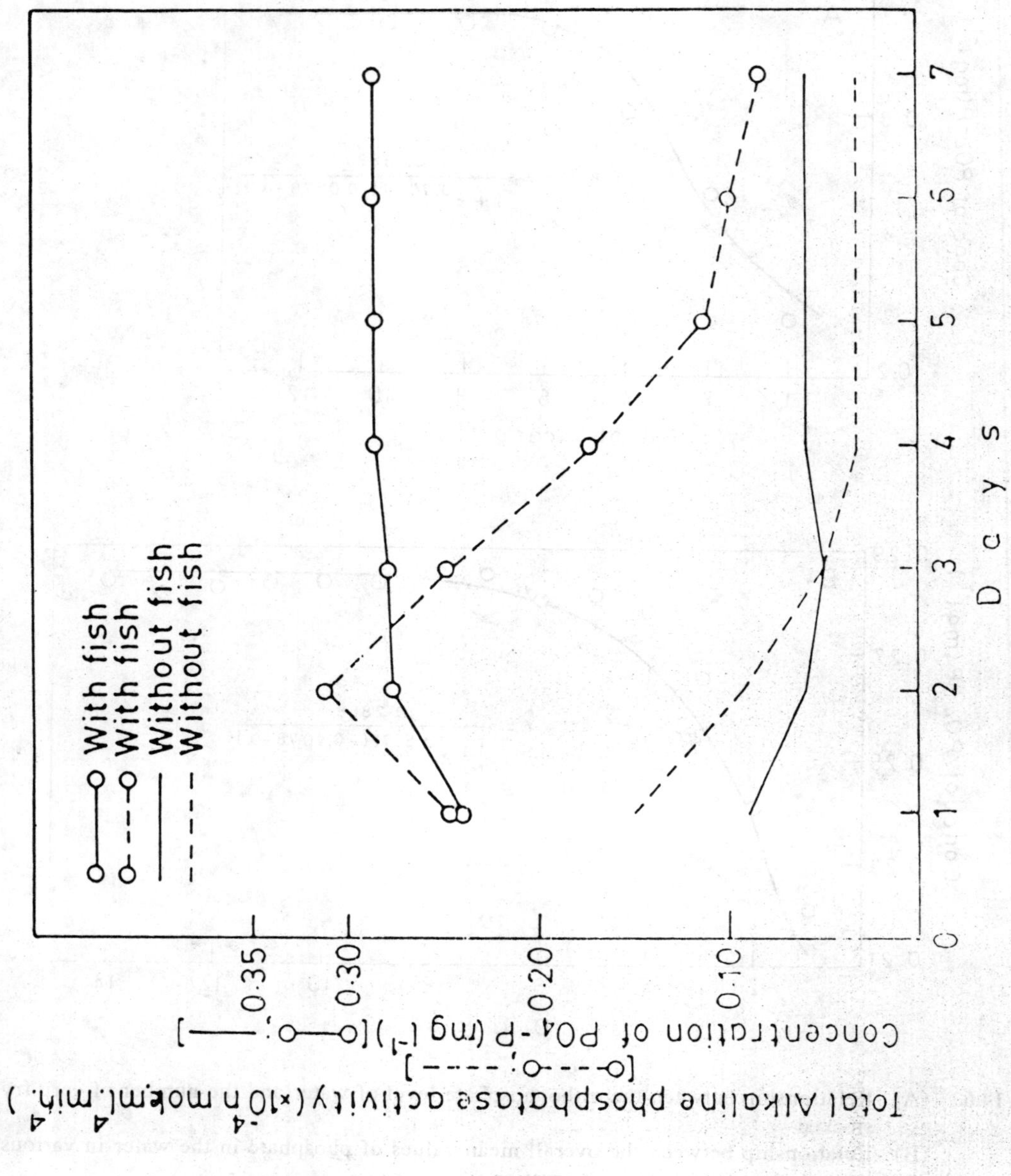

Fig. 14.7. Overall mean values of alkaline phosphatase activity and phosphate level of water for various treatments are plotted against days.

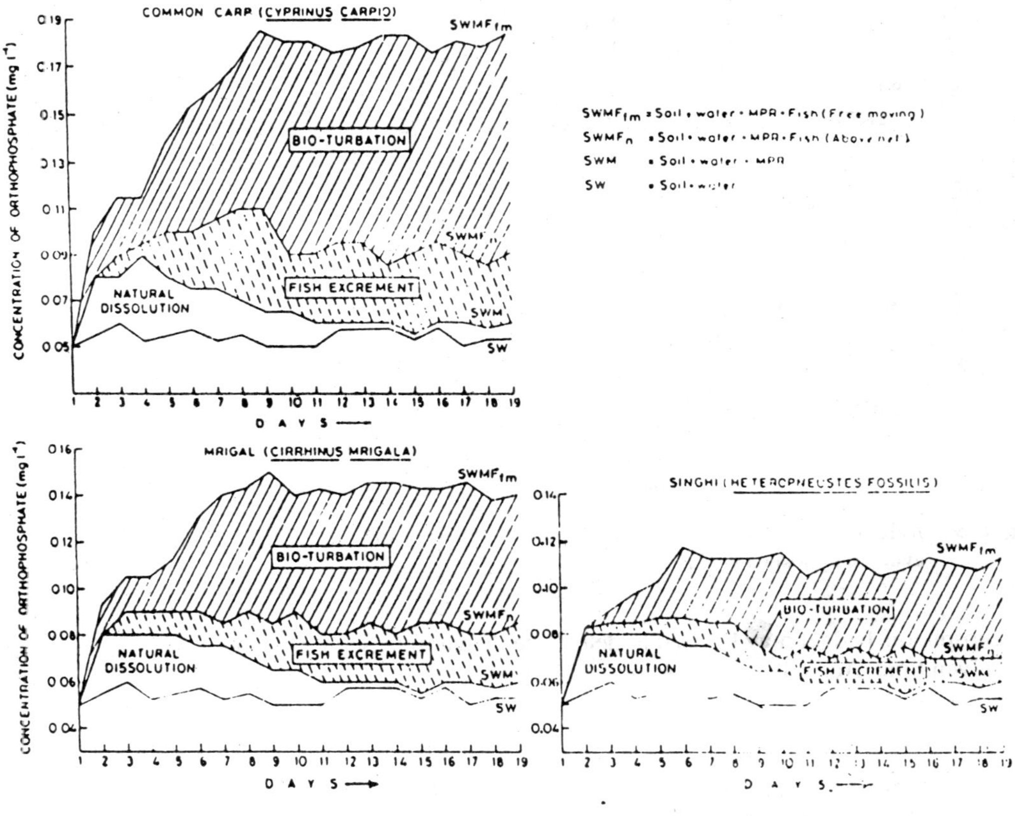

Fig. 14.8. Responses of orthophosphate of water to various treatments for three fishes.

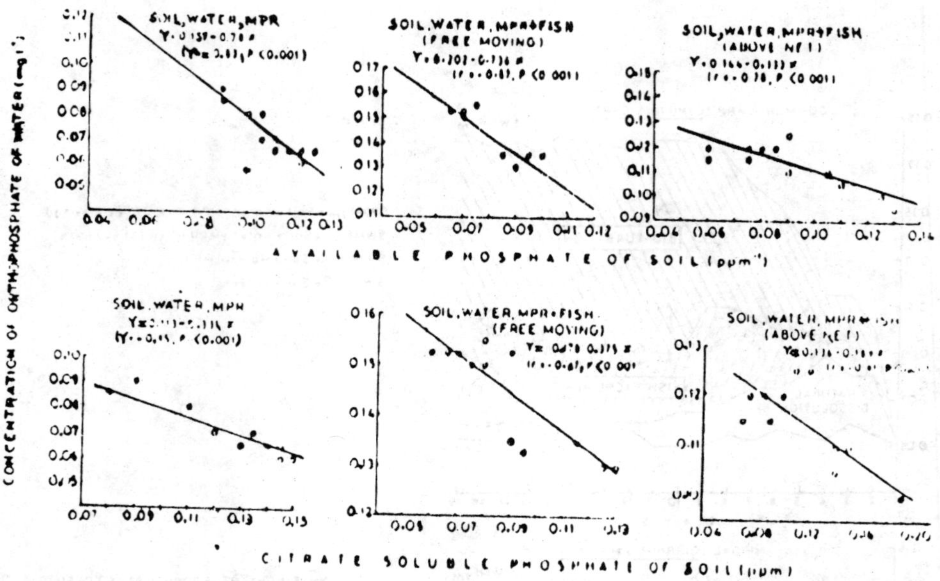

Fig. 14.9. Relationship between orthophosphate level of water and citrate-soluble phosphate and available phosphate of bottom soil.

Phosphorus budget may be used to determine the net release of phosphorus from the sediment at a given time. According to Cooke *et al.* (1977) :

$\Sigma P_{ext} \quad = \Sigma P_{out} = P_{net}$

$\Delta P_{lake} \quad = P_{net} = P_{int}$

ΣP_{ext} = Sum of all phosphorus income from external sources

ΣP_{out} = Sum of phosphorus in outflows

P_{net} = Net phosphorus income or loss

ΔP_{lake} = Change in phosphorus content of the lake over the studied interval.

P_{int} = Internal phosphorus load

Phosphorus and Eutrophication

Eutrophication remains one of the World's principal problems of water quality. The term eutrophication describes the biological reaction of aquatic systems to nutrient enrichment, the eventual consequence of which is the development of primary production to nuisance proportions (Marsden 1989). Undoubtedly, the main cause is excessive loading of phosphorus and nitrogen resulting in high algal biomass, dominance by cyanobacteria and loss of macrophytes.

A majority of the phosphate data for freshwaters are reported as total-P and inorganic soluble phosphate, orthophosphate. The total phosphate concentration of most uncontaminated surface waters are between 10-50 μg p l^{-1}. Phosphate levels are generally lowest. Lakes in mountaineous regions of crystalline rocks show low phosphate value, whereas, high values were recorded in low land waters derived from sedimentary rock deposits. Again, lakes rich in organic matters tend to show higher total phosphorus concentration. A few sedimentary coastal areas are rich in phosphate rock and lakes from drainage from these deposits have abnormally high phosphorus content.

Total-P concentration often occur in extremely low concentration i.e. <1 μg l^{-1} in non polluted water to extremely high values of >200 mg l^{-1} in saline lakes. The total phosphate cocentration of most natural waters are between 10-50 μg p l^{-1} (Wetzel 1975). Vollenweider (1968) categorised water bodies with total-P concentration of 30-100 μg P l^{-1} as eutrophic.

Generally there are two phases in eutrophic water, a macrophyte dominated stage and a phytoplankton dominated stage. Each state is presumed to be stable within a range of nutrient concentrations and a reduction in nutrient concentration alone is often insufficient to switch from phytoplankton to macrophyte state (Timms and Mass 1984).

In an extremely detailed treatment relating phosphorus and nitrogen to the productivity of lakes, Vollenweider (1968) demonstrated that the amount of total phosphorus generally increases with lake productivity (Table 14.3). Though there are a number of exceptions to this relationship, the general principle is useful in the context of eutrophication.

Nutrient loading

The concept of nutrienet loading as a factor controlling lake productivity or trophic state has been one of the most productive theories in limnology, and has stimulated much discussion and

TABLE 14.3. *General relationship between epilimnetic phosphorus and lake productivity*

General Level of Lake Productivity	Change in Alkalinity in Epilimnion During (mval l^{-1})	Total Phosphorus (μg l^{-1})
Ultra-oligotrophic	< 0.2	< 5
Oligo-mesotrophic	0.6	5-10
Meso-eutrophic	0.6-1.0	10-30
Eutrophic		30-100
Hypereutrophic	> 1.0	> 100

research. Originally proposed by Vollenweider (1968) and later modified by him (1975, 1976), this concept has had a great impact on all subsequent eutrophicatioin research and lake management.

There are several approaches to predict the load of nutrients that would become critical. The critical level or load (L_C) if exceeds would lead from oligotrophy to eutrophy. There has been an improvement of methods for estimating the critical loading values. In the simplest model, mean depth was the only factor for consideration. Subsequently, shoreline, volume, area, water retention time and flushing rate were taken into consideration. Vollenweider (1976) contributed much to the development of the nutrient loading concept of lakes. Phosphorus sedimentation, lake dimension, residence time of incoming phosphorus in relation to the hydraulic residence time and annual water budgets are important factors in eutrophication which Vollenweider considered for his improved equation. Vollenweider's refined equation is widely applicable to define the annual critical loading value in miligrams of phosphorus per square meter according to :

$$L_c = (10 \text{ to } 20) \times q_s \left(1 + \sqrt{Z/q_s}\right)$$

In the equation q_s is the so called hydraulic load calculated by dividing the total discharge of water into the lake by lake area and is, therefore, expressed in terms of meters per year.

Based on nutrient loading OECD (1982) categorised two basic loading tolerance time (1) permissible limit and (2) dangerous level. Permissible level is defined as the maximum allowable for a lake to remain oligotrophic in the long time whereas, dangerous loadings are sufficiently great to cause a lake to become eutrophic.

On the basis of basic loading tolerance line Xiangsan *et al.* (1990) have classified a number of Chinese lakes. The plot of phosphate load of these lakes as a function of mean depth hydraulic residence time (Fig. 14.10) shows that all the Chinese lakes investigated are above the excessive loading curve.

There are many good review articles on different types of eutrophication models such as empirical phosphorus loading models (Reckhow 1979), phosphorus-chlorophyll models (Nicholls and Dillon 1978) simple dynamic model (Rainey 1967), Eawag model (Gachter *el al.* 1983), glumose model (Jorgensen 1976), Salmo model (Benndorf *et al.* 1981) and other models developed by Kamp-Nielsen (1985), Jorgensen (1983) and Straskraba and Gnauck (1985). The benefits of using empirical models in limnology in general was discussed by Peters (1986). Chapra (1980) proposed to link a number of sub-models to predict the trophic state of lakes from the phosphorus loading (Fig. 14.11). The Organisation for Economic Cooperation and Development (OECD) had conducted an intensive study to define the relationship between the nutrient load and the water quality of lakes. This study included about 200 lakes distributed throughout Western Europe, North America, Japan and Australia. Models describing the dependence of lake trophic state on external phosphorus loading are of two general types : empirical, steady-state models derived from statistical treatment of data from large number of

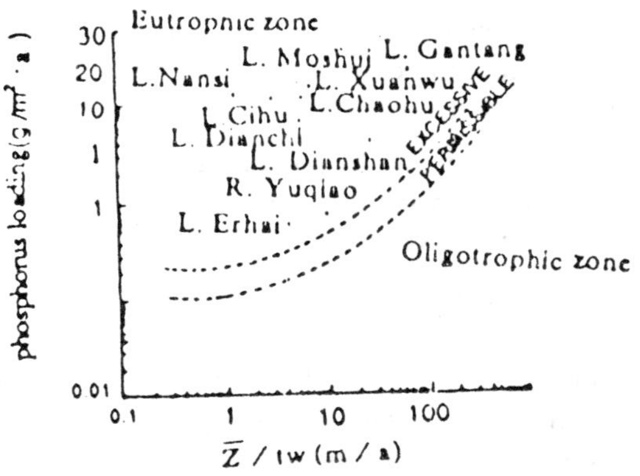

Fig. 14.10. The relationship between P loading and lake trophic state response of some lakes in China.

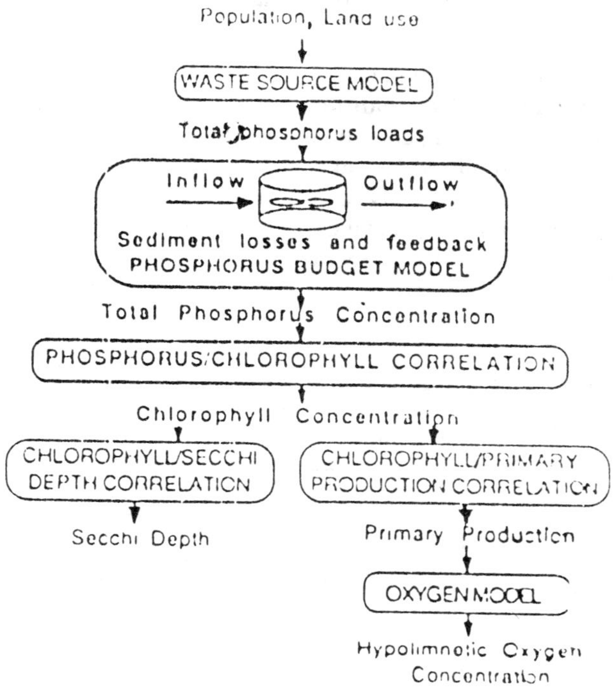

Fig. 14.11. Schematic diagram of P loading model in lakes.

lakes and theoretical, dynamic models based on detailed mathematical dęscription of kinetics of nutrient and population dynamics. Dynamic theoretical models are more suited for prediction of rates of response to perturbations than the empirical steady state models. Simple model with flexible structure generally have advantage over complex models (Ahlgren *et al.* 1988).

Vollenweider and Kerekes (1982), Vollenweider and his co-workers (Oecd, 1982) developed a simple relationship between phosphorus loading and phosphorus concentration in lake, the retention time and average depth of the lake as follows :

$$\text{T-P} = (L_p/q_s)\,(1 + \sqrt{tw}\,)$$

where, T-P = average annual in-lake T-P concentration (mg 1^{-1})
L_p = annual areal phosphorus loading (mg/m^2.a)
q_s = annual areal water loading (m/a = z/tw)
tw = hydraulic residence time (a)
z = mean depth (m).

The idea that large lakes are less susxceptible to pollutants than smaller ones because of dilution effect, was later modified on the basis of data collected from stratified lakes that rarely circulate (Cole 1988).

Vollenweider (1968, 1969, 1972) attempted to evaluate the probable recovery time of a lake to attain a new steady state under conditions of reduced inputs. Based on a combined consideration of numerous influencing conditions very approximately, provisional loading rates for nitrogen and phosphorus required for lakes to maintain a steady-state can be estimated (Table 14.4). Rapid eutrophication is likely to occur as the loading rate to or below those of the lakes prior to accelerated inputs, for lakes of average size and average hydrological replenishment timę, the period needed for recovery will be on the order of 2-10 years (Wetzel 1975).

TABLE 14.4. *Permissible loading level for total nitrogen and total phosphorus (g m^{-2} year^{-1})*

Mean Depth (m)	Permissible Loading		Dangerous Loading	
	N	P	N	P
5	1.0	0.07	2.0	0.13
10	1.5	0.10	3.0	0.20
50	4.0	0.25	8.0	0.50
100	6.0	0.40	12.0	0.80
150	7.5	0.50	15.0	1.00
200	9.0	0.60	18.0	1.20

Restoration

Unabetting eutrophication resulting from sewage and other pollutants associated with human activities in phosphorus back from seas has stimulated research on environmental protection. In eutrophic water bodies reduction of phosphate input from point sources and various non-point sources is considered important for restoration programme (Forsberg and Wallsten 1986; Cullen and Forsberg 1988). One approach is to reduce or eliminate the use of high phosphate detergents. Another scheme is to modify sewage before it can enter a water body. This method has been successful where ferric chloride precipitate about 90 per cent of the phosphate from the sewage as ferric phosphate (Thomas 1969). Spatzierer *et al.* (1985) also emphasized the role of chemical precipitation with salts of ferrous iron. Aluminium salts in sewage effluents are also being used in the chemical precipitation process. A third method is to ensure that domestic wastes do not reach the lake. Considerable success in this regard has been made in some European lakes and in Lake Washington, USA (Edmondson and Lehman 1981).

Internal loading usually refers to the autochthonous input of phosphorus for which sediment acts as an important source. The effect of internal loading was clearly demonstrated in some polluted shallow lakes when external sources of phosphorus are reduced or eliminated (Ryding and Forsberg 1976). The release of phosphate from the sediment occur despite dissolved oxygen tensions of at least 60 per cent of saturation in the bottom water (Riemann 1977). In two eutrophic lakes in Ohio 65-100 per cent of the phosphate increase during summer stratification could be attributed to internal loading (Cooke *et al.* 1977). In highly eutrophic lakes, sediment P release compensates for reductions in external loading, and phytoplankton biomass is insensitive to such change in lake P concentration. In lakes with annual total phosphorus concentration greater than 100 mg m^{-3}, few improvements have been achieved with at least 60 per cent the reduction in nutrient loading (Marsden 1989). In highly eutrophic Lake Trummen, reduction in the external load of phosphorus did not bring the desired improvement in water quality until the layers of phosphorus rich sediments were removed (Bjork 1972). Cooke *et al.* (1986) obtained similar results by removing nutrient rich sediments by dredging, or by the use of chemicals where nutrients are fixed in sediments.

Quantitative aspect of P cycle

Phosphorus in the atmosphere

The role of atmosphere in the phosphorus cycle is poorly understood. Since it does not exist in the form of stable gaseous compounds, phosphorus in the atmosphere is either adsorbed on particulate matter in the dust and exhaust fumes or dissolved in sea-spray. Dust and sea-spray appear to be the major source of phosphorus in atmosphere. The fall out of phosphorus (dry deposition and precipitation) has been estimated to within the range of 3.6 to 9.2 Tg PY^{-1} for terrestrial ecosystem, 0.054 to 0.140 Tg P y^{-1} for freshwater ecosystem and 2.6 to 3.5 Tg P y^{-1} for marine ecosystem. Thus the total fallout from the atmosphere is 6.3 to 12.8 Tg P y^{-1} (Pierrou 1976). Many scientist have shown that precipitation can contribute a major source of nutrients in

TABLE 14.5. *Atmospheric phosphorus loading*

Country	mg/P/m^2 year	References	
New Zealand	41	Bargh	1977
Canada	27	Schindler & Nighswander	1970
Canada	24-53	Schindler *et al.*	1976
Canada	75	Dillon & Rigler	1975
Canada	77	Dillon	1975
USA	10	Murphy & Doskey	1976
USA	102	Sonzogni & Lee	1974
USA	70	Brakke	1977
USA	41	Duffy *et al.*	1978
USA/Europa	10-100	Wetzel	1975
USSR	11-15	Evdokimova *et al.*	1976
Finland	14	Haapala	1977
Finland	6	Jansson	1977
Danmark	10-40	Harremoes	1977
Sweden	20	Ahl & Oden	1975
Norway	34	Berge *et al.*	1979

many lakes of the world (Table 14.5). The influx of phosphorus to the atmosphere due to high temperature of combustion of organic matter is about 0.8 Tg P y^{-1} of which 0.07 Tg P y^{-1} is the result of burning of coal (Bertine and Goldberg 1971).

Terrestrial phosphorus

The terrestrial phosphorus is primarily derived from various sources such as transfer of phosphorus from dead organic matter to soil, atmospheric fallout, industries, etc. The transfer of phosphorus from the terrestrial biomass to soil as dead organic matter has been estimated to be 136.4 Tg P y^{-1}; out of which 133.3 Tg P y^{-1} is from plant sources and only 3.1 Tg P y^{-1} from animal material (Pierrou 1976). The natural input of phosphorus in the soil from the atmospheric fallout due to the seaspray is estimated to be 3.6-9.2 Tg P y^{-1} (Pierrou 1976); man made annual contribution in the form of superphosphate is 9.93 Tg P y^{-1} (FAO, 1975) and 1.1 Tg as human excreta. Hutchinson (1952) estimated that guano deposit contribute about 0.1 Tg P annually to terrestrial phosphorus.

Phosphorus is also utilized by a large number of plants and animals. The uptake of phosphorus only by plants from soil as calculated by Bazilevich (1974) is 178 Tg P y^{-1}, and including the freshwater ecosystems, this estimate is 236.8 Tg P y^{-1} (Stumm 1973). The absorption of phosphorus to the terrestrial biota from aquatic ecosystem is 0.065 Tg P y^{-1} and from foodstuffs and pharmaceutical products is 0.063 Tg P y^{-1} (Pierrou 1976).

Most part of the phosphorus in the soil is immobilized and inaccessible to plants due to the formation of complexes with iron, aluminium and calcium. According to Phillips & Webb (1971), soluble phosphorus immobilised by reactions with soil components within 2-3 cm from the fertilizer granule. Other factors responsible for the loss of availability of soluble phosphorus to terrestrial plants are leaching of soluble phosphorus (2.5-12.3 Tg P y^{-1}) and erosion of phosphorus rich soils (Pierrou 1976). However, some soil components, such as humic acids increase the solubility and thus availability of phosphorus. The bacterial action is also important in making phosphorus available to plants.

Sediment phosphorus

The parent rock material from which soil evolves by weathering is variable in phosphorus content and this variability increases with the thickness and heterogenity of the stratified soil layer. Basic igneous rocks contain relatively little phosphorus as apatite. Phosphorus percentage of other rocks are : sandstone - ca 0.02, gneiss - 0.04, unweathered loss - 0.07, andesite - 0.16, and diabase - 0.03. Limestone containing ca 1.3 per cent P and rare deposits of rock phosphate (10 to 15 per cent), in which biotic accumulations either from organisms themselves or from quano were concentrated, are largely of sedimentary origin (Hutchinson 1950).

In freshwater ecosystem the amounts of phosphorus (introduced to sediments are estimated to be of the order of 1 Tg P y^{-1}, while the amounts released from sediments per year are estimated to be less than 1 Tg P y^{-1} (Pierrou 1976). Emery *et al.* (1955) calculated that the amount of phosphorus deposited in ocean sediments is 13 Tg p y^{-1}. The amounts of phosphorus released from ocean sediments is not known but is probably relatively small.

Aquatic phosphorus

The phosphorus content of ground water is generally low; average concentrations are 20 µg P l^{-1}. Even in areas of soils containing high phosphorus content the values are low due to (1) high insolubility of phosphate-containing minerals, (2) the scavenging of surface phosphate concentrations by biotic removal and (3) soil percolation (Wetzel 1975). In general the regional climatic characteristics of surface waters are closely related to the soil characteristics of their drainage basins. The quantities of phosphorus entering surface drainage are influenced by the amount of phosphorus by soils, topography, vegetative cover, quantity and duration of runoff flow, land use and pollutions (Wetzel 1975). The topography of the land in the catchment basin influences the extent of erosion and subsequent export of nutrients. Moreover, the relative erosion is influenced markedly by the type of vegetation.

Urbanization results an increase in phosphorus, discharged to surface water in approximately direct proportion to population densities (Wetzel 1969). Losses of phosphorus from heavy residential fertilizations, storm, sewer, drainage and leaves increase phosphorus input to surface drainage (Wetzel 1975).

The amounts of phosphorus in flowing waters are generally than 100 mg m^{-3} total-P (Vollenweider 1968). The flux of phosphorus in freshwater system transported by river runoff has been estimated at between 1.9 and 2.0 Tg P y^{-1} (Stumm 1973; Gulbrandsen and Robertson 1973). Emery *et al.* (1955) estimated the total phosphorus transport by rivers including sediments to be 14 Tg P y^{-1}. According to Pierrou (1976) the total phosphorus transport by rivers is in the order of 17.4 Tg P y^{-1} of which 3.7 Tg P y^{-1} is the dissolved particulate phosphorus and 13.7 Tg P y^{-1} is the sediment loading.

Uptake of phosphorus by phytoplankton

The uptake of phosphorus by phytoplankton in ocean has been variously calculated as 1300 Tg P y^{-1} (Emery *et al.* 1955) and 990 Tg P y^{-1} (Stumm 1973). A similar estimate about 1000 Tg P y^{-1} can be made for the amount of phosphorus deposited in ocean detritus (Pierrou 1976). The large storage of phosphorus in the ocean (Table 14.6) and the large internal circulation within the ocean can absorb relatively large addition of phosphorus without causing marked effect on the concentration of phosphorus in water. Stumm (1973) stated that even small increase in phosphorus concentration will increase the fraction of the ocean bottom covered by anoxic waters. According to Broecker (1974) the ocean may balance the oxygen decrease by changing the dynamics of loss and gain of phosphorus between water and sediments. The reaction time is in the order of 10-100 ky.

TABLE 14.6. *Different sources of phosphorus (Tg P)*

Biomass :	
Human	< 1
Terrestrial	1305
Marine	128
Freshwater	< 1
Waters :	
Fresh	90
Marine	120,000-128,000
Soil :	160,000
Rocks :	
Total solid sphere	1.1×10^{13}
Mineable	6500-59,000

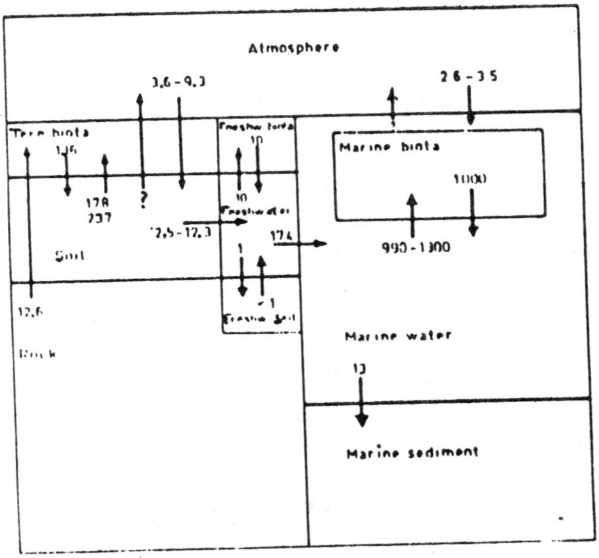

Fig. 14.12. A brief outline of global flows of phosphorus.

Global P cycle

A simplified model of the global phosphorus cycle (Pierrou 1976) revealed that the turnover rate of phosphorus is regulated by the rate of diaegnesis of phosphorus containing sediments into phosphate rock (Fig. 14.12); the process takes 0.1 - 1 Gy (Broecker 1974) implying that a period of more than 1 Gy is required for one global cycle of phosphorus to be completed.

General comments

Phosphorus is essential for all living organisms. Geologic in origin, phosphorus is a macro-nutrient but its availability is extremely low. Phosphorus often acts as limiting element in oligotrophic system but may cause hazards in eutrophic lakes.

It appears that phosphorus plays an important role in regulating productivity in oligotrophic systems. Mud may act as sink for phosphorus in eutrophic system and as a source of P in oligotrophic systems.

The total phosphorus concentration of most natural waters is between 10-50 $\mu g\ P\ l^{-1}$, and water bodies with total-P concentration of 30-100 $\mu g\ P\ l^{-1}$ are considered as eutrophic. Based on nutrient loading, there are two basic loading tolerance limits : permissible limit and dangerous level. On this basis, most of the Chinese lakes are above the excessive loading curve.

Models describing the dependence of lake trophic state on external phosphorus loading are of two general types : empirical steady-state models derived from statistical treatment of data from large number of lakes and theoretical, dynamic model based on detailed mathematical description of kinetics of nutrients and population dynamics. Dynamic theoretical models are more applicable for prediction of rates of response to purturbations than the empirical steady-state model. Simple model with flexible structure generally have advantages over complex models.

Unabetting eutrophication resulting from sewage and other pollutants associated with human activities has stimulated much discussion and research on environmental protection. Reduction of phosphate input from point sources and various non point sources is considered important for restoration programme. One approach is to reduce or eliminate the use of high phosphate detergents. Another method is the chemical treatment of sewage before it can enter a water body. Removal of nutrient rich sediment by dredging or by the use of chemicals where nutrients are fixed in sediments are used for environmental protection.

REFERENCES

Aaronson, S. 1971. The synthesis of extracellular macromolecules and membranes by a population of the phytoflagellate *Ochromonas danica*. Limnol. Oceanogr. **16** : 1-9.

Aaronson, S. and Patni, N.J. 1976. The role of surface and extracellular phosphatases in the phosphorus requirement of *Ochromonas*. Limnol. Oceanogr. **21** : 838-845.

Ahl, T. and Oden, S. 1975. Narsaltkallor - en oversikt. Nordforsk, Helsinki, **1** : 99-133.

Ahlgren, I., Frisk, T. and Kamp-Nielsen, L. 1988. Empirical and theoretical models of phosphorus loading, retention and concentration vs. lake trophic state. Hydrobiologia **170** : 285-303.

Alexander, G.R. 1978. The ratiande for a ban on detergent phosphate in the Great Lakes Basin. In : Phosphorus in the Environment, its chemistry and biochemistry. A Ciba Foundation Symposium. Elsevier, Amsterdam, pp. 269-284.

Andersen, J.M. 1974. Nitrogen and phosphorus budgets and the role of sediments in six shallow Danish Lakes. Arch. Hydrobiol;. **74** : 528-550.

Andersson, G., Berggren, H., Cornberg, G. and Gelin, C. 1978. Effects of planktivorous and benthivorous fish on organisms and water chemistry in eutrophic lakes. Hydrobiologia **59** : 9-15.

Andersson, G., Granelli, W. and Stenson, J. 1988. The influence of animals on phosphorus cycling in lake ecosystems. Hydrobiologia **170** : 267-284.

Antia, N.J. and Watt, A 1965. Phosphatase activity in some species of marine phytoplankters. *J. Fish Res. Bd. Can.* **22** : 793-799.

Arson, O.M., Solomatina, V.D. and Romanenko, V.D. 1984. Role of phosphorus in aquous medium in the regulation of bioenergetic processes in fish. Gidrobiol Zh. **20** : 53-57.

Bargh, B.J. 1977. Output of water, suspended sediment and phosphorus and nitrogen forms from a small forested catchment. New Zealand Forestry Science **7** : 162-171.

Barlow, J.P. and Bishop, L. W. 1965. Phosphate regeneration by zooplankton in Cayuga lake. Limnol. Oceanogr. **10** : 15-24.

Bazilevich, N.I. (1974). Energy flow and biogeochemical regularities of the main world ecosystems. In : A. J. Cave (ed.) Proceeding of the first International Congress of Ecology. Structure, functioning and Management of Ecosystems, Wegeningen : Pudoc, pp. 182-186.

Benndorf, J., Koschel, R. and Recknagel, F. 1981. Erste Ergebnisse der Okosystemanalyse des Pelagials des Stechlinsees mit der Modellmethodse. Limnologica (Berlin) **13** : 195-202.

Berge, D., Rognerud, S. and Johannessen, M. 1979. Videreutvikling av fosforbelastningsmodellen for store sjiktede innsjoer. NIVA-arbok. Oslo, 39-42.

Berman, T. 1969. Phosphatase release of inorganic phosphorus in Lake Kinneret. Nature **224** : 1231-1232.

Berman, T. 1970. Alkaline phosphatases and phosphorus availability in Lake Kinneret. Limnol. Oceanogr. **15** : 663-674.

Berman, T. and Moses, G. 1972. Phosphorus availability and alkaline phosphatase activities in two Israeli fish ponds. Hydrobiologia **40** : 487-498.

Berner, R.A. 1980. Early diagenesis. A theoretical approach. Princeton series in geochemistry. Princeton Univ. Press. Princeton 241 p.

Bertine, K.K. and Goldberg, E.D. 1971. Fossil fuel combustion and the major sedimentary cycle. Science **173** : 233-235.

Bjork, S. 1972. Swedish lake restoration programme gets results. Ambio. **1** : 153-165.

Boavida, M.J. and Heath, R.T. 1984. Are the phosphatases released by *Daphnia magna* components of its food? Limnol. Oceanogr. **29** : 641-645.

Bostrom, B. 1984. Potential mobility of phosphorus in different types of lake sediments. Int. Revue ges. Hydrobiol. **69** : 457-474.

Bostrom, B., Jansson, M. and Forsberg, C. 1982. Phosphorus release from lake sediments. Arch. Hydrobiol. Beih. Ergebn. Limnol. **18** : 5-59.

Bostrom, B., Ahlgren, I. and Bell, R. 1985. Internal nutrient loading in a eutrophic lake, reflected in seasonal variations of sediment parameters : Verh. Int. Verein. Limnol. **22** : 3335-3339.

Bostrom, B., Andersen, J.M. Fleischer, S. and Jansson, M. 1988. Exchange of phosphorus across the sediment-water interface. Hydrobiologia **170** : 229-244.

Boyd, C.E. 1982. Water quality management for pond fish culture. Elsevier, Amsterdam Oxford New York, 318 p.

Brakke, D.F. 1977. Rainwater : nutrient additions to a hypereutrophic lake, Hydrobiologia **52** : 159-163.

Broecker, W.S. 1974. Chemical Oceanography. Harcourt Brace Jovanovich, New York, 214 p.

Brunskill, G.J., Povoledo, D., Graham, B.W. and Stainton M. 1971. Chemistry of surface sediments of sixteen lakes in the experimental lakes Area, north western Ontario. J. Fish. Res. Bd. Canada. **28** : 277-294.

Chapra, S.C. 1980. Application of the phosphorus loading concept to the Great Lakes. In : Loehr, C., Martin, C.S. and Rast, W. (eds.), Phosphorus management strategies for lakes. Ann Arbor Science Publishers, Ann Arbor, pp. 135-152.

Chrost, R.J., Siuda, W. and Halemejko, G. 1984. Longterm studies on alkaline phosphatase activity (APA) in a lake with fish aquaculture in relation to lake eutrophication and phosphorus cycle. Arch. Hydrobiol. **70** : 1-32.

Cole, G.A. 1988. Textbook of Limnology. Waveland Press Inc., Illinois, 401 pp.

Cooke, G.D., McComas, M.R., Waller, D.W. and Kennedy, R.H. 1977. The occurrence of internal phosphorus loading, in two small, eutrophic, glacial lakes in northeastern Ohio. Hydrobiologia. **56** : 129-135.

Cooke, G.D., Welch, E.B., Peterson, S.A. and Newroth, P.R. 1986. Lake and reservoir restoration. Butterworth Publ., Boston.

Cullen, P. and Forsberg, C. 1988. Experiences with reducing point sources of phosphorus to lakes. Hydrobiologia. **170** : 321-336.

Davis, R.B. 1974. Stratigraphic effects of tubificids in profoundal lake sediments. Limnol. Oceanogr. **19** : 466-488.

Davis, R.B., Thurlow, D.L. and Brewster, F.E. 1975. Effects of burrowing tubificid worms on the exchange of phosphorus between lake sediment and overlying water. Verh. Int. Verein. Limnol. **19** : 382-394.

DE Pinto, J. U., Edzwald, J.K., Switzenbaum, M.S. and Young, T.C. 1980. Phosphorus removal in lower Great Lakes municipal treatment plants. US Envir. Prot. Agency Rep. no. 600/2-80-177. Cincinnati, Ohio.

Dillon, P.J. 1975. The phosphorus budget of Cameron Lake, Ontario : The importance of flushing rate to the degree of eutrophy of lakes. Limnol. Oceanogr. **19** : 28-39.

Dillon, P. J. and Rigler, F.H. 1975. A simple method for predicting the capacity of a lake for development based on lake trophic status. J. Fish. Res. Bd. Can. **32** : 1519-1531.

Doremus, C. and Clesceri, L.S. 1982. Microbial metabolism in surface sediments and its role in the immobilization of phosphorus in oligotrophic lake sediments. Hydrobiologia **91** : 261-268.

Drake, J.C. and Heaney, S.I. 1987. Occurrence of phosphorus and its potential remobilization in the littoral sediments of a productive English lake. Freshwater Biology. **17** : 513-523.

Duffy, P.D., Schreiber, J.D., McClurkin, D.C. and McDowell, L.L. 1978. Aqueous and sediment phase phosphorus yields from five Southern Pine Watersheads. J. Envir. Qual. **7** : 45-50.

Edmondson, W.T. and Lehman, J.T. 1981. The effect of changes in the nutrient income on the condition of Lake Washington Limnol. Oceanogr. **26** : 1-29.

Einsele, W. 1936. Uber die Beziehungen des Einsenkreislaufs zum Phosphatkreislauf im eutrophen see., Arch. Hydrobiol. **29** : 664-686.

Emery, K.O., Orr, W.L. and Rittenberg, S.C. 1955. Nutrient budgets in the ocean. In : Essays in the Natural Sciences in Honor of Captain Allan Hancock, University of Southern California Press, Los Angeles, CA., 299-309 pp.

Enell, M. and Lofgren, S. 1988. Phosphorus in interstitial water : methods and dynamics. Hydrobiologia **170** : 103-132.

Evdokimova, T.J., Grishina, L.A., Vasilyevskaya, V.D., Samvilova, E.M. and Batriskaya, T.L. 1976. Biogeochemical cycles of elements in some natural zones of European USSR. In Svensson, B. H. and Soderlund, R. (eds.), Nitrogen, phosphorus and sulphur-global cycles, - SCOPE Report 7, Ecol. Bull Stock holm, **22** : 135-155.

Fao. 1975. Production Yearbook, 1974. Vol. 28 Rome : FAO 328 pp.

Ferrante, J.G. 1976. The role of zooplankton in the intrabio-coenotic phosphorus cycle and factors affecting phosphorus excretion in a lake Hydrobiologia **49** : 203-214.

Forsberg, C. and Wallsten, M. 1986. Fox formsatting inom Horbyans avrinningsomrade. Vatten **42** : 310-315.

Francko, D.A. 1983. Size-fractionation of alkaline phosphatase activity in lake water by membrane filtration. J. Fresh. Ecol. **2** : 305-309.

Franko, D.A. 1984. Phytoplankton metabolism and cyclic nucleotides. II. Nucleotide-induced perturbations of alkaline phosphatase activity. Arch. Hydrobiol. **100** : 409-421.

Fuller, W.H. 1972. Phosphorus cycle. In : Fairbridge, R.W. (ed.). The Encyclopedia of Geochemistry and Environmental Sciences. Vol. IV A. Von Reinhold Company, New York, 946-951 pp.

Gachter, R., Imboden, D., Buhrer, H. and Stadelmann, P. 1983. Mogliche Massnahmen zur Restaurierung des Sempacherseas. Schweiz. Z. Hydrol. **45** : 246-266.

Gardner, W.S., Nalepa, T.F., Quigley, M. A. and Malezyk, J.M. 1981. Release of phosphorus by certain benthic invertebrates. Can J. Fish. aquat. Sci. **38** : 978-981.

Glew, R.H. and Heath, E.C. 1971. Studies on the extracellular alkaline phosphatase of *Micrococcus sodonensis*. I. Isolation and characterization. J. Biol. Chem. **246** : 1556-1565.

Golterman, H.L. 1982. Loading concentration models for phosphate in shallow lakes. Hydrobiologia, **19** : 169-174.

Graneli, W. 1979. The influence of *Chironomus plumosus* larvae on the exchange of dissolved substances between sediment and water. Hydrobiologia **66** : 149-159.

Groot, C.J. de and Golterman, H.L. 1991. Phosphorus compounds in sediments : Organic aspects. Third International Workshop on Phosphorus in Sediments (Sept. 30 - Oct. 3, 1991), Woudschoten/ Utrecht, the Nederlands, 20 p.

Gulbrandsen, R.A. and Robertson, C.E. 1973. Inorganic phosphorus in sea water. In : Griffith, E.J., Beeton, A., Spencer, J.M. and Mitchell, D.T. (eds.), Environmental phosphorus Handbook. John Wiley, New York, 117-140 pp.

Gunatilaka, A 1984. Observations on phosphorus dynamics and orthophosphate turnover in à tropical lake - Parakrama Samudra, Sri Lanka. Verh. Int. Verein. Limnol. **22** : 1567-1571.

Haapala, K. 1977. Luftburen fororeningstilforsel. Vattenstyrelsens observationer 1971-1976. Nordforsk, Helsinki **2** : 293-316 pp.

Hakansson, L. and Jansson, M. 1983. Principles of lake sedimentology. Springer Verlag, Berlin, 316 pp.

Haaremoes, P. 1977. Betydningen af forurening fra regnafstromning for valg afurbane aflobssystemer-en oversigt over nordisk litterature og vurolering af status. Nordforsk, Helsinki, **2** : 293-316.

Halemejko, G.Z. and Chrost, R.J. 1984. The role of phosphatases in phosphorus mineralization during decomposition of lake phytoplankton blooms. Arch. Hydrobiol. **101** : 489-502.

Hargrave, B.T. and Geen, G.H. 1968. Phosphorus excretion by zooplankton. Limnol. Oceanogr. **13** : 332-342.

Healey, F.P. 1973. Characteristics of phosphorus deficiency in *Anabaena* J. Phycol. **9** : 383-394.

Healey, F.P. 1985. Interacting effects of light and nutrient limitation on the growth rate of *Synechococcus linearis* (Cyanophyceae). J. Phycol. **21** : 134-146.

Healey, F.P. and Hendzel, L.L. 1979. Fluorometric measurement of alkaline phosphatase activity in algae. Freshwat. Biol. **9** : 429-439.

Hecky, R.E. and Kilham, P. 1988. Nutrient limitation of phytoplankton in freshwater and marine environments : A review of recent evidence on the effects of enrichment. Limnol. Oceanogr. **33** : 796-822.

Herodek, S. 1984. The eutrophication of Lake Balaton : measurements, modeling and management. Verh. Int. Verein. Limnol. **22** : 1087-1091.

Hieltjes, A.H.M. and Lijklema, L. 1980. Fractionation of inorganic phosphates in calcareous sediments. J. envir. Qual. **9** : 405-407.

Holdren, G.C. and Armstrong, D.E. 1980. Factors affecting phosphorus release from intact lake sediment cores. Env. Sci. and Tech. **14** : 79-87.

Holtan, H., Kamp-Nielsen, L. and Stuanes, A.O. 1988. Phosphorus in soil, water and sediment : an overview. Hydrobiologia **170** : 19-34.

Hrbacek, J., Dvorakova, M., Korinek, V. and Prochazkova, L. 1961. Demonstration of the effect of the fish stock on the species composition on zooplankton and the intensity of metabolism of the whole plankton association. Verh. Int. Verein. Limnol. **14** : 192-195.

Huber, A.L., Gabrielson, J.O., Dolin, P.J. and Kidby, D.K. 1983. Decomposition of *Cladophora.* III. Heterotroph populations and phosphatase activity associated with *in vitro* phosphorus mineralization. Bot. mar. **26** : 181-188.

Hurlbert, S.H., Zedler, J. and Fairbanks, D. 1972. Ecosystem alteration by mosquito-fish (*Gambusia affinis*) predation. Science **175** : 639-641.

Hutchinson, G.E. 1950. The biogeochemistry of vertebrate excretion. Bull. Amer. Mus. Nat. Hist. **96** : 554 pp.

Hutchinson, G.E. 1952. The biochemistry of phosphorus. In : Wolterink, L. F. (ed.) The Biology of Phosphorus, Michigan state college press, 1-35 pp.

Jackson, J.A. and Schindler, D.W. 1975. The biogeochemistry of phosphorus in an experimental lake environment : evidence for the formation of humic-metal-phosphate complexes. Verh. Int. Verein. Limnol. **19** : 21-22.

Jana, B.B. and Das, S.K. 1992. Bioturbation induced changes of fertilizer value of phosphate rock in relation to alkaline phosphatase activity. Aquaculture **103** : 321-330.

Jana, B.B. and Sahuy, S.N. 1993. Relative performance of three bottom grazing fishes (*Cyprinus carpio, Cirrhinus mrigala, Heteropneustes fossilis*) in increasing the fertilizer value of phosphate rock. Aquaculture (in press).

Jana, B.B., Das S.K. and Chakraborty, D. 1992. Use of common carp (*Cyprinus carpio* L.) in increasing the fertilizer value of phosphate rock. Journal of Applied Ichthyology **8** : 222-227.

Jansson, M. 1976. Phosphatases in lake water : Characterization of enzymes from phytoplankton and zooplankton by gel filtration. Science **194** : 320-321.

Jansson, M. 1977. Vattenbalans Och Kemiska Budget-berakningar for Slugsjoen 1971-1975. Koukkelprojektets rapport 5. Inst. of Limnology, Univ. Uppsala, 3-45 pp.

Jansson, M., Olsson, H. and Broberg, O. 1981. Characterization of acid phosphatases in the acidified Lake Gardsjon, Sweden. Arch. Hydrobiol. **92** : 377-395.

Jansson, M., Olsson, H. and Pettersson, K. 1988. Phosphatases; Origin characteristics and function in lakes. Hydrobiologia **170** : 157-175.

Jones, J.G. 1972a. Studies on freshwater bacteria : Association with algae and alkaline phosphatase activity. J. Ecol. **60** : 59-75.

Jones, J.G. 1972b. Studies on freshwater micro-organisms : phosphatase activity in lakes of differing degrees of eutrophication. J. Ecol. **60** : 777-791.

Jonge, V.N. de., Engelikes, M.E. and Bakker, J.F. 1991. Bioavailability of phosphorus compounds in sediments of Western Dutch Wadden Sea. Third International Workshop on Phosphorus in sediments, Woudschoten/Utrecht, The Nederlands, 25 pp.

Jorgensen, S.E. 1976. A eutrophication model for a lake. Ecol. Modelling. **2** : 147-165.

Jorgensen, S.E. 1983. Eutrophication models of lakes. In : Jorgensen, S.E. (ed.), Application of ecological modelling in environmental management. Elsevier, Amsterdam, 5 p.

Jorgensen, S.E. and Revsbech, N.P. 1985. Diffuse boundary layers and the oxygen uptake of sediments and detritus. Limnol. Oceanogr. **30** : 111-122.

Jorgensen. S.E., Hoi, I. and Jacobsen, O.S. 1973. A prognosis for a lake. Vatten. **29** : 382-404.

Jorgensen, S.E., Kamp-Nielsen, L. and Jacobsen, O.S. 1975. A submodel for anaerobic mud-water exchange of phosphate. Ecol. Modelling, **1** : 133-146.

Kamp-Nielsen, L. 1974. Mud-water exchange of phosphate and other ions in undisturbed sediment cores and factors affecting the exchange rates. Arch. Hydrobiol. **73** : 218-237.

Kamp-Nielsen, L. 1985. Modelling of eutrophication processes. Proceedings of the International Congress on Lake Pollution and Recovery, Rome, April 1985.

Kitchell. J.F., Koonce, J.F. and Tennis, P.S. 1975. Phosphorus flux through fishes. Verh. Int. Verein. Limnol. **19** : 2478-2484.

Kitchell, J.F., O'Neill, R.V., Webb, D., Galepp, G.W., Bartell, S.M., Koonce, J.F. and Ausmus, B.S. 1979. Consumer regulation of nutrient cycling. Bio-Science. **29** : 28-34.

Knoblauch, A.R. 1982. Estimation of phosphorus exchange rates in the upper part of the Wahnbach Reservoir using balance calculations. Archiv Für Hydrobiologie Ergebnisse der Limnologie. **18** : 69-77.

Kormondy, E.J. 1978. Concepts of ecology. Prentice Hall of India Pvt. Ltd., New Delhi, 238 pp.

Krezoski, J.R., Mozley, S.C. and Robbins, J.A. 1978. Influence of benthic macroinvertebrates on mixing of profundal sediments in Southeastern Lake Huron. Limnol. Oceanogr. **23** : 1011-1016.

Kuenzler, E.J. and Perras, J.P. 1965. Phosphatases of marine algae. Biol. Bull. Woods Hole. **128** : 271-284.

Lamarra, V.A. 1975. Digestive activities of carp as a major contributor to the nutrient loading of lakes. Verh. Int. Verein. Limnol. **19** : 2461-2468.

Lang, C. 1990. Quantitative relationship between oligochaete communities and phosphorus concentrations in lakes. Fresh-water Biology. **24** : 327-334.

Lappalainen, K.M. 1982. Convection in bottom sediments and its role in material exchange between water and sediments. Hydrobiologia **86** : 105-108.

Lean, D.R.S. 1973 : Movements of phosphorus between its biologically important forms in lake water. J. Fish. Res. Ed. Can. **30** : 1525-1536.

Lean, D.R.S. and Rigler, F.H. 1974. A test of the hypothesis that abiotic phosphate complexing influences phosphorus kinetics in epilimnetic lake water. Limnol. Oceanogr. **19** : 784-788.

Lee, G.F., Sonzogni, W.C. and Spear, R.D. 1977. Significance of oxic vs anoxic conditions for Lake Mendota sediment phosphorus release. In : Golterman, H.L. (ed.), Interactions between sediments and freshwater. Dr. W. Junk B.V., The Hague, 294-306 pp.

Lehman, J.T. 1980. Nutrient cycling as an interface between algae and grazers in freshwater communities. In : Kerfoot, W.C. (ed.) Evolution and ecology of zooplankton communities. The University Press of New England, Hanover, (N.H.), Lond, 251-263 pp.

Lijklema, L. 1980. Interaction of orthophosphate with iron (III) and aluminium hydroxides, Env. Sci. and Techn. **14** : 537-540.

Lijklema, L. 1985. Internal loading, mechanisms and assessment of magnitude. In : Proceedings of the International Congress on Lakes Pollution and Recovery, Rome.

Lijklema, L., Gelencser, P. and Szilagyi, F. 1983. Sediments and sediment-water interaction. In : Somlyody, L., Herodek, S. and Fischer, J. (Eds.), Eutrophication of shallow lakes : Modelling and management. Internatl. Institute for Applied systems Analysis, Laxemburg, Austria, 81-100 pp.

Lin, C.K. 1977. Accumulation of water soluble phosphorus and hydrolysis of phosphatases by *Cladophora glomerata* (Chlorophyceae). J. Phycol. **13** : 46-51.

Lofgren, S. and Ryding, S.O. 1985. Apatite solubility and microbial activities as regulators of internal loading in shallow, eutrophic lakes. Verh. Int. Verein. Limnol. **22** : 3329-3334.

Lynch, M. and Shapiro, J. 1981. Predation, enrichment and phytoplankton community structure. Limnol. Oceanogr. **26** : 86-102.

Marsden, M.W. (1989) : Lake restoration by reducing external phosphorus loading : the influence of sediment phosphorus release. Freshwater Biology **21** : 139-162.

Matisoff, G., Fisher, J.B. and McCall, P.L. 1981. Kinetics of nutrient and metal release from decomposing lake sediments. Geochim. Cosmochim. Acta. **45** : 2333-2347.

Matisoff, G., Fisher, J.B. and Matis, S. 1985. Effects of benthic macroinvertebrates on the exchange of solute between sediments and freshwater. Hydrobiologia **122** : 19-33.

Merritt, R.W., Cummins, K.W. and Burton, T.M. 1984. The role of aquatic insects in the processing and cycling of nutrients. In : Resh, V.H. and Rosenberg, D.M. (eds.) The ecology of aquatic insects, Praeger, New York 134-163 pp.

Mortimer, C.H. 1941. The exchange of dissolved substances between mud and water in lakes. I.J. Ecol. **29** : 280-329.

Mortimer, C.H. 1942. The exchange of dissolved substances between mud and water in lakes. II. J. Ecol. **30** : 147-201.

Murphy, T.J. and Doskey, P.V. 1976. Input of phosphorus from precipitation to Lake Michigan. J. Great Lakes Res. **2** : 60-70.

Nakashima, B.S. and Leggett, W.C. 1980. The role of fishes in the regulation of phosphorus availability in lakes. Can. J. Fish. aquat. Sci. **37** : 1540-1549.

Neame, P.A. 1977. Phosphorus flux across the sediment water interface. In : Golterman, H.L. (ed.), Interactions between sediments and freshwater, W. Junk, The Hague, 307-312 pp.

Nicholls, K.H. and Dillon, P.J. 1978. An evaluation of phosphorus-chlorophyll-phytoplankton relationship in lakes. Int. Revue ges. Hydrobiol. **63** : 141-154.

Odum, E.P. 1983. Basic ecology. Saunders College Publishing, Holt-Saunders, Japan, 613 pp.

Oecd 1982. Eutrophication of waters. Monitoring, assessment and control. OECD, Paris, 154 pp.

Ohle, W. 1937. Kolloidgele als Nàhrstoffregulanten der Gewässer. Naturwissenschaften **25** : 471-474.

Olsson, H. 1990. Phosphatase activity in relation to phytoplankton composition and pH in Swedish lakes. Freshwat. Biol. **23** : 353-362.

Osborne, P.L. 1981. Phosphorus and nitrogen budgets of Barton Broad and predicted effects of a reduction in nutrient loading on phytoplankton biomass in Barton, Sutton and Stalnam Broads, Norfolk, United Kingdom. Int. Revue Gao. Hydrobiol. **66** : 171-202.

Overbeck, J. and Babenzien, H.D. 1964. Uber den Nachweis von freien Enzymen in Gewasser. Arch. Hydrobiol. **60** : 107-114.

Patni, N.J., Dawale, S.W. and Aaronson, S. 1977. Extracellular phosphatases of *Chlamydomonas reinhardi* and their regulation. J. Bact. **130** : 205-211.

Persson, L. 1983. Food consumption and the significance of detritus and algae to intraspecific competition in roach *Rutilis rutilis* in a shallow eutrophic lake. Oikos. **41** : 118-125.

Peters, R.H. 1986. The role of prediction in limnology. Limnol. Oceanogr. **31** : 1143-1159.

Peters, R.H. and Rigler, F.H. 1973. Phosphorus release by *Daphnia*. Limnol. Oceanogr. **18** : 821-839.

Peters, J. and Van Liere, L. 1985. Dredging and ground-water movements, effects on phosphorus release from sediments. Verh. Int. Verein. Limnol. **22** : 3340-3344.

Petr, T. 1977. Bioturbation and exchange of chemicals in the mud water interface. In : Golterman, H.L. (ed.), Interactions between sediments and fresh water, Dr. W. Junk B.V., The Hague, 216-226 pp.

Pettersson, K. 1980. Alkaline phosphatase activity and algal surplus phosphorus as phosphorus-deficiency indicators in Lake Erken. Arch. Hydrobiol. **89** : 54-87.

Pettersson, K. and Bostrom, B. 1986. Phosphorus exchange between sediment and water in Lake Balaton - In : Sly, P.G. (ed.) - Sediments and water interactions Springer-Verlag, New York, 427-435 pp.

Phillips, A.B. and Webb, J.R. 1971. Phosphorus fertilizers. In : Olson, R.A., Amy, T.J., Hanway, J.J. and Kilmer, V.J. (eds.), Fertilizer Technology and Use, 2nd edn. Soil. Science society of America. Inc., Madison, WI, 271-301 pp.

Pierrou, U. 1976. The Global phosphorus Cycle. In : Svensson, B.H. and Soderlund, R. (eds.), Nitrogen phosphorus and Sulphur-Global Cycles. Scope Report 7. Ecol. Bull., Stockholm, **22** : 75-88 pp.

Rainey, R.H. 1967. Natural displacement of pollution from the Great Lakes. Science, **155** : 1242-1243.

Rebsbech, N.P., Sorensen, J., Blackburn, T.H. and Lomholt, J.P. 1980. Distribution of oxygen in marine sediments measured with microelectrodes. Limnol. Oceanogr. **25** : 403-411.

Reckhow, K.H. 1979. Empirical lake models for phosphorus : development, applications, limitations and uncertainty. In : Scavia, D. and Robertson, A. (eds.) Perspectives on lake ecosystem modelling. ann Arbor Science Publishers, Ann Arbor., 193-221 pp.

Reichardt, W. 1971. Catalytic mobilization of phosphate in lake water and by Cyanophyta. Hydrobiologia **38** : 377-394.

Reichardt, W., Overbeck, J. and Steubing, L. 1967. Free dissolved enzymes in lake waters. Nature **216** : 1345-1347.

Riemann, B. 1977. Phosphorus budget for a non-stratified Danish lake and horizontal differences in phytoplankton growth. Arch. Hydrobiol. **79** : 357-381.

Rigler, F.H. 1956. A tracer study of the phosphorus cycle in lake water. Ecology **37** : 550-562.

Rigler, F.H. 1961. The uptake and release of inorganic phosphorus by *Daphnia magna* Straus. Limnol. Oceanogr. **6** : 165-174.

Rigler, F.H. 1964. The phosphorus fractions and the turnover time of inorganic phosphorus in different types of lakes. Limnol. Oceanogr. **9** : 511-518.

Rigler, F.H. 1973. A dynamic view of the phosphorus cycle in lakes. In : Griffith, E.J., Beeton, A., Spencer, J.M. and Mitchell, D.T. (eds.), Environmental phosphorus Handbook. J. Wiley and Sons NY : 539-568 pp.

Ripl. W. and Lindmark, G. 1979. The impact of algae and nutrient composition on sediment exchange dynamics. Arch. Hydrobiol. **86** : 45-65.

Rippey, B. 1977. The behaviour of phosphorus and silicon in undisturbed cores of Lough Neagh sediments. In : Golterman, H.L. (ed.), Interactions between sediments and fresh-water, W. Junk B.V., The Hague, 348-353. pp.

Ryding, S.O. 1981. Reversibility of man induced eutrophication. Experiences of a lake recovery study in Sweden. Int. Revue Ges. Hydrobiol. **66** : 449-503.

Ryding, S.O. 1985. Chemical and microbiological processes as regulators of the exchange of substances between sediments and water in shallow eutrophic lakes. Int. Revue Ges. Hydrobiol. **70** : 657-702.

Ryding, S.O. and Forsberg, C. 1976. Six polluted lakes : a preliminary evaluation of the treatment and recovery process. Ambio **4** : 151-156.

Ryding, S.O. and Forsberg, C. 1977. Sediments as a nutrient source in shallow polluted lakes. In : Golterman, H.L. (ed.), Interactions between sediments and freshwater, Dr. W. Junk B.V., The Hague, 227-234 pp.

Schaffner, W.R. and Oglesby, R.T. 1978. Phosphorus loadings to lakes and some of their responses - I. Limnol. Oceanogr. **23** : 120-134.

Schindler, D.W. and Nighswander, J.E. 1970. Nutrient supply and primary production in Clear Lake, eastern Ontario. J. Fish. Res. Bd. Can. **27** : 2009-2036.

Schindler, D.W., Newbury, R.W., Beaty, K.G. and Campbell, P. 1976. Natural water and chemical budgets for a small precambrian lake basin in central Canada. J. Fish. Res. Bd Can. **33** : 2526-2543.

Serruya, C., Edelstein, M., Pollingher, U. and Serruya, S. 1974. Lake Kinnert sediments : Nutrient composition of pore water and mud water exchanges. Limnol. Oceanogr. **19** : 489-508.

Shapiro, J. and Carlson, R. 1982. Comment on the role of fish in regulation of phosphorus availability in lakes. Can. J. Fish. Aquat. Sci. **39** : 364.

Sonzogni, W.C. and Lee, G.F. 1974. Nutrient sources for Lake Mendota-1972. Trans. Wis. Acad. Sci. Arts. Lett. **62** : 133-164.

Sonzogni, W.C., Chapra, S.C., Armstrong, D.E. and Longan, T.J. 1982. Bioavailability of phosphorus inputs to lakes. J. Environ. Qual. **11** : 555-563.

Spatzierer, G., Ludwig, C. and Matsche, N. 1985. Biological phosphorus removal in combination with simultaneous precipitation. Wat. Sci. Tech. **17** : 163-176.

Starkel, W.M. 1985. Predicting the effect of macrobenthos on the sediment water flux of metals and phosphorus. Can. J. Fish. Aquat. Sci. **42** : 95-100.

Stauffer, R.E. 1985. Relationship between phosphorus loading and trophic state in calcareous lakes of southeast Wisconsin. Limnol. Oceanogr. **30** : 123-145.

Stauffer, R.E. and Armstrong, D.E. 1986. Cycling of iron, manganese, silica, phosphorus, calcium and potassium in two stratified basins of Shagawa Lake, Minnesota, Geochimica et Cosmochimica Acta. **50** : 215-229.

Steiner, M. 1938. Zur Keantnis der Phosphat-kreislaufes in Seen. Naturwissenschaften **26** : 723-724.

Stewart, A.J. and Wetzel, R.G. 1982. Phytoplankton contribution to alkaline phosphatase activity. Arch. Hydrobiol. **93** 265-271.

Straskraba, M. and Gnauck, A.H. 1985. Freshwater ecosystems. Modelling and simulation. Developments in environmental modelling, 8 Elsevier, Amsterdam, 309 pp.

Stumm, W. 1973. The acceleration of the hydrogeochemical cycling of phosphorus. Water Res. **7** : 131-144.

Syers, J.K., Harris, R.F. and Armstrong, D.E. 1973. Phosphate chemistry in lake sediments. J. Env. Qual. **2** : 1-14.

Taylor, W.D. and Leen, D.R.S. 1981. Radiotracer experiments on phosphorus uptake and release by limnetic microzooplankton. Can. J. Fish. aquat. Sci. **38** : 1316-1321.

Thomas, E.A. 1969. Die Elimination der Phosphor verbindungal aus dem Abwasser. Berichte der ATV **23** : 253-275.

Timms, R.M. and Mass, B. 1984. Prevention of growth of potentially dense phytoplankton populations by zooplankton grazing in the presence of zooplanktivorous fish in a shallow wetland system. Limnol. Oceanogr. **29** : 472-486.

Tiwari, B.K. and Mishra, R.R. 1982. A study on biological activity measurements and heterotrophic bacteria in a small freshwater lake. Hydrobiologia **94** : 257-267.

Van Liere, L. and Mur, L.R. 1982. The influence of groundwater movement on the phosphorus release from sediments, as measured in a continuous flow system. Hydrobiologia **92** : 511-518.

Van Raaphorst, W. and Brinkman, A.G. 1984. The calculation of transport coefficients of phosphate and calcium fluxes across the sediment - water interface, from experiments with undisturbed sediment cores. Water Sci. Tech. **17** : 941-951.

Vollenweider, R.A. 1968. The scientific basis of lake eutrophication, with particular reference to phosphorus and nitrogen as eutrophication factors. Tech. Rep. DAS/DSI/68.27. OECD, Paris, 159 pp.

Vollenweider, R.A. 1969. Moglichkeiten und Grenzen elementarer Modelle der Stoffbilanz von Seen. Arch. Hydrobiol. **66** : 1-36.

Vollenweider, R.A. 1972. Input-output models. Mimeographed report, Burlington, Ontario, Can. Cent. Inland Waters, 40 pp.

Vollenweider, R.A. 1975. Input-output models. With special reference to the phosphorus loading concept in limnology. Schweiz. Z. Hydrol. **37** : 53-84.

Vollenweider, R.A. 1976. Advances in defining critical loading levels for phosphorus in lake eutrophication Mam. 1st. ital. Idrobiol. **33** : 53-83.

Vollenweider, R.A. and Krekes, J.J. 1982. Eutrophication of waters. Monitoring, assessment and control. OECD, Paris, 154 pp.

Walther, K. and Fries, L. 1976. Extracellular alkaline phosphatase in multicellular marine algae and their utilization of glycerophosphate. Physiol. Pl. **36** : 118-122.

Wetzel, R.G. 1969. Factors influencing photosynthesis and excretion of dissolved organic matter by aquatic macrophytes in hard-water lakes. Verh. Int. Verein. Limnol. **17** : 72-85.

Wetzel, R.G. 1975. Limnology. W.B. Saunders Company, Philadelphia, London, Toronto, 743 pp.

Wetzel, R.G. 1981. Longterm dissolved and particulate alkaline phosphatase activity in hardwater lake in relation to lake stability and phosphorus enrichments Verh. Int. Verein. Limnol. **21** : 369-381.

Wisniewski, R.J. and Planter, M. 1985. Exchange of phosphorus across sediment water interface (with special attention to the influence of biotic factors) in several lakes of different trophic status. Verh. Int. Verein. Limnol. **22** : 3345-3349.

Wright, D.I. and Shapiro, J. 1984. Nutrient reduction by biomanipulation : An unexpected phenomenon and its possible cause. Verh. Int. Verein. Limnol. **22** : 518-524.

Wynne, D. 1981. The role of phosphatases in the metabolism of *Peridinium cinctum*, from Lake Kinneret. Hydrobiologia **83** : 93-99.

Xiangsan, J., Hongliang, L., Qingying, T., Zongshe, Z. and Xuan, Z. 1990. Eutrophication of lakes in China. Proceedings of the 4th International Conference on the conservation and management of lakes, Hangzhou, 652 pp.

Chapter 15

Interrelationships of Phyto- and Zooplankton in Indian Wetlands

Ashok K. Pandit

Centre of Research for Development, University of Kashmir, Srinagar-190 006, Jammu & Kashmir, India

Abstract. The paper reports on a comparative study of phytoplankton and Zooplankton interrelationships, particularly of seasonal succession, species composition, food-chain and prey-predator relationship, Biomass and production, in two wetlands of India.

Key words : *Wetlands, Phytoplankton, Zooplankton, Species composition, Seasonal succession, Trophic relations, Eutrophication, Biomass, Production.*

Introduction

Depending upon the nutrient availability, light penetration, grazing pressure and other associated factors, plankton population shows great variation in the wetlands. Since the wetlands are dominated by profuse growth of macrophytes, the productivity of plankton in typical wetlands is relatively low compared to natural lakes and ponds. For instance, the phytoplankton production of the waters of Kolleru wetland, situated in Andhra Pradesh between the deltas of Godavari and Krishna rivers, ranged between 1.60-4.94 g O_2 m^{-2} day^{-1}, being comparatively low when compared to that of 3.64-23.20 g O_2 m^{-2} day^{-1} of the fish tanks (Seshavatharam and Venu, 1982). The distribution, density and biomass of zooplankton in wetland ecosystems is, however, determined by the availability of food sources in the form of phytoplankton and microbe-organic detritus complex as well as the associated environmental factors, and hence a close relationship exists between the phyto- and zooplankton communities in these productive ecosystems. Though some information is available on the distribution, density and biomass of phyto- and zooplankton in lakes and ponds, yet very little is known about these aspects in wetlands except some work carried out in wetlands of Kashmir (Kaul *et al*, 1978; Kaul and Pandit, 1982; Pandit and

Kaul, 1982). The aim of the present study is, therefore, to elucidate the phyto- and zooplankton interrelationships in the subtemperate wetlands of Kashmir Himalaya and its comparison with the tropical wetlands as typified by the chaurs of Bihar.

Seasonal Succession of Phyto- and Zooplankton

The studies indicate that, in general, two peaks of zooplankton growth are obtained, one in late spring to early summer and the other in autumn. The first peak, shortly after the spring phytoplankton peak in the sub-temperate wetlands and late winter maximum in tropical wetlands, which in turn start gradually declining, indicate the basic source of food being very much conducive for the growth of zooplankton. The sudden drop of algal cells in summer appears to be related to excessive grazing of phytoplankton by large number of herbivorous zooplankton which subsequently switch over to its feeding on huge quantities of bacteria. Besides, many planktonic algae are unsuitable as food for many herbivorous zooplankters and certain large sized zooplankters are secondary consumers with bacteria and planktogenic detritus forming a large proportion of their diet. This, for obvious reasons, complicates the interaction between the phyto- and zooplankton. Thus, in a balanced ecosystem a certain ratio prevails between the biomass of the producers and primary consumers, the rise in primary production causing an increase in zooplankton biomass also.

Trophic Relations in the Plankton Community

Food and Feeding Habits of Zooplankton

Protozoa : Studies on the food habits of the most common species of Protozoa in wetlands of Kashmir reveal *Actinosphaerium eichhornii* and *Acanthocystis chaetophora* to be predatory types capable of capturing many prey species of various sizes with the help of axopods, while giant amoeba, *Pelomyxa palustris,* prefers only small ciliates and zooflagellates. *Amoeba* sp., *Paramecium caudatum, P. aurelia, Stentor roseli* and vorticellids exhibit an omnivorous feeding and their food consists of a variety of algae and other protozoans, besides bacteria and organic detritus. Majority of the protozoans namely, *Centropyxis, Arcella, Difflugia, Euglypha, Wailesella* and *Lesquereusia* among rhizopods, *Porodon, Pseudoporodon, Holophrya, Colpoda, Colpidium, Euplotes, Opisthotrichia, Climacostomum* and *Bursaria* among ciliates and *Monas* and *Paramastix* among zooflagellates, are distinguished as primary consumers as their feeding involves both herbivorous and detritivorous types. The amount of algae, bacteria and detritus grazed, however, vary from species to species, depending upon the food preference of a particular consumer species coupled with the availability of food resources. Among the protozoans, only *Paramecium bursaria* exhibits a holophytic type of nutrition by entering into a symbiotic association with green *Chlorella.*

Rotifera : Among the Rotaria only *Polyarthra vulgaris, Epiphanes* sp., *Keratella cochlearis* and *Trichocera* sp. are the regular herbivores with the last two species often seen drilling into an algal filament to suck out its contents. *Euchlanis diliata* is an omnivore and feeds on *Pandorina,* diatoms and protozoans which form its favourable food. Certain rotifers like *Syncheata pectinata,*

Asplanchna priodonta and *Gastropus stylifer* are raptorial predators, sucking out the prey with the help of a piston. Among these *Asplanchna priodonta* shows a distinct preference for *Brachionus* over other available rotifers. Large *Cephalodella*, besides being highly predaceous over protozoans and other available rotifers, often exhibits what is called "cannabolism", by devouring small individuals of its own type. Majority of the filter-feeding rotifers (*Brachionus* sp., *Keratella valga, K. quadrata, Monostyla* sp., *Lecane* sp., *Notholca acuminata, Phliodina* sp., *Colurella* sp., *Platyias* sp., *Diplois* sp., *Mytilina* sp., *Pompholyx* sp. and *Filina* sp.) employ the corona for rafting the prey, comprising protozoans, zooflagellates and bacteria, into the mouth. These rotifers are categorized as secondary consumers on the basis of food habits. The higher percentage of rotifers in fall of spring in wetlands may be attributed to the availability of rich protozoan food which in turn subsists on the higher amounts of organic matter providing the basic source of food, especially in highly eutrophicated wetlands of Kashmir like Malgam and Mirgund.

Crustacea : The food of filter-feeding cladocerans (water-fleas) consists of both bacteria and a variety of algae usually held in suspension (seston)- Pandit (1980) observed that these organisms, under laboratory conditions, were seen to reject blue-green algae in favour of other algal food stuffs. Ostracods, again graze on algae preferably of oligotrophic nature. Both the groups, being largely herbivorous, are distinguished as primary consumers. While Cladocera and Ostracoda show some degree of uniformity in their food habits, the copepods exhibit a great diversity in their feeding behaviour. Among the latter group of animals only *Eucyclops agilis* seems to be a pure herbivore while *Bryocamptus* sp. and *Canthocamptus* sp. evince omnivorous feeding. Laboratory experiments have shown *Cyclops vernalis* and *C. bicuspidatus* consuming 6 or even more prey animals every day, depending upon the size of the latter. In one set of experiments to confirm the food selection of *C. bicuspidatus* 75 per cent of the prey animals were *Ceriodaphnia* sp. and *Simocephalus* sp. and the remaining 25 per cent were *Diaptomus* sp. (Pandit, 1980). In plankton samples, immature *Daphnia* were occasionally found grasped and partly consumed by *C. bicuspidatus,* though no predation on *Daphnia* was seen to occur under laboratory experimentation. *Diaptomus* sp., which formed the prey for *C. bicuspidatus,* were in turn seen preying upon *C. vernalis*, rotifers and ostracods, besides their own detritus feeding nauplii, hence showing cannabolism. The food of *C. vernalis* in summer is provided by most common water-fleas viz., *Bosmina, Cerio-daphnia, Simocephalus, Scapholebris, Alona* and *Chydorus*, besides *Daphnia,* nauplii of diaptomids and rotifers. A small portion of algae has been reported to occur constantly in the intestine of this species (Pandit, 1980). Further observations showed rotifers, chironomids, cladocerans and protozoans in addition to algae making up the variety of potential food sources for other cyclopoids in wetlands of Kashmir.

Planktonic Food Chains and Food Web

The food habits of various zooplankton species inhabiting the typical wetlands of Kashmir (Nowgam, Malgam, Haigam, Mirgund and Hokarsar) are summarised elsewhere in details by Pandit (1980) and Pandit and Kaul (1982). According to different trophic levels, in the food-chain, the zooplankton inhabiting the wetlands can be classified into : (i) primary consumers which feed directly on plant material (algae and bacteria), (ii) secondary consumers which in turn graze upon the primary consumers, and (iii) tertiary consumers consuming secondary consumers. Even omnivores, acting as both primary and secondary consumers, are distinguishable.

The diversity of zooplankton communities coupled with the variety of potential foods, comprising phytoplankton, zooplankton, detritus and bacteria suggest much complex and variable nature of trophic relations in the food-chains that lead to the building of a series of linear food-chains each consisting of a number of steps (Figure 15.1) and a generalized type of food web prevalent in the wetlands (Figure 15.2).

The constituent links of the trophic chains, according to Pandit (1980) and Pandit and Kaul (1982), exhibit two main features :

(i) that each link in the food chain has a food source restricted to biomass than the density,

(ii) that the accelerated eutrophication causes a shortening of food chain since primary consumers largely switch from planktonic algae to bacteria and detritus, and

(iii) that the grazing of a particular species over the other of lower trophic level being primarily a matter of size than density.

Trophic Structure of Plankton Community

Trophic structure of plankton community in Kashmir wetlands revealed upright pyramids of number and biomass (Figures 15.3 and 15.4). The pyramids for various wetland sites, however, differed on quantitative basis. A comparison of biomass at various trophic levels in the pyramid gives an idea of the relative levels of production in different wetlands. Thus, the ratio of standing crop between trophic level is :

Wetland Site	H/P (%)	C/O+H (%)
Nowgam	15.24	15.87
Malgam	13.49	9.36
Haigam	16.43	7.85
Mirgund	25.13	4.94
Hokarsar	34.21	8.93

The turn over rates for the producers and the primary or secondary consumers, differ a great deal, which makes the comparison of the production estimated from biomass at various trophic levels so difficult. According to Nauwerck (1963), the phytoplankton is replaced as many as 120 times a year, whilst the value for zooplankton is 8-10 times. In the study on Kashmir

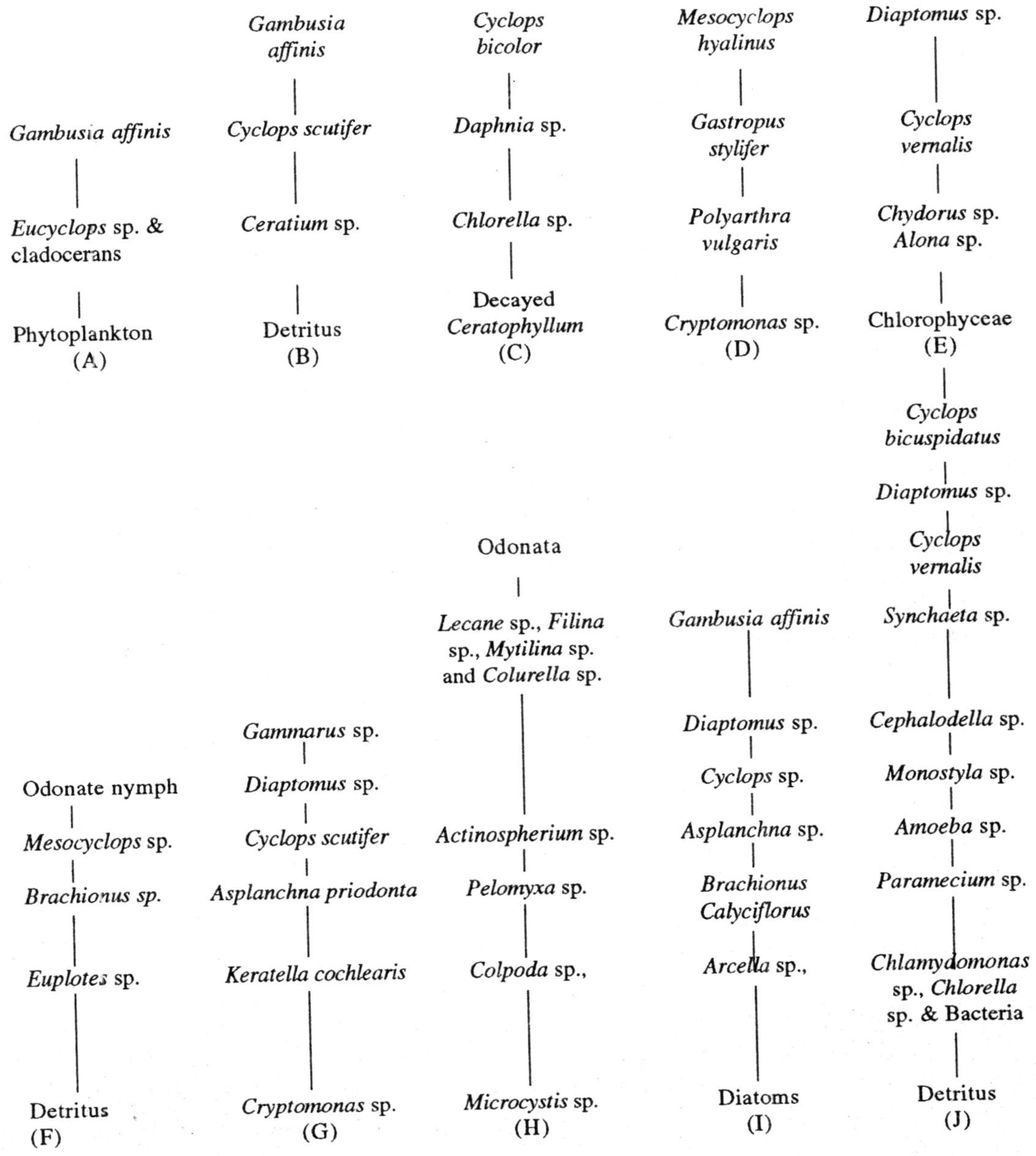

Fig. 15.1. A series of food chains in Kashmir wetlands based on trophic relations of plankton (Pandit, 1980; Pandit and Kaul, 1982).

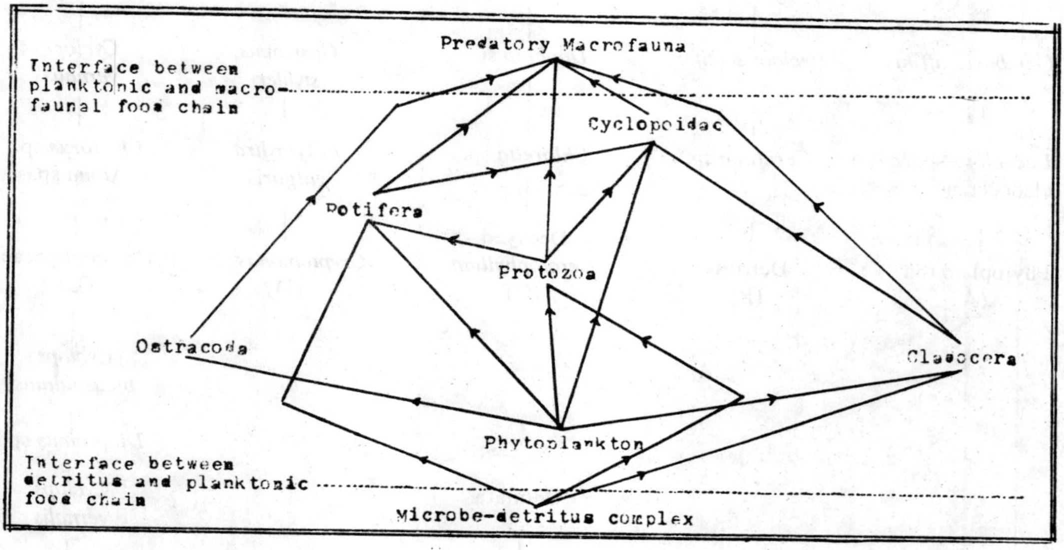

Fig. 15.2. A typical food web of plankton community in Kashmir wetlands showing the possible flow of materials (Pandit, 1980).

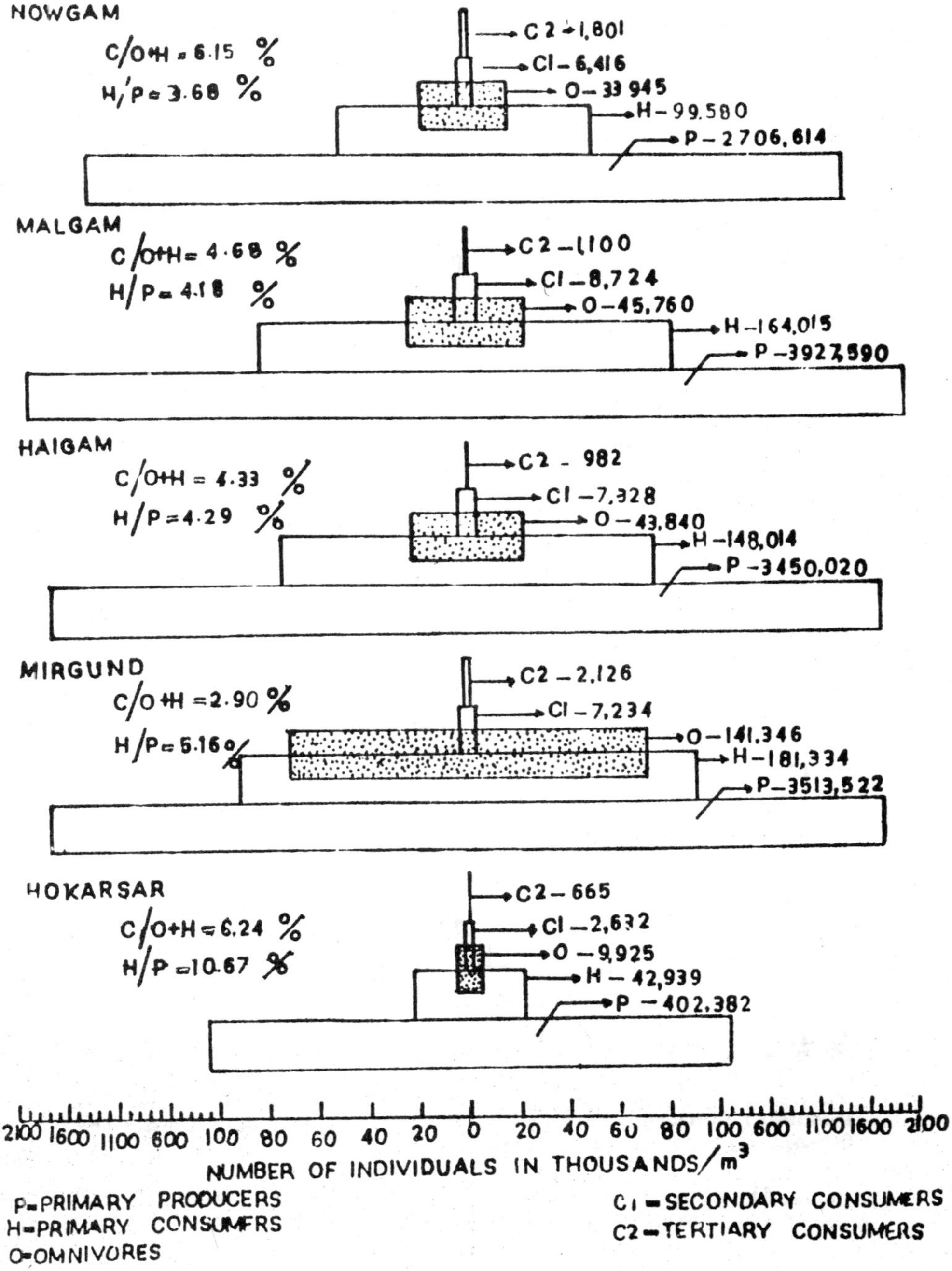

Fig. 15.3. Pyramids of number (based on average density) of plankton in five wetlands of Kashmir (Source : Pandit, 1980; c.f. Pandit and Kaul, 1982).

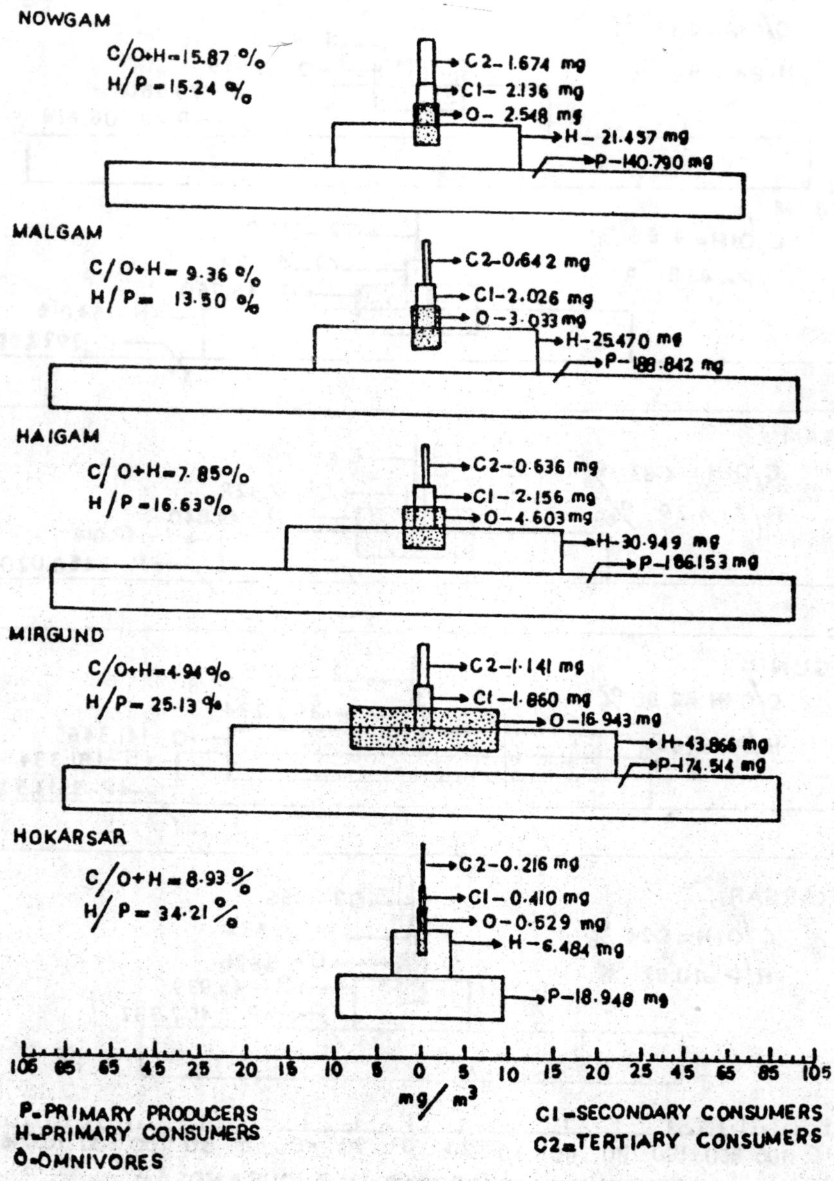

Fig. 15.4. Pyramids of biomass (based on mean biomass) of plankton in five wetlands of Kashmir (Source : Pandit, 1980; c.f. Pandit and Kaul, 1982).

wetlands, however, the phytoplankton density and biomass increased more rapidly as compared to herbivores and a considerable part of it (70-80%) was left unutilized as the residue of primary production. It resulted in a pyramidal structure since the microscopic filterfeeders (water-fleas, rotifers and protozoans) that dominate in eutrophic wetlands graze on only small quantities of primary production with bacteria and fine tripton particles being their main diet. The pyramidal structure is evident even with regard to density and biomass within different taxonomic groups but for rotifers showing the predominance of carnivores over the herbivores and omnivores. In both Protozoa and Crustacea, however, there is a striking predominance of herbivores in comparison to omnivores and carnivores (Pandit, 1980; Pandit and Kaul, 1982).

Eutrophication Changes the Structure of Plankton Community

Whatever may be the mode of origin of wetlands, from the moment of its birth, fertility and the productivity of these ecosystems begin to increase under the impact of certain natural forces bringing in nutrients. The influence of human activities (anthropogenic pressures) superimposing on natural forces have accelerated the rate of eutrophication of these and other biotopes in the immediate vicinity of human settlements. In this way many shallow lakes have also been influenced and got finally converted into wetlands, though with varying degree of eutrophication.

The discharge of organic wastes and nutrients into lentic waterbodies like lakes and wetlands creates a situation different from that of lotic habitats. During the ice-free period, the conditions favour dilution and mixing and the nutrients in the effluents cause eutrophication, leading to an increase in primary production. This is distinguished from natural eutrophication only by its rapidity. The increased supply of nutrients to the waters in due course leads to changes in the species composition and a rise in production particularly towards margins. Thus, a comparison of planktonic communities in different wetlands indicates various degrees of eutrophication. The differences are well illustrated by the strongly eutrophied Mirgund and Malgam sites, moderately eutrophied Nowgam and Haigam sites and the least eutrophied Hokarsar. This is because of the fact that plankton is very sensitive indicator of pollution than the chemical parameters as a small increase in nutrients can evoke an immediate reaction which gets easily manifested in the larger changes in the species composition and the abundance of phytoplankton and thus affects the zooplankton, the immediate consumers of phytoplankton.

Changes in the Species Composition and Indicator Species

A perusal of the data collected on wetlands of Kashmir indicate that, in general, Cyanophyceae, Chlorophyceae (excluding *Sphaerocystis* sp. and desmids), Euglenophyceae, few species of Bacillariophyceae and *Peridinium* sp. prefer eutrophic waters. In such shallow feeding niches are more often found the strongly eutrophic crustaceans viz., *Bosmina longirostris, Chydorus* sp.; *Cyclops vernalis* and *Eucyclops agilis.* While majority of the protozoans and rotifer species are restricted to sites rich in organic detritus, certain rotifers namely *Keratella cochlearis, Brachionus calyciflorus, Epiphanes* sp., and *Asplanchna priodonta* alongwith the protozoans like *Arcella* sp. and *Centropyxis* sp. can tolerate severe pollution and hence exhibit their abundance at the sewage outfalls. Such sites are often dominated by certain blue-greens like *Oscillitoria, Anabaena, Gomphosposphaera* and *Microcystis*. Among the phytoplankton, majority of desmids in addition to *Sphaerocystis schroeteri,* Chrysophyceae (*Dinobryon sertularia* and *D. sociale*), Dinophyceae (*Ceratium* sp.) and certain diatoms (*Diatoma balfouriana, Astronella formosa, Cymatopleura solea, Cyclotella stelligera, Coconeis* sp., *Ceratoneis* sp. and *Synedra acus*) and among the zooplankton only 6 crustaceans viz., *Sida crystillina, Pseudosida bidentata, Daphnia middendorffiana, Cyclops bicuspidatus* and *Diaptomus* sp. are, on the other hand, more characteristic of oligotrophic conditions. *Bosmina coregoni*, though an indicator of oligotrophy, actually prefers eurytopic conditions. *Diaptomus* sp., distinguished as a strong indicator of oligotrophy, is very rare in wetlands and has been occasionally found in the gut contents of a carnivorous fish (*Gambusia affinis*) and in a free state in Hokarsar wetland.

Pandit (1980) further observed that in highly eutrophicated Mirgund wetland the number of phytoplankton species (179) had increased and that of zooplankton (97) decreased while opposite was true for the less eutrophicated Hokarsar site having 156 species of phytoplankton and 100 species of zooplankton. Pronounced eutrophication, thus, seemingly results in a reduction in the average number of zooplankton and a corresponding increase in the phytoplankton species, this being especially true for the final stages of stagnation. The drop in the number of zooplankton species and the consequent decrease in the zooplankton species diversity as a result of pronounced eutrophication points to the existence of strongly limiting factors which disturb the stability of plankton community and cause a simultaneous increase in biomass.

Changes in the Plankton Biomass and Production

The phytoplankton is opined to be rath

r a sensitive indicator of even slight eutrophication as compared to the chemical properties of water. Even if the changes in the nutrient content are very low, their combined influence may evoke a reaction in the phytoplankton. The studies carried on Kashmir wetlands further suggest that certain planktonic algae either invade the community at a very early stage of eutrophication. Remarkable spatial differences, chiefly caused by the nutrient load of the incoming waste waters, are often observed in phytoplankton biomass. A comparison of different wetlands in regard to primary production at phytoplankton level, thus, shows Malgam to be most productive (189 mg m^{-2} mean biomass) and Hokarsar least productive (19 mg m^{-2} mean biomass). Mirgund, Haigam and Nowgam are the intermediates between the two extremes (Table 15.1).

The rise in primary production leads to increased biomass of herbivorous zooplankton, though the phytoplankton biomass increases more rapidly with eutrophication than the zooplankton biomass. In the final phases of eutrophication however, the effects are reversed (Jumppanen, 1976; Pandit, 1980). Similar results have also been obtained by other investigators (Shuskina, 1966; Shcherbakov, 1967; Patalas, 1968 and Widuto, 1977) while working on other waterbodies.

TABLE 15.1. *Total number of species, average density (x 10^4 indiv. m^{-3}) and mean biomass (mg m^{-3}) of phytoplankton and zooplankton communities in five wetlands of Kashmir (Pandit, 1980)*

Parameters	Nowgam	Malgam	Haigam	Mirgund	Hokarsar
*Phytoplankton**					
Number of Species	189	177	183	179	156
Density	270.66	392.75	345.00	351.35	40.23
Biomass	140.80	188.84	186.15	174.51	18.95
*Zooplankton***					
Number of Species	109	109	118	97	100
Density	14.17	21.98	20.01	33.20	5.61
Biomass	27.81	31.17	38.34	63.81	7.64

* Comprises Cyanophyceae, Chlorophyceae, Bacillariophyceae, Euglenophyceae, Xanthophyceae, Chrysophyceae and Dinophyceae.

** Comprises Protozoa, Rotifera and Crustacea.

In general, the Kashmir wetlands show an exceedingly greater ratio, the phytoplankton biomass being greater (4-5 times) than that of zooplankton, a situation quite contrasting to the one obtained by Nauwerck (1963) according to whom the phytoplankton biomass never exceeds zooplankton biomass in the eutrophic Lake Erken. Since nutrient value of planktonic algae decrease in eutrophied biotopes, bacteria and detritus form a large proportion of omnivorous filter-feeders. Chrysomonads, important as food, also replaced by green and blue-green algae, is one of the important reasons for the replacement of indicators of oligotrophy like *Sida crystillina, Pseudosida bidentata, Bosmina coregoni* and *Daphnia middendorffiana* by Rotifera, Cladocera and Copepoda, the former species being encountered only in small numbers.

Ratio between the Phytoplankton and Zooplankton

In a balanced ecosystem a certain ratio prevails between the phytoplankton and the zooplankton. Thus, in tropical wetlands (chaurs) of Bihar (Table 15.2), the proportion of phyto- and zooplankton obtained on the basis of populations have been calculated to be 99.2 per cent and 0.8 per cent in January and 95 per cent and 5 per cent in August and on the basis of species composition it came to be 89.7 per cent and 10.3 per cent in February and 83.9 per cent and 16.1 per cent in August respectively (Rai and Dutta Munshi, 1982). Preliminary observations on Pichhola-Fatehsagar wetland complex (Udaipur, Rajasthan) reveal that in all 93 phytoplankton (algal) species and 57 zooplankton (Protozoa, Rotifera, Cladocera, Copepoda and Ostracoda) species have been recorded, thereby depicting a proportion of 62 per cent and 38 per cent respectively (Vyas, 1989). In contrast, Kashmir wetlands evinced still higher proportions of zooplankton, comprising 14.17-28.73 per cent by weight and 5.3-12.25 per cent by density as

TABLE 15.2. *Density (N), number of species (S) and species diversity (H) of phyto- and zooplankton communities in Chaurs of Bihar (Rai and Datta Munshi, 1982)*

Parameters	Jan.	Feb.	Aug.	Sep.	Oct.	Nov.	Dec.
*Phytoplankton**							
N (indiv. l^{-1})	28,619	16,051	8,288	14,026	23,037	24,176	19,050
S	85	96	36	71	77	87	84
H	5.08	5.27	3.37	3.98	4.52	5.06	5.05
*Zooplankton***							
N (indiv. l^{-1})	442	191	440	603	723	610	545
S	11	11	7	12	13	14	12
H	3.25	3.75	2.57	3.52	3.55	3.67	3.69

* includes Cyanophyceae, Chlorophyceae and Bacillariophyceae.
** includes Rotifers and Crustacea (Cladocera and Copepoda).

TABLE 15.3. *Relation between phyto- and zooplankton in Kashmir wetlands (Pandit, 1980)*

Wetlands	Biomass					Density	
	% Phytoplankton (P)	% Herbivorous Zooplankton (H)	P/H	% Phytopl-ankton	% Zoopl-ankton	% Phytopl-ankton	% Zoopl-ankton
Nowgam	86.78	13.22	6.56	83.50	16.50	95.02	4.98
Malgam	88.12	11.88	7.47	85.83	14.17	94.70	5.30
Haigam	85.74	14.26	6.01	82.92	17.08	94.52	5.48
Mirgund	79.91	20.09	4.00	73.23	26.77	91.37	8.63
Hokarsar	74.50	25.50	2.92	71.27	28.73	87.75	12.25

against 71.27-85.83 per cent by weight and 87.75-94.7 per cent by density of phytoplankton (Table 15.3). Similarly the proportions of 57.4 per cent and 42.6 per cent were obtained for phytoplankton and zooplankton respectively on the basis of total number of species recorded in these wetlands. The ratio between the phytoplankton (P) and herbivorous zooplankton (H) biomass in five different wetlands of Kashmir has been further calculated as :

Wetland Site	*P/H ratio*
Nowgam	6.56
Malgam	7.47
Haigam	6.01
Mirgund	4.00
Hokarsar	2.92

As evident from the above table, the highest ratio is recorded for Malgam, followed by Nowgam, Haigam and Mirgund, decreasing to the lowest for Hokarsar. The values suggest the phytoplankton biomass increasing more rapidly with eutrophication as compared to zooplankton but during the final stages of eutrophication, as evinced by the strongly eutrophied Mirgund site, the relationship between the biomasses is quite reversed. Pandit (1980) also believed that the ratio between the phyto- and zooplankton biomass shows rather wide temporal variations, being higher in spring and autumn than in summer and winter. The low proportion of zooplankton in tropical wetlands of Bihar compared to Kashmir wetlands is probably due to the fact that protozoans are not taken into account, otherwise the system shows some similarities though under different climatic conditions (Table 15.2).

The fact that during initial stages of eutrophication the phytoplankton biomass grows more rapidly than that of herbivores and during the final stages the relationship between the biomasses is quite reversed, in all likelihood, is due to the rapid turn over of algal populations and the bacterial production, as a comparatively small phytoplankton biomass seems to be sufficient to maintain a large biomass of primary consumers in less eutrophic biotopes. In the highly eutrophic wetlands, however, a considerable part of phytoplankton biomass remains unutilized as a residue of primary production which then is decomposed by bacteria and utilized by detritivores, thus leading to shortening of food chains. The role of planktonic algae in the zooplankton diet is further diminished by a notable decline in the proportion of primary consumers and that of omnivores and detritivores showing a corresponding increase in the zooplankton biomass, as is clearly indicated by the distribution of zooplankton by food source (Table 15.4). In the strongly eutrophied Mirgund site, the primary consumers comprised 68.74 per cent by weight and 54.61 per cent by density as compared to the least eutrophied Hokarsar site where the corresponding values were as high as 84.68 per cent and 76.46 per cent respectively. A similar but less pronounced change may be discerned in Malgam, Nowgam and Haigam wetlands also. Since primary consumers mainly use the phytoplankton as food, it appears that only a part of the net primary production is utilized by the zooplankton in stringly eutrophied biotopes. An important part (70-80 per cent) of unedible algae presumably sediments and is decomposed by bacteria and then utilized by detritivores. This may result in a sharp increase in decomposition and also shortening of food chain, since primary consumers largely switch from planktonic algae to bacteria and detritus. This is well supported by the fact that microscopic filter-feeders, dominating in eutrophic wetlands, grazed only small quantities of phytoplankton while bacteria and fine tripton particles constituted their main diet. Thus, it

TABLE 15.4. *Distribution of zooplankton in Kashmir wetlands by food habits (Pandit, 1980; Pandit and Kaul, 1982).*

Wetlands	Herbivores (H) %		Omnivores and Detritivores (O) %		Carnivores (C_1) %		Carnivores (C_2) %	
	Density	Biomass	Density	Biomass	Density	Biomass	Density	Biomass
Nowgam	70.25	77.14	23.95	9.16	4.53	7.68	1.27	6.02
Malgam	74.62	81.71	20.91	9.73	3.97	6.50	0.50	2.06
Haigam	73.95	80.71	21.90	12.00	3.66	5.62	0.50	1.66
Mirgund	54.61	68.74	42.57	26.56	2.18	2.91	0.64	1.79
Hokarsar	76.96	84.68	17.67	6.92	4.69	5.37	1.18	2.83

seems that at critical times of the year, the food web of an extremely eutrophied wetland may largely consist of producers and decomposers. In contrast, in a well balanced ecosystem, greater part of energy fixed by primary producers flows to the metabolism and net production of primary consumers and then to secondary consumers, an assumption well supported by H/P and C/O+H ratio's and supported by Jumppanen (1976). Thus, the disturbances occurring in production as a result of vigorous eutrophication first appear at the level of primary consumers and presumably also at that of secondary consumers, but gradually extend throughout the whole food chain and the ecosystem. The lower consumption of planktonic algae by zooplankton in the eutrophied wetlands may be attributed to the unfavourable structure of the plankton community (blue-green algae being rejected by water-fleas) and the gradual disappearance of larger herbivorous crustaceans. It is further envisaged that larger zooplankton species especially *Daphnia* are more active in the decrease of the algal biomass than the smaller species (Anonymous, 1992). The ecological niche ordinarily harbouring such animals as a consequence remain unoccupied and may at times disturb the equilibrium of the ecosystem, besides the energy flow as also suggested by Jumppanen (1976) and Pandit (1980). They opined that the larger planktonic herbivores compose a weak but rather important link in the food chain in the process of rapid eutrophication, being comparatively sensitive to chemical changes in addition to the composition and production of phytoplankton. The very same reason explains the numerical surge of large cladocerans (e.g. *Daphnia* sp.), preceded by a short lived increase of small cladocerans (e.g. *Bosmina longirostris, Alona* sp. and *Chydorus* sp.) and a marked increase in the number of *Cyclops vernalis*. Such a shift is presumably a predator avoidance mechanism helping the small crustaceans to escape predation by fish (*Gambusia affinis*, fry and fingerlings of *Cyprinus carpio*) and other predators (Brynildson and Kempinger 1973), Anderson (1974) and Richards *et al* (1975). This hypothesis gains further support from the recent studies (long term observations) carried on the relations of the biomass of phytoplankton, Cladocera and Copepoda in two reservoirs of Czechoslovakia (Anonymous, 1992). From these observations it follows that higher fishstock induces higher share of cyclopoid copepods in the biomass of crustacean zooplankton in the growing season. For this phenomenon at least two assumptions can be presented. One obvious is that copepods escape more rapidly to the fish impact than do the cladocerans. The other possible explanation is that the larger cladocerans induce size distribution of phytoplankton which is not suitable as food for the development stages of cyclopoid copepods. The first very hypothesis, therefore, reasonably explains the greater percentage compositions of copepods in the total zooplankton population and biomass during autumn, a period when the fish impact is the most intense especially in wetlands of Kashmir (Pandit, 1980; Anonymous, 1992). In other studies, Brook (1969) observed a shift in dominance, from large-bodied to small-bodied planktonic herbivores (as a result of size dependant predation), being very much indicative of higher degree of eutrophication. The mass occurrence of *Daphnia pulex*, tolerant to pollution, at few stations in Kashmir wetlands is due to the fact that predaceous animals and fish feeding on zooplankton usually avoid polluted biotopes (Pandit, 1980). Similar results are also obtained by Jumppanen (1976).

Predator – Prey Relationships

It is likely that predator-prey relationship, competition and permanence of waterbody are the most important factors influencing the distribution pattern of certain ecologically important species like *Cyclops vernalis, C. bicuspidatus* and *Diaptomus* sp. and the dimensions of food web. Since *C. bicuspidatus* preys heavily on *Diaptomus* sp. both seem to be positively associated and, therefore, both decline. Such a decline may also be attributed to their oligotrophic nature (Carl, 1940; Tundisi and Tundisi, 1976; Anderson and Green, 1975-76). The disappearance of *Diaptomus* sp. may further be attributed to the heavy predation of *Diaptomus* nauplii by *C. vernalis* and adults by *C. bicuspidatus*, freshwater shrimps (*Gammarus* sp.), small mosquito fish (*Gambusia affinis*) and fry of carp (*Cyprinus carpio*), and its cannabolism. The dominance of *C. vernalis* (an inhabitant of shallow waters unlike *C. bicuspidatus* and *Diaptomus* sp. prefering deep waters) in wetlands may further be due to little predation by *Diaptomus* sp. (Anderson, 1970). This also supports the view point held by Anderson (1970, 72) that *C. bicuspidatus* declines in presence of more abundant *C. vernalis*. Thus, *Diaptomus* sp. – *C. bicuspidatus* association again forms a weaker link in the food chain of wetlands (Pandit, 1980, Pandit and Kaul, 1982), the most important food links showing the possible flow of materials being depicted in a complex food web (Fig. 15.5).

Summary and Conclusions

In wetlands the profuse growth of macrophytes is mainly responsible for the fixation of bulk of organic carbon, and hence the productivity of plankton in these ecosystems is relatively low compared to lakes and ponds. The distribution, density and biomass of zooplankton in wetlands is, however, determined by the availability of food sources in the form of phytoplankton and microbe-organic detritus complex as well as the associated environmental factors, and hence a close relationship exists between the phyto– and zooplankton communities in the so called

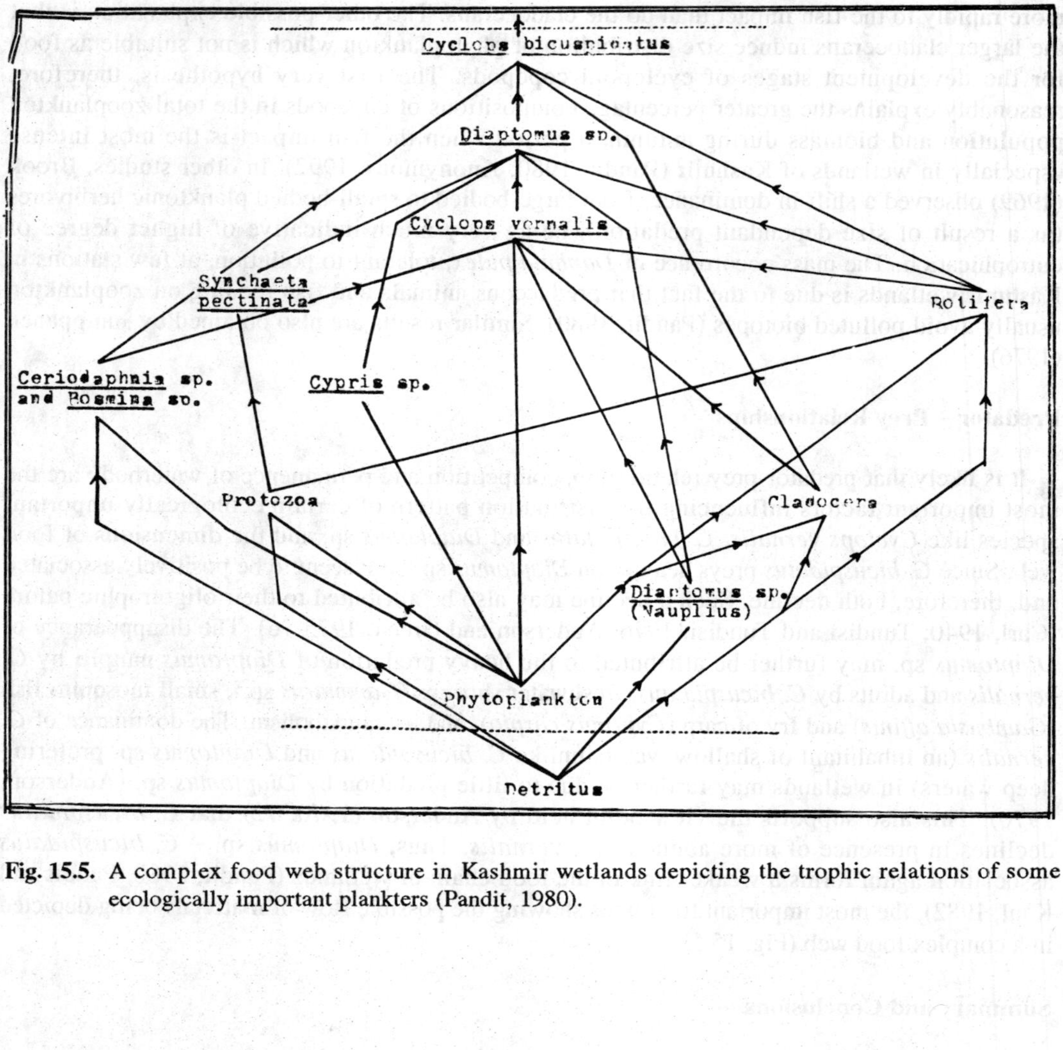

Fig. 15.5. A complex food web structure in Kashmir wetlands depicting the trophic relations of some ecologically important plankters (Pandit, 1980).

productive ecosystems. Though the knowledge on the interrelationships between the phyto- and zooplankton in wetland ecosystems is generally lacking or rather very little, yet and attempt is made to draw a comparison between Kashmir wetlands and chaurs (wetlands) of Bihar on the basis of these aspects. The studies on the micro-cosm community in different wetlands, under varying climatic conditions, indicate that, in general, two peaks of zooplankton growth are obtained, one in late spring to early summer and the other in fall. The first peak, shortly after the spring phytoplankton peak in sub-mediterranean type of wetlands of Kashmir and late winter maximum in tropical wetlands of Bihar, which in turn start gradually declining, indicate the basic source of food being very much conducive for the growth of phytoplankton. The sudden drop in algal cells in summer appears to be related to excessive grazing of phytoplankton by large number of herbivorous zooplankton which subsequently switch over to its feeding on huge quantities of bacteria. Besides, many planktonic algae are unsuitable as food for many herbivorous zooplanktons and certain large sized zooplankters are secondary consumers with bacteria and planktogenic detritus forming a large proportion of their diet. This, for obvious reasons, complicates the interaction between the phyto- and zooplankton. Thus in a balanced ecosystem a certain ratio prevails between the biomass of the producers and primary consumers, the rise in primary production causing an increase in zooplankton biomass also. In tropical wetlands the proportions of phyto- and zooplankton recorded on the basis of population have been calculated to be 99.2 per cent and 0.8 per cent in January and 95 per cent and 5 per cent in August and on the basis of species composition it turned to be 89.7 per cent and 10.3 per cent in February and 83.9 per cent and 16.1 per cent in August respectively. In contrast, Kashmir wetlands registered higher proportions of zooplankton, comprising 14. 17-28.73 per cent by weight and 5-12 per cent by density as against 71.27-85.83 per cent by weight and 88-95 per cent by density of phytoplankton.

The trophic structure of plankton communities further indicate that during initial stages of eutrophication the phytoplankton biomass increases more rapidly as compared to zooplankton but during the final stages of eutrophication the relationship between the biomasses is quite reversed. Since the primary consumers mainly use the phytoplankton as food, it appears that only a part of net primary production is utilized by zooplankton in strongly eutrophied biotopes and a considerable part (70-80%) of the former is, however, left unutilized as a residue of primary production which then is decomposed by bacteria and utilized by detritivores, thus leading to shortening of food chains. This is well supported by the fact that microscopic filter-feeders, dominating in eutrophic wetlands, grazed only small quantities of phytoplankton while bacteria and fine tripton particles constituted their main diet.

Acknowledgements

This paper was presented as an 'invited lecture' at an International Seminar on Wetland Ecology and Management, Bombay Natural History Society Ecological Research Centre, Bharatpur (Rajasthan), February 23-25, 1990. The researches on Kashmir wetlands were possible through a subvention from the University Grants Commission, New Delhi, India, which is gratefully acknowledged.

REFERENCES

Anderson, R.S. 1970. Predator-prey relationships and predation rates for crustacean zooplankters from some lakes in Western Canada. *Canadian J. Zool.* **48** : 1229-1240.

Anderson, R.S. 1972. Zooplankton composition and change in an alpine lake. *Verh. Internat. Verein. Limnol.* **8** : 264-268.

Anderson, R.S. 1974. Crustacean plankton communities of 340 lakes and ponds in and near the National Parks of the Canadian Rocky Mountains. *J. Fish. Res. Board Can.* **31** : 855-869.

Anderson, R.S. and Green, R.B. 1975. Zooplankton and phytoplankton studies in the western lakes, Alberta, Canada. *Verh. Internat. Verein. Limnol.* **19** : 571-579.

Anderson, R.S. and Green, R.B. 1975. Zooplankton and phytoplankton studies in the western llakaes, Alberta, Canada. *Canadian Wildl. Serv.* Occasional Papeer No. **27** : 40 pp.

Anonymous, 1992. *32nd Annual Report for the Year 1991.* Hydrobiological Institute, Czechoslovak Academy of Sciences, Ceske Budejovice, Czechoslovakia.

Brook, J.L. 1969. Eutrophicationa nd changes in the composition of zooplankton. In : *Eutrophication -- Causes, Consequences and Correctives.* (Ed. Rohlich, G.A.) National Academy of Sciences, Washington, D.C., pp. 236-255.

Brynildson, O.M. and Kempinger, J.J. 1973. Production, food, and harvest of trout in Nebish lake, Wisconsin. *Wis. Dept. Nat. Resour. Tech. Bull.* **65** : 1-20.

Carl, C.C. 1940. The distribution of Cladocera and free-living Copepoda in British Columbia. *Ecol. Monogr.* **10** : 55-110.

Jumppanen, K. 1976. Effects of waste waters on a lake ecosystem. *Ann. Zool. Fennici.* **13** : 85-138.

Kaul, V., Fotedar, D.N., Pandit, A.K. and Trisal, C.L. 1978. A comparative study of plankton populations of some typical fresh waterbodies of Jammu and Kashmir State. In : *Environmental Physiology and Ecology of Plants* (Eds. D.N. Sen and R.P. Bansal), B. Singh M.P. Singh, Dehra Dun, India, pp. 249-269.

Kaul, V. and Pandit, A.K. 1982. Biotic factors and food chain structure in some typical wetlands of Kashmir. *Pollution Research,* **1** : 49-54.

Nauwerck, A. 1963. Die Beziehungen Zwischen Zooplankton and Phytoplankton in See Erken. *Symbolae Bot. Upsaliensis.* **17** (5) : 1-163.

Pandit, A.K. 1980. *Biotic factor and food chain structure in some typical wetlands of Kashmir.* Ph.D. Thesis, University of Kashmir, Srinagar - 190 006, Jammu & Kashmir, India.

Pandit, A.K. and Kaul, V. 1982. Trophic structure in some typical wetlands. In : *Wetlands-Ecology and Management,* Vol. II. (Eds. B. Gopal, R.E. Turner, R.G. Wetzel and D.F. Whigham), International Scientific Publications and National Institute of Ecology, Jaipur, India. pp. 55-82.

Patalas, K. 1968. Crustacean Plankton of some Masurian lakes characterized by different fish yield. *Pol. Arch. Hydrobiol.* **15** : 289-304.

Rai, D.N. and Datta Munshi, J.S. 1982. Ecological characteristics of 'Chaurs' of North Bihar. In : *Wetlands-Ecology and Management,* Vol. II. (Eds. B. Gopal, R.E. Turner, R.G. Wetzel and D.F. Whigham), International Scientific Publications and National Institute of Ecology, Jaipur, India. pp. 89-95.

Richards, R.S., Goldman, C.R., Frantz, T.C. and Wickwire, R. 1975. Where have all the *Daphnia* gone? The decline of a major cladoceran in Lake Tohoe, California-Nevada. *Verh. Internat. Verein. Limnol.* **19** : 835-842.

Seshavatharam, V. and Venu, P. 1982. Some observations on the ecology of Kolleru Lake. In : *Wetlands-Ecology and Management*, Vol. II. (Eds. B. Gopal, R.E. Turner, R.G. Wetzel and D.F. Whigham), Intern. Sci. Publ. and National Institute of Ecology, Jaipur, India. pp. 35-44.

Shcherbakov, A.P. 1967. Glubokoe Lake, hydrobiological description. *Moskva Izdat* ''Nauka''.

Shushkina, E.A. 1966. The estimation of zooplankton production in different lakes. *Verh. Internat. Verein. Limnol.* **16** : 419-424.

Tundisi, T.M. and Tundisi, J.G. 1976. Plankton studies in a lacustrine environment. *Oecologia* (Beri). **25** : 265-270.

Vyas, L.N. 1989. State of art of Pichhola-Fatehsagar wetland complex. In : *Wetlands, Mangroves and Biosphere Reserves* (Proceedings of the Indo-US Workshop, New Delhi, 4-7 January, 1989), Ministry of Environment and Forests, Government of India. pp. 66-81.

Widuto, J. 1977. Zooplankton in the Kortowskil Lake after 18 years of experimental restoration in the lake. *Pol. Arch. Hydrobiol.* **24** (1) : 91-106.

Chapter 16

Production and Seasonality of Dominant Macrophyte – *Eichhornia Crassipes* and Associated Epiphytic Macro-Invertebrates of a Freshwater Body

M.P. Sinha, Ratna Sinha, M. Phillip AND S.K. Bose

Department of Zoology, Ranchi University, Ranchi-834 008, India

Abstract. The temporal variation in standing crop and production of *Eichhornia crassipes* and population density, species diversity, evenness and equitability of associated macroinvertebrate fauna have been dealt with. Five months from February to June were a period of rapid growth for the weed during which standing crop increased considerably reaching a maximum upto 9828.60 g/m^2. The maxima of productivity was observed in April being 158.75 gC/m^2/month. The epiphytic fauna was considerably lower than macrobenthic fauna. Insects in general were the dominant group followed by the oligochaets. The maximum percentage of oligochaets, insects, chironomids, gastropods was 56.50 per cent, 48.50 per cent, 64.60 per cent and 28.40 per cent respectively in different months during the period of study. The population density has been further analysed for species diversity (Shanon-Weaver, Margalef and Simpson index), evenness and equitability. The expected number of species was more in February, March and April and less in rest months than the observed value of the same while the expected values of species diversity was more in these months than observed values.

Key words : *Eichhornia,* Epiphytic, macroinvertebrates, Temporal variation, Species diversity, Evenness, Equitability.

Introduction

The most fundamental dimension of an aquatic biotope is ecological production (Whittaker, 1961) which prepares the base of all types of productive utilization of water. The primary producers of an aquatic ecosystem include both microphytes and macrophytes which when occur in great abundance produce adverse impacts on the ecosystem. Microphytes by producing algalblooms have been reported to be responsible for fish mortality apart from many other limnological effects. Similarly among macrophytes the free floating plants are troublesome aquatics in tropical countries because of their rapid adventive spread in warm climate. Their growth and dry matter production shows marked variation under comparable climatic condition because of the physico-chemical complexes of environment of water bodies which have direct bearing on their growth rate and productivity (Sinha and Sahai, 1978).

The *Eichhornia crassipes* a floating hydrophyte, introduced in India in 1896 from Brazil as an ornamental flower owing to its intense growth and capable of making the water infertile, has become a problem of aquatic habitats. The hydrophyte enhances water loss many times through evapotranspiration, its thick cover which floats over water surface reduces the photic zone and thus the productivity and energetics of the system as well as the cover also checks the gaseous exchange at air water interface. Considerable literature on different aspect of *Eichhornia* is available. In India, observations on the biomass and production of ephemeral plant communities started by the work of Bellamy (1967) followed by contributions from Gopal (1967), Das (1968, 1969) Sahai and Sinha (1970). Chaudhury *et al.* (1976) and Mishra *et al.* (1968) reported temperature as a favouring growth factor for *Eichhornia.* Michael (1976) studied its growth and management. Trivedy *et al.* (1978) reported ecological observations on floating island of the weed. The growth along with production aspect has been dealt with in detail by Haniffa and Pandian (1978), while Ayyappan *et al.* (1986) have described the various aspects of decomposition and the related implication on the water body. Many more works are being carried out on nutrient uptake and on trophic significance (Sen *et al.*, 1993) of *Eichhornia crassipes,* but surprisingly there is great paucity of knowledge on the weed associated fauna particularly of Indian waters, while plenty of literature is available on weed fauna of western countries.

Aquatic invertebrates living on macrophytes (epiphytic macroinvertebrates, Hutchinson, 1967) are important and integral parts of over all macroinvertebrate fauna of an aquatic biotope which take shelter in the leaf and stems of weeds. Early studies of animal population in various aquatic habitat have usually included some references to the animals present on the vegetation (Carpenter, 1927; Whitehead, 1935; Moon, 1936; Jones, 1949; Smyly, 1957). There are evidences from the studies of vegetation in ponds and lakes, that different plants shelter different animals (Krecker, 1939; Andrews and Hasler, 1943; Edmondson, 1944; Bryden, 1952; Smyly, 1952; Rosine, 1955; Gerking, 1957). The density of epiphytic organisms has been reported to be related to the amount of surface area available and the degree of leaf dissection (Mrachek, 1966; Krull, 1970; Biochino and Biochino, 1980; Dvorak and Best 1982; Schramm *et al.*, 1987). The temporal variation in the abundance of epiphytic invertebrates have also been documented (Rosine, 1955; Mrachek, 1966; Soszka, 1975; Voights, 1976; Smock and Stoneburner, 1980) and the epiphytic invertebrate fauna has been found as a food source for fish (Pieczynski, 1973; Werner *et al.*, 1977).

Very little is known about the epiphytic fauna so far Indian scenario is concerned. Apart from a few informations (Rai and Dutta Munshi, 1979; Kumar and Mitra, 1986; Sinha *et al.*, 1993) as this important ecological aspect is virtually least studied, the present project was taken up to have some basic information pertaining to epiphytic community composition, density and diversity attached to very common and dominant aquatic weed *Eichhornia crassipes.* Hence the production and seasonality in both the hydrophyte as well as the epiphytic macroinvertebrates have been studied.

The Study Area

The present study was carried out on Ranchi lake situated in the South-East corner of the town. The lake covers its origin partly to the removal of certain soluble minerals, such as felspar and partly to erosion by running water. The lake covers an area of about 55 acres in the rainy season which gradually decreases to 54.10 acres during summer season. Due to the current steps taken by local administration to beautify the lake, its area has been decreased to 53.5 acres (Fig. 16.1 A and B). The average depth is 25 ft. in rainy season while only 18 feet in summer. The lake has three small islands in it and so far the bottom is concerned, mud is fine clay with insignificant traces of silica. The whole surface at times possesses the small islands of the floating weed when dispersed by wind action.

Material and Methods

Sampling : Monthly sampling was done for one year (Oct. 1990 to Sept. 1991) preferably in the morning hours. For the present study, the samples of *E. crassipes* were collected by using a detachable wooden frame (50 cm × 50 cm) from the sampling site selected along the littoral zone of the lake having homogenous spreading of *Eichhornia crassipes.* The wooden frame was placed on the weed and pushed into water to cover the weed. The whole plant material and associated fauna were carefully removed from the quadrate and kept in sufficiently large polythene bags to bring them to the laboratory. Five replicates of each sample was done.

Processing of Eichhornia crassipes : The epiphytic macroinvertebrates were carefully removed and after removal of soil and other associated unwanted materials standard methods were followed (Westlake 1963, 1965) to assess the fresh weight and biomass. The epiphytic fauna removed *Eichhornia* mass of each sample was shaken well to remove the water on plant. After removal of water fresh weight was taken by physical balance. A fixed weight of each sample was dried at 110°C for 24 hours. The weight of the dried sample taken and calculated for the entire sample in terms of g dry weight per square metre. Similarly for organic carbon and inorganic carbon the method followed by Singh and Sahai (1979) was followed.

The productivity of *E. crassipes* was determined by light and dark bottle method (Wetzel, 1965, 1969). The production was expressed as g oxygen/m^2/month, g C/m^2/month and carbohydrate/m^2/month and also the calorific values were calculated to assess the energetics.

Processing of epiphytic macroinvertebrate fauna : The epiphytic macroinvertebrate organisations were carefully removed from *Eichhornia crassipes* manually and were preserved in 70 per cent ethanol. The epiphytic macroinvertebrate fauna bearing portions (which were the

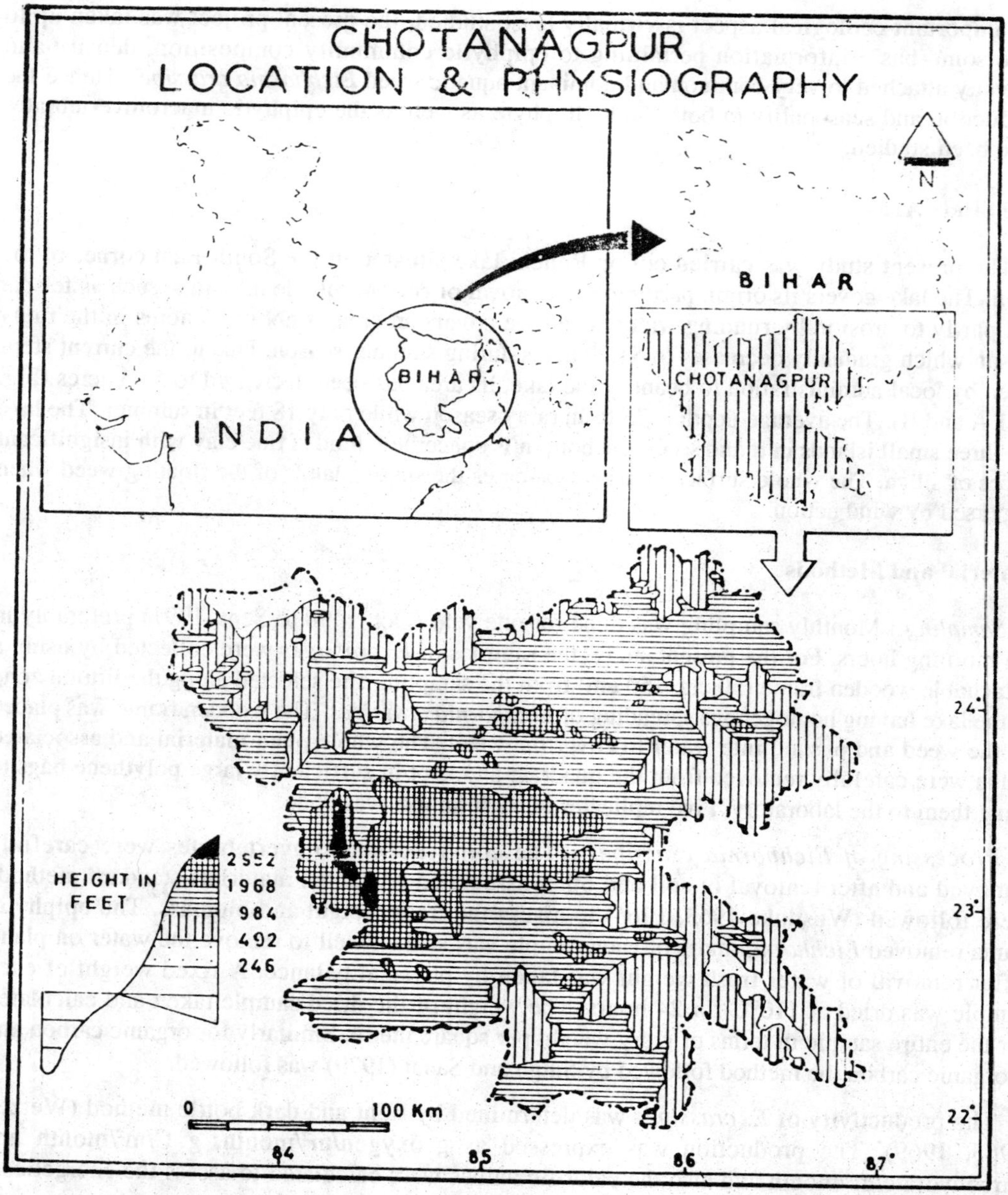

Fig. 16.1A. Map showing details of location and morphometry of study area of Chota Nagpur.

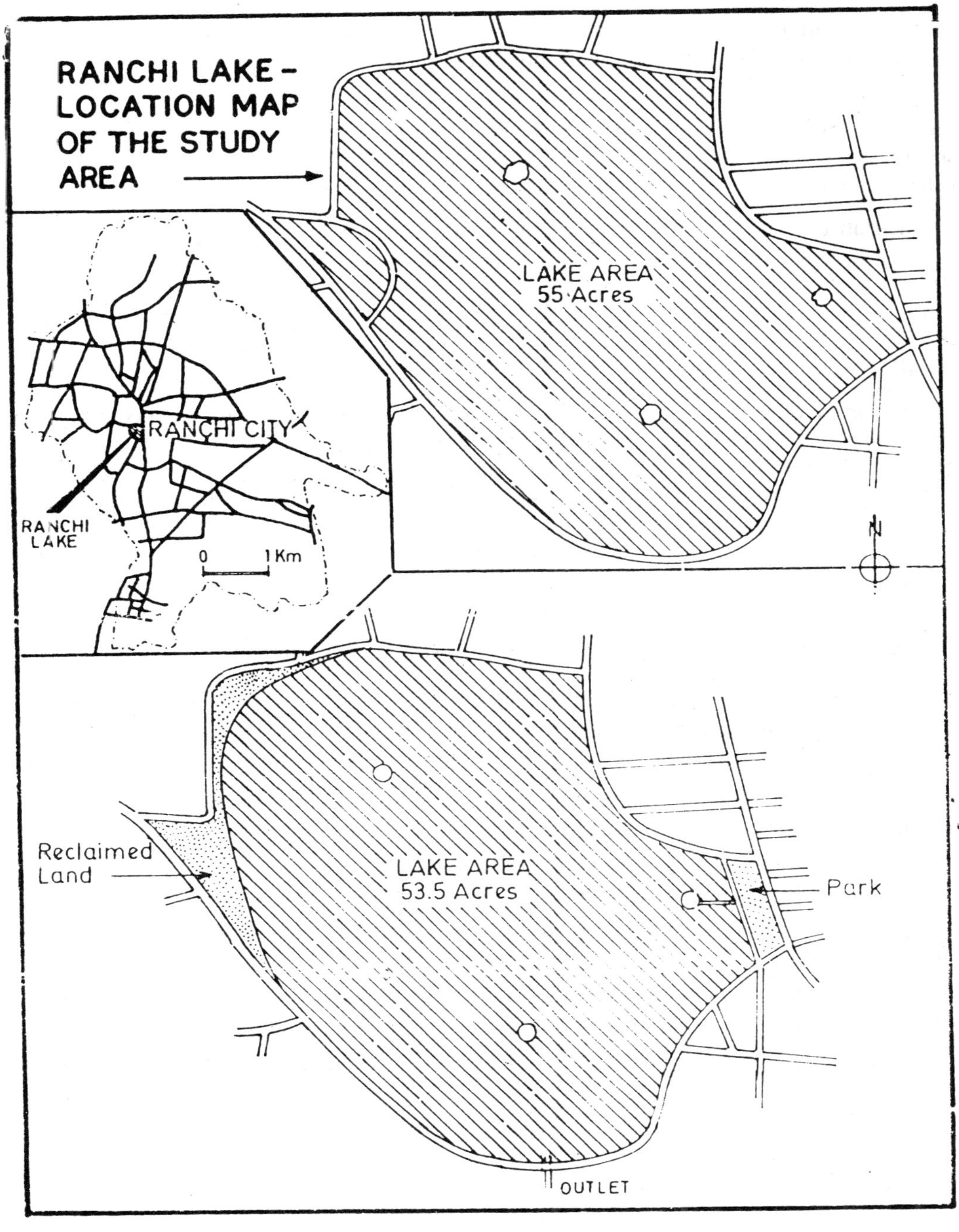

Fig. 16.1B. Map Showing details of location and morphometry of study area of Ranchi Lake.

submerged parts) were cut off from the *E. crassipes* plant and population density was calculated which was subsequently converted into epiphytic population density per square metre.

The population density values were used for calculating the species diversity following Shanon-Weaver (1949), Simpson (1949), Margalef (1958), and the evenness component of species diversity was calculated following the formula of Pielou (1969), while equitability was calculated following Lloyd and Ghelardi (1964).

The statistical analysis of data when required was done following Zar (1974).

Results

A. *The weed*

Standing crop and biomass

The portion of water body under investigation covered by *E. crassipes* varied considerably in terms of percentage area during the one year period. The percentage area occupied by *E. crassipes* cover did not only depend on the growth of the weed but also on the wind velocity and movement of surface water under influence of wind. At times the weed got distributed throughout the surface area and again packed in one portion of the water body. The considerable variation in occupied area by *E. crassipes,* as presented in Table 16.1, ranged from 27.50 per cent to 72.52 per cent in October 1990 and May 1991 respectively. In most of the months more than 50 per

TABLE 16.1. *Seasonal changes in standing crop, biomass, organic carbon and inorganic carbon of Eichhornia crassipes* (1990-91).

Months	Eichhornia cover in %	Standing crop g/m^2	Biomass g/m^2	Organic Carbon g/m^2	Inorganic Carbon g/m^2
October	27.50	4585.50 ± 11.08	318.990	229.274	89.716
November	35.32	4362.10 ± 19.16	311.578	228.490	83.088
December	39.27	5766.20 ± 21.27	417.552	328.076	89.476
January	60.24	7899.50 ± 18.72	493.512	345.458	148.054
February	51.62	8764.20 ± 17.61	547.762	408.331	139.431
March	61.24	9625.20 ± 19.25	624.440	470.889	153.511
April	69.70	9828.60 ± 23.27	634.763	491.429	143.334
May	72.52	9543.70 ± 27.70	616.363	457.301	159.062
June	52.10	9275.40 ± 16.20	635.163	453.687	181.476
July	56.27	8686.20 ± 11.62	526.436	368.505	157.931
August	35.82	7899.20 ± 9.97	482.266	357.542	124.724
September	27.52	7426.40 ± 16.25	453.401	304.780	148.620

cent of the water body was covered with *E. crassipes.* From October 1990 to May 1991 there was a gradual increase in *E. crassipes cover,* excepting February where there was a slight decrease and there after it revealed gradual decrease upto the end of investigation in September 1991.

The standing crop of *E. crassipes* as g/m² was found minimum in November 1990 and was observed to have gradual rise in the following months up to April when it was maximum (9828.60 ± 23.27 g/m²). After this period of maxima there was a gradual decrease in the standing crop reaching to 7426.40 ± 16.25 g/m² as minimum in September 1991 (Table 16.1). Out of the twelve months of investigation the standing crop of *E. crassipes* was more than 9000 g/m² during four months i.e. from March to June, it was more than 8000 g/m² in February and July and more than 7000 g/m² in January, August and September. By the data recorded, it is clear that standing crop values are more in summer (March to May) and less in post monsoon period.

The biomass values (Table 16.1) also revealed the same trend of fluctuation as the standing crop values. The minimum value was recorded in November 1991 (311.578 g/m²) and maximum in June 1991 (635.163 g/m²). The higher values of biomass ranging between 616.363 g/m² to 635.163 g/m² was restricted to summer months i.e. from March 1991 to June 1991. There was a change of just double amount in biomass so far the minimum and maximum values are concerned.

Organic and Inorganic Carbon

The organic carbon content was found to be minimum in November 1990 (228.490 g/m²) and thereafter it revealed a gradual rising trend in successive months reaching to maxima of the order of 491.429 g/m² in April 1992. After April there was a steady fall in organic carbon value till the end of the present investigation in the month of September 1991 (304.780 g/m²). Though the organic carbon values were maximum during the summer months like the biomass values yet the higher and lower values of organic carbon did not correspond to those of biomass. Biomass was maximum in June 1991 (635.163 g/m²) while the organic carbon content was found maximum in April 1991. Hence, it appears that the ratio of organic carbon and inorganic carbon content does not remain the same throughout.

The inorganic carbon content showed a variation between 89.476 g/m² (the minima in December 1990) to 181.746 g/m² (the maxima in June 1991. From October 1990 to December 1990 the values were very low (89.716, 83.088 and 89.476 g/m² respectively) while it rose to the tune of 124.724 (August 1991) to 181.476 (June 1991) in the following months. The inorganic carbon content values also did not correspond to the fluctuations in the biomass values.

As presented in Table 16.2 the data on the trend of variation in standing crop in successive months show that there was a decrease of 223.40 g/m² in October-November period but after that there was continuous increase in standing crops of *E. crassipes* till March-April, followed again by decrease upto the end. *E. crassipes* in the water body increased its standing crop continuously from November 1990 to April 1991. The maximum increase was recorded in December 1990 - January 1991 (2133.30 g/m²). The maximum decrease in standing crop was

TABLE 16.2. *Monthly variation in standing crop, biomass, organic and inorganic carbon content of Eichhornis crassipes (1990-91)*

Months	Standing crop g/m²/month	Biomass g/m²/month	Organic Carbon g/m²/month	Inorganic Carbon g/m²/month
Oct 1990 - Dec 1990	– 223.40	– 7.412	– 0.784	– 6.628
Nov 1990 - Dec 1990	+ 1180.70	+ 105.974	– 99.586	+ 6.388
Dec 1990 - Jan 1991	+ 2133.30	+ 75.960	+ 17.382	+ 58.578
Jan 1991 - Feb 1991	+ 864.70	+ 54.250	+ 62.873	– 8.623
Feb 1991 - Mar 1991	+ 861.00	+ 76.678	+ 62.558	+ 14.080
Mar 1991 - Apr 1991	+ 1.021	+ 10.323	+ 20.540	– 10.177
Apr 1991 - May 1991	– 284.90	– 18.400	– 34.128	– 15.728
May 1991 - Jun 1991	– 268.30	+ 18.800	– 3.614	+ 22.414
Jun 1991 - Jul 1991	– 589.20	– 108.727	– 85.182	– 23.545
Jul 1991 - Aug 1991	– 787.00	– 44.170	– 10.913	– 33.207
Aug 1991 - Sep 1991	– 472.80	– 28.865	– 52.812	+ 23.897

recorded in July-August (789.00 g/m²). Similarly there was increase in biomass from November 1990-December 1990 to March-April 1991 and again in May-June 1991 while a decrease (18.400 g/m²) was observed in April-May 1991 period. Maximum decrease in biomass was noticed in June-July (108.17 g/m²). The maximum gain in biomass in successive month was 105.974 g/m²) between November-December.

The trend of enhancement in organic carbon content was recorded between November-December 1990 to March-April 1991. Excepting these months the organic carbon content always decreased in the following month in comparison to the present month. The maximum gain in a following month in organic carbon content of *E. crassipes* was 62.873 g/m² in January-February 1991 and the maximum loss was 2.821 g/m² in August-September 1991.

Unlike the trends of increase and decrease in standing crop, biomass and organic carbon the inorganic carbon content of *E. crassipes* was irregular. The value of inorganic carbon content increased in November-December (6.388 g/m²), December '90-January 1991 (58.518 g/m²), February-March 1991 (14.080 g/m²) April - May 1991 (15.728 g/m²), May - June 1991 (22.414 g/m²) and August - September 1991 (23.896 g/m²) while decrease during rest period, the maxima being 33.20 g/m² in July-August 1991 and minima 06.628 g/m² in October-November 1990.

The ratio between standing crop and biomass of *E. crassipes* was maximum in December 1990 (7.241) and minimum in July 1991 (6.060). The data recorded revealed that the ratio remains roughly of a definite tune.

TABLE 16.3. *The standing crop, biomass, organic and inorganic carbon content in terms of percentage*

Months	Biomass in relation to Std. crop	Organic carbon in relation to Std. crop	Inorganic carbon in relation to Std. crop	Organic carbon as biomass	Inorganic carbon as biomass	Inorganic carbon as organic carbon
October	6.956	4.999	1.956	71.874	28.125	39.130
November	7.142	5.238	1.904	73.333	26.641	36.363
December	7.241	5.689	1.551	78.571	21.428	27.272
January	6.247	4.373	1.874	69.999	30.000	42.857
February	6.249	4.659	1.590	74.545	25.454	34.146
March	6.487	4.892	1.595	75.409	24.583	32.600
April	6.458	4.979	1.458	77.409	22.580	29.166
May	6.458	4.791	1.666	74.193	25.806	34.782
June	6.847	4.862	1.956	71.428	28.571	40.100
July	6.060	4.242	1.818	70.299	29.897	42.857
August	6.205	4.526	1.578	74.137	25.862	34.883
September	6.108	4.104	2.001	67.220	32.778	48.763

The ratio between organic carbon and the standing crop was again highest in December 1990 (5.689) and lowest in September 1991 (4.104). There was no particular trend of increase or decrease in this ratio.

So far relation between inorganic carbon and biomass is concerned, it was maximum in September 1991 being 2.001 when the ratio between organic carbon and standing crop was found minimum. The lowest value of this ratio was recorded to be 1.458 in April 1991.

The organic carbon in relation to biomass was highest in April (77.419) and lowest in September (67.220) while the inorganic carbon content in relation to biomass was highest in September 1991 (32.778) while lowest in December 1990 (21.428). The inorganic and organic carbon content ratios ranged between a maximum of 42.857 (January 1991) to a minimum of 27.272 (November 1990).

Production

Monthly variation in production of *E. crassipes* was estimated in terms of gram carbon/m^2 (g C/m^2), gram oxygen/m^2 (gO_2/m^2) and gram hydrocarbon/m^2 (g HC/m^2) and corresponding energetics was calculated based on gram oxygen/m^2/month value. As is evident from the data presented in Table 16.4, the production was most in June 1991 (158. 75 g C/m^2, 422. 275 g O_2/m^2

TABLE 16.4. *Seasonal variation in production of E. crassipes in terms of gC/m²/month, gO$_2$/m²/month and gCH/m²/month during 1990-91*

Months	Production as gC/m²/month	Production as gO$_2$/m²/month	Production as gCH/m²/month	Energy as Cal/m²/month based on Oxygen
October	57.25	152.285	161.844	560.256
November	56.27	149.678	158.287	550.665
December	104.25	277.305	293.255	1020.205
January	123.25	327.845	346.702	1206.141
February	136.75	363.755	384.677	1338.254
March	156.00	414.960	438.828	1526.637
April	158.67	422.075	446.338	1552.813
May	154.00	409.640	433.202	1507.065
June	158.75	422.275	446.563	1553.549
July	131.50	349.790	369.909	1286.877
August	120.50	320.530	338.966	1179.229
September	114.37	303.240	321.722	1115.620

and 446.563 g HC/m²) while least in November 1990 (56.27 g C/m², 149.678 g O$_2$/m² and 158.287 g HC/m²). A rising trend in production was noticed from November 1990 to June 1991 with slight decrease in May 1991 in comparison to the previous month. After June 1991 the production of *E. crassipes* decreased till the end of study. A corresponding rise and fall in calorific values was recorded (Table 16.4).

B. *The weed fauna*

Species composition

The epiphytic fauna on *E. crassipes* recorded during the study belonged to Nematoda, Oligocaeta, Hirudinea, Insecta, Hydracharina and Gastropoda. All together 23 species of epiphytic macroinvertebrates constituted the weed fauna. As presentd in Table 16.5, Nematoda was represented by *Rhabditis* sp. while Oligochaeta by *Nais communis, Dero indica, Stylaria fossularis, Aulophorus hymanae, Chaetogaster langi, Branchiodrillus semperi, Aelosoma bengalensis.* Hirudinea was represented by single species - *H. javonica.* Among insects Chironomus larvae, *Ceratopogen* sp., *Caenis perpersilla, Eretes stictius, Hydrophilus indicus* were encountered while *Ceriodaphnia* sp., *Cyclops* sp. and *Cypris* sp. were the representative of crustacean planktonic forms. *Hydrachna kloomi* was the only representative of Hydracharina. Gastropoda was represented by *Pila globosa, Indoplanorbis exustas* and *Lymnaea luteola.*

TABLE 16.5. *Species composition of epiphytic macroinvertebrate community with trend of seasonality during 1990-91*

Group/Species	Max. Population per month	Min. Population per month	% F.O.
1. **Nematoda**			
Rhabditis sp.	12//Fev 1991	5/ Jan 1991	33.333
2. **Oligocheata**			
Nais communis	63/Apr 1991	3/Dec 1990	91.166
Dero indica	24/Apr 1991	2/Feb 1991	75.000
Stylaria fossularis	5/Feb 1991	1/Dec 1990	33.333
Aulophorus hymanae	16/Apr 1991	3/Nov 1990	91.166
Chaetogaster langi	10/Mar 1991	5/Feb 1991	16.666
Branchiodrilus sempere	32/Apr 1991	4/Nov 1990	66.666
Aelosoma bengalensis	35/Apr 1991	2/Oct 1990	75.000
3. **Hirudinea**			
Hirudinaria javonica	3/May '91	1/Aug '91	25.000
4. **Diptera**			
Chironomus larvae	216/May 1991	36/Feb & Mar 1991	100.00
5. **Insects other than Diptera**			
Ceratopogen sp.	19/Jul 1991	4/Aug 1991	41.166
Caenis perpusilla	26/Mar 1991	5/Jan 1991	100.000
Laccophilus chinensis	51/Mar 1991	5/Jan 1991	41.166
Eretes sticticus	22/Apr 1991	6/Feb 1991	50.000
Hydrophilus indicus	26/Mar 1991	5/Apr 1991	16.666
6. **Crustacea**			
Ceriodaphnia sp.	24/May1991	3/Dec 1990	50.000
Cyclops sp.	44/May 1991	5/Oct 1990	25.000
Cypris sp.	30/Dec 1990	3/Aug 1991	50.000
7. **Hydracarina**			
Hydrachna kloomi	68/Apr 1991	1/Nov1990	41.666
8. **Gastropoda**			
Pila globosa	2/Oct 1990	1/Dec 1990	41.666
Indoplanorbis exustus	17/Apr 1991	1/Nov 1991	50.000
Lymnaea luteola	35/Jan 1991	3/Sep 1991	66.666

F.O. = Frequency of occurrence

Out of the species, only chironomus larvae and *C. perpusilla* were found throughout the period of study with frequency of occurrence 100 per cent. *Nais communis* and *A. hymanae* the two oligochaetes had 91.166 per cent of frequency of occurrence next to the two insects. Except *D. indica* (75.00 per cent), *B. semperi* (33.66%), *A. bengalensis* (75.00%) among oligochaetes and *Lymnaea luteola* (66.66%) among Gastropods, rest all species had 50 per cent or less frequency of occurrence during the investigation which reflects their intermittent presence in the fauna (Table 16.5).

Population density and dynamics

On the whole the peak of the epiphytic fauna was observed in April 1991 when the density was 501/m² and lowest was 75/m² in November 1990. A bimodal peak pattern was observed for the fauna, the first and prominent peak was in April which after decreasing to minimum in July (184/m²) formed the second peak in September (266/mm²). Hence, the two lower population densities were observed in November 1990 (75/m²) and July 1991 (184/m²). Out of 23 species encountered during the study 17 was the highest number of species found at a time (April 1991), while minimum 7 species were observed in June 1991. The average individual number per species was highest (30.714) in June 1991 and lowest (6.818) in November 1990.

Among Oligochaetes *Nais communis, Aulophorus hymanae,* and *Aelosoma bengalensis* were the dominant forms. In general the two peaks of seasonality revealed by oligochaetes was observed in April (170/m²) and July (104/m²). The maximum and minimum population density of different species and the corresponding number per square metre have been shown in Table 16.5, and percentage composition has been presented through histograms of Fig 16.2. (A-H). Oligochaeta constituted maximum 56.5 per cent of total fauna in July 1991 and shared a minimum of 7.4 per cent in December 1990.

Rhabditis sp. of Nematoda and *Hirudinaria javonica* of Hirudinea and *Hydrachna kloomi* were least in the samples both from population density and frequency of occurrence view point. These forms could not constitute even one per cent of the fauna throughout the year of study.

The most dominating form of the weed-fauna was chironomus larvae 216/m² which formed 49.4 per cent to total fauna in May 1991 was the first peak value of its population density while the second peak was formed by 155/m² population density sharing 59.0 per cent of the total fauna. The overall peak pattern of the weed-fauna was controlled by the group which dominated the samples both by regular presence with higher number. Insects other than chironomids abundant in March 1991 having 118/m² population density which shared 44.7 per cent of the fauna. The occurrence of insect species was higher in warmer months of the year while in rest of months they were of occasional occurrence. Crustacean planktonic forms constituted maximum 19.7 per cent in November 1990 and minimum 2.7 per cent in February 1991 apart from remaining absent in a few months.

Both qualitatively and quantitatively, after Chironomids and Oligochaetes the gastropods were the third important group 28.4 per cent and 3.6 per cent of total weed-fauna in January 1991 and May 1991 respectively. Among gastropods *Lymnaea luteola* was the dominant species.

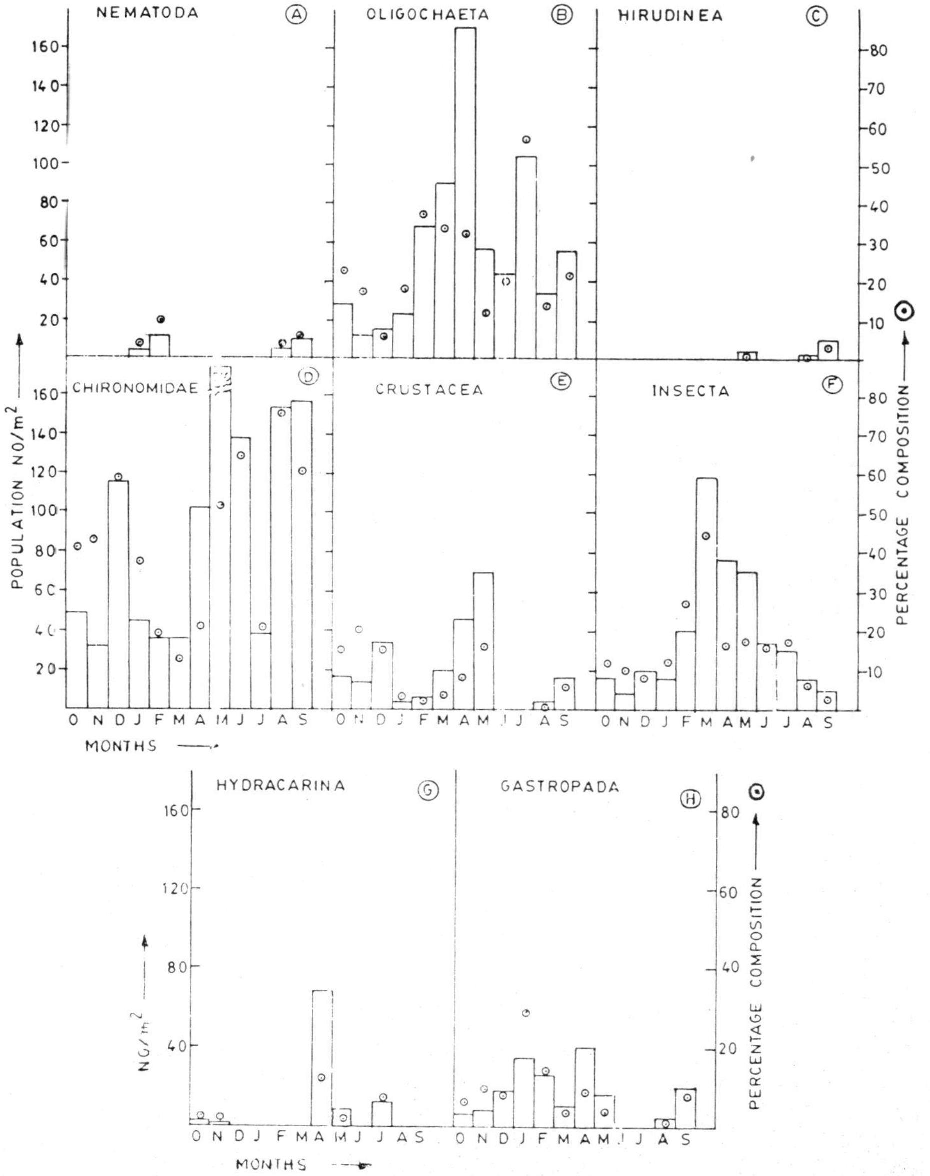

Fig. 16.2. Seasonal variation in population density (histograms) and percentage composition of different groups (A to H) indicated as circles with dott in the centre.

Species diversity

The species diversity of weed-fauna was calculated by formula of Shanon-Weaver (1949), Margalef (1949) and Simpson (1949). The species diversity by Shanon-Weaver's formula varied between a maximum of 3.6864 (April 1991) and a minimum value 1.7416 was recorded in August 1991. The values were more than 3 in February, March and April while more than 2 in rest all months except August. The highest value was observed when both the number of species and individual were maximum but the minimum value was noted when the number of species was relatively higher and number of individuals lower. The species diversity values were found different for the samples in which individual numbers were different but the number of species was the same (Fig. 16.3C).

Similar to the higher value of species diversity by Shanon-Weaver formula, by Margalef formula higher value was recorded in April being 2.5737 but the lower value 1.1171 was noted in June 1991 in contrast to the previous result. The species diversity value for most of the samples by Margalef method was more than 2. The three lower values recorded 1.1171, 1.2422, and 1.9700 for the month of June, July and September respectively (Fig. 16.3B). The lower values were observed when the number of species was low in the sample unlike to the Shanon-Weaver method.

By Simpson's formula the higher value was recorded to be 1.498 in October 1990 and lower value 0.513 in August (Fig. 16.3A). The range of variation in species diversity was very low in this method in comparison to the previous two methods.

The evenness component of species diversity and the equitability in the sample were calculated following Pielou (1969) and Lloyd and Ghelardi (1964). The evenness value ranged between 0.4705 (minimum, August 1991) to 0.298 (maximum, July '91) (Fig. 16.4). The range of variation in evenness value during the period of study was comparatively low. The equitability, however, showed greater range of variation, the minimum 0.307 in August 1991 exactly corresponding to the period of highest and lowest values of evenness. The higher values of equitability were recorded in February (1.133), March (1.071), April (1.117) apart from the highest value in July 1991.

Discussion

Hydrophytic plants, in general, are important biotic component of any aquatic system and play an important role by harbouring a considerable variety and amount of epiphytic fauna apart from their own contribution in ecological production, as well as also become a problem for the system in question if grow and spread in abundance. The data obtained during the present study on the growth of *E. crassipes* point out that five months from February to June 1991 was a period of rapid growth during which its standing crop increased considerably. Haniffa and Pandian (1978) termed this period as long-phase of accelerating growth. After the growth phase and during monsoon months there was a gradual decreasing phase. Westlake (1963) opined *E. crassipes* to be exceptionally productive as might be expected from the fact that it grows in warmer climatic

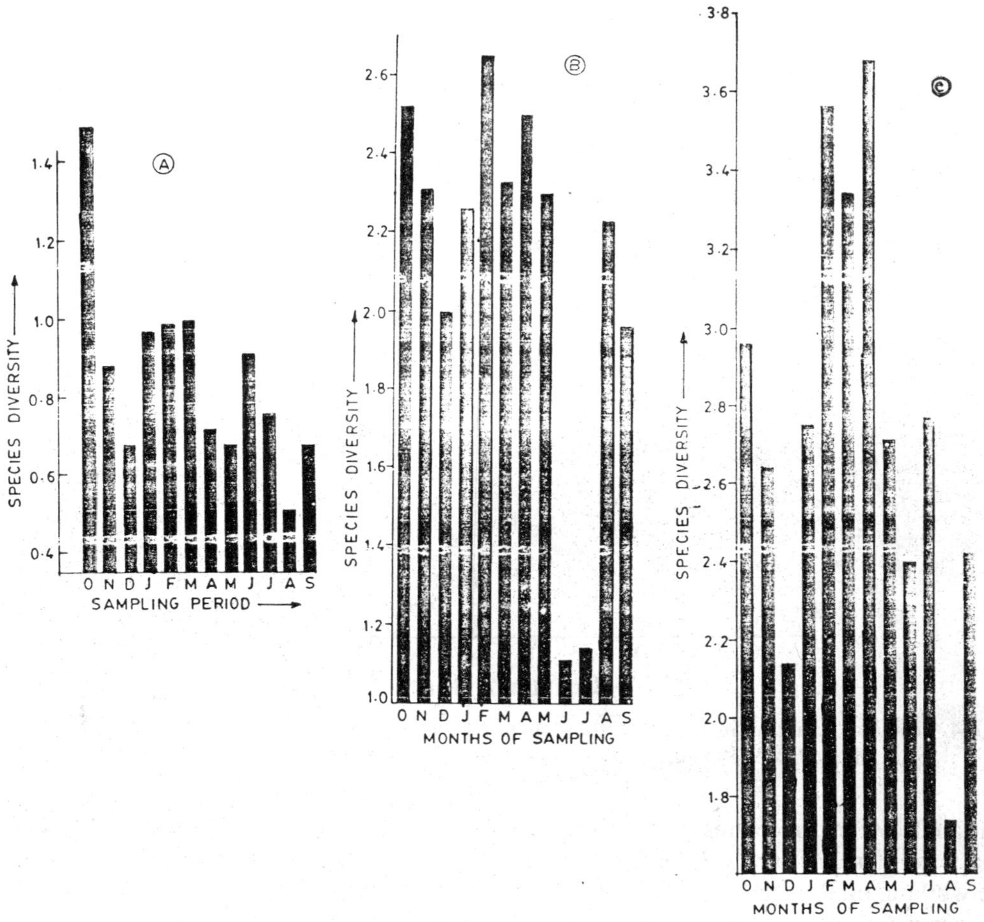

Fig. 16.3. Species diversity of epiphytic macroinvertebrate fauna following Simpson-A, Margalef-B and Shanon-Weaver-C.

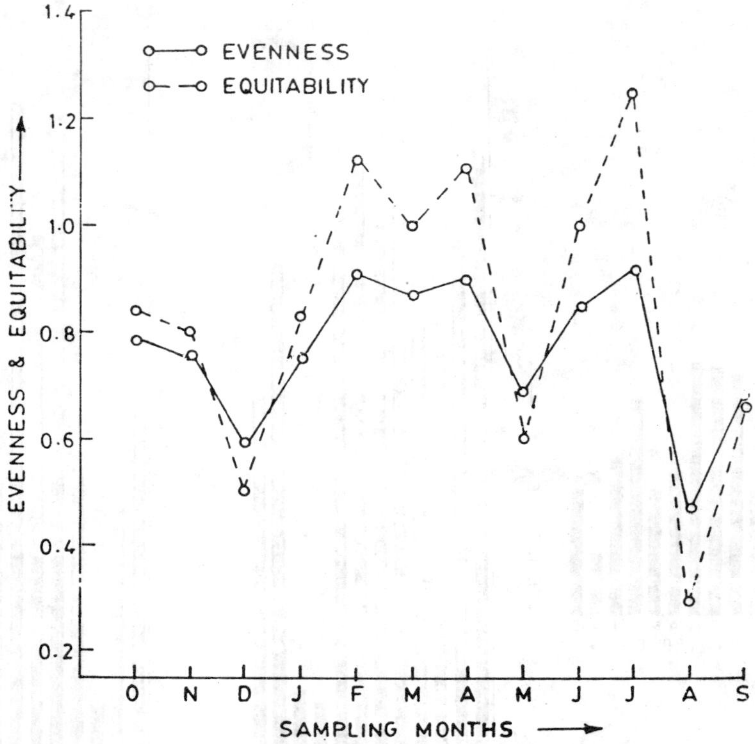

Fig. 16.4. Seasonal variation in evenness component of species diversity and equitability of epiphytic macroinvertebrate fauna attached to *E. crassipes*

condition. However, in contrast to the present finding Sinha and Sahai (1978) reported maximum biomass value in January (723.0 ± 23.0 g/m²) and minimum in September (462.0 ± 40.0 g/m²) while working on Ramgarh Lake, Gorakhpur. They reported relatively lower biomass values in warmer months which contradicts their own earlier reports.

A maximum value of *E. crassipes* dry weight production has been reported to be a tune of 0.72 kg/m² for Ramgarh Lake at Gorakhpur (Sahai and Sinha, 1970) and 0.63 kg/m² for Jalwania pond (Singh and Sahai, 1979), while the maximum of the same in the present observation was 634.763 g/m², a considerably lower than that reported by Penfound and Earle (1947) who observed it to be of the tune of 1.56 kg/m² for several water bodies of Mississippi.

The standing crop in the present study was maximum in April being 9828.60 g/m², a considerably lower value as reported by Singh and Sahai (1979).

The total production of *E. crassipes* has been observed to be 723 ± 23 g/m² (Sinha, 1969; Sinha and Sahai, 1978), and 2067 g/m² by gopal *et al.* (1978) from water bodies of Jaipur, while in the present case the value never came to be more than 634.76 g/m^2, a considerably lower value than the reports of Gopal *et al.* This may be attributed to less suitable growth conditions in the experimental water body for the weed.

The authors observed gross production maxima as 158.75gC/m^2/month or 422.275 gO_2/m^2/month or 446.563 gHC/m^2/month, while Das (1968) reported its production as 3.5 to 9.5 g/m^2/day, Sinha (1969) observed 3.8 g/m^2/day, which show the similar range of production. The review of literature reveals (Gopal *et al.*, 1978) that the gross production of *E. crassipes* are usually higher than the microphytic production the same water body. Sumitra (1971) and Haniffa and Pandian (1978) generally observed higher gross production for macrophytes in south Indian ponds so long their biomass was less than 10.00 g/m^2. The variety of substances (probiotic or antibiotic; Talling, 1975) present in the littoral region of tropical reservoir produce adverse impact on the production of macrophytes (Westlake, 1976; Haniffa and Pandian, 1978) by accelerating or inhibiting the photosynthetic rate. This factor seems to be operative in the present water body as the highest production was not observed when the value of standing crop and biomass was the most. However, higher gross production rate was correlated with the higher growth rate of the weed, showing the high photosynthetic efficiency during this period. The production values declined after June 1991 considerably. Sumitra (1971), Sreenivasan (1971), Natarajan and Pathak (1980) have reported the higher photosynthetic efficiency in summer months and a sharp seasonality in production value maximum being in warmer months, which are in agreement with the present findings.

The aquatic invertebrates living on the macrophyte in general, have been reported to be many times more abundant than invertebrates living in the bottom sediments (Richardson, 1921; Needham, 1929; Gerking, 1957). The present finding, however, contradicts the above observation as the number of epiphytic fauna was recorded considerably lower than the benthic fauna. This may be due to the morphometry of the hydrophyte, which although provides more surface area but support comparatively lower density of epiphytic fauna. There are evidences to conclude that different vegetations in ponds and lakes shelter different animals (Krecker, 1939; Edmondson, 1944; Bryden, 1952; Smyly, 1952; Rosine, 1955). The quantitative investigations by Krecker and Rosine showed in addition, that plants with very finely divided leaves possessed larger and more varied animal population than those with simple entire leaves. The basis of difference of epiphytic fauna on different aquatic plants has been suggested by Harrod (1964) is the characteristics of the plants themselves rather than the physical condition of the environment, since all the plants are growing under the same environmental conditions. The positive correlation of higher epiphytic population density with increasing standing crop value of *Eichhornia crassipes* clearly reflects that the surface area plays an important role in sustaining the population. Many authors like Krecker (1939), Andrews and Hasler (1943), Gerking (1957), Mrachek (1966), Krull (1970), Biochino and Biochino (1980), Dvorak and Best (1982), Schramm *et al.* (1987) have also observed that in general, number and weights of macroinvertebrates per unit weight of macrophyte are positively related to the amount of surface area available and the degree of leaf dissection. According to Harrod (1964) apart from the surface area, two other factors, the periphyton on the plants and the chemical nature of the plants are important for qualitative and quantitative changes in epiphytic population. Though no detailed qualitative and quantitative study was attempted on periphyton on the surfaces of *E. crassipes* during the present study yet the casual study on this aspect showed common occurrence of several species of green algae, diatoms rotifers and some crustaceans, which serve as food for the epiphytic fauna.

The chemical nature of plant might make it a favourable or unfavourable habitat for the animals (Harrod, 1964). So far the chemical nature of *E. crassipes* is concerned it is exceptionally rich in Potassium (Dey and Sarmah, 1982) and has high carotene contents in its leaves (Dey and Sarmah, 1982). Riechert and Trede (1977) while working on water hyacinth as feed of some weedifagous fishes reported the presence of raphides and oxalates as problematic materials due to which it is not taken raw by the fishes and many other animals excepting a few which are used for its biological control. Hence it seems possible that the chemical contents of *E. crassipes* do not have any impact on attracting and harbouring the epiphytic fauna while other weeds may have.

In a comparative study of population density epiphytic fauna on different macrophytes Schramm *et al.* (1987) reported *E. crassipes* consistently supporting high density and biomass. They found a positive relationship between abundance of invertebrates and the amount of colonizable surface are per unit weight of macrophyte, which is in conformity with the present finding.

Andrews and Hasler (1943) and Mrachek (1966) observed higher population densities of oligochaetes on *E. crassipes*. They found, in general, that insects were abundant but comprised a relatively low proportion of the total invertebrates collected from the submergent and emergent macrophytes. In the present study insects in general were the dominant group and the oligochaets ranked second from quantity view point. Our finding is in agreement with Schramm *et al.*, (1987) who observed abundance of insects on *E. crassipes* constituting 10 to 25 per cent of total population. O'Hara (1968) found insects sharing 15 per cent of total number of invertebrates on *E. crassipes* in southern Florida. As found in other investigations (Andrews and Hasler, 1943; Rosine, 1955, Gerking, 1957; Mrachek, 1966; Pieczynski, 1973; Sozzka, 1975; Martin and Shireman, 1976; Schramm *et al.*, 1987; Sinha *et al.*, 1993) Chironomidae was the most abundant taxon, the highest percentage value reaching upto 75.30 per cent of total epiphytic population.

In other studies in Florida, Gastropoda were 28 per cent of the invertebrates collected on *E. crassipes* (O'Hara, 1968) and 15 per cent of total population on the same hydrophyte (Martin and Shireman, 1976). We observed a high of 28.40 per cent and low of 1.50 per cent gastropods on *E. crassipes* during the study.

Like variation in abundance of different epiphytic groups on *E. crassipes* the diversity has also been reported to vary considerably. The review of literature reveals that virtually nothing has been done on species diversity aspect of epiphytic invertebrates found on *E. crassipes* or other hydrophytes.

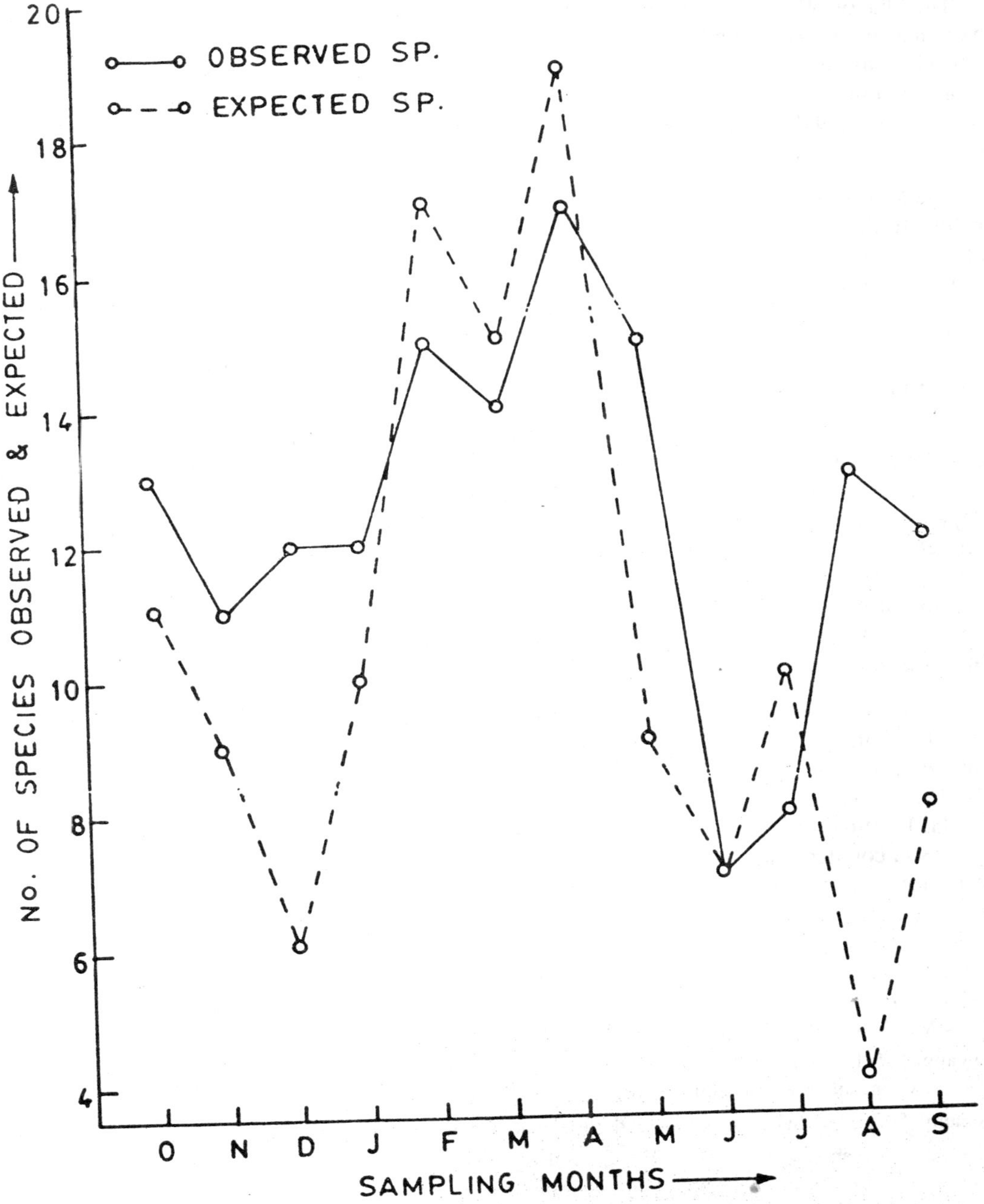

Fig. 16.5. Seasonal variationin number of species observed in the samples and expected of epiphytic macro invertebrates attached to *E. crassipes*.

The Shanon-Weaver species diversity index ranged from 1.7416 to 3.6864. By application of explanation by Wilhm and Dorris (1966), the habitat was moderately polluted throughout the period of investigation except February, March and April when the species diversity index was more than three and hence the habitat can be said unpolluted. But by application of scale of Staub *et al.*, (1970) habitat was moderately polluted in August (1.7416), light polluted during rest all months except February, March and April when it was slightly polluted.

The Margalef diversity index varied considerably from 1.1171 to 2.6550. The higher diversity values so derived reflect the suitability of habitat for the organisms in one hand while on the other the higher values have been reported to be related with longer food chain and complex food web which does not appear in the present study as the higher values are near five or around five.

The Simpson's diversity index ranged from 0.513 to 1.498. By this range of diversity, following Mac Arthur's explanation (1965 and 1972) it can be said that the habitat has diversified resources with lower diversity of their utilization by an average species.

As reflected by the variation in equitability values, due to the "inequitability" in the distribution of individuals and the species of the weed-fauna, the samples of weed-fauna community in the months of February, March, April and July 1991 have a species diversity "appropriate" to a community of 113.33 per cent, 107.14 per cent, 111.76 per cent and 125.00 per cent as many species as actually occur respectively. This value varied only from 30.76 per cent to 100.00 per cent for the samples of other months. Thus these values of equitability are the measure of the species relationships existing at that one place of sampling and time in one hand while on the other these throw light on over all estimate of apportioning of species (Lloyd and Ghelardi, 1964).

On the basis of the expected number of species and the actually recorded number of species (Fig. 16.5), the samples of February, March, April and July 1991 of 15, 14, 17 and 8 species have more diversity, judged by Shanon-Weaver function, than would a hypothetical community of 17, 15, 19 and 10 species respectively which confirmed Mac Arthur's model (see Lloyd and Ghelardi, 1964), while for the samples of the rest months with their respective number of species encountered have no more diversity calculated by Shanon-Weaver function than their respective expected number of species of Mac Arthur's model (1957).

REFERENCES

Andrews, J.D. and Hasler, A.D. 1943. Fluctuation in the animal populations of the littoral zone in Lake Mendota. *Trans. Wisc. Acad. Sci. Arts. Lett.* **35** : 175-185.

Ayyappan, S.J., Olah, J., Raghavan, S.L., Sinha, V.R.P. and Purushotaman, C.S. 1986. Macrophyte decomposition in two tropical lakes. *Arch. Hydrobiol.* **106** (2) : 19-231.

Bellamy, D.F. 1967. Succession and depth time scale in ephemeral swamp ecosystem. *Trop. Ecol.* **8** : 67-73.

Biochino, A.A. and Biochino, G.I. 1980. Quantitative estimation of phytophilous invertebrates. *Hydrobiol. J.* **15** : 74-76.

Bryden, R.W. 1952. Ecology of *Pelmatohydra oligactis* in Kirkpatricks Lake, Tennessee. *Ecol. Monogr.* **22** : 45-68.

Carpenter, K.E. 1927. Faunistic ecology of some Cardiganshire streams. *J. Ecol.* **15** : 33-54.

Choudhary, H., Murthy, D.S., Dey, R.K. and Reddy, V.G.K. 1976. Role of Chinese carp *C. idella* in biological control of noxic aquatic weeds in India. In : Aquatic Weeds in India (Eds. C.K. Varshney and I. Rozoska) Dr. W. Junk, B.V. The Hague.

Das, R.R. 1968. Growth and distribution of *Eichhornia crassipes* and *Spirodela polyrhiza.* Ph.D. Thesis, B.H.U., Varanasi.

Das, R.R. 1969. A study of reproduction in *Eichhornia crassipes. Trop. Ecol.* **10** : 195-198.

Dey, S.C. and Sarmah, S. 1982. Prospect of water hyacinth *Eichhornia crassipes* feed to cultivable fishes - a preliminary study with *Tilapia mossumbica* Peters. *Matsya* **8** : 40-44.

Dvorak, J. and Best, E.P.H. 1982. Macroinvertebrate communities associated with the macrophytes of Lake Vechten : structural and functional relationships. *Hydrobiologia.* **95** : 115-126.

Edmondson, W.T. 1944. Ecological studies of sessile Rotatoria. Part-I. Factors affecting distribution. *Ecol. Monogr.* **14** : 31-66.

Gerking, S.D. 1957. A method of sampling the littoral macrofauna and its application. *Ecology.* **38** : 219-226.

Gopal, B. 1967. Contribution of *Azolla Pinnata* R. Br. to the productivity of temporary ponds at Varanasi. *Trop. Ecol.* **8** : 126-130.

Gopal, B., Sharma, K.P. and Trivedy, R.K. 1978. Studies on ecology and production in Indian freshwater ecosystems on primary producer level with emphasis on Macrophytes. pp. 349-376. In : *Glimpses of Ecology* (Eds. J.S. Singh and B. Gopal) International Scientific Pub., Jaipur, India.

Haniffa, M.A. and Pandian, T.J. 1978. Morphometry, primary production and energy flow in a tropical pond. *Hydrobiologia* **59** (1) : 23-48.

Harrod, J.J. 1964. The distribution of invertebrates on submerged aquatic plants in a chalk stream. *J. Anim. Ecol.* **33** (2) : 335-341.

Hutchinson, G.E. 1967. *A treatise on limonolgy-II, Introduction to lake biology and the limnoplankton.* Wiley, N.Y.

Jones, J.R.E. 1949. An ecological study of the river Rheidol, North Cardiganshire, Wales. *J. Anim. Ecol.* **18** : 67-88.

Krecker, F.H. 1939. A comparative study of the animal population of certain submerged aquatic plants. *Ecology* **20** : 553-562.

Krull, J.N. 1970. Aquatic plant-macroinvertebrate association and water fowl. *J. Wildl. Manage.* **34** : 707-718.

Lloyd, M. and Ghelardi, R.J. 1964. A table for calculating the equitability component of species diversity. *J. Anim. Ecol.* **33** (2) : 217-225.

Mac Arthur, R.H. 1965. Patterns of species diversity. *Biol. Rev.* **40** : 510-533.

Mac Arthur, R.H. 1972. *Geographical Ecology. Patterns in the distribution of species.* Harper and Row. N.Y. pp. 269.

Margalef, R. 1958. Information theory in ecology. *Gen. Syst.* **3** : 36-71.

Martin, R.G. and Shireman, J.V. 1976. A quantitative sampling method for hydrilla-inhabiting macro-invertebrates. *J. Aquat. Plant Manag.* **14** : 16-19.

Mitchael, D.S. 1976. The growth and management of *Eichhornia crassipes* and *Salvinia* sp. in the native environment and alien situation. In : *Aquatic Weeds in India* (Eds. C.K. Varshney and I. Rozaska). Dr. W. Junk. B.V. the Hague.

Mishra, B.K., Sahu, A.K. and Pani, K.C. 1988. Recycling of aquatic weed water hyacinth and animal wastes in the rearing of Indian major carps. *Aquaculture.* **68** : 59-64.

Moon, H.P. 1936. The shallow littoral region of a bay at the north-west end of Windermere. *Proc. Zool. Soc. London* 491-515.

Mrachek, R.J. 1966. Macroscopic invertebrates on the higher aquatic plants at Clear Lake, Iowa. *Iowa Acad. Sci.* **73** : 168-177.

Natarajan, A.V. and Pathak, V. 1980. Bioenergetic approach to the productivity of man-made lakes. *J. Inland Fish. Soc. India* **12** (1) : 1-18.

Needham, P.R. 1929. Quantitative studies of the fish food supply in selected areas; a biological survey of the Erie-Niagara system. *Ann. Rept. N.Y. St. Conserv. Deptt. Suppl.* **18** : 220-232.

O'Hara, J. 1968. Invertebrates type found in water hyacinth. *Quart. J. Fla. Acad. Sci.* **30** : 73-80.

Penfound, W.T. and Earle, T.T. 1948. The biology of water hyacinth. *Ecol. Monogr.* **18** : 447-472.

Pieczynski, E. 1973. Experimentally increased fish stock in the pond type lake Warniak. XIII. Number and biomass of the fauna associated with macrophytes. *Ekol. Polska.* **21** : 595-610.

Pielou, E.C. 1966. The measurement of diversity in different type of biological collections. *J. Theor. Biol.* **13** : 131-144.

Rai, D.N. and Datta Munshi, J.S. 1979. The influence of thick floating vegetation (Water hyacinth, *Eichhornia crassipes*) on the physico-chemical environment of freshwater wetland. *Hydrobiologia.* **65** : 65-69.

Richardson, R.E. 1921. The small bottom and shore fauna of the middle and lower Illinois River and its connecting lakes, Chillicothe to Grafton. *Bull. Ill. Nat. Hist. Surv.* **13** : 363-522.

Riechert, C. and Trede, R. 1977. Preliminary experiments on utilization of water Hyacinth by grass carp. *Weed Res.* **17** (6) : 357-360.

Rosine, W.N. 1955. The distribution of invertebrates on submerged aquatic plant surfaces in Muskee Lake, Colorado. *Ecology* **36** : 308-314.

Sahai, R. and Sinha, A.B. 1970. A contribution to the ecology of Indian aquatics-I. Seasonal changes and biomass of water hyacinth (*Eichhornia crassipes* Mart) Solms. *Hydrobiol.* **35** : 376-382.

Schramm, H.L., Jirka, K.J. and Hoyer, M.V. 1987. Epiphytic macroinvertebrates on dominant macrophytes in two control Florida Lakes. *J. Freshwat. Ecol.* **4** (2) : 151-162.

Shannon, C.E. and Weaver, W. 1949. The mathematical theory of communication. Urbana III : Univ. of Illinois Press. 117 p.

Singh, S.B. and Sahai, R. 1979. Seasonal changes in biomass of *Eichhornia crassipes* (Mart) Solms in 'Jalwania' pond of Gorakhpur. *Indian J. Ecol.* **6** (1) : 30-34.

Simpson, E.H. 1949. Measurement of diversity. *Nature* **6** (1) : 30-34.

Sinha, A.B. 1969. *Investigation on the Ecology of Ramgarh Lake, Gorakhpur.* Ph.D. Thesis, Gorakhpur Univ.

Sinha, A.B. and Sahai, R. 1978. Studies on the factors affecting growth and productivity of free floating macrophytes in Ramgarh Lake, Gorakhpur. pp. 377-382. In : *Glimpses of Ecology* (Eds. J.S. Singh and B. Gopal). International Scientific Pub. Jaipur.

Sinha, M.P., Sinha Ratna and Mehrotra, P.N. 1993. Seasonal variation and species diversity in epiphytic oligochaet fauna attached to water hyacinth of a tropical freshwater body at Ranchi. *J. Freshwater Biol.* Communicated.

Sen, N.S., Naqvi, A.H., Appa Rao, N.V. and Ahmad, M.F. 1993. Distribution growth and trophic significance of water hyacinth *Eichhornia crassipes* (Kunth). pp. 93-106. In : *Advances in Limnology* (Ed. H.R. Singh). Narendra Pub. House, Delhi.

Smock, L.A. and Stoneburner, D.L. 1980. The response of macroinvertebrates to aquatic macrophyte decomposition. *Oikos.* **35** : 397-403.

Smyly, W.J.P. 1952. The Entomostraca of the weeds of a moorland pond. *J. Anim. Ecol.* **21** : 1-11.

Smyly, W.J.P. 1957. Distribution and seasonal abundance of Entromostraca in moorland ponds near Windermere. *Hydrobiologia* **11** : 59-72.

Soszka, G.J. 1975. Ecological relation between invertebrates and submerged macrophytes in the lake littoral. *Ekol. Polska.* **23** : 393-415.

Sreenivasan, A. 1971. Shallow variation in primary productivity of three tropical ponds. *Hydrobiologia* **37** (3-4) : 553-564.

Staub, R., Appling, J.W., Hofsteiler, A.M. and Hass, I.J. 1970. The effect of industrial wastes of Memphis and Shelby country on primary plankton producers. *Bioscience* **20** : 905-912.

Sumitra, V. 1971. Seasonal variation in primary productivity of three tropical ponds. *Hydrobiologia* **38** : 395-408.

Trivedy, R.K., Sharma, K.P., Goyal, P.K. and Gopal, B. 1978. Some ecological observations on floating islands. *Hydrobiologia* **60** (2) : 187-190.

Voights, D.K. 1976. Aquatic invertebrate abundance in relation to changing marsh vegetation. *Am. Midl. Nat.* **95** : 313-322.

Werner, E.E., Hall, D.J., Laughlin, D.R. Wagner, D.J., Wilsmann, L.A. and Funk, F.C. 1977. Habitat partitioning in a freshwater fish community. *J. Fish. Res. Board. Can.* **34** : 360-370.

Westlake, D.F. 1963. Comparison of plant productivity. *Biol. Rev.* **38** : 385-425.

Westlake, D.F. 1965. Some basic data for investigation of the productivity of aquatic macrophytes. In : *Primary Productivity in Aquatic Environment* (Ed. C.R. Goldman). *Mem. Inst. Hydrobiol.* **18** (Suppl.). Univ. Calif. Press, Berkeley.

Westlake, D.F. 1976. Primary production. In : *The Function of Freshwater Ecosystem* (Eds. E.D. Le Gren and R.H. Lowe Mc Connoll). I.B.P. Hand Book 22, Cambridge Univ. Press, London.

Whitehead, H. 1935. An ecological study of the invertebrate fauna of chalk stream near Great Driffield, Yorkshire. *J. Anim. Ecol.* **4** : 58-78.

Whittaker, B.A. 1961. Estimation of net primary production of forest and shrub communities. *Ecology* **42** : 177-180.

Wilhm, J.L. and Dorris, T.C. 1966. Species diversity of benthic macroinvertebrates in a stream receiving domestic and oil refinery effluent. *Amn. Midl. Nat.* **76** : 427-449.

Zar, J.H. 1974. *Biostatistical Analysis.* Prentice-Hall, new Jersey. 251-262.

Chapter 17

Interference and Exploitation Competition in Larval *Bufo marinus*

Ross A. Alford

Department of Zoology, James Cook University, Townsville, Qld 4811, Australia

Abstract. Anuran larvae have recently been used extensively to explore how competition operates. Intraspecific competition among tadpoles can be either exploitation competition (via the consumption of limited resources) or interference competition, which occurs through behavioural interference or allelopathy. These forms of competition depend on density and resource levels in different ways, and thus may have different consequences for population regulation. This experiment demonstrated that both forms of competition can occur in laboratory populations of larval cane toads, *Bufo marinus*. Doubling tadpole density while keeping feeding rate per individual constant affected growth rate similarly to halving feeding rate with constant density. The effects of density and relative feeding rate on log-transformed body size at Day 14 of the experiment were statistically independent. This suggests that interference and exploitation competition have additive effects on larval growth rates in this species.

Key words : *Bufo marinus, Larva, competition, feeding and growth rate.*

Introduction

Competition appears to be important in the ecology of tadpoles of many anuran species. Intraspecific competition can have major effects in both the field (Wilbur, 1972, 1976; Morin, 1983; Smith, 1983, 1990; Hearnden, 1991) and in the laboratory (references below). Intraspecific competition can be caused by exploitation (the depletion of resources), or by interference. In interference competition some tadpoles either reduce the access of others to preferred resource (Savage, 1952, Steinwascher, 1978a, b), increase levels of stress (Adolph 1931; Gromko *et al.* 1973; John and Fenster 1975; Mishra and Dash 1984; Mahapatro and Dash 1987) or release allelopathic chemicals or symbiotes (Richards 1958; Rose 1959, 1960; Licht, 1967; Steinwascher, 1979a, b; Beebe 1991, Griffiths *et al.* 1993). Effects similar to those of interference competition may be caused by fouling of the water in laboratory experiments (Wilbur, 1977b; Dash and Mahapatro 1990; Murray 1990).

In many laboratory experiments, the negative effects of density have been produced by a combination of interference and exploitation mechanisms (Wilbur 1977b; Steinwascher, 1979b; Hota and Dash 1981; Semlitsch and Caldwell, 1982; Berven and Chadra, 1988; Kehr 1989). Interference can occur in the field as well as in the laboratory; Petranka (1989) showed that localised allelopathic effects can occur in natural ponds when animals are present at high densities.

Many recent studies have used tadpoles to examine general models of the operation of interspecific and intraspecific competition (Alford in press). These usually have concentrated on exploitation competition. The intensity of competition is often varied by rearing tadpoles at a variety of densities on identical food supplies. This varies the rate of food supply per individual and hence the degree of exploitation competition. Unfortunately, without control treatments that vary density while holding rate of supply/individual constant, such experiments cannot distinguish effects due to resource limitation from effects due to interference. In general, interference competition, which may be due to any of the inhibitory effects outlined above, can increase the effects of density observed in such experiments. If the results of such experiments are to be generalised to models of the behaviour of populations in the field, it is necessary to distinguish between interference and exploitation competition, as these forms of competition are likely to respond differently to changes in resource levels (Steinwascher 1979a).

An example of this sort of experiment is that of Wilbur (1977a), who investigated the effects of density on laboratory populations of larval *Bufo americanus*. They were raised at six densities with the same ration of food supplied to all densities. The tadpoles responded strongly to density, but the effects of interference and exploitation could not be separated. Dash and Hota (1980), Sokol (1984), and Tejedo and Reques (1992) performed similar experiments using larval *Rana tigerina, Litoria ewingi*, and *Bufo calamita*, respectively, and found similar patterns.

In order to separate the effects of exploitative and interference competition, Wilbur (1977b) raised larval *Rana sylvatica* in the laboratory at four densities crossed with four rates of food supply. Unfortunately, his results were difficult to interpret, possibly because of fouling by food at low densities and high feeding rates. Steinwascher (1979b), as part of a larger design, manipulated density and relative feeding rates of larval *Rana utricularia.* He suggested that the relative importance of interference and exploitation competition change with alterations in resource level. Berven and Chadra (1988) used a similar design to separate the influences of egg size, density, and food level on tadpoles of *Rana sylvatica.* They found that size at metamorphosis decreasd and length of larval period increased with decreasing food/tadpole. They found no effect of density independent of food level; interference competition was apparently minimal or absent. Murray (1990) found a similar result with larval *Rana sylvatica.* Kehr (1989) experimentally manipulated food level and density of *Hyla pulchella* tadpoles in two experiments in aquaria and could evidence for both interference and exploitation competition. These experiments have shown that both forms of competition can operate in laboratory experiments but leave their relative intensities difficult to evaluate.

The Cane Toad, *Bufo marinus*, is an introduced pest in Australia and is the subject of a major biological control programme (Cohen and Alford 1993). Field work (Hearnden 1991) has demonstrated that growth and survival rates of tadpoles are affected strongly by the density of conspecifics. Experiments conducted to date have not separated interference and exploitation effects. Interference might be important in *B. marinus* as tadpoles often form dense aggregates (Tyler 1989; Hearnden 1991). The experiment reported on in this paper was designed to determine whether intraspecific competition among larval *Bufo marinus* was solely interference competition, exploitation competition, or a combination of both forms. If competition included both interference and exploitation, I wanted to examine the relative intensities of the two forms.

Materials and methods

A single clutch of *Bufo marinus* eggs was collected at Mount Margaret Dam, near Townsville, Queensland, Australia (19° 20'S, 146° 37'E), on 11 February 1988. They were returned to the laboratory, and hatched on 13 and 14 February. The experiment was set up on 16 February. Hatchlings were assigned at random to experimental units. Ten hatchlings included in the randomisation were weighed; their mean wet mass was 6.3 mg (S.D. = 1.1, N = 10).

The experiment was a 3 X 3 factorial design with eight replicates. Three feeding levels (1X, 2X, and 4X) were crossed with three initial densities (1, 2, and 4 tadpoles per container). One replicate of each treatment was included in each of eight spatial blocks. Treatment codes combine the initial density (ID) and feeding regimen, i.e. 1-2X means one tadpole fed at the 2X rate.

The rearing containers used were 440 mL capacity polypropylene containers, 112 mm in diameter at the base tapering to 96 mm at the rim, and 55 mm deep. Animals were fed, water was changed and the containers scrubbed and rinsed clean at 3-4 day intervals. Water was drawn from the James Cook University Biological Sciences recirculating freshwater aquarium system. Survival was monitored daily.

Food was a 3 : 1 mixture by mass of ground lucerne (alfalfa) pellets and Tetra-Min^{tm} Tropical Fish Flakes passed through a 250 μm sieve. The initial feeding rate was 1X = 0.5 mg/day on Day 0 of the experiment, 16 February. On Day 11 (27 February) this was changed to X = 1.0 mg/day. On Day 14 (1 March) all animals were weighed and the experiment was terminated, as mortality in the highest ID/lowest food treatment had reached 75 per cent.

Animals were weighed to the nearest 0.1 mg by placing them on a piece of window screening, blotting all water from the screening, then dropping them from the screening into a tared container of water on an electronic balance. This procedure produces wet weights that are repeatable to within ± 0.5 mg and minimises handling of tadpoles.

TABLE 17.1. *Summary of the results of the experiment on Day 14. Exponential growth constants calculated using k = [(mean log mass) - log (mean initial mass)]/age, where age was 14 days. Using mean log(mass) is equivalent to using log(geometric mean mass) as suggested by Alford and Jackson (1993).*

Initial density	Feeding regimen	Relative feeding rate	Mass (mg) mean	Std. Dev.	N	Exponential growth constant (k)	Percent surviving
1	1X	1	33.8	4.8	7	0.1194	87.5
	2X	2	40.3	6.0	8	0.1318	100.0
	4X	4	56.7	8.0	8	0.1563	100.0
2	1X	0.5	20.9	9.1	9	0.0783	56.2
	2X	1	24.2	8.9	14	0.0911	87.5
	4X	2	37.4	11.0	12	0.1238	75.0
4	1X	0.25	22.4	4.1	8	0.0896	25.0
	2X	0.5	18.1	6.3	19	0.0698	59.4
	4X	1	21.4	9.3	23	0.0804	71.9

The data were analysed in two ways. Wet weight at day 14 of the experiment was analysed using a factorial analysis of variance (ANOVA). For the ANOVA, the feeding rate treatments were recoded into relative feeding rates (RFRs; Table 17.1). This resulted in an unbalanced design with three levels of RFR (0.5X, 1X, 2X) crossed with three levels of ID (1, 2, and 4 individuals/container) and fully crossed with eight replicates, which were considered random. The design was unbalanced because two of the nine possible combinations of ID and RFR did not occur (1 individual with a RFR of 0.5X; 4 individuals with a RFR of 2X). Two of the original treatments (1-4X and 4-1X, RFRs 4 and 0.25 respectively) were omitted from this analysis because they would have unbalanced it further. Percent surviving to Day 14 in each container was analysed using a Kruskal-Wallis nonparametric one-way ANOVA (SAS Institute Inc., 1987). For this analysis, each combination of ID and absolute feeding rate was regarded as a single treatment.

For use in interpretation, growth constants (k) of the exponential equation $mass_t = a \cdot e^{kt}$ were calculated using the geometric mean of mass at day 14 (Alford and Jackson 1993). This assumes that tadpoles were following the exponential curve, an assumption which seems reasonable (Alford and Jackson 1993).

Results

The results of the experiment are summarised in Table 17.1. Mean mass at Day 14 varied from 56.7 mg (a ninefold increase over the mean hatchling weight) in treatment 1-4X to 18.1 mg (less than a threefold increase over man hatchling mass) in treatment 4-2X. Percent surviving to Day 14 varied significantly among treatments (Kruskal-Wallis test, $\chi^2 = 39.513$, 8 d.f., $P < 0.0001$). The data (Table 17.1) indicate that within levels of initial density, survival tended to decrease with decreasing food levels. It also tended to decrease with increasing initial density, and reached its lowest level in the lowest-food, highest density treatment (4-1X). Survival to Day 14 in this treatment was so low that the resulting increases in food level per individual were apparently beginning to cause an acceleration in individual growth rates (Table 17.1).

The ANOVA of mass at Day 14 (Table 17.2) showed that both relative feeding rate and initial density affected mass at day 14. The absence of any significant interactions suggests that these factors acted independently. Fig. 17.1 can aid in interpreting these effects. The ordinate in this figure is plotted on a log scale so that equivalent increases or decreases in relative growth rate will produce equivalent changes in the elevations of points regardless of their position along the growth curve.

The absence of interactions in the ANOVA is reflected in the similar patterns of change of mass at day 14 with RFR within levels of ID across all treatments except 4-1X. Within ID 2, decreasing the RFR from 2 to 1 and 1 to 0.5 decreased mass at Day 14 by 35 per cent and 14 per cent, respectively. In ID 1, decreases of RFR from 4 to 2 and 2 to 1 decreased mass at Day 14 by 29 per cent and 16 per cent respectively. In ID 4, decreasing RFR from 1 to 0.5 decreased mass at Day 14 by 15 per cent. Changes caused by altering ID while keeping RFR constant

TABLE 17.2. *Results of three-way ANOVA of log-transformed mass at Day 14 of the experiment, with replicate considered as a random factor. Treatments 4-1X and 1-4X omitted because they increase the lack of balance in the design and because low survival in 4-1X confounded survival with growth results. Mean squares are Type IV (partial) mean squares, and error terms are composites because of unbalanced data (SAS Institute Inc. 1987); this causes non-integer degrees of freedom.*

	Hypothesis term		Error term		F	P (F)
effect	M.S.	d.f.	M.S.	d.f.		
Relative feeding rate (F)	0.9475	2	0.1340	18.6	7.069	0.0052
Initial density (D)	0.5931	2	0.0477	11.4	12.443	0.0014
F X D	0.0081	2	0.0926	11.7	0.087	0.9169
Replicate (R)	0.1846	7	0.1245	2.5	1.482	0.4302
F X R	0.1184	14	0.0662	4.2	1.790	0.2977
DXR	0.0487	14	0.0813	7.8	0.599	0.8069
F X D X R	0.0908	11	0.2429	39.0	0.374	0.9586
Error	0.2429	39				

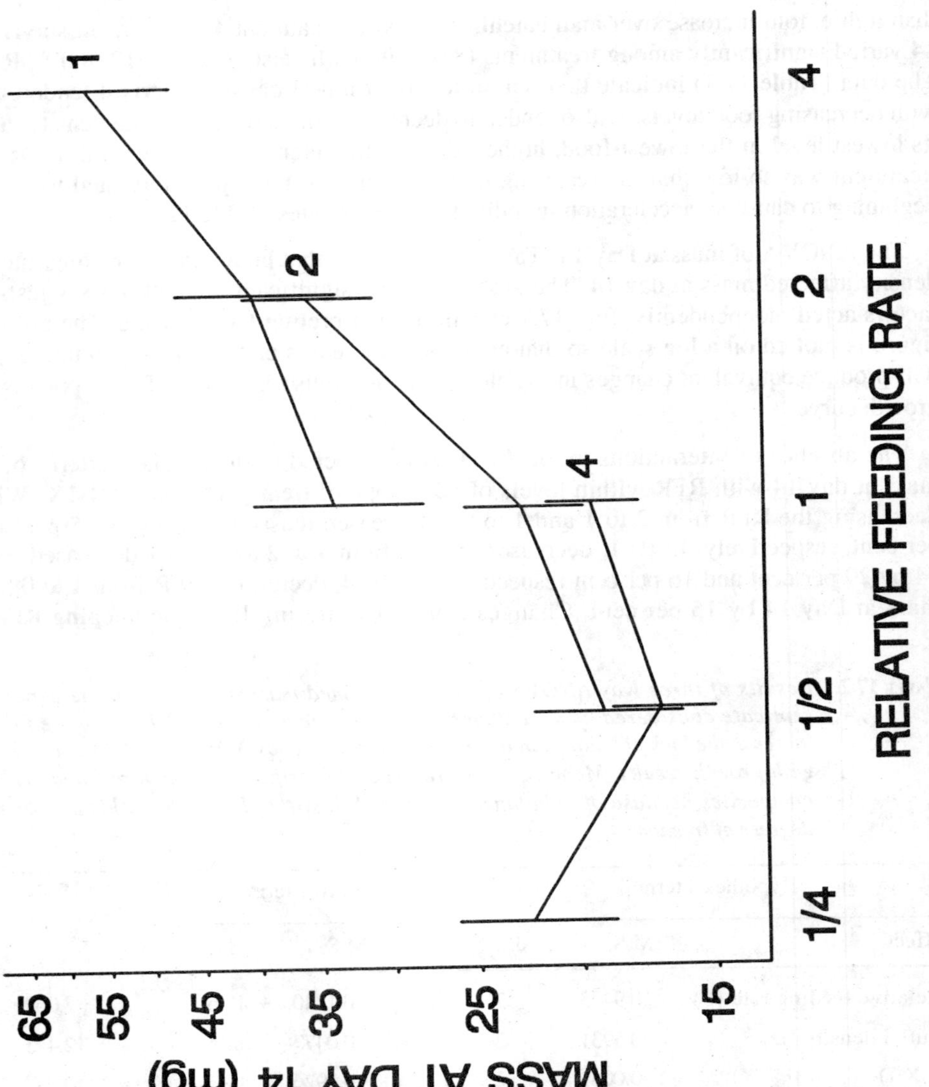

Fig. 17.1. Mean mass at Day 14 ± 95 per cent LSD confidence interval (SAS Institute Inc., 1987) plotted against relative feeding rate for each treatment combination in the experiment. Relative feeding rates of 1/4 and 4 were not included in the ANOVA (Table 17.2) as they would have further unbalanced the design. Solid lines connect means for animals reared at the same density; numbers at the end of each series indicate the initial density of that series. The scale of the y-axs is logarithmic so that equivalent proportional changes in relative growth rate produce equal vertcal displacements.

followed a similar pattern. Within RFR 2, doubling the ID from 1 to 2 individuals decreased mass at day 14 by 7 per cent. In RFR 1, doubling the ID from 1 to 2 and 2 to 4 individuals decreased mass at Day 14 by 28 per cent and 12 per cent, respectively. In RFR 1/2, doubling the ID from 2 to 4 individuals decreased mass at Day 14 by 13 per cent. These decreases are calculated using the arithmetic means of mass; to compare how changes in RFR and ID affected competition, it is more meaningful to examine changes in the growth rate constant, *k*.

Examining the patterns of changes in *k* for initial densities of 1 and 2 animals/container reveals very similar patterns in the effect of doubling the feeding rate/individual on proportional growth rate. Within initial density 2, for example, halving the RFR from 2.0 to 1.0 decreased *k* by a factor of 26 per cent. Halving it a second time from 1.0 to 0.5 decreased the *k* by 14 per cent. Within initial density 1 a similar pattern occurred : halving the RFR from 4.0 to 2.0 decreased *k* by 16 per cent, and halving it a second time decreased *k* an additional 9 per cent. The pattern within initial density 4 is different; decreasing the RFR from 1.0 to 0.5 decreased *k* by 13 per cent, but decreasing the RFR a second time, from 0.5 to 0.25, actually resulted in *k* increasing by 28 per cent.

The effects of doubling the density while maintaining a constant RFR were similar in three of the four situations where comparisons can be made. Doubling the initial density from 2 to 4 individuals at RFRs of 0.5 and 1, and from 1 to 2 at a RFR of 2, all had similar effects on growth rate, decreasing *k* by 11 per cent, 12 per cent, and 6 per cent, respectively. The effect of doubling the initial density from 1 to 2 within the RFR of 1 was considerably greater, decreasing *k* by 24 per cent.

Discussion

The results of this experiment allow a clear separation of the effects of interference and exploitation competition. They do not appear likely to have been caused in any part by pollution from excess food, a flaw that appears in several earlier papers with similar aims. The only complication was caused by the relatively high mortality rates of tadpoles in the 4-1X treatment. Survival to Day 14 in this treatments was only 25 per cent. It is clear from Figure 17.1 that in this treatment the high mortality had increased the food ration per surviving individual. This lead to increased growth rates and greater masses at Day 14 in this treatment than in the other treatments with initial densities of four individuals. It does not appear that mortality in any other treatments was high enough to lead to similar effects. On average, halving the RFR led to a 16 per cent (S.D. = 6.3, N = 5) reduction in the value of *k*, the exponential growth rate parameter. Doubling the density while maintaining a constant RFR led to a 13 per cent (S.D. = 7.6, N = 4) decrease in *k*. The mean decreases are very similar, and do not differ significantly ($t = 0.51$, 7 d.f., $P = 0.629$). It thus appears that interference and exploitation competition had approximately equivalent effects on the growth rates of *B. marinus* larvae under the conditions of this experiment.

This result differs from those of most others who have attempted to separate these sources of competition. Berven and Chadra (1988) and Murray (1990) found little evidence for interference competition in any form. These experiments were conducted on tadpoles of *Rana.* spp. (Kehr's

(1989) results suggested that both forms of competition were important in *Hyla pulchella*, but because he conducted separate experiments on interference and exploitation it is difficult to interpret the relative intensities of the two types of competition. Wilbur (1977b) found evidence for the operation of both types of competition in *Rana sylvatica*, but his experiment was compromised by severe pollution by excess food, resulting in decreases in growth rate with increases in RFR in many treatments. Steinwascher (1979b) found that both exploitation and interference competition occurred among tadpoles of *Rana utricularia*, and that the intensity of interference effects depended strongly on the degree of food limitation experienced by the larvae.

The equivalence of effects of interference and exploitation competition found in this experiment, and the insensitivity of interference effects to resource levels, suggest that interference effects may be very important in the biology of larval *Bufo marinus*. This species often occurs at very high densities in the field (Hearnden 1991), and depending on the degree to which interference effects persist in field environments, interference may be a major factor controlling populations of this species. I would suggest that future studies of intraspecific competition in anuran larvae should (1) clearly distinguish between effects of interference and exploitation competition, (2) attempt to examine the effects of both types of competition separately, (3) attempt to determine the mechanism(s) which cause interference, and (4) if interference has more than one cause, attempt to separate the contribution of each cause to the overall effects of interference.

Acknowledgements

This research was funded by the Council of Nature Conservation Ministers and by Australian Research Grants Scheme grant A 18715284. J. Wolstenholme provided valuable technical assistance. The manuscript was improved by comments from Reid N. Harris and Stephen J. Richards.

REFERENCES

Adolph, E.F. 1931. The size of the body and the size of the environment in the growth of tadpoles. *Biol. Bull.* **61** : 350-375.

Alford, R.A. and Jackson, G.D. 1993. Do cephalopods and the larvae of other taxa grow asymptotically? *Am. Nat.* 141 : 717-728.

Beebe, T.J.C. 1991. Purification of an agent causing growth inhibition in anuran larvae and its identification as a unicellular unpigmented alga. *Can. J. Zool.* **69** : 2146-2153.

Berven, K.A. and Chadra, B.G. 1988. The relationship among egg size, density, and food level on larval development in the wood frog (*Rana sylvatica*). *Oecologia* **75** : 67-72.

Cohen, M.P. and Alford, R.A. 1993. Growth, survival, and activity patterns of recently metamorphosed *Bufo marinus*. *Wildlife Research*. **20** : 1-13.

Dash, M.C. and Hota, A.K. 1980. Density effects on the survival, growth rate, and metamorphosis of Rana tigerina tadpoles. *Ecology*. **61** : 1025-1028.

Dash, M.C. and Mahapatro, B.K. 1990. Effect of food quality and quantity on growth and metamorphosis of *Bufo stomaticus* larvae. *J. Ecobiol.* **2** : 67-76.

Griffiths, R.A., Denton, J. and Wong, A.L.C. 1993. The effect of food level on competition in tadpoles : interference mediated by protothecan algae?. *J. Anim. Ecol.* **62** : 274-279.

Gromko, M.H., Mason, F.S. and Smith-Gill, S.J. 1973. Analysis of the crowding effect in *Rana pipiens* tadpoles. *J. Exp. Zool.* **186** : 63-72.

Hearnden, M.N. 1991. *The Reproductive and Larval Ecology of* Bufo marinus (*Anura : Bufonidae*). Ph.D. Thesis, James Cook University, Townsville, Australia, 245 pp.

Hota, A.K. and Dash, M.C. 1981. Growth and metamorphosis of *Rana tigrina* larvae : effects of food level and larval density. *Oikos* **37** : 349-352.

John, K.R. and Fenster, D. 1975. The effects of partitions on the growth rates of crowded *Rana pipiens* larvae. *Am. Midl. Nat.* **93** : 123-130.

Kehr, A. 1989. Factores dependientes de la densidad y su influencia sobre el crecimiento individual de los estados larvales de *Hyla pulchella pulchella* (Amphibia Anura). *Limnobios* **2** : 757-761.

Licht, L.E. 1967. Growth inhibition in crowded tadpoles : intraspecific and interspecific effects. *Ecology.* **48** (5) : 736-745.

Mahapatro, B.K. and Dash, M.C. 1987. Density effect on growth and metamorphosis of *Bufo stomaticus* larvae. *Alytes* **6** : 88-98.

Mishra, P.K. and Dash, M.C. 1984. Metamorphosis of *Polypedates maculatus* (Gray 1830) : an analysis of crowding effect. *Alytes* **3** : 163-172.

Morin, P.J. 1983. Predation, competition, and the composition of larval anuran guilds. *Ecol. Monogr.* **53** : 119-138.

Murray, D.L. 1990. The effects of food and density on growth and metamorphosis in wood frogs (*Rana sylvatica*) from central Labrador. *Can. J. Zool.* **68** : 1221-1226.

Petranka, J.W. 1989. Chemical interference competition in tadpoles : does it occur outside laboratory aquaria? *Copeia* **1989** : 921-930.

Richards, C.M. 1958. The inhibition of growth in crowded *Rana pipiens* tadpoles. *Physiol. Zool.* **31** : 138-151.

Rose, S.M. 1959. Failure of survival of slowly growing members of a population. *Science.* **129** : 1026.

Rose, S.M. 1960. A feedback mechanism of growth control in tadpoles. *Ecology* **41** (1) : 188-199.

SAS Institute Inc. 1987. *SAS/STAT Guide For Personal Computers, Version 6 Edition.* SAS Institute Inc., Cary, NC, USA. 1028 pp.

Savage, R.M. 1952. Ecological, physiological, and anatomical observations on some species of anuran tadpoles. *Proc. Zool. Soc. London* **122** : 467-514.

Semlitsch, R.D. and Caldwell, J.P. 1982. Effects of density on growth, metamorphosis, and survivorship in tadpoles of *Scaphiopus holbrooki. Ecology* **63** : 905-911.

Smith, D.C. 1983. Factors controlling tadpole populations of the chorus frog (*Pseudacris triseriata*) on Isle Royale, Michigan. *Ecology* **64** : 501-510.

Smith, D.C. 1990. Population structure and competition among kin in the chorus frog (*Pseudacris triseriata*). *Evolution* **44** : 1529-1541.

Sokol, A. 1984. Plasticity in the fine timing of metamorphosis in tadpoles of the hylid frog, *Litoria ewingi*. *Copeia* **1984** : 868-873.

Steinwascher, K. 1978a. The effect of coprophagy on the growth of *Rana catesbeiana* tadpoles. *Copeia* 1978 : 130-134.

Steinwascher, K. 1978b. Interference and exploitation competition among tadpoles of *Rana utricularia*. *Ecology* **59** : 1039-1046.

Steinwascher, K. 1979a. Competitive interactions among tadpoles : Responses to resource level. *Ecology* **60** (6) : 1172-1183.

Steinwascher, K. 1979b. Host-parasite interaction as a potential population-regulating mechanism. *Ecology* **60** : 884-890.

Tejedo, M. and Reques, R. 1992. Effects of egg size and density on metamorphic traits in tadpoles of the natterjack toad (*Bufo calamita*). *J. Herpetol.* **26** : 146-152.

Tyler, M.J. 1989. *Australian Frogs.* Viking O'Neil, Ringwood, Victoria, Australia, 220 pp.

Wilbur, H.M. 1972. Competition, predation, and the structure of the *Ambystoma-Rana sylvatica* community. *Ecology* **53** : 3-21.

Wilbur, H.M. 1976. Density-dependent aspects of metamorphosis in *Ambystoma* and *Rana sylvatica*. *Ecology* **57** : 1289-1296.

Wilbur, H.M. 1977a. Density-dependent aspects of growth and metamorphosis in *Bufo americanus*. *Ecology* **58** : 196-200.

Wilbur, H.M. 1977b. Interactions of food level and population density in *Rana sylvatica*. *Ecology* **58** : 206-209.

Chapter 18

Floodplain Fisheries of India

Y.S. Yadava*

Central Inland Fisheries Research Institute, Barrackpore-743 101, West Bengal, India

Abstract. Floodplain lakes in the form of *jheels, beels, mans* and pats form one of the most lucrative source of fisheries in the states of Eastern Uttar Pradesh, Northern Bihar, West Bengal, Assam, valley districts of Manipur, Tripura and foothills of Arunachal Pradesh and Meghalaya. The production potentialities of such lakes is a silver lining in the realm of inland fisheries. If managed scientifically, the floodplain lakes would constitute the frontline areas of inland fisheries from where the country would meet its demand of *c.* 12 million tons of fish by the year 2000 A.D. However, a combination of anthropogenic activities have reduced the production levels of many of these lakes through siltation, habitat destruction, heavy weed infestation and by isolating them from the seasonal floods and natural sources of fish stock. The present review attempts to portray feasible technologies for rehabilitation of the dwindling fisheries and measures for sustaining an optimum fish production from the floodplain lakes.

Key words : *Floodplain, Resources, Ecological dynamics, Developmental measures*

Introduction

The inland fish production in India has increased from a meagre 0.277 million t to about 1.2 million t in the last four decades. India, with its present population of 800 millions, which is likely to cross 1 billion mark by the turn of the century will need *c.* 12 million t of fish, a half of which is to be contributed by the inland sector. To meet the estimated demand, a 6 fold increase in the present catch will be required against a growth of *c.* 4 times achieved since the inception of the five year plans (Srivastava, 1983). For boosting the inland fish production during the ensuring decade, ambitious plans are being contemplated to develop capture fishery resources through their conservation and judicious management.

***Present Address :** Dr. Y.S. Yadava, Advisor (Fisheries), North Eastern Council Secretariat, Ministry of Home Affairs, Shillong-793 001, Meghalaya, India.

Rivers and their floodplains, canals, reservoirs, lagoons and backwaters constitute the multifaceted arenas of inland capture fisheries from where the country would meet its demand by 2000 A.D. The floodplain lakes which form an integral component of the Ganga and the Brahmaputra basins are the thrust areas offering enormous scope to augment their fish production and add to the country's fish basket. The present review besides elucidating the resource potentialities and ecological dynamics of the floodplain lakes, highlights appropriate technologies for their speedy and sustainable development.

Evolutionary status and resource potentialities

Floodplain lakes in the form of *jheels, beels, mans* and pats form one of the most lucrative source of fisheries in the States of Eastern Uttar Pradesh, Northern Bihar, West Bengal, Assam, valley districts of Manipur, Tripura and foot-hills of Arunachal Pradesh and Meghalaya (Table 18.1).

The origin of floodplain lakes is as kaleidoscopic as the often changing tortuous course of the major rivers and their tributaries. The Brahmaputra and Barak basins in the north eastern region

TABLE 18.1. *The floodplain lakes resources and their distribution in India.*

States	Distribution (district-wise)	River basin	Local name	Area (ha)
Arunachal Pradesh	East Kameng, Lower Subansiri, East Siang, Dibang valley, Lohit, Changlang and Tirap	Kameng, Subansiri Siang, Dibang, Lohit, Dihing and Tirap	Beel	2500
Assam	Brahmaputra & Barak valley districts	Brahmaputra & Barak	Beel	1,00,000
Bihar	Saran, Champaran, Saharsa, Muzaffarpur, Darbhanga, Monghyr & Purnea	Gandak and Kosi	Mans & Chaurs	40,000
Manipur	Imphal, Thoubal & Bishnupur	Iral, Imphal & Thoubal	Pat	16,500
Meghalaya	West Khasi Hills, East and West Garo Hills	Someshwari & Jinjiram	Beel	375
West Bengal	24-Parganas, Hooghly and Nadia	Ganga (Hooghly) & Matlah	Beel	42,000

lie in a zone of acute seismic activity. Frequent earthquakes due to crustal instability have induced local and sudden changes in the basement levels, resulting later in pertinent water shed changes of the fluvial environment (Gansser, 1964; Jhingran *et al.*, 1976; Das 1983). This adverse feature coupled with heavy rainfall have resulted in formation of either typical oxbow type, lake-like or true tectonic depressions in the region. Similar scenario prevails in Arunachal Pradesh and Meghalaya too. *Beels* in these states have evolved in foothill (plains) regions where the rivers from the eastern Himalayas and the Patkai ranges respectively invade the flood plains. The *pats* of Manipur owe their origin to the stream meanders of the Iral, Imphal and Thoubal rivers in the mature valley of Imphal.

North Bihar in its entirety may be treated as a vast inland delta, as all the principal rivers emerging from the central Himalayas flow into the Ganga on the South. The Ghangra, the Gandak, the Kamla and the Kosi bring heavy silt and detritus load from the Himalayas. In this process, the rivers have oscillated from east to west. Both the Gandak and Kosi rivers have changed their course ever so often, leaving behind varying stretches of water mostly in the form of true oxbow lakes (Shetty and Malhotra, 1983).

Physiographically, the floodplain lakes can be grouped into two main categories. Those retaining continuity with the river/tributary through some channel, either throughout the year or atleast during rainy season, are termed as live. Others, which are completely cut off, are known as dead and these mostly receive water from the catchment area. The Assam and Manipur lakes are generally sprawling water sheets, covering several hundred hectares. Most of them have functional link with their riverine source and expand tremendously during monsoon months. The lakes in other states are mostly shallow and small in size.

Ecological Dynamics

The dynamics of floodplain lake morphometry, as effected by the incursion of river water, varying degrees of precipitation and nature of catchment are highly complex and present contrasting pictures of their area and depth within a year. These water bodies expand and contract. The channel which brings and drains most of the inflow and outflow of lakes having a riverine connection, principally acts as a *spill way* to limit retention of excess water from storage. Such lakes represent a combination of lotic and lentic habitat, becoming at times a natural lake ecosystem. However, it never arrives at that relative state, because it is always subject to manipulations of outflow and water level changes. Being shallow, the atmospheric temperature has a direct bearing on the water temperature and small reflexivity and penetration of incident radiation deep into the water are responsible for the physical micro-climate of the lakes.

The limnological features indicate the possibility for a better yield of fish (Yadava, 1988). Greater percentage contact of water with the sediments, deeper euphotic zone and longer sunshine hours culminating in a warmer temperature regime exhibit intermediate to high productivity levels in the lakes. The physical and chemical characteristics of the lake water and soil also support such preposition.

The form of many large and small lakes favour extensive development of marginal and submerged vegetation. With optimum values of light quality and quantity, temperature and alkalinity, the macrophytes play a major role in governing the biotic components and primary production in the lakes. The deposits of the decaying weeds at the bottom contribute to the richness of the bottom soil in terms of detritus and benthic organisms. Primary production has been considered as an indicator of fish production and a harvest of 1.2 per cent as fish is considered excellent (Odum, 1960). Significantly, in floodplain lakes a conversion of 0.24-1.21 per cent of primary production as fish recorded (Dey and Lahon, 1979; Pathak *et al.*, 1985; Yadava, 1987; Yadava *et al*, 1987) depict encouraging values of conversion efficiency. The relatively high efficiency is also reflected in the fish production (377 kg ha^{-1} yr^{-1} from Dhir beel, Assam - capture based and > 1000 kg ha^{-1} yr^{-1} from Kulia beel, West Bengal - management based) of the lakes.

The floodplain lakes generally possess high potential for *in situ* fish production and where the topography allows, provide a collection sink for the fish production on neighbouring flooded catchments. In contrast to average annual fish yield of *c.* 5-75 kg ha^{-1} yr^{-1} of the open water lakes and reservoirs recorded (Sreenivasan, 1965; Holt, 1966; Jhingran and Tripathi, 1969), the average annual yield from floodplain lakes is > 160 kg ha^{-1} yr^{-1} and if small scale subsistence fishing is taken into account, yields higher than this can be expected.

In live floodplain lakes, migratory propensity of the fishes, especially the commercial groups is of special significance. Adults and juveniles of Indian major carps depict such migration during monsoon for seeking breeding and feeding pastures in the freshly inundated floodplains. In closed lakes, the fauna comprises catfishes and trash/weed fishes. Devoid of carps, the productivity of such lakes is low. The situation in contexts of fish yield in live lakes is typical in that the recruits come to the fishery from the river. Majority of them enter the fishable stock in their very first year of life, an average when they are only 6-8 months old. Although, the population of such species are exposed to regular capture, the stock is supplemented through fresh recruitment from year to year. The diminishing mesh size of the nets used in the fishery and supplemented by the low water column throughout the winters, alarmingly, do cause complete annual removal of the fish stock. Quantitatively, the fish yield and fish production in such lakes, therefore, almost rate equal, especially at annual level. Further, in these fisheries, discarding of fish is not practised at any stage and all fish, small and big, are utilized for human consumption.

A rich assortment of traditional fishing gears is used by the floodplain lake fishermen. Passive gears, such as set barriers *Banas* (Yadava and Choudhury, 1986), dip nets and traps capture fish as they are migrating to or away from the lakes. Active gears, like drag, seine, gill, frame trawls or scoop nets are used mainly during trough season, when as an norm maximum fishes are captured. Certain types of fish husbandry methods (*Katal* fishing - Yadava *et al.*, 1981) using branches or vegetation placed in the water are extensively used. The diverse fish population of the floodplain lakes offer a multi-species fishery, but as their abundance is distributed lognormally (Welcomme, 1979), only a few species predominate.

Present State of Affairs

Traditionally, the fish harvest of the floodplain lakes have been considered as a renewable natural resource, available for maximum exploitation and able to provide a source of state revenue. However, a combination of the processes of river bed evolution and extensive works for irrigation, drainage and flood control have resulted in the disappearance of many of the original features. The consequences loom higher on both the socio-economic and environmental skyline. Within their restricted area, the floodplain lakes illustrate the on-going transformation, that is generated both natural and human desired. Major factors responsible for the deterioration are :

- habitat destruction, especially siltation in the connecting channels and severing of link with the mainstream,
- extensive use of lake margins for agricultural purpose during lean season, leading to cultural eutrophication of the lake proper,
- large scale appropriation of nutrients by the thick mantle of undesirable macrophytes, especially water hyacinth,
- isolation from the seasonal floods and natural stock of juvenile and adult fish. Indiscriminate destruction of seed and brood fish of commercial species during recruitment phase,
- dumping of industrial effluents and washings of agricultural wastes from the catchment areas abetting in speedy eutrophication,
- conflicting patterns of land use, water management and deforestation in the catchment area,
- inadequate sector management policies like ineffective ownership and leasing system, regulation, control and implementation of fishing activities etc., and
- poor socio-economic status of the fishing community

Appropriate Technologies

The National Commission on Flood Control in its recommendations submitted on March 21, 1980 has emphatically laid stress on utilizing the floodplain lakes as exclusively fishery resources. The concentrated research efforts during the last one decade have resulted in the generation of suitable technologies for development of floodplain lake fishery. The developmental programmes designated for this sector are broad ranging, keeping in views the magnitude and multitude of problems encountered.

With the advent of culture-based capture fisheries, as is happening even in the seas, culture and capture are merg in into one integrated whole. The present strategy for optimum exploitation of floodplain lake fisheries revolves around the concept of keeping the deeper central portions as exclusive zones for fisheries. Desilting of connecting channels and construction of perimeter bundhs are pre-requisite for such exercises. Most of the lakes are dendritic and have pockets retaining water throughout/part of the year. By way of micro dams/sluices, these pockets can be converted to tanks. Intensive aquaculture practices are suggested for such areas. It is also stressed to adopt integrated fish farming rather than composite fish culture alone. A multicommodity farming system is more advantageous to the farmer than the monocropping system, if the commodity mix fits well into the available resources and needs. The integration of pig and duck husbandry with fish culture is gaining very popularity due to high returns.

In recent years, new areas of culture system have emerged. Of these, mention may be made of pen and cage culture which have attained various operational levels in South-East Asia and Far East as well as in some of the East-European countries and Norway. In India, pen culture has been field-tested for fry and fingerling rearing in reservoirs (Abraham, 1983) and to culture table size fish in beels (Yadava, 1983; Chitranshi, 1988). *Beels* and *mans* provide ideal habitat for pen and cage culture operations. Pen culture of *Cyprinus carpio* in an indigenously designed bamboo screen pen yielded a production of 16 kg 10 m^{-2} 5 months in an Assam *beel.* With 80 per cent survival, the cost of production worked out to be Rs. 3.84 kg^{-1} of fish. In a *man* in Bihar, Indian major carps raised for 6 months in a split bamboo pen covering 0.1 ha yielded a fish crop of 4 t ha^{-1} with 76.2 per cent return on investment. Nursery rearing of carps obviates the necessity of constructing cost intensive nursery farms. Necessary techniques have been demonstrated in Poongar Swamp adjoining Bhavanisagar reservoir in Tamil Nadu.

Rearing of fishes in cages offer exciting avenues of culture in open water bodies. Productions varying from 0.190 kg $cage^{-1}$ to 4,832 kg $cage^{-1}$ in 90 days from Assam beels and 2.2 kg to 12.0 kg m^{-2} from Bihar *mans* have been achieved using air-breathing fishes. However, in terms of economics, the practical utility of the increased production at higher densities of fish in cages is to be weighed against the additional cost of inputs for raising fish at higher density (*i.e.* artificial feed, infection, environmental impact, mortality, etc.) and the reduced return to the extent of price being dependent on the time, size and number of the fishes actually harvested.

Infestation of submerged and free floating weeds have reached dynamic proportions in most of the floodplain lakes. The weeds have hastened the process of swampification of most of these potential water bodies. The submerged weeds (*Hydrilla, Vallisneria, Najas* etc.) can be biologically controlled, but the floating weeds, chiefly water hyacinth has very little use except its eradication. While manual/mechanical methods of removal may be put to use in small lakes (< 100 ha), the technology for larger ones demand a combination of both mechanical and chemical. Technology demonstrated in Takmu *pat* (500 ha), in Manipur resulted in removal of water hyacinth which had formed dense mats upto one meter in height. Application of 2,4-D sodium salt formulation (80 per cent a.i.) achieved a weel-kill to the extent of 90 per cent of the water hyacinth. Paraquat herbicide (gramaxone) was used to kill the grass grown over the weed mat. A commercial wetting agent (Dedanol) was mixed with the herbicidal spray solution to facilitate adhesion and spread of the solution on waxy and hairy leaf surface of the weeds. The cost of labour, chemicals and field contigencies worked out to Rs. 517. 15 ha^{-1} for clearance of water hyacinth in Takmu *pat*.

Biological control of submerged vegetation would necessitate introduction of indigenous and exotic herbivores like *Puntius pulchellus, Puntius dobsonii and Ctenopharyngodon idella* (grass carp). The former two species are endemic to south India and would have to be introduced in a large scale for effective utilization. The floodplain lakes have to be assiduously segregated before release of exotic carps. The lakes having no connection with the riverine source may be considered for stocking of exotic carps. If stocked in others, their escape in the drainage may endanger the indigenous forms.

A proper understanding of the complex relationship of food chain and patterns of energy flow in floodplain lakes will help in formulating policies for stock manipulation. In general, there are two main routes through which energy flows in the aquatic ecosystem-grazing chain and detritus chain. The energy of producers can be utilized either directly by consumers through grazing chain or the unused energy can get deposited as organic detritus and utilised through detritus chain. Taking into consideration the vast load of detritus available in the system and also the eutrophic character of the lakes, we may expect a continuous high rate of production by stocking of detritus-oriented fishes. *Cirrhinus mrigala, C. reba, Labeo rohita, L. bata* etc. are effective detritivores and considered suitable.

The simple exploitation of the floodplain lakes by capture fishery does not realise the full potential productivity of the system. Besides the reliance on traps, snares and lures, the *Banas Katal* fishing are widely practised. *Banas* fishing is generally in vogue in live lakes and ensures a good profit. With more effective management practises (Yadava and Choudhury, 1986), these methods can form the back bone of capture fishery in the floodplain lakes.

From all facts it is now possible to record with a degree of certainty that the live floodplain lakes themselves, in the process of inundation, serve as the spawning grounds of Indian major carps. The breeders from the riverine source in response to the annual floods enter the easily accessible shallow areas, which make available to the larve ample nursery space and food in the inundable margins. The development of commercial fisheries in the lakes would involve a recurring expenditure on the collection of spawn and its transport for stocking. Autostocking of such lakes, therefore, is the only viable alternative for their quick and profitable development. For this, the desilting of connecting channels and their effective management is a pre-requisite. The operation of sluice gates, wherever, installed is to be oriented in such a way so that it facilitates entry of brood fish and juveniles into the lake proper. This would entail suitable legislation to overcome the conflict between agriculture and fisheries.

The sustained fisheries development of the floodplain lake requires both micro-and macro-planning approach. As the micro-planning approach is essentially project oriented and involves investment in a carefully planned manner, all important aspects concerned with project cycle, such as the identification, financing, appraisal and subsequently implementation will be required. While micro-planning techniques and project-oriented approach to development would need to continue, greater attention is to be devoted to macro-planning *i.e.* the sector approach. As such, the major issues to be tackled under the sector development are the:

Fig. 18.1. A panoramic view of a floodplain lake in the Brahmaputra valley in Assam

Fig. 18.2. A haul of major carps and feather backs from a floodplain lake

Fig. 18.3. Khora jal - A dip net popularly used in the floodplain lakes of the north-east

Fig. 18.4. Pen culture in floodplain lakes - A bamboo screen pen installed in Muktapur *man* in Muzaffarpur, Bihar

Fig. 18.5. Cage culture in floodplain lakes - A 2 × 1 × 1 m split bamboo cage for raising airbreathing fishes in weed-choked *beels* in Assam

(1) Co-operative development
(2) Credit and subsidy schemes,
(3) Changes in lease period policy
(4) Technology transfer,
(5) Marketing,
(6) Human resource development
(7) Insurance schemes, and
(8) Socio-economic considerations

Of the above issues, the changes in lease period policy demands immediate action. Where the lakes are developed for increased production under a long-term programme of management and maintenance, it is inappropriate that they would continue to be considered as a source of state revenue with the value of the leases rising with the increased size of the harvests. In these circumstances the lease rent becomes simply a tax on development and will be evaded and lead to conflicts of control, bad reporting of catches etc.

Conclusion

The floodplain lake fisheries in India have a long history of development in association with the riparian settlements and in most cases the resources can be considered to be heavily exploited. A *single window concept,* having well-planned integrated programme of action has to be created in order to acheive the main target of sustainable development and optimum exploitation of the floodplain lake fisheries.

REFERENCES

Abraham, M. 1983. Observation on rearing of hatchlings in pens. *Prog. Natl. Sem. Cage Pen Culture* : 21-25.

Chitranshi, V.R. 1988. Pen and cage culture of fishes in Ox-bow lakes. In : Summer Institute on *Resource Management and Conservation of Inland Capture Fisheries of India,* CICFRI, Barrackpore, 4-23 July, 1988.

Das, P.K. 1983. A model for upper tertiary sedimentation of Arunachal Pradesh with special reference to Subansiri district. *J. Assam Sci. Soc.,* **25** (2) : 66-69.

Dey, S.C. and Lahon, B. 1979. Observations on the interrelationship of fish population in an ox-bow lake of Assam, India. *Proc. Ind. Sci. Congr.* **67** : 159 (Abstr).

Gansser, A. 1964. *Geology of the Himalayas.* Interscience Publishers. pp. 220-229.

Holt, S.I. 1966. The contribution of freshwater fish production to human nutrition and well being In : *Biological basis of freshwater fish production.* (Ed., S.D. Gerkins), Blackwell Scientific Publication. pp 455-467.

Jhinngran, A.G., Thakur, V.c. and Tandon, T.K. 1976. Structure and tectonics of the Himalayas. In : *Sem. Himalayan Geol.,* **I** : 1-39.

Jhingran, V.G. and Tripathi, S.D. 1969. A review of the measures adopted for the development of the fisheries of reservoirs in India. In : *Proc. Sem. Ecol. Fish. Freshwater Reservoirs, Barrackpore,* pp. 637-651.

Odum, E.P. 1960. The role of tidal marshes in estuarine production. *New York State Conservationist.* **I** : 1-4.

Pathak, V, Saha, S.B. and Bhagat, M.J. 1985. Patterns of energy utilization and productivity in beel ecosystem. *J. Hydrobiol.,* **1** (2) : 47-52.

Shetty, H.P.C. and Malhotra, J.C. 1983. A report on the survey of North Bihar in relation to effects of Gandak and Kosi river valley projects on the fisheries of the area. CIFRI, Barrackpore. *Survey Report,* **7** : 1-30.

Sreenivasan, A. 1964. The limnology, primary production and fish production in a tropical pond. *Limnol. Oceanogr.,* **9** (3) : 391-396.

Srivastava, U.K. 1983. Data base for planning and control. In : *Inland Fish Marketing in India.* IIM, Ahmedabad. **8** : 1-170.

Welcomme, R.L. 1979. *Fisheries ecology of flood plain.* Longman Ltd., 317 p.

Yadava, Y.S. 1988. Riverine flood plain fishery of the Brahmaputra basin. In : Summer Institute on *Resource Management and Conservation of Inland Capture Fisheries of India. CICFRI, Barrackpore.* 4-23 July, 1988.

Yadava, Y.S. and Choudhury, M. 1986. 'Banas' fishing in beels of Assam. *J. Bombay Nat. Hist.,* **83** (2) : 452-456.

Yadava, Y.S. Choudhury, M. and Kolekar, V. 1981. 'Katal' fishing - A special device for catching fish in beels of Assam. *J. Inl. fish. Soc. India.,* **13** (1) : 81-86.

Yadava, Y.S. Choudhury, M. Kolekar, V. and Singh, R.K. 1983. Pen farming-Utilizing marginal areas of beels in Assam for carp culture. *Proc. Natl. Sem. Cage. Pen Culture;* 55-58.

Yadava, Y.S., Singh, R.K., Choudhury, M. and Kolekar, V. 1987. Limnology and productivity of Dighali beel (Assam). *Trop. Ecol.,* **28** (2) : 137-146.

Chapter 19

Energetics of the Larval Anurans : A Review of Strategies Employed

Ashok Kumar Hota

School of Life Sciences, Sambalpur University, Jyotivihar-768 019, Burla, Orissa, India

Abstract. When the different energy components of anuran larva are regressed with body mass by log-log approximation, the exponent, *b,* varies within a range. This suggests that energy components of anuran larvae can be scaled to a constant power of body mass. The various energetic efficiencies are independent of larval body mass but depends on the metabolic state of the individuals. Anuran larvae have unusually high assimilation efficiencies and low maintenance cost, except for *Polypedates maculatus,* which permits rapid growth or high accumulation of energy reserves. The strategy of anuran larvae is to optimize the use of available resources of the aquatic habitat before being transformed into terrestrial juveniles.

Key words : *Complex life cycles, size constrains, energy partitioning, energetic efficiency, metamorphic efficiency, optimisation resource use, production maximization.*

Introduction

Anurans have complex life cycles featuring a dramatic shift in habitat which includes an abrupt ontogenetic change in an individuals morphology, physiology and behaviour. They have developing aquatic larva that eventually metamorphoses into a small version of the terrestrial adult, morphologically quite different from the aquatic larva (Just *et al.*, 1981; Wassersug, 1984). The larval and adult anurans are ecologically independent so as not to compete for the same resources. Such complex life cycle, forms are often viewed as adaptations to exploit transient opportunities for growth and dispersal (Wilbur, 1984). The organism irreversibly switches its adaptations from one of these functions to the other through the process of metamorphosis. Thus, it implies that larval anurans must metamorphose into juveniles so that they can migrate to new habitats to facilitate breeding.

In the consideration of amphibian life history, much importance has been given to larval growth and metamorphosis (Wilbur and Collins, 1973; Wassersug, 1975; Dash and Hota, 1980; Wilbur 1980; 1984). The adaptive significance of the larval stage of anurans is believed to reside in their ability to exploit transient opportunities for rapid growth in temporary ponds. The growth opportunity, together with the mortality, risk, influences the timing of anuran metamorphosis i.e. a shift in the utilization of the resources of different habitat (Wernes, 1986). Thus, very correctly, Salthe and Meecham (1974) refer the larval stage of anurans as the "energy gathering stage." Crump (1981) suggests that energy accumulation may be an important determinant in determining the timing of anuran metamorphosis. Pandian and Marian (1985a) demonstrated that the feeding rate of larval anuran, interacting with a variety of environmental factors, is a good quantitative predictor of growth and developmental rates and hence energy content at the time of metamorphosis and larval duration. Greater energy accumulation at the terminal stages of larval anuran is advantageous for the organism as it enhances metamorphic efficiency (Pandian and Marian, 1985b). According to Seale (1987), the physiological, ecological and biochemical components of amphibian energetics are complex and incompletely understood. The adapted complex life cycle pattern of anurans to exploit aquatic and terrestrial food sources increases the energy resource base of the organisms. The foraging activity of the terrestrial adult anurans influences the energetic parameters of their aquatic larva (Dash, 1991). These studies have become conventional wisdom in understanding the energetic strategies of anuran larvae. Yet many questions have been left to be answered. Whether there is any intra-specific similarity between the energetic parameters of anuran larvae? Whether energetics consideration can offer some insight into interspecific variations in the size at metamorphosis? This article, therefore, addresses itself in answering these questions by considering the energetic parameters of anuran larvae in general.

Energy Partitioning

Animal bioenergetics concerns with the rates of energy expenditure, the losses and gains, and the efficiencies of energy transformation as functional relations of the whole organisms confirming to the laws of thermodynamics. The relationship between energy assimilation, utilization and expenditure are neatly summarized by the energy budget equation developed for conventional bioenergetic scheme (Grodzinski and Wunder, 1975) :

$$A = P + R = C - FU$$

where A = assimilated energy, P = energy of production (since larval anurans are non reproductive, it is production due to growth only), R = respiratory energy, C = consumed energy and FU = faecal and urinary energy. It is evident from the equation that consumption (C) exceeds assimilation by the amount of energy lost through faeces and urines. Thus, for determination of energy budget, separate estimate of each energy component is required. Many studies of larval anuran energetics have estimated these energy components to draw the required energy budget. According to Seale (1987) many interdependent factors influence the proportion of energy partitioned for maintenance and production of amphibians and the process of energy partitioning varies among species and life stages and with change in the environment. For example, both rates of growth and maximum body sizes attained by larval anurans are plastic (Smith Gill and Berven 1979, Dash and Hota 1980, Wilbur 1980, Smith-Gill 1983, and Travis 1983). Unfortunately, quantitative energy budgets for larval anurans have been established only for a very few species.

From the available literature, it is evident that two major approaches have been used for energy partitioning in larval anurans. Assimilative energy has been calculated from laboratory estimates of consumption-excretion and respiration-production method. Consumption excretion values are higher than respiration-production values (Table 19.1). According to Seale (1987), variations in metabolism, behaviour and environmental interactions for different amphibians could cause 50 per cent or larger deviations in the relative amount of energy devoted to each energy component. The potential source of such discrepencies in estimating assimilative energy in anuran larvae probably arise in estimating the maintenance energy or the energy cost of respiration. In practice, measurements of respiration have either been obtained for starved animals (Mohanty and Dash, 1988), or been obtained for post absorptive animal (Hota, 1984) in small metabolic chambers. Metabolism of vertebrate animals in small metabolic chambers is lower than animals in larger cages or in the field (Smith, 1976). In post feeding animals, the process of deamination and other conversion processes involved in digesting, metabolizing and assimilating food (the specific dynamic action proposed by Broody 1945) are associated with elevated metabolic rates (Jobling, 1981). Crump (1979) points out that in most amphibian energetic studies, the measurement of the energy content of nitrogen wastes, mucus and gut lining secretions leads to an error. Further, the energy budget equation applies strictly to rates of transfer among components in the steady state, a condition rarely satisfied by larval anurans. At the metamorphic climax stage, production becomes negative. Prior to the climax stage in many species internal reorganisation occurs for the final metamorphic transformation (Hota and Dash, 1986). Therefore, larval anurans probably are never in steady state with respect to energy use. Probably, all these facts together, explain the

TABLE 19.1. *Differences in energy flow measurements of P+R Vrs. C-FU for larval anurans. Consumption values are higher than respiration production values.*

Anuran larva	$\frac{(P+R)}{(C\text{-}FU)} \times 100$	Source
Rana tigerina	90.89	Monanty and Dash (1988)
Rana tigrina	51.31	Hota (1984) (Unpublished)
Bufo melanostictus	33.17	Hota (1984) (Unpublished)
Bufo stomaticus	71.80	Mohapatro and Dash (1992)
Polypedates maculatus	96.66	Dash and Mishra (1989)

discrepencies in the estimate of assimilative energy through consumption-excretion method and respiration-production method.

Constrains of Size on Energy Components

Numerous physiological and morphological parameters are scaled to body size in accordance with the allometric equation:

$$Y = aW^b \text{.........(1)}$$

where, Y is a parameter, W is body mass and *a* and *b* are fitted constants (Huxley 1932, Zeuthen 1953, Hemmingsen 1960, Gunther 1975, Kleiber 1975). Values of Y increase in a curvilinear fashion with increasing W when b is positive and decrease curvilinearly with increasing W when b is negative. This relationship is conveniently transformed into a straight line equation of the form:

$$\text{Log } Y = \log a + b \text{ low } W \text{ ....(2)}$$

Mass specific metabolic rates (Cm^3 O_2/g/h) decline during growth phase and it is a potential source of error in estimating metabolic loss. However, the relationship between body mass (W) and metabolic rate (M) is expressed as the allometric equation:

$$M = aW^b \text{ ......(3)}$$

The exponent *b*, in anuran tadpoles varies with a number of factors like temperature, trophic state, time of day and time of feeding and it ranges from 0.83 to > 1 (Guimond and Hutchison 1968, Feder 1981, Bradford 1983). This indicates that larval anurans have the ability of metabolic compensation for increasing body size. According to the large scale interspecific 'empirical law' (Kleiber, 1975) or 'general biological rule' (Schmidt-Nielsen, 1979) relating basal or standard metabolic rate and body size of individuals, the exponent *b* has a vlue 0.75 such that equation 3 can be written as:

$$M = aW^{0.75}$$

This indicates that metabolic rate is directly proportional to a given power of body mass and the constant of proportionality is independent of body mass within a group of "metabolically similar animals." The regular relationship between metabolic rate and the value 0.75 for the exponent *b* is not easily explained and there is no consensus concerning the precise theoritical significance of this value (Bratholomew, 1977).

Humpreys (1979) has analysed 235 energy budgets from literature to determine the relationship between production and respiration in animal populations. It is not obvious why production should be proportional to respiration. But through such analysis it has been possible

TABLE 19.2. *Regression equations of component of energy budgets of larval anurans scaled to body mass basing on allometric equations $Y = aW^b$. Each energy component (Y) is measured in terms of energy time^{-1} on a whole animal basis*

	Units of measurement	Mass	a	b	Coeff. det. (R^2)	Source
Consumption (C)						
Rana tigrina	K. Joules. day^{-1}	g	0.52	0.57	0.95	Hotas, 1984
Bufo melanostictus	K. Joules. day^{-1}	g	0.31	0.54	0.90	Hota, 1984
B. Stomaticus	K. Joules day^{-1}	g	--	0.50	0.95	Mohapatra and Dash 1992
Polypedates maculatus	K. Joules day^{-1}	g	--	0.59	0.97	Dash and Mishra 1989
Faecal Energy (FU)						
R. tigrina	K. Joules. day^{-1}	g	0.013	0.56	0.94	Hota, 1984
B. melanostictus	K. Joules day^{-1}	g	0.72	0.49	0.99	Hota, 1984
P. maculatus	K. Joules day^{-1}	g	--	0.64	0.97	Dash and Mishra, 1989
Maintenance Energy (R)						
R. tigrina	K. Joules day^{-1}	g	0.16	0.56	0.91	Hota, 1984
B. melanostictus	K. Joules day^{-1}	g	0.11	0.59	0.94	Hota, 1984
B. stomaticus	K. Joules day^{-1}	g	--	0.50	0.95	Mohapatra and Dash, 1992
P. Maculatus	K. Joules day^{-1}	g	--	0.58	0.97	Dash and Mishra, 1989
Production Energy (P)						
R. tigrina	K. Joules day^{-1}	g	0.21	0.56	0.92	Hota, 1984
B. melanostictus	K. Joules day^{-1}	g	0.31	0.57	0.88	Hota, 1984
B. stomaticus	K. Joules day^{-1}	g	—	0.58	0.95	Mohapatra and Dash, 1992

to identify several groups of metabolically similar animals in animal populations. Lavinge (1982) suggest that some basis of general patterns exists between various components of energy budget in animal populations. Such general patterns can be utilized for large scale interspecific comparisons for generating reasonable hypotheses.

Table 19.2 gives a comparison of the regression equations of the energy components of four species of anuran larva scaled to their body mass. The exponent value *b*, fitted by log-log approximation varies between 0.50 to 0.60 with an exception of the faecal energy of *Polypedates maculatus* (b = 0.64). In all the four species, the energy components have been estimated in laboratory glass aquaria. The same mixed food (mixture of boiled *Amaranthus sp* leaves, boiled egg yolk and cooked minced goat meat) has been given *ad libitum* in alternate days. Therefore, the estimate of energy component in these four species is comparable. Results of previous studies indicate that the observed variations in the allometric relationship of energy components can be explained in relation to life history considerations (Mc Neill and Lawton, 1971), habitat type and trophic type (Hamprey 1979). The tadpoles of *R. tigrina* prefer to metamorphose at a larger size within 29-31 days of hatching and are very active. The tadpoles of *Bufo* sp. are usually small, slow moving and metamorphose within 14-16 days. In presence of sufficient mixed food, tadpoles of *Bufo* species are predated by tadpoles of *Rana tigrina* if cultured together (Hota and Dash, 1983). Whereas, the metamorphic size of the tadpoles of *P. maculatus* is intermediate between *Rana* and *Bufo* species. They are very active and attain the metamorphic climax stage within 30-33 days of hatching. The swift swimming activity of the tadpoles of *Polypedates* may be responsible for a greater proportion of energy loss as faecal energy indicated by the regression equations. It might therefore, be deduced the all energy components in the biogenetic scheme should be scaled to body mass and in an interspecific comparison over a wide range of body size, that each component might be scaled to a constant power of body mass. This is consistent with Kleiber's (1975) suggestion that body size does not affect the efficiency of food utilization.

Energetic Efficiencies

A variety of ratios are calculated involving rates of consumption, production and respiration (Odum, 1971). As discussed previously, the energy components of the energy flow equations of larval anurans can be scaled to a constant power of body mass - ranging between 0.50 to 0.60. This indicate that there is striking similarities between different energy components of larval anurans. From this relationship a variety of other similarity criteria can be estimated as proposed by Stahl (1963) for metabolically similar animals (Table 19.3). Such an analysis indicate that the ecological efficiencies of larval anurans are independent of their body mass instead the energy components are proportional to each other i.e. $P \propto R \propto A \propto C$. According to Dash (1991), the ecological efficiency of larval anurans are dependent on their developmental stages. The entire larval period of anurans is clearly divisible into three distinct stages like (I) premetamorphic stage with faster growth and little development (II) prometamorphic stage with continuous growth and differential development of parts and (III) metamorphic climax stage with degrowth and reorganisation of organ systems for transformation (Etkin 1964).

Stage III is the fasting stage. During this stage, complete resurption of the long tail serves as the energy source for the fasting individuals. Stage II is the developmental stage of larval anurans. Development of hind limb makes the larva to become an active swimmer. Besides the differential development of parts, in some larval anurans like *R. tigrina,* the phenomena of ''first metamorphosis'' of gut reorganisation from herbivorous to carnivorous type occurs in this stage (Varute, 1970). But, the phenomena of first metamorphosis is not found in *R. clamitans* (Jenssen, 1967), in *B. melanosticus* (Hota and Dash, 1986) and in *B. stomaticus* (Dash and Mohapatra, 1990). Whereas, stage I of larval anuran is the growth phase. From this theoritical consideration it appears that stage II is loaded with internal and external activities. Therefore, a bulk share of the assimilated energy has to be spent on maintenance to sustain the activity. Stage I is rather sedantary (limb less) and, therefore, much of the assimilated energy is conserved

TABLE 19.3. *Additional similarities between variious energy components as proposed by Stahl (1963)*

Relationship		Ratio
1. Production/Consumption	PC	$\frac{a_1 W^{0.6}}{a_2 W^{0.6}} = a^3$
2. Assimilation/Consumption	A/C	$\frac{a_4 W^{0.6}}{a_2 W^{0.6}} = a^5$
3. Production/Assimilation	P/A	$\frac{a_1 W^{0.6}}{a_4 W^{0.6}} = a^6$
4. Respiration Assimilation	R/A	$\frac{a_7 W^{0.6}}{a_4 W^{0.6}} = a^8$
5. Gross efficiency	P/C × 100	$= 100\ a^3$
6. Production efficiency	P/A × 100 = P/P+R =	$\frac{P}{C\text{-}FU} = 100\ a^6$

TABLE 19.4. *Energetic efficiencies (%) of larval anurans*

Species	Premetamorphosis (Stage I)			Prometamorphosis (Stage II)			Source
	A/C	R/A	P/A	A/C	R/A	P/A	
Rana tigerina	85	60.99	22.68	82	66.10	27.54	Mohanty and Dash (1988)
Rana tigrina	71.66	41.98	7.91	71.98	42.32	5.96	Hota (1984)
Rana tigrina	88.32	68.54	4.21	89.18	71.43	1.93	Hota (1986)
(Instead of mixed food, live *B. melanostictus* tadpoles were given as food)							
Bufo stomaticus	70	62.54	17.27	72	51.32	15.08	Mohapatra and Dash (1992)
Bufo melanostictus	76.92	52.56	18.44	70.02	36.28	2.92	Hota (1984)
Polypedates maculatus	89	65.40	7.30	94	74.30	10.70	Dash and Mishra (1989)

in the form of growth. Or in other words, premetamorphic stage of larval anuran is the real energy gathering stage.

It is evident from Table 19.4 that larvae of different species showed different strategies for adopting energetic efficiencies. For *Bufo sp.* limbless premetamorphic stage is the energy gathering period. Assimilative efficiency is maximum in the tadpoles of *P. maculatus*. Food habits in mammals directly influence the assimilation efficiencies (Mc Nab 1980), so also the assimilation efficiencies of *R. tigrina* are influenced by food habit. Mohanty and Dash (1988) used mixed food but restricted the tadpoles to 250 ml of active space only. Hota (1984) used sufficient mixed food and reared them in 2L of active space. Hota (1986) allowed maximum foraging activity for *R. tigrina* tadpoles by giving live *B. melanostictus* tadpoles as their food. Accordingly, *R. tigrina* tadpoles employed different strategies of energy partitioning (Table 19.4). In a given set of experimental condition, the assimilative efficiency of larval anuran is independent of size and developmental state. Partitioning of assimilation energy for maintenance and production depends much on the metabolic state of the larval anurans.

Larval anurans depend on accumulated reserves to supply the energy required for morphological transformation of the fasting tadpoles during metamorphic climax stage (Dodd and Dodd, 1976; Crump, 1981). Pandian and Marian (1985b) gave the concept of metamorphic efficiency (Me). They measured '*Me*' for *Rana tigrina* larvae using the equation:

$$\text{Me (\%)} = \frac{\text{Energy content of froglet}}{\text{Energy content of tadpoles at metamorphic climax stage}} \times 100$$

It ranges from 64.2 per cent to 85.2 per cent. It is also unfluenced by food quality, temperature, tadpole densities and water quality. Field estimate of metamorphic efficiency of *Rana catesbeiana* tadpoles ranges from 80 to 90 per cent (Seale, 1987). Variability in metamorphic efficiency has been attributed to the feeding relationship of larval anurans (Pandian and Marian 1985a). The feeding rates of larval anurans, interacting with a variety of factors, larval density, foraging activity, food levels and temperature, are good quantitative predictor of growth and developmental rates and hence energy content at the onset of metamorphosis and the larval duration. Hence, the efficiency of energy accumulation (Production efficiency) by the aquatic larval anuran is an important factor in determining the abundance of terrestrial adult population.

Energy Budget

Energy transfer between components of the energy budget for the larval anurans are indicated as a percentage of initial consumption (Table 19.5). It is obvious from the comparison that the pattern of gross energy allocation does not differ much between species, excepting the fact that *P. maculatus* spends about 63 per cent of ingested energy for respiration and about 22 per cent for

TABLE 19.5. *Cumulative Energy budget of the larval period of four species of anurans. Values are expressed as KJ end^{-1}. Nos. in the parenthesis are percentage of consumption*

Energy values of larval Tissues are KJ/g dry wt.	Cumulative energy vlaues			
	1	2	3	4
Components of energy budget	*Rana tigrina*	*Bufo melanostictus*	*Bufo stomaticus*	*Polypedates Maculatus*
Energy values of larval tissues (KJ/g dry wt.)	19.121	23.221	22.835	15.914
Consumption (C)	36.112 (100)	3.640 (100)	3.052 (100)	19.394 (100)
Rejecta (FU)				
Faeces (F)	9.665 (26.75)	1.042 (28.63)	0.880 (28.83)	0.310 (1.60)
Urine (U)	2.552 (7.07)	0.050 (1.37)	0.038 (1.25)	4.240 (21.86)
Respiration (R)	10.837 (30.01)	0.686 (18.85)	1.198 (39.25)	12.217 (63)
Production (P)	1.423 (3.94)	0.159 (4.37)	0.312 (10.22)	1.676 (8.64)

urinary excretion. But when the absolute values of cumulative energy budgets are compared (Table 19.5), it is found that the energy components of the *R. tigrina* larva are approximately 10 times greater than those of the *Bufo* species. There is no significant difference in the initial weight of hatchlings of *R. tigrina* and *Bufo* species, whereas the transformation size of *R. tigrina* larva is ten times greater and the larval period is twice that of the larva of *Bufo* species. So the differences in the cumulative energy budget of these species are reflected in the larval transformation size.

Adult *Polypedates* are very active and are smallest of all the four species considered here. Their tadpoles are also very active and are of intermediate sized larva compared to *Rana* and *Bufo* species. Therefore, the proportion of energy spent by them in respiration is much higher in comparison to other species of larval anurans.

Metamorphosing anurans are extremely succeptible to predation (''Wassersug and Sperry 1977). Rapid growth of larval anurans minimizes the predation risk. Also, larger juveniles with more lipid accumulation are more likely to survive the first winter and reproduce earlier than smaller ones (Collins 1979). Tadpoles of many anuran species form dense aggregate to elevate ambient temperatures for accelerating development and growth rates (Wassersug 1975). But, oxygen availability imposes an important limitation (Mohanty and Dash, 1986). The tendency of *Rana tigrina* tadpoles is to maximise production with a optimum use of the available resources.

Bufonids transform at much smaller sizes with a very little energy reserves compared to similar sized ranids and hylids. They have relatively high mortality rates in the larval habitat and low mortality rates in the adult habitat as compared to the ranids and hylids. (Werner, 1986). In the terrestrial phase the bufomids possess a variety of strong toxins that especially effective against predators like birds and mammals (Cei *et al* 1972, Arnold and Wassersug 1978), which may result in lower mortality rate in the terrestrial habitat relative to ranids and hylids. Such a phenomena suggests that *Bufo* species have partially conquerred the problem of predation risk on terrestrial habitat and can emerge at a small size with a very little energy reserves dispensing with the advantage of being transformed as a larger juveniles. Further, *Bufo* species lay their eggs in temporary ditches with very little resource to utilize. Therefore, they adopt the strategy to transform earlier with a little energy reserves to take advantages of higher growth rate available on terrestrial habitat.

If resources are limiting, energy acquisitions in the field are density dependent. Tadpoles can separate available food within their guts for several hours and graze continuously and non-selectively hence, they may alter their own food supply by controlling the energy flow and nutrient cycling (Seale, 1980). This suggests that tadpoles can regulate the rate of supply of their own food source in the field by altering the growth and energy acquisition. By comparing the total energy deposited in eggs by adult anurans and the energy content of emerging froglets, it emerges that the metamorphosing animals exported about hundred times more energy to the terrestrial ecosystem than was deposited in eggs (Seale, 1987). So the environmental conditions of the aquatic habitat determines the energy acquisition of the larval anurans to be exported to terrestrial habitat. In the pond studied by Seale (1980), metamorphosing anurans represented a large energy gain to the amphibian community from the aquatic ecosystems yet an insignificant proportion of the total energy fixed by primary production in the pond. Hence, the complex life history strategies of anurans is advantageous to the amphibian community in view of energy acquisition.

Conclusions

From the review of situations, it appears that larval anurans adopt strategies for maximizing energy conversion efficiency from optimum resource levels. The complex life cycle of anurans increase the available energy resource base of species. Patterns of energy acquisition and mobilisation differs among different species of anuran larvae particularly the energy allocated for maintenance and production. Available active space also influences the energy partitioning of larval anurans. The general strategy of anuran tadpoles is to adjust metabolism physiologically to environmental conditions. The energy thus saved is diverted for somatic tissue production or energy gathering.

REFERENCES

Arnold, S.J. and Wassersug, R.J. 1978. Differential predation on metamorphic anurans by garter snakes (*Thamnophis)*: Social behaviour as a possible defense. *Ecology* **59** : 1014-1022.

Bartholomew, G.A. 1977. Energy metabolism. pp. 57-110. In : *Animal Physiology : Principle and Adaptations.* 3rd edn. (ed. M.S. Gordon). Mac Millan Publ. Co., New York.

Bradford, D.F. 1983. Winterkill, oxygen relation and energy metabolism of a submerged dormant amphibian *Rana muscosa. Ecology* **64** : 1171-1183.

Broody, S. 1945. *Bioenergetics and Growth.* Hafner Publ. Co., New York 1023 p.

Cei, J.M., Erspamer, V. and Roseghini, M. 1972. Biogenic amines. pp. 233-243. In : *Evolution in the Genus Bufo* (ed. W.F. Blair). Univ. Texas Press, Austin.

Collins, J.P. 1979. Intrapopulation variation in the body size at metamorphosis in the bull frogs, *Rana catesbeiana. Ecology* **60** : 738-749.

Crump, M.L. 1979. Intrapopulation variability in energy parameters of the salamander *Plethodon cinereus. Oecologia* **38** : 235-247.

Crump, M.L. 1981. Energy accumulation and amphibian metamorphosis. *Oecologia* **49** : 167-169.

Dash, M.C. 1991. Energetics of amphibian tadpoles and its significance to frog culture. *S.U.J.Sc. Tech.* **10** : 89-93.

Dash, M.C. and Hota, A.K. 1980. Density effects on the survival growth rate and metamorphosis of *Rana tirgina* tadpoles. *Ecology* **61** : 1025-1028.

Dash, M.C. and Mishra, P.K. 1989. Energy cost of metamorphosis in a tropical tree frog, Polypedates maculatus (Gray). *J. Ecobiol.* **1** : 131-136.

Dash, M.C. and Mohapatra, B.K. 1990. Effect of food quality and quantity on growth and metamorphosis of *Bufo stomaticus* larvae. *J. Ecobiol* . **2** : 67-76.

Dodd, M.H.I. and Dodd, I.M. 1976. The Biology of metamorphosis. pp. 467-576. In : *Physiology of the Amphibia* Vol. 3 (ed. B. Lofts). Ac Press, New York.

Etkin, W. 1964. Metamorphosis. pp. 427-468. In : *Physiology of the amphibia* (ed. J.A. Moore). Ac. Press, New York.

Feder, M.E. 1981. Effect of body size, trophic state, time of day and experimental stress on oxygen consumption of anuran larvae : An experimental assessment and evaluation of the literature. *Comp. Biochem. Physiol.* A. **70** : 497-508.

Grodzinski, W. and Wunder, B.A. 1975. Ecological energetics of small mammals pp. 173-204. In : *Small Mammals. Their Productivity and Population Dynamics* (IBP No. 5) (eds. F.B. Golley, K. Petrusewicz and K. Ryzkowski) Camb. Univ Press. London.

Guimond, R.W. and Hutchison, V.H. 1968. The effect of temperature and photoperiod on gas exchange in the leopard frog *Rana pipiens. Comp. Biochem. Physiol.* **27** : 177-195.

Gunther, B. 1975. Dimensional analysis and theory of biological similarity. *Physiol. Rev.* **55** : 659-699.

Hemmingsen, A.M. 1960. Energy metabolism as related to body size and respiratory surfaces and its evolution. *Rep. Steno. Memo. Hosp. Norol. Insulin Lab.* **9** : 1-110.

Hota, A.K. 1984. Growth, production and energetics of the tadpoles of *Rana tigrina* (Daud) and *Bufo melanostictus* (Bloch and Schneider). Ph.D. Thesis, Sambalpur University.

Hota, A.K. 1986. Energy accumulation and metamorphosis of *Rana tigrina* larvae. *Proc. First World Conference on Trade in Frog Legs vis-a-vis Environmental Consideration* (The marine product export development authority, India) p. 67.

Hota, A.K. and Dash, M.C. 1983. Evidence of interspecific predation among larval anurans : Predation of *Rana tigrina* larvae on *Bufo melanostictus* larvae. *Biol. Bull. India* **5** : 54-55.

Hota, A.K. and Dash, M.C. 1986. Growth and metamorphosis of anuran larvae : effect of diet and temperature. *Alytes.* **5** : 165-172.

Humpreys, W.F. 1979. Production and respiration in animal populations. *J.Anim. Ecol.* **48** : 427-453.

Huxley, J.S. 1932. *Problems of relative growth.* Lincoln Mac Veigh, New York.

Jenssen, T.A. 1967. Food habits of the green frog *Rana clamitans* before and during metamorphosis. *Copeia* 1967 : 214-218.

Jobling, M. 1981. The influence of feeding on the metabolic rate of fishes : A short review. *J. Fish Biol.* **18** : 385-400.

Just, J.J., Kraus-Just, J. and Check, D.A. 1981. Survey of chordate metamorphosis pp. 265-326. In : *Metamorphosis : A Problem in Developmental Biology* (eds. L.I. Gilbert and E. Frieden). Plenum, New York.

Kleiber, M. 1975. *The Fire of Life : An Introduction to Animal Energetics.* R.E. Krieger Publ. Co.

Lavinge, D.M. 1982. Similarity in energy budgets of animal populations. *J. Anim. Ecol.* **51** : 195-206.

Mc Nab, B.K. 1980. Food habits, energetics and population biology of mammals. *Am. Nat.* **116** : 106-124.

Mc Neill, S. and Lawton, J.H. 1971. Annual production and respiration in animal population. *Nature* **225** : 472-474.

Mohanty, S.N. and Dash, M.C. 1986. Effects of diet and aeration on the growth and metamorphosis of *Rana tigrina* tadpoles. *Aquaculture* **51** : 89-96.

Mohanty, S.N. and Dash, M.C. 1988. Larval energetics of *Rana tigerina* (Daud). *Ind. J. Exp. Biol.* **26** : 88-91.

Mohapatro, B.K. and Dash, M.C. 1992. Energy budget of *Bufo stomaticus* tadpoles. *J. Ecobiol.* **4** : 297-302.

Odum, E.P. 1971. *Fundamentals of Ecology.* 3rd edn. W.B. Saunders Co. Toronto.

Pandian, T.J. and Marian, M.P. 1985a. Predicting anuran metamorphosis and energetics. *Physiol. Zool.* **58** : 538-552.

Pandian, T.J. and Marian, M.P. 1985b. Time and energy costs of metamorphosis in the Indian Bullfrog. *Rana tigrina. Copeia* **1985** : 653-662.

Salthe, S.N. and Meecham, J.S. 1974. Reproductive and courtship patterns. pp. 310-519. In : *Physiology of Amphibia* Vol. 2 (ed. B. Lofts). Ac. Press, New York.

Schmidt-Nielsen, K. 1979. *Animal Physiology : Adaptation and Environment.* 2nd edn. Cambridge Univ. Press, Cambridge.

Seale, D.B. 1980. Influence of amphibian larvae on primary production, nutrient flux and competition in a pond ecosystem. *Ecology* **61** : 1531-1550.

Seale, D.B. 1987. Amphibia. In *Animal Energetics* Vol. 2 (eds. T.J. Pandian and F.J. Veernberg) Ac Press, New York.

Smith, G.C. 1976. Ecological energetics of three species of ectothermic vertebrates. *Ecology* **57** : 252-264.

Smith-Gill, S.J. 1983. Developmental plasticity : Developmental conversion versus phenotypic modulation. *Am. Zool.* **24** : 47-55.

Smith-Gill, S.J. and Berven, K.A. 1979. Predicting amphibian metamorphosis. *Am. Nat.* **113** : 563-585.

Stahl, W.R. 1963. Similarity analysis of physiological systems. *Perspectives in Biological Medicine* **6** : 291-321.

Travis, J. 1983. Variation in growth and survival of *Hyla gratiosa* larvae in experimental enclosures. *Copeia* 1983 : 232-237.

Varute, A.T. 1970. B-glucuronidase in the alimentary canal of herbivorous and carnivorous anuran tadpoles and adults. *Comp. Biochem Physiol.* **33** : 143-148.

Wassersug, R.J. 1975. The adaptive significance of the tadpole stage with comments on the maintenance of complex life cycles in anurans. *Am. Zool.* **15** : 405-417.

Wassersug, R.J. 1984. Why tadpoles love fast food. *Nat. Hist* **93** : 60-68.

Wassersug, R.J. and Sperry, D. 1977. The relationship of locomotion to differential predation on *Pseudacris triseriata* (Anura : Hylidae). *Ecology* **58** : 830-839.

Werner, E.E. 1986. Amphibian metamorphosis : Growth rate, Predation risk and optimal size at transformation. *Am. Nat.* **128** : 319-341.

Wilbur, H.M. 1980. Complex life cycles. *Ann. Rev. Ecol. System.* **11** : 67-93.

Wilbur, H.M. 1984. Complex life cycles and community organisation in amphibians. pp. 195-224. In : *A New Ecology* : *Novel Approaches to Interacting Systems* (eds. P.W. Price, C.N. Slobgodchikoff. and W.S. Gaud). Willey, New York.

Wilbur, H.M. and Collins, J.P. 1973. Ecological aspects of amphibian metamorphosis. *Science* 182 : 1305-1314.

Zeuthen, E. 1953. Oxygen-uptake as related to body size in organisms. *Q. Rev. Biol.* **28** : 1-12.

Chapter 20

Effect of Environmental Factors on the Pineal Gland Function of Certain Tropical Animals

N. Saxena, M.P. Sinha AND P.N. Mehrotra

Department of Zoology, Ranchi University, Ranchi-834 008, India

Abstract. The pineal gland mediates the environmental information to the reproductive axis of the animal. The Indian palm squirrel, *Funambulus pennanti* and the Indian garden lizard, *Calotes versicolor* showed a marked seasonal sexual cycle. Pinealectomized animals were unable to respond to environmental cues and presented an active testes throughout the year. Long photoperiod stimulated the gonads of *F. pennanti* and led to a retardation of testicular regression. Short photoperiod inhibited the gonads of the squirrel and accelerated the gonadal regression. A constant light condition stimulated the gonads of both animals but had an opposite effect on the pineal activity. However, continuous darkness inhibited the reptilian as well as mammalian gonad. The pineal 5-methoxyindoles suppress gonads and the time of administration is critical in causing this effect. The biochemical activity of the pineal exhibited a distinct circadian and circannual cyclicity.

***Key words** : Pineal, Environment, Testes, Squirrel, Lizard.*

Introduction

Studies indicate that the pineal gland is involved in long term adaptation of function especially of reproductivity to environmental lighting conditions in seasonal breeders (Follet and Follet, 1981; Kappers and Pevet, 1979; Oksche and Pevet, 1981; Reiter, 1978; 1981). Among the environmental factors photoperiod is one of the most important factors influencing the reproduction. The pineal has been considered to be mediating between external light factors and activity of hypothalamo-hypophysial-gonadal axis (Bhatnagar *et al.*, 1990; Chan *et al.*, 1990; de Vlaming and Vodicnik, 1977; Fenwick, 1970; Rhodes, 1989; Scalera *et al.*, 1990). Seasonal changes in temperature (Peter and Hontela, 1978) and rainfall (Lofts, 1978; Sundararaj and Garg, 1981) influence the breeding activity thereby controlling reproduction. Thus, the pineal can be said to be perceiving, differentiating and integrating all types of environmental information.

The more severe and drastic the environmental conditions, the greater the necessity for the synchronization of reproduction with the environment. The animal needs to adapt itself to the environmental hazards in order to cope with these variations.

Effects of environmental factors upon the control mechanisms of neuroendocrine system have been studied in temperate zone mammals for which photoperiod is the most important factor controlling the animal's reproductive system. But the tropical animals experience completely different conditions. They are exposed to a diverse set of changing environmental conditions including humidity, temperature, rainfall, photoperiod, daylength and food. An attempt has been made elucidate the role of different environmental factors in regulation of reproduction of the tropical animals, Indian palm squirrel and Indian garden lizard.

Material and Methods

I. Seasonality in Pineal and Testicular Cycle in Relation to Various Ecofactors

(1) Annual Pineal and Gonadal Cycle in the Indian palm squirrel, *Funambulus pennanti.*

Ten adult males of *Funambulus pennanti* (100-120 grams in body weight) were obtained in the second week of every month. The animals were acclimatized to laboratory conditions for a period of two weeks. They were housed in wire net cages and exposed to normal environmental conditions. They were supplied with food and water *ad libitum.* After two weeks all the animals were weighted and then sacrificed by cervical dislocation. Their testes were removed. weighed on a microelectrical balance and fixed in Bouin's fluid for histological studies. The heads of all animals were fixed in Bouin's fluid. After a period of two weeks the pineal glands were dissected out from each head and weighed on a microelectrical balance. The data were analysed statistically with the help of analysis of variance (ANOVA) and correlation analysis (Bruning and Kintz. 1977).

(2) Annual pineal and gonadal cycle in the Indian garden Lizard. *Calotes versicolor.*

Six adult males of *Calotes versicolor* were obtained in the laste week of every month. Testes of one side, vas deference and kidney were removed, weighed and fixed in Bouin's fluid for histological studies. Testis of the other side was processed for cholesterol estimation by the method of Sackett (1925). The renal sex segment, cell height and seminiferous tubule diameter was measured with an ocular micrometer. Heads of all the animals were fixed in Bouin's fluid. After two weeks the pineal glands were dissected out and weighed on a microelectrical balance. The data was analysed statistically by Student's 't' test, ANOVA and correlation analysis.

II. Effect of Pinealectomy on the gonads of Funambulus pennanti, during different months of the year

Fourteen adult males were obtained in the second week of every month and acclimatized for a period of two weeks, in a room fully exposed to natural environmental conditions. They were housed in wire net cages (25" × 25" × 30" in size), throughout the experimentation. They were provided with food and water *ad libitum*. They were divided into the following two groups of seven animals each.

Group I : The animals were pinealectomized by the method of Haldar-Misra (1986).

Group II : The animals were sham-operated.

After sixty days, the body weights of all the animals were noted and then they were sacrified by cervical dislocation. The experiment was performed throughout the year in order to get the data for each month. Testes were weighed and immediately fixed in Bouin's fluid. The brains of pinealectomized animals were checked histologically to ensure the completeness of pinealectomy.

III. Effect of Photoperiod on the Gonadal Function

(1) Effect of long photoperiod (14 L : 10 D) on the gonads of the Indian palm squirrel, *Funambulus pennanti*

Twentyfour adult males of *F. pennanti* were obtained in the first week of September (inactive testes) and acclimatized to lab. conditions for period of two weeks. The animals were maintained under the same conditions as mentioned in Experiment II. After two weeks they were divided into the following four groups of six animals each :

Group I : Animals were pinealectomized and kept under a long photoperiod condition of 14 L : 10 D.
Group II : Animals were sham-operated and kept under a long photoperiod condition of 14 L : 10 D.
Group III : Animals were pinealectomized and kept under natural environmental conditions.
Group IV : Animals were sham-operated and kept under natural environmental conditions.

Groups III and IV served as controls. After sixty days all animals were weighed and sacrificed by cervical dislocation. Their gonads were dissected out and weighed on a microelectrical balance. The data was analysed by Student's 't' test.

(2) Effect of Short Photoperiod (10 L : 14 D) on the gonads of the Indian palm squirrel, *Funambulus pennanti*

The experiment was started in the testicular regressive phase (August) and continued upto March (testicular active phase) to observe the effect of short photoperiod (10 L : 14 D) on gonadal activity after they had recrudesced. One hundred and twelve adult males of *Funambulus pennanti* were divided into the following two groups :

Group I : Fifty six squirrels were kept under a short photoperiod condition of 10 L : 14 D.

Group II : Fifty six squirrels were kept under natural environmental conditions.

Group II served as controls for Group I. Eight animals were sacrificed from each group after a period of 4, 8, 12, 16, 20, 24 and 28 weeks. An initial sacrifice of 8 animals was done at the time of starting the experiment. Testes were dissected out, weighed and fixed in Bouin's fluid for histological observations. Epididymis were removed and weighed on a microelectrical balance. Data was analysed with the help of Student's 't' test.

(3) Effect of Continuous light on Gonads

(a) *Funambulus pennanti*

The experiment was performed during the testicular regressive phase (August-October). Adult males were obtained and acclimatized for two weeks in a room fully exposed to ambient environmental conditions. Sixteen animals were divided into the following two groups of eight animals each :

Group I : The animals were exposed to a continuous light condition.
Group II : The animals were exposed to natural environmental conditions.

Group II served as controls for Group I. After completion of 90 days, all animals were weighed and sacrificed. Their testes were removed and fixed in Bouin's fluid for routine histological observations. Epididymis were dissected out and then weight. The pineal glands were disssected out and weighed on a microelectrical balance. Data was analysed by Student's 't' test.

(b) *Calotes versicolor*

Fifty adult lizards were obtained in the first week of February. After acclimatizing them to laboratory conditions for two weeks, they were divided into two groups of twenty four animals each. Animals of Group I were exposed to a constant light condition while those of Group II were kept under natural day length (NDL). Group II served as controls for Group I. Eight animals from each group were weighed individually and then sacrificed by decapitation after 20, 40 and 60 days. Testes and vas deferens were removed, weighed and fixed in Bouin's fluid for routine histology. Pineal glands were dissected out, weighed on a microelectrical balance and then processed for the estimation of total protein by technique of Lowery *et al.*, (1951). Seminiferous tubule diameter was measured with the help of ocular micrometer. Data was analysed by Student's 't' test.

(4) Effect of constant Darkness on Gonads

(a) *Funambulus pennanti*

Sixty adult males were divided into two groups of thirty animals each. Animals of Group I were subjected to a continuous dark condition (DD) while those of Group II were maintained under NDL and served as controls. The animals were sacrificed from each group on day 15, 30 and 60 after onset on experiment and their body weights noted. Testes were weighed and fixed in Bouin's fluid for histological observations. Pineal glands were removed, weighed on a microelectrical balanced and kept frozen till the biochemical assay of protein.

(b) *Calotes versicolor*

Experiment was done during August-October with fifty adult males. Animals of Group I were kept under continuous darkness (DD) while those of Group II were kept under NDL and served as controls. The animals were sacrificed after 20, 40 and 60 days. The experimental procedures were similar to those of Expt. 3 (a).

IV. Environment and Pineal 5-Methoxyindoles

A. Evening injections :

(a) Sixteen adult lizards (*C. versicolor*) were obtained in June (gonad active) and November (gonad inactive). They were acclimatized to lab. conditions for two weeks and then divided into two groups of 8 animals each. Animals of Group I were injected intraperitoneally (i.p.) with 10 μg of 5-methoxytryptamine (MT, Sigma Chemical Co., USA) in 0.1 ml of normal saline between 1500-1600 hrs. for 30 continuous days. Animals of Group II were injected with 0.1 ml of normal saline. Similar treatment was repeated during both phases. After 30 days, blood was collected for estimation of plasma cholesterol. Testis of one side, kidney and vas deferens were dissected out, weighed and fixed in Bouin's fluid. The other testis was processed for the estimation of cholesterol. The renal sex segment (RSS) cell height and seminiferous tubule diameter was measured. Data was analysed by Student's 't' test.

(b) One hundred and fifty adult males of *Calotes versicolor* were obtained in April, acclimatized and divided into five groups of 30 animals each. Animals of Groups I, II, III and IV were injected, between 1650-1700 hrs, with 1, 10, 20 and 50 μg of melatonin/animals/day, respectively. Group V served as control and were injected with normal saline. Six animals from each group were sacrificed by decapitation after 5, 10, 20, 40 and 60 days. Testes were removed, weighed and processed for cholesterol estimation. Data was analysed by Studient's 't' test.

(c) Thirty two adult males of *Funambulus pennanti* were obtained in June (gonad active), acclimatized to lab. conditions and divided into following four groups of eight animals each :

Group I : Animals were pinealectomized and injected subcutaneously with 10 μg 5-methoxytryptophol (ML)/animals/day.

Group II : Animals were sham-operated and injected subcutaneously with 10 μg ML/animal/day.

Group III : Animals were pinealectomized and injected subcutaneously with 0.1 ml normal saline/animal/day.

Group IV : Animals were sham-operated and injected subcutaneously with 0.1 ml normal saline/animal/day.

Injections were given between 1650-1700 hrs for 45 continuous days. The animals were sacrificed by cervical dislocation. Their testes and epididymis were dissected out and weighed on a microelectrical balance. Data was analysed statistically by Student's 't' test.

(d) Sexually active males of *Funambulus pennanti* were obtained in May and injected subcutaneously (between 1650-1700 hrs) with 10 μg 5-methoxytryptophan (MTP)/ animals/day for 60 continuous days to both sham-operated and pinealectomized animals.

(e) Adult pinealectomized and sham-operated males of *Funambulus pennanti* were injected subcutaneously (between 1650-1700 hrs) with 10 μg MIAA/animal/day for 60 continuous days during March-May. After completion of experiment, the animals were weighed and sacrificed by cervical dislocation. Testes were removed, weighed, fixed and processed for histological observations.

(f) Adult pinealectomized and sham-operated males of *Funambulus pennanti* was injected (between 1650-1700 hrs) subcutaneously with 1, 10 and 50 μg of melatonin (aMT) and 5-methoxytryptamine (MT)/animal/days for 20, 40 and 60 continuous days during the active phase (April-May), regressive phase (September-October) and inactive phase (October-November). After the required duration the animals were weighed and sacrificed. Testes were removed, weighed and fixed in Bouin's fluid for histological observations.

B. **Morning Injections**

(a) Twenty one adult males of *Calotes vesicolor* were obtained in May (active phase) and acclimatized to lab. conditions. The following three groups of seven animals each were made :

Group I : Animals were injected intraperitoneally with 10 μg aMT/animal/day.
Group II : Animals were injected intraperitoneally with 10 μg MT/animal/day.
Group III : Animals were injected intraperiotoneally with normal saline.

All injections were given between 0650-0700 hrs for 45 continuous days after which the animals were weighed and decapitated. Blood was collected for cholesterol estimation in plasma. Testis of one side was removed, weighed and fixed in Bouin's fluid for histological studies. Testis of other side was processed for total cholesterol etimation.

(b) Twenty one adult males of *Funambulus pennanti* were obtained in December and acclimatized to laboratory conditions for two weeks. They were divided into following three groups of 7 animals each :

Group I : Animals were injected subcutaneously with 10 μg aMT/animal/day
Group II : Animals were injected subcutaneously with 10 μg MT/animal/day.
Group III : Animals were injected subcutaneously with normal saline.

After 45 days of continuous administration, between 0650-0700 hrs. all animals were weighed and sacrificed. Their gonads were removed, weighed and fixed in Bouin's fluid for routine histology.

V. Effect of Environment on Pineal Biochemistry

A. Circannual Experiments

(a) *Calotes versicolor*

Adult males of *C. versicolor* were obtained in the last week of every month and acclimatized for 2 weeks in a room fully exposed to environmental conditions. They were divided into two groups of eight animals each. Protein content was measured in the pineal glands of animals of Group I. Pineal cholesterol estimation was done of Group II animals. Data was statistically analysed by ANOVA.

(b) *Funambulus pennanti*

Sixteen adult males were obtained in the last week of every alternate month and divided into two groups of 8 animals each :

Group I : Protein content of the pineal gland was measured by the method of Lowry *et al.*, (1951).

Group II : Cholesterol level of the pineal gland was measured following the method of Sackett (1925).

Data was statistically analysed by ANOVA.

B. Circadian Experiments

(a) *Calotes versicolor*

Adult males of *C. versicolor* were obtained in December (pineal active, gonad inactive) and May (pineal inactive, gonad active). They were exposed to ambient environmental conditions for two weeks. The animals were sacrificed at intervals of 4 hours, during the 24 hour scale, starting at 1000 hrs. The pineal glands were removed, weighed quickly on a microelectrical balance and kept frozen till the biochemical assays of protein and cholesterol by the above mentioned methods.

(b) *Funambulus pennanti*

Circadian experiments were performed during the months of December (testes recrudescing, pineal active) and April (testes active, pineal inactive). The animals were sacrificed at intervals of 4 hrs during the 24 hour scale, starting at 1000 hrs. Body weights of all animals were noted. The pineal glands were removed, weighed quickly and kept frozen till the biochemical assay of protein and cholesterol.

TABLE 20.1. *Annual variation in the weights of testes (g/100 g body weight) and pineal gland (mg ′ 100 g body weight) of the Indian palm squirrel Funambulus pennanti.*

Month	Testes	Pineal Gland
June	0.80 ± 0.02	1.00 ± 0.075
July	0.79 ± 0.01	1.12 ± 0.05
August	0.62 ± 0.02	1.19 ± 0.03
September	0.50 ± 0.03	1.60 ± 0.06
October	0.37 ± 0.04	2.00 ± 0.07
November	0.22 ± 0.02	2.50 ± 0.10
December	0.55 ± 0.02	2.71 ± 0.05
January	0.81 ± 0.04	3.12 ± 0.12
February	0.78 ± 0.01	2.00 ± 0.10
March	0.80 ± 0.02	1.25 ± 0.03
April	0.81 ± 0.01	1.12 ± 0.02
May	0.80 ± 0.03	0.75 ± 0.02

TABLE 20.2. *Annual variation in the daylength (hrs) and pineal gland weights (mg/100 g body wieght) of male* Calotes versicolor *and* Lonchura punctulata.

Month	Daylength	*Calotes versicolor*	*Lonchura punctulata*
June	13.44	1.18 ± 0.10	1.05 ± 0.040
July	13.33	1.08 ± 0.06	2.01 ± 0.060
August	12.85	1.20 ± 0.06	2.94 ± 0.050
September	12.35	1.56 ± 0.10	4.07 ± 0.043
October	11.59	2.22 ± 0.08	3.92 ± 0.030
November	11.07	3.73 ± 0.08	4.13 ± 0.050
December	10.72	3.83 ± 0.06	4.65 ± 0.040
January	10.92	3.61 ± 0.06	4.77 ± 0.040
February	11.48	3.43 ± 0.19	4.50 ± 0.050
March	12.17	2.63 ± 0.17	4.14 ± 0.017
April	12.77	1.96 ± 0.16	2.01 ± 0.040
May	13.15	1.52 ± 0.13	1.50 ± 0.060

Results and Discussion

I. Annual Pineal and Testicular Cycle in Relation to Various Ecofactors

F. pennanti

Results are presented in Table 20.1 analysis of variance of the data presented a significant variation in the testes weight ($F = 35.50$, $p < 0.001$) throughout the year. High testes weight was observed during March-July when daylength and temperature were increasing and humidity decreasing. A decrease in testes weight was observed after July. This period was accompanied with decreasing daylength, temperature and humidity. Lowest testicular weight with arrest of spermatogenesis was recorded in November. Testicular recrudescence started in December and active testes was found in January. Table 20.1 also indicates the annual variation in the pineal gland weight of *F. pennanti* in relation to environmental factors. A significant annual variation ($F = 135.46$, $p < 0.001$) was observed in the pineal gland weight of the squirrel. Results of correlation and regression analysis indicated that temperature, humidity and daylength significantly contribute to the pineal activity variation. In case of testes, temperature and daylength significantly contribute to the variability in its activity.

C. versicolor

A marked annual variation was observed in the testes, pineal, accessory sex organ weight, renal sex segment (RSS) cell height, seminiferous tubule diameter and testicular cholesterol level of the animals (Figs. 20.1 & 20.2; Tables 20.2-20.3). Increase in the gonadal activity coincided with increasing daylength. High values were noted during summer months when the days were long. A decreasing trend was noted with the decrease in daylength and low values were observed when days were short. An opposite trend was observed for the pineal gland weight and activity of the animals. Results of correlation analysis indicated a significant positive correlation between

TABLE 20.3. *Summary of the analysis of variance (ANOVA) of the data on the annual changes in testes, pineal gland and accessory sex organs on Calotes versicolor*

Parameter	F	Level of significance
Testes weight	853.63	$p < 0.001$
Seminiferous tubule diameter	154.46	$p < 0.001$
Testicular cholesterol	164.90	$p < 0.001$
Pineal gland weight	76.62	$p < 0.001$
Pineal protein	93.41	$p < 0.001$
Pineal cholesterol	319.76	$p < 0.001$
Vas deferens weight	60.59	$p < 0.001$
Kidney weight	285.34	$p < 0.001$
Renal sex segment cell height	389.22	$p < 0.001$

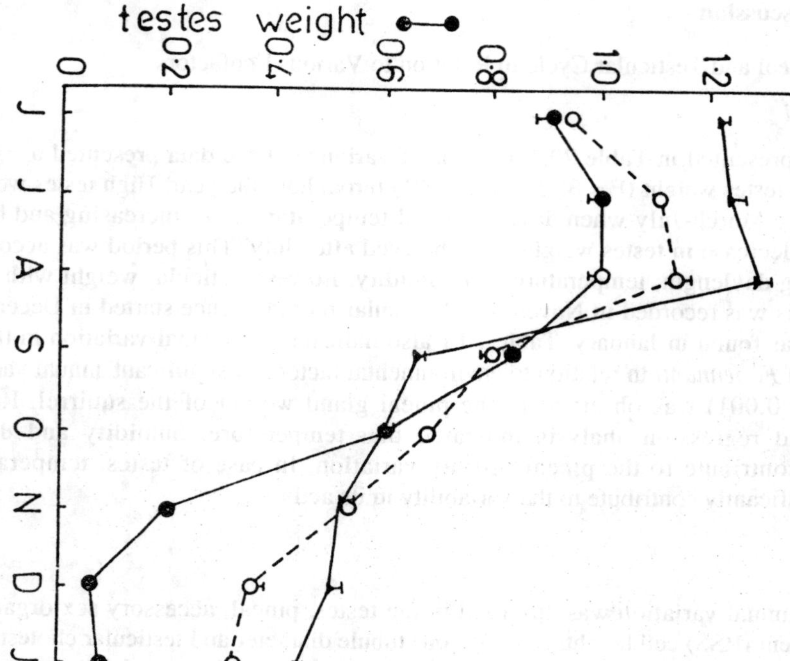

Fig. 20.1. Annual changes in the testes weight (g/100 g body weight), seminiferous tubule diameter (µm) and testes cholesterol level (mg/100 mg tissue) of the Indian garden lizard, *Calotes versicolor*.

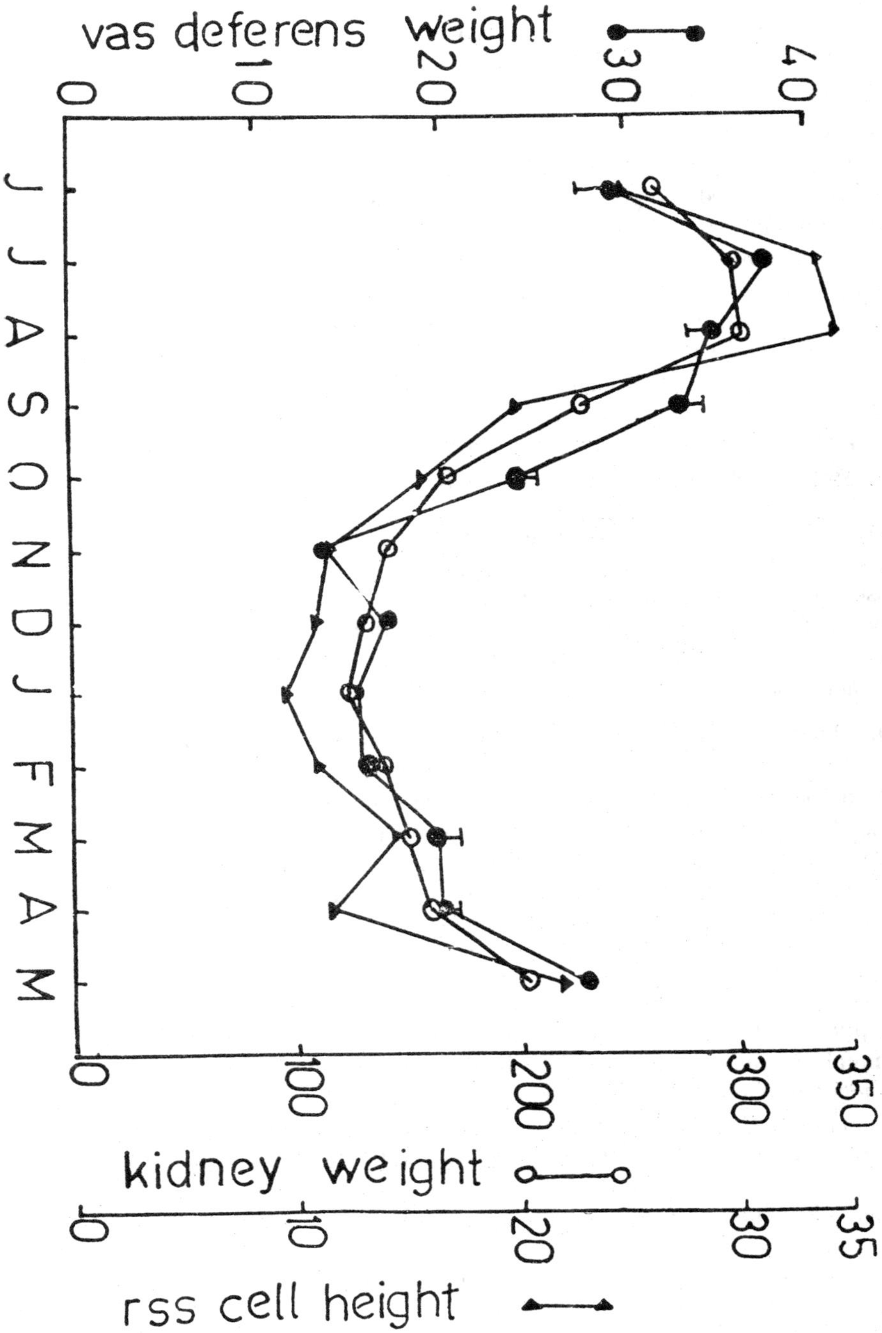

Fig. 20.2. Annual changes in the vas deferens weight mg/100 g body weight), kidney weight (mg/100 g body weight) and renal sex segment (RSS) cell height (μ) of male Indian garden lizard, *Calotes versicolor*.

testes and daylength ($r = 0.72 \pm 0.21$; $p < 0.01$). A significant negative correlation was established between pineal and daylength ($r = -0.94 \pm 0.10$; $p < 0.001$) as well as between testes and pineal ($r = -0.57 \pm 0.25$; $p < 0.05$).

The results clearly show the sexual cycle of both the animals to be related to the environment. The present observations on the Indian palm squirrel can be related to those of Prasad *et al.* (1966) who described two peaks of pregnancies, one in March and the other in July. Considering the temporal relationship between the seasonal sexual cycle and annual changes in environmental factors, clear observation can be made. It can be said that besides daylength, rainfall either directly by increasing humidity or indirectly by acting on the feeding processes participates to the induction of sexual quiescence in *F. pennanti*. It has been reported that in mammals, rainfall indirectly affects the reproductive activities by causing changes in food availability and quality (Perry and Rowlands, 1973; Delany and Happold, 1979). Food quality which depends on other environmental factors may be a potential cue for timing the annual reproductive cycle because it is well known that the consumed plant can modify the annual gonadal cycle of some herbivores or carnivores (Desjardins, 1979; Pevet, 1987).

II. Effect of Pinealectomy during Different Seasons

Results of Experiment II are presented in Fig. 20.3. Analysis of variance of the data shows a highly significant variation in the testes weight in relation to pinealectomy ($F = 126.22$, $p < 0.001$) as well as phase ($F = 35.4$, $p < 0.001$). The effect of pinealectomy also indicated a significant variation in different phases ($F = 9.20$, $p < 0.001$).

It has been hypothetized that besides photoperiod, the pineal would be able to mediate other environmental factors such as temperature, humidity, food etc. separately or in association depending on the ecological situation which the concerned species inhabits (Pevet, 1979, 1985; Pevet and Haldar-Misra, 1982 a, b, Vivien-Roels and Pevet, 1983; Brainard *et al.*, 1984). The present results also show that the pineal gland is necessary for the expression of the seasonal sexual cycle. The significant inverse relationship between the annual changes of the pineal gland and testes weight suggest that other environmental factors might also be integrated via the pineal gland.

There is substantial evidence showing that in animals under natural photoperiods the function of the pineal gland may vary with season. Biochemical data provide information related to circannual fluctuations of serotonin in the bovine pineal organ (Philo and Reiter, 1980). Environmental photoperiod determines the structure (Quay, 1956) and metabolism (Quay 1963; Wurtman *et al.*, 1963) of the mammalian pineal and hence the ability of the pineal to influence the reproductive system depends on the photoperiod (Vivien-Roels and Pevet, 1983).

III. Effect of Photoperiod

The results of Experiment (1) are presented in Table 20.4. It is clearly evident that the sham-operated animals indicated a retardation of testicular regression whereas the pinealectomized

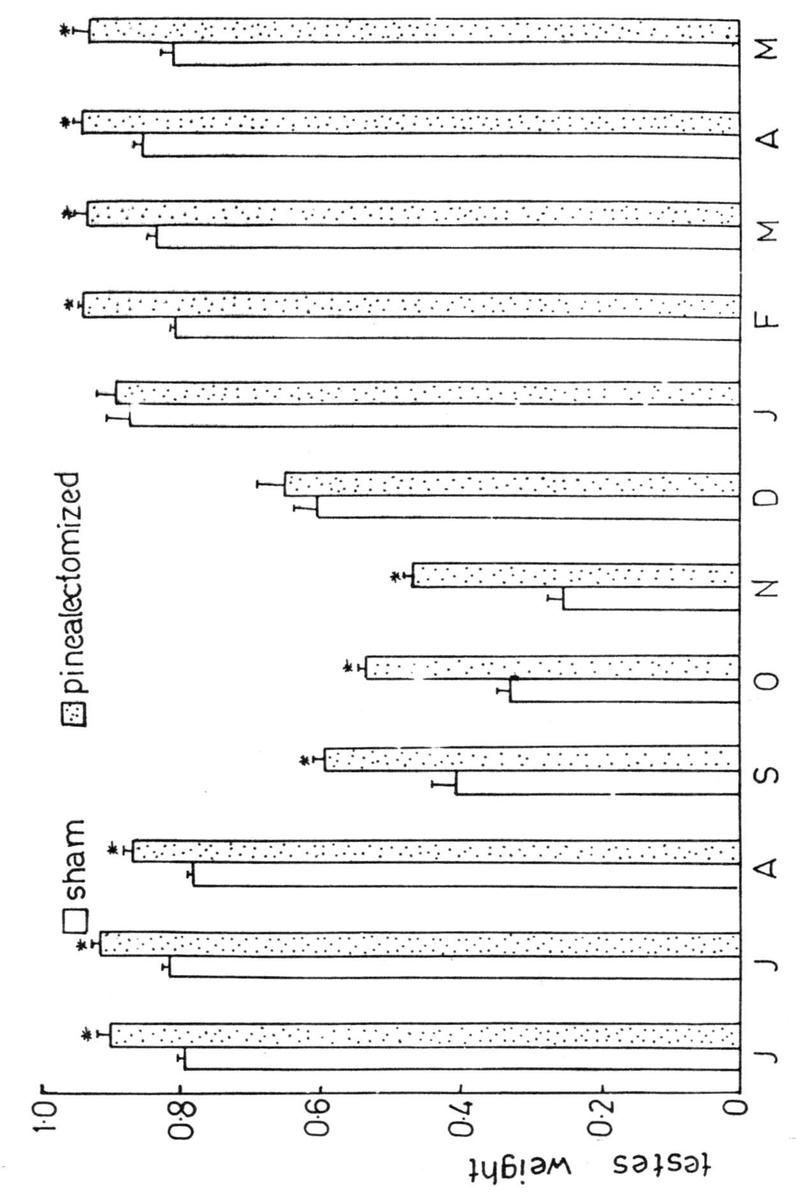

Fig. 20.3. Effect of pinealectomy on the testes weight (g/100 g body weight) of *Funambulus pennanti,* during the different phases of its annual gonadal cycle. Significance of difference from control-*, $p < 0.001$.

animals did not show any significant difference in their testes weight from those of controls under NDL. This proves that in the absence of the pineal gland the animals are unable to read the environmental information. The experimental results also show that in this species long photoperiod is stimulatory to the gonads and the pineal gland is absolutely necessary for the perception of this environmental information.

Photoperiod regulates the indolamine synthesis (Cardinali, 1981) and hormonal effects of the pineal gland on the neuroendocrine reproductive axis (Reiter, 1980). The photic information perceived by retina gets transduced into a neural message and reaches the pineal gland. In the pineal gland, the sympathetic nerve endings bring about the release of nor-epinephrine which interacts with the β-receptors linked to an adenylate cyclase cyclic AMP system (Klein *et al.*, 1981). This causes aMT production which is released into the blood. Low levels of NAT activity (Klein *et al.*, 1981) and aMT content (Lynch, 1971; Rolag *et al.*, 1980) and depressed circulating melatonin titers (Wilkinson *et al.*, 1977) have been reported. During the dark period of the light-dark cycle, melatonin production is high (Panke *et al.*, 1978; Brown *et al.*, 1981). Superior cervical ganglionectomy of the animal abolishes this indolamine rhythm. All these conditions are involved in the regulation of reproduction in animals. Almost all the biological rhythms are synchronized to environmental lighting. The daily changes in pineal HIOMT activity and NE content are totally dependent upon the environmental lighting condition. A diurnal rhythm in HIOMT activity occurs in the rat (Quay, 1974). These rhythms are disturbed and even blocked by superior cervical ganglionectomy which blocks the connection of sympathetic nerves to the pineal gland. It has also been observed that in the golden hamster pinealectomy prevents gonadal regression normally brought about by blinding or experimental or natural short photoperiod (Czyba *et al.*, 1964; Hoffman and Reiter, 1965).

The results of Experiment III (2) are presented in Fig. 20.4. A significant reduction in the testes and epididymis weight of the animals exposed to 10 L : 14 D was observed in October, February and March when compared with their respective controls under NDL. Histologically, the testes of animals under short photoperiod showed a complete arrest of spermatogenic activity in March while the testes of animals under NDL were fully active with an increased tubular diameter and bunches of spermatozoa.

TABLE 20.4. *Effect of long photoperiod (14 L : 10 D) on the testes weight (g/100 g body weight) of sham-operated (SO) and pinealectomized (Px) Indian palm squirrel,* F. pennanti *during September-November*

	Normal Daylength (NDL)	14 L : 10 D
SO	0.30 ± 0.01	0.45 ± 0.01*
Px	0.52 ± 0.02	0.50 ± 0.02

Significance of difference from NDL control-*, $p < 0.001$.

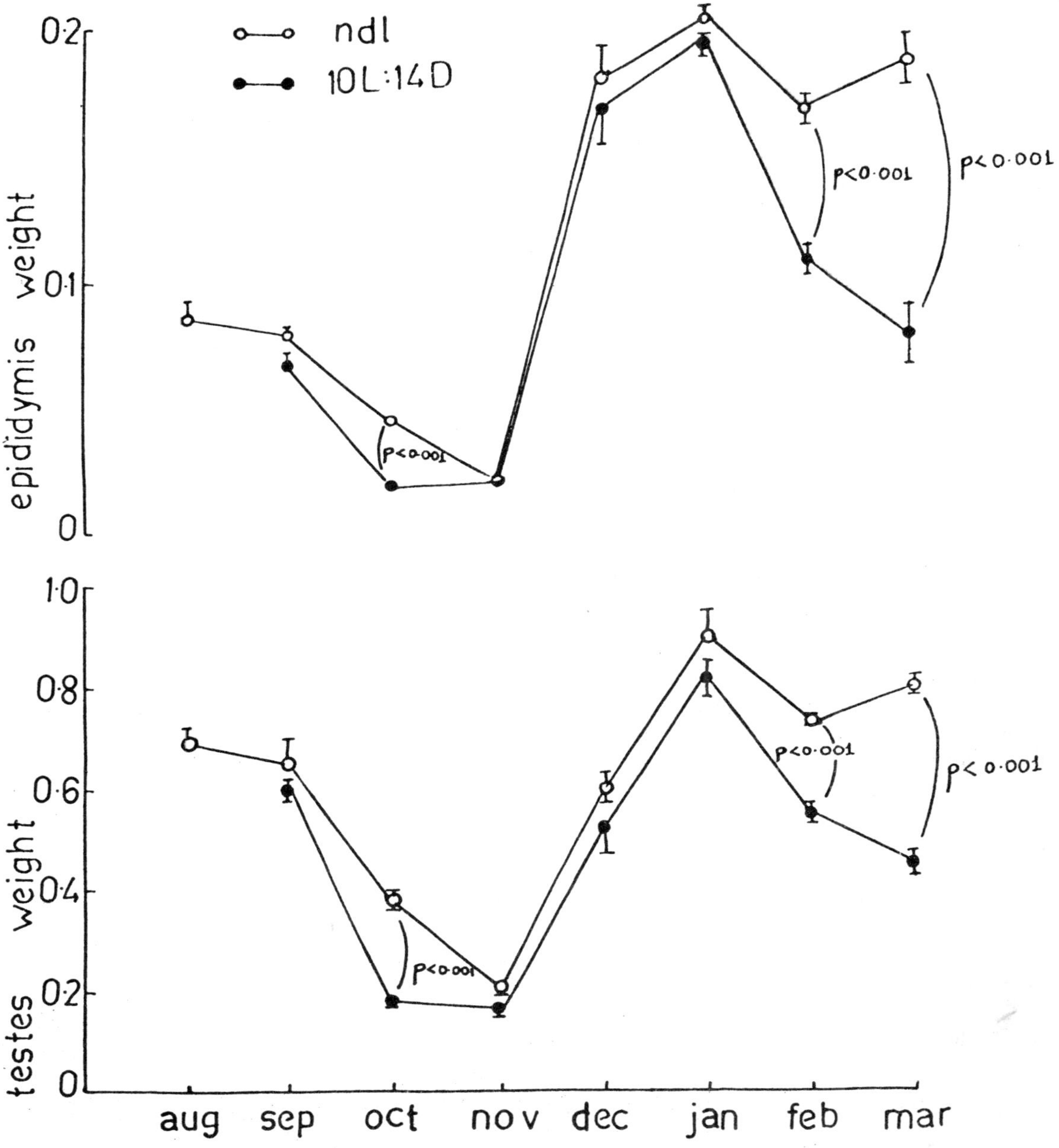

Fig. 20.4. Effect of short photoperiod (10 L : 14 D) on the testes weight (g/100 g body weight) of *F. Pennanti* during August - March.

Since the activity of the pineal gland is directly dependent upon photic information reaching it via the lateral eyes (Reiter, 1974) its role in long term adaptation of reproduction to environmental lighting has been investigated in highly photoperiodic species like hamster (Reiter 1974, 1978, 1981; Hoffmann, 1979) and the vole (Farrar and Clark, 1976; Thorpe and Herbert, 1976). Epididymis being under the direct control of testicular androgens showed a similar effect as that on testes.

A 10 L : 14 D condition accelerates gonadal regression of the Indian palm squirrel in September and October while it maintained the inactive testes (during the inactive phase) for a much longer duration (60 days) than that occurring in nature (30 days). However, testicular recrudescence occurs even under a short photoperiodic conditions thus confirming that testicular recrudescence is a spontaneous phenomenon in this animal. The testes of animal experiencing normal daylength showed a faster gonadal recrudescence than those of squirrels in 10 L : 14 D. An effect of short photoperiod, on the gonads of *F. pennanti,* was again evident after the testicular recrudescence was completed in January. A reduced testicular weight was observed after 4 and 8 weeks of short photoperiodic exposure from January, when compared with that of animals under NDL. Histological observations showed that the testes of animals exposed to 10 hrs of light also showed a reduced spermatogenic activity in March. This was in contrast to the animals under NDL as their testes were fully active. Thus, short photoperiod does not affect a recrudescing testes but it delays gonadal activation after testicular recrudescence has taken place. Thus, after a period of photorefractoriness, the testes again become photosensitive in the month of February.

It can be concluded that gonadal atrophy in squirrel is induced by exposure to short photoperiods. A short photoperiod has no effect on testicular recrudescence (which appears to be endogenous) but after this phenomenon is completed the gonad again becomes sensitive to photoperiod and a short photoperiod is again effective in inhabiting gonadal function. Thus, the function of the pineal gland, in this animal, might be changing throughout the various seasons. The pineal appears to function as a synchronizer of the stages of the annual reproductive cycle with the appropriate season of the year and its presence is assential for the transaction of the photoperiodic effects in this squirrel.

Results of Experiment III 3 (a) show that a constant light condition for 90 days was stimulatory to the gonads and epididymis of *F. pennanti*. The testes and epididymal weights of animals exposed to constant light were significantly higher than that of control animals kept under natural environmental conditions (Table 20.5). Testes of animals under constant light showed acceleration of spermatogenesis whereas animals under NDL indicated an inactive state. The pineal gland weight of animals under constant light was significantly lower ($p < 0.001$) than that of animals under NDL.

The effect of photoperiod on the reproductive status of the animal depends on its ability to differentiate the length of one photoperiod from that of the other (Rudeen and Reiter, 1980).

In the case of *C. versicolor,* no significant effect was observed on testes and pineal gland after exposure to LL for 20 days (Table 20.6). The pineal gland weight and its protein content were

TABLE 20.5. *Effect of constant light condition on the pineal, testes and epididymal weight of Indian palm squirrel,* Funambulus pennanti. *Relative weight of tissue-mg/100 g body weight for pineal gland and g/100 g body weight for testes and accessory sex organs* ·

	Pineal Gland	Testes	Epididymis
NDL Control	2.70 ± 0.10	0.45 ± 0.03	0.06 ± 0.005
Constant light	2.00* ± 0.15	0.68** ± 0.04	0.09** ± 0.003

NDL-Normal daylength; *, $p < 0.005$; **, $p < 0.001$.

also significant reduced after 40 and 60 days of exposure to LL. After 60 days of exposure the testes and vas deferens weight and seminiferous tubule diameter were significantly higher as compared to the controls under NDL. These results show that LL inhibits the pineal gland activity of this tropical lizard. The results also suggest that the activity of the testes and the pineal gland of the Indian garden lizard may be inversely related to each other. A LL condition, for 70 days, in mouse has been found to cause a significant reduction in pinealocyte size and Golgi complex size. The vas differens weight showed an increase after 40 and 60 days of exposure to LL and this further confirms the stimulation of gonads. In rodents a LL environment causes pineal gland reduction in weight (Fiske *et al.*, 1962; Quay, 1961), lipid (Quay, 1961) and serotonin content (Quay and Halevy, 1962).

The results of the experiment of exposure of *F. pennanti* to constant darkness are presented in Table 20.7. The table shows that after 15 days no significant effects was visible on the gonads as well as the pineal gland of the animals. After 30 days of exposure, a significant reduction was observed in the testes weight. A corresponding significant increase was noted in the pineal gland weight and protein content. After 60 days the testes of animals under constant darkness exhibited a significantly lower weight than that of controls. The pineal weight and protein level of animals under darkness revealed a significant increase in comparison to controls. The testes of animals under DD indicated a suppression of spermatogenesis while animals under NDL exhibited active testes with presence of spermatozoa.

Results of Experiment III 4 (b) are presented in Table 20.8 and indicate an inhibition of gonad and vas deferens after 40 and 60 days of exposure. A corresponding stimulatory effect was evident on the pineal gland. The present study provides evidence for the fact that continuous darkness leads to alterations in the activity of the pineal gland and consequently the gonads of the animals. It has been reported that pinealectomy alters the ability of a number of animals to respond appropriately to the changes in the light dark environment (Thorpe and Herbert, 1976; Turek and Campbell, 1979).

Again, in the case of *F. pennanti* it was observed that animals maintained a steady weight after 15 days. A significant decline in the body weight was evident after 30 days of exposure to

TABLE 6. *Effect of continuous illumination (LL) on the pineal gland, vas deferens and testes of the Indian garden lizard,* Calotes versicolor.
Relative weights of pineal gland, vas deferens-and testes-g/100 mg body weight.

Group	Pineal gland		Vas deferens weight	Testes	
	Weight	Protein (µg/100 mg tissue		Weight	Seminiferous tubule diameter (µm)
20 Days					
NDL	3.04 ± 0.20	126.21 ± 7.20	14.20 ± 2.15	89.60 ± 5.60	158.60 ± 6.21
LL	3.00 ± 0.12	120.10 ± 5.45	18.00 ± 2.00	100.24 ± 7.00	162.42 ± 12.12
40 Days					
NDL	2.64 ± 0.15	121.00 ± 4.62	18.10 ± 1.01	105.41 ± 4 .61	164.00 ± 7.11
LL	2.01[c] ± 0.09	103.36[a] ± 5.20	23.12[c] ± 0.80	137.21[d] ± 4.40	198.30[b] ± 10.37
60 Days					
NDL	2.00 ± 0.10	112.62 ± 8.72	22.00 ± 0.90	156.70 ± 6.10	188.41 ± 6.00
LL	1.80[d] ± 0.03	86.50[a] ± 5.40	27.16[c] ± 1.31	190.00[d] ± 7.93	216.90[c] ± 6.20

NDL–Natural daylength
Significance of difference from NDL–a, $p < 0.05$; b, $p < 0.02$, c, $p < 0.01$; d, $p < 0.001$.

DD. A further decrease was noted after 60 days. It can be suggested that the animals in dark are under a condition of stress which might be affecting their feeding habits resulting into change in body weight. It may also be suggested that darkness produces an inhibitory influence on the body weight of squirrels. This finding corroborates the previous study on Djungarian hamster (Hoffmann, 1973) which shows that the seasonal rise in body weight is enhanced by exposing them to long photoperiods in mid winter and that the stimulatory effect of light is suppressed by aMT. The results with *F. pinnanti* and *C. versicolor* suggest that darkness inhibits gonadal activity of these tropical species and the pineal gland appears to be involved in this phenomenon.

Evening injections of MT of *C. versicolor* during June-July significantly reduced gonadal and accessory sex organ weight and activity. A significant reduction was observed in the seminiferous tubule diamer, RSS cell height and plasma and testicular cholesterol levels. No

TABLE 20.7. *Effect of continuous darkness on the pineal protein, pineal, testicular and body weight of Indian palm squirrel,* Funambulus pennanti.

	Pineal Protein (μg/100 mg tissue)	Pineal Weight (mg/100 g body wt.)	Testes Weight (g/100 g body wt.)	Body Weight (g)
15 Days				
Control	125.62 ± 3.24	3.01 ± 0.02	0.99 ± 0.05	124.00 ± 2.23
Dark	129.00 ± 6.74	2.90 ± 0.09	0.97 ± 0.02	120.00 ± 1.94
30 Days				
Control	100.21 ± 2.64	2.54 ± 0.06	0.96 ± 0.04	126.00 ± 1.52
Dark	119.76 ± 5.20**	3.06 ± 0.02***	0.81 ± 0.03*	112.00 ± 3.52**
60 Days				
Control	86.02 ± 6.26	2.00 ± 0.07	0.88 ± 0:02	121.00 ± 1.26
Dark	115.50 ± 3.76***	3.22 ± 0.01***	0.74 ± 0.01***	108.00 ± 1.16***

Significance of difference from control-*, $p < 0.01$; ** $p < 0.005$.; ***, $p < 0.001$.

significant effect was observed during November-December (Tables 20.9 and 20.10). Results of evening injections of aMT to the Indian garden lizard indicated that only 10-15 μg aMT injected for 40-60 days causes significant reduction in testicular activity. ANOVA indicates significant variation in reduction in testes weight and cholesterol in relation to dose (wt-F = 5.16, $p < 0.001$; cholesterol-F = 2.83, $p < 0.05$) as well as duration (wt-F = 9.50, $p < 0.001$; cholesterol-F = 4.51, $p < 0.05$) of aMT administration.

Injections of ML, during the evening hours resulted in a significant decrease in the testes weight of sham-operated as well as pinealectomized animals (Table 20.11). Table 20.12 shows that MTP administration to *F. pinnanti* during evening also led to a significant reduction in the testes weight of both sham-operated ($p < 0.005$) as well as pinealectomized squirrels ($p < 0.001$). A significant increase was noted in the pineal gland weight of sham-operated animals injected with MTP ($p < 0.001$). Evening injections of MIAA, during March-May, was unable to cause any significant effect on the testes of the Indian palm squirrel (Table 20.13).

TABLE 20.8. *Effect of constant darkness on the pineal gland, vas deferens and testes of* C. versicolor. *Relative weights of pineal gland, vas deferens–mg/100 mg body weight and testes–g/100 g body weight.*

Group	Pineal		Testes		Vas deferens weight
	Weight	Protein (μg/100 mg tissue)	Weight	Seminiferous tubule diameter (μm)	
20 Days					
NDL	1.22 ± 0.04	70.20 ± 3.60	1.04 ± 0.02	296.40 ± 6.70	42.56 ± 1.20
LL	1.21 ± 0.08	67.00 ± 5.20	1.02 ± 0.01	287.00 ± 8.00	40.45 ± 2.00
40 Days					
NDL	1.34 ± 0.07	75.61 ± 4.60	0.96 ± 0.01	267.30 ± 4.10	35.00 ± 1.20
LL	1.90 ± 0.02[c]	99.52 ± 2.20[c]	0.72 ± 0.006[c]	249.20 ± 3.20[a]	30.00 ± 0.20[b]
60 Days					
NDL	150 ± 0.04	82.62 ± 1.90	0.79 ± 0.008	241.60 ± 3.12	30.12 ± 1.60
LL	3.10 ± 0.02[c]	107.76 ± 3.21[c]	0.61 ± 0.002[c]	2.00.73 ± 4.00[c]	22.11 ± 0.36[c]

NDL-Natural daylength
Significance of difference from NDL control-a, $p < 0.001$; b, $p < 0.005$; c, $p < 0.001$.

TABLE 20.9. *Effect of 5-methoxytryptamine (MT) injections on testes, kidney and vas deferens of* Calotes versicolor.

	Jun-July		Nov-Dec	
	Control	MT	Control	MT
Testes				
Weight (g/100 g b. wt.)	0.98 ± 0.046	0.72 ± 0.02**	0.04 ± 0.005	0.04 ± 0.002
Seminiferous tubule diameter (μm)	240.50 ± 8.620	172.22 ± 6.00**	154.20 ± 7.610	150.42 ± 10.10
Kidney				
Weight (g/100 g b. wt.)	0.20 ± 0.014	0.14 ± 0.011*	0.13 ± 0.007	0.13 ± 0.01
RSS cell height (μ)	31.20 ± 2.000	18.00 ± 0.400*	16.00 ± 0.500	15.00 ± 0.06
Vas Deferens				
Weight (g/100 g b. wt.)	0.04 ± 0.004	0.03 ± 0.001*	0.01 ± 0.004	0.01 ± 0.001

Significance of difference from control-*, $p < 0.005$, **, $p < 0.001$; b. wt. = body weight.

TABLE 20.10. *Effect of 5-methoxytryptamine (MT) injections on the cholesterol levels of blood plasma and testes of* Calotes versicolor.

	Jun-July		Nov-Dec	
	Control	MT	Control	MT
Plasma (mg/100 ml)	211.21 ± 10.42	167.30 ± 4.61*	104.44 ± 6.70	100.61 ± 10.25
Testes (mg/100 mg tissue)	1.45 ± 0.02	0.89 ± 0.01**	0.80 ± 0.06	0.70 ± 0.03

Singnificane of difference from control-*, $p < 0.005$; **, $p < 0.001$.

TABLE 20.11. *Effect of evening injections of 5-methoxytryptophol (ML) on the testes and epididymal weight (g/100 g body weight) of pinealectomized (Px) and sham-operated (SO)* F. pennanti, *during the active phase of the annual gonadal cycle*

	Testes	Epididymis
Px Control	0.89 ± 0.005	0.26 ± 0.005
SO Control	0.83 ± 0.01	0.20 ± 0.004
Px + ML	0.79 ± 0.004*	0.19 ± 0.006*
SO + ML	0.78 ± 0.004*	0.14 ± 0.004*

Significance of difference from control : *, $p < 0.001$.

TABLE 20.12. *Effect of evening injections of 5-methoxytryptophan (MTP) on the weight of testes (g/100 g body weight) and pineal gland (mg/100 g body weight) of* Funambulus pennanti, *during the active phase.*

	Testes	Pineal Gland
Px Control	0.27 ± 0.02	--
SO Control	0.82 ± 0.02	1.90 ± 0.02
Px + MTP	0.69 ± 0.02*	--
SO + MTP	0.70 ± 0.01*	2.30 ± 0.02*

Px-Pinealectomized, SO--Sham-operated
Significance of difference from control-*, $p < 0.001$.

Injections of the 5-methoxyindoles during morning hours were without any effect on the mammalian as well as the reptilian gonad. Results of MT and aMT injection to *C. versicolor* during morning hours had no significant effect on testes weight and biochemistry (Table 20.14). Plasma cholesterol level of treated animals was also not significantly different from controls. Histologically, testes of saline injected controls as well as aMT and MT injected animals indicated an active state with bunches of spermatozoa.

Again in the case of *F. pinnanti* the morning injections of MT as well as aMT indicated no significant effect on the testes of the animal (Fig. 20.5).

Above mentioned results clearly indicate the effects of 5-methoxyindole administration to be dependent on the time of the day when they are administered. Evening injections in all cases, except for MIAA, resulted in inhibition of the gonadal activity both in the case of squirrel as well as the lizard whereas the same injections during morning hours were ineffective in altering gonadal activity. This demonstrates that the environment plays a very crucial role in the expression of the effects of these 5-methoxyindoles on the reproductive organs of the various tropical animals. The results also indicate that the duration of 5-methoxyindole administration is also critical. Administration for a short duration of 20 and 40 days may not affect the

TABLE 20.13. *Effect of subcutaneous evening injections of 5-methoxyindole-3-acetic acid (5-MIAA) on the testes and epididymal weight (g/100 g body weight) of sham-operated (SO) and pinealectomized (Px)* Funambulus pennanti, *during the sexually active phase*

Group	Testes	Epididymis
SO Control	0.84 ± 0.012	0.19 ± 0.005
Px Control	0.92 ± 0.004*	0.26 ± 0.007*
SO + 5-MIAA	0.82 ± 0.01	0.19 ± 0.013
Px + 5-MIAA	0.91 ± 0.01	0.25 ± 0.01

Significance of difference between SO and Px control-*, $p < 0.001$.

TABLE 20.14. *Effect of morning injections of melatonin (aMT) and 5-methoxytryptamine (MT) on the testes weight (g/100 g body weight), testicular cholesterol (mg/100 mg tissue) and plasma cholesterol level (mg/100 ml) of* C. versicolor *during the active phase*

	Testes weight	Testes Cholesterol	Plasma Cholesterol
Control	0.90 ± 0.01	2.15 ± 0.05	199.00 ± 2.66
aMT	0.91 ± 0.01	2.09 ± 0.08	204.85 ± 5.93
MT	0.89 ± 0.02	2.00 ± 0.05	206.20 ± 4.65

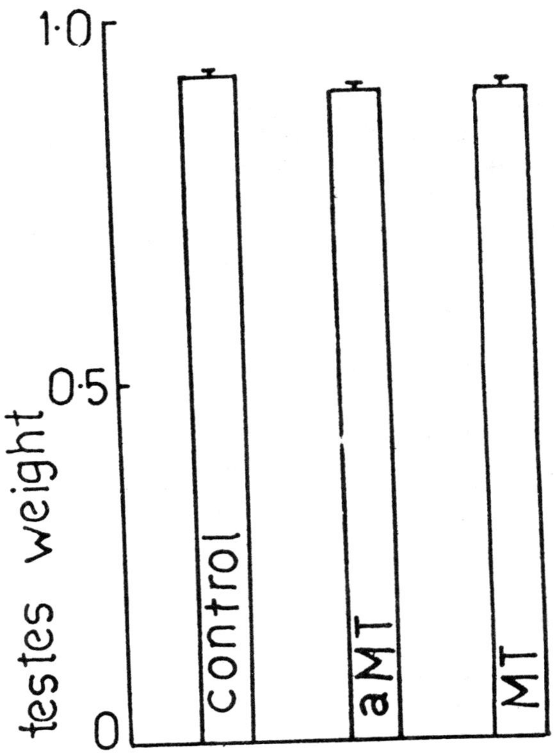

Fig. 20.5. Effect of morning injections of melatonin (aMT) and 5-methoxytryptamine (MT) on the testes weight (g/100 g body weight) of *F. pennanti* duirng the active phase.

reproductive activity. A longer duration may be required to produce an effect. Therefore, it was interesting to note that although the evening injections suppressed gonadal activity in the cases studies (except MIAA) aMT and MT administration via injections during morning hours could not produce any effect and the active gonad, in all these cases, failed to respond to the exogenous indolamines. This enables us to suggest that there exists a rhythm in the sensitivity to exogenous 5-methoxyindoles and the 5-methoxyindoles administered just before the onset of the dark period (i.e. during evening hours) are capable of producing an effect. There is an endogenous peak release of different pineal compounds in different animals. In Syrian hamsters and *Peromyscus leucopus* the timing of injection has a pronounced effect on the reproductive organs (Tamarkin *et al.*, 1977). Unlike the present results morning injections affect the gonads of Turkish hamsters (Carter *et al.*, 1982). The present animals exhibit a rhythm in the sensitivity to the exogenous 5-methoxyindoles.

Results of the circannual study of pineal protein and cholesterol level of the Indian garden lizard, *Calotes versicolor* are presented in Table 20.15. A marked seasonal variation was observed in both the parameters. High values were observed during winter months and low during summer. The peak in pineal protein content was observed in February whereas for cholesterol it was found to be present in January. ANOVA indicated a significant variation in the pineal protein ($F = 93.41$, $p < 0.001$) and cholesterol levels ($F = 319.76$, $p < 0.001$).

Circannual studies on the pineal biochemistry of Indian palm squirrel showed a marked annual variation in the pineal gland protein ($F = 82.46$, $p < 0.001$) and cholesterol content ($F = 128.40$, $p < 0.001$). The results are presented in Fig. 20.6. The table shows the highest value to be

TABLE 20.15. *Annual changes in the pineal protein (µg/100 mg tissue) and cholesterol (mg/100 mg tissue) levels of male Indian garden lizard,* Calotes versicolor

Month	Protein	Cholesterol
June	46.50 ± 1.25	
July	50.00 ± 1.23	1.00 ± 0.046
August	64.66 ± 1.40	
September	75.66 ± 1.89	3.01 ± 0.033
October	99.33 ± 2.24	
November	123.00 ± 3.93	4.61 ± 0.043
December	128.50 ± 4.69	
January	126.50 ± 3.99	4.87 ± 0.170
February	129.16 ± 3.35	
March	100.83 ± 4.01	3.20 ± 0.030
April	83.66 ± 2.66	
May	61.00 ± 3.01	2.51 ± 0.029

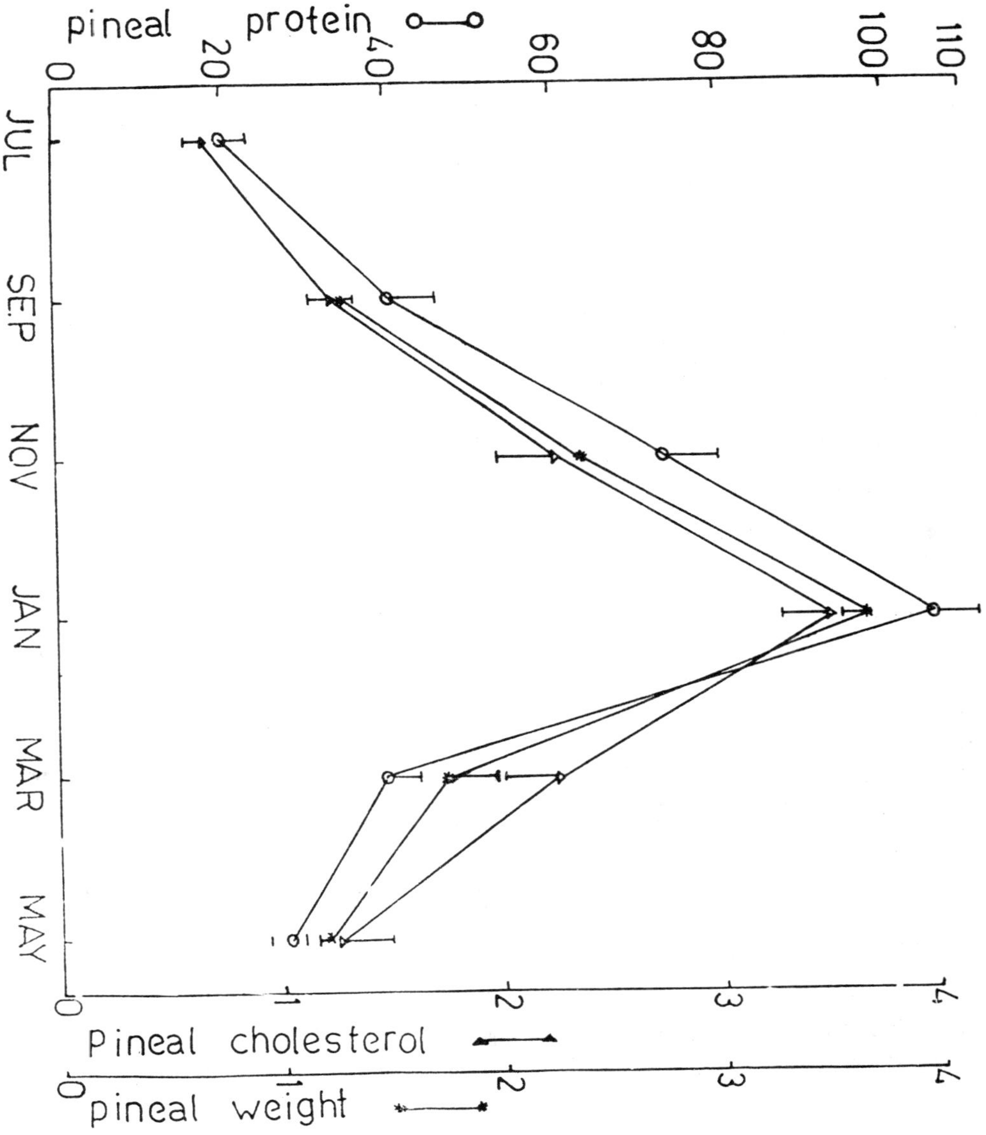

Fig. 20.6. Annual changes in the pineal protein (µg/100 mg tissue), pineal cholesterol (mg/100 mg tissue) and pineal weight (mg/100 g body weight) of *F. pennanti*.

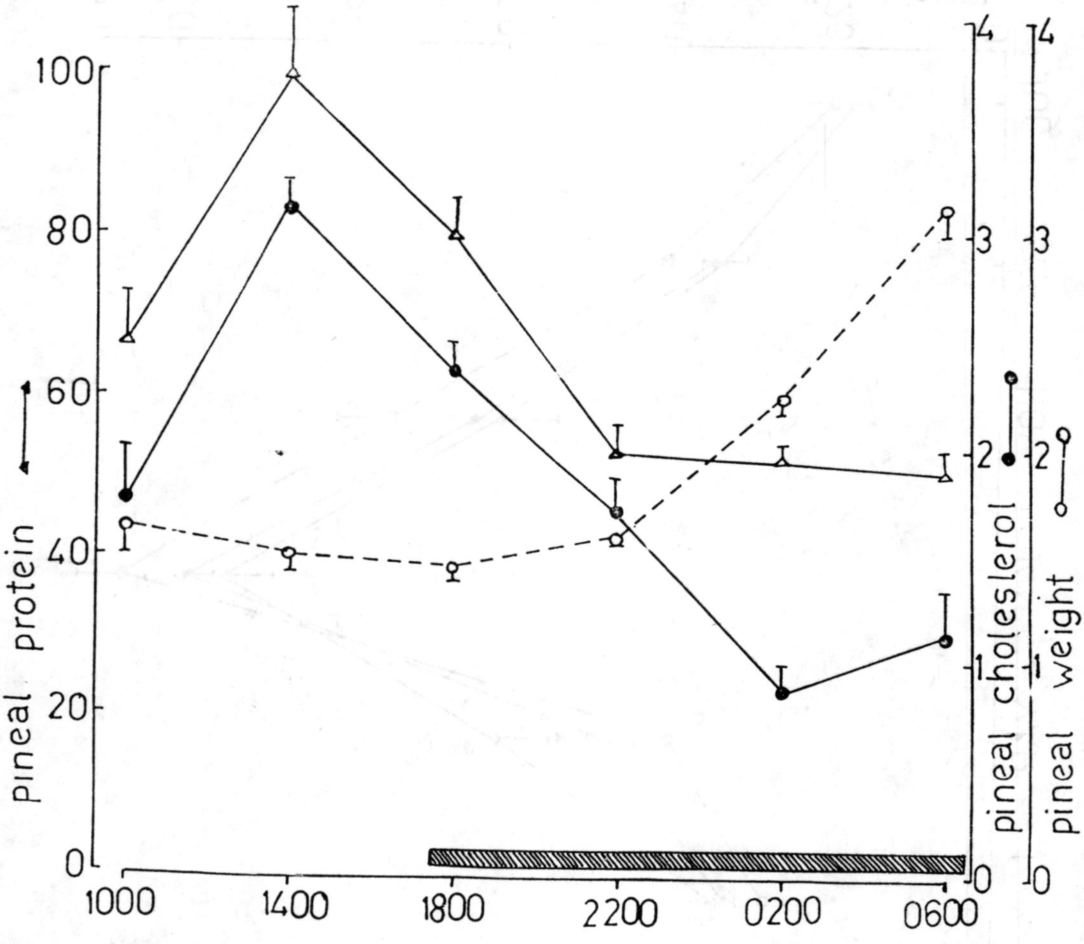

Fig. 20.7. Circadian changes in the pineal protein (μg/100 mg tissue), pineal cholesterol (mg/100 mg tissue) and pineal wieght (mg/100 g body weight) of *F. pennanti*, during December.

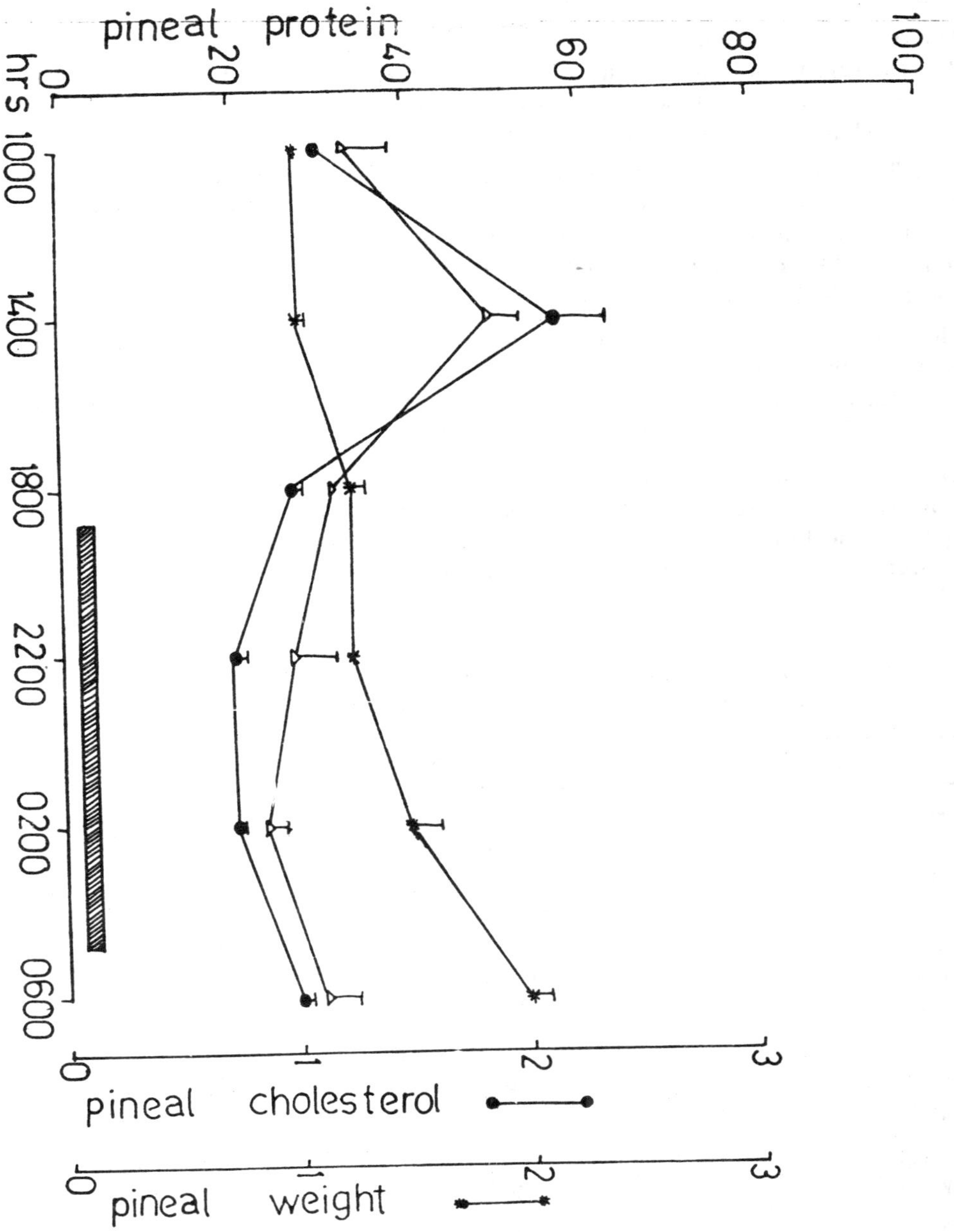

Fig. 20.8. Circadian changes in the pineal protein (μg/100 mg tissue), pineal cholesterol (mg/100 mg tissue) and pineal weight (mg/100 g body weight) of *F. pennanti* during April.

presented in January and low values during May-July.

The results of circadian study, in case of *C. versicolor* are presented in Tables 20.16 and 20.17. The results indicated a marked daily variation. A similar pattern of rhythmicity was observed for both the constituents. The peak was always found to be present in the afternoon at 1400 hrs and the lowest value after midnight at 0200 hrs.

Figs. 20.7 and 20.8 presents the circadian changes in the pineal protein and cholesterol levels of *F. pennanti*, during December and April. A similar pattern of rhythmicity was observed in the protein and cholesterol levels, the peak being at 1400 hrs and the lowest value after midnight at 0200 hrs, During the month of April the peak in pineal protein and cholesterol level was seen at 1400 hrs and the minimum value at 0200 hrs.

The results of the biochemical study of the pineal gland showed a clear relation of the pineal activity with the daily pattern of the light as well as with the season. Biochemical mechanisms link the pineal gland to adaptive and other responses involving other regions of the body, and are correlated to the daily pattern of light.

The activity of the pineal gland varies with changes in environmental conditions where an animal lives (Vivien-Roels and Pevet, 1983). Thus, the pineal gland has an important endocrine and physiological significance and it shows great sensitivity to environmental control. The pineal gland secretes aMT which regulates the gonadal function of mammals (Reiter, 1980). Melatonin

TABLE 20.16. *Circadian changes in the protein content of the pineal gland (μg/100 mg tissue) of male* C. versicolor *and* L. punctulata *during December and May*

	C. versicolor	*L. punctulata*
December		
10.00 hrs	125.66 ± 0.71	124.83 ± 1.49
14.00 hrs	130.83 ± 1.19	140.00 ± 1.00
18.00 hrs	122.00 ± 1.00	119.66 ± 1.08
22.00 hrs	120.16 ± 0.54	110.50 ± 1.33
02.00 hrs	113.00 ± 1.15	106.74 ± 0.83
06.00 hrs	122.16 ± 1.07	117.75 ± 0.96
May		
10.00 hrs	59.00 ± 1.46	58.66 ± 1.60
14.00 hrs	65.00 ± 1.03	65.33 ± 0.98
18.00 hrs	57.00 ± 1.46	55.66 ± 1.11
22.00 hrs	53.50 ± 1.08	49.50 ± 1.43
02.00 hrs	46.83 ± 0.70	42.50 ± 1.25
06.00 hrs	53.66 ± 1.05	53.33 ± 1.14

TABLE 20.17. *Circadian changes in the cholesterol content of the pineal gland (mg/100 mg tissue) of male* C. versicolor *and* L. punctulata *during December and May*

	C. versicolor	*L. punctulata*
December		
10.00 hrs	4.18 ± 0.06	5.02 ± 0.063
14.00 hrs	4.82 ± 0.04	5.70 ± 0.05
18.00 hrs	4.30 ± 0.03	4.78 ± 0.031
22.00 hrs	3.75 ± 0.05	4.57 ± 0.03
02.00 hrs	3.10 ± 0.04	4.02 ± 0.04
06.00 hrs	4.05 ± 0.03	4.45 ± 0.04
May		
10.00 hrs	1.98 ± 0.07	1.71 ± 0.018
14.00 hrs	2.60 ± 0.07	1.79 ± 0.03
18.00 hrs	2.15 ± 0.06	1.74 ± 0.01
22.00 hrs	1.71 ± 0.03	1.61 ± 0.02
02.00 hrs	1.15 ± 0.01	1.22 ± 0.05
06.00 hrs	1.88 ± 0.04	1.55 ± 0.05

is found in the pineal gland of a large number of animals and in the serum, urine, nervous tissue, retina, Harderian gland and intestine of chickens, rats, sheep and human beings (Arendt *et al.*, 1975; Lynch *et al.*, 1975; Pang *et al.*, 1974, 1977; Raikhlin *et al.*, 1975). Plasma concentration of testosterone which is produced by testes also undergoes seasonal and circadian changes (Webley *et al.*, 1985). There are reports relating to the pineal gland as a function of the season (Petit and Vivien-Roels, 1977; Rollag *et al.*, 1978; Vivien-Roels and Petit, 1975).

In these Indian tropical animals, circadian studies were performed to study the effects of light and darkness whereas the circannual variations were made to correlate them with the various stages of the reproductive cycle. The results show that the biochemical activities of the pineal gland are under the sensitive influence of the environment. In animals of temperate regions photoperiod and temperature are the most important variables for the reproductive activity of animals (Assenmacher and Farner, 1978). But animals of tropical zones experience entirely different environmental conditions and for them annual cycles in rainfall and humidity control the reproductive processes of animals. The pineal gland is involved in this phenomenon (Goldman and Darrow, 1983; Gwinner, 1981; Lofts, 1975; Vivien-Roels *et al.*, 1979).

In both the species a marked pattern of cyclicity was observed for the pineal biochemical constituents. As there was a decrease in the temperature and daylength from September to January, a gradual increase in the biochemical constitutions was observed. During this period the humidity was high. Again with the increasing daylength the protein and cholesterol levels started decreasing and remained low in May-July when the days were long and temperature and rainfall high. Therefore, it may be suggested that for this species the environmental light and temperature affect the level of these two pineal constituents.

Earlier reports indicate alteration in pineal protein content and the environmental factors are involved in this (Nir *et al.*, 1969). Continuous light causes changes in pineal RNA (Nir *et al.*, 1969). *Calotes versicolor* and *Funambulus pennanti* also exhibited distinct changes in the pineal protein and cholesterol levels, with alterations in natural environmental factors. The circadian experiments also revealed a significant variability. The content in squirrel decreased with onset of dark period and remained low during night. As a functional point of view the protein and peptides have been considered as possible pineal hormones (Thieblot and Blaise, 1967). There are diverse opinions regarding the physiological significance of lipids which may be involved primarily in general metabolism, membrane structure and function within the organ (Wolfe, 1965). Therefore, it appears that the secretary activity of the pineal gland is related to the environment for the control of its reproduction.

In natural environments, the animal is never exposed to a single environmental factor at a time but to a combination of all the factors which may act in association to synchronize the reproductive activities of the animal. Since the animals depend on the climatic factors, any deviation from their (climatic factors) natural period of occurrence will markedly affect their breeding cycles. If this keeps continuing the species might, one day, become extinct.

Therefore, occurrence of any climatic variable for a prolonged period may disturb the cycle of food availability for the animal e.g. the rainfall of July and September makes a difficult environment for *Funambulus pennanti* since during this period the fields are flooded with rain water and the animals are unable to find fruits, nuts, seeds etc. This brings the animal under a condition of stress. If this continues for longer durations, the breeding activity of the animals is also affected accordingly. This will prove hazardous for the survival of the offspring and eventually the species as a whole. Similarly, an opposite condition of drought can also adversely affect the reproductive activities of the animals. Any pressure of the environment may prove hazardous for the animals.

It has been suggested that the most important function of the pineal is to help the species to adapt to geographical and mateorological conditions. The environment to which the animal is exposed regulates its reproduction and the pineal gland is involved in this phenomenon.

Acknowledgement

Grateful acknowledgements are due to CSIR, New Delhi for financial support.

REFERENCES

Arendt, J., Paunier, L. and Sizonenko, P.C. (1975). Melatonin radioimmunoassay. J. Clin. Endocr. Metab., **40** : 347-350.

Bhatnagar, K.P., Heiko, D.F. and Stephan (1990). The megachiropteran pineal organ : A comparative morphological and volumetric investigation with special emphasis on the remarkably large pineal of *D. praedatrix*. J. Anat., **168** (O) : 143-166.

Brainard, G.C., Vaughan, M.K. and Reiter, R.J. (1984). The influence of artificial and natural short photoperiod on male Syrian hamsters : Reproductive effects. Int. J. Biometeorology, **28** : 317-326.

Brown, G., Grota, L. and Niles, L. (1981). Melatonin : Origin, control of circadian rhythm and site of action. In : Melatonin-Current Status and Perspectives. Birau, N. and Schloot, W. (eds.). Pergamon Press, New York.

Bruning, J.L. and Kintz, B.L. (1977). Computational Handbook of Statistics. Scott. Foresman and Co. Palo, Alto, Calif.

Carter, D.S., Hall, V.D., Tamarkin, L. and Goldman, B.D. (1982). Pineal is required for testicular maintenance in the Turkish hamster (*Mesocricetus brandti*). Endocrinology, **111** : 863-871.

Chan, Y.S., Yuen-Ming, C. and Pang, S.F. (1990) Patterns of pineal melatonin secretion in rabbits : Diurnal variation of basal and pulsatile release. Neuroendocrinol. **51** (2) : 147-155.

Czyba, J.C., Girod, C. and Durand, N. (1964). Sur l' antagonisme epiphysohypophysoire et les variations saisonniers de al spermatogenese chez le Hamster dore (*Mesocricetus auratus*). C.R. Soc. Biol. (Paris), **158** : 742.

DeLany, M.J. and Happold, D.C.D. (1979). Ecology of African Mammals (Longman, London).

Desjardins, C. (1979). Potential sources of variation affecting studies on pituitary gonad function. In : Animals models for research on contraception and fertility. Harper & Row, Hagerstown, pp. 13-32.

deVlaming, V.L. and Vodicnik, M.J. (1977). Effects of pinealectomy on pituitary gonadotrops, pituitary gonadotropin potency and phypothalamic gonadotropin releasing activity in *Notemigonus crysoleucas*. J. Fish. Biol., **10** : 73-86.

Farrar, G.M. and Clark, J.R. (1976). Effect of chemical sympathectomy and pinealectomy upon gonads of voles (*Microtus agrestis*) exposed to short photoperiod. Neuroendocrinology, **22** : 134-143.

Fenwick, J.C. (1970). The pineal organ. In : Fish Physiology Hoar, W.S. and Randall, R.J. (eds.), Academic Press, **4** : 91-108.

Fiske, V.M., Pound, J. and Putnam, J. (1962). Endocrinology, **71** : 130-133.

Goldman, B.D. and Darrow, J.M. (1983). The pineal gland and mammalian photoperiodism. Neuroendocrinology, **37** : 386-396.

Gwinner, E. (1981). Annual rhythms : Perspectives. Handbook of Behavioural Neurobiology. In : Biological Rhythms. Aschoff, J. (ed.), **4** : 381-389.

Haldar-Misra, C. (1986). Methods of pinealectomy in some vertebrates. Ind. J. Exptl. Biol. **24** : 319-322.

Hoffman, K. (1973). The influence of photoperiod and melatonin on testis size, body weight and pelage colour in the Djungarian hamster (*Phodopus sungorus sungorus*). J. Comp. Physiol. **85** : 267-282.

Hoffman, K. (1979). Photoperiod, pineal, melatonin and reproduction in hamsters. In : The pineal gland of vertebrates including man. Ariens-Kappers, J. and Pevet, P. (eds.). Elsevier, Amsterdam, pp. 397-415.

Hoffman, R.A and Reiter, R.J. (1965). Pineal gland. Influence on gonads of male hamsters. Science, **148** : 1609-1611.

Klein, D.C., Auerbach, D.A., Namboodiri, M.A.A. and Wheler, G.H.T. (1981). Indole metabolism in the mammalian pineal gland. In : The pineal gland. Reiter, R.J. (ed.) Vol. I, pp. 199-228.

Lofts, B. (1975). Animal photoperiodism. In : The Institute of Biology's Studies in Biology, Arnold, E. (ed.), Vol. 25, pp. 1-64.

Lofts, B. (1978). In : Environmental Endocrinology, Assenmacher, I. and Farner, D.S. (eds.) Springer-Verlag, 37-43.

Lowry, O.H., Rosenbrough, N.J., Farr, A.L. and Randall, R.J. (1951). Protein measurements with the folin phenol reagent. J. Biol. Chem., **193** : 265-275.

Lynch, H.J. (1971). Diurnal oscillations in pineal melatonin content. Life Sci., **10** : 791.

Lynch, H.J. , Wurtman, R.J., Koskowitz, M.A., Archer, M.C. and Ho, M.H. (1975). Daily rhythm in human urinary melatonin. Science. N.Y., **187** : 169-171.

Nir, I. Hirschmann, N.Mishkinsky, J. and Sulman, F.G. (1969). The effect of light and darkness on nucleic acids and protein metabolism of the pineal gland. Life Sci. **8** : 279-287.

Oksche, A. and Pevet, P. (1981). The Pineal Organ. Photobiology, Biochemistry and Endocrinology. Amsterdam Elsevier/North Holland. pp. 1-366.

Pang, S.F., Ralph, C.L. and Reilly, D.P. (1974). Melatonin in chicken brain : its origin, diurnal variation and regional distribution. Gen. Comp. Endocrinol. **22** : 499-506.

Pang, S.F., Brown, G.M., Grota, L.J., Chambers, J.W. and Rodman R.L. (1977). Neuroendocrinology. **23** : 1.

Panke, E.S., Reiter, R.J., Rollag, M.D. and Panke, T.W. (1978). Pineal serotonin N-acetyltransferase activity and melatonin concentrations in prepubertal and adult Syrian hamsters exposed to short daily photoperiods. Endocr. Res. Commun. **5** : 311.

Perry, J.S. and Rowlands, I.W. (1973). The environment and reproduction in mammals and birds. Blackwell Scientific Publications, Oxford.

Peter, R. and Hontela, A. (1978). Environmental Endocrinology. Assenmacher, I. and Farner, D.S. (eds.) pp. 20-25.

Petit, A. and Vivien-Roels, B. (1977). Les rhythmes circadiens et circannuels du taux de 5-hydroxytryptamine (serotonine) Chez *Lacerta muralis Laurenti* (Reptiles-Lacertiliens). Arch. Biol. **88** : 217-234.

Pevet, P. (1979). Secretary processes in the mammalian pinealocytes under natural and experimental conditions. In : The Pineal Gland of Vertebrates including Man. Kappers, J.A. and Pevet, P. (eds.). Prog. Brain Res., Elsevier, Amsterdam. **52** : 149-194.

Pevet, P. (1985). In : The Pineal Gland, Current State of Pineal Research. Mess, B., Ruzsas, C.S., Tima., L. and Pevet, P. (eds.). Akademia Kiado, Budapest and Elsevier Science Publisher, Amsterdam, pp. 163-186.

Pevet P. (1987). Environmental control of the annual reproductive cycle in mammals. Role of the Pineal Gland. In Comparative Physiology of Environmental Adaptations, Pevet, P. (ed.), 3, 8th ESCP Conf. Strasbourg 1986, Karger Basel, pp. 82-100.

Pevet, P. and Haldar-Misra, C. (1982 a). Effect of 5-methoxytryptamine on testicular atrophy induced by experimental or natural short photoperiods in the golden hamster (*Mesocricetus auratus*). J. Neural Transm. 55 : 69-84.

Pevet, P. and Haldar-Misra, C. (1982 b). Daily 5-methoxytryptamine injections inhibit short-day-induced testicular atrophy in the golden hamster. J. Neural Transm. **55** : 95-99.

Philo, R. and Reiter, R.J. (1980). A circannual rhythm in bovine pineal serotonin. Experientia. **36** : 664-665.

Prasad, M.R.N., Dhariwal, G.K., Seth, P., Reddi, A.N., Shivashankar, A.K. and Uberai, N.K. (1966). Biology of reproduction in the Indian palm squirrel, *F. pennanti*. In : Comp. Biol. Reproduction in Mammal. (Symp. Zool. Soc., Zool. Soc. London/Academic Press, London). No. **15** : 353-364.

Quay, W.B. (1956). Volumetric and cytologic variations in the pineal body of *Peromyscus leucopus* (Rodentia) with respect to sex, captivity and daylength. J. Morph. **98** : 471.

Quay, W.B. (1961). Reduction of mammalian pineal weight and lipid during continuous light. Gen. Comp. Endocrinol. **1** : 211-217.

Quay, W.B. (1963). Circadian rhythm in rat pineal serotonin and its modifications by estrous cycle and photoperiod. Gen. Comp. Endorinol. **3** : 473.

Quay, W.B. (1974). Pineal Chemistry, in Cellular and Physiological Mechanisms. Charles C. Thomas Publ., Springfield, III, Illinois.

Quay, W.B. and Halevy, A. (1962). Physiol. Zool. **35** : 1-7.

Raikhlin, N.T., Kvetnoy, I.M. and Tolkachev, V.N. (1975). Melatonin may be synthesized in enterochromaffin cells. Nature, Lond. **255** : 344-345.

Reiter, R.J. (1974). Circannual reproductive rhythms in mammals related to photoperiod and pineal function. A review, Chronobiologia. **1** : 365-394.

Reiter, R.J. (1978). Interaction of photoperiod, pineal and seasonal reproduction as exemplified by findings in the hamster. Prog. Reprod. Biol. **4** : 169-190.

Reiter, R.J. (1980). The pineal and its hormones in the control of reproduction in mammals. Endocr. Rev. **1** : 109.

Reiter, R.J. (1981). The Pineal Gland. Reproductive Effects. Boca Raton : CRC Press, Vol. II.

Rhodes, D.H. (1989). The influence of multiple photoperiods and pinealectomy on gonads, pelage and body weight in male meadow voles, *Microtus pennsylvanicus*. Comp. Biochem. Physiol. A. Comp. Physiol. **93** (2) : 445-450.

Rollag, M.D., O'Callaghan, P.L. and Niswender, G.D. (1978). Biol. Reprod. **18** : 279.

Rollag, M.D., Panke, E.S. and Reiter, R.J. (1980). Pineal melatonin content in male hamsters throughout the seasonal reproductive cycle. Proc. Soc. Exp. Biol. Med. **165** : 330.

Sackett, G.E. (1925). In : Practical Clinical Biochemistry. Varley, H. (ed). William Heinemann Medical Books Ltd. p. 309.

Scalera, G., Carlo, B., Pier, C.M., Giorgio, T. and Carlo, P. (1990). Circadian rhythms of drinking, feeding and body weight in pinealectomized rats. Neuroendocrinology Lett. **11** (6) : 325-332.

Sundararaj, B.I. and Garg, S.K. (1981). Pineal Function. Elsevier, North Holland Biomedical Press.

Tamarkin, L., Hollister, C.W., Lefebvre, N.G. and Goldman, B.D. (1977). Melatonin induction of gonadal quiescence in pinealectonized Syrian hamsters. Science, **198** : 953-955.

Thieblot, L. and Blaise, S. (1967). Principe anti-gonadotrope de la glande pineale. Rev. Roum. Endocrinol. **4** : 269.

Thorpe, P.A. and Herbert, J. (1976). Studies on the duration of the breeding season and photorefractriness in female ferrets pinealectomized or treated with melatonin. J. Endocrinol. **70** : 255-262.

Turek, F.W. and Campbell, C.S. (1979). Photoperiodic regulation of neuroendocrine-gonadal activity. Biol. Reprod. **20** : 32-50.

Vivien-Roels, B. and Petit, A. (1975). Dosage spectro-fluorimetrique de la serotonine (5-hydroxy-tryptamine = 5-HT) epiphysaire chez les Reptiles. Etude des variations du Taux de 5-HT en fonction du Cycle saisonneir chez Testudo hermanni. C.R. Acad. Sci. **280** : 467-470.

Vivien-Roels, B. and Pevet, P. (1983). The pineal gland and the synchronization of reproductive cycles with variations of the environmental climatic conditions with special reference to temperature. In : Pineal Research Review, New York, Alan R. Liss Inc. **1** : 91-143.

Vivien-Roels, B., Arendt, J. and Bradtke, J. (1979). Circadian and circannual fluctuations of pineal indolamines (serotonin and melatonin) in *Testudo hermanni*, Gmelin (Reptilia, Chelonia) I under natural conditons of photoperiod and temperaturjse. Gen. Comp. Endocrinol. **37** : 197-210.

Webley, G.E, Pope, G.S. Johnson, E. (1985). Seasonal changes in the testes and accessory reproductive organs and seasonal and circadian changes in plasma testosterone concentration in the male grey squirrel, *Sciurus carolinensis.* Gen. Comp. Endocrinol., **59** : 1-15.

Wilkinson, M., Arendt, J., Bradtke, J. and de Ziegler, D. (1977). Determination of a dark induced increase of pineal N-acetyltransferase activity and simultaneous radioimmunoassay of melatonin in pineal serum and pituitary tissue of the male rat. J. Endocr. **72** : 243.

Wurtman, R.J., Axelrod, J. and Phillips, L.S. (1963). Melatonin synthesis in the pineal gland : Control by light. Science, **142** : 1071.

Wolfe, D.E. (1965). The epiphyseal cell : an electron-microscopic study of its intercellular relationships and intracellular morphology in the pineal body of the albino rat. Progr. Brain Res. **10** : 332-376.

Chapter 21

Ecological Studies of Orissan Forest from Eastern Ghats

S. Basu, H.P. Dani* AND D.P. Pati

School of Life Sciences, Sambalpur University Burla-768 019, India

Abstract. An investigation was carried out to study the phytosociology, population structure, productive pattern and soil microbiological characteristics in three forest sites (Udiapur natural teak forest, Podhal reserve forest and Parapit mixed forest) of Phulbani district of Orissa, representing a part of the Eastern Ghat mountain range of India. Analysis of tree vegetation revealed *Tectona grandis, Shorea robusta* and *Madhuca indica* as the dominant species in the three forest sites. Densities of trees were 1570, 4070 and 4350/ha in Udaipur, Parapit and Podhal respectively. Total basal area was maximum (74.83 m^2/ha) in Podhal and minimum (19.97 m^2/ha) in Parapit. Tree species diversity (H) in three sites ranged from 1.913-3.355 with maximum in Parapit and minimum in Udaipur. But the dominance index for tree vegetation was least in Parapit and highest in Podhal. In all the sites, herbaceous diversity was maximum during rainy and minimum during summer season, whereas reverse trend was marked in case of species dominance index. Population analysis of dominance tree species revealed pyramidal structure in Podhal and Parapit. Inverted pyramidal population structure in Udiapur reflected poor regenerating potential due to biotic interference and adverse biodaphic conditions. The forest sites exhibited a annual production of 7.24 (Parapit) to 17.70 t/ha (Udaipur) to which the leaf litter was observed to be the major contributor. Besides, forest floor herbaceous and shrub layer also contributed significantly (26 to 40 per cent) to the total production. Annual turnover of litter in the sites ranged from 69 to 93 per cent with turnover time of 1.07 to 1.44 years.

Soil metabolic study in the sites indicated seasonal trends with peak soil respiration during rainy season. From the annual mean rate of soil respiration (199-337 mg $CO_2/m^2/hr$) contribution of microbial component was calculated. Annual microbial respiration amounted to 237 (in Parapit) to 399 g CO_2-C/m^2 (in Podhal). Total microbial biomass amounted to 0.229-0.869 mg C/g soil and this constituted 11 to 14 per cent of the total carbon input from the producer compartment. The percent of soil organic C reflected in microbial biomass was in the range of 1.06 to 1.28 per cent. Soil microfungal biomass constituted 70-83 per cent of the total microbial biomas. Annual microbial activity per unit microbial biomass was

* Boudh Panchayat College, Boudh.

comparatively higher in Parapit and Udaipur (2.21-2.34) and lower in Podhal (1.56). This reflected the relative functional efficiency of soil microbial component in different sites. Soil microbial component utilised 59 to 76 per cent of the total available substrate for maintenance and the rest for growth. This sustained a microbial turnover of 5 to 12 times per year with the generation time of 30 to 80 days and microbial productivity of 146 to 732 g C/m²/yr in mixed open canopy forest of parapit and mature sal forest of Podhal respectively.

Key words : *Phytosociology, Species diversity, Population, Production, Soil microbial biomass, Metabolism.*

Introduction

Ecological importance of the forest for maintenance of quality environment needs no further elaboration. On global basis, forest occupies a geographical area of 4000 million ha of which 70 per cent is in the tropic. Tropical forests are characterised as highly complex ecosystems which in the course of uninterrupted evolution over millions of years, have achieved a remarkable balance between plants, animals and the physical environment (Richards, 1977). The stable nature of these systems in mainly attributed to the high level of species diversity. Tropical forests contribute around 69 per cent of the earth's biological productivity (Brunig, 1974) which is due to the high rates of carbon and nutrient turnover in soil. In spite of the several important functional attributes, annually, an estimated 11.1 million ha of tropical forests and woodlands are destroyed or seriously degraded through agricultural expansion, logging and fuelwood collection (Parrotta, 1992).

The Indian subcontinent has a climate most suited to the forest (Puri, 1960). Hence, agriculture and forestry should develop side by side contributing to the sustenance of productive and protective nature of the environment. The complementary nature of the two are available from the Himalayan ecosystem studies (Singh *et al.*, 1984).

The Eastern Ghats are the important mountain ranges on the eastern coast of Peninsular India. The range has confluence with the Western Ghats at its southern extremity in Nilgiri regions. Goegraphers consider the Khondmal hills in the Boudh-Phulbani district as forming the northern extremity of the Eastern Ghats. The range supports mostly forest vegetation of either dry or moist deciduous type. Elements of sal and teak interrupted by the mixed type vegetation are the characteristic feature of Eastern Ghats. Although synecological information about the floristic elements of Eastern Ghats are available from Meher-Homji (1977) and Legris and Meher-Homji (1982), detailed structural and functional aspects of forest vegetation of northern extremity of Eastern Ghats has not received much attention. During recent years, a rapid increase in human settlements with subsequent urbanisation in this area has resulted in deforestation and conversion of natural forest of savanna/scrubland. The conversion of the forest has led to several changes in soil properties (Basu and Behera, 1993). Unless the forests of the area are managed properly, the area will soon be facing several environmental problems of severe dimensions. The present paper deals with the phytosociological informations, productive pattern and soil microbiological characteristics of three forest sites of Phulbani district of Orissa in order

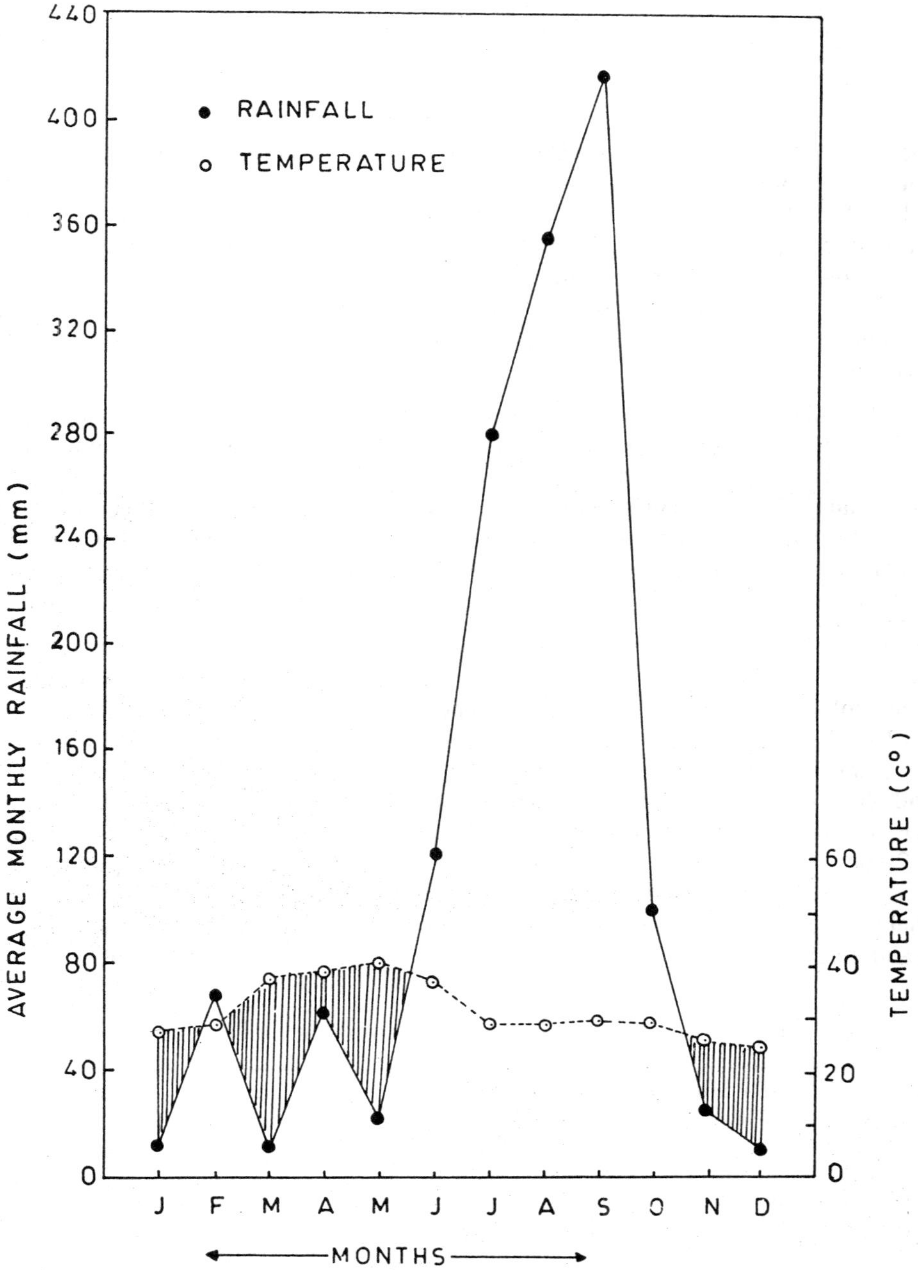

Fig. 21.1. Ombrothermic diagram of Phulbani district.

to make an assessment of the ecological status of these forests.

Sites

The three different forest sites (1) Udaipur natural teak forest (2) Podhal reserve forest and (3) Parapit mixed forest selected for the present study were located between 19°34' to 24°54' N latitude and 83°30' to 84°48' E longitude. The Udaipur forest site is very close to the river Tel, a tributary of the river Mahanadi. The Podhal reserve forest is one of the oldest and largest natural forest located close to a stream and represents a riverine forest. The forest of Parapit is a disturbed forest site where afforestation by the forest department of Government of Orissa has been taken up. The site at many places is comparable with a scrubland because of human interference.

The climate of the area is tropical monsoonal with three distinct seasons : summer (March-May), rainy (June-October) and winter (November-February). The average annual rainfall of the sites is around 1400 mm, of which 70 per cent falls during rainy season. Monthly mean maximum and minimum temperature varied from 24°-39°C and 8°-24°C respectively with May as the hottest and December as the coldest month. Ombrothermic diagram of the area has been represented in Fig. 21.1. As revealed from the figures the wet period extends from June to mid October and for the rest of the period of the year, the area experiences a dry climate.

The soil of Udaipur forest is alluvial in nature having relatively more amount of silt content. The Podhal forest have grey soil, which is derived from the Charnockite type of parent rock. The grey colour of the soil is due to the presence of magnesium in it. The soil being relatively clayey has high water retaining capacity. The soil of parapit is of sandy loamy texture. Sandy soils are derived from gneisses of granites and mainly composed of quartz and clay. Some of the soil physico-chemical characteristics of the study sites are presented in Table 21.1. From the table it is evident that sand constituted 70 to 77 per cent of soil. Percentage of silt ranged from 15 to 20 per cent with the minimum value in Parapit and maximum in Udaipur natural forest. The clay content varied from a minimum of 8 per cent (Parapit) to a maximum of 12 per cent (Podhal forest). The soil water holding capacity was minimum on Parapit mixed forest (40 per cent) and maximum in Podhal reserve forest (51%). Except Udaipur, in the other two sites, the

TABLE 21.1. *Soil physico-chemical characteristics of different study sites.*

Sites	Sand + Silt + Clay	Bulk density (g/cm^3)	Water holding capacity (%)	pH	Organic carbon (%)	Total nitrogen (%)	C : N ratio
Udaipur	70 + 20 + 10	1.5	42	7.8	6.2	0.47	13
Podhal	70 + 18 + 12	1.7	51	6.8	8.1	0.74	11
Parapit	77 + 15 + 8	1.4	40	6.3	4.7	0.24	19

TABLE 21.2. *Analysis of tree vegetation of different forest sites.*

Species	Frequency	Density (No/ha)	Abundance	Total basal area (m^2/ha)	IVI
		UDAIPUR			
Acacia leucophloea Willd.	20	20	1.0	0.535	6.9
Bombax ceiba Linn.	10	10	1.0	0.963	4.6
Butea monosperma (Lamk) Taub.	80	300	3.75	3.818	44.5
Cassia fistula Linn.	20	20	1.0	2.215	9.9
Lagerstroemia parviflora Roxb.	30	40	1.3	0.154	9.7
Pterocarpus marsupium Roxb.	60	100	1.7	12.230	42.3
Shorea robusta Gaertn.	20	30	1.5	1.237	8.8
Soymida febrifuga Adr. Juss.	20	20	1.0	2.104	9.8
Tectona grandis Linn. f.	100	900	9.0	27.510	129.5
Terminalia arjuna Bedd.	20	40.	2.0	4.581	15.30
Terminalia tomentosa W.et. A.	10	10	1.0	0.336	3.5
Zizyphus xylopyrus Willd.	40	80	2.0	0.254	14.8
		PODHAL			
Adina Crodifolia Hook. f.	70	90	1.28	10.30	20.74
Anogeissus latifolia Wall.	80	440	5.5	2.18	18.66
Buchanania lanzan Spr.	60	110	1.83	0.28	7.12
Brideliaretusa Spr.	30	30	1.0	1.01	4.11
Careya arborea Roxb.	30	30	1.0	1.93	5.27
Cassia fistula Linn.	30	30	1.0	1.93	5.27
Cleistanthus collinus Benth.	50	50	1.0	0.08	4.79
Dalbergia paniculata Roxb.	30	30	1.0	3.43	7.39
Dillenia pentagyna Roxb.	30	30	1.0	0.59	3.59
Dendrocalamus strictus Nees.	30	30	1.0	5.36	9.96
Diospyros melanoxylon Roxb.	40	80	2.0	5.15	11.59
Emblica officinalis Gaertn.	30	40	1.33	0.34	3.49
Elaeodendron glaucum Pers.	40	40	1.0	0.19	3.05
Flemingia chappar Pers.	40	60	1.5	–	4.19
Grewia hirsuta Vahl.	30	40	1.33	–	3.03
Gardenia latifolia Ait.	30	30	1.0	–	2.80

(Table 21.2 Contd)

Species	Frequency	Density (No/ha)	Abundance	Total basal area (m^2/ha)	IVI
Holarrhena antidysenterica Wall.	60	90	1.5	0.08	6.39
Indigofera pulchela Roxb.	20	20	1.0	0.06	1.53
Kydia calycina Roxb.	40	40	1.0	–	3.73
Lagerstroemia parviflora Roxb.	40	40	1.0	0.19	3.19
Lannea coromandelica (Houtt.) Merr.	30	30	1.0	2.38	5.98
Madhuca indica Gmel	50	90	1.8	7.15	15.18
Ougeinia oojeinensis (Roxb) Hochr.	30	50	1.66	1.43	5.17
Phoenix acaulis Buch.	60	120	2.0	–	6.98
Pterocarpus marsupium Roxb.	70	120	1.71	2.86	1.48
Schleicsera oleosa (Lour.) Oken.	30	30	1.0	0.86	2.83
Semecarpus anacardium Linn. f.	30	40	1.33	0.07	3.13
Shorea robusta Gaertn.	100	2100	21.0	11.45	71.61
Terminalia tomentosa W. et A.	70	240	3.43	15.46	31.09
Woodfordia fruticosa (L) Kurtz.	60	90	1.5	–	6.29
Xeromphis spinosa (Thunb.) Keoy.	40	40	1.0	0.07	2.83
Zizyphus xylocarpus Willd.	40	50	1.25	–	3.96
		PARAPIT			
Acacia catechu Willd.	10	50	5.0	0.99	12.09
Acacia leucophloea Willd.	10	10	1.0	0.47	2.99
Adina cordifolia Hook. f.	20	20	1.0	0.38	3.81
Aegle marmelos Corr.	15	15	1.0	0.95	5.41
Alangium salvifolium (L.f) Wang.	5	20	4.0	0.03	0.92
Anogeissus latifolia Wall.	45	90	2.0	0.28	8.168
Azadirachta indica Juss.	20	30	1.5	0.05	2.97
Bauhinia purpurea Linn.	25	30	1.2	0.85	6.28
Bombax ceiba Linn.	25	25	1.0	4.47	19.87
Bridelia retusa Spr.	20	25	1.25	0.08	2.88
Buchanania lanzan Spr.	35	50	1.43	0.13	5.23
Butea monosperma (Lamk) Taub.	15	15	1.0	0.02	1.4

(Table 21.2 Contd)

Species	Frequency	Density (No/ha)	Abundence	Total basal area (m^2/ha)	IVI
Careva arborea Roxb.	15	15	1.0	0.42	3.42
Cassia fistula Linn.	15	15	1.0	0.02	1.90
Chloroxylon swietenia D.C.	20	25	1.25	0.12	3.05
Cleistanthus collinus Benth.	100	400	4.0	0.36	24.63
Diospyros melanoxylon Roxb.	75	100	1.66	0.18	14.53
Diospyros montana Roxb.	60	225	3.75	0.18	14.11
Emblica officinalis Gaertn.	30	40	1.33	0.20	4.08
Gardenia latifolia Ait.	35	40	1.14	–	4.38
Garuga pinnata Roxb.	15	15	1.0	0.96	3.69
Grewia tiliaefolia Vahl.	15	15	1.0	0.04	1.98
Holarrhena antidysenterica Wall.	100	400	4.0	0.26	24.29
Ixora parviflora Vahl.	55	110	2.0	–	8.66
Lagerstroemia parviflora Roxb.	30	40	1.33	0.12	4.42
Lannea coromandelica (Houtt) Merr.	15	20	1.33	0.94	5.57
Limonia acidissima Linn.	15	15	1.0	0.07	2.08
Madhuca indica Gmel.	30	40	1.33	0.95	25.83
Morinda tinctoria Roxb.	20	25	1.25	0.71	5.28
Phoenix acaulis Buch.	30	50	1.66	–	4.34
Schleicsera oleosa (Lour) Oken.	15	20	1.33	0.57	4.13
Semecarpus anacardium Linn.	20	25	1.25	0.06	2.28
Shorea robusta Gaertn.	50	110	2.2	0.54	10.28
Sterculia urens Roxb.	10	10	1.0	0.84	3.61
Tamarindus indica Linn.	10	10	1.0	3.18	13.19
Tectona grandis Linn. f.	20	100	5.0	0.18	10.54
Terminalia tomentosa W.et A.	40	75	1.87	0.37	7.51
Woodfordia fruticosa Kurtz.	75	75	1.0	–	8.98
Xeromphis spinosa (Thunb) Keoy.	15	20	1.33	–	1.96
Zizyphus xylocarpus Willd.	50	75	1.5	–	6.93

pH of the soil was in acidic range. The amount of organic carbon varied from a minimum of 4.7 per cent (parapit mixed forest) to a maximum of 8.1 per cent (Podhal reserve forest). Total soil nitrogen ranged from 0.24 per cent in Parapit mixed forest to 0.74 per cent in Podhal reserve forest. The range of C : N ratio was 11 to 19 with minimum ratio in Podhal and maximum in Parapit forest.

Phytosociology

In order to get a correct ecological picture of the area, detailed phytosociological analysis of the vegetation were made setting 20 random quadrats (10 × 10 m) in each of the sites. Vegetational data were quantitatively analysed for frequency, density and basal area (Misra, 1968). Relative frequency, relative density and relative basal area were analysed following Phillips (1959). The Importance Value Index (IVI) for individual species was calculated as the sum of the relative frequency, relative density and relative dominance (Curtis, 1959).

Phytosociological analysis of the vegetation of these sites are presented in Table 21.2. On the basis of the Importance Value Indices (IVI), *Tectona grandis, Shorea robusta and Madhuca indica* were the dominant species in Udaipur, Podhal and Parapit forest sites respectively. The main associates of *T. grandis* were *Butea monosperma* (IVI = 44.5) at Udiapur, *Terminalia tomentosa* (IVI = 31.09) at Podhal and *Cleistanthus collinus* (IVI = 24.63) at Parapit mixed forest site. Considering the major contribution of different tree species to the total IVI, the study sites are classified into the following vegetational types.

Udaipur : *Toctona-Butea-Pterocarpus*

Podhal : *Shorea-Terminalia-Adina*

Parapit : *Madhuca-Cleistanthus-Holarrhena*

Total density, total basal area of tree species and contribution of some dominant species to total basal area have been mentioned in Table 21.3. Total density showed a range of 1570 to 4350/ha. Singh and Misra (1978) reported a tree density range of 936 to 1174 no./ha in a mixed deciduous forest of Uttar Pradesh. Thus the tree density values observed in the present study are in much higher range than that of Singh and Misra (1978). Total basal area varied from 19.97 to 74.83 m^2/ha with minimum in Parapit and maximum in Podhal forest. Total basal area estimated in the present study is comparable with the values obtained that of Tewari (1982) and Sexena and Singh (1982) for some Central Himalayan forests. However, total basal area estimate was observed to be more than that of Bandhu (1970), Singh (1974), Singh and Misra (1978) and Rout and Gupta (1989) for some deciduous forests of India and less than the basal area estimate of tropical rain forest (Singh *et al.*, 1984).

The relationship between the density and basal area showed a negative trend, confirming observations of Chaturvedi and Singh (1987) for some Central Himalayan pine forest. In the present study, low total basal area coincieded with relatively higher tree density in Parapit, which

TABLE 21.3. *Total density, total basal area of tree species and contribution of some dominant species to total basal area in different forest sites.*

Sites	Total density (no/ha)	Total basal area (m^2/ha)	Species	Contribution %
Udaipur	1570	55.94	*Tectona grandis*	49
			Pterocarpus marsupium	22
			Butea monosperma	7
Podhal	4350	74.83	*Terminalia tomentosa*	21
			Shorea robusta	15
			Adina cordifolia	14
			Madhuca indica	10
			Dendrocalamus strictus	7
Parapit	4070	19.97	*Bombax ceiba*	17
			Tamarindus indica	12
			Acacia catechu	4

was due to relatively more tree individuals in lower girth classes. On the contrary, in Udaipur in spite of lower tree density, higher total basal area was observed because of distribution of more tree individuals in higher girth classes. Since low tree density and high total basal area is the characteristic featuré of relatively mature/climax forest (Adhikari *et al.*, 1991), Udaipur forest may be considered close to mature/climax stage In Udaipur, maximum contribution to total basal area came from *Tectona grandis* whereas in Parapit and Podhal, the maximum contribution came from *Bombax ceiba* and *Terminalia tomentosa*. However, in Podhal and Parapit, no single species contributed significantly to total basal area.

Phytosociological analysis of forest floor herbaceous vegetation with respect to summer, rainy and winter seasons are presented in Table 21.4. Intercomparison of the number of herbaceous species revealed the presence of maximum number of species during the rainy season. *Atylosia scarabaeoides* was observed to be the dominant herbaceous species during summer and rainy seasons in Udiapur and Parapit. However, in Podhal, the dominant herbaceous species was *Hemidesmus indicus* during summer and *Costus speciosus* during rainy.

Species diversity and dominance

Species diversity (H) of tree and herbaceous vegetation of the three sites was calculated following Shannon and Weaver (1963) and concentration of dominance (cd) was calculated following Simpson (1949) (Table 21.5). Tree diversity was maximum in Parapit (H = 3.355) and minimum in Udiapur (H = 1.913). The dominance index for the tree vegetation however was

TABLE 21.4. *Lists of herbaceous species and their IVI values in different study sites.*

Species	Summer	Rainy	Winter
	UDAIPUR		
Achyranthus aspera Linn.	--	2.62	4.62
Ageratum conyzoides Linn.	--	11.92	24.78
Atylosia scarabaeoides (linn.) Benth.	156.74	28.63	42.72
Cassia occidentalis Linn.	--	8.44	--
Cassia tora Linn.	--	14.48	--
Chrysopogon aciculatus Trin.	--	2.28	--
Cleome viscosa Linn.	--	8.98	12.34
Cynodon dactylon (Linn.) Pers.	--	12.72	46.24
Eragrostis viscosa (Retz) Trin.	59.96	8.34	28.39
Euphorbia hirta Linn.	--	16.73	18.28
Evolvulus alsinoides Linn.	--	19.28	28.37
Hemidesmus indicus (Linn.) Schultz.	--	22.72	--
Justicia simplex D. Don.	--	21.28	--
Malvastrum tricuspidatum A. Gray.	--	12.36	2.28
Phyllanthus niruri Linn.	--	7.78	--
Plumbago zeylanica Linn.	--	18.48	2.34
Sida cordifolia Lamk.	83.62	13.86	28.15
Spermacocea hispida Linn.	--	6.23	--
Sphaeranthus indicus Linn.	--	1.28	--
Tephrosia purpurea Pers.	--	16.48	18.24
Tridax procumbens Linn.	--	8.28	12.34
Triumfetta rhombiodea Jacq.	--	5.82	--
Vendellia crustacea Linn.	--	18.28	7.43
Vicoa indica (Willd) DC	--	13.17	23.27
	PODHAL		
Adiantum caudatum Linn.	--	1.28	--
Andrographis paniculata Linn.	--	4.28	16.23
Ageratum conyzoides Linn.	--	8.26	21.18
Asparagus racemosus Willd.	37.97	21.36	4.28

(Table 21.4 Contd)

Species	Summer	Rainy	Winter
Atylosia scarabaeoides Linn.	114.16	8.23	32.43
Costus speciosus Smith.	--	35.28	--
Curculigo orchioides Gaertn.	--	2.32	--
Curcuma angustifolia Roxb.	--	32.28	--
Cynodon dactylon (Linn) Pers.	--	21.33	32.37
Cyperus rotundus Linn.	--	8.23	12.23
Dioscorea bulbifera Linn.	--	12.48	--
Dioscorea pentaphylla Linn.	--	8.23	--
Eragrostis viscosa (Retz). Trin.	--	14.34	16.38
Evolvulus alsinoides Linn.	--	21.46	24.32
Heliotropium indicum Linn.	12.14	2.28	18.37
Hemidesmus indicus (Linn.) Schultz.	136.18	8.28	6.23
Heteropogon contortus (Linn.) P.	–	18.42	32.72
Ischaemum angustifolium Hack.	–	8.82	14.23
Plumbago zevlanica Linn.	–	15.34	18.48
Pteris sp.	–	3.85	–
Selaginella sp.	–	2.34	–
Spermacocea hispida Linn.	–	8.34	6.32
Sphaeranthus indicus Linn.	–	6.28	2.18
Tephrosia purpurea Pers.	–	4.28	9.16
Tridax procumbens Linn.	–	2.38	6.28
Triumfetta rhomboidea Jacq.	–	8.23	–
Vicoa indica (Willd) DC	–	12.28	26.72
	PARAPIT		
Achyranthus aspera Linn.	--	4.62	1.23
Andrographis paniculata Linn.	--	8.63	12.37
Ageratum conyzoides Linn.	--	11.92	16.82

(Table 21.2 Contd)

Species	Summer	Rainy	Winter
Asparagus racemosus Willd.	48.37	8.44	2.37
Atylosia scarabaeoides Linn.	129.82	28.98	37.86
Cassia occidentalis Linn.	--	14.48	18.61
Cassia tora Linn.	--	18.72	23.62
Chrysopogon aciculatus Trin.	--	7.28	8.85
Cleome viscosa Linn.	--	6.73	8.63
Curculigo orchioides Gaertn.	--	2.28	2.32
Cynodon dactylon (Linn) Pers.	--	18.34	24.48
Dioscorea bulbifera Linn.	--	1.28	--
Eragrostis viscosa Trin.	59.9	12.36	14.51
Euphorbia hirta Linn.	--	7.78	4.28
Evolvulus alsinoides Linn.	--	28.48	12.37
Hemidesmus indicus (Linn.) Schultz.	22.83	6.86	18.13
Justicia simplex D. Don	--	6.36	9.23
Oldenlandia corymbosa Linn.	--	2.82	--
Ocimum basilicum Linn.	--	6.48	6.72
Phyllanthus niruri Linn.	--	8.28	12.46
Plumbago zeylanica Linn.	--	18.17	14.42
Sida cordifolia Lamk.	39.29	5.28	4.28
Spermacocea hispida Linn.	--	1.28	--
Tephrosia purpurea Pers.	--	24.72	19.63
Tridax procumbens Linn.	--	4.37	8.32
Triumfetta rhomboidea Jacq.	--	17.63	--
Vendellia crustacea Linn.	--	3.71	--
Vicoa indica (Willd.) DC	--	14.28	18.73

TABLE 21.5. *Diversity and dominance indices of tree and herb species in different forest sites.*

Sites	Species diversity (H)		Concentration of dominance (cd)	
	Tree	Herb	Tree	Herb
Udaipur	1.913	2.097	0.238	0.390
Podhal	2.819	2.051	0.872	0.368
Parapit	3.355	2.389	0.044	0.276

observed to be least in Parapit and highest in Podhal. The indices were observed to be higher than the average values for tropical forest (cd = 0.06, Knight, 1975). In all the three sites, herbaceous diversity was maximum during rainy and minimum during summer season, whereas the reverse trend was marked in case of species dominance index. Annual mean herbaceous diversity was maximum in Parapit (H = 2.389) and minimum in Podhal forest (H = 2.051). But the annual mean dominance was estimated to be maximum in Udaipur and minimum in Parapit. The study in general indicated the trend of inverse relationship between the diversity and dominance supporting the views of Mc Naughton (1967), Singh and Misra (1969), Murthy and Pathak (1972) and Joshi and Behera (1991). The relationship between the dominance and the total basal area of three sites showed a positive correspondence with each other. Since, basal area is an index of biomass, it can be concluded that increase in species dominance results in an increase in stand biomass in tropical forest. This can be attributed to the absence of species competition in the least diversified tropical forest showing maximum dominance index.

Dominance-diversity curves have been repeatedly used to interpret the community organisation in terms of resource share and niche space (Whittacker, 1969, 1975; Saxana and Singh, 1982). The dominance-diversity curves (Fig. 21.2) for tree species in three different study sites were found to approach normal distribution model of Preston (1948) with less number of species in higher IVI range. In Udaipur, the top niche was occupied by *Tectona grandis. Butea monosperma* and *Pterocarpus marsupium* occupied the intermediate niche. In the Podhal and Parapit forests, *Shorea robusta* and *Madhuca indica* occupied the top niche. The intermediate niche was occupied by *Terminalia tomentosa Adina cordifolia* and *Anogeissus latifolia* at Podhal and *Cleistanthus collinus* and *Holarrhena antidysenterica* at Parapit.

Figure 21.3 illustrates one of the example of dominance-diversity curve for herbaceous species during summer, rainy and winter season in forest. During summer, the low diversified herbaceous community approached geometric series. But during rainy season, with the increase in herbaceous diversity the dominance-diversity curve approached log normal distribution. Thus along with the variation of herbaceous diversity, there occurs a variation in dominance-diversity curve for herbaceous community on the tropical forest floor.

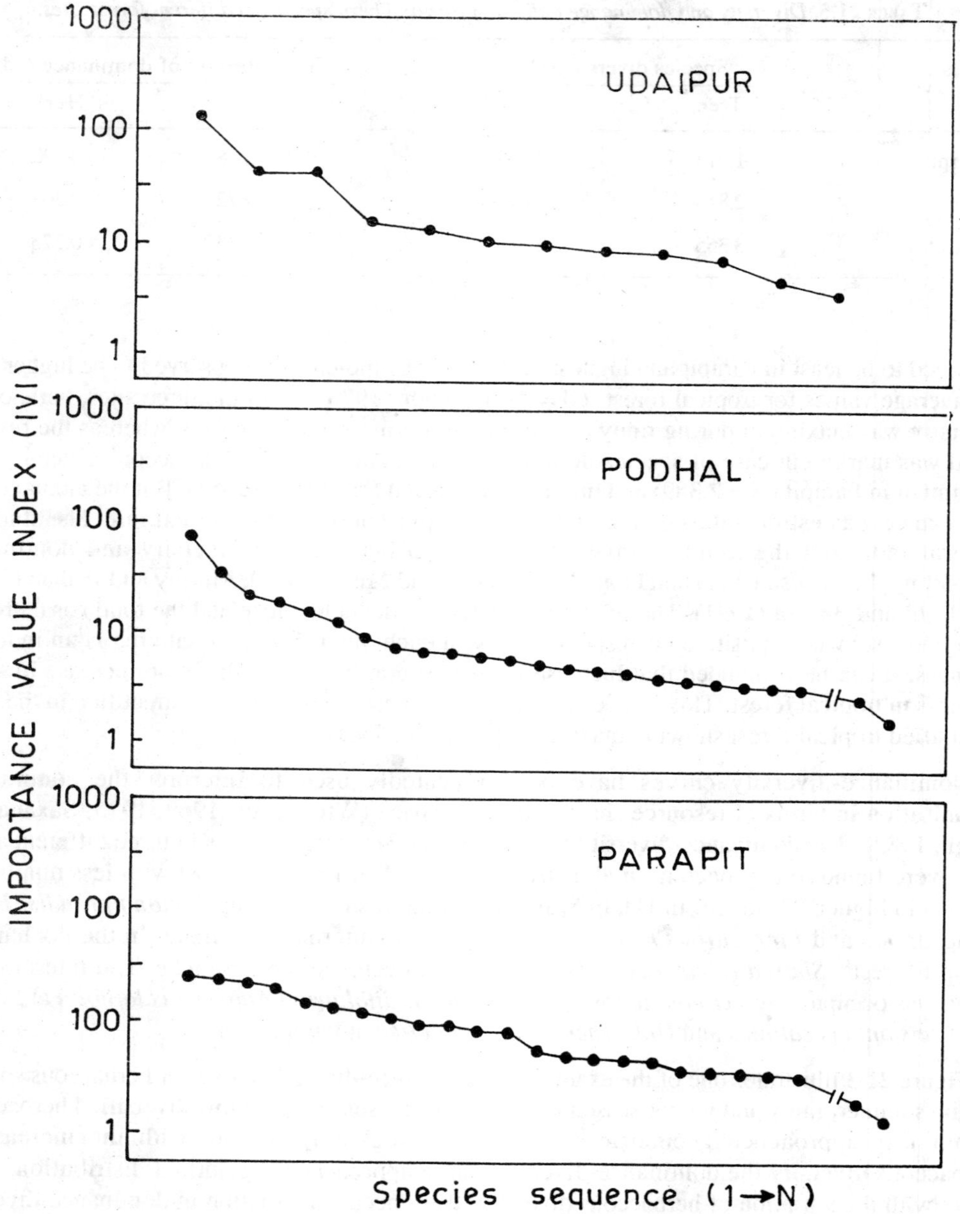

Fig. 21.2. Dominance-Diversity curve for tree species in three different forest sites

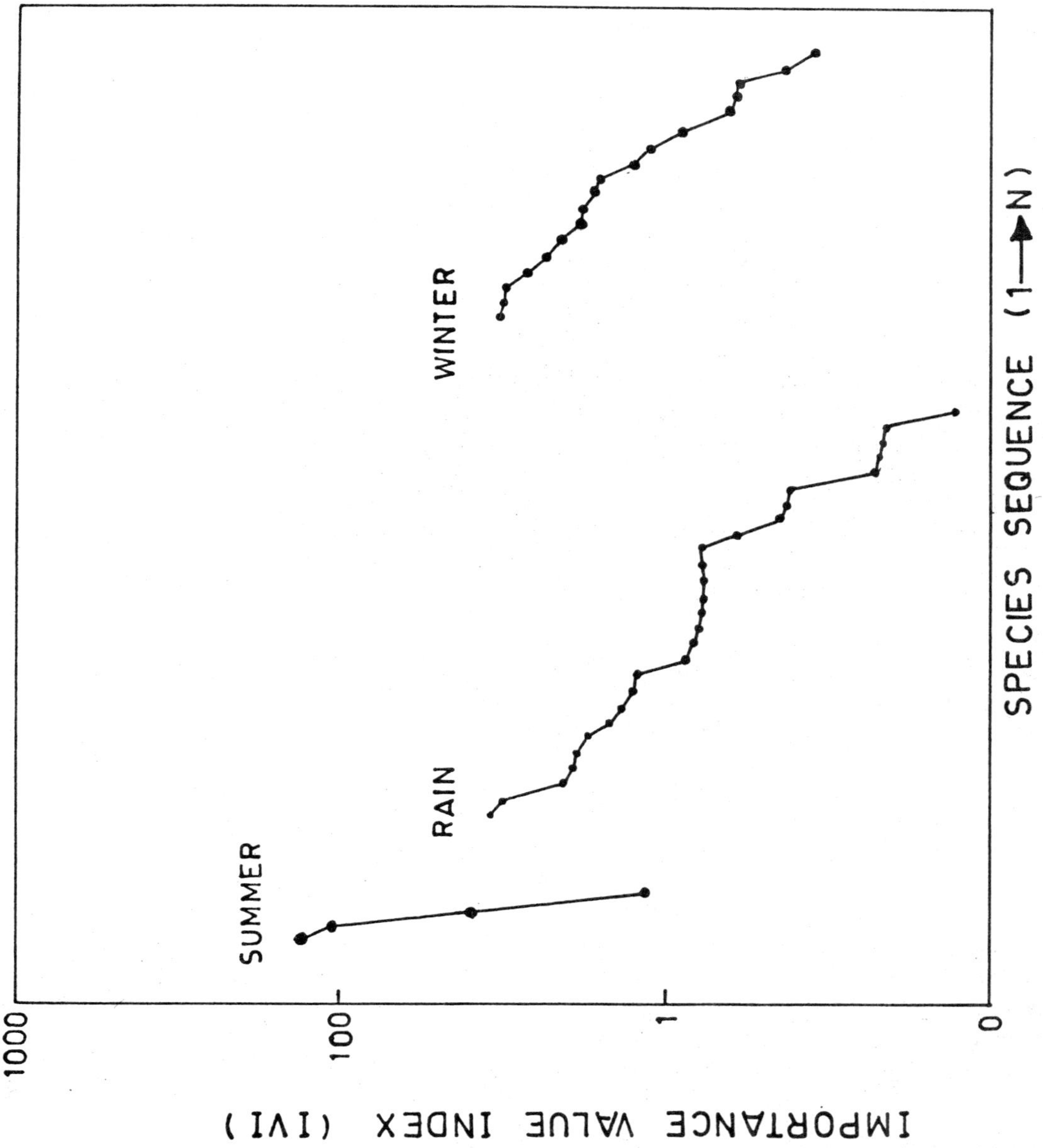

Fig. 21.3. Dominance-Diversity curve for herbaceous species during different seasons in Podhal forest

TABLE 21.6. *Vegetational Biomass of different forest sites.*

Sites	Tree biomass (t/ha)	Herb biomass (t/ha)	Shrub biomass (t/ha)	Total biomass (t/ha)
Udaipur	133.13	1.84 ± 0.11	2.80 ± 0.42	137.77
Podhal	185.43	1.96 ± 0.21	2.55 ± 0.45	189.94
Parapit	98.43	1.91 ± 0.16	1.00 ± 0.36	101.34

Biomass

Biomass estimate is also needed for a thorough understanding of the structure and dynamic functions of forest ecosystem (Ovington, 1962). From the total basal area, tree biomass was calculated for different forest sites (Table 21.6) following the equation of Brown *et al.* (1988; quoted by Anderson and Ingram, 1989).

$$Y = 34.4703 - 8.0671\,D + 0.6589\,D^2$$

Y = biomass (in Kg)

D = diameter at breast height (in cm)

Relatively low tree biomass in Parapit forest may be attributed to the presence of tree individuals confined to lower girth class. The biomass estimates of the present study are lower than that of tropical rain forest and Himalayan forest but within the range of Singh (1974), Chaturvedi and Singh (1987) and Singh and Yadava (1991). The total biomass comprising of tree, herb and shrub biomass was in the range of 101.34 to 189.94 t/ha.

Population

Analysis of population structure of different tree species in a forest community provides information about their status and biotic interactions. Besides, the analysis can also be extended for the evaluation of regenerating capacity of the forest. On the basis of the cbh measurement, the population structure of trees was prepared recognising girth classes (Ralhan *et al.*, 1982). The total number of individuals belonging to A-G girth classes was calculated species-wise. From these data, percent density for each girth class was calculated for various tree species as :

$$\text{Percent density} = \frac{\text{No. of individuals in each girth class}}{\text{Total no. of individuals in all the girth classes}} \times 100$$

Percent density distribution of some important tree species in different girth classes has been

TABLE 21.7. *Percent density distribution of different tree species in different girth classes*

Tree species	Udaipur							Podhal							Parapit						
	A	B	C	D	E	F	G	A	B	C	D	E	F	G	A	B	C	D	E	F	G
Shorea robusta	18	–	–	82	–	–	–	50	23	12	12	2	1	–	45	45	–	10	–	–	–
Terminalia tomentosa	–	–	–	–	100	–	–	20	9	6	48	17	–	–	50	30	–	20	–	–	–
Tectona grandis	15	16	29	31	4	5	–	–	–	–	–	–	–	–	–	100	–	–	–	–	–
Buchanania lanzan	–	–	–	–	–	–	–	75	–	–	25	–	–	–	–	87	13	–	–	–	–
Butea monosperma	–	–	–	–	58	42	–	–	–	–	–	–	–	–	–	–	–	–	–	–	–
Anogeissus latifolia	–	–	–	–	–	–	–	–	68	27	5	–	–	–	20	80	–	–	–	–	–
Pterocarpus marsupium	–	–	–	20	80	–	–	70	15	8	7	–	–	–	–	–	–	–	–	–	–
Madhuca indica	–	–	–	20	80	–	–	5	–	–	16	44	–	35	–	–	–	–	–	–	–
Diospyros melanoxylon	–	–	–	–	–	–	–	40	60	–	–	–	–	–	60	20	20	–	–	–	–
Holarrhena antidysenterica	–	–	–	–	–	–	–	44	51	5	–	–	–	–	54	46	–	–	–	–	–
Cleistanthus collinus	–	–	–	–	–	–	–	40	60	–	–	–	–	–	42	58	–	–	–	–	–
Adina cordifolia	–	–	–	–	–	–	–	–	9	21	39	31	–	–	–	–	80	20	–	–	–

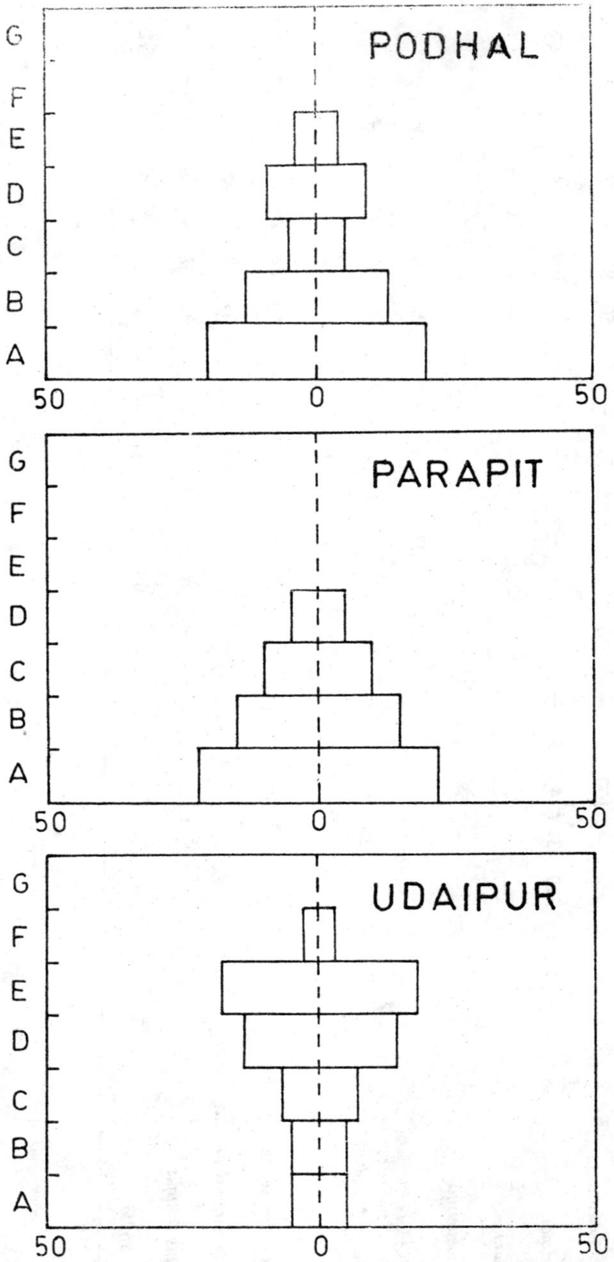

Fig. 21.4. Sum total population structure of tree species in different forest sites

presented in Table 21.7. The study indicated that in most of the species, either there is absence or poor representation of lower and higher girth classes. These types of incomplete population structure can be due to (1) failure of germination and (2) biotic interference. Absence of *Anogeissus latifolia* seedling in Podhal was due to the failure of germination because of the inhibitory effect of profuse germination of *Shorea robusta* (Legris and Meher-Homji, 1982). In the study sites, grazing was observed to be one of the important biotic factor, which affected the population structure of *Shorea robusta, Terminalia tomentosa Dalbergia sisoo, Zizyphus sp.* etc.

Sum total population structure of all tree species with respect to three different forests (Fig. 21.4) revealed upright pyramidal structure in Parapit, and to some extent in Podhal with maximum percent density of girth class A. In these two sites, a gradual decline in percent density of individuals from girth class A to E was marked. This decline explained normal mortality of individual during their transformation from lower to higher girth classes. But in Udaipur the structures were comparable with more or less inverted pyramidal type, which indicated relatively poor survival of tree individual in lower girth classes. This reflected poor regenerating potential of Udaipur forest.

Production

Productivity of forest ecosystem includes annual increments of boles, branches and its parts, leaves and underground parts of tree species. Besides, the forest floor herbaceous vegetation also contributes to the total production of forest ecosystem (Singh and Singh, 1980; Chaturvedi and Singh 1987). Because of the higher turnover of the herbaceous litter, herbaceous stratum is also expected to have a major influence on the nutrient dynamics in the forest ecosystem. Besides herb, shrub species present on the forest floor also do contribute significantly to the forest stand production. In deciduous forest, tree species have the characteristics of shedding their leaves, twigs, flowers, cones and fruit annually which is termed as litter fall. On the basis of data presented by Bray and Gorham (1964) annual litter fall is considered to represent one third of total annual net primary production in forest ecosystem. In the Present study, litter production, shrub and herbaceous production was investigated.

Litter production

To estimate the monthly rate of litterfall, 10 litter traps (1 × 1 × 0.25 m) were randomly placed in each site and litter materials fallen in these traps were collected at monthly intervals. Litter materials after collection from the traps were sorted into two categories (1) leaf (2) twig and miscellaneous (which includes flower, fruit and bark materials). Leaf litter were further separated into different species-wise (considering dominant tree species of the site). Different categories of litter after collection were over dried at 85°C for 24 h and weighed to determine their dry mass.

Tectona grandis was estimated to contribute 31 to 40 per cent of the total litterfall in Udaipur and Parapit forests. In Podhal, *Shorea robusta* was the major contributor contributing 42 per cent to total litter production. In all the sites, leaf litter was observed to be a major constituent of total litter production. Leaf litter was estimated to constitute 70-80 per cent of the total litter. Leaf litter as a significant percentage of total litter has also been reported by various workers in Indian forests (Chaturvedi and Singh, 1987; Rout and Gupta, 1990).

When litter production data of individual tree species along with their IVI was taken into consideration, it was observed that except Parapit, the major contribution to litter production was made by the species exhibiting maximum IVI. In Parapit, the dominant contributor to litter production was *Tectona grandis*. But *T. grandis* was not the dominant species on the basis of IVI. Thus in this site, there was no correspondence between synecological characteristics and litter production.

Annual litter production of Parapit, Podhal and Udaipur forests were 4.33, 9.72 and 13.06 t/ha respectively. The ratio of litter production : biomass reflects the production potentialities of biomass in different sites. The ratio for Udaipur, Podhal and Parapit forests were 0.098, 0.052 and 0.044 respectively. This indicated more production per unit bomass in Udaipur forest.

Hebaceous and Shrub production

Herbaceous and shrub production were estimated by hervesting peak herbaceous and shrub foliar biomass laying 10 random quadrats. The collected materials were then over dried to determine dry biomass.

Table 21.8 represents herbaceous and shrub production in different study sites. The range of above ground herbaceous production estimated in the present study (1.84-1.96 t/ha/yr) is comparatively lower than that of Joshi *et al.* (1990) for a regenerating forest of this region. Maximum herbaceous production was estimated in Podhal and minimum in Udaipur. After the establishment period of vegetation ground vegetation production declines logarithmically in relation to an increase in tree cover (Ford and Newbould, 1977). This justifies the lower range of above ground herbaceous production in the matures forest of Udaipur. However, the range was observed to be higher than that of Singh and Singh (1980) and Rout and Gupta (1990). Shrub production showed a range of 1.00-2.80 t/ha with maximum production in Udaipur and minimum in Parapit mixed forest. In Udaipur and Podhal the amount of shrub production was found to be more than that of herbs. Total herbaceous and shrub production in different sites

TABLE 21.8. *Peak herbaceous and shrub production in different forest sites.*

Sites	Herbaceous production (t/ha)	Shrub production (t/ha)	Total (t/ha)
Udaipur	1.84 ± 0.11	2.80 ± 0.42	4.64
Podhal	1.96 ± 0.21	2.55 ± 0.45	4.51
Parapit	1.91 ± 0.16	1.00 ± 0.36	2.91

TABLE 21.9. *Above ground peak herbaceous and shrub (foliar) biomass as percentage of tree biomass*

Sites	Herb	Shrub	Herb + Shrub
Udaipur	1.38	2.80	3.49
Podhal	1.06	2.55	2.96
Parapit	1.94	1.00	2.43

TABLE 21.10. *Above ground herbaceous and shrub (foliar) production as percentage of litter production*

Sites	Herb	Shrub	Herb + Shrub
Udaipur	14.08	21.43	35.52
Podhal	20.17	26.24	46.41
Parapit	44.11	23.09	67.16

varied from a minimum of 2.91 t/ha/yr (in Parapit) to a maximum of 4.64 t/ha/yr (In Udaipur).

Peak herbaceous biomass as percentage of tree biomass varied from 1.06 to 1.94 per cent (Table 21.9). The corresponding range for shrub biomas was 1.00 to 2.80 per cent. Both herb and shrub together constituted a maximum of 3.49 per cent of above ground tree biomass in Udiapur.

Herbaceous production as percentage of litter production amounted to a minimum of 14.08 per cent in Udaipur and maximum of 44.11 per cent in Parapit (Table 21.10). Shrub production in different forest sites amounted to 21.43 per cent (Udaipur)-26.24 per cent (Podhal) of total litter production. Herbaceous and shrub production together as percentage of annual litter production varied from a minimum of 35.52 per cent (in Udaipur) to a maximum of 67.16 per cent (in Parapit).

A comparative account of above ground herbaceous and shrub biomass with tree biomass and with litter production in different forest sites revealed that although forest floor herbaceous and shrub vegetation in terms of biomass constitute very little of total forest biomass, contributes significantly to annual net primary productivity of tropical forest.

Turnover of litter

The turnover rate (K) of litter was calculated directly following Jenny *et al.* (1949) and Olson (1963) as $K = A/(A + F)$, where 'A' is the annual increment of the litter (i.e. litterfall on each site) and 'F, is the biomass of litter at steady state. Turnover time is the reciprocal of the turnover rate, expressed as $t = 1/K$. In the present study, standing crop of partially decomposed litter at the end of annual cycle of litter production was considered to represent the 'F'.

TABLE 21.11. *Production and turnover of litter in different study sites*

Sites	Average annual litter production (t/ha/yr)	Partial decomposed litter during December (t/ha)	Turnover rate (K)	Turnover time (yrs)
Udaipur	13.06	1.01	0.93	1.07
Podhal	9.72	0.98	0.91	1.09
Parapit	4.33	1.89	0.69	1.44

Turnover rate and turnover time of litter for different study sites are mentioned in Table 21.11. The turnover rate of 0.69-0.93 indicated that in these sites, annually 69 to 93 per cent of the total litter produced is subjected to decomposition. Comparatively lower turnover rate in Parapit can be explained on the basis of relatively poor microbiological status (Basu *et al.*, 1992; Basu, 1992). However, higher turnover rate in other two sites confirms the view of Golley *et al.* (1975) about the dynamic nature of the tropical forest.

Soil metabolism

Soil metabolism refers to the sum total activity of all the soil dwelling organisms resulting in the evolution of carbon dioxide. The study of CO_2 evolution i.e. soil respiration is one of the best approaches for the evaluation of the activity of soil inhabiting organisms in relation to carbon and energy flow in terrestrial ecosystem (Singh and Gupta, 1977). Among the soil organisms, microorganisms are the major contributor to the soil respiration (Reichle *et al.*, 1975). Study of Behera *et al.* (1990) have revealed a major contribution of microorganisms to total soil metabolism in a tropical deciduous forest from Orissa.

In the present study, alkali abşorption technique (Wilkamp, 1966) was employed to study soil metabolism in three different study sites. In all the sites, maximum rate of soil respiration was observed during rainy and minimum rate during the summer season (Table 21.12). This reflected the favourable effect of soil moisture and moderate temperature on microbial mineralization of organic matter (Behera and Pati, 1986). The mean annual rate of soil respiration was highest in Podhal and the lowest in Parapit. The parapit site being a scrub forest has the eroded soil at several places with very thin litter accumulation (Dani, 1992). This may be one of the possible reason for the lowest annual mean soil respiration rate in Parapit forest. On the contrary, Podhal forest represented a riverine forest with optimum soil moisture throughout the year and organic carbon level with relatively better edaphic condition. This explained the highest annual mean respiration rate in Podhal.

In Indian deciduous tropical forest soil root contributes around 50.5 per cent to total soil respiration (Behera *et al.*, 1990). Taking this estimate into consideration, microbial respiration

TABLE 21.12. *Seasonal trends of soil respiration (mg/CO_2/m^2/hr)* in different forest sites

Sites	Rainy	Winter	Summer	Mean
Udaipur	439 ± 14	322 ± 12	172 ± 12	311
Podhal	459 ± 38	372 ± 26	181 ± 13	337
Parapit	328 ± 21	157 ± 17	113 ± 17	199

was calculated in different study sites. Microbial respiration rate varied from a minimum of 99 mg (Parapit) to 167 mg CO_2/m^2/hr (Podhal). The rate for Udiapur soil was 154 mg CO_2/m^2/hr. Annual microbial respiration for Udaipur, Podhal and Parapit were 368, 399 and 237 g CO_2-C/m^2. This indicated comparatively better microbial mineralization in Podhal and Udaipur forests than the Parapit forest.

Total microbial biomass

Quantitative estimation of microorganisms in terms of biomass is one of the ways of evaluating their role in the decomposition process. Microbial biomass, a labile fraction of soil organic matter and repository of plant available nutrients (Jenkinson and Ladd, 1981; Smith and Paul, 1990) is considered as an important functional index of soil subsystem (Powlson *et al.*, 1987; Insam and Domsch, 1988). The measurement of microbial biomass provide a valuable tool for analysing and predicting long term effects of changes in soil condition.

For the analysis of microbial biomass, field moist soil samples were stored at normal temperature (28 ± 2°C) for a week to settle down the respiration. Soil microbial biomass C was determined using the $CHCl_3$ fumigation incubation method of Jenkinson and Powlson (1976) except that liquid $CHCl_3$ was used instead of vapour and CO_2-C evolved from fumigated soil during 10-20 days was taken as control (Merckx *et al.*, 1985; Srivastava and Singh, 1988). Microbial biomass C was calculated by dividing the flush of CO_2-C (CO_2-C) evolved from fumigated soil during 0-10 days minus that evolved from same soil during 10-20 days) by a k_c (fraction of biomass C evolved as CO_2-C during the first 10 days of incubation) value of 0.45 (Jenkinson and Ladd, 1981). All estimates are expressed on oven dry weight basis.

Total soil microbial biomass during different months of sampling in study sites showed a range of 0.229 (in Parapit) to 0.869 mg/g dry soil (in Podhal) which is comparable with that of Singh *et al.* (1989) and Srivastava and Singh (1989, 1991) for some tropical forests from India.

Average microbial biomass varied from a minimum of 0.342 mg C/g dry soil (in Parapit forest) to a maximum of 0.816 mg C/g dry soil (in Podhal forest) (Table 21.13). The percent of soil organic carbon reflected in microbial biomass were in the range of 1.06 to 1.28. A wide spectrum of ratio ranging from 0.27 to over 7.0 per cent (Anderson and Domsch 1980; Brookes *et al.*, 1984; Mc Gill *et al.*, 1986; Mulongoy, 1986; Anderson and Domsch, 1989; Insam *et al.*, 1989) have been reported. This wide spectrum of ratios can be explained due to the difference in

TABLE 21.13. *Average microbial biomass in different forest sites*

Sites	Microbial biomass C (mg/g dry soil)	Microbial biomass C / Soil organic C %
Udaipur	0.570 ± 0.216	1.06
Podhal	0.816 ± 0.048	1.28
Parapit	0.342 ± 0.103	1.28

soil type, overlying vegetation cover and analytical methods (Anderson and Domsch, 1989).

From the data of average microbial biomass and annual microbial respiration, microbial activity per unit microbial biomass (g CO_2-C/g C microbial biomass/yr) was calculated for the three study sites. The activity was 2.21 in Parapit, 1.56 in Podhal and 2.34 in Udaipur. Thus the microbial activity per unit microbial biomass was the lowest in Podhal mature forest. This reflected the relative efficiency of microbial function in three different sites. The more efficiently the soil micro-organisms function, less carbon is lost through respiration (Insam, 1990). Thus the microbial functional efficiency was maximum in Podhal. This may be explained further on the basis of the concept of economic metabolism proposed by Anderson and Domsch (1990) i.e. in the more complex and mature forest ecosystem, microbial communities channelise relatively more substract C for their biomass synthesis which ultimatley lead to more biomass build up in soil.

Fungal biomass

Fungi are commonly the largest component of the soil biomass (Gray and Williams, 1971; Anderson and Domsch, 1975; Toth and Hammer, 1977). Amount different fungal structures, the hyphal structure is the most important structure because it takes active part in exploiting the decomposing substrate. So Measurement of hyphae is considered to provide in index of microfungal activity in soil (Behera, 1980). Soil fungal biomass is estimated using microscopic measurement of hyphae in the agar film (Jones and Mollison, 1948). Soil samples were randomly collected from 0-20 cm soil depth in each of the sites. For each soil sample, five agar films were prepared and ten random observations per film were made. Hyphal length and diameter were converted to biovolume using the formula :

$$\text{Biovolume} = \pi r^2 l$$

Where r = radius of hyphaen

l = length of hyphaen (Parkinson *et al.*, 1971)

Considering average fungal protoplasmic value of 1.1 and dry matter of 21 per cent (Bakken and Olsen, 1983), biovolume data were converted into dry fungal biomass.

TABLE 21.14. *Microfungal biomass in different forest sites*

Sites	Microfungal biomass (mg/g dry soil)	Mean microfungal biomass (g/m^2)	$\frac{\text{Microfungal biomass C}}{\text{Soil organic C}}$ %
Udaipur	0.936 ± 0.353	258 ± 97	0.78
Podhal	1.376 ± 0.268	430 ± 84	0.97
Parapit	0.545 ± 0.070	170 ± 22	0.92

The average microfungal biomass showed a range of 0.545 to 1.376 mg/g dry soil with the maximum biomass in Podhal and minimum in the Parapit site (Table 21.14). The range of mean microfungal biomass is higher than that of Satpathy *et al.* (1982), Pati (1984) and Behera *et al.* (1991) for different tropical grassland and forest soils. On area basis, microfungal biomass as percentage of soil organic C ranged from a minimum of 0.78 (in Udaipur) to a maximum of 0.97 (in Podhal).

A comparison of mean soil fungal biomass carbon of different sites with that of total soil microbial biomass indicated that the former constituted 70-83 per cent of the latter. Several workers (Clark and Paul, 1970; Satchell, 1971; Tesarova and Repova, 1984; Anderson and Domsch, 1975, 1978) have reported the dominance of fungal biomass in different soil types of the world. Thus the observation of present study corroborates the views of the above workers.

Annual substrate requirement for microbial maintenance and growth

Annual substrate requirement for microbial maintenance was calculated following the equation of Marr *et al.* (1963) :

$$ax = Y\frac{ds}{dt}$$

Where a = specific maintenance requirement
= 0.001/ha (Babiuk and Paul, 1970),
Y = yield coefficient = 0.5 (Hisset and Gray, 1976)
X = microbial biomass (viable)
$\frac{ds}{dt}$ = substrate required for maintenance.

A viable proportion of total microbial biomass pool is taken into account for the analysis of microbial growth and productivity because most of the microbial cells in soil remian in dormant/physiologically inactive stage which require very little energy to maintain their cell integrity. About 26 to 34 per cent (X = 30 per cent) of total microbial population are viable in

TABLE 21.15. *Substrate utilisation pattern and soil microbial productivity in three different forest sites*

Sites	Total substrate available* ($gC/m^2/yr$)	Substrate required for microbial maintenance ($gC/m^2/yr$)	Substrate available for microbial growth ($gC/m^2/yr$)	Microbial generation time (day)	Microbial productivity ($gC/m^2/yr$)
Udaipur	1396	827	569	30	569
Podhal	1773	1340	433	38	732
Parapit	817	562	255	80	146

*Calculated as sum of litter production, herbaceous and shrub production and rhizo deposition.

tropical grassland and forest soils (Pati and Behera, 1989; Basu *et al.*, 1991). Annual substrate requirement for microbial maintenance and growth are mentioned in Table 21.15. Annual maintenance requirement as percentage of total substrate input amounted to 59 per cent in Udaipur, 69 per cent in Parapit and 76 per cent in Podhal. Amount of substrate available for supporting microbial growth was calculated after subtracting annual maintenance requirement from the total substrate input. Number of microbial generation in different study sites was calculated following the equation of Gray and Williams (1971) i.e.

$$Y = (S + xR) = xR$$

where Y = yield coefficient = 0.5

S = substrate available for growth

X = microbial biomass (viable)

R = number of microbial generation/yr.

Calculated values of R were 4.55/yr in Parapit, 9.75/yr in Podhal and 12.06/yr in Udaipur. From the values of R, generation time and microbial productivity for three different sites were calculated (Table 21.15). In Parapit, the generation time was maximum (80 days) as a result of which microbial productivity was calculated to be minimum (145.92 $gC/m^2/yr$). With relatively less amount of substrate input from the overlying vegetation compartment, the soil of the site is thus yet to attain stability to support better microbial growth and productivity.

Conclusion

The climatic climax forest vegetation of Eastern Ghats have several interesting characteristics such as presence of several endemic floristic elements, co-existence of both sal and teak elements in certain areas representing tension belt (Legris and Meher Homji, 1982), variance of forest types ranging from tropical dry deciduous to semi evergreen/evergreen types etc. In general, having rich biodiversity the forests of Eastern Ghats have several ecological implications including climatic control over Eastern and Peninsular India. In spite of this, the forest cover of the area is being reduced progressively leading to the problem of acclerated soil erosion, reduced infiltration and consequently low recharge of ground water. This disruption is being reflected in slow regeneration of natural forest. The problem is being further aggravated due to increased biotic pressure arising out of overstocking, unauthorised logging, mining, overall impact of urbanisation and introduction of commercial monoculture. The conversion of dense natural forest to savanna and other cultural zones is causing a decline in the soil organic matter-microbial status and consequent soil fertility resulting in the development of impoverished soil patches (Basu and Behera, 1993). On the other hand, reduced or non-availability of major and minor forest products has its negative impact on the life support system and sustainable economy of the indigeneous people leading to the widespread marginalisation and migration throwing the people ultimately to a poverty and exploitation trap (Dani, 1992). Hence this necessitates the immediate formulation of proper scientific policy for maintenance and management of Eastern Ghats forest range.

Acknowledgement

Authors thank Dr. N. Behera, Reader, School of Life Sciences for guidance and Head, School of Life Sciences for necessary facilities. Financial assistance from Department of Science, Technology and Environment, Government of Orissa in form of a research project is gratefully acknowledged. SB is grateful to CSIR, New Delhi for the award of Senior Research Fellowship to carry out the work. Authors express their gratitude to all the forest officials of Boudh Forest Division for their help and cooperation during the field work.

REFERENCES

Adhikari, B.S. Rikhari, H.C. Rawat, Y.S. and Singh, S.P. 1991. High altitude forest : Composition, diversity and profile structure in a part of Kumaun Himalaya. *Trop. Ecol.* **32** (1) : 86-97.

Anderson, J.P.E. and Domsch, K.H. 1975. Measurement of bacterial and fungal contributions to respiration of selected agricultural and forest soils. *Can J. Microbiol.* **21** : 314-322.

Anderson, J.P.E. and Domsch, K.H. 1978. Mineralization of bacteria and Fungi in Chloroform-Fumigated soils. *Soil. Boil. Biochem.* **10** : 207-213.

Andersons, J.P.E. and Domsch, K.H. 1980. Quantities of plant nutrients in the microbial biomass of selected soils. *Soil. Sci.* **130** : 211-216.

Anderson, T.H. and Domsch, K.H. 1989. Ratios of microbial biomass carbon to total organic carbon in arable soils. *Soil. Boil. Biochem.* **21** : 471-479.

Anderson, T.H. and Domsch, K.H. 1990. Application of eco-physiological quotients (q CO_2 and qD) on microbial biomasses from soils of different cropping histories. *Soil Biol. Biochem.* **22** (2) : 251-255.

Anderson, J.M. and Ingram, J.S.I. 1989. *Tropical Soil Biology and Fertility : A handbook of methods.* C.A.B. International. 171 p.

Babiuk, L.A. and Pual, E.A. 1970. The use of floorescien isothicocyanate in the determination of the bacterial biomass of grassland soil. *Can J. Microbiol.* **16** : 57-62.

Bakken, L.R. and Olsen, R.A. 1983. Bouyant densities and dry-matter contents of mircoorganisms : Conversion of a measured biovolume into biomass. *Appl. Environ. Microbiol.* **45** : 1188-1195.

Bandhu, D. 1970. A study of the productive structure of northern tropical dry deciduous forest near Varanasi. I. Stand structure and nonphotosynthetic biomass. *Trop. Ecol.* **11** (1) : 90-104.

Basu, S. 1992. *Microbial biomass and soil metabolism in some tropical forest soils.* Ph. D. Thesis, Sambalpur University, Orissa, India. 161 p.

Basu, S. Joshi, S.K. Pati, D.P. and Behera, N. 1991. Soil respiration in relation to microbial biomass in a tropical deciduous forest floor from India. *Rev. Ecol. Biol. Sol.* **28** (4) : 377-386.

Basu, S. Pati, D.P. and Behera, N. 1992. Microfungal biomass in some tropical forest soils of Orissa, India. *For. Ecol. Manage.* **55** : 333-339.

Basu, S. and Behera, N. 1993. The effect of tropical forest conversion on soil microbial biomass. *Biol. Fertil. Soils.* Vol. 16 (in press).

Behera, N. 1980. *Ecology of soil fungi in some tropical soil of Orissa, India.* Ph. D. Thesis, Sambalpur University, Orissa, India. 137 p.

Behera, N and Pati, D.P. 1986. Carbon budget of a protected tropical grassland with reference to primary production and total soil respiration. *Rev. Ecol. Biol. Sol.* **23** (2) : 167-181.

Behera, N. Joshi, S.K. and Pati, D.P. 1990. Root contribution to total soil metabolism in a tropical forest soil from Orissa, Indian. *For Ecol. Manage.* **36** : 125-134.

Behera, N. Pati, D.P. and Basu, S. 1991. Ecological studies of soil microfungi in a tropical forest soil of Orissa, India. *Trop. Ecol.* **32** (1) : 136-143.

Bray, J.R. and Gorham, E. 1964. Litter production in forests of world. *Adv. Ecol. Res.* **2** : 101-157.

Brookes, P.C. Powlson, D.S. and Jenkinson, D.S. 1984. Phosphorus in the soil microbial biomass. *Soil Biol. Biochem.* **16** : 169-175.

Brunig, E.F. 1974. Okosysteme in den tropen umschau. **74** : 405-410.

Chaturvedi, O.P. and Singh, J.S. 1987. The structure and function of pine forest in Central Himalaya. I Dry matter dynamics. *Ann. Bot.* **60** : 237-252.

Clark, F.E. and Paul, E.A. 1970. The microflora of grassland. *Adv. Agron.* **22** : 375-435.

Curtis, J.T. 1959. *Vegetation of Wisconsin. An Ordination of Plant Communities.* (Madison : Wisconsin Press).

Dani, H.P. 1992. *Ecological studies on a tropical deciduous forest from Orissa, India.* Ph. D. Thesis, Sambalpur University, Orissa, India. 173 p.

Ford, E.D. and Newbould, P.J. 1977. Stand structure and dry weight production through the sweet chestnut (*Castanea sativa* Mill) Coppice cycle. *J. Ecol.* **58** : 275-296.

Golley, F.B. Mc Ginnis, J.T. Clements, R.G. Child, G.I. and Duerer, M.J. 1975. Mineral cycling in a tropical moist forest ecosystem. Univ. of Georgia Press, Athens.

Gray, T.R.G. and Williams, S.T. 1971. Microbial Productivity in soil. In : *Microbes and Biological Productivity.* Symp. Soc. Gen. Microbiol. **21** : 255-286.

Hisset, R. and Gray, T.R.G. 1976. Microsites and Time Changes in Soil Microbe Ecology. pp. 23-40. In : *The Role of Terrestrial and Aquatic organisms in Decomposition Processes* (Ed. J.M. Anderson & A. Macfadyen). Blackwell Sci. Publ., Oxford.

Insam, H. and Domsch, K.H. 1988. Relationship between soil organic carbon and microbial bomass on chronosequences of reclamation sites. *Microb. Ecol.* **15** : 177-188.

Insam, H.Parkinson, D.and Domsch, K.H. 1989. The influence of mciroclimate on soil microbial biomass levels. *Soil. Biol. Biochem.* **21** : 211-221.

Insam, H. 1990. Are the soil microbial biomass and basal respiration governed by the climatic regime. *Soil. Biol. Bicohem.* **22** (4) : 525-532.

Jenkinson, D.S. and Ladd, J.N. 1981. Microbial Biomass in Soil : Measurement and Turnover. In : *Soil Biochemistry* (Ed. E.A. Paul & J.N. Ladd). Marcel Dekker, New York.

Jenkinson, D.S. and Powlson, D.S. 1976. The effects of biocidal treatments on metabolism in soil. V.A. method for measuring soil biomass. *Soil. Biol. Biochem.* **8** : 209-213.

Jenny, H. Gestel, S.P. and Bingham. 1949. Comparative study of decomposition rates of organic matter in temperate and tropical region. *Soil. Sci.***68** : 419-432.

Jones, P.C.T. and Mollison, J.E. 1948. A technique for the quantitative estimation of soil microorganisms. *J. Gen. Microbial.* **2** : 54-69.

Joshi, S.K. Pati, D.P. and Behera, N. 1990. Primary production of herbaceous layer in a tropical deciduous forest in Orissa, India. *Trop. Ecol.* **31** (2) : 73-83.

Joshi, S.K. and Behera, N. 1991. Qualitative analysis of vegetation from mixed tropical forest of Orissa, India. *Ind. For.* **117** (3) : 200-206.

Knight, B.H. 1975. A phytosociological analysis of species rich tropical forest on Barro, Colorado Insland, Panama. *Ecol. Monogr.* **45** : 259-289.

Legris, P. and Meher-Homji, V.M. 1982. *The Eastern Ghat Vegetation and bioclimatic aspect.* Proc. Sem. Res. Dev. Envi. in the Eastern Ghats. 24-27.

Marr, G.A. Nilson, E.H. and Clark. D.J. 1963. The maintenance requirement of *Escherichia coli Ann. New York. Acad. Sci.* **102** : 536-548.

Mc Gill, W.B. Cannon, K. R. Roberton, J.A. and Cook, F. 1986. Dynamics of soil microbial biomass and water soluble organic C in Breton L after 50 years of cropping to two rotations. *Can J. Soil. Sci.* **66** : 1-19.

Mc Naughton, S.J. 1967. Relationship among functional properties of California grassland. *Nature* **216** : 168-169.

Meher-Homji, V.M. 1977. Tropical dry deciduous forest of Peninsular India. *Feddes Report.* **88** (1-2) : 113-134.

Merckx, R. Den Hartog, A. and Van Veen, J.A. 1985. Turnover of root derived material and related microbial biomass formation in soils of different texture. *Soil. Biol. Biochem.* **17** : 565-569.

Misra, R. 1968. *Ecology workbook.* Oxford and IBH Publ. Co., Calcutta. 244 p.

Mulongoy, K. 1986. Microbial biomass and maize nitrogen uptake under a *Psophocarpus palustris* live mulch grown on a tropical alfisol. *Soil Biol. Biochem.* **18** : 395-398.

Murthy, N.S. and Pathak, S.J. 1972. Dominance, diversity and net production in grassland at Rajkot. India. *Jap. J. Ecol.* **222**: 47-49.

Olson, J.S. 1963. Energy storage and balance of Producers and decomposers in ecological systems. *Ecology.* 322-331.

Ovington, J.D. 1962. Quantitative Ecology and the woodland ecosystem concept. *Advan. Ecol. Res.* **1** : 103-193.

Parkinson, D. Gray, T.R.G. and Williams, S.T. 1971. *Methods to study ecology of soil microorganisms*. IBP Handbook No. 19. Blackwell Sci. Publ. Oxford. 116 p.

Parrotta, J.A. 1992. The role of plantation forests in rehabilitating degraded tropical ecosystems. *Agril. Eco. Environ.* **41** : 115-133.

Pati, D.P. 1984. *Studies on soil microbial population and soil metabolism of some tropical grasslands of Orissa, India.* Ph. D. Thesis, Sambalpur University. 167 p.

Pati, D.P. and Behera, N. 1989. Evaluation of two techniques for estimating bacterial population in tropical soil. *Sambalpur Univ. J. Sci. Tech.* **9** : 39-44.

Phillips, E.A. 1959. *Methods of vegetation study.* A. Holtdryden Book. Henry Holt and Co. Inc. New York, 107 p.

Powlson, D.S. Brookes, P.C. and Christensen, B.T. 1987. Measurement of soil microbial biomass provides an early indication of changes in the total soil matter due to straw incorporation. *Soil. Biol. Biochem.* **19** : 159-164.

Preston, F.W. 1948. The commonness and rarity of species. *Ecology*. **29** : 254-283.

Puri, G.S. 1960. Indian Forest Ecology. Vol. I & II. Oxford Book Co. New Delhi.

Ralhan, P.K. Saxena, A.K. and Singh, J.S. 1982. Analysis of forest vegetation at and arond Nainital in Kumaun Himalaya. *Proc. Ind. Natl. Sci. Acad.* **48** : 122-138.

Reichle, D.E. Mc Brayer, J.F. And ausmus, B.S. 1975. Ecological energetics of decomposer invertebrates in deciduous forest and total soil respiration budget. pp. 283-292. In : *Progress in Soil Zoology*. (Ed. J. Vanek) Academia Publ. House of Czekoslovak, Acad. Sci. Prague.

Richards, P.W. 1977. Tropical forests and Woodland : An overview. *Agro-Ecosystems.* **3** : 225-238.

Rout, S.K. and Gupta, S.R. 1989. Analysis of forest vegetation of Morni Hills in North East Haryana. *Proc. Natl. Acad. Sci.* 117-126.

Rout, S.K. and Gupta, S.R. 1990. Forest floor, litterfall and nutrient return in sub tropical ecosystems of Siwaliks in Northern India. II. Litterfall pattern and nutrient turnover rates. *Flora* **184** : 405-421.

Satchell, J.E. 1971. Feasibility study of an energy budget for Meathop Wood. pp. 619-630. In : *Productivity of forest ecosystem, ecology and conservation* (Ed. P. Duvigneaud). UNESCO, Paris.

Satpathy, B. Behera, N. and Dash, M. C. 1982. Microbial population, biomass and activity in some tropical soils of Orissa, India. *Ind. Biol. Bull. India.* **4** (3) : 150-157.

Saxena, A.K. and Singh. J.S. 1982. A phytosociological analysis of woody species in forest communities of a part of Kumaun, Himalaya, *Vegetatio.* 50 : 3-22.

Shannon, C.E. and Weaver W. 1963. *The mathematical theory of communication.* Univ. Illinois Press. Urbana. 117 p.

Simpson, E.H. 1949. Measurement of diversity. Nature. **163** : 688.

Singh, E.J. and Yadava, P.S. 1991. Above ground biomass and net production in two oak species growing in the natural forest of Manipur. India. *Trop. Ecol.* **32** (1) : 117-126.

Singh, J.S. and Misra, R. 1969. Diversity, Dominance, Stability and net production in the grasslands at Varanasi, India. *Can J. Bot.* **47** : 425-427.

Singh, J.S. and Gupta, S.R. 1977. Plant decomposition and soil respiration in terrestrial ecosystems. *Bot. Rev.* **43** : 449-528.

Singh, J.S. Singh, S.P. Saxena, A.K. and Rawat, Y.S. 1984. India's silent valley and its threatened rain forest ecosystems. *Environ. Conserve.* **11** : 223-233.

Singh. J.S. Raghubanshi, A.S. Singh, R.S. and Srivastava, S.C. 1989. Microbial biomass acts as a source of plant nutrients in dry tropical forest and savanna. *Nature.* 338 : 499-500.

Singh, K.P. and Misra, R. 1978. Structure and Functioning of natural, modified and silvicultural ecosystems in Eastern Uttar Pradesh. Final. Tech. Rep. (1975-76) MAB Res. Project LBHU. 161 p.

Singh, K.P. and Singh, R.P. 1980. Biomass and net production of herbaceous layer in tropical dry deciduous forest of Varanasi. *Trop. Ecol.* **21** : 48-57.

Singh, R.P. 1974. *A Study of primary productivity and nutrient cycling in Chakia forest, Varanasi,* Ph. D. Thesis, B.H. U. Varanasi. 179 p.

Smith, J.L. and Paul, E.A. 1990. The significance of Soil Microbial Biomass Estimations p. 357-396. In : Soil Biochemistry (Ed. J.M. Bollag & G. Stotzky). Marcel Dekker, New York.

Srivastava, S.C. and Singh, J.S. 1988. Carbon and Phosphorus in the soil biomass of some tropical soils of India. *Soil Biol. Biochem.* **20** : 743-747.

Srivastava, S.C. and Singh, J.S. 1989. Effect of cultivation on microbial biomass C and N·and dry tropical forest soil. *Biol. Fertil. Soils.* **8** : 343-348.

Srivastava, S.C. and Singh, J.S. 1991. Mircobial C, N and P in dry tropical forest soils : effects of alternate land uses and nutrient flux. *Soil. Biol. Biochem.* **23** (2) : 117-124.

Tesarova, M. and Repova, A. 1984. Microbial biomass of a grassland soil as estimated by the fumigation and direct counting method. *Folia Microbiol.* **29** : 1989-1992.

Tewari, J.C. 1982. *Vegetational analysis along altitudinal gradients around Nainital.* Ph. D. Thesis. Kumaun Univ. Nainital.

Toth, J.A. and Hammer, I. 1977. Quantitative microbiological studies on the soil of "Sikfourt Project" *Acta Biol. Bebrecina,* Bebrecen. Hungaria. **14** : 33-44.

Whittaker, R.H. 1969. Evolution of diversity in plant communities. *Brook Heaven. Symp. Biol.* **22** : 178-196.

Whittaker, R.H. 1975. *Communities and ecosystems.* Mac Millan, New York.

Witkamp, M. 1966. Rates of CO_2 evolution from the forest floor. *Ecology.* **47** : 492-494.

Chapter 22

Eco-Development : A Comparative Study of Two Villages of Rural Orissa, India

Aditya Kishore Dash[1] AND **Arttabandhu Mishra**[2]

[1]*Department of Environmental Science Sambalpur University, Jyoti Vihar-768 019, Orissa, India;* [2]*School of Life Sciences, Sambalpur University, Jyoti Vihar-768 019, Orissa, India*

Abstract. The paper reports on the participation of village folk of two villages of Orissa, Kesharpur (Nayagarh), and Kudamunda (Sambalpur), to conserve degraded forests on economic and ecological considerations. Kudamunda represents a *Shorea robusta* and Kesharpur a *Diopsyros ferrea* dominated plant community. Summer sampling indicated the presence of 42 plant species of which ten species were common in both the sites with various degrees of importance from the point of view of occupation of community space and contribution to the importance value. It was also found that from the conserved communities removal of plants does not take place. Kudamunda plant community where conservation efforts began seven years later than Kesharpar exhibited more growth. Despite ten common species both the communities were analyzed to be ecologically dissimilar.

Evaluating the extent of awareness among the people regarding an integrated view of eco-development it was found that a very small percentage of people are aware. Levels of public awareness were found to be more in Kesharpur than in Kudamunda.

Almost all the species found in the area were useful. Looking at the economic importance it was found that 20 species are used as major and minor timber, 16 species have food and five have fodder value, 30 species are used as medicine and 10 other as gum, resign, oil or dye producing plants. Three plants are found to produce fibre. People in both the areas were observed to be well organized to conserve and utilize their resources, but Kudamunda requires some external supports which Kesharpur has.

Key words : Eco-development, Regenerating forest, Diversity, Frequency, Density, Abundance, Evenness, Economic Utility.

Introduction

Present concern throughout the world regarding the depleted resources essential for survival and development, decreases in productivity; pollution changing climate, drought, cyclone and floods, is being described as eco-degradation. Referring to the Human Environment conference at Stockholm (1972) we visualise a process of change in the minds of people, in government and non-government agencies about eco-development. These essentially mean a process of development (Satisfaction of the basic survival and cultural needs) with ecological considerations questioning the limits to economic growth. Over the last two decades many examples of people's action in this regard are in record, most of which emphasize protection of natural forests or plantations but the real focus of eco-development is in the integration of available natural resources of an area for the bare basic needs of the people. It is easy to integrate the needs and resources theoretically but in practice it demands specific skills, finance and political will on one hand and management skills and behavioural adjustments on the other thereby demanding a process of new learning and unlearning leading to real eco-development (Mishra & Dash 1981). To achieve this goal, people with proper nonviolation and education have to form organisations that would develop the resource base with constant inputs from academic, social and government agencies to solve the intrinsic and extrinsic problems. It is observed that with the erosion of local control and traditional management of natural resources (Jodha 1990, Chambers *et al.* 1989) loss of biological productivity and biodiversity has led to acute scarcity of water, biomass and energy. This causes failure of the current approach to rural development. Bhat *et al.* (1983) argue in terms of an integrated participatory and village ecosystem approach for micro-level planning for a sustainable development through their experiment in two village ecosystems. Agarwal and Narain (1989) elaborate the need for an ecologically sound rural development (ecosystem specific) and discuss the factors and forces in detail with appropriate models. Ganapathy (1990) indicates the changing pattern of man forest interaction and its implication on ecosystem management, emphasizing the reforestation of most deciduous zone to meet the local demands. Reddy (1990) in his study on modelling village agricultural systems emphasizes plant biomass as the only renewable resource of the village ecosystem to cope up with the present and future energy requirements and argues that an understanding of the logic of village agricultural system should be the basis of rural development, Udhaya Shankar (1990) in a study on the impact of deforestation on the socio-economic condition of village life and environment finds the negative impact on irrigation, fishing, fuel collection, fodder production, green manure, wood based industries and such other aspects leading to wide spread poverty. Aiyer (1990) discusses the development of viable models of agro-ecosystems to compromise with forest ecosystems, emphasizing the management of domestic animals, horticulure, medicinal plants along with the preservation of forests, trees and wild animals.

Considering the importance of these studies, the study under discussion was planned for comparing two village areas of Sambalpur and Nayagarh districts, where forest protection by communities are taken as an important activity of eco-development. Forest protection by communities is widely seen in many areas of Orissa in particular and the country in general of which these two are samples. Singh and Singh (1992) and Sashikant *et al.* (1991) have given an indicative discussion of community forest management in Orissa with particular case studies of Nayagarh, Dhenkanal and Bolangir. Pati *et al.* (1993) present a description of the community protected forest of Kudamunda which is one of the fields of this study as well.

The study sites

Kesharpur is located between 85-86°N Lat. and 20-21°E Long., southwest of Bhubaneswar at a distance of 14 km and outleast of Nayagarh district in Orissa state. The people of Kesharpur village have protected the hill namely Binjagiri. By the late sixties, Binjagiri was completely denuded, creating problems for the locality in forest product collection and sand casting of their crop lands. During 1976, people of Kesharpur took a decision protect a patch of Binjagiri, though half-heatedly to start with, a NSS camp in 1978 with plantation programme, by the then DFO, Nayagarh and local MLA, motivated the villagers of Kesharpur lately being involved whole-heatedly in forest protection, Meanwhile, other hills, like Malati hill near village Manapur were also protected, and plantations were undertaken in other villages. In 1982, a workshop was organised under the auspices of NSS in three villages – Gami, Nagamunduli and Kesharpur. It was attended by representatives of 22 villages of the area. The workshop led to formation of "Brikshya O' Jeevar Bandhu Parishad" (Council of Friends for trees and animals) a voluntary organisation consisting of members from these 22 villages.

Regeneration from available root stock and plantation are the two major objectives taken up by this council of friends for trees and animals. The total area of the hill Binjagiri is approximately 360 hectares, and that legal status of the forest is that of protected forest, locally known as Khesra forest. The flora and fauna of Binjagiri hill were virtually undisturbed till as recently as 1940. Around seven streams use to flow from the hill the Kusumi river. Almost all the villages have fish ponds. The ponds are managed by the village Council. They are mainly used for bathing and more significantly for pisciculture. Village common land is cultivated by the village Council on share cropping basis. The existence of an informal village council and their prior experience in management of other village resources i.e. schools, village land, village ponds etc. provide a strong base for inclusion of management of forest resources also in their sphere of activities. The village Council have formed a set of rules, defining the rights and duties of villagers. Kesharpur has also another significant special rural for the trees in river bank. The wood is to be equally shared between the care-taker and the villagers. Various forest officials have co-operated with the villagers at various occasions. The support by 'BOJBP' has been most crucial for this forest protection movement.

Now the Binjagiri looks green with a large number of trees. At present, 'BOJBP' is working in 255 villages through its sister organisations. Approximately 39,000 acres of forest area is being managed as common property resources (CPR) in these 255 villages. The villages present a clear picture of an eco-development culture although they have many more things to do to attain perfection. They do not do it because they know ecology but they understand their own needs and the resources around them are managed according to the needs.

Kudamunda is situated between 84-85°N Lat. and 21-20°E Long., about 30 km north-eastern direction from Sambalpur on road and is in the Rengali block of Sambalpur district of Orissa. Situated on the foot hills of "Jharghat Carpati" hill on one side and the Phuljari Khaman and Sangramal reserve forest on the other side, Kudamunda till 1950 was a small forest village on the catchment of river Mahanadi. After Hirakud Dam and canal system, the irrigation canals reached the vicinity of Kudamunda and impact of deforestation came in. Around 60's people of Kudamunda realised the problem of deforestation and started protecting a small patch of natural forest for a decade or so. But due to intra-village tension and intervention of nearby villagers it was cutaway. Again during 1983 the youth club of Kudamunda (Janasebaka Youth Club) initiated the protection of forest and maintained it for last ten years.

On one side of the village, over a patch of more than one hundred acres of land is covered with a regenerated forest that has been protected by the villagers. The other side is surrounded by the water systems and crop fields. About 500 people (70 families) are residing in the village, most of which are economically backward and illiterate mainly Scheduled Tribes. No amenities like electrification are available to the village. Transportation system is available to the village but one has to walk 5 kms to reach the village from road. Regarding the management they have formulated a village forest protection committee and have employed watchman paid from the village fund gathered by household subscriptions. Besides the forest, the villages have in their plant to develop irrigation, get electrification and develop the road to their village.

Unlike Kesharpur, the Kudamunda group has not succeed in motivating or organising large number of villagers but their initiatives has been well taken by more than ten villages of the area where groups have started protecting patches of natural forest. Their forest product needs are collected from the areas other than the conserved patch which they have not touched even for pruning, thinning or gap-filling of which they were aware. Besides they have not taken up any other eco-development activities, not because they do not need them but their initiatives have not been supported by any external agencies or individuals like that of Kesharpur. Very recently the Orissa Environmental Consciousness Society (Sambalpur University) and another NGO called MSS (Manav Adhikara Seva Samiti) have started extending their support to the village level organisation of the area.

The regenerating forest community : An analysis

To study the regenerated forests at Kudamunda and Kesharpur, quadrats of 10 × 10 m were chosen at random. In each quadrat, plants within the height of 10′, 10-20′ and more than 20′ were recorded. Kesharpur being a site of longer duration of regeneration, 59 per cent of the quadrats did not show trees with height of more than 20′ where as Kudamunda a site of shorter duration of regeneration. more than 83 per cent of the quadrats did not show trees with a height of more then 20′. Similarly, trees with the height range of 10-20′ were absent in 50 per cent of the quadrats at Kudamunda and 23 per cent of the quadrats at Kesharpur where efforts of conservation began in 1976 but at Kudamunda it started during 1983. In both the sites ten species were found to be common.

TABLE 22.1. *Density of some dominant tree species in the study sites*

Kudamunda	No./Hectara	Kesharpur	No./Hectare
S. robusta	2500	*D. sylvatica*	1800
B. latifolia	2300	*A. arabica*	400
T. tomentosa	1900	*A. chinensis*	800
D. melanonxylen	1100	*C. fistula*	600
A. cordifolia	1200	*S. malbarica*	400
T. belerica	1200	Kakhada	600
M. latifolia	1000	*A. acuminata*	500
P. lactifera	900	*A. leucophloea*	700
C. collinus	800	*I. parviflora*	700
L. parviflora	500	*B. monosperra*	500

Considering the frequency of plants in different quadrats it was found that *P. zeylanica, A. paniculata* and *S. robusta, A. latifolia* and *T. tomentosa* were present in all the quadrats whereas *P. dactylifera, D. melanoxylen, A. cordifolia, H. antidysenterica* and *L. parviflora* were found in 90 per cent of the quadrats at Kudamunda. Five plants show a 90 per cent frequency at Kudamunda but none of the plants show such high frequency at Kesharpur. Tree species showing highest density at Kudamunda and Kesharpur were *S. robusta* and *D. ferrea* respectively. Importance value indices (IVI) for different species in both the sites were calculated following Odum (1971). At Kudamunda *S. robusta* exhibited the maximum IVI. The average number of the dominant tree *D. sylvatica* at Kesharpur was 1800/hect and at Kudamunda *S. robusta* was 2500/hect (Table 22.1).

In the height range of 0-10' *T. tomentosa* and *D. ferrea* were the dominant species at Kudamunda and Kesharpur respectively, but in the height range of 10-20' and above, *S. robusta* and *D. sylvatica* were dominant at Kudamunda and Kesharpur respectively (Fig. 22.1 & 22.2, Table 22.2). The dominance diversity pattern of trees look different at both the study sites. At Kudamunda the two dominant species occupy separate niche space from the rest but at Kesharpur the niche space is equally shared by all the trees ad evident from the perfect concave curve. In the layer of 0-10' at Kudamunda *T. tomentosa* is not segregated from the other species sharing the space equally but at Kesharpur *D. feerea* occupies the top niche totally by itself segregating the rest, indicating a marked difference of vegetational behaviour (Fig. 22.3).

Phytosociological characteristics of ten common species for both the sites were synthesized and represented in Fig. 22.4 and Table 22.3. The relative Density, Frequency, Abundance and the Importance Value Indices of all the common species of both the sites show that none of them are

TABLE 22.2. *Species sequence in order of ni/N values (10^{-3})*

Kesharpur (within 10′)	Abbre-viation	R.F.	R.A.	R.D.	ni/N	Kudamunda (within 10′)	Abbre-viation	R.F.	R.A	R.D.	ni/N
D. ferrea	Df	38.6	63.0	99.2	200.8	*T. tomentosa*	Tt	46.0	58.5	85.8	190.3
R. malabarica	Rm	33.8	32.8	44.5	110.6	*D. melanozylen*	Dm	41.4	49.4	65.2	156.0
Z. oenoples	Zo	28.9	24.0	28.3	81.2	*A. cardifolia*	Ac	41.4	42.8	62.7	146.9
F. cataphrata	Fc	24.1	23.6	41.4	79.1	*P. zeylanica*	Pz	46.0	23.5	34.4	103.9
Takualai	Tl	28.9	21.4	25.3	75.6	*p. dectylifera*	Pd	41.4	20.5	27.1	89.0
H. indicus	Hi	28.9	19.7	23.3	71.9	*A. latifolia*	Al	32.2	25.7	26.4	84.3
P. heyneanum	Ph	33.8	13.2	18.2	65.2	*z. oenoplea*	Zo	36.8	13.7	16.1	66.6
G. tiliaefolia	Gt	19.3	24.4	19.2	62.9	*W. floribunda*	Wf	36.8	11.2	16.2	61.2
C. baccifera	Cb	19.3	19.2	15.1	53.6	*A. pamiculata*	Ap	46.0	5.0	0.73	51.73
C. species	Cs	14.4	20.5	12.1	47.0	*G. turgida*	Gt	13.8	15.0	6.6	35.4
D. sylvatica	Ds	19.3	88.7	69.9	177.9	*s. robustaz*	Sr	46.0	125.0	183.4	354.4
A. arabica	Aa	39.8	55.9	40.3	136.0	*B. Latifolia*	Bl	46.0	106.5	157.7	310.2
A. chinensis	Ac	43.4	25.1	44.5	113.0	*T. tomentosa*	Tt	46.0	58.5	85.8	190.3
C. fistula	Cf	38.6	23.1	36.4	98.1	*D. melanozylen*	Dm	41.4	49.4	65.2	156.0
S. malabarica	Sm	38.6	21.2	33.4	93.2	*A. cordifolia*	Ac	41.4	42.8	62.7	146.9
Kakhada	K	33.8	24.2	33.4	91.4	*T. balarica*	Tb	23.0	44.0	64.5	131.5
A. acuminate	Aa	28.9	26.5	31.4	86.8	*M. latifolia*	Ml	32.2	36.4	33.0	101.6
A. leucophloea	Al	33.8	21.3	29.3	84.4	*P. dectylifera*	Pd	41.4	20.5	27.1	89.0
I. parviflora	Ip	33.8	16.9	31.4	82.1	*C. collinus*	Cc	27.6	32.5	28.6	88.7
B. monosperma	Bm	28.9	24.0	28.3	81.2	*L. parviflora*	Lp	41.4	18.8	24.9	85.1

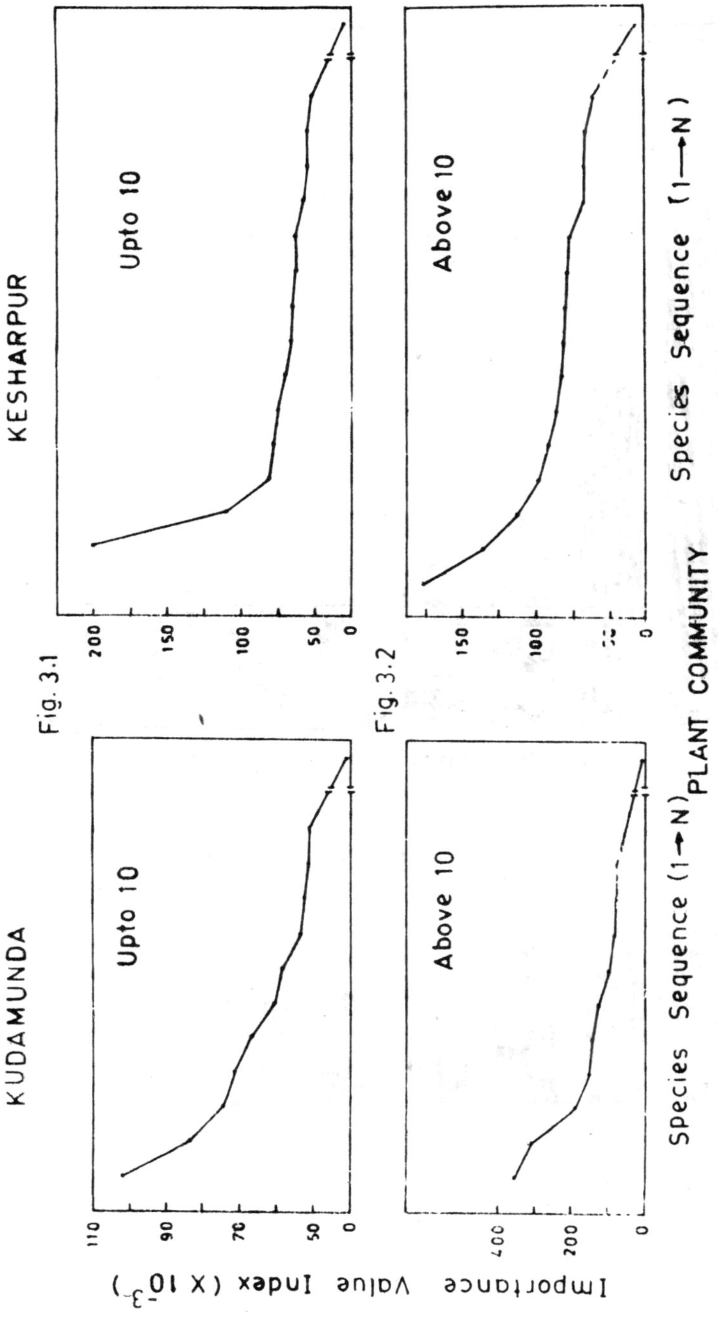

Fig. 22.1. Dominance diversity curve of Kudamunda and Kesharpur

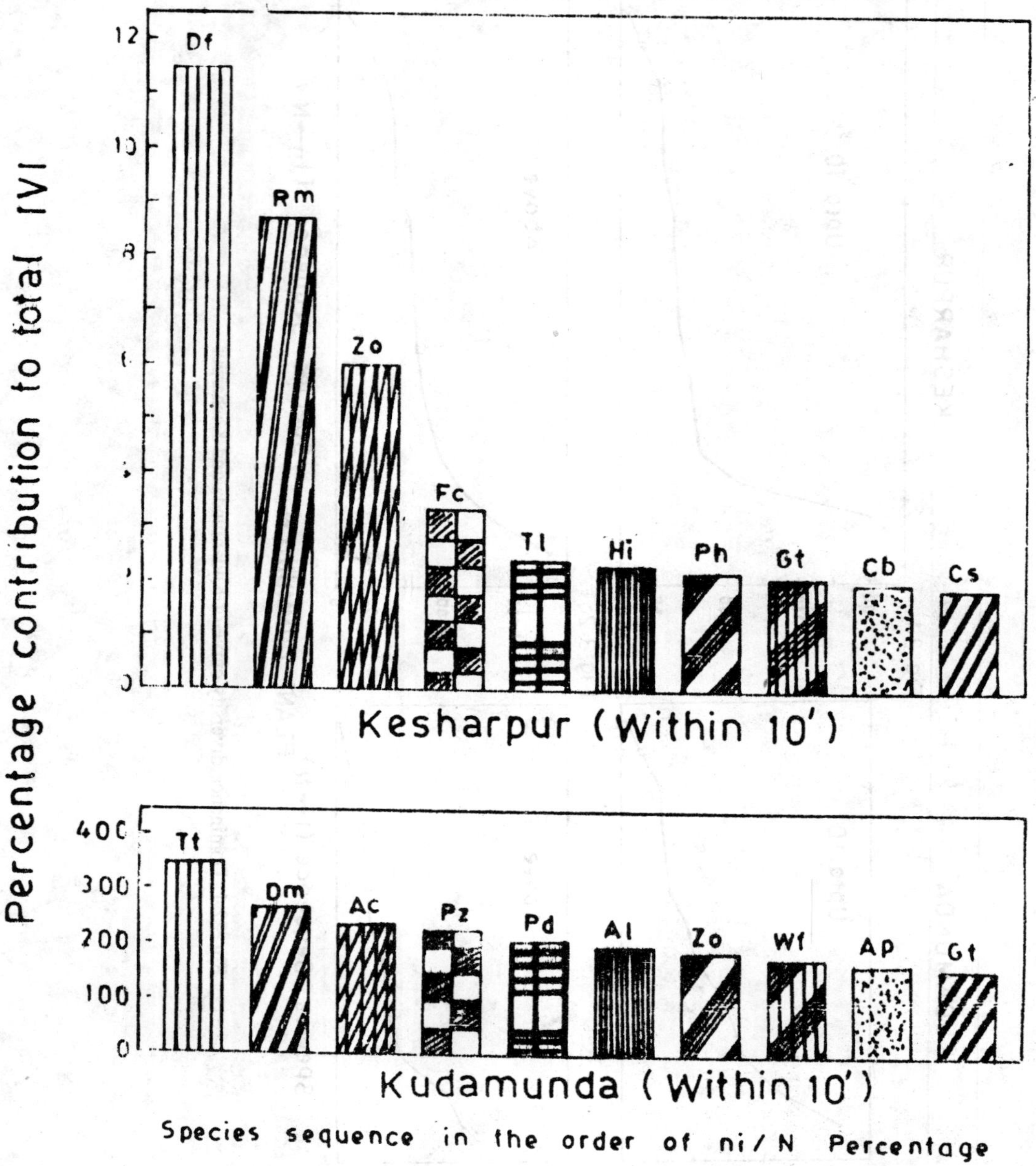

Fig. 22.2. Contribution of some dominant species to total IVI in different species

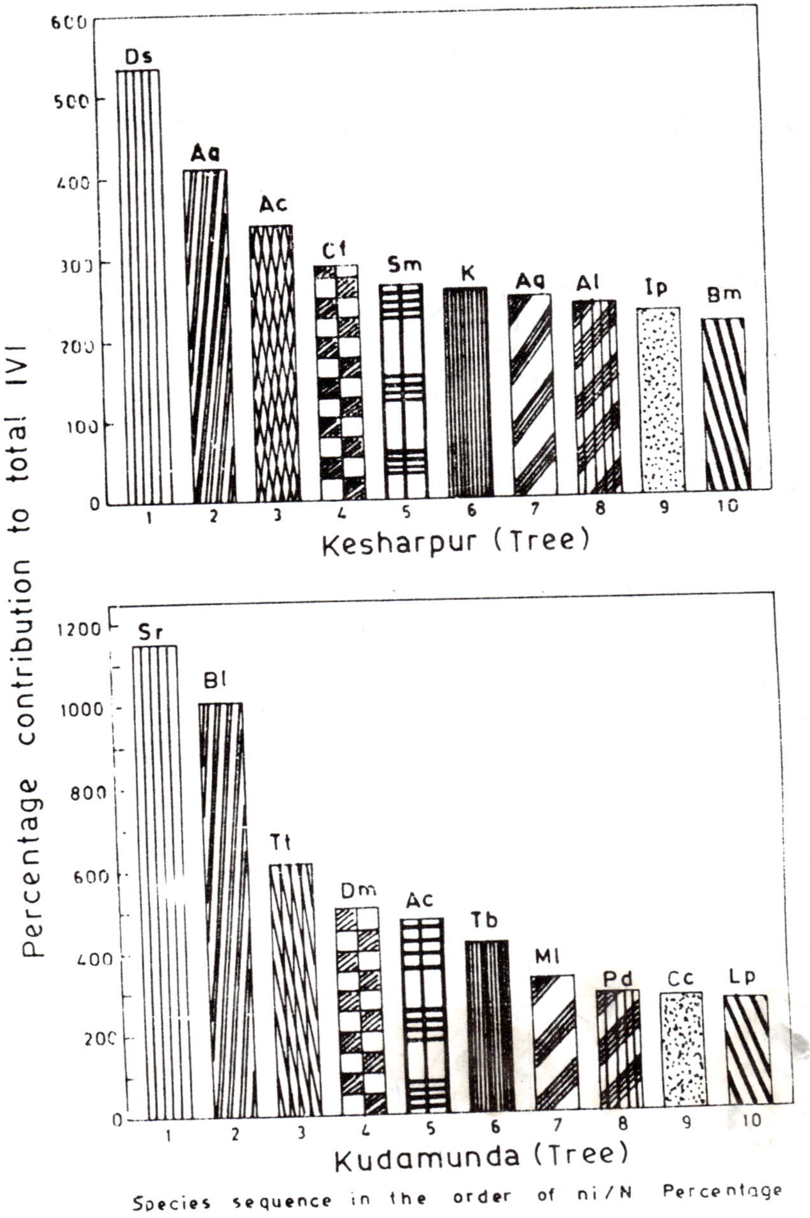

Fig. 22.2 a. Contribution of some dominant species to total IVI in different species

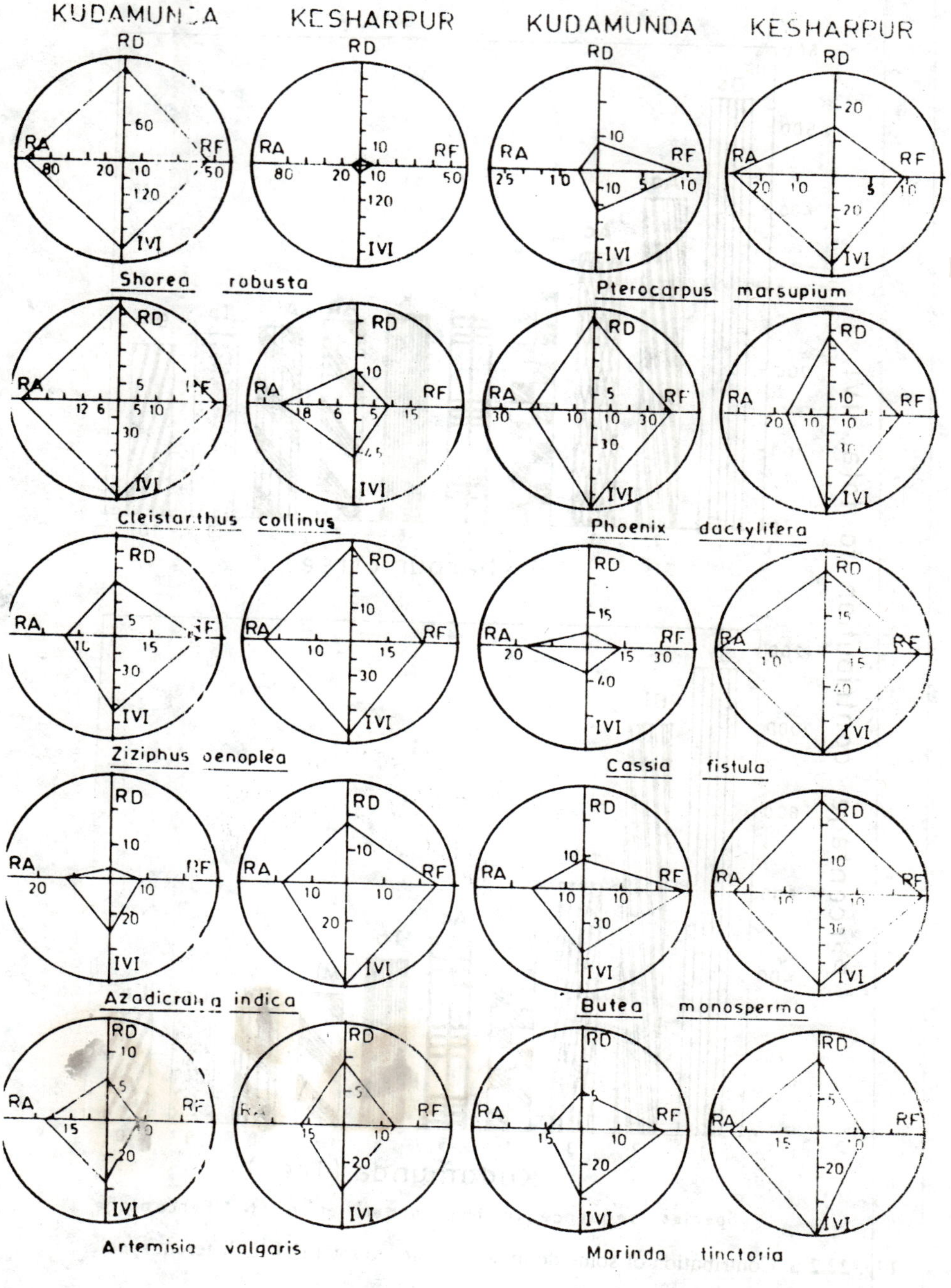

Fig. 22.3. Contribution of common species of the total ni/N% of the community

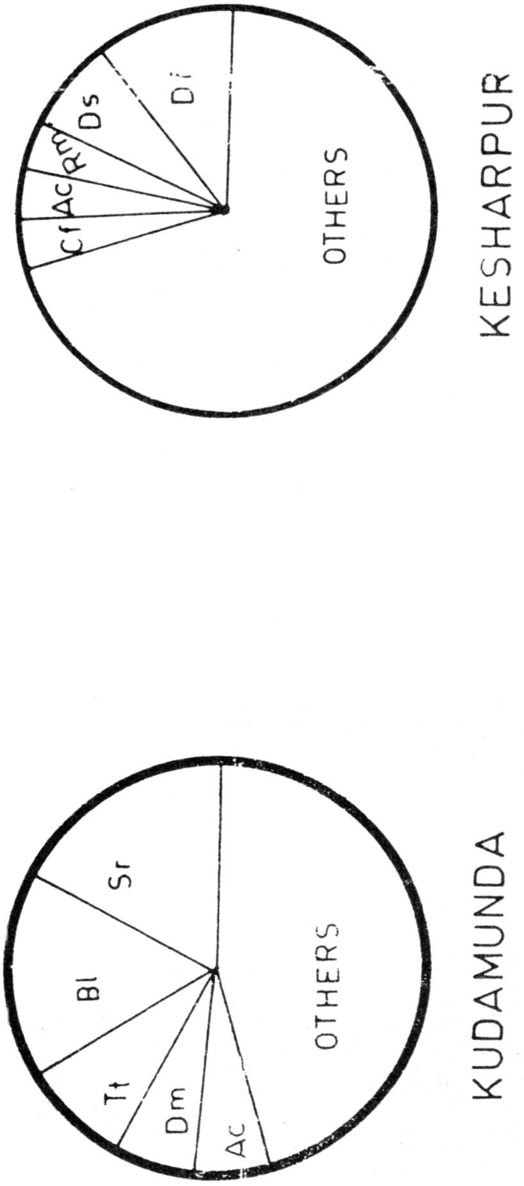

Fig. 22.4. Contribution of five dominant species to the total IVI

TABLE 22.3. *Indices of some common species in the two communities* ($\times 10^{-3}$)

Species	Kesharpur				Kudamunda			
	RF	RA	RD	ni/N	RF	RA	RD	ni/N
S. robusta	9.6	5.1	2.0	16.7	46.0	106.5	157.7	310.2
P. marsupium	9.6	28.3	14.1	52.0	9.2	5.0	7.3	21.5
C. collinus	9.6	25.7	10.1	45.4	27.6	32.5	28.6	88.7
P. dectylifera	38.6	15.4	24.3	78.3	41.4	20.5	27.1	89.0
Z. oenoplea	28.9	24.0	28.3	81.2	36.8	13.7	16.1	66.6
C. fistula	38.6	23.1	36.4	98.1	13.8	16.6	7.3	37.7
A. indica	24.1	17.4	17.2	58.7	9.2	12.5	3.6	25.3
B. monosperma	28.9	24.0	28.3	81.2	27.6	14.1	12.4	54.1
A. vulagaris	14.4	15.4	9.1	38.9	9.2	22.5	6.6	38.3
M. tinctoria	14.4	34.3	11.1	59.8	23.0	8.0	5.8	36.8

RA : Relative Abundance, RD : Relative Density, RF : Relative frequency, IVI (ni) : Important Value Index

comparable to each other and show differences except two species like *P. dectylifera* and *A. vulgaris.* Considering five dominant secies of both the sites it was found that their contribution to the total IVI at Kesharpur is little more than 25 per cent whereas at Kudamunda it is more than 50 per cent (Fig. 22.4). The similarity index between both the sites was calculated to be 0.312 which is very low and this leads to the conclusion that the sites are not similar to each other with respect to the plant community. Shannon Weaver's (1949) index of general diversity (H) was calculated to be 1.5383 for Kesharpur and 1.3536 for Kudamunda and the Evenness indices are 0.9476 and 0.1611 respectively. This indicates that more evenly distributed plant community exist at Kesharpur than at Kudamunda although diversity indices are comparable.

The utility of the available species in both the study sites is presented in Table 22.4 which indicates that efforts of conservation of forests by communities is economically rewarding. If the cost of the products is equated with that of the efforts it is more obvious.

Eco-development : An evaluation

It is widely agreed that the basic cause of poverty is the scarcity of biotic resources to meet the day to day needs. Due to the ecodegradation the scarcity situation is aggravated. In an effort to

TABLE 22.4. *Utility of the available plant species in two study sites*

Sl. No.	Species (Local name in Parentheses)	Economic importances of plants and their uses
	KESHARPUR	
1.	*Acacia arabica* (Babala)	Minor timber
2.	*Acacia leucopholoea* (Gohira)	Fencing, minor timber
3.	*Aegle marmilos* (Bela)	Food, medicine
4.	*Alangium chinensis* (Ankula)	Timber, medicine
5.	*Anogeissus accuminata* (Fasi)	Timber for major use
6.	*Bambusa vulgaris* (Baunsa)	Construction material
7.	*Bridelia retusa* (Kasi)	Timber for major use
8.	*Cipadessa baccifera* (Nahalabeli)	Tooth brushing, minor timber
9.	*Cycas species* (Araguna)	Food, beautification
10.	*Dalbergia sissoo* (Shisu)	Major timber medicinal use
11.	*Diospyros ferrea* (Guakoli)	Food
12.	*diospyros sylvatica* (Kalachua)	Timber, gum
13.	*Euphorbia species* (Lankagaba)	Medicinal use
14.	*Flacourtia cataphrata* (Bhainchakoli)	Food
15.	*Gmelina arborea* (Gamhari)	Major timber
16.	*Grewaia teliaefolia* (Dhamana)	Strong timber for Agro implements
17.	*Hemidesmos indicus* (Sugandhi)	Medicinal
18.	*Ixora parviflora* (Telakuruan)	Oil extract and timber
19.	*Unidentified sp.* (Kakhada)	Used as minor timber
20.	*Lannea grandis* (Mai)	Used as timber, gum, medicine, fodder
21.	*Largerstomia flosreginae* (Patuli)	Fuel, medicinal and minor timber
22.	*Mangifera indica* (Amba)	Food, wood for plank and fuel.
23.	*Mimusops hexandra* (Khirakoli)	Edible berry
24.	*Pterospermum heynaenum* (Giringa)	Strong timber
25.	*Randiana malbarica* (Firika)	Thorny bush used for fencing material.
26.	*Unidentified sp.* (Panijamuna)	Creeper, used in traditional medicine.
27.	*Satnalia malbarica* (Simuli)	Light timber, fibre used in bed and pillow, medicinal use
28.	*Strychnos potatorum* (Katakala)	Minor timber, medicinal use
29.	*Strychnos nuxvomica* (Kochila)	Minor timber, important medicinal plant.
30.	*Stribulus asper* (Sahada)	Minor timber, medicinal plants.
31.	*Unidentified sp.* (Takualai)	Creeper used as fibre in fencing and house roof.
32.	*Vitex leucoxylon* (Chirgudia)	Creeper producing fibre

(Contd. Next Page)

Sl. No	Secies (Local name in Parentheses)	Economic importance of plants and their uses
	KUDAMUNDA	
1.	*Adina cordifolia* (Kuruma or Haldi)	Used as furniture timber
2.	*Andrographis paniculata* (Bhuilimba)	Medicinal plant
3.	*Anoegisus latifolia* (Dhanra)	Used as timber for minor purposes
4.	*Antidesma ghaesembilla* (Jamurla)	Fruit used as food
5.	*Buchanania latifolia* (Char)	Fruit edible, seed used like cashew, wood useful.
6.	*Carey arborea* (Kumbhi)	Used as a fibre producing tree timber value.
7.	*Carthamus tinctorius* (Kumbhi)	Timber, oil yeilding seeds oil used as medicine
8.	*Caseoria graveolens* (Beniman)	Tribals use as medicine
9.	*Chloroxylen swetenia* (Bheru)	Strong wood, used in agro implements and construction medicinal use.
10.	*Diospyros melanoxylon* (Kendu)	Fruit edible, leaves for beedi rolling, wood for agro implements and construction.
11.	*Embelia tsjeriam* (Baibidang)	An established medicinal plant
12.	*Emblica officinalis* (Ainla)	Very important medicinal plant
13.	*Gardenia gummifera* (Chauladhua)	Used as medicinal plant
14.	*Gardenia turgida* (Kurdu)	Timber
15.	*Holarrhena antidysenterica* (Kurei)	Very important medicinal plant bark for dysentry, seeds for many uses.
16.	*Unidentified sp.* (Khasi)	Timber, minor.
17.	*Lagerstoemia parviflora* (Senha)	Used as medicinal plant
18.	*Madhuca latifolia* (Mahula)	Multipurpose use, Timber, food, fodder, oil
19.	Odina wodier (Kekot)	Used as timber
20.	*Ougeinia delbergioides* (Bandhana)	Very strong timber used in Carts and plough
21.	*Olex scandans* (Bhadbhadria)	Leaves used as food and medicine
22.	*Plumbago zeylanica* (Chitapaur)	Used as a medicinal herb
23.	*Unidentified sp.* (Rohel)	Used as small timber and medicinal plant.
24.	*Semecarpus anacardium* (Bhalia or Bheluan)	Marker's nut, timber and medical use
25.	*Soymida fabrifuga* (Rohen)	Timber and medicinal use
26.	*Syzygium cumini* (Jamun)	Edible fruits. Soil conserving roots. Timber durable to water exposure
27.	*Terminalia bellirica* (Bahada)	Medicine, Timber
28.	*Terminalia chebula* (Harida)	Medicine, Timber
29.	*Terminalia tomentosa* (Asana or Sahaj)	Timber for household use
30.	*Unidentified sp.* (Telko)	Use unknown
31.	*Woodfordia floribunda* (Dhatuk)	Flowers used to extract dye. Medicinal.
32.	*Zizyphus xylophyra* (Ghantu or Ghantala)	Strong timber used mainly in houses and Ogroimplements, Charcol.

evaluate Eco-development in both the study areas, a sample of 50 persons at Kesharpur and 40 persons at Kudamunda were chosen at random and in course of discussion eighteen questions were asked. The response to each question was recorded after the interview and an analysis was made to know the extent to which eco-development has helped the people to satisfy their basic needs.

We observed that the family size in both the areas are different. Families consisting of two to three members were found among 30 per cent of the respondents at Kesharpur but none at Kudamunda. At Kudamunda 20 per cent of the respondents and none at Kesharpur had four member families. Sixty percent of the respondents of Kesharpur and 37.5 per cent of Kudamunda had 5-6 member families. Ten per cent of them in Kesharpur and 12.5 per cent of them at Kudamunda had eight member families. Eleven member families, were seen among 12.5 per cent of the respondents at Kudamunda but none at Kesharpur.

Education has always been considered as an index of development. The highest qualification among the respondents is class VIII in Kesharpur and class IX in Kudamunda although in both the areas beyond sample educated persons are there. About 50 per cent of the sample interviewed at Kesharpur read upto Class II and III whereas in Kudamunda it is about 37.5 per cent. Twenty per cent are illiterate at Kesharpur and 37.5 per cent in Kudamunda. Considering other aspects of education in both the areas it seems that Kesharpur is more educated than Kudamunda.

As per the information collected in both the areas about food and nutrition it is seen that 50 per cent of the total respondents in Kesharpur depend upon their own land for food collection, whereas 12.5 per cent do so at Kudamunda. In Kudamunda 50 per cent of the population depend upon their own land along with other sources for food whereas 30 per cent at Kesharpur do so. Landless people depending only on other sources are 37.5 per cent for Kudamunda and 20 per cent for Kesharpur.

Considering protein consumption as the index of better nutrition the respondents were asked how many times in a month and year they take animal protein. At Kudamunda 12.5 per cent of the respondents replied that they take animal proteins twice a year. A some percentage of respondents replied in each case for twelve times, thirteen times, twenty five times and thirty six times at Kudamunda whereas 37.5 per cent said that they take such proteins twice a month. At Kesharpur 20 per cent said that they take thrice and another 20 per cent went upto 24 times a year, 30 per cent went to four times and other 30 per cent for 13 times a year. This shows that people at Kudamunda are poorly fed and at Kesharpur most of them belong to the middle nutrition class. The animal proteins were dominantly fish at Kesharpur but meat at Kudamunda.

Regarding fuel wood requirement almost all the villagers depend upon the village forest in both the areas. The village forest being the only source for them and wood being the major fuel appears to be the cause for deforestation but in reality it is not the case, as the amount and type of wood used for fuel can never clear a forest.

For health service almost all the villages in both the areas depend upon the government dispensary along with traditional practices. They either collect the medicine from the village practitioner or go to nearby areas where such facility is available.

Fodder requirement is one of the important factors in the rural areas. Different persons collect fodder from different sources. About 40 per cent of the total population in Kesharpur and 75 per cent in Kudamunda collect the fodder from their own land and grazing land, rest 60 per cent of Kesharpur and 25 per cent of Kudamunda either collect it from the forest land or from own land or purchase it.

Fertilizer to maximise crop production is a major rural need. In general there is a trend to go for synthetic fertilizer with or without cowdung, compost or green manure. It is found that 20 per cent of the respondents of Kesharpur use cowdung and 20 per cent use synthetic fertilizer where as another 20 per cent use both and the rest 40 per cent do not add any such fertilizer. But in Kudamunda 37.5 per cent use cowdung but none of them use synthetic and cowdung together as fertilizer whereas 25 per cent do not add any form of fertilizer.

There is no production of fibre of clothes in villages. They entirely depend on the market for clothes, ropes used in beds and agricultural operations.

Transportation facilities in both the areas is poor. At Kesharpur 40 per cent of the people depends on bicycle, 20 per cent on motor cycle and rest 40 per cent on public transport. In Kudamuna 50 per cent depends upon bicylce and rest 50 per cent are pedestrians. All of them depend on public transport. It shows that in both the areas transportation is a problem.

Fore entertainment, 30 per cent of the respondents at Kesharpur use radio, visit cinema and 30 per cent use T.V. and 10 per cent go to the cinema hall and 30 per cent of the people depend on traditional village entertainment. In Kudamunda 37.5 per cent of the respondents have their own traditional ways whereas 25 per cent talked of video and films respectively. Only 12.5 per cent remain content with their radio sets. At Kesharpur unlike Kudamunda radio and video are not given importance. Kudamunda being without electricity therefore, no question of T.V. arises.

Asking the question about the hike of prices the respondents came up with different responses. At Kesharpur 50 per cent of the people said that, the natural resources are very much limited which is decreasing from day to day as the consumption rate is increasing. The other 50 per cent said that, there is an artificial price rise of the commodities. At Kudamunda, 65.5 per cent respondents thought that due to availability of limited resources and highest consumption rate, price rise is caused; 12.5 per cent of them said that they do not know any thing about it, which indicates lack of information and awareness.

Asking the question about the agencies who are responsible for large scale depletion of the natural resources various responses were received. At Kesharpur 60 per cent of the respondents made the village people responsible for it. The other 40 per cent said that is the government. In Kudamunda 50 per cent of the people blamed the villagers and other 50 per cent blamed the government agencies for the cause of resources depletion.

Species common to both sites

Sl No.	Name of the species	Uses
	Species common to both sites	
1.	*Azardictus indica* (Nimba)	Very important medicine, timber resin, gum.
2.	*Artemisia vulgaris* (Panairi)	Medicinal, Snake repellent.
3.	*Betua monosperma* (Palasa)	Used for planks, Charcol used in industry, bark for fibre, leaves for plates and thatching, medicinal.
4.	*Cassia fistula* (Sunari)	Minor timber, medicinal
5.	*Cleistanthus collinus* (Karla or Karada)	Bark used for tanning, Fruit used as piscicide, minor timber.
6.	*Morinda tinctoria* (Bija or Piasal)	Medicinal use, Furniture timber, agroimplements
7.	*Pterocarpus marsupium* (Bija or Piasal)	Important timber, Medicinal use.
8.	*Phoenix dactylifera* (Khajuri)	Edible fruit and juice.
9.	*Shorea robusta* (Sala)	Timber, plank, bark for tanning, leaves widely used. Seeds for oil, gums as resign, Fuel. Medicine.
10.	*Zizyphus oenoplea* (Kanteikoli)	Edible berry, Fencing material

About the effects of deforestation people replied differently and the answers centered around air pollution; rainfall, imbalance in atmospheric gases and scarcity of forest products, questions, asked regarding the solution different answers were recorded. Out of 40 per cent of the total respondents, 50 per cent at Kudamunda and 50 per cent at Kesharpur replied that, the only solution is peoples' awareness. Plantation and village level organisation were emphasized by 25 per cent of them at Kudamunda, 10 per cent and 12.5 per cent of the respondents at Kesharpur and Kudamunda respectively thought of fencing the conserved forests and 12.5 per cent of them Kudamunda could not participate in the discussion. Natural regeneration and plantation as simultaneous effects were emphasized by 20 per cent of Kesharpur respondents but none of Kudamunda. At Kesharpur 10 per cent of the people thought to control the artificial rise of prices whereas 20 per cent of them at Kesharpur thought that there is no solution.

Asking the question on the hike in transport fare different people replied differently. The answer were pointed at increase of petrol and diesel price and motor parts, high population growth, lack of information and artificial rise etc.

Coming to the question of counteracting such problems 40 per cent of the respondents at Kesharpur said that economic consciousness is necessary and 62.5 per cent at Kudamunda answered similarly. 40 per cent in Kesharpur said that there should be peoples' resistance and non-co-operation where as 20 per cent of them could not think of anything. In Kudamunda 25 per cent replied for resistance and 12.5 per cent could say nothing about it.

This also indicates that despite efforts to conserve forests and getting organised people in general lack a satisfactory level of awareness, required for an ecological way of life, but in comparison to other areas where no efforts of eco-development has been taken, the people in these two areas are definitely more aware.

At Kesharpur the people have started managing other resource components like croplands, fishponds and grazing lands in certain villages which indicate that gradually on integrated approach to eco-development can be taken up whereas in Kudamunda area, people have perceived such possibilities but due to lack of proper organisation and training they are yet to begin. Intervention of external agencies has completed more than ten years for Kesharpur area but it is in its initial stage for Kudamunda. With an extension of support services for five to ten years Kudamunda can also show signs of perceptive eco-development.

Conclusion

As emphasised earlier, eco-development is a process of meeting all the needs of people of the concerned area. In the modern context, varieties of needs have increased to an extent that, only local resources can't meet them. Moreover, utilising the local resources demands certain skills which may not be available in all the areas. Despite this problem of modern needs and amenitities, certain basic needs like the staple food, cooking energy, traditional medicines, portable water, fodder and fertilizer can always be made available locally. To achieve this an integrated approach to utilize the natural resource base is always required. Therefore, resource bases like forests, grazing lands, crop lands, streams and water bodies should be integrated, keeping in view the quantitative needs of the local population in this regard. Areas which claim to implement eco-development action programmes through the exotic tree planting are very often rejected by the local people. Fortunately, Kesharpur and Kudamunda are two pockets of optimism like many of them in Orissa and India Kesharpur has gone one step ahead over Kudamunda by undertaking plantation, managing village ponds and creating village fund because it started earlier. A number of supporting agencies are regularly helping them.

Both Kudamunda and Kesharpur have the irrigation potentials from the water systems present in and around their village. Both the areas have immense potentialities to plan for integrated production systems, to produce more biomass from grazing lands, livestock, wasteland, crop land, forest and water systems provided the sector approach is changed and the exploitation of the rural elites and the rich is countered by the poor, for the management of the common property resources of both the areas. Information at Kesharpur and Kudamunda regarding the detail social structure is not collected but it is certain that without an understanding of the long term goals of eco-development cannot be achieved.

Details of management of both the forests is beyond the scope of the present study but factors like legality, income generation for the poor villagers from the forest, equity in distribution of the benefits, management and settlements of conflicts arising from time to time and designing a sustainable productive system for satisfaction of the needs of every family of the village are questions which could not be taken up in this study. Perfect model for eco-development is neither available at these study sites nor any where else reported so for.

Organizational activity at Kesharpur is indicative of the fact that a simple task of forest protection can take the people's organisation a long way which has not began at Kudamunda. The interest and inter relationships that the people of Kudamunda have in sure to go ahead if they get support services like the people of Kesharpur. However, in both the areas, integrated models of eco-development can be visualised provided planning and implementation are sincerely taken up.

REFERENCES

Agarwal, A. and Narain, S., 1989. Village Ecosystem planning Towards Green village 9-16 Center for science and Environment, New Delhi.

Aiyer, Q.S., 1990. Development of viable models of Agro Ecosystems to compromise with forest ecosystem in the tribal hamlets along the Western Ghats. *Environmental Research* I, 106.

Bhat, P.R., Bhat, D.M. Ravindranath, N.H. and Gatil, M. 1983. Microlevel planning for sustainable development of two village ecosystem in Uttara Khanda district.

Chambers, R., Saxena, N.C. and Shan, T. 1989. *To the Hands of poor water and trees*, Oxford and IBH Publishing Co. Pvt. Ltd., New Delhi.

Ganapathy, P.M., 1990. Studies on the changing pattern of man-Forest Integration and its Implication of Ecology and Management. *Environmental Research* I. 28.

Jodha, N.S., 1990. Rural common property resources : Contributions and Crises, *Westland News,* Society for promotion of West Land Development, New Delhi.

Mishra, A. and Dash, M.C. 1981. Education for survival and Development *Proc. Int. Sem. on Env.* Edn. New Delhi, Dec. 1981, Indian Environmental Society.

Odum, E.P., 1971. *Fundamentals of Ecology.* W.B. Saunders company Phildelphia. London, Toronto P.P. 574.

Pati, S., Panda, R. and Rai, A., 1993 Comparative Assessment of Forest protection by Communities. *Proc. Workshop. Joint Forest Management* BBSR. 1993 May.

Reddy, N.A.K. 1990. Studies on model village Agricultural Eco-systems. *Environmental Research* I. 66.

Sashikant, Neera, M. Singh and Kundan, K. Singh. 1991. ''Community *Based Forest Management Systems* (Case studies from Orissa)'' Indian Institute of Forest management, Bhopal, Swedish Inter. Dev. Authority. New Delhi. Iso/Swed forest New Delhi.

Singh, Neera, M. and Singh, Kundan, K., 1992. Forest protection by communities. The Balangir Experience. Personal Communication.

Udhayashankar, S., 1990. A Coordinated study of the Background and Impact of Deforestation on the socio-economic condition of village life and Environment in Dakshina Kannada District. *Environmental Research* I. 105.

Chapter 23

Mangrove Ecosystem and its Biodiversity in Orissa Coast with Special Reference to Bhitarkanika

C.S. Kar[1] AND G.R. Satpathy[2]

[1]Bhitarkanika National Park Forest Department Govt. of Orissa; [2]School of Life Sciences Sambalpur University Jyotivihar, Orissa

Abstract. This paper presents part of an investigation involving biological diversity of Bhitarkanika mangrove forest ecosystem of Orissa coast. A study has been made about the traditional uses of mangrove forest products in Orissa, pattern of exploitation and management of mangrove forests during the 'Zamindari' period, as well as present threats to this fragile ecosystem. It also deals with ecological interaction between mangroves and wildlife in general and the mass nesting of olive ridley sea turtles in particular and suggests conservation measures for their protection and management.

Key words : *Wildlife, Biodiversity, Mangroves, Conservation.*

Introduction

Mangroves are essentially tropical or sub-tropical plant communities adopted for survival in saline habitats and interfaces where land and the sea meet. Mac Nae (1968) named mangroves as mangal vegetation. These systems usually act as a buffer ecotonal zone between land and oceans and are capable of tolerating violent seasonal winds. They play an extremely important role in maintaining high productivity and rich biotic diversity of coastal waters and are of interest from economic, scientific as well as wildlife management point of view. The mangroves in Indian coasts are largely confined to the Gulf of Kutch in Gujarat; Bombay region; in the delta system of Ganges and Bramhaputra (the Sunderbans); in the delta system of Brahmani, Baitarani and Dhamra river system of Orissa (Bhitarkanika) and in the Mahanadi delta of Orissa; in the mouths of Godavari and Krishna rivers in Andhra Pradesh; in the Cauvery delta and point Calimere region of Tamil Nadu. Besides, mangrove forests are also present in the Indian Bay islands of Andaman and Niçobar.

The world's total mangrove area which spans over 30 countries, including those for the various island nations is about 99,300 sq.km. The total area of the mangroves in India, including the area permanently under water, is reckoned at about 6,740 sq.km. This covers about 7 per cent of the world's mangroves (Annon, 1987). Orissa at present has a total of about 195 sq.km of mangroves of which about 125 sq.km. lies mostly in the erstwhile zamindari areas of Kanika and Kujang. No detailed systematic study on the structure and function of these systems have been made. However, some recent attempts to characterize the floristics of Orissan mangrove forests have been made (Banerjee 1984, 1985, 1986, 1987; Banerjee and Dash 1972; Banerjee and Rao 1985; Carter and Narayanswamy 1922; Champion and Seth 1968; Choudhury 1984; 1986, 1987, 1990; Dash and Kar 1987 a, b. 1990; Haines 1921-25; Kar 1988; Majumdar and Banerjee 1985; Mishra and Panigrahi 1987; Mooney 1950; Panda 1987; Panigrahi *et al.* 1964; Pattnaik and Choudhury 1988; Samal and Pattnaik 1988).

Study Site, Materials and Methods

The study site Bhitarkanika mangrove forest, declared as a wildlife sanctuary during April 1975 and as a proposed National Park during October 1988, is located in the undivided Cuttack and Balasore (now Kendrapara and Bhadrak) districts of Orissa, in the deltaic regions of Brahmani, Baitarani and Dhamra river system. The area lies between 20°30'N to 21°C'N latitude and 86°45'E to 87°6'E longitude (Fig. 23.1).

The Forest Survey of India (FSI) assessments using satellite data records about extent of mangrove forest cover of Orissa was compared. Extensive survey and ground status truthing was made to determine the present status and causes of mangrove forest depletion in Orissa. An assessment was also made about the status of some dominant wildlife species occurring in the area more particularly about wildlife species which are at the apex of coastal estuarines mangrove ecosystem whose survival is entirely dependent on the mangrove habitat.

Mangroves of Orissa

Orissa situated between latitudes 17.5°N to 23°N and longitudes 81.5°E to 87.5°E has a 480 km coastline covering the undivided coastal districts of Balasore, Cuttack, Puri and Ganjam from north to south. Although no detailed survey of the former area and extent of the mangroves of the Orissa coast has been made but it is believed that the total land area under mangroves has been reduced considerably due to human activities in recent times. Blasco (1977) pointed out that the area under mangroves has been reduced from 12,000 has to 5,000 ha. In the past Mahanadi delta was thought to be the only mangroves in Orissa. The other coastal stretches like the Brahmani-Baitarani-Dhamra delta, and the coastal distributary Devi-Budhabalanga estuarine mangroves formations and possibly the Subarnarekha stretch were ignored (Annon, 1987) largely because of the inaccessibility of these areas.

The mangroves in the 670 km² Bhitarkanika Wildlife Sanctuary of Cuttack district include 170 km² of protected forests notified under Indian Forest Act, 1972 in the deltaic region of river

Bramhani-Baitarani (Fig. 23.1). to a lesser extent mangroves also occur in the Balasore district, in the wheeler island (Nadiachadha); in some newly formed mudflats near Palmyras Point, at confluence point of several rivers, in the Hukitola island near False Point. The mangroves of Chilika and Kujang areas have been worse affected by human activities in the last few decades. At present only small patches of denunded mangroves occur in the Kujang coast and in the coastine of Balasore district in between Dhamra and Chandanipal and in Balaramgudi area near Chandipur.

According to remote sensing data, in Orissa the area under mangrove forests in 1975 was 234 sq/km which was reduced to 214.58 sq/km in 1984. As such about 2 sq.km of mangrove forests per year is being lost in Orissa. The Forest Survey of Indian (FSI) assessments of mangrove forest cover using satellite data records total mangrove forest area of Orissa to be 199 sq.km (FSI, 1987) and 195.sq.km (FSI, 1991). The space borne remote sensing and these state of Forest Reports (SFRs) have effectively focussed the attention of planners and administrators at speedy depletion/ degradation of mangrove forest resources including forest encroachments which is adversely affecting the wildlife habitat in coastal areas.

Biotic Diversity

Flora

The mangrove vegetation of Orissa comprises more than thirty tree species comprising of both exclusive and non-exclusive mangroves. The common species alongwith their local vernacular names include *Aegialitis rotundifolia* (Banarua), *Aegiceras majus* (Kharasi), *Avicennia alba* (Dhala Bani), *Avicennia marina* (Singala Bani or Nadaka Bani), *Avicennia officinalis* (Bada Bani), *Brownlowia tersa* (Nati sundari), *Bruguirera parviflora* (Bandari), *Bruguiera gymnorhiza* (Bandari), *Bruguiera cylindrica* (Bandari), *Bruguiera Sexangula* (Bandari), *Ceriops decandra* (Ganani), *Ceriops tagal* (Garani), *Cynometra ramiflora* (Singada), *Excoecaria agallocha* (Guan), *Heritiera fomes* (Sundari), *Haritiera littoralis* (Bada sundari or Dhala Sundari), *Heritiera kanikansis* (Kanika Sundari), *Kandelia candel* (Sindhuka), *Lumnitzera racemosa* (Churanda), *Rhizophora apiculata* (Rai), *R. mucronata* (Rai), *R. stylosa* (Rai), *Sonneratia apetala* (Keruan), *Sonneratia caseolaris* (Orua), *Sonneratia griffithi* Chakada), *Xylocarpus gangeticus* (Pita Korua), *Xylocarpus granatum* (Sisumara), *Xylocarpus mekongensis*, and *Xylocarpus moluccensis* (Pitamari). The palm *Phoenix paludosa* (Hentala) is widespred throughout the mangrove forests of Orissa and the leaves provide ideal nest building materials for the estuarine crocodile, *Crocodylus porosus.* Recently another palm species *Nypa fruticans* (Golpatta in Bengali) has been reintroduced into the Bhitarkanika forests. *Acanthus illicifolius* (Harkanch) is widespred in all mangrove forest areas of Orissa.

Among non-exclusive mangroves common tree species met within Bhitarkanika are *Amoora cuculata* (Uanra), *Cerbera manghas* (Pani Amba), *Dolichandrone spathacea* (Gosinga), *Intsia bijuga* (Ma-Sitha), *Tamarix indica* (Jagula), *Tamarix dioica, Tamarix ericoides, Thespesia populnea* (Habali) and *Thespesia populneoides.*

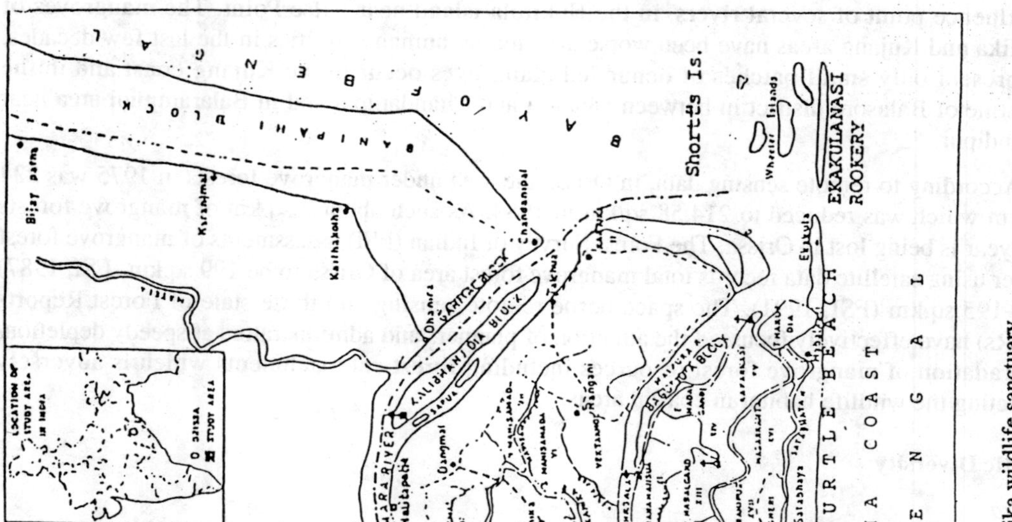

Fig. 23.1. Map of Bhitarkanika wildlife sanctuary.

Other important non-exclusive mangroves include *Caesalpinia nuga* and *Caesalpinia cristata* (Nentei Kanta), *Crinum defixum* (pani Kenduli), *Clerodendrum inerme* (Chiani), *Dalbergia spinosa* (Goira Kanta), *Derris scandens* (Kata Nai), *Derris uliginosa, Flagellaria indica* (Bahumruga), *Heliotropium curasavicum, Hibiscus tiliaceus* (Bania), *Merope angulata* (Bana lembu), *Myriostachya wightiana* (Nalia ghasa). Like *Phoenix paludosa,* the aquatic fern *Achrostichum aureum* (Kharakhari) also provide good nesting materials for the salt water crocodile *Crocodylus porosus. Finlysonia obovata* (Lata Rai) is abundantly found in less saline area viz; in the creeks of Khola, Mahisamada, Suajora, Ghanagalia, Babargalia etc. on one or both banks. *Suaeda nudiflora, Suaeda maritima* and *Suaeda monoecia* (Giria saga), Occur along the tracts of this region particularly where the soil is clayey loam. *Portresia coartata* (Dhani ghasa or Tanti Dhana) is also found extensively on river banks of the sanctuary.

Important mangrove associates occurring in Bhitarkanika mangroves include *Capparis horrido* (Mendhi or Asadhua), *Pandanus fascicularis* (Luni Kia), *Phragmitis* karka (Nala), *Pongamia pinnata* (Karanja), *Salvadora persica* (Miriga), *Sapium indica* (Batula), *Tylophora asthmatica* (Anantamula) and *Tylophora tenuissima.*

The characteristic beach vegetation includes 'Kansari Lata' *Ipomea pes-caprae* (Linn), `Guda Kanka', *Spinifex squarrosa, Launaea pinnatifida* (Cass), *Hydrophylax maritina* (Linn) and some beach grasses such as *Cyperus arenarius, Indigofera enncaphylla* etc. During the past 20 years coastal plantations of *Casuarina equisitefolia* have been made all along the sandy tracts of the coast as a cyclone preventive measure.

Mangroves Fauna

There are very few animals which are absolutely exclusive to the mangroves and some animals spend at least a part of their life cycle in the mangrove ecosystem. Some of the animals are so suited to the mangroves in fact, that they can be considered as characteristic of the mangrove habitat. The animals found in the mangrove community, according to their habitats can be categorised as mud dwellers, the aquatic inhabitants restricted to the water ways, forest dwellers found in the forests, and canopy dwellers found in tree conopy.

The mangrove mud is a very special habitat which may further provide a variety of microhabitats. The mud-dwelling animals show fascinating unusual features of structure, function and behaviour which are characteristics of the mangrove environment. The mud dwellers include the amphibious fish (*Periophthalmus* sp.) and a veriety of crabs and molluscs. These mud dwellers are abundantly found along all river banks and mud flats of the Bhitarkanika mangrove area. About eight species of crabs are found including the commercially important mangrove crab *Scylla serrata.* Fishermen use iron hooks or baited disc-net traps to catch this species.

King crabs or the Horse-shoe crabs (*Tachypleus gigas and Carsinoscorpius rotandicauda*) which are relic of past ages and which came on earth 400 million years back during the upper Silurian age, still find their way to Bhitarkanika mud flats for their mating and laying eggs where their pre-historic trilobitic larvae thrived in the pristine and primitive Bhitarkanika mangrove atmosphere. King crabs are slaughtered widely for the extraction of the highly priced chemical, 'Lectine', and have become threatened globally. The other interesting crab is the Hermit crab (*Clibarnius padavensis*), which does not have a shell of its own and occupies dead gastroped shells. As it grows larger it changes to larger shells. Among the crustaceans, the tiger prawn, *Peneaus monodon* in the main economic attraction of the tidal creeks. The post larval stages of the species are nourished and take shelter in the high nutrient nursery of the mangrove till they reach the adult stage. The content of the growth promoting harmone, the gibberellic acid of mangrove leaves, also adds to the attraction.

The aquatic inhabitants of the Bhitarkanika mangrove environment include planktons, shrimps, prawns, other neritic fauna, fishes, terrapins, brackish water and marine turtles, monitor lizard, saltwater crocodiles and the dolphins. Innumerable diesel drop like phytoplankton colonies which are the developmental stages of a benthic algae (*Derbasia marine*) are seen in winter months in the creeks. Due to the presence of Noctiluca and other such luminescent planktons one would witness bioluminescence over water in the creeks, rivers and estuary during the night time which all most looks like a fairy land. Luminiscent fire flies also add romance to Bhitarkanika rivers and creeks during the night time. Phytoplanktons are the autotrophs which are consumed by zooplanktons, the primary heterotrophs which in turn are consumed by higher consumers like fish, turtles and water birds ending in the crocodile at the pinnacle of the aquatic food-chain.

Some of the aquatic inhabitant of Bhitarkanika mangroves can be considered as mangrove visitors like the brackish water turtle (*Pelochelys bibroni*) which migrate to the Gahirmatha coast during the breeding season for sharing the same nesting beach with the olive ridley sea turtle *Lepidochelys olivacea* for laying eggs. Due to increasing fishing activities this species alongwith *Batagur baska* is becoming rare in the rigion. Gahirmatha coast of the Bhitarkanika sanctuary is considered as the largest nesting ground of the olive ridley sea turtle *Lepidochelys olivacea* in the world. The sanctuary is also the home of the endangered salt water or estuarine crocodile (*Crocodylus porosus*) which hold world's largest wild salt water crocodile of 7.5 meter length (Guines Book of World Record-1991) and some white specimens of this species. More than 1100 captive reared young Crocodiles (*C. porosus*) have been released into different river systems of the sanctuary as a part of the GOI/FAO/UNDP sponsored "grow-rear and release'' programme to build up the depleted population in nature. At present more than 500 crocodiles of different age groups are bring reared at Dangmal Salt Water crocodile Research and Conservation Centre. Besides, several of the captive hatched and released crocodiles have started laying eggs, thus producing about 20 nests per annum during the laying season in May.

The forest dwellers or the terrestrial mangrove inhabitants include ants, spiders, Indian Flap Shelled turtle (*Lissemys punctata*), yellow monitor and common land monitor lizard (*Varanus flavescens* and *V. bengalensis*), water monitor lizard (*Varanus salvator*), Common garden lizard (*Calotes versicolor*), Common skink (*Mabuya carinata*), Snake skink (*Riopa punctata*), Common Indian house gecko (*Hemidactylus flaviviridis*), Indian chameleon (*Chamaeleon zeylanicus*), Snakes such as King Cobra (*Ophiophagus hannah*), Python (*Python molurus*), Banded Krait

(*Bungarus fasciatus*), Russels viper (*Viper russeli*), Common Rat Snake (*Ptyas nueous*). Painted bronze back (*Dendrelaphis pictus*), Common Indian bronze back or tree snake (*Dendrelaphis tristis*), Common Indian Krait (*Bungarus caerulens*), Blind Snake (*Typhlina bramina*), Checkered Keelback (*Xenochrophis Piscator*), etc.

Mammals of the Bhitarkanika mangroves include primates such as Rhesus monkey (*Macaca mulata*), Carnivores such as Panther (*Panthera pardus*), Leopard Cat (*Felis bengalensis*), *Fishing cat* (*Felis viverrina*), Jungle cat (*Felis chaus*), Small Indian Civet cat (*Viverricula indica*), Common palm civet or toddy cat (*Paradoxurus hermaphroditus*), Common mangoose (*Herpestes edwardsi*), Small Indian mangoose (*Herpestes citticollis*), Jackal (*Canis aureus*), Indian fox (*Vulpes bengalensis*), Striped hyaena (*Hyaena hyaena*); Pigs such as Wild Pig (*Sus scrofa*); Rodents such as Inidan porcupine (*Hystrix indica*), mole rat (Bandicota bengalensis), Long tailed tree mouse (*Vandeleuria oleracea*); Deers such as spotted deer (*Axis axis*), and Sambar (*Cervus unicolor*); Mustelids such as common otter (*Lutra lutra*); Smooth Indian otter (*Lutra perspicillata*) and Bats such as Bearded sheath tailed bat (*Taphozous mehanopogon*), the short nosed fruit bat (*Cynopterus sphinz*), Indian Flying Fox (*Pteropus siganteus*) etc. Aquatic mammals are represented by two species of dolphins and the Finless or little Indian porpoise. The dolphins are found in coastal waters, in the rivers viz; Bramhani, Baitarani, Dhamra and particularly at the confluence point of these large rivers.

The tree canopy of the Bhitarkanika mangrove ecosystem have certain distinctions. The woody limbs and the leaves, flowers and fruits of which it is composed are found in no other type of forest. The only elements of any significance are the insects and birds. Honey bees are particularly abundant during the flowering seasons of mangrove trees, being specifically attracted to the plentiful nectar. The quality of honey differs greatly according to the trees. Lage quantities of about 3000 Kg to 5000 Kg of honey are produced particularly during February to May. Near the crocodile rehabilitation centre, in a roughly 1 km^2 area of Bhitarkanika forest (Baga gahana also known as 'Chadeigahana') more than 50,000 migratory birds assemble each season (mid May to mid December) and make their nests. The birds include open billed storks, white ibis, cormorants, herons, egrets and painted storks. The heronry is also visited by the endangered spotted billed Pelicans. The scavenger Bengal vulture (*Vulture bengalensis*) make their nests at the top of tall trees at Khola mouth. They can be seen in large numbers at Gahirmatha coast during winter for scavenging upon the dead olive ridleys which reach the coast due to the off shore fishing activities. Some of the rare birds like Brahminy kite (*Haliastur indus*) and the White bellied Sea Eagle (*Halicetus leucogaster*) also nest on tall trees in the banks of Bhitarkanika rivers. Most of the mangroves tree canopy dwellers are migratory and extremely mobile creatures, and only a very few are actually mangrove inhabitants. Spoonbills and the black necked storkes have been sighted on a few occasions in the Gahirmatha coast and Bhitarkanika forest area. The most notable tree canopy dwellers of thre mangrove forest are the Kingfishers which can be regarded as exclusively to mangroves and are highly mangrove dependent. Six species of King fishers have been recorded in this area. Other migratory birds such as Barheaded geese, Brahminy ducks and waders also visit the area during winter. Shore associated birds include sand pipers and gulls. Thousands of black and brown headed sea gulls visit the Gahirmatha coast coinciding with the peak nesting season of the olive ridley sea turtles at the above rookery during winter and early summer months. These birds are secondary predators of the olive ridley eggs but are important predators of the hatchilings. Large number of crows and kites also visit the Gahirmatha rookery during arribadas of olive ridleys and during the peak hatchling emergence period. Tree canopy dwellers also include the rhesus monkeys. The mangroves provide all the habitat requirements of some species of threatened, rare and endangered species of wildlife and their conservation is highly dependant on the existence of mangroves. In the recent times, the mangrove vegetation in Orissa is depleting in a faster rate and this in turn has resulted in the depletion and disappearance of many endangered and vulnerable wildlife species as a loss of habitat. The 'global status of mangrove ecosystems (Saenger *et al.* 1983) presented a list of plants and animals: at risk'' which included 2 species of molluses, 7 species of reptiles, 65 species of birds and 15 species of mammals. Since, some of the most endangered species of wildlife such as the salt water crocodile (*C. porosus*), estuarine turtle (*P. bibroni*), Batagur terrapin (*Batagur baska*), Olive ridley sea turtle (*L. olivacea*), Green sea turtle (*Chelonia mydas*), Hawksbill sea turtle (*Eretomochelys imbricata*) Leatherback sea turtle (*D. coriacea*) etc. are found in the coastal-estuarine mangrove ecosystem of Bhitarkanika, protection of the habitat in this last remaining mangrove areas of Orissa (Kar and Dash, 1984) should be the principal tool from wildlife management point of view.

Mangroves as a resource

Mangroves have triditionally been exploited in Orissa as forests. Most mangrove areas in the state have received practically no attention in terms of rational exploitation. As a result, mangroves have completely disappeared from some places and in other areas they are being depleted and degraded at an alarming rate. The only mangrove area which received some sort of protection and management in Orissa has been the Kanika forests. When Kanika was 'Zamindari' area, the total mangrove forest area was divided into different forest blocks such as protected forest and protected reserve forest blocks. The local population were paying 'kara' (revenue) for collection and use of mangrove wood for the construction of houses, fencing, for fire wood and as wood for burning bricks. Timber of mangrove forests were also used for construction work such as bridges, jetties, dock pillings, boat building, beams and poles for buildings, flooring, fence posts, furniture, etc. Leaves of *Phoenix paludosa* (hentala) was extensively used for thatching the houses in Kanika and Kujang and it is still continuing in Kanika. Mangroves were also used as poles for fish trap. Tannin and dyes were extracted from *Rhizophora, Ceriops,* etc. and were used as dyes for net and line preservation. At present, this practice has been largely replaced by the extensive use of nylon and plastic nets. Pneumatophores were used as floats in fishing nets. Leaves of *Avicennia* sp. were used as cattle feed. The wood of Sundari (Heritiera sp.) was locally used for construction of plough. During Kanika Raj the exploitation was permitted only in different protected forest blocks on a rotation basis while no exploitation was permitted in the protected reserve forest blocks which served as 'core areas'. Licences were also given seasonally for fishing, Collection of animal and forest products such as shrimps, prawns, crabs, honey and sea turtle eggs. In the early 1970's, due to rapid immigration of human population into the area and ever increasing demand of forest products, this management system of the mangroves proved unsatisfactory, as it was difficult to control and some areas were either clear felled or converted into human habitation, agricultural fields, fish ponds, salt pens, etc. and some areas were left only with less valuable species. Regeneration was thus a problem. Moreover, regeneration has been poor after large scale clearfelling in some areas. Extensive plantation of non-mangrove species in some mangrove areas have also been done in recent years.

Threats to Mangroves

The mangroves in Orissa are dwindling fast due to human encroachments. The encroachments for human settlements, conversion of mangrove forest areas into agricultural fields and fire wood collection etc. are severe outside the sanctuary limits but have also occurred within the Bhitarkanika sanctuary area in the recent past. In some areas the refugees rehabilitation programme and construction of saline embankments for conversion of land to agriculture and prawn ponds have resulted in the depletion of mangrove forest from within and in the vicinity. In the process of coastal erosion two forest blocks have been completely lost and the remnants of the mangrove trees along Gahirmatha coast can be seen even now and particularly during the low tides. Mangroves have also disappeared from the river banks in most places resulting in the erosion of river banks. The effects of pollution and oil on Orissan mangroves have not been studied.

In Bhitarkanika mangroves fishing is done by diverse manners. One such method is the construction of bunds at the upper end of the creeks to a height of upto the level of mid tide. When the water level increases during the high tide, the bunds will be completely submerged. Fishes usually have the tendency to move towards the shore. During low tide fishes are trapped inside the embankment and those trying to escape by leaping, are collected by nets. Such bunds are usually illegally constructed in the Bhitarkanika mangroves particularly in the Adbhutia, Habalganda, Sunei Rupei Jora, etc. The bunds constructed in this fishing method effect the pattern of the current leading to siltation which in turn gives rise to poor nutrient condition of the soil and causes the adjacent mangrove vegetation to perish. Further more, since this fishing method is more intensively carried out in mangroves of the forest areas close to the coast where the tidal amplitude is high, the bunds built also acts as barrier for the effective dispersal of viviparous mangrove seedlings. This prevents further colonisation of mangrove vegetation in the area and may partly account for sea erosion.

Fishing of most juvenile marine fishes and prawns using stake nets and nets of small mesh sizes is affecting the future of large scale fishery in the area. The stake-nets when set up in the water column act as barrier for the movement of fishes, crabs and prawns and other marine species and they avoid to enter into the mangrove ecosystem.

When villages and agricultural fields are situated close to the rivers and mangrove forests, pesticides and fertilizer find their way into the ecosystem there by causing serious food chain contamination through biological magnification.

Oil is especially dangerous to mangroves, because when deposited on air breathing roots (pneumatophores), it cuts off oxygen supply to the roots. Increasing number of mechanised vessels, construction of ports, harbours and oil drilling operations along the coast may become a serious threat to the existance of mangroves in the foreseeable future.

Ecological Links of Mangrove Forests

Mangrove forests serve as a link between terrestrial and marine ecosystems. They receive a continuous supply of inorganic nutrients from the adjacent land mass and release dead organic matters to the sea through rivers and estuary.

The huge Maipura estuary near Gahirmatha rookery which is nourished by two major river systems of the state of Orissa (vis; Bramhani, Baitarani and their tributaries) receiving much of the litter and organic matters from the mangrove forest ecosystems of Bhitarkanika Sanctuary, possibly provides a fertile breeding ground for the planktons which in turn is linked with the food chain of crustaceans and pelagic fishes on which the turtles feed. The detritus food chain of a mangrove forest is complex and no study has been made in Orissan mangroves to understand the foodweb relationships, nutrient cycling, energy flow and regenerative power of these systems. A large number of hatchling olive ridley turtles at Gahirmatha coast, Orissa, soon after they enter into the sea are drifted back by the action of the tides into the estuary and rivers and spent about a week in the mangrove environment.

Turtle Nesting Sites

Throughout the world mass nesting beaches of the olive ridleys are found very closely associated with the presence of mangrove forests in the vicinity (Table 23.1). Olive ridleys prefer to nest in large numbers on sandbars near river mouths (Loveridge, 1946). Since the coastline of Orissa is interrupted by numerous small and large rivers (viz; Rushikulya, Devi, Prachi, Kadua Nadi, Kushabhadra, Gobari, Hansua, Baunsagarh, Maipura, Dhamra, Budhabalanga, Subarnarekha etc. from South to North and their tributaries), the nesting of olive ridleys is comparatively more at the sandspits at these river mouths. Some of these nesting beaches are far away from human habitation and the texture of beach sand is perhaps more suitable for digging the nests, providing oxygen requirements and maintaining other microclimatic conditions necessary for the development of eggs. Two large mass nesting beaches of the olive ridleys in Orissa located on the northern and southern extreme end of Cuttack district–the rookery at Gahirmatha and another near Devi river estuary with minor arribadas occurring in between and in the Hukitola island, together host one of the largest aggregations of olive ridleys in the world and certainly the largest in India. Table 23.2 shows the estimated number of *Lepidochelys olivacea* at major nesting sites of the world. On the mass nesting beaches the olive ridley females, instead of scattering over wide areas, congregate to form huge arribads of thousand and thousands of individuals and emerge on to the beaches for nesting enmass. The mass mating, mass nesting places and the migratory route of the Olive ridley sea turtles have been traced in the Bay of

TABLE 23.1. *Some environmental parameters of olive ridley concentration localities (Hughes, 1972)*

Locality	Salinity	Range of sea surface, temp. (°C)	Annual rain fall (mm)	Mangroves along shore lines.	Neritic fauna	Organic content	Tides	Spring tidoi range (m)	Relative abundance of olive ridleys.
1. Orissa, India	34.00	25-28	1000	Abundant	Abundant	High	Semi-diurnal	2-4	Abundant
2. Surinam	34.00	26-27	1000	Abundant	Abundant	High	Semi-diurnal	2-4	Abundant
3. Pacific S. Mexico	33.00	27-28	1000	Abundant	Abundant	High	Semi-diurnal	2-4	Abundant
4. West Africa	35.00	25-27	1000	Abundant	Abundant	High	Semi-diurnal	2-4	Present
5. North Australia	35.00	27-29	1000	Abundant locally	Present	high	Semi-diurnal	4	Present
6. South China Sea	33.00	25-28	1000	Abundant	Present	–	Mixed	2-4	Present
7. N. Mocambique and Tanzania	35.00	25-28	500	Abundant locally	Abundant locally	High in inshore locally	Semi-diurnal	2-4	Present
8. West Malgasy	36.00	24-28	1000	Abundant locally	Abundant locally	High in inshore locally	Semi-diurnal	2-4	Present

*Data drawn from numerous sources
Tidal data from Pattullo (1966)

TABLE 23.2. *Estimated number of female olive ridley sea turtle, Lepidochelys olivacea at some major nesting sites of the world. (Modified after ross, 1982).*

Nesting Locations.	Approximate number of female/ year (thousands)	Reference.
Mismaloya, Mexico	20-50*	Marquez 1976
Tlacoyunque, Mexico	20-50*	Marquez 1976
Chacahu, Mexico	20-50*	Marquez 1976
Escobilla, Mexico	50-100*	Marquez 1976
Nancite, Costa Rica	200	Hughes & Richard 1974
Ostional, Costa Rica	100	Hughes & Richard 1974
Eilanti, Surinam	1	Schulz 1975
West Africa	--	Carr 1952
Mozambique	1?	Hughes 1974 a,b
Malagassy Republic	1?	Hughes 1974 a,b
Hawkes Bay, Pakistan	1?	Salm 1976
Coromandel Coast, India	10?	Valliappan & Whitaker 1975
Orissa Coast, India	150-600	Present study, Kar 1988.
Burma	--	Smith 1933
North East Australia	1?	Bustard 1972
Trenggnu, Malaysia	1	Moll, Pers. Comm.

– No data

*Estimated numbers upto 1974. These numbers are known to be greatly reduced in 1977-79 and thereafter.

Bengal as shows in Fig. 23.2. The huge mass arrival at Gahirmatha coast involving an estimated number of 3 to 4 lakhs females in each arribadas is unparallel and unique in comparison to any nesting site in the world in general and Indian in particular. Fig. 23.3 shows the estimated sizes of olive ridley arribadas for the past 16 years (1976-1991) at Gahirmatha coast, Orissa. Each mass nesting period lasts for about a week. The first arribada in Gahirmatha coast usually occurs during late December to early February and the second arribada during March to Mid April. At the above rookery, smaller arribadas may occur during the months of April and May. Besides, sporadic nesting continues in all most all the months of the year. The above mass nesting sites in the Cuttack district of Orissa receive about six lakh olive ridley nesting and more than six crore of eggs annually. Assuming the male : female ratio as 1 : 1, 10 lakh turtles visit Orissa coastal waters annually. Each adult turtle, on an average, weigh 40 kg and thus more than 40 thousand

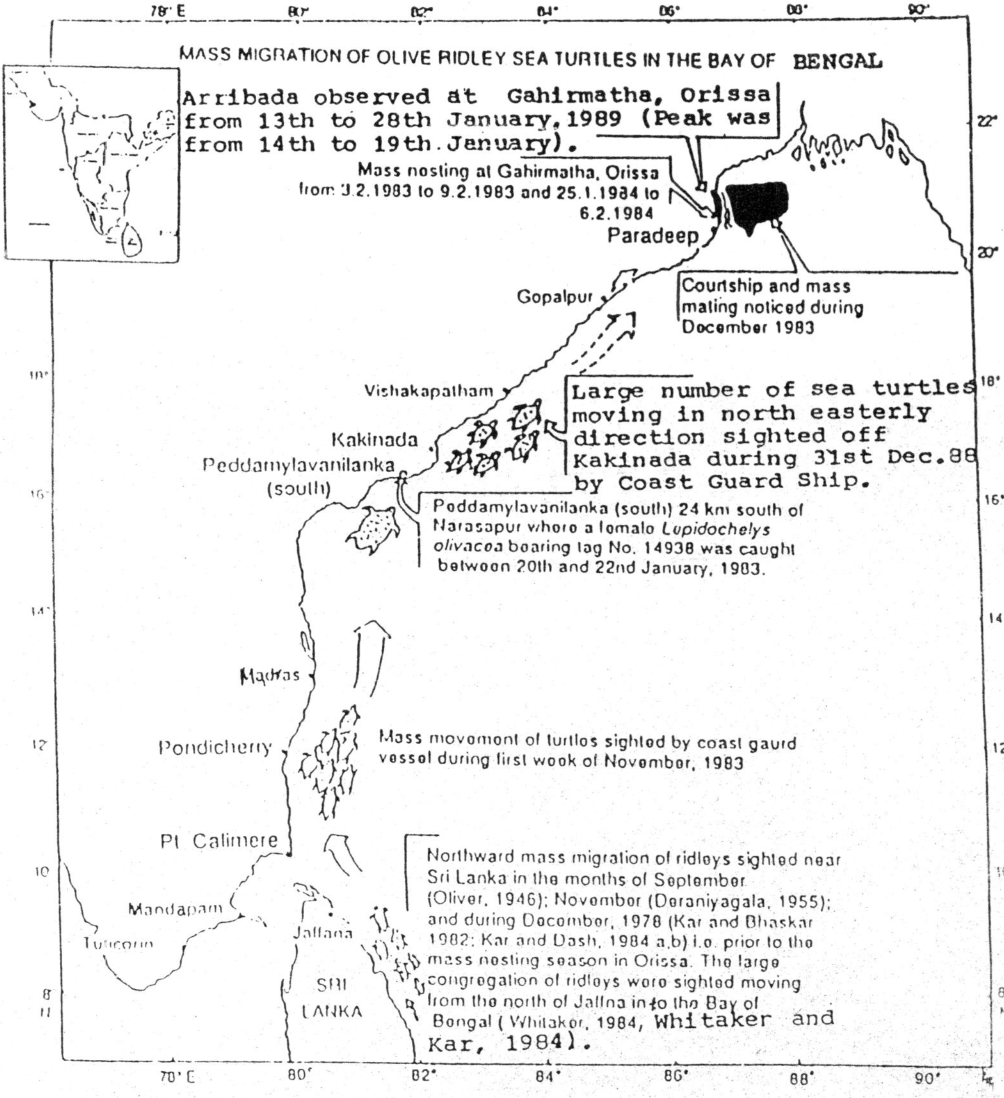

Fig. 23.2. Mass mating, mass nesting places and the migratory route of the olive Ridley sea turtles in the Bay of Bengal.

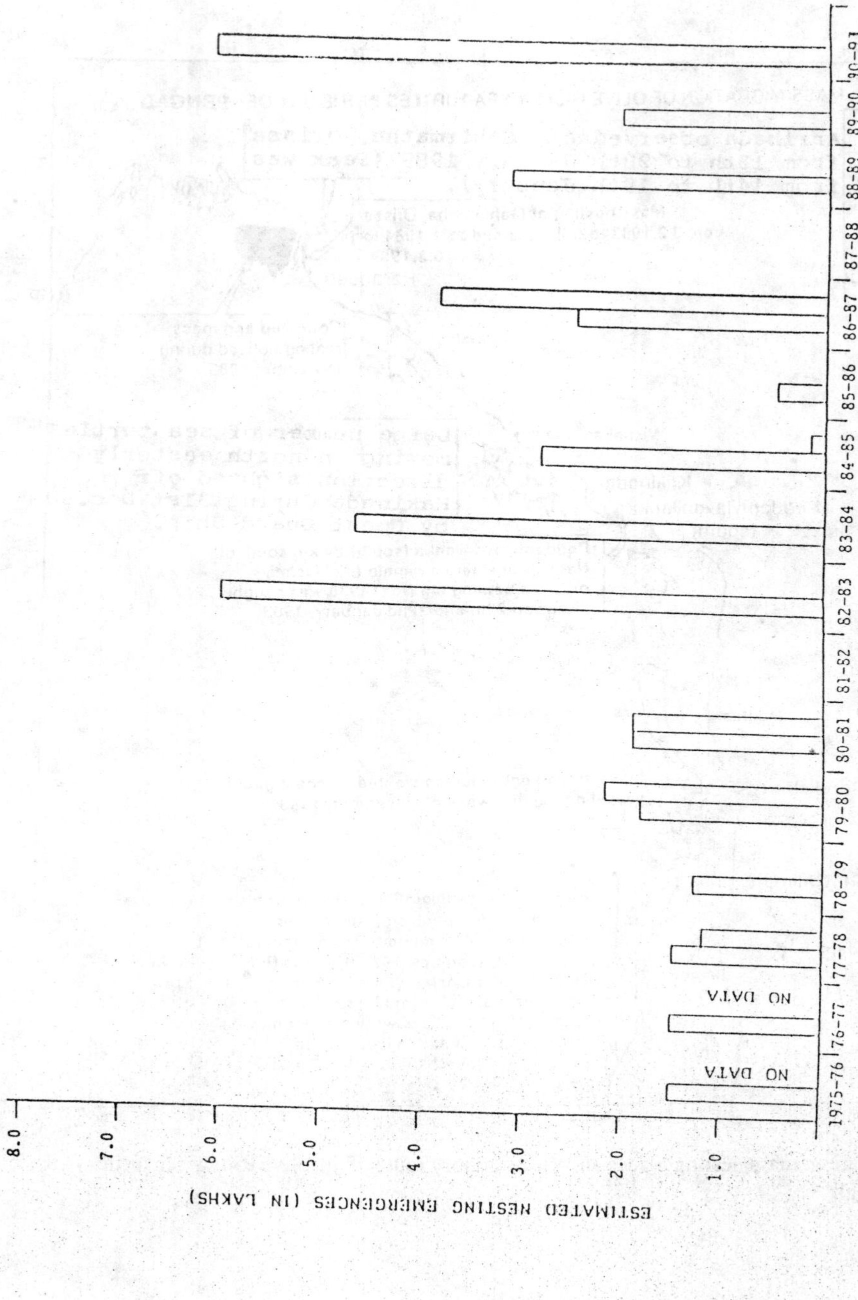

Fig. 23.3. Estimated sizes of olive ridley arribadas at Gahirmatha coast, Orissa during the study period (1976-91)

tons of turtle biomass visit the Orissa coast every year which is a huge resource in terms of protein food. Therefore, its conservation and management which in turn is associated with the conservation and management of the mangrove ecosystem, are essential to protect the species and simultaneously to get some benefit for human welfare (Kar 1988; Dash and Kar 1990).

Protection Measures taken for Conservation of Sea Turtles

The mass nesting ground was brought to the notice of F.A.O., Specialist Dr. H.R. Dustard during 1974. After this a special conservation and research project was made operative under the assistance from FAO/UNDP and Government of India till 1980 and subsequently as a Centrally Sponsored Project till date. The entire mass nesting area has been notified as part of the Bhitarkanika Wildlife Sanctuary, gazetted during April, 1975 and also as a part of proposed Bhitarkanika National Park notified during October, 1988. The congregation zone from Dhamra mouth to Mahanadi mouth upto 10 kms. of the territorial waters has been proposed as Gahirmatha Marine Sanctuary.

''Operation Gee turt'' has been initiated by patrolling of the coastal waters with the help of coast guard ship and patrol boats to prevent smuggling of adult turtles by mechanised fishing boats. Tight vigilance is maintained to apprehend smuggling of adult turtles by roadways, railways, fish landing stations and check gates. This has given dividents. Sumgglings of turtles from Orissa has been totally stopped by land route since 1984. Round the year research and monitoring operation is continuing to observe and record beach eco-system used by turtles as nesting ground.

A close coordinated operation is carried out with coast guard personnel and officers of Fisheries Department to prevent death of turtles. In spite of all these arrangements the incidental kill of adult turtles is on the increase. To reduce the incidental death the following conservation measures are suggested.

Conservation Measures Suggested for sea Turtles

1. The territorial waters in the Exclusive Economic zone (EEZ) from Hukitola island to shortt's island should be declared as a fishing free zone and a Marine Sanctuary for effective protection of turtles.
2. No fishing activity should be allowed, within 10 kms. zone from shore and strict implementation of OMFR Act, (1980) should be done till the area is properly notified as Marine Sanctuary.
3. Turtle Excluder Devise/Trawling Efficiency Devise (TED) should be made compulsory in all trawler nets to exclude sea turtles and still allow the trawlers to be able to catch shrimps.

Conservation Measures Suggested for Protection and Management of Wild Life and the Mangroves in Orissa

1. Survey of the exact area and extent of existing mangrove forests and the areas in which mangroves existed in the recent past in Orissa should be made and listed in an inventory and be declared as protected areas. Mangrove plantations should be made in those areas where mangroves have been eliminated in the recent past.
2. The status of the Bhitarkanika Wildlife Sanctuary should be elevated to the level of a National Park or Biosphere Reserve extending its limit northwards to include Wheeler islands and the Shortt's island and southwards to include Hukitola island to include the nesting beaches of sea turtles and the mangrove forests.
3. The Bhitarkanika and adjacent mangrove areas may be divided into three zones :

   – *a core area or preservation zone* set aside for environmental reasons. These natural areas should be retained in order to provide scientific reference sites and maintaining genetic reservoirs. To maintain the ecological balance of esturine areas, mangrove forests along the coasts and adjoining mouths of major river systems should be closed from development of fish ponds, salt pens and other activities. The area to be preserved should cover at least 2-3 kilometer stretch of mangroves along the coast line and on both sides of the mouth of the rivers fronting the sea. Further, mangrove forests on small islands which serve as major ecological compact of the island's ecosystem should in no case be disturbed.

   – *a buffer zone or forest reserve zone* managed for forests as well as forestry and wildlife products on a sustained yield basis.

   – *a reclamation zone* where the land is used for pond culture (fishes, crabs, prawns), agriculture, saltpens or other uses.
4. Exploitation of mangrove forests along the coast line and river banks which act as natural buffer against shore erosion, strong winds and floods should be left untouched and where mangrove plants have been eliminated plantation programme of mangrove plants should be initiated to compensate for areas lost or adversely affected. Planting along the river banks may not be necessary where a seed source is maintained or created. On the other hand, mangrove plantations in between the river bank and embankment margins would provide not only bank stability but a protection buffer to the exposed inner mangrove system. Regardless of location, swamplands, which are covered with virgin mangrove forest and dense young growth should be totally preserved since these are important in maintaining the ecological balance in the mangrove ecosystem.
5. Extensive social forestry plantations be made in the buffer zones of the sanctuary and other non-conventional energy sources should be made available to provide an alternative source of fire wood to the local people to reduce pressure on mangroves in the core areas. Degraded mangrove forests should be restored by large scale plantation and rigid protection.

6. Plantation of mangrove species in private lands of the sanctuary should be encouraged.
7. Quantitative informations on the physical, biological and socio-economic aspects of the mangrove system be collected and a comprehensive data base be developed. It should include the structure and function of mangroves, their flora and fauna, quantification of yields from various uses both actual and potential, and the socio-economic structure of mangrove dependent human populations.
8. During construction of roads through mangrove areas essential precautions should be taken to minimise interference with natural water flow in these systems. Whenever possible roads should be made parallel to surface flow patterns, including tides, and not perpendicular. Unavoidable obstructions however, should contain suitable culverts as closely as possible to the original major flows with a total flow cross section sufficient to accommodate 50 year flood levels.
9. Bhitarkanika mangrove ecosystem is interesting and beautiful and in recent years have become an important part of a tourism resource base in Orissa. However, any major infrastructure for tourism such as access roads, accommodations and other services may degrade the very resource being used. Human visitation into the sanctuary should be strictly controlled particularly during the nesting seasons of crocodiles, birds and the mass nesting season of the olive ridley sea turtles.
10. The licence system adopted by the forest department have provisions for issuing permits for the harvest of mangrove leaves as cattle feed also allowing certain areas to be used for cattle grazing. However, this may be done only in buffer zones under strict vigilance and the core areas should not be subjected to any kind of grazing.
11. Fishing inside the rivers of the Bhitarkanika sanctuary should be completely stopped for protection of the crocodiles and construction of bunds in the creeks be stopped for effective dispersal of viviparous mangrove seedlings. Prohibition for the usage of stake nets in the Patsala river and the estuarine region will minimise Juvenile fishery harvest and thereby improving the yeild of marine fishery. Mesh size regulation will help to prevent accidental capture of hatchling sea turtles and illicit fishing of Juvenile fishes.
12. All sorts of coastal developments including establishment of military base, oil drilling operations, construction of new ports fishing harbours, Jetties and industrial complex, sand mining along the coast; construction of dams, reservoirs and saline embankments which has its impact on fresh water discharges, salinity, siltation, tidal amplitude etc.-ultimately affect the mangrove ecosystem. Therefore, these should be carefully planned by a team of experts from various fields. The effects of these must be considered at the feasibility/ planning phase of the projects.
13. Preliminary surveys have been made for extraction of oil near the Mahanadi delta. When oil or gas are extracted from within the mangrove areas or in the vicinity, the greatest damage may come from pipelines and roads. Unless properly designed and located, these structures may alter drainage patterns apart from directly clearing the mangrove forests in some areas. Therefore, precautions should be taken during planning stage and after words

to minimize the risk of oil leakage from pipes, tanks and wells because of the risk of damage to the flora and fauna of the region.

14. Pesticides and fertilisers may find their way into the neighbouring mangroves and cause serious food chain contamination through biological magnification. This effect should be taken into account in any planning for conversion to agriculture.
15. Attention should be given to include the subject of mangroves in the educational curriculum of schools and colleges, Audiovisual materials be made to communicate the value of mangroves to the students and the general public.
16. An extensive programme to create awareness of the value of mangroves and mangrove conservation be made among the decision makers and the local people who interact with the ecosystem.

Conclusion

Although Bhitarkanika has lost several species it still retains a diverse flora and fauna of great scientific and educational value. Wetlands of Bhitarkanika provide critical habitats for waterfowl and other birds and it serves as a refuge for wildlife. It is also a site for recreation, tourist attraction, education and research. There are thickly populated villages within the proposed national park. With different government agencies working and so many different demands made upon such a small area, conflicts of interest are inevitable. However, such conflicts must be resolved if Bhitarkanika's mangrove forest and Wildlife are to survive and Bhitarkanika is to continue as a wildlife conservation site of international importance.

Acknowledgement

The authors wish to thank Professor M.C. Dash for his helpful suggestions and encouragements and to the Forest Department, Govt. of Orissa for providing facilities for the above work.

REFERENCES

Annon. 1987. *Mangroves in India–Status Report,* Govt. of India, Ministry of Environment and forests. 150p.

Banerjee, L.K. 1984. Vegetation of Bhitarkanika Sanctuary in Cuttack District, Orissa, India, *J. Econ. Tax. Bot.* **5** (5) : 1065-1079.

Benerjee, L.K. 1985. Comparative study on Mangroves of the Sundarbans and that of the Mahanadi delta in Eastern India. p. 527-530. In : *The mangroves : Proc. Nat. Symp. Biol. Util. Cons. Mangroves,* Nov. 1985. Shivaji : Univ. Kolhapur.

Banerjee, L.K. 1986. New distributional records of some mangrove species from Orissa coast. *J. Bombay Nat. Hist. Soc.* **83** (1) : 271-73.

Banerjee, L.K. 1987. Ecological Studies on the mangals in the Mahanadi estuarine delta, Orissa : India. *Trop. Ecol.* **28** : 117-125.

Banerjee, L.K. and G.C. Das. 1972. New distributional records from Orissa Coast, *Bull. Bot. Surv. India.* **14** (1-4) : 184-186.

Banerjee, L.K. and T.A. Rao. 1985. *Mangals of Mahanadi delta,* Cuttack district Orissa State, India.

Blasco, F. 1975. *The Mangroves of India. Trav. Inst. Fr. Pondicherry. Institute Francis Pondicherry, India (French and English Versions).* p. 175.

Blasco, F. 1977. Outlines of ecology, botany, forestry of the mangals of the Indian Subcontinent pp. 241-260. In : *Wet coastal ecosystems* (Ed. V.J. Chapman).

Bustard, H.R. 1972. *Sea Turtles : Their natural history and conservation.* Londons, Collins.

Carr. A.F. 1952. *Handbook of turtles.* Ithaca, New York. comstock Bulb. Association. p. 542.

Carter, H.G. and V. Narayanaswamy. 1922. Lists of Plants from Barkuda and adjacent island in Chilika Lake. *Mem. Asia. Soc. Bengal.* **7** : 257-309.

Champion. H.G. and S.K. Seth. 1986. *A revised survey of the forest types of India,* New Delhi.

Choudhury, B.P. 1984. A glimpse into the Vegetation of Bhitarkanika Wildlife Sanctuary in the State of Orissa, India. *Bot Reptr.* **3** (2) : 121-124.

Choudhury, B.P. 1986. Nomenclature and distribution of some mangrove elements in the Bhitarkanika. *NWCSO News Letter* (No. 1 and 2) 58-60.

Choudhury, B.P. 1987. Socio-economic aspects of Bhitarkanika wildlife Sanctuary in the State of Orissa. *Orissa Review.* **43** (9) : 45-47.

Choudhury, B.P. 1990. The unique Mangrove Forest of Bhitarkanika in the State of Orissa. *Orissa Review,* April, 1990. pp. 34-39.

Dash, M.C. and Kar, C.S. 1987 a. Conservation and management of Sea turtle resource in Orissa coast. p. 303-316. In : *Environment and Ecotoxicology.* Academy of Environmental Biology (Eds.).

Dash, M.C. and Kar, C.S. 1987 b. Mangrove ecosystem resource conservation in Orissa coast. *IX International Symp. on Tropical Ecology–Ecological Management of Tropical Ecosystems and International Conference on Rehabilitation of disturbed Ecosystem : A Global Issue,* held at Dept. Bot. B.H.U. Varanasi, Dec. 11-16 1987.

Dash, M.C. and Kar, C.S. 1990. *The Turtle Paradise–Gahirmatha–An ecological Analysis and Conservation Strategy.* Interprint, New Delhi, p. 295.

Haines, H.H. 1921-1925. *The Botany of Bihar and Orissa:* 6 part London.

Hughes, D.A. and J.D. Richard. 1974. The nesting of the pacific ridley turtle, Lepidochelys on plays Nancite, *Costa Rica Marine Biol.* **24** : 97-107.

Hughes, G.R. (1974 a,b) : *The sea turtles of South East Africa. I. Status, morphology and distributions. II. The biology of the Tongaland Loggerhead turtle Caretta caretta L. with comments on the leatherback turtle Dermochelys coriacca L. In the study region,* Invest. Rep. Nos. 35, 36. South regions. Invest Rep. Nos 35, 36 South African Asso. Mar. Biol. Res. Oceanographic Res. Institute, Durbans South Africa.

Kar, C.S. 1988. Ecological Studies on the pacific Olive Ridley Sea turtle, *Lepidochelys olivacea* (Eschscholtz, 1829) in Orissa. Ph. D. thesis, Sambalpur University.

Kar, C.S. and Dash, M.C. 1984. Conservation and Status of Sea turtles in Orissa. Proc. of the workshop on sea turtle conservation, 27-29, Feb. 1984. Madras, In : *C.M.F.R.I. Spl. Publ.* No. 18, pp. 39-107.

Loveridge, A. 1946. *Reptiles of the pacific world* pp. 20-21. Mc Millan Co. New York.

Macnae, W. 1968. A general account of the fauna and flora of mangrove swamps and forests in the Indo-West Pacific Regions. *Adv. Mar. Biol.* **6** : 75-270.

Macnae, W. 1974. Mangrove Forest and fisheries. Indian Ocean Programme. *Rome, Indian Ocean fishery Commission.* Publ. No. **34** : p. 35.

Majumdar, N.C. and Banerjee, L.K. 1985. A new Species of Heritier Dry and (Sterculiaceae) from Orissa. *Bull. Bot. Surv. India,* **27** (1-4) : 150-151.

Marquez. R. 1976. Reservas naturales para La conservation de Las tortugas marinas en Maxico. INS/SI i. 83, p. 23.

Mishra, S.C. and G. Panigrahi. 1987. Studies on the mangrove flora of Orissa with particular reference to the Rhizophoraceae R. Br. *J. Econ. Tax. Bot.* Vol. II (1), pp. 121-131.

Mooney, H.F. 1950. *Supplement to the Botany of Bihar and Orissa,* Ranchi.

Panda, P.C. 1987. Floristics in Orissa–Past, Present and Future. *Orissa, Review,* January 1987. Vol. XLIII, No. **6** : 29-32.

Panigrahi, G. Choudhury, S.D.C.S., Raju and Deka, G.K. 1964. A contribution to the botany of Orissa. *Bull. Bot. Surv. India.* **6** : 237-266.

Pattnaik, H. and Choudhury, B.P. 1988. Present status and future development of mangrove vegetation in Bhitarkanika. In : *Proc. INDO-US Workshop on Wetlands, Mangroves, and Biosphere Reserves New Delhi; India* : 89-95.

Rao, T.A. and A.K. Mukherjee. 1970. Some interesting plant records from the Orissa Coast. *J. Bombay Nat. Hist. Soc.* **56** (3) : 659-660.

Rao, T.A. and Mukherjee. A.K. 1970. Ecological aspects along the shore of the Burabalanga tidal estuary, Balasore district, *Proc. Ind.Ac. Sci.* **76** (B) : 201-206.

Rao, T.A. and Banerjee, L.K. 1982. Tidal mangrove of the Mahanadi delta Utkal Coast, India. *J. Bombay. Nat. Hist. Soc.* 219-226.

Ross, J.P. 1982. Historical decline of Loggerhead, Ridley and Leatherback Sea Turtles. p. 189-195. In : *Biology and Conservation of sea turtle.* (Ed. K. Bjorndal) Smithsonian Institution Press. Washington, D.C.

Salm, R.V. 1976. Marine turtle management in Seychelles and Pakistan, *Environmental Conservation.* **3** : 267-68.

Samal, R.C. and J.M. Pattnaik. 1988. Management Strategy for Bhitarkanika Mangrove forest. *Proc. Indo-US Workshop on Wetlands, Mangrove and Biosphere Reserves.* New Delhi, India. pp. 96-99.

Saenger, P. Hegerl. E.J. and Davie. J.D.S. 1983. Global status of mangrove ecosystems. IUCN Commission on Ecology papers Number 3. *The Environmentalist.* Vol. **3** (1983). Supplement No. 3.

Samantari, B. and Pattnaik, H. 1962. Flora of Orissa. P. 160-176. In : *Orissa-Past and Present* (Ed. P. Parija).

Schulz, J.P. 1975. *Sea turtles nesting In : Surinam. Zoologische Verhendelingen (Leiden).* **143** : 1-143.

Sidhu, S.S. 1963. Studies on the mangroves of India. 1. East Gadavari region, Indian Forester. **89** (5) : 337-351.

Smith, M.A. 1933. *The Fauna of British India including Ceylon and Burma.* pp. 1-185.

Valliappan, S. and Whitaker, R. 1975. Ridley sea turtle on the Coromandel coast of India. *Herpetological Review.* **6** : 42.

Chapter 24

Sustainable Development in the Himalaya—a Nebulous Goal

K.G. Saxena AND **K.S. Rao**

G.B. Pant Institute of Himalayan Environment and Development Kosi-Almora-263 643, India

Abstract. Although a realization towards sustainability has become profound in scientific, social and political concerns for development, more so for the so-called 'ecologically fragile zones' like Himalaya, it remains uncertain to ascertain as to what extent sustainability has been, or likely to be, altered, and the rate thereof. While the trends of unsustainability, whether it concerns biological or social or industrial facet of development, in the past have been well documented, characterization of elements rendering sustainability, practical constraints in implementing the development approaches argued to be sustainable and the benefits likely to accrue if the 'known approaches' are successfully applied suffer from elements of ambivalence and dualism. Industrial development, at the same time conserving the environment and preserving the sociol-cultural setup, are desired, but mechanisms to achieve the dual purposes are not known. Need of 'area specific integrated development programme' is realized, but the actions converge towards patch-up integration of conventional sectoral development programmes, rather than any innovative endeavours for designing prescriptions of integrated resource management in a true sense. Preservation of traditional culture is advocated simultaneously with changes accompanying drastic improvement in the comforts of hill people attuned to vagaries of the difficulties in mountains for generations. Traditional privileges of using forest resources are withdrawn despite a strong realization of acute crisis of forest based subsistence needs of the rural poor. Gigantic hydroelectric projects are undertaken, simultaneously, expressing the need of development from indigenous resources and indigenous talent at micro (village) level, restrictions on the expansion of cropland are imposed, accepting the physical constraints in increasing crop production and also to make an 'area' self-sufficient for its food requirements. If the goal is to achieve sustainability in a true sense, concerted efforts to remove confusion emanating from the 'dualism' and 'ambivalence' in perceptions and needs of development are prerequisites. The conceptual basis of sustainable development is not likely to find a foothold unless a consensus, from both social and political ends, on the nature and pace of development is achieved.

Key words : *Sustainable Development, Himalaya, Tihri Dam, Rehabilitation*

Introduction

Sustainable development is a globally acceptable goal. There are conflicting opinions on the feasibility of achieving this goal. A fact that hundreds of millions are still deprived of even the basic essentials of life sustenance and a majority of human beings are 'underdeveloped' or 'developing' ones, renders a pessimistic thinking that it is perhaps impossible rather than difficult to achieve this goal. On the other hand, scientific and technological developments known to have solved or eased problems in the past make one to side an optimistic view. Sustainability though known from the early stages of human civilization, has been emphasized only during the last couple of decades. It has received recognition more as a subject of economic versus ecological needs of development than a problem to work at for finding solutions. Unsustainable trends of development are well known but the answers on how could they be replaced by the sustainable ones are only little known. Planning and management which could avoid adverse effects on ecological processes without putting a brake on human aspirations for economic and social improvement are well conceptualized but not available in acceptable and applicable forms (Redcliftt 1987). The need is to identify new patterns/trends of development which are sustainable and also a change in socio-political attitude favouring adoption of the new patterns/ trends. This would necessitate a radical change in the concern for development, from 'development exclusively of ourselves and present generation' to 'development of all including ourselves and for many generations including the present one'.

On account of being monetarily poor and lack of formal education, rural poor are mostly neglected in development policy formulation and programme execution process. Alien interventions, including those by Government, could be considered to accompany opposite effects. In one respect, they bring betterment in ways which would have not been possible through indigenous mechanisms, while in the other respect, they increase dependence on such interventions and thereby suppress the indigenous instruments of development. Development targeted interventions are often conceived as impositions serving interests of the people other than those inhabiting the area where project is executed. This paves a way for emergence of regional conflicts which defeat the very object of development programmes. Experiences in the past suggest that people's participation in development planning can ensure long term ecological stability as well as a sustained socio-economic progress (Stadel 1991). Complicated problems are faced in accelerating the pace of development in areas such as Himalaya where the development of the people is as important as ecology of the region on account of its far reaching impacts. Natural resources on which the very existence of mankind depends, are limited. There are technological limitations in harnessing these resources to meet the development needs. Sustainable development envisages optimal resource use incorporating ecological principles of conservation and recycling, with a vision of progress of human beings for generations (Anonymous 1987). Multipurpose river valley projects in the Himalaya have drawn considerable attention on their likely impacts. Some issues relevant to the philosophy of sustainable development concerning Tihri dam project in Garhwal Himalaya are addressed in this article.

Sustainable Development - Tihri Dam Project Context

Spatial variability in the quality as well as quantity of natural resources is a prominent feature of India. Often poorly developed regions passess a rich resource capital while more developed ones suffer from scarcity. Resource use patterns and management practices do vary. In the Himalaya, conventional development oriented interventions are recent introductions and thereby new to the traditional societies. Utilization of untapped natural resources for development is acceptable to all. These are drawbacks in planning projects for such uses, misconceptions about their likely sociol-economic and ecological impacts and decisions on how the accrued benefits should be distributed which create controversies. A major share of the likely benefits viz. electricity potential and irrigational facilities from river valley projects to people in the plains directly affects the sentiments of the local people and prevails over a realization for the benefits which would have not been available had the project been not executed. Though there are many ways of defining sustainable development, all converge on the objects viz. adequate economic returns, maintenance of natural resources and productivity indefinitely, minimal adverse environmental impacts, optimal production with minimal external inputs, satisfaction of human needs and a provision for the social needs (Davidson and Dence 1988; Herman 1990). While microlevel development interventions accompany slow changes, influencing smaller areas, and thereby not readily detectable, major projects such as river valley hydroelectric projects accompany readily perceptible changes affecting larger areas. The rationale behind all such projects is that monsoon run-off, which otherwise goes waste, should be used for power generation and irrigation. The potential of water flows and energy is known to the hill people (Kothyari *et al.,* 1991) but traditionally they designed its uses for small scale cottage industries and subsistence needs. Criticism essentially centres on the technology introduced for water resource utilization and our social economic and ecological capacity to sustain the technology and its impacts.

Project Planning

Four fold increase in the installed capacity from 600 MW in 1972 to 2400 MW in 1986 indicates the limitations of planners to for see the magnitude of resource utilization. Maximum possible installed capacity is a limit set by the natural processes under the influence but not control of man. Feasibility of exploiting the maximum potential depends upon the technological potential to harness the resource and economic potential to install the technology. Opportunity of exploiting the water resources in the Tihri area was known as early as in 1949 but the project could commence only in 1978. Still the information on water resource potential in the Himalaya are in the form of coarse approximations rather than real measurements. Unless an authentic inventory of the resource potential is created, decision making process would continue to be blamed for lacunae in project conceptualization and planning. Arguments are also put forth in favour of micro-hydel projects at a number of sites instead of a mega project and run-of-river technology instead of dams. However, evidences in support of superiority of these alternatives over the dam project are lacking.

Geological Uncertainties

The Himalayan region is considered to be geological active and to consist of poorly consolidated and highly folded sediments. The risks of geological origin such as earthquakes, landslides, rockfalls and spatial dynamics of river courses are difficult to predict precisely for their occurrence in space or in time. As the knowledge system is enriched more with likely impacts of the catastrophic events than with alternative remedial measures, opposition to the project, often taking pretext of the environment, are obvious. Spatial or temporal scale of observing these catastrophic events are too coarse to satisfy the requirements of decision makers formulating the development projects. An earthquake defeating the goals of development envisaged in hydroelectric project is much uncertain as the possibility of project delivering goods anticipated in its plan. When arguments against a development project are emphasized at a later stage of project execution, they lead to conflicting stands difficult to reconcile. The success of project execution agency is judged from the completion of project and so they ought to ensure about the safety measures. A sincere consideration to the unpredictable events do matter but before project execution.

Life Span of Reservoir

Life span of the reservoir is the most crucial determinant of economic viability of dam based projects. Accuracy in calculating life of reservoir depends upon the quality of data on sedimentation, silt load, run-off and rainfall regime. Estimates on sedimentation vary from 8.1 ha m/100 sq.km/year to 16.5 ha m/100 sq.km/year (Table 24.1). Similarly, serious drawbacks in available statistics on land use (Saxena *et al.*, 1993) which regulates water flows do exist. It is basically because of inconsistencies in data that estimates on life span of the reservoir vary from 62 years to 160 years and benefit : cost ratio of the project from 1.28 to 3.96. A reappraisal of the project plan in view of these inconsistencies is important but if the process is time consuming, delay in an economy like ours accompanies significant increase in project cost. During a period of 7 years, the cost of Tihri dam project increased by more than two times, from Rs. 1365 crores

TABLE 24.1. *Variability in data on sedimentation*

Data source	Estimate on sedimentation (ha m/100 km^2/yr)
Project Authorities (1969)	8.10
Central Water Commission (1972)	12.50
Ganga Discharge Organization (1986)	16.53
Central Water Commission (1987)	14.86
Project Authorities (1991)	14.50

in 1983 to Rs. 3465 in 1990. Delay in project completion could manifest the risk of economic unsustainability to the project.

Catchment Area Development

Catchment area development is an action which appeals the interests of the project authorities and the people inhabiting the region as well. Focus in this component is to minimize soil erosion for maximizing the life of the reservoir and productive potential of the land. Integrated development under the disciplines viz. forestry, agriculture, soil conservation, animal husbandry has been envisaged. A real measure of the effectiveness of integrated development is lacking. Increase in allocation to this component of the project (1.5% of the total provision in 1983 to 2.5% in 1990) indicates a positive response to environmental concerns. However, diversion of funds by itself shall not serve the purpose unless the integration process is clearly understood and applied. Integrated development is a conceptual thesis of avoiding sector problems. A perfect integrated development programme would satisfy economic, social and ecological needs together. However, clear prescriptions of such programmes are lacking, with a result that it is the 'patch-up' of management of different resource sectors than their real integration which becomes apparent.

Compensatory Afforestation

Compensatory afforestation has been made obligatory to account loss in forest resources as a result of submergence. In case of Tihri dam project, the land for afforestation could be made available far away in Bundelkhand region. Thus corrective action meets the land use requirements in a national and not the regional perspective. If forests are considered to regulate the environmental quality, compensatory afforestation needs to be carried out in the vicinity of the project area. Such need is felt but legal constraints of land ownership come in way of taking environmentally sound decisions. Perhaps the pressure on forests because of activities such as construction of new roads, broadening of existing roads, accompanying the mega projects are more severe than the direct influence in the form of submergence. Forests are closely linked to the life style of the hill people. Sustainability of hill agriculture depends upon the forest based organic inputs and fuelwood constitutes the major source of energy. Providing alternatives to these essential needs has not received proper attention in development planning. Institutional mechanisms which make people to feel that a share of the benefits from the project would be invested for their well being, can stimulate people's participation in the development process.

Rehabilitation

Mass displacement of human beings is perhaps the most sensitive issue of the mega scale hydroelectric project. It is practically impossible to rehabilitate people in areas resembling their native places in all respects. Although the new Tihri town is quite near to the old one, there is a difference of over 500 m in elevation levelgs of the two locations. Rural population is being rehabilitated in the foot hill where social and environmental set ups are quite different from those in the hills. Many of these impacts argued to be adverse are difficult to overcome and therefore compromises are essential. Conflicts concerning rehabilitation could be avoided if the effected population is exposed to both positive and negative impacts before project is executed. Direct benefits like employment opportunities obviously reach to a minority. There are also practical difficulties to ensure equal access to rehabilitatory packages for obvious reasons. The conflict between those who enjoy the benefits and those who are kept away from the benefits, within the society emerges. This culminates in a feeling that development interventions ignore the values which sustained local people and their environment for generations. It is perhaps basically this feeling which gets expressed in the form of people's movements which in turn are exploited for political mileage.

Discussion and Conclusion

All our efforts tend to ensure that the markers of development invariably follow progressive upward trend immune to externalities. Since resources are limited with uneven distribution but the demands are many, competition for use of resources for alternative demands and competition among distinct regions to develop are obvious. Thus the two objects of development *viz.* progressiveness and stability appear more contradictory than complementary. Stability reflects sustainability but distinction is often made for contextual meaning of these terms. Conway (1985), in the context of agricultural systems, implied stability to refer constancy in crop yields in the face of small disturbances while sustainability to major natural or economic disturbances. The object of achieving sustainability is invariably subdued by the aspirations of achieving higher levels of development. In India, potential of electricity generation by water based technology is concentrated in the Himalayan region. Thus there is no alternative but to execute such projects in the Himalayan region to satisfy industrial requirements of the country. It is not that people in the Himalaya do not want development elsewhere in the plains but certainly not at the cost of their own development. Supply of electricity for industrial development, improved irrigational facilities and drinking water needs elsewhere in the plains should be considered as important as providing irrigational facilities of hill farmers and drinking water to hill people. If expansion of cropland in the hills is restricted and traditional rights of forest resource use are curtailed for achieving ecological balance, appropriate intensive farming packages should be offered to the hill farmers. Climatic constraints and geological risks cannot be manipulated and therefore development in the Himalaya must take care of these attributes. Interventions providing contingencies to these risks should be designed. Preservation of Himalayan ecology could be ensured only when the development of the Himalyan people is taken care of. The problem is that ambivalent views on and actions for development prevail. Environmentally sound economic development is argued for but the requisite actions are not known. We express a concern for development emanating from indigenous talent, resources and technology but are unable to contain reliance on external resources and technology.

Development is a never ending process of change realized by human beings for having advanced to a better state of living. Sustainable development is a conceptual framework of integrating economic growth and environmental and social concerns and to make this framework operational is perhaps a question for the future (Mykeltun 1990). The advocates of sustainable development foresee a possibility of development by way of unfolding the realities and necessarily not by industrial development or economic growth, particularly when the latter accompanies social disintegration and environmental degradation. The decisions in favour or against the continuation of development projects are argued with preconceived notions for or against industrial development and environmental conservation. Controversies pertaining to economic growth versus environmental preservation, microlevel development projects versus megalevel ones, labour intensive production systems versus capital intensive production systems and a change from traditional isolated and closed societies to well connected and open societies should be resolved giving due appreciation to the vision of development as understood by the hill people themselves. When it is accepted that technological and economic potential to evade the geological risks and climatic extremes are limited and people have habitated such environments for generations, it is better that the people themselves decide how they want to develop.

Acknowledgements

We are thankful to Prof. A.N. Purohit, Director, G.B. Pant Institute of Himalayan Environment and Development for stimulating discussions. The views expressed in this article are to the authors and need not necessarily of the Institute.

REFERENCES

Anonymous, 1987. *Our Common Future*. The World Commission on Environment and Development, Oxford University Press, Oxford.

Conway, G. 1985. Agroecosystem analysis. *Agricultural Administration*, **20** : 31-35.

Davidson, A. and Dence, M. (Eds.). 1988. *The Brundtland Challenge and the Cost of Inaction*. The Institute for Research on Public Policy, Halifax.

Herman, E.D. 1990. Toward some operation principles of sustainable development. *Ecological Economics*, **2** : 1-6.

Kothyari, B.P., Rao, K.S., Saxena, K.G., Kumar, T. and Ramakrishnan, P.S. 1991. Institutional approaches in development and transfer of water-harvest technology in the Himalaya. In : *Advances in Water Resources Technology*. (Ed. G. Tsakiris). A.A. Balkema, Rotterdam.

Mykeltun, J. (Ed.). 1990. *Sustainable Development, Science and Policy the Conference Report*. Norwegian Research Council for Science and the Humanities, Oslo.

Redcliftt, M. 1987. *Sustainable Development. exploring the Contradictions*. Methuen, London.

Saxena, K.G., Rao, K.S. and Purohit, A.N. 1993. Sustainable forestry-prospects in India. *Journal of Sustainable Forestry* (in press).

Stadel, C. 1991. Environmental stress and sustainable development in the tropical Andes. *Mountain Research and Development*, **11** : 213-223.

Chapter 25

Ensheathed Cyanobacteria as Indicator of Polluted Waters

Siba Prasad Adhikary

P.G. Department of Botany, Utkal University, Bhubaneswar 751 004, Orissa

Abstract. Survey of industrial waste water receiving areas for the occurrence of various algal forms was carried out. All these industries discharge large quantities of untreated or partially treated liquid wastes containing objectionable amounts of various organic wastes and inorganic salts to the surrounding area including rice-fields. A variety of cyanobacterial species predominantly ensheathed forms belonging to the genera *Chroococcus, Gloeothece, Gloeocapsa, Lyngbya, Plectonema* and *Calothrix* occur in the waste water receiving areas through out the year. These organisms also retained their sheath in the culture vessels in the laboratory. Chemical nature of the isolated sheath materials showed that it was composed of predominantly carbohydrate and protein. The formation of thick and rigid sheath around cell wall of these organisms in polluted waters indicate that the ensheathed cyanobacteria which normally rarely grow in unpolluted water bodies can be used as the test system (indicator) or industrial effluent polluted water bodies.

Key words : *Industrial effluent, Cyanobacteria, sheath, Pollution indicator.*

Introduction

Water bodies of most of the developing countries are polluted mainly due to the discharge of industrial waste waters containing various toxic substances. These partially treated or untreated wastes degrade the water quality and affect adversely on the aquatic organisms. Usually pollution of industrial effluents has been determined by analysis of the chemical constituents of the wastes. In the recent years biological parameters and microbial assays have been employed to assess the pollution load of various effluents (Hynes, 1966). Palmer (1980) suggested the use of algae as the test organism and drawn up a list of algae inhabiting in polluted waters. In the recent years biological studies of algae growing in various industrial waste water polluted habitats of India has also been carried out (Rana *et al.*, 1971, Agarwal and Kumar, 1978, Somasekhar and Ramaswamy, 1983).

The Molasses distillery of Aska, Fertilizer factory and Thermal power station of Talcher and Spoinge iron factory of Palasponga, located in the Orissa state discharges untreated liquid wastes into the surrounding area since last decade. Ecological studies were carried out on the occurrence of algal forms in the waste water polluted and unpolluted areas of these industries. The results showed that only few ensheathed forms of cyanobacteria belonging to *Chroococcus, Gloeothece, Calothrix* and organisms belonging to LPP group occurred in the industrial effluent receiving areas. With a view to understand the higher tolerance capacity of ensheathed cyanobacteria than sheathless forms in the field, growth response of ensheathed *Gloeothece* PCC 6909 and sheathless *Anacystis nidulans* CCAP 1405/1 to various concentrations of these industrial effluents were studied in the laboratory.

Materials and Methods

The molasses distillery, Aska (19°16' N, 84°55'E) discharges large quantity of untreated liquid wastes into the surrounding rice-fields and finally to Rusikullya river since 1970. The reddish effluent is highly acidic and contain objectionable amounts of total solids and chlorides (Table 25.1). The fertilizer factory, Talcher (20°57'N, 85°15'E) manufacturing urea and ammonia since 1979 discharges untreated effluent to Nandira river through a 3 km long open canal amidst crop fields. The black coloured effluent is highly alkaline and contain objectionable amounts of ammoniacal nitrogen and oil (Table 25.1). Five km down in the course of Nandira river, few open canals carrying white coloured waste waters of the Talcher thermal power station, a coal based power plant, joins the river. The effluent is rich with solid particles and high amount of oil further degrade the water quality of river Nandira. The soponge iron factory of Palasponga (21°35'N, 86°42'E) manufactures sponge iron using coal and iron ore as raw materials since 1983. The waste water discharged out is black in colour and loaded with dust particles of iron ore, coal, sponge iron and oiled (Table 25.1).

Soil samples, algal patches and the effluent were collected at regular intervals through out the year during 1987-1988. Physico-chemical analysis of the effluent was carried out following the method given by A.P.H.A. (1971). The soil samples were kept in BG 11 medium with or without combined nitrogen (Rippka *et al.*, 1979) at room temperature under fluorescent light. The algae appeared in the cultures and algae samples collected from the spot were identified following Desikachary (1959), Phillipose (1967) and Thampson (1959). Two cyanobacterial species, the ensheathed *Gloeothece* sp. (PCC 6909) and sheathless *Anacystis nidulans* (CCAP 1405/1) were grown in presence of varying concentrations of the neutralized industrial effluents (1, 5, 10, 15, 20, 25, 30, 50 and 100% v/v) prepared with BG 11 medium. Growth experiments were conducted in 30 ml capacity cotton stoppered test tubes containing 10 ml of medium and the cultures were incubated at 26 ± 1°C under 2,400 lux light intensity upto 20 days. Growth was recorded in terms of O.D. of the homogenized culture suspension at 760 nm as per the procedure followed earlier (Adhikary, 1983). Growth response of the test organisms in presence of varying concentrations of industrial effluents was calculated taking the growth on control cultures as 100 per cent.

TABLE 25.1. *Physico-chemical characteristics of the effluent of molasses distillery of Aska, fertilizer factory and thermal power station of Talcher and sponge iron factory of Palasponga of Orissa state. Waste waters were collected from the effluent discharge points of the industries during December 1988 and analysed. All the values except pH are in mg/litre.*

Characteristics	Effluent from			
	Molasses distillery	Fertilizer factory	Thermal power station	Sponge iron factory
Colour	Reddish brown	Black	Lime white	Black
Odour	Burnt sugar	Ammoniacal	Burnt ash	Burnt ash
pH	4.8	10.5	8.0	6.5
Temperature (°C) at discharge point	85	48	55	35
Total solids	5225	2260	3140	2890
Dissolved oxygen	0.5	2.2	1.0	0
Total nitrogen	86	1760	16	11
Phosphate	9.3	4.5	2.8	2.6
Sulphate	208	169	102	176
Chloride	754	360	215	312
Calcium	68	152	11	126
Magnesium	16.4	14.3	0.6	18.5
Iron (total)	5.6	Trace	Trace	286
Phenol	Nil	156	24	Nil
Oil a nd grease	12	30	42	27

Results and Discussion

Various algal forms occurred in the effluent polluted and unpolluted fields around the industries. The effluent of molasses distillery, rich with various nutrients such as nitrogen, phosphorus, calcium and iron (Table 25.2) encouraged growth of various algal forms (Table 25.2). However, the heterocystous filamentous cyanobacterium *Calothrix marchica* Lemm. var. *intermedia* Rao showing thick sheath layers around the trichome occur profusely in the effluent affected fields. Though 15 species of microalgae were recorded from the sponge iron factory area, mostly ensheathed forms of cyanobacteria viz. *Gloeothece, Lyngbya, Calothrix* and few green algal forms were found growing in the effluent polluted fields (Table 25.4). To the country none of the organisms except a species of *Gloeothece* was observed in the Nandira river after receiving the waste water from the Talcher fertilizer factory which manufactures urea and

TABLE 25.2. *Occurrence of different algal forms in the molasses distillery effluent affected and unaffected rice-fields of Aska, Ganjam district.*

Organism	Effluent affected field	Effluent unaffected field
Chroococcus minutus	+	-
Gloeocapsa sp.	+	-
Merismopedia glauca	+	-
Oscillatoria princeps	+	-
Oscillatoria cruviceps	+	+
Oscillatoria formosa	-	+
Phormidium ambigucum	+	-
Spirulina sp.	+	-
Anabaena variabilis	-	+
Anabaena sp.	+	+
Nostoc tenue	-	+
Nostoc punctiforme	-	+
Cylindrospermum muscicola	-	+
Calothrix marchica	+	-
Scytonema schmidlei	+	-
Westiellopsis prolifica	-	+
Scenedesmus quadricauda	+	-
Pediastrum duplex	+	+

ammonia. Due to extreme alkaline pH of the waste water and objectionable amounts of total solids, ammoniacal nitrogen and oil, almost all the species which occurred in the river before receiving the industrial wastes were affected (Table 25.3). In the down stream of Nandira, after receiving the waste water from Talcher thermal power station, only few ensheathed cyanobacterial species viz. *Gloeothece* sp. and *Chroococcus minutus* appeared in the river course (Table 25.3). These results showed that only the species *Gloeothece, Gloeocapsa, Chroococcus, Calothrix* and organisms belonging to LPP group which have well defined sheath around the cells/trichome can tolerate in the industrial effluent affected areas.

Two different cyanobacterial species, the ensheathed *Gloeothece* PCC 6909 and sheathless *Anacystis nidulans* CCAP 1405/1 were grown in presence of various concentrations of the neutralized distillery, fertilizer factory, thermal power station and sponge iron factory effluents to understand the tolerance of ensheathed cyanobacterial forms in the industrial effluent polluted areas. The results showed that *Gloeothece* sp., which possess a distinct sheath around

TABLE 25.3. *Occurrence of different algal forms in the course of Nandira river before and after receiving wastewater from Talcher fertilizer factory and Talcher thermal power station, Dhenkanal district.*

Organism	Before receiving effluent from industries	After receiving effluent from Fertilizer factory	After receiving effluent from Thermal
Chroococcus minutus	-	-	+
Gloeothece sp.	-	+	+
Oscillatoria formosa	+	-	-
Coelosphaerium sp.	+	-	-
Cylindrocystis sp.	-	-	+
Scenedesmus quadricauda	-	-	+
Melosira granulata	+	-	-
Spirogyra sp.	+	-	-
Ulothrix sp.	+	-	-
Hydrodictyon reticulatum	+	-	-

the cells tolerated considerably higher concentrations of the neutralized effluents in comparison to that of the tolerance capacity of *Anacystis nidulans* (Fig. 25.1 to 25.4).This shows the protective nature of extracellular sheath to the organisms to such adverse conditions which are not condusive for growth of sheathless forms. These results showed that the ensheathed forms of cyanobacteria which rearely occur in unpolluted water bodies (Palmer, 1980) can be used as the test system (indicator) of industrial effluent polluted water bodies.

Experiments were also conducted to find out the chemical nature of the isolated sheath fractions of *Gloeothece* PCC 6909 grown in presence or absence of lower concentrations of the industrial effluents. Freshly cultivated cells were suspended in Tris/HCl buffer (pH, 8.0) mixed with a pinch of DNAase and subjected to sonication at 4°C. Crude sheath fractions were separated by centrifugation at 3020 × g, 30 min. Discontinuous sucrose gradient (10 ml of 60% and 5 ml each of 55, 50, 45 and 40%, w/v) were loaded with the sediment of last centrifugation and run in a swing out rotor at 16,300 × g, 4°C, 4 h. Sheaths were removed from 60 per cent sucrose and further purified on the same type of gradient. Residual contaminating cell wall components were removed by treatment with lysozyme (5 mg in 25 ml ammonium acetate buffer, pH 6.8, 37°C, 12 h) followed by extraction with the detergent sodium dodecyl sulfate (4%, in Tris/HCl buffer, pH 7.8, 100°C, 15 min). The yield of sheath material from BG 11 medium and from the distillery, fertilizer factory, thermal power station and sponge iron factory waste water (10%, v/v) supplemented cultures of *Gloeothece* were 8.2, 10.6, 8.0, 8.2 and 9.1 per

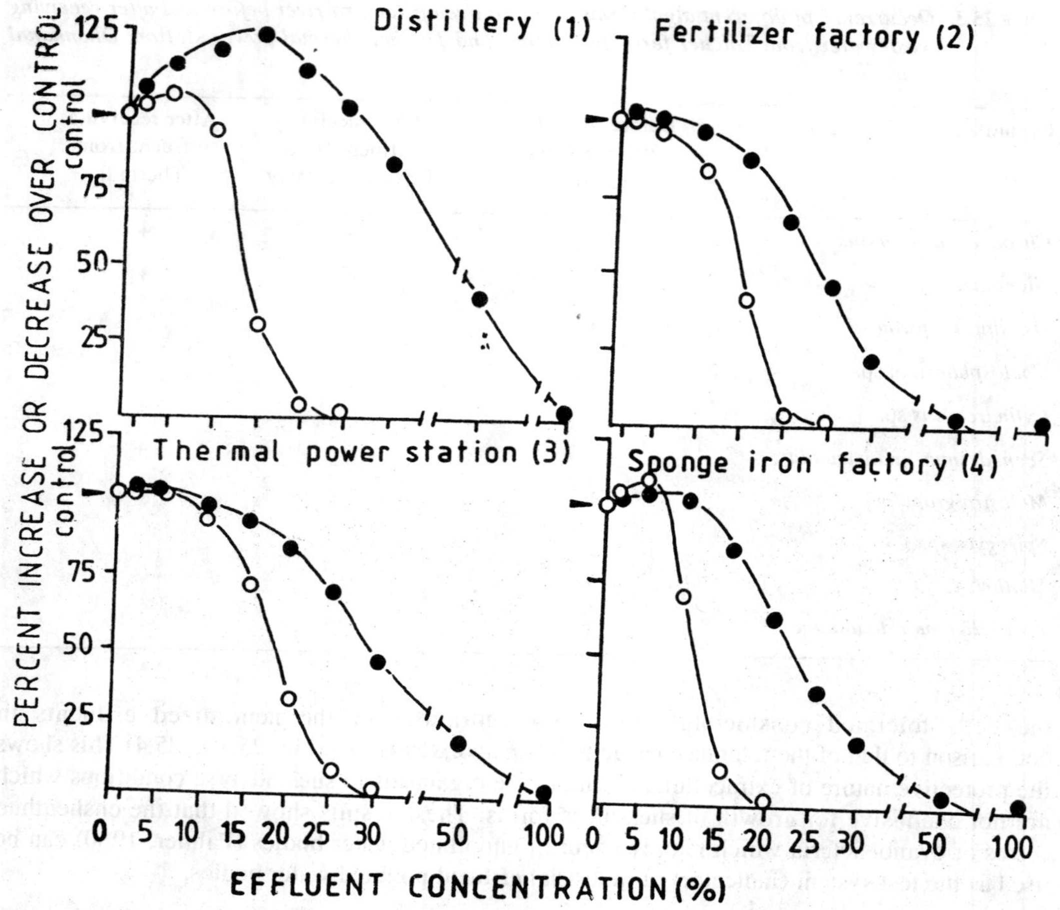

Fig. 25.1 to 25.4. Growth (per cent increase or decrease over control) of the sheathless *Anacystis nidulans* (CCAP 1405/1) and ensheathed *Gloeothece* sp. (PCC 6909) to different concentrations of the industrial effluents of (1) Molasses distillery, (2) Fertilizer factory, (3) Thermal power station and (4) Sponge iron factory.
O : *Anacystis nidulans*
● : *Gloeothece* sp.

TABLE 25.4. *Occurrence of different algal forms in the Sponge Iron Factory (OSIL, Orissa) effluent affected and unaffected fields of Palasponga, Keonjhar district.*

Organism	Effluent affected field	Effluent unaffected field
Aphanothece saxicola	-	+ +
Gloeothece sp.	+ +	+
Oscillatoria anomalis	+	+
Lyngbya major	+ +	-
Nostoc piscinale	-	+ +
Anabaenopsis circularis	+	-
Anabaena variabilis	-	+
Anabaena torulosa	-	+ +
Nodularia spumigena	+	-
Calothrix marchica	+	+
Calothrix castelli	+ +	-
Scenedesmus dimorphus	+	-
Pinnularia subcapitata	+	-
Surrella linearis	+	-
Pediastrum sp.	+	+

+ : occur; ++ : occur abundantly

cent respectively on a cell dry weight basis. Chemical analysis of the sheath fractions were carried out as per the procedure followed earlier (Adhikary *et al*., 1986). The results showed that isolated sheath fractions of the cyanobacterium from cultures supplemented with or without industrial effluents were composed of mostly neutral sugars followed by protein (Table 25.5). All the fractions contained predominantly glucose, galactose, mannose, 6-deoxy rhamnose, 2-0-methyl pentose and an unknown sugar residue. In addition, presence of uronic acid was detected in the sheath of the organism grown with industrial wastes. Absence of muramic acid, diaminopimelic acid and B-hydroxy fatty acids showed that these were free from cell wall contaminants. Quantitative difference was observed in the composition of sheath materials of the organism grown in presence of various industrial waste waters (Table 25.5). There are reports (Tease and Walker, 1987) that alteration of chemical composition of sheath materials affect the capacity of extracellular sheath to bind to heavy metals. Thus it is necessary

TABLE 25.5. *Chemical composition (% of dried material) of the isolated sheath fractions (after lysozyme and SDS treatment)* of Gloeothece PCC 6909 grown in presence or absence of 10% (v/v)** effluents of molasses distillery of Aska, Fertilizer factory and thermal power station of Talcher and sponge iron factory of palasponga.*

Constituent	Sheath fraction of *Gloeothece* PCC 6909				
	Grown in BG 11 medium	Grown in BG 11 medium + 10% effluent of			
		Molasses distillery	Fertilizer factory	Thermal station	Sponge iron factory
Sugar					
Unknown sugar-x	2.4	3.9	3.1	2.5	3.0
6-deoxy Rhamnose	4.2	4.6	3.9	4.8	3.6
2-0-methyl pentose	0.8	1.0	1.3	1.6	0.9
Xylose	0	0.2	0.7	0	0.3
Mannose	6.3	7.8	6.2	6.0	6.4
Galactose	7.5	10.5	9.3	8.3	8.6
Glucose	10.2	14.3	11.8	10.5	10.5
Total sugars	31.4	42.3	36.3	33.7	33.3
Total amino acids	8.4	6.2	13.6	10.2	10.4
Glucosamine	--	--	--	--	--
Muramic acid	--	--	--	--	--
Diaminopimelic acid	--	--	--	--	--
Uronic acid	--	1.8	1.3	1.4	2.1
Phosphate	0.26	1.2	0.39	0.96	1.4
Total fatty acids	0.8	0.34	0.52	0.62	0.67

*Lysozyme : 5 mg in 25 ml ammonium acetate buffer, pH 6.8; 37°C, 12 h.
SDS : 4% (w/v) in Tris/HCl, pH 7.8; 100°C, 15 min.
**10% (v/v) proportions of the effluent of the industries were prepared in BG 11 medium (Rippka *et al.* 1979).

to work further to understand whether sheath materials developed around the organisms growing in polluted waters can bind more heavy metals and decrease the pollution load of water bodies.

Acknowledgements

Thanks are due to the Head of the Department of Botany, Utkal University, Bhubaneswar for providing facilities, to Dr. S. Ayyappan, Central Institute of Fresh water Aquaculture, Kausalyaganga for his assistance in the analytical portion of the work and to the Department of Science and Technology, Government of Orissa for financial assistance.

REFERENCES

Adhikary, S.P. 1983. Growth measurements by monitoring light scattering of a filamentous blue-green alga which does not give uniform and stable suspension in culture vessels. *Z. Allg. Mikrobiol.* **23** : 475-483.

Adhikary, S.P., Weckesser, J., Jurgens, U.J., Golekei, J.R. and Borowiak, D. 1986. Isolation and chemical characterization of sheath from the cyanobacterium *Chroococcus* minutus SAG B. 41.79. *J. Gen. Microbial.* **132** : 2595-2599.

A.P.H.A. (American Public Health Association) 1971. *Standard methods for examination of water and waste water.* 13th Ed. APHA, NY, USA.

Agarwal, M. and Kumar, H.D. 1978. Physico-chemical and phicological assessment of two mercury polluted environments. *I.J. Env. Health,* **20** : 141-155.

Desikachary, T.V. 1959. *Cyanophyta,* ICAR, New Delhi, 686 p.

Palmer, C.M. 1980. *Algae and water pollution.* Castle House Publications Ltd. England, pp.31-78.

Phillipose, M.T. 1969. *Chloroccocales,* ICAR, New Delhi, 365 p.

Rana, B.C., Gopal, T. and Kumar, H.D. 1971. Studies on biological effects of industrial wastes on the growth of algae. *Env. Health* **13** : 138-143.

Rippka, R., Deruelles, J., Waterbury, J.R. and Stanier, R.Y. 1979. Generic assignments, strain histories and properties of pure cultures of cyanobacteria. *J. Gen. Microbiol* **111** : 1-61.

Somasekhar, R.K. and Ramaswamy, S.N. 1983. Algal indicators of paper mill waste waters. *Phykos,* **22** : 161-166.

Tease, B.E. and Walker, R.W. 1987. Comparative composition of sheath of the cyanobacterium *Gloeothece* ATCC 27152 cultured with or without combined nitrogen. *J. Gen. Microbiol.* **133** : 3331-3339.

Thampson, R.H. 1959. *Algae* pp. 155-170. In : *Freshwater Biology* (Ed. W.T. Edomondson), John Wiley, New York.

Chapter 26

Herbicide Toxicity : *in vitro* Effects of some Herbicides on Chloroplasts Isolated from *Cyperus rotundus* L.

T.V. Rao AND Tribhuban Panda

Biochemistry and Biophysics Laboratory School of Life Sciences,
Sambalpur University Jyotivihar-768 019, Orissa

Abstract. Mechanism of action of herbicides applied in paddy cultivation was investigated in isolated chloroplasts of *Cyperus rotundus*. Action of these chemicals on electron transport chain was analysed by the primary photochemical reactions. Electron transport in the entire chain and PSI segment were probed by measuring MV reduction with water-DPC and DCPIP-ascorbate as electron donor respectively. Results of these experiments suggest that the herbicides probably act on the PSII segment. This finding was then corroborated by the study of PSII function by DCPIP photo reduction with various electron donors. The measurement of Chl a fluorescence reveals that it is substantially enhanced with addition of herbicides. From the excitation and emission spectra energy transfer values were computed which indicate the action of herbicides on the PSII side of the chain. Data from fluorescence polarisation and absorption spectra show that the herbicides have clear role on the membrane disorganisation.

Key words : arozin, oxadiazon, butachlor, herbicide, chloroplast, PSII, Cyperus rotundus, fluorescence polarisation, MV reduction.

Abbreviation

CDEA : α chloro diethyl acetamide; chl. a : Chlorophyll a; DCMU : 3-(3, 4-dichloro-phenyl)-1, 1-dimethly urea; DCPIP : 2, 6-dichlorophenol indophenol; EDTA : Ethylenediamine tetraacetic acid; EPTC : Ethylene-N, N-di-n-propyl thiocarbamate; MH : 1, 2-dihydro-3, 6-pyridazinedione; MV : Methyl viologen; PSI : Photosystem I; PSII : Photosystem II; Q_A : Primary electron acceptor quinone of PSII; Q_B : Secondary electron acceptor quinone of PSII.

I-

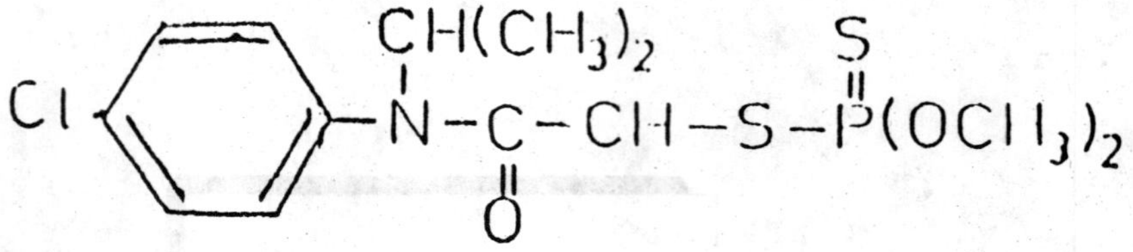

(Arozin)

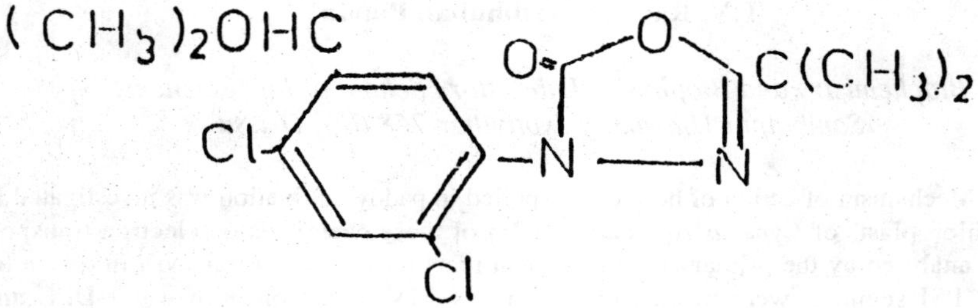

(Oxadiazon)

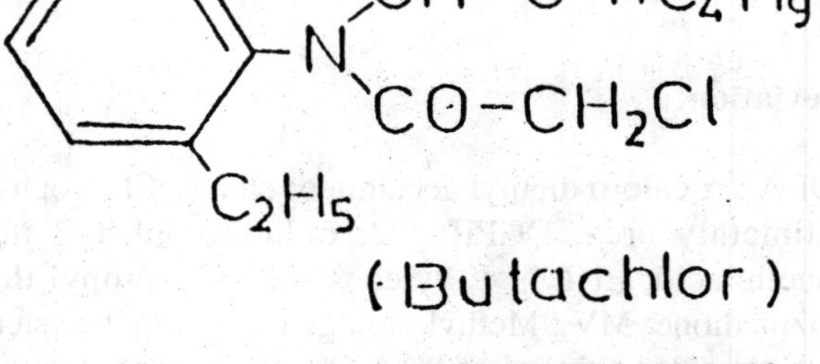

(Butachlor)

Fig. 26.1. Chemical structure of herbicides a: arozin, b: oxadiazon, c: butachlor

Introduction

In modern agriculture practice herbicides are used for eradication of weeds or retarding their growth selectively. Herbicide application holds an important place in agronomics for its cost effectiveness. But beneath this advantage lies the toxic effect of these chemicals in physiology of the plants. They have got diverse mode of action in plants. Chemicals like s-triazine, pyrazon, substituted ureas, bromacil etc. disturb or block Hill reaction (Paton and Smith 1965). Sodium arsenite, picloram, dichlobenil, dinoseb etc. affect vital steps in respiration; specifically phosphorylation, allowing respiration to free wheel. Some of them damage the permeability properties of the membranes (Magalhes and Ashton 1969). 2, 4-D is known to dislocate normal translocation of food materials in plants and to cause changes i.e. nucleic acid and protein metabolism (Chang and Foy 1971). It also activates ethylene production causing damage to plants. Herbicides like MH, CDEA, Kill plants primarily by dislocating normal mitotic division in seedlings (Jaworski 1969). Carbamates like EPTC and bensulide inhibit cell elongation and expansion during the process of germination (Cutter *et al* 1968).

Literature on the molecular mechanism of herbicide action show that they act at specific target site (Vasiler *et al* 1988, Gardner 1989). Many herbicides like parquat block photosynthetic electron transport in plants by preventing oxidation of the primary acceptor Q_A of PSII by the secondary acceptor Q_B (Barber and Falco 1989). Compounds of diverse chemical nature such as DCMU, atrazin, dinoseb, inhibit the electron flow in the thylakoid membranes as well as bind to the same protein known as "herbicide binding protein" (Naber and Van Rensen 1987, Chow *et al* 1989, van der Bolt *et al* 1992). This is a protein of Q_B in PSII complex of 32 kilo Daltons molecular mass. Inhibition of electron flow in chloroplast are investigated by the assay of chloroplast fluorescence which is the result of accumulation of energy (Yamamoto *et al* 1981). Butochlor an α chloroacetamide is being used as common weedicide in rice cultivation in Orissa for about 20 years. Recently two other chemicals oxadiazon and arozin (Fig 26.1.) have been introduced as more potent weed killers than butachlor. Although their efficiency has been reported yet little is known about their mechanism of action. In this context we have taken these three chemicals to investigate their action on leaf chloroplasts of a dicot weed *Cyperus rotundus*. The major objective of our work has been their molecular mechanism of action on isolated chloroplasts. For this we have used fluorescence and absorbance studies and assay of electron flow by DCPIP Hill reaction and methyl viologen reduction in isolated chloroplasts as parameters of investigation.

Materials and Methods

Tubers of *Cyperus rotundus* were collected from the Botanical Garden of Sambalpur University. They were germinated in pots containing garden loam. Within 12-15 days of germination the leaves were harvested for experiments.

Herbicide preparation

Arozin, oxadiazon and butachlor were gifted by Agricultural Research Station Barmunda,

Satupali, Sambalpur. Herbicides were diluted to required dilution by glass distilled water.

Preparation of Chloroplast

Leaves of *Cyperus rotundus* were cut into pieces and homogenized in mortar and pestle with ice cold isolation medium containing 0.25 M sucrose, 0.01 M EDTA-Na_2, and 0.1 M phosphate buffer (pH 7.8). After homogenization the homogenate was squeezed through 16 layers of cheese cloth. The filtrate was centrifuged at 500 xg for 1 min. The supernatant was centrifuged at 2000 xg for 10 min. and the pellet was suspended in a small volume of isolation medium. All operation were conducted at 4°C under safe green light.

DCPIP Photoreduction

Measurement of DCPIP photoreduction of isolated chloroplasts was carried out following the procedure described by Biswal and Mohanty (1976). The reaction mixture containing chloroplasts was illuminated for 30 sec. with saturating white light (0.008 w cm^{-2}) coming from a 150 watt projection lamp. The incident light was passed through a water filter to cut off infra-red light. The final volume of the reaction mixture was 1.5 ml and it contained chloroplasts equivalent to 8.5 mg of chlorophyll, 15 M DCPIP, 10 mM KCl, 0.1 mM $MgSO_4$ and 10 mM sodium phosphate buffer of pH 6.8. The photoreduction of the dye was measured at 600 nm in a Shimadzu UV-VIS double beam spectrophotometer model 150-02.

MV-Reduction

For MV-reduction assay chloroplast were isolated in a medium containing sucrose 0.4 M, KCl 10 mM, $MgCl_2$ 5.0 mM, 20 mM tris-HCl buffer of pH 7.6 at 4°C. The reaction mixture in final volume of 2ml contain KCl 10 mM, $MgCl_2$ 5.0 mM, Tris-HCl buffer 50 mM of pH 8.0, MV 0.5 mM and chloroplasts equivalent to 8.0 µgm of chlorophyll. Ascorbate and DCPIP were used at the concentration of 2.5 mM and 5.0 mM respectively. MV reduction was measured with the help of an oxygen electrode (Hansatech) as moles of oxygen consumed/mg of chlorophyll/hr.

Fluorescence Assey

Room temperature flourescence excitation spectra of chloroplast suspension were recorded with a fluorescence spectrophotometer (Hitachi, Model 680-40) for emission at 685 and 735 nm. The width of excitation and emission slits were fixed at 5 nm. Chloroplast equivalent to 5 mg/ml were taken in the cuvette for all experiments.

Emission spectra under similar condition was measured by the method of (Panda *et al* 1987) with 450 nm as excitation wave length.

The vertical and horizontal polarization of fluorescence was measured at 685 nm and 735 nm and 620 nm excitation wave length (Yamamoto *et al* 1981). Polarization (P) was calculated according to the following equation :

$$P = \frac{I_{vv} - \left[\dfrac{I_{vh} x I_{hv}}{I_{hh}} \right]}{I_{vv} + \left[\dfrac{I_{vh} x I_{hv}}{I_{hh}} \right]}$$

where

I = Intensity of fluorescence
v = Vertical
h = Horizontal

Results and Discussion

Herbicides such as DCMU are known to bind with electron transport chain having inactive binding site in the acceptor site of SPII. They strongly inhibit electron transport chain in SPII even at low concentration (Purcell *et al* 1991). In this investigation electron transport measurement in chloroplast of *Cyperus rotundus* by reduction of methyl viologen shows substantial decline from control with arozin treatment (Table 26.1). The decline (82%) of the MV reduction is comparable with 2 ppm DCMU treatment (85%) taken as reference. The rate of MV

TABLE 26.1. *Methyl viologen reduction in isolated chloroplasts of C. rotundus. Experimental condition as given in "material and methods"*

Condition	Herbicide treatment	μ moles of O_2 consumed/mg chl/hr	% loss in MV reduction
(a) Entire chain			
H_2O - > MV	None	62	0.00
H_2O - > MV	Arozin (5 ppm)	11	83.30
H_2O - > MV	DCMU (2 ppm)	9	85.59
DPC - > MV	Arozin (5 ppm)	16	74.30
DPC - > MV	DCMU (2 ppm)	15	75.99
(b) PSI Section			
DCPIP - > Mv	None	670	0.000
DCPIP - > Mv	Arozin (5 ppm)	665	0.746
DCPIP - ASC - > Mv	DCMU (2 ppm)	660	1.490

TABLE 26.2. *DCPIP photoreduction in chloroplast of C. rotundus. Details of the experiment as described in "material and methods.*

Concentration of Herbicide (ppm)		H_2O —> DCPIP		DPC —> DCPIP	
		μ moles of DCPIP reduced/ mg chl/hr	% change from control	μ moles of DCPIP reduced/ mg chl/hr	% change from control
	0	23.00	0.00	26.0	0.00
Arozin	1	19.00	17.39	21.5	17.31
	2	12.70	44.80	16.9	34.10
	4	2.10	90.87	6.3	75.80
	5	1.00	95.70	2.5	90.39
DCMU					
	2	1.05	95.5	2.1	90.87

TABLE 26.3. *Chlorophyll a fluorescence, fluorescence polarisation and energy transfer parameters in the chloroplasts isolated from C. rotundus. Other experimental details as given in "material and methods.*

Treatment	F[1]685	F735	F685/F735	P 685	[2]E 475/E6[3]00	E600/E685
Control	37.7	4.86	7.77	0.0714	5.33	0.057
Arozin	59.5	7.46	7.97	0.040	2.80	0.022
Oxadiazon	62.0	7.70	8.05	0.043	2.60	0.016
Butachlor	59.0	7.30	8.02	0.041	2.50	0.020

1 - Fluorescene intensity in arbitrary unit, 2 - Fluorescence polarisation, 3 - Energy transfer

reduction with DPC as electron donor is significantly inhibited (75%) in presence of arozin, whereas with DCPIP-Ascorbate couple as electron donor the rate was not significantly impaired either with arozin or DCMU. These results indicate the poisoning effects of the herbicides on the electron flow in the SPII segment. The lack of inhibitory effect of arozin in the PSI sector (DCPIP-Ascorbate as electron donor) depicts the similarity with DCMU (Naber and van Rensen 1987) in its action. Further analysis of the PSII by DCPIP photoreduction (Table 26.2) confirms resemblance of its action with DCMU. These findings point out that arozin mainly affect the PSII functioning.

Fluorescence provides information on the changes in organisation of the membrane specifically pigment-proteins associated with energy transfer (Bohme *et al* 1981, Gruszecki *et al*

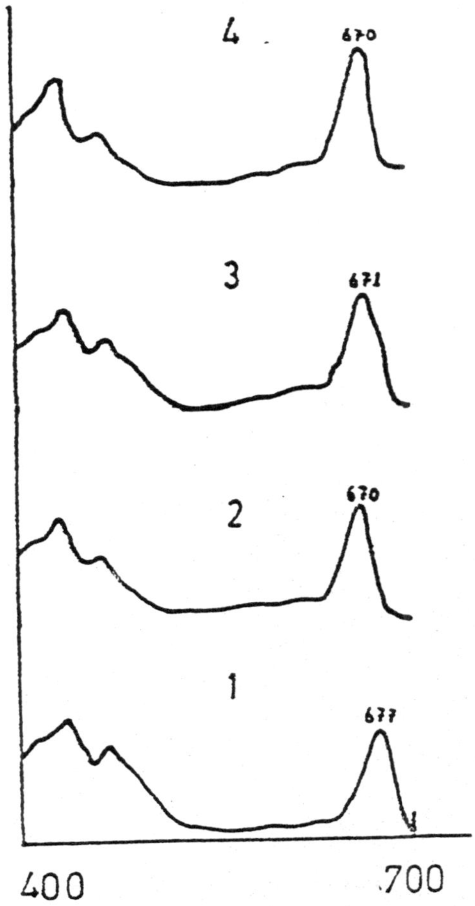

Fig. 26.2. Fluorescence excitation spectra of isolated *C. rotundus* chloroplasts. 1: control, 2: 5 ppm arozin, 3: 6 ppm oxadiazon, 4: 6 ppm butachlor. Other experimental condition as in "material and methods".

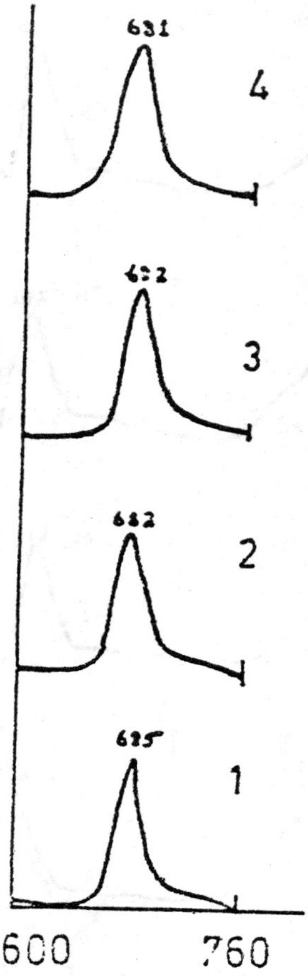

Fig. 26.3. Flourescence emission spectra of isolated *C. rotundus*. 1: control 2: 5 ppm arozin, 3: 6 ppm oxadiazon, 4: 6 ppm butachlor. Other experimental condition as in "material and methods".

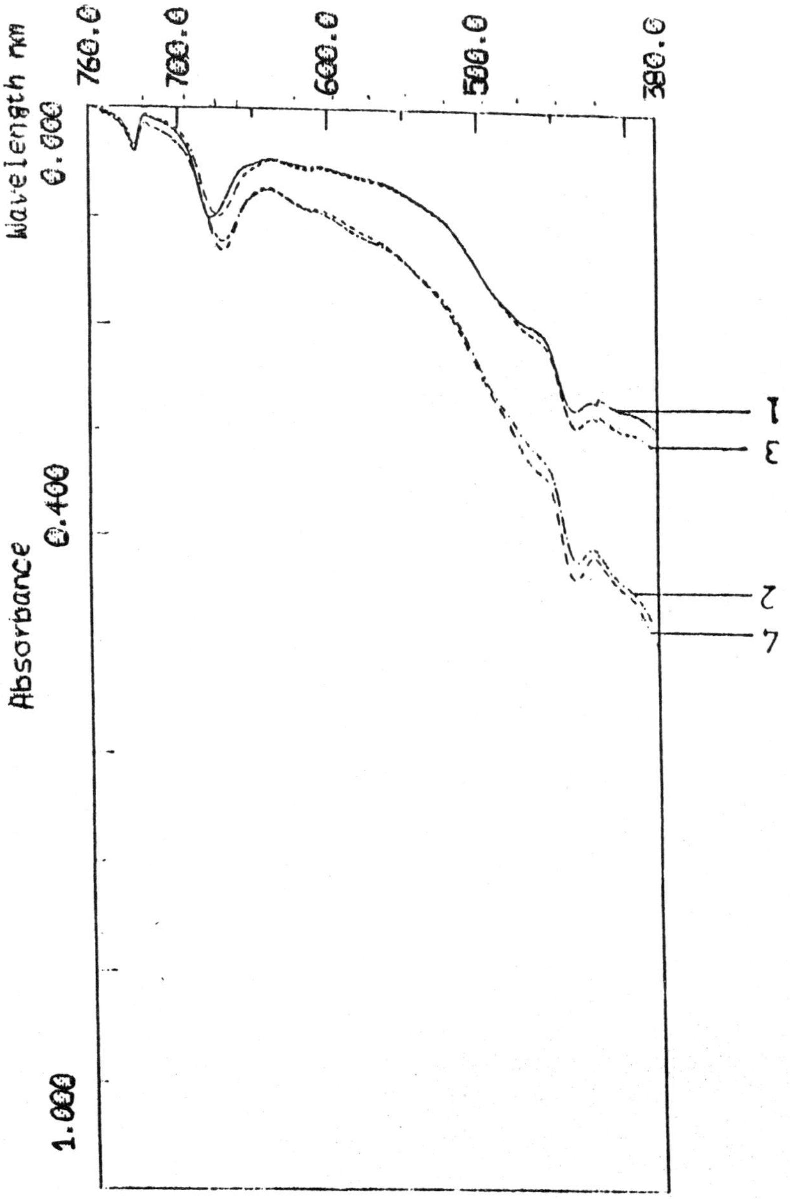

Fig. 26.4. Absorption spectra of isolated *C. rotundus* chloroplasts. 1: control, 2: 5 ppm arozin, 3: 6 ppm oxadiazon, 4: 6 ppm butachlor. Other experimental condition as in "material and methods".

1991). Damage to the component of energy transfer is reflected in the chl a fluorescence characteristics. In the present investigation these characteristics are summarised in Table 26.3. The intensity of fluorescence register an increase of about 56 per cent from the control value at both 685 nm and 735 nm in the three herbicides namely arozin, oxadiazon and butachlor. Rise in the fluorescence intensity may reflect molecular disorganisation due to loss in coupling between LHCP and the reaction centre on the other hand it may also reflect an increase in the level of disorganised chlorophyll in the membrane (Bohme *et al* 1981). The ratio of F_{685} to F_{735} denotes the relative states of the photosystems *i.e.* F_{685} for PS II and F_{735} for PS I. The decline in this ratio (Table 26.3) suggests a possible increase in PS II fluorescence *vis a vis* the decrease in fluorescence intensity in PS I. However, the data on primary photochemical reaction (Tables 26.1 and 26.2) indicate the damage to PS II by the herbicide arozin which corroborates the findings with fluorescence.

The evidence of excitation energy transfer from carotenoids to chlorophyll is the contribution of the light absorbed by carotenoids in chlorophyll fluorescence. The carotenoid band in the chlorophyll excitation spectrum can be quantitatively characterised by the ratio of the intensities in the excitation spectrum at 475 nm to the intensity at 600 nm. At shorter wave length (475 nm) both carotenoids and chlorophyll and at longer wave length (600 nm) only chlorophyll are responsible for the change in absorption hence reflecting the change in the excitation spectrum (Gruszecki, 1991). The energy transfer changes (Table 26.3) for the excitation and emission spectra (Figs. 26.2 and 26.3) show that E_{475} to E_{600} ratio declines by about 50 per cent in all these herbicides. Further, E_{600} to E_{685} ratio declines more significantly by about 65 per cent of the control value under the same treatment. Since the fluorescence at 685 nm corresponds the PS II segments. The decrease in this ratio points at the possible disorganisation of PS II unit in the chloroplast.

The efficiency of energy transfer depends on the natural direction of emission and absorption moment in donor and acceptor molecule respectively. Energy transfer path between initial absorption and reaction centre is governed by the orientation of various pigments in the photosynthetic system. In this investigation P_{685} value on herbicide treatment declines significantly from the control value suggesting a change in the PS II conformation. Yamamoto *et al* (1981) have reported decrease in the polarization in case of DCMU treated pea chloroplast. Additional information from the absorption spectra (Fig. 26.4) depicts possible changes in the interaction between chlorophyll binding proteins and reaction centre resulting in the blue shift.

From these informations it is possible to predict that herbicides arozin, oxadiazon and butachlor might be interacting with PS II segment of electron transport chain in *Cyperus rotundus* Chloroplast. This has to be analysed further with more sophisticated investigation for confirmation.

Acknowledgement

The work was partly sponsored by UGC in the form of JRF to one of the author (TP). We express our gratitude to Prof. U.C. Biswal for extending help in the investigation.

REFERENCES

Barber, J. Mazur and Falco, S.C. 1989. The development of herbicide resistance crops. *Annu. Rev. Plant Physiol. Plant Mol. Biol.* **40** : 441-470.

Biswal, U.C. and Mohanty, P. 1976. Aging induced changes in photosynthetic electron transport of detached barley leaves. *Plant Cell Physiol.* **17** : 323-331.

Bohme, H., Kunert, K.J. an Boget, P., 1981. Site of herbicide action on photosynthesis : A fluorescence assay study. *Weed Sci* **29** : 371-373.

Change, I.K. and Foy, C.L., 1971. Effect of picloram Qa mitochondrial swelling and ATPase. *Weed Sci.* **19** (1) : 53-57.

Chow, W.S., Osmond, C.B., and Huang, L.K., 1989. Photosystem II function and herbicide binding site during photoinhibition spinach chloroplast *in vivo* and in *vitro. Photosynth. Res.* **21** : 17-26.

Cutter, E.G., Ashton, F.M., and Donna, H., 1968. The effect of bensulide on the growth, anatomy and morphology of oat roots. *Weed Res.* **8** (4) : 346-352.

Gardner, G., 1989. A sterio chemical model for the active site of photosystem II herbicides. *Photochem and Photobiol*, **49** (3) : 331-336.

Gruszecki, W.I., K. Veevanjaneyulu, Zelent, B., and Leblane, R. M., 1991. Energy transfer process during senescence : fluorescence and Photoacoustic studies of intact pea leaves. *Biochem. Biophys.* Acta. **1056** : 173-180.

Jaworski, E.G., 1969. ''Chloroacetamide'' : *Degradation of herbicides* (Ed. P.C. Keerney and D.D. Kaufman). Marcel Dekkar Inc. New York.

Magalhes, A.A., Ashton, F.M., 1969. Effect of dicamba on oxygen uptake and cell membrane. Permeability in leaf tissue of *Cyperus rotundus*. L. *Weed Res.* **9** (1) : 48-52.

Naber, J.D. and Van Rensen, J.S., 1987. Determination of the exchange parameters of herbicides on Q_B Proteins of photosystsm II. *Progress in Photosynth. Res.* III (Ed. J. Biogens), ISBN 9024734525.

Panda, S.K., Mishra, A.K., Biswal, U.C., 1987. Manganese induced modification in the membrane lipid peroxidation and change in chlorophyll fluorescence *Phytochemistry* (Pergamon) **26** : 3217-3219.

Paton, D.J., and Smith, J.F., 1965. Influence of ioxynil in the photoreduction of endogenous plastoquinone by isolated chloroplast of *Vicia fabia. Weed Abst.* **14** : 227 (1577).

Prucell, M., Leroux, G.D. and Carpentier, R., 1991. Interacton of the electron donor diphenyl carbazide with the herbicide binding niche of photosystem II. *Biochem. Biophys. Acta.* **1058** : 374-378.

Var der Belt, F., Vermas, W., 1992. Photoinactivation of photosystem II as studied with site directed D_2 mutant of *Cyanobacterium synechocystis* Sp. Pcc 6803. *Biochem. Biophys. Acta.* **1098** : 247-254.

Vasiler, I. R., Matorin, D.N., Lyadsky Vladimir, V., and Venediktov Pavel, S., 1988. Multiple action site for photosystem II herbicide as revealed by delayed fluorescence. *Photosynth. Res.* **15** : 33-39.

Yamamoto, Y., Ford, R.C., Barber, J., 1981. Measurement of degree of chlorophyll fluorescence polarisation in relation to the regulation of excitation energy transfer between photosystem I and II in pea chloroplasts. *Biochem. Biophys. Acta.* **637** : 224-230.

Chapter 27

Impact of NPK Fertilisation on Population Biology of Tropical Earthworms in Cane Sugar Field

S.K. Sahu[1], P.K. Dash[2] AND U. Nag[3]

[1,3] Department of Environmental Sciences, Sambalpur University, Jyotivihar-768 019, Orissa
[2] Department of Botany, Panchayat College, Bargarh, Orissa

Abstract. Studies were carried out in a cane sugar field, a part of which was fertilised with 740 kg NPK/ha. The fertilisation was found to cause about 30 per cent decrease in population density and 16 to 36 per cent decrease in reproductive rate of earthworms. The information is very important in agroecosystems where there is high fertiliser input for enhancing production.

Key words : *Earthworm, NPK, fertiliser, population dynamics, reproductive strategy.*

Introduction

Role of earthworm in agriculture is known since the time of Darwin (1881). The natural observations of Darwin have been greatly supported by the scientific studies made in different agroecosystems of the world which show that : (i) earthworms represent a dominant fraction of the soil invertebrate biomass (> 80%) (Dash, 1978; Lavelle, 1983); (ii) they affect soil physical structure through borrowing and casting (Dash, 1978; Lee, 1985); (iii) they process around 20 to 25 per cent of the net primary production (Dash and Patra, 1977; Senapati and Dash, 1983, 1984); (vi) they accelerate the rate of decomposition and mineralisation of soil organic matter by about 25 to 40 per cent (Syers *et al.*, 1979; Senapati and Dash, 1984, Dash and Senapati, 1985) and (v) they also contribute about 50 to 200 kg of nitrogenous materials/acre/yr through urines, mucus, dead tissues and worm casts (Dash and Patra, 1977, 1979; Senapati and Dash, 1984; Dash and Senapati, 1985; Lavelle *et al.*, 1989).

Inorganic fertilisers have played a pivotal role in modern agriculture. Increased use of fertilisers to boost the productivity and high net : gross ratios in crop is now-a-days a common practice. Further in the areas of assured irrigation fertiliser is overused specially in case of cereals, cotton and sugarcane. It is therefore essential to assess the impact of fertiliser application on beneficial organisms like earthworms. Although some information are recorded on the effect of fertilisers on earthworm populations in different agroecosystems (Heath, 1962; Zajonc, 1975; Nowak, 1976; Gerard and Hay, 1979; Edwards and Lofty, 1982; Reddy and Goud, 1987; Brussard and De Ridder, 1990), no detail information is available on their effect on population dynamics, reproductive strategy, biomass and secondary production, and particularly in cane sugar field. Keeping this in view the present investigation was undertaken as a part of comprehensive programme to study the impact of NPK fertilization on population biology of tropical earthworms in cane sugar field.

Materials and Methods

Study site and climate

Studies were carried out in two irrigated cane sugar plots situated at Tora of newly formed Bargarh district in Orissa state. The study site is about 45 km away from Sambalupur University campus and located at 21°25'N latitude and 83°52'E longitude with an elevation of about 160 m above mean sea level.

A pasture field of 25 m × 25 m size was selected as the study site. Out of two plots of 10 m × 10 m size selected in the field, plot I served as control where no fertiliser was applied, whereas plot II was the experimental plot applied with fertiliser (three times during the studies) at the rate of 400 kg N + 170 kg P_2O_5 + 170 kg K_2O/ha as recommended for agriculture. The control and experimental plots were equally divided into 100 subplots each of one square metre area. The field sampling and experiments were conducted during October 1990 to November 1991.

CO 62175 variety of sugarcane having crop duration of about 12 to 13 months were planted in the control (plot I) and experimental (plot II) plots. Plantation was done in the October III week, 1990 by ridge and furrow method putting budded sets cut from fresh cane stalk in furrows. Harvest was done in the November IV week, 1991. The study plots come under the command area of Hirakud dam. Supply of irrigated water through canal system was made from the date of plantation (October III week, 1990) and continued upto one month before harvest (October IV week, 1991) at regular interval.

The study site lies in the arid-humid agroclimatic belt (Lenka, 1972). The climatological data of the study site are given in Table 27.1. Solar radiation, rain fall and relative humidity data for the study site were obtained from the nearest meteorological station at Chakuli. Soil temperature and soil moisture were recorded by soil thermometer and oven drying respectively. Soil of this area belongs to sandy loam type. Maximum and minimum soil temperature at 20 cm depth were 33.5°C (May II, 1991) and 21.0°C (Jan. IV, 1991) in plot I and 34.1°C (May II, 1991) and 21.8°C (Jan. II, 1991) in plot II, respectively. Maximum and minimum soil moisture (g% at 20 cm depth) were 24.8 (Aug. II. 1991) and 15.1 (May IV, 1991) in plot I and 24.3 (Aug. II, 1991)and 13.8 (May IV, 1991) in plot II, respectively (Table 27.1).

TABLE 27.1. *Climatological data for the study site*

Month (Week) year			Solar radiation (10^3 kg/m²/ month)	Rainfall (mm/ fortnight)	Air temp. (°C)	Relative humidity (%)	Soil temp (°C)		Soil moisture (g%)	
							plot I (Control)	Plot II (Expt. al)	Plot I (Control)	Plot II (Expt. al)
Oct.	(II)	1990	490.5*	128.4**	25.8	84.3	25.3	25.5	20.9	20.7
	(IV)			2.2**	23.8	73.0	24.9	25.1	21.6	21.0
Nov.	(II)		445.8	43.8	23.4	71.9	24.7	25.3	20.9	20.0
	(IV)			0.0	22.6	71.0	24.1	24.5	20.6	19.7
Dec.	(II)		425.6	0.0	20.6	71.0	22.2	22.9	20.2	19.1
	(IV)			0.0	20.2	67.5	22.5	22.6	19.0	19.0
Jan.	(II)	1991	473.7	24.6	19.2	71.0	21.1	21.8	19.0	18.4
	(IV)			4.4	19.1	69.1	21.0	21.9	19.4	18.2
Feb.	(II)		498.8	0.0	22.6	64.1	23.3	24.0	19.0	18.0
	(IV)			10.8	24.1	65.8	25.7	26.2	18.7	17.8
Mar.	(II)		625.5	35.8	24.4	75.3	27.1	27.3	17.0	16.6
	(IV)			2.0	27.1	66.6	27.9	28.1	17.2	16.1
Apr.	(II)		654.3	10.0	27.1	64.9	27.7	28.2	16.0	15.1
	(IV)			0.0	29.9	59.8	30.2	30.7	15.6	15.1
May.	(II)		700.7	0.0	33.3	52.9	33.5	34.1	15.2	14.0
	(IV)			45.2	33.2	61.6	33.2	33.6	15.1	13.8
Jun.	(II)		463.4	89.2	29.8	72.9	31.4	31.7	16.8	15.9
	(IV)			88.2	29.4	75.7	31.0	31.1	20.4	19.3
Jul.	(II)		463.3	58.4	28.2	81.3	28.3	28.5	20.5	19.9
	(IV)			376.6	26.1	87.5	25.6	25.7	24.7	23.9
Aug.	(II)		416.6	321.2	25.9	89.9	25.6	25.6	24.8	24.3
	(IV)			163.8	25.9	87.3	25.8	26.0	23.7	23.2
Sep.	(II)		428.2	126.4	26.7	88.1	26.1	26.6	23.0	22.6
	(IV)			159.4	27.4	84.2	26.6	27.2	21.7	21.0
Oct.	(II)		490.5	16.8	26.0	82.1	25.6	26.3	21.2	20.7
	(IV)			0.0	24.4	75.0	24.9	25.7	21.0	20.5
Nov.	(II)	1991	445.8	7.2	22.0	78.6	23.3	24.0	20.3	20.3
	(IV)			6.2	19.8	88.2	23.0	23.6	20.8	20.0

*Monthly data **Fortnightly data

Table 27.2. *Dynamics of total density (nos/m²/fortnight) of different earthworms in control plot (plot I).*

Month (week) Year	D.willsi	L. mauriti	O. surensis	D.calebi	D. bolaui	Total worms
Oct. (II) 1990	75	16	5	4	4	104
(IV)	86	12	5	5	5	113
Nov. (II)	91	17	7	5	4	124
(IV)	97	22	11	8	3	141
Dec. (II)	105	26	12	12	4	159
(IV)	107	33	16	16	3	175
Jan. (II) 1991	110	40	18	15	3	186
(IV)	115	46	15	12	3	191
Feb. (II)	128	35	11	8	2	184
(IV)	132	25	9	8	1	175
Mar. (II)	99	23	7	5	0	134
(IV)	86	15	5	5	0	111
Apr. (II)	60	16	3	3	0	82
(IV)	28	8	3	3	0	42
May. (II)	15	7	1	1	0	24
(IV)	15	6	1	1	0	23
Jun. (II)	16	6	1	1	0	24
(IV)	31	8	2	1	1	43
Jul. (II)	42	8	3	2	4	59
(IV)	59	12	3	3	6	83
Aug. (II)	76	14	5	5	8	108
(IV)	94	15	5	5	14	133
Sep. (II)	106	20	7	7	17	157
(IV)	151	21	8	7	15	202
Oct. (II)	143	23	8	9	11	194
(IV)	92	27	11	9	8	147
Nov. (II) 1991	78	28	15	12	7	140
(IV)	71	30	16	12	7	136

TABLE 27.3. *Dynamics of total density (nos/m²/fortnight) of different eartworms in experimental plot (plot II).*

Month (Week)		Year	D. willsi	L. mauritii	O. surensis	D. calebi	D. bolaui	Total worms
Oct.	(II)	1990	75	15	4	5	4	103
	(IV)		75	12	5	4	4	100
Nov.	(II)		79	14	4	5	1	103
	(IV)		78	16	6	7	1	108
Dec.	(II)		82	18	5	9	0	114
	(IV)		83	20	6	11	1	121
Jan.	(II)	1991	81	25	8	8	1	123
	(IV)		86	32	6	5	0	129
Feb.	(II)		90	28	4	3	0	125
	(IV)		78	16	2	4	0	100
Mar.	(II)		71	11	2	3	0	87
	(IV)		58	12	1	2	0	73
Apr.	(II)		49	9	1	2	0	61
	(IV)		25	5	1	2	0	33
May.	(II)		11	7	0	1	0	19
	(IV)		8	4	0	0	0	12
Jun.	(II)		13	1	0	0	0	14
	(IV)		24	4	1	1	0	30
Jul.	(II)		39	5	1	2	1	48
	(IV)		57	7	2	2	4	72
Aug.	(II)		70	8	2	3	4	87
	(IV)		78	12	2	4	5	101
Sep.	(II)		88	11	2	3	7	111
	(IV)		96	14	2	3	8	123
Oct.	(II)		82	14	2	4	5	107
	(IV)		67	15	3	4	3	92
Nov.	(II)	1991	55	17	3	5	1	81
	(IV)		45	19	3	5	1	73

Sampling and analysis

Fortnightly five random samples each of 25 cm × 25 cm × 40 cm size were taken from the control (plot I) and experimental (plot II) Plots (Dash and Senapati, 1980). The sampling dates were fixed between 12th to 14th (II week) and between 26th to 28th (IV week) in each month. Hand sorting and wet sieving methods were used for the extraction of worms and cocoons (Dash and Senapti, 1980; Senapati and Dash, 1981). Preservation and analysis of worms and cocoons were made as per the methods standardised by Dash and Patra (1977), Senapati *et al*., (1979), Dash and Senapati (1980) and Senapati and Dash (1980, 1981). Rate of reproduction was calculated as suggested by Sahu and Senapati (1988).

Results

Population density and dynamics

Five species of earthworms (*Drawida willsi,* Michaelsen; *Lampito mauritii, Kinberg*; *Octochaetona surensis* (Michaelsen); *Drawida calebi*, Gates and *Dichogaster bolaui* (Michaelsen) were available in both the study plots. Total number of earthworms (non/m²/fortnight) ranged form 23 to 202 in plot I and from 12 to 129 in plot II (Table 27.2 and 27.3). Population turnover value (maximum population : minimum population) were 8.8 and 10.8 in plot I and II, respectively indicating greater population fluctuation in experimental plot. Average fortnightly population were 121/m² and 84/m² in plot I and II respectively (Table 27.2 and 27.3). Comparison of total population in plot I and II showed significant difference at 0.05 level ($t = 2.98$, $df = 54$) (Table 27.4). In plot I, *D. Willsi, L. mauritii, O. surensis, D. Calebi* and *D. bolaui* constituted 67.8 per cent, 16.5 per cent, 6.6 per cent, 5.0 per cent and 4.1 per cent whereas in plot II they constituted 73.8 per cent, 15.5 per cent, 3.6 per cent, 4.7 per cent and 2.4 per cent of the total population, respectively. *D. willsi and L. mauriti* constituted more than 80 per cent of the total worm number and biomas in both the plots. The present report emphasizes activity of these two dominant species.

Total earthworm population (nos/m²/fortnight) of *D. willsi* ranged from 15 (May II, 1991) to 151 (Sep. IV, 1991) and 8 (May IV, 1991) to 96 (Sep. IV, 1991) in plot I and II respectively whereas *L. mauritii* population ranged from 6 (May IV, 1991) to 46 (Jan. IV, 1991) in plot I and 1 (Jun. II, 1991) to 32 (Jan. IV. 1991) in plot II (Table 27.2 and 27.3). Population turnover value were 10.1 and 12.0 for *D. willsi* and 7.8 and 32.0 for *L. mauritii* in plot I and plot II respectively. Average population (nos/m²/fortnight) of *D. willsi* and *L. mauritii* were 82 and 20 in plot I and 62 and 13 in plot II respectively. This showed 24 per cent decrease in *D. willsi* and 35 per cent decrease in *L. mauritii* population in fertiliser applied plot (plot II). Comparison of the total *D. willsi* and *L. mauritii* population in plot I and II also showed significant difference at 0.05 level (Table 27.4).

Population of *D. willsi* constituted 33.3-48.2 per cent of juveniles, 25.0-34.3 per cent of immatures and 17.9-40.0% of adults in plot I (Fig. 27.1). In plot II the population comprised

TABLE 27.4. 't' test of total and different earthworms between control (plot I) and experimental plot (plot II).

Earthworm species	Plot	Sampling occasion	Mean ± Standard deviation	Calculated value of 't'
Total earthworms	I	28	121.21 ± 55.68	2.98*
	II	28	83.96 ± 35.79	
D. willsi	I	28	82.43 ± 38.13	2.31*
	II	28	62.45 ± 25.94	
L. mauritii	I	28	19.96 ± 10.57	4.39*
	II	28	13.25 ± 7.31	
O. surensis	I	28	7.61 ± 5.08	4.65*
	II	28	2.78 ± 2.08	
D. calebi	I	28	6.47 ± 4.31	2.79*
	II	28	3.82 ± 2.57	
D. bolaui	I	28	4.64 ± 4.79	2.80*
	II	28	1.82 ± 2.35	

*P < 0.05

25.0-37.5 per cent of adults (Fig. 27.2). Juvenile, immature and adult worms of *L. maruiti* constituted 33.3-47.8 per cent, 25.0-38.1 per cent and 20.0-42.0 per cent in plot I and 33.3-50.0 per cent, 20.0-41.7 per cent and 20.0-100.0 per cent in plot II, respectively (Fig. 27.3 and 27.4) Average percentage occurrence of juvenile, immature and adult age groups of *D. willsi* were 44.8, 33.1 and 22.1 in plot I and 41.5, 33.5 and 25.0 in plot II and of *L. mauritii* were 41.9, 34.2 and 24.0 in plot I and 38.0, 31.8 and 29.9 in plot II respectively. The ratio of non-adults (juveniles and immatures) to adults were 3.5 and 3.0 for *D. willsi* and 3.2 and 2.3 for *L. mauritii* in plot I and II respectively indicating 14 to 28 per cent change in nonadult to adult ratios of both the worms in fertiliser applied plot over the control plot.

Reproductive strategy

Cocoons (nos/m²/fortnight) of *D. willsi* ranged from 6-86 and 5-50 in plot I and II respectively whereas *L. mauritii* cocoons ranged from 6-43 in plot I and 0-25 in plot II (Table 27.5 and 27.6). *D. willsi* showed two cocoon peaks (Feb. IV and Sep. IV 1991 in plot I and II) and *L. mauritii* showed single cocoon peak (Jan. IV, 1991 in plot I and II) in both the plots. Average rate of reproduction was calculated to be 2.39 and 2.0 for *D. willsi* and 1.53 and 0.98 for *L. mauritii* in plot I and II respectively indicating 16 per cent (*D. willsi*) and 36.0 per cent (*L. mauritii*) reduction in the rate of reproduction at plot II in comparison to plot I. 't' test also indicates that

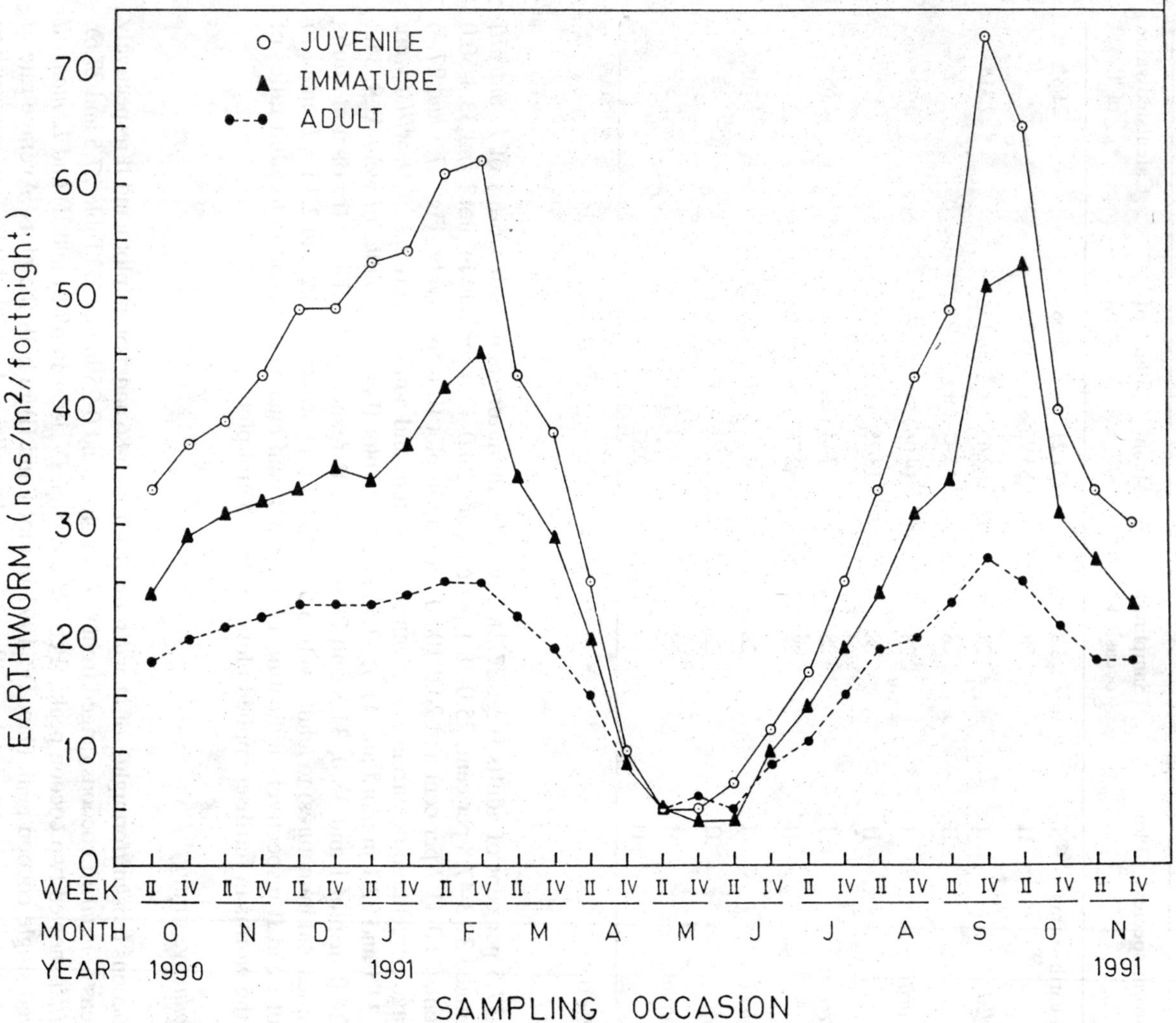

Fig. 27.1. Fortnightly dynamics of different age class density of *D. willsi* in control cane sugar plot (plot I)

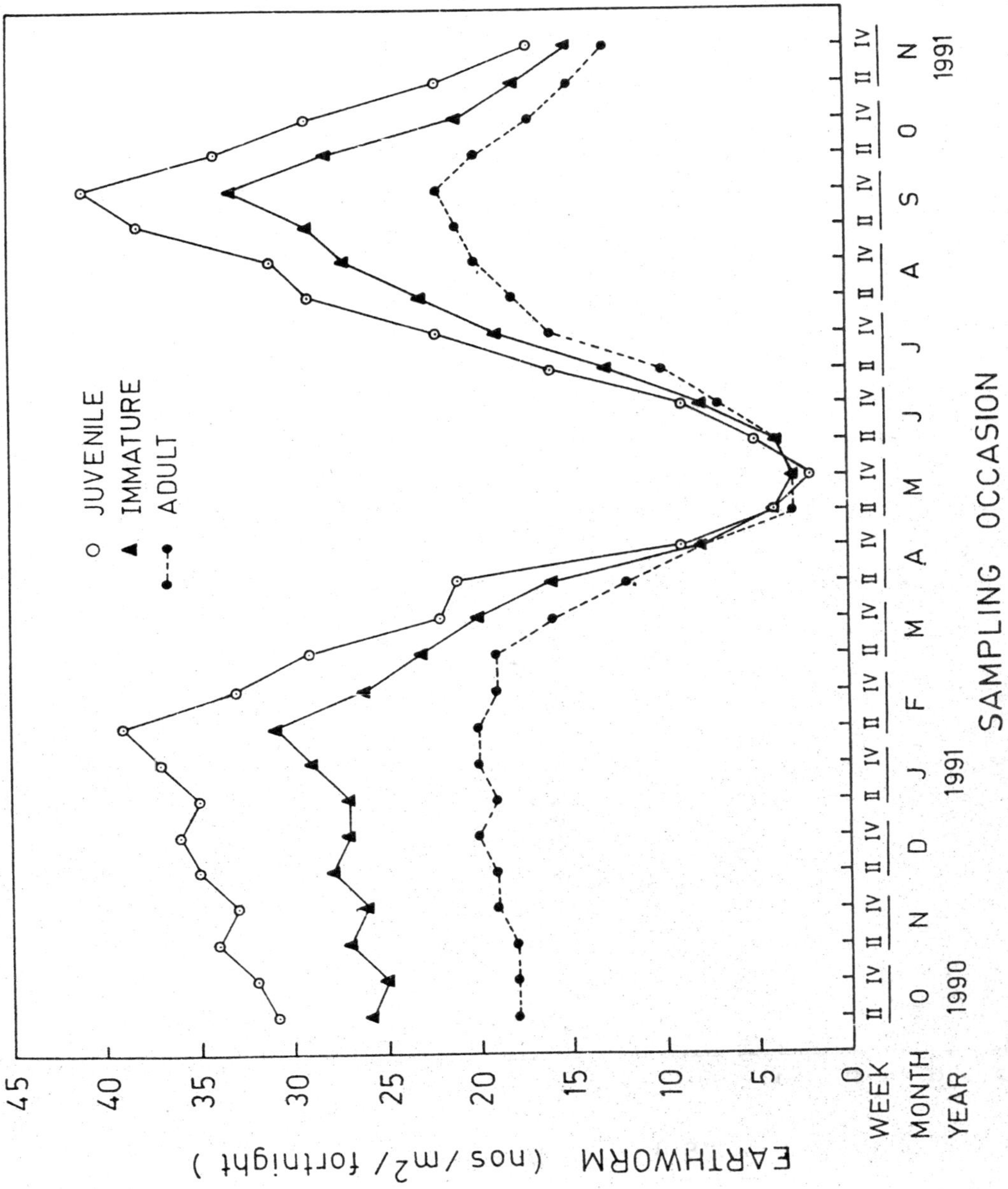

Fig. 27.2. Fortnightly dynamics of different age class density of *D. willsi* in experimental cane sugar plot (Plot II)

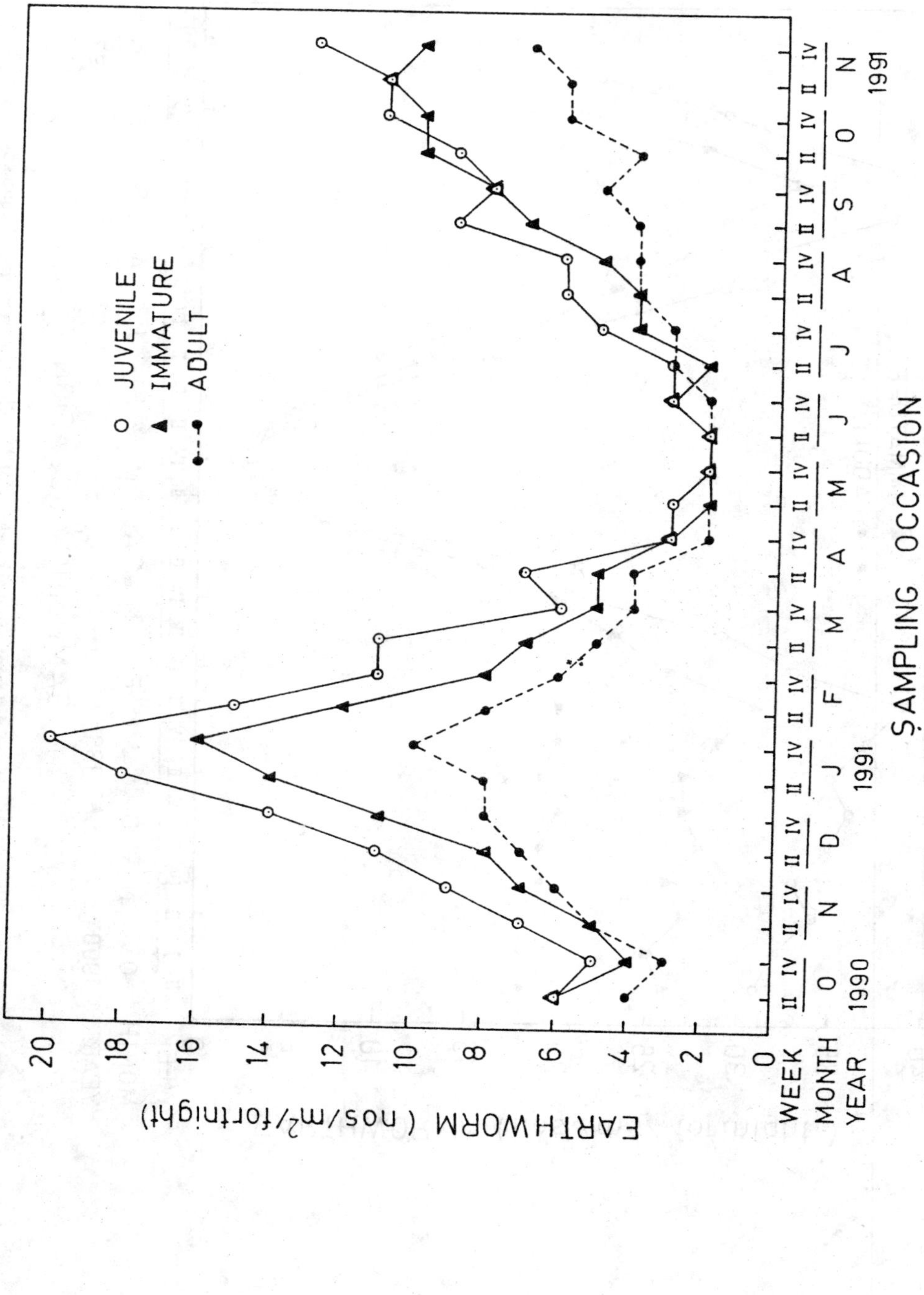

Fig. 27.3. Fortnighly dynamics of different age class density of ***L. mauritii*** in control cane sugar plot (plot I)

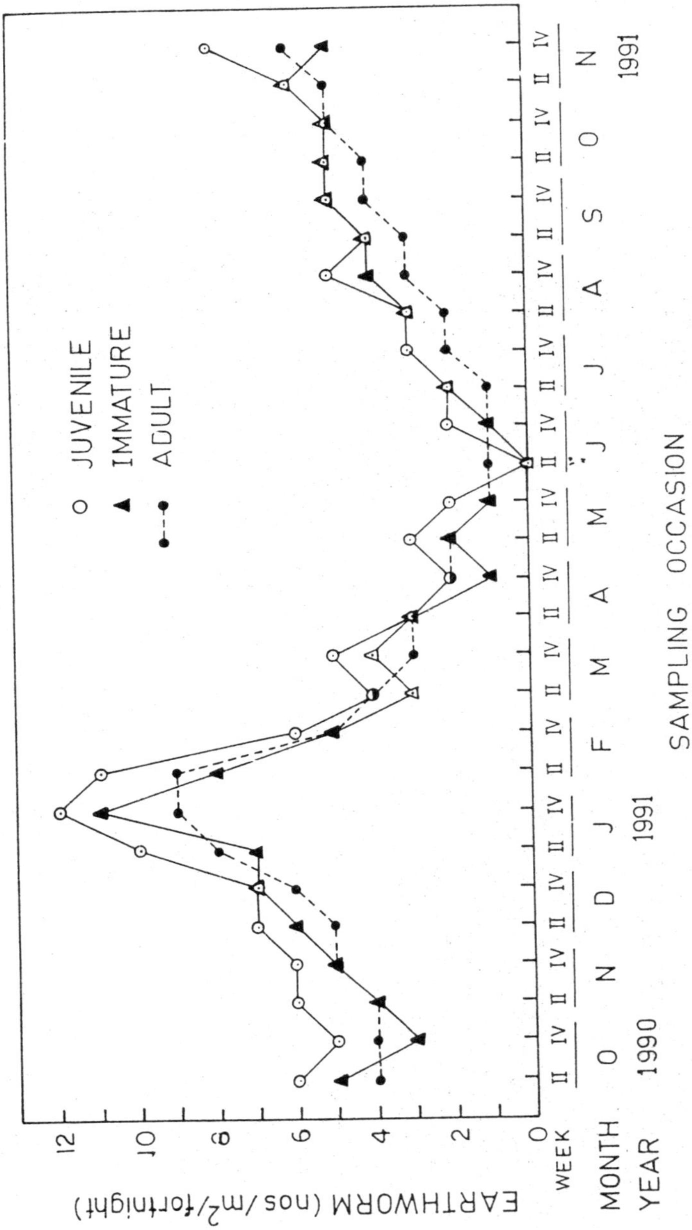

Fig. 27.4. Fortnightly dynamics of different age class density of *L. mauritii* in experimental cane sugar plot (plot II)

TABLE 27.5. *Dynamics of cocoons, adults (nos/m²/fortnight) and rate of reproduction of D. willsi in control and experimental plot.*

Month (Week) Year	Control plot (plot I)			Experimental plot (plot II)		
	Total cocoons (C)	Total adults (A)	Rate of reproduction [C × (SI/IP)]/A	Total cocoons (C)	Total adults (A)	Rate of reproduction [C × (SI/IP)]/A
Oct. (II) 1990	39	18	2.17	37	18	2.06
(IV)	52	20	2.60	39	18	2.17
Nov. (II)	55	21	2,62	38	18	2.11
(IV)	59	22	2.68	39	19	2.05
Dec. (II)	63	23	2.74	40	19	2.11
(IV)	62	23	2.70	43	20	2.15
Jan. (II) 1991	66	23	2.87	43	19	2.26
(IV)	66	24	2.75	45	20	2.25
Feb. (II)	70	25	2.80	47	20	2.35
(IV)	76	25	3.04	42	19	2.21
Mar. (II)	59	22	2.68	38	19	2.00
(IV)	46	19	2.42	33	16	2.06
Apr. (II)	33	15	2.20	25	12	2.08
(IV)	13	9	1.44	11	8	1.38
May. (II)	6	5	1.20	5	3	1.67
(IV)	6	6	1.00	5	3	1.67
Jun. (II)	8	5	1.60	6	4	1.50
(IV)	16	9	1.78	14	7	2.00
Jul. (II)	22	11	2.00	19	10	1.90
(IV)	36	15	2.40	30	16	1.88
Aug. (II)	43	19	2.26	39	18	2.17
(IV)	52	20	2.60	41	20	2.05
Sep. (II)	66	23	2.87	45	21	2.14
(IV)	86	27	3.19	50	22	2.27
Oct. (II)	72	25	2.88	43	20	2.15
(IV)	55	21	2.62	34	17	2.00
Nov. (II) 1991	46	18	2.56	26	16	1.73
(IV)	39	18	2.17	20	13	1.54

SI-Sampling interval in days, IP-Incubation period of cacoon in days

TABLE 27.6. *Dynamics of cocoons, adults (nos/m²/fortnight) and rate of reproduction of L. mauritii in control and experimental plot.*

Month (Week) Year	Control plot (plot I)			Experimental plot (plot II)		
	Total cocoons (C)	Total adults (A)	Rate of reproduction [C * (SI/IP)]/A	Total cocoons (C)	Total adults (A)	Rate of reproduction [C * (SI/IP)]/A
Oct. (II) 1990	13	4	1.63	12	4	1.50
(IV)	11	3	1.83	10	4	1.25
Nov. (II)	16	5	1.60	10	4	1.25
(IV)	19	6	1.58	12	5	1.20
Dec. (II)	22	7	1.57	12	5	1.20
(IV)	29	8	1.81	14	6	1.17
Jan. (II) 1991	33	8	2.06	20	8	1.25
(IV)	43	10	2.15	25	9	1.39
Feb. (II)	26	8	1.63	19	9	1.06
(IV)	19	6	1.58	10	5	1.00
Mar. (II)	19	5	1.90	6	4	0.75
(IV)	10	4	1.25	6	3	1.00
Apr. (II)	13	4	1.63	4	3	0.67
(IV)	6	2	1.50	2	2	0.50
May. (II)	6	2	1.50	2	2	0.50
(IV)	3	2	0.75	0	1	0.00
Jun. (II)	3	2	0.75	0	1	0.00
(IV)	3	2	0.75	2	1	1.00
Jul. (II)	6	3	1.00	2	1	1.00
(IV)	10	3	1.67	4	2	1.00
Aug. (II)	10	4	1.25	4	2	1.00
(IV)	13	4	1.63	6	3	1.00
Sep. (II)	13	4	1.63	6	3	1.00
(IV)	16	5	1.60	10	4	1.25
Oct. (II)	16	4	2.00	10	4	1.25
(IV)	19	6	1.58	11	5	1.10
Nov. (II) 1991	19	6	1.58	11	5	1.10
(IV)	19	7	1.36	11	6	0.92

SI-Sampling interval in days, IP-Incubation period of cocoon in days.

TABLE 27.7. *'t' test for rate of reproduction of earthworms between control (plot I) and experimental plot (plot II).*

Earthworm species	Plot	Sampling occasion	Mean	Standard deviation	Calculated value of 't'
L. mauritii	I	28	1.53 ±	0.36	5.71*
	II	28	0.98 ±	0.36	
D. willsi	I	28	2.39 ±	0.30	3.47*
	II	28	2.0 ±	0.06	

*P < 0.05

there was significant difference in the rate of reproduction of *D. willsi* and *L. mauritii* worm in two plots ('t' = 3.47. and 5.71 df = 54, $p < 0.05$) (Table 27.7).

Discussion

Population density and dynamics

The effect of inorganic fertilisers on earthworm population seems to be rather variable. Some authors claim that they are beneficial to earthworms (Jacob and Weigland, 1952; Watkin, 1954; Heath, 1962; Zajonc, 1970; Gerard and Hay, 1979; Edwards and Lofty, 1982; Reddy and Gound, 1987). However, harmful effects on earthworms have also been reported from the works of Richardson (1938), Huhta *et al.* (1967), Edwards and Lofty (1975), Zajonc (1975), Nowak (1976), Gerard and Hay (1979), and Brussard and De Ridder (1990). But information on the effects of fertilisers on structure and seasonal dynamics of earthworm populations are scanty (Lee, 1985). The present work is the first such report on the effect of fertilisers on earthworm structure and dyanmics in cane sugar field from India indicating about 30 per cent decrease in worm population. Further, higher population turnover value in fertiliser applied plot (plot II) exhibits greater instability in earthworm population.

Reproductive strategy

Reproductive strategy of very few species of earthworms are known from both temperate and tropical agroecosystems (Satchell, 1967; Lavelle, 1978; Hartenstein *et al.*, 1979; Senapati *et al.*, 1979; Dash and Senapati, 1980; Pani, 1987; Sahu *et al.*, 1988; Sahu, 1989). But there is great paucity of information on the interaction of agrochemicals and reproductive activity of earthworms. Few investigations that have been made on the effects of different stress on the reproductive biology of earthworms are by Griffiths *et al.*, (1967), Senapati and Dash (1981), Lofs-Holmin (1982), Pradhan and Mishra (1986), Senapati *et al.*, (1991). However, no information is available on the effects of fertilisers on reproductive strategies of earthworms. The present study indicates about 16 per cent (*D. Willşi*) and 36 per cent (*L. mauritii*) reduction in the rate of reproduction of dominant earthworms.

The decreased population density and reproductive rate of earthworms in plot II indicate that earthworms are at greater stress in fertiliser applied plot than in the control plot. The information is thus very important in agroecosystems where there is high fertiliser input for enhancing production.

Acknowledgements

The authors express their sincere thanks to Dr. P.C. Mishra, Reader and Head, Department of Environmental Sciences, Sambalpur University for providing laboratory facilities and Dr. B.K. Senapati for his encouragements.

REFERENCES

Brussard, W.C.L. and De Ridder, J.A. 1990. Long term effects of nitrogen fertilisers on grassland earthworms (Oligochaeta : Lumbricidae), their relation to soil acidification. *Agric. Ecosystems and Environment* **30** : 71-80.

Darwin, C. 1881. *Formation of the Vegetable Moulds through the Action of Worms with Observations on their Habits*. John Murray, London p. 298.

Dash, M.C. 1978. Role of earthworms in the decomposer system. pp. 399-406. In : *Glimpses of Ecology* (Eds. J.S. Singh and B. Gopal). Int. Scientific Publ., New Delhi.

Dash, M.C. and Patra, U.C. 1977. Density, biomass and energy budget of a tropical earthworm population from a grassland site in Orissa, India. *Rev. Ecol. Biol. Sol.* **14** : 461-471.

Dash, M.C. and Patra, U.C. 1979. Wormcast production and nitrogen contribution to soil by a tropical earthworm from grassland site in Orissa, India. *Rev. Ecol. Biol.* Sol. **16** : 79-83.

Dash, M.C. and Senapati, B.K. 1980. Cocoon morphology, hatching and emergence pattern in tropical earthworms. *Pedobiologia* **20** : 317-324.

Dash, M.C. and Senapati, B.K. 1985. Vermitechnology : Potentiality of earthworms for vermicomposting and vermifed. p. 61-69. In : Proc. Natl. Soil Biol. Symp. on *Current trends in Soil Biology* (Eds. M.M. Mishra and K.K. Kapoor) Hissar.

Edwards, C.A. and lofty, J.R. 1975. The influence of cultivation on soil animal populations. pp. 399-408. In : *Progress in Soil Biology* (Ed. J. Vanek). Academic Publ. House, Prague.

Edwards, C.A. and Lofty, J.R. 1982. *Biology of earthworms*. Chapman and Hall, London, p. 333.

Gerard, B.M. and Hay, R.K.M. 1979. The effect on earthworms of ploughing, tined cultivation, direct drilling and nitrogen in a barley monoculture system. *J. Agric. Sci. Cambridge* **93** : 147-155.

Griffiths, D.C., Raw, F. and Lofty, J.R. 1967. The effects of soil fauna of insecticides tested against wire worms (Agriotes sp.) in wheat. *Ann. Appl. Biol.* **60** : 479.

Hartenstein, R., Neuhauser, E.F. and Kaplan, D.L. 1979. Reproductive potential of the earthworm *Eisenia foetida. Oecologia* **43** : 329-340.

Heath, G.W. 1962. The influence of ley management on earthworm populations. *J. Br. Grassl. Soc.* **17** : 237-244.

Huhta, V., Karpinnen, E., Nurminen, M. and Valpas, A. 1967. Effect of silvicultural practices upon arthropod, annelid and nematode populations in coniferous forest soil. *Ann. Zool. Fenn.* **4** : 87-143.

Jacob, A. and Weigland, K. 1952. Transformations of the mineral nitrogen of fertilisers in the soil *Z. Pfl Ernahr. Dung.* **59** : 48-60.

Lavelle, P. 1978 Les vers de terre, de la Savana de Lamto (Cote d'Ivoire) : peuplements, populations et functions dans I' ecosystems. *Publ. Lab. Zool. E.N.S.* **12** : 1-301.

lavelle, P. 1983. The soil fauna of tropical savannas : The earthworms. pp. 485-504. In : *Tropical Savannas* (Ed. P. Bourliere). Elsevier Scientific Publ. Comp. Amsterdam (The Netherlands).

Lavelle, P., Barois, I., Martin, A., Zaidi, Z. and Schaefer, R. 1989. Management of earthworm populations in Agroecosystems. : A possible way to maintain soil quality? pp. 109-122. In : *Ecology of arable land* (Eds. M. Clarholm and L. Bergstrom). Kluwer Academic Publishers.

Lee, K.E. 1985. *Earthworms, their ecology and relationships with soils and land use*. Academic Press, New York.

Lenka, D. 1972. Agroclimatic zones and potential evapotranspiration in Orissa. *Prakruti-U.U.J. (Sci.)* **9** : 53-61.

Lofs-Holmin, A. 1982. Measuring cocoon production of earthworms *Allolobophora caliginosa* (Sav.) as a method of testing sublethal toxicity of pesticides. *Swedish J. Agric.* Res. **12** : 117-119.

Nowak, E. 1976. The effect of fertilization on earthworms and other soil fauna. *Pol. Ecol. Stud.* **2** (4) : 195-207.

Pani, S.C. 1987. *Aspects of ecological studies on tropical earthworms in irrigated agricultural system of Orissa, India.* Ph. D. thesis submitted to Sambalpur University, Orissa, India, p. 232.

Pradhan, S.C. and Mishra, P.C. 1986. Changes in population and metabolism of earthworms following application of carbary insecticide to a grassland site. pp. 111-121. In : *Verms and Vermitechnology.* Proc. Nat. Sem. Org. Waste Utiliz. Vermicomp. (Eds. M.C. Dash, B.K. Senapati and P.C. Mishra). Five Star Press, Burla, Orissa.

Richardson, H.C. 1938. The nitrogen cycle in grassland soils with special reference to the Rothamsted park grass experiments. *J. Agric. Sci. Camb.* **28** : 73-121.

Reddy, M.V. and Goud, A.N. 1987. The effect of fertilisers on the population density of earthworm, *Malabaria* sp. in rice (*Oryza sativa*) nurseries. *J. Environ. Biol.* **8** (2) : 191-194.

Sahu, S.K. 1989. *The ecological implications of interspecific variations in some tropical earthworms from pasture ecosystems of Orissa, India.* Ph. D. thesis submitted to Sambalpur University, Orissa, India, p. 289.

Sahu, S.K. and Senapati, B.K. 1988. Alternative proposals for quantification of reproduction in a tropical earthworm. *Trop. Ecol.* **29** : 6-14.

Sahu, S.K., Mishra, S.K. and Senapati, B.K. 1988. Population biology and reproductive strategy of *Dichogaster bolaui* (Michaelsen) in two tropical agroecosystems. *Proc. Ind. Acad. Sci. (Anim. Sci.)* **97** (3) : 239-250.

Satchell, J.E. 1967. Lumbricidae. pp. 259-322. In : *Soil Biology* (Eds. A. Burges and F. Raw). Academic Press, London.

Senapati, B.K. and Dash, M.C. 1980. Effect of formalin preservation on weight of the tropical earthworms. *Rev. Ecol. Biol. Sol.* **17** : 371-377.

Senapati, B.K. and Dash, M.C. 1981. Effect of grazing on the elements of production in the vegetation of oligochaete components of a tropical pasture land. *Rev. Ecol. Biol. Sol.* **18** : 487-505.

Senapati, B.K. and Dash, M.C. 1983. Energetics of earthworm populations in tropical pastures of India. *Proc. Ind. Acad. Sci. (Anim. Sci.)* **92** : 315-322.

Senapati, B.K. and Dash, M.C. 1984. Functional role of earthworms in the decomposer subsystems. *Trop. Ecol.* **25** (2) : 54-73.

Senapati, B.K. Dash, H.K.. and Dash, M.C. 1979. Seasonal dynamics and emergence pattern of a tropical earthworm *Drawida calebi* (Oligochaeta). *Int. J. Inver. Rep.* **1** : 271-277.

Senapati, B.K., Biswal, J., Sahu, S.K. and Pani, S.C. 1991. Impact of malathion on *Drawida willsi* , Michaelsen, a dominant earthworm in Indian rice fields. I. Survival, Population dynamics reproductive strategy and secondary production. *Pedobiologia* **35** : 117-128.

Syers, J.K., Sharply, A.M. and Keeney, D.R. 1979. Cycling of nitrogen by surface casting earthworms in a pasture ecosystem. *Soil Biol. Biochem.* **11** : 181-185.

Watkin, B.R., 1954. The animal factor and levels of nitrogen. *J. Br. Grassl. Soc.* **9** : 35-46.

Zajonc, I. 1970. Earthworm synusial of meadow stratocenoses and choriocenoses. *Acta. Zooltech. Univ. Agr. Nitra* **21** : 203-211.

Zajonc, I. 1975. Variations in meadow associations of earthworms caused by the influence of nitrogen fertilizers and liquid manure irrigation. pp. 497-503. In : *Progress in Soil Zoology* Proc. 5th Int. Colloq. Soil Zool. (Ed. J. Vanek). Prague.

Chapter 28

Some Aspects of Cadmium Toxicity on the Vitellogenesis in a Hillstream Teleost *Garra mullya* (Sykes)

Ehsan A. Khan and M.P. Sinha

Department of Zoology Ranchi University, Ranchi-83008, Bihar, India

Abstract. Effects of sublethal cadmium toxicity on vitellogenesis in a hillstream teleost *Garra mullya* (Sykes) have been described. Results indicated that cadmium produces inhibitory effects on protein synthesis mechanism of liver which is the site of vitellogenin (yolk protein) synthesis as was evident by significant fall in total protein, RNA and DNA in liver as well as reduced hepatosomatic index (HSI) in Cd-treated fish. Reduced protein mobilisation also resulted in fall of serum protein and ovary protein as well as gonosomatic index (GSI) of Cd-exposed fish. Prolonged sublethal treatment of cadmium chloride also resulted in significant reduction of oocyte diameter and total egg count per fish. Data of energy content (calorific value) of liver, ovary and muscle during vitellogenic growth of fish revealed a reduction in energy mobilisation in all three tissues of Cd-treated specimens. Cadmium caused a net reduction of calorific content by 38.5 per cent in liver and 16.5 per cent in ovary. The depletion of calorific content of muscle during vitellogenic growth was reduced by 5.6 per cent in Cd-treated fish when compared to control. Results indicated inhibitory role of cadmium chloride during vitellogenic growth of *Garra mullya*.

Key words : *Cd-toxicity, Garra mullya, liver, ovary, vitellogenesis.*

Introduction

Effects of pollutants, specially, heavy metals, on the aquatic organisms in general and fish in particular, are the area of special interest as most of the pollution causing effluents are discharged into water-bodies like ponds, lakes and rivers. Various genera of fishes found in the rivers and streams form major source of supplementary food for the rural mass who practice capture fisheries in rivers and streams. Heavy metal form several industrial, mining and other sources enormously contribute to the pollution problem in rivers and streams resulting into adverse impact on biota including fishes.

Fish population is generally considered very sensitive to all kinds of environmental changes to which it is exposed as they are exclusively aquatic with external mode of fertilization (most fresh water teleosts). Though certain stages in the life cycle of marine and fresh water fishes are more susceptible to environmental and pollutional stress (Von Westernhagen, 1968, 1970; Von Westernhagen, 1988). Rosenthal and Alderdice 1976; yet the effects on physiological and metabolic processes related with vitellogenic growth of the fish are not less significant (Mount, 1968; Hiltibran, 1971; Bengtesson, 1974; Benoit, 1975; Alderdice *et al.*, 1979 a, b; Rombough and Garside, 1982). The review of wide range of literature available on the effect of different pollutant on fish eggs and larvae (Von Westernhagen, 1988) reveals differences in susceptibility between the early developmental stages like embryos and larvae of cod, *Gadus merhua*, herring, *Clupea harengus*, plaice, *Pleuronectes platessa* (Kuhnhold 1972), Black Sea flounder, (Mazmanidi and Bazhasvili 1975) and several other marine species (Wilson, 1972).

Heavy metal toxicity has been found to cause inhibition of growth in Atlantic salmon (*Salmo salar*) when exposed to low level of toxicity (0.47 μg cd/l) while significant reduction in viable hatch has been noticed at comparatively higher level between 300 and 800 μg cd/l (Rombough and Garside, 1982). Similarly, exposure of mature fish to micro quantities of zinc, cadmium, copper and mercury has been reported to lead to the reduction upto 80 per cent in the eggs produced on some fresh water species. (Von Westernhage, 1988).

The basic requisite for successful reproduction is the production of eggs in sufficient quantities so as to preserve the species. Toxic effects of various heavy metals appear to affect this basic requirement by inhibiting metabolic activities at cellular levels (Nilsson, 1970; Hiltibran, 1971; Sax and Sax, 1974). So far Cd-toxicity is concerned, Banerjee *et al.* (1978) noticed significant changes in haematological parameters and growth as well as Cd-deposition in tissues of air breathing fish *Clarias batrachus and non air breathing fish Tilapia mossumbica*. Inhibition of egg-growth and survival due to cadmium toxicity has been reported in *Salmo gairdneri* (Beattie and Pascoe, 1978); in herring (Von Westernhagen *et al.*, 1974), in flounder (Von Westernhagen and Dethlefsen, 1975) and in garpike eggs (Von Westernhagen *et al.* 1975). But no such report is available on hillstream fish forms.

Egg yolk is the most important material which meets the nutritional requirement of developing embryos in oviparous animals including teleost fishes. In most fish the process of yolk deposition in the oocytes (vitellogenesis) is a seasonal phenomenon. Accordingly, the onset of breeding season in oviparous vertebrates in general and fish in particular is characterised by initiation of biochemical processes related to egg production. The key product in vitellogenesis is a lipophasphoprotein called vitellogenin which is synthesized in liver under the influence of estrogen produced by ovary (Plack *et al.* 1971; Emmersen and Petersen, 1976; Emmersen and Emmersen, 1976; Nath and Sundararaj, 1981 a, b). After its synthesis the vitellogenin is transported by blood to the ovary where it is cleaved into final egg yolk proteins-phosvitin and lipovitelling and deposited in occytes as yolk granules (Tata, 1976; Ng and Idler, 1983). In accordance with this fundamental process, the initiation of breeding season in fish is marked by variations in the cellular constituents of liver, blood serum, gonad and to some extent muscle which are the results of variations in external (environmental) as well as internal (sex steroid) factors (Diana and Mackay, 1979 Emmersen and Emmersen, 1976; Medda *et al.* 1980). As the process of vitellogenin synthesis sets in the liver the elevated levels of serum components particularly phosphorous, calcium, protein have been recorded in fish (Follett and Redshaw, 1968; Plack *et al.* 1971; Emmersen and Petersen, 1976; Emmersen and Emmersen, 1976; Nath and Sundararaj, 1981 a, b; Okuzawa *et al.* 1986; Barannikova *et al.* 1989; Quinitio *et al.*1989).

As the production of mature and viable eggs in fish in mediated by complex biochemical and metabolic processes via liver-blood-ovary axis, any change in the external environmental condition, pollutional or other wise, is most likely to act through this axis. A perusal of the available literature, however, indicate an utter paucity of research on the effects of heavy metal toxicity on the biochemical processes related to vitellogenic growth of fresh water fishes. Heavy metals like Zinc, cadmium and copper at low concentration (Cu, 3.7-31.0 µg/l; Cd, 0.6-60.0 µg/l) caused significant decrease in the spawning activity and egg numbers per female (Eaton, 1973). Zinc toxicity has been found in minnow, *Phoxinus phoxinus* (Bengtsson, 1974) and in *Brachydanio rerio* (Speranza *et al.* 1977). Effect of Cd-toxicity on Indian fresh water fishes are limited to few species and only related to histological and hematological aspects (Banerjee *et al.* 1978; Wani and Latey, 1982, 1983).

Further, relatively few works appear to have been devoted to sublethal effects. Notable among them was by Rosenthal and Alderidic (1976). According to Rosenthal and Alderdice (1976), sublethal effects may be defined as "Those responses to environmental changes- histological, morphological, physiological, or ethological – that may be induced in one stage of development but expressed at a later stage of development in terms of reduced survival potential". Thus, the term sublethal for developing fish eggs applies to the different processes and stages of development of vitellogenesis which may result in production of vitellogenesis which may result in production of non-viable egg and/or reduction in the number of viable eggs.

An individual's sublethal response always excludes immediate death, through its life expectancy may be shortened due to impairment of its morphological, histological or physiological set-up (Vonn Westgernhagen, 1988).

The present study was designed specially to study the sublethal effects of cadmium (Cd) toxicity on hill stream teleost (*Garra mullya* (Sykes). This species is a teleost fish belonging to order cypriniformes and it is very common in the rivers and streams of Chotanagpur plateau. It is also greatly relished as food in the surrounding rural areas and dominates the fish catches of local fishermen. Chotanagpur plateau is one of the richest mineral belts in the country having vast deposits of coal, iron, copper, manganese and even uranium. Mining potential of this area has resulted in the concentration of large, moderate and smaller industries in this plateau region which often discharge their effluents containing heavy metals in surrounding rivers and streams.

The objective of the present communication is to provide some information pertaining to the sublethal effects of cadmium on the basic egg producing phenomenon "vitellogenesis" in a hill stream teleost *Garra mullya* from biochemical view point.

TABLE 28.1. *Physico-chemical profile of aquaria-water during the period of experiment*

Parameters	March	April	May	June	M	SD	V	CV	SE	M at 95% CL
Temp°C	23.5	24.7	26.2	25.5	24.975	1.003	1.006	4.016	0.501	24.975 ± 1.594
pH	7.4	7.5	7.2	7.2	7.325	0.129	0.016	1.773	0.064	7.325 ± 0.203
Total hardness (ppm)	66.0	69.0	61.0	72.0	67.000	4.062	16.500	6.062	2.031	67.000 ± 6.462
Total alkalinity (ppm)	105.0	107.0	116.0	112.0	110.000	4.301	18.500	3.910	2.150	110.000 ± 6.841
Dissolved oxygen (ppm)	6.3	6.5	7.2	7.5	6.875	0.491	0.241	7.153	0.245	6.875 ± 0.779
Free carbon dioxide (ppm)	0.5	0.5	1.0	1.0	0.750	0.250	0.062	33.333	0.125	0.750 ± 0.397

M = Mean, SD = Standard Deviation, V = Variance, CV = Co-efficient of variation,
SE = Standard error, M at 95% Cl = Mean at 95% confidence level.

Material and Methods

Mature specimens of *Garra mullya* were collected from the Suvarnarekha river at Getalsud, 30 km west of Ranchi town (23°20' latitude and 85°30' longitude) with the help of local fishermen. Fishes were brought to the laboratory and mature specimens of similar weight (40 ± 3g) and length (15.5 ± 1.00 cm) were sorted out. The fishes were kept in aquaria for a week in normal tap water for acclimatization. The 96 hour LC_{50} value of cadmium chloride for the experimental fish was determined by Probit analysis (Finney, 1971; Fisher and Yates, 1974), and was found to be 8.6 mg/l. The water used in the experiment had the Physico-Chemical characteristics as given in Table 28.1.

Experiment plant

Ovary of *Garra mullya* at the collection site enters the preparatory phases in February and spawning begins by mid-July (Khan and Mehrotra, 1991). Ovary remains in the maturing phase during April and eggs become ripe and ready for spawning by June end. As the objective of the present study was to find out the degree of biochemical effect on vitellogenesis, the fishes were kept in Cd-containing water from beginning of March to June end. Care was taken that fishes used for experiment were mature and of similar length and weight. As the LC_{50} value was 8.5 mg/l for cadmium chloride, sublethal dose was considered to be safe at 4.00 mg/l for producing effect during prolonged period. Experiment was started on 1st March, 1991 and continued till June 30, 1991. About eighty specimens were used each for control and Cd-treatment. Fishes were lodged in four aquaria of similar size, each for control and treated, as keeping them in one aquaria would have caused crowding condition. All the aquaria had sufficient aeration as well as light, and water was changed weekly, Specimens were fed on *Hydrilla* Leaves and some bryophytes and lichens sticked to small stone chips.

Test fishes were lodged in control as well as $CdCl_2$ containing waters on 1st March and dissection for tissue analysis and eggs counts were carried out after 30 days in each month from March to June. Control groups were maintained under identical conditions without $CdCl_2$. Average mortality during the experiment was roughly around 5.00 per cent.

Tissue Analysis

After every 30 days, beginning from March 1, the liver, ovary and muscle of both control and treated fish were dissected out and weight. Gonosomatic index (GSI *i.e.* ovary wt/100 g body Wt) and hepatosomatic index (HSI *i.e.* liver wt/100 g body wt) were determined for each fish. About ten to fifteen fishes were sacrificed each month. Blood was collected by heart puncture. Ovary, liver and muscle samples were kept in fish saline (5.5 g NaCl, 0.14 g KCl and 0.12 g $CaCl_2$ per litre distilled water) until processed for estimation of various biochemical constituents. Few ovaries were also preserved in 10 per cent formaldehyde solution for measuring oocyte diameter and total egg counts.

Extraction of DNA, RNA, and total protein in tissues and protein in blood serum from the homogenates by the method of Munro and Fleck (1966) as modified by Abalain *et al.* (1980). The DNA and RNA contents were measured by determining the optical density at 260 nm on a UV-spectrophotometer with calf thymus DNA and yeast RNA as standards. After extraction, the total protein contents of tissues and blood serum were estimated by the method of Lowery *et al.* (1951) by measuring optical density on spectrophotometer at 620 nm with bovine serum albumin as standard.

The water contents of ovary, liver and muscle were determined by drying for 48 hrs. at 105°C in an electric oven. Ash content was measured by igniting the samples until tissues were completely burnt. Fat content was determined by using the method of Floch *et al.* (1957). Carbohydrate was estimated by difference (Eliassen and Vahl, 1982 a).

Energy Content

Energy Content of tissue were calculated assuming calorific values of 9.5, 5.7 and 4.0 $Kcal^{-1}$ for fat, protein and carbohydrate respectively (Kleiber, 1975).

Oocyte diameter and total egg Count

At least five ovaries were used each month to determine the total number of eggs per fish.

Each preserved ovary was weighed individually and the number of eggs from appropriate weighted sub-samples were counted under a dissecting microscope. The means of the total number of eggs and growing oocytes in the sub-samples were determined to calculate the total number of eggs per ovary (Abidin, 1986).

For the measurement of oocyte diameter, only those eggs were considered in which some degree of yolk deposition was noticed and it was assessed by colour, shape and general appearance, particularly yolk material under microscope (Jacob and Nair, 1983). As first sampling was done during maturing stage (March end), majority of eggs were yolk laden to different extents. Hence the measurement of mean oocyte diameter gave an overall size of the developing oocytes. For measurement of diameter, occular micrometer was used (in × 100). Random samples of 1000 eggs (ova) from each ovary of different regions (anterior, middle and posterior) were taken so as to eliminate sampling error resulting from regional differences.

Statistical analysis of all the data was carried by students. 't' test (Bringing and Kintz, 1977).

Results

Sublethal treatment of cadmium chloride to the experimental fish produced a very significant effect during the prolonged treatment. As the objective of the study was to find out the toxic effects on liver-blood-ovary axis in relation to vitellogenesis, biochemical constituents like total protein of liver, ovary and blood serum, liver-RNA, Liver-DNA and gonosomatic index (GSI) as well as hepatosomatic index (HSI) were determined. These parameters provided an overall metabolic state of the fish in relation to the progress of vitellogenic process.

Biochemical effects

Results of the experiment reveal that Cd-treatment begin to produce significant effect in vitellogenin synthesis in liver from the very first month (March) of treatment as is evident from the data of total liver protein, liver-RNA liver-DNA and hepatosomatic index (Fig. 28.1 A-D). These parameters recorded decrease in the liver of treated fish in comparison to control. An analysis of the data by student t-test indicated significant change in liver protein from March itself ($p < 0.005$). The change was more significant in the following months April, May and June ($p < 0.001$). As the RNA-content is prerequisite for protein synthesis and the ratio of RNA to DNA (RNA per unit DNA) is a better index of metabolic activity (Bulow, 1970), the contents of RNA and DNA in liver were determined (Fig. 28.1 C, D). Results indicated highly significant decrease in RNA content of treated fishes during first three months ($p < 0.001$). But the level of significance was low in June ($p < 0.005$) when the protein mobilisation in liver attains decreasing trend due to near completion of vitellogenesis in oocytes. Corresponding change in the liver-DNA was also noticed and this change was also significant being $p < 0.025$ in March and $p < 0.001$ in following months. Variation in the HSI during initial period (March) was statistically non-significant while it was highly significant from April onwards ($p < 0.001$).

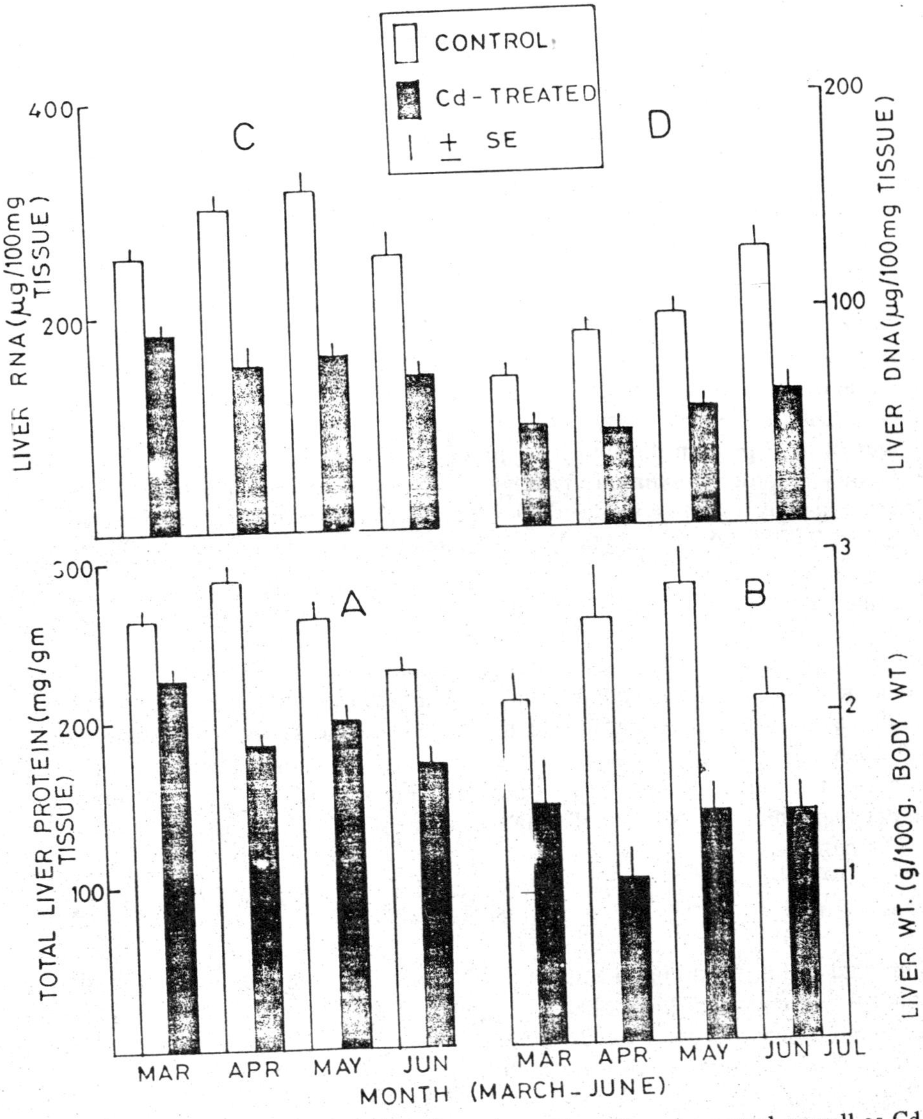

Fig. 28.1. Hepatosomatic index (HSI) and biochemical profile of liver in control as well as Cd-treated specimens between March and June.

A. Total protein content of liver.
Student's t-test values : March = $p < 0.005$;
April-June = $p < 0.001$.

B. HSI
March = Non-significant; April-June = $p < 0.001$.

C. RNA content of Liver
March-May = $p < 0.001$; June = $p < 0.005$.

D. DNA content of Liver
March = $p < 0.025$; April-June = $p < 0.001$.

As the vitellogenin synthesis proceeds in liver, it gets transferred to the blood serum for transportation to ovary. Cd-treatment resulted in the significant decrease in the total blood serum protein (Fig. 28.2, B). As the uptake of liver protein (vitellogenin) by blood is preceded by its synthesis in liver, the variations of serum protein was less significant during initial months being $p < 0.05$ in March and $p < 0.005$ in April. However, during May and June serum protein recorded highly significant decrease ($p < 0.001$) in the Cd-treated specimens in comparison to control groups. Variation of serum protein was in accordance with the change of liver protein, RNA and DNA.

Ovarian up-take of vitellogenin from blood serum is reflected by total protein content of the ovary, gonosomatic index (GSI) and diameter of the developing oocytes (Fig. 28.2 A, C, D). Initial rate of vitellogenin uptake by ovary was slow as was indicated by non-significant change of ovary protein, GSI and oocyte diameter in March and to some extent in April. Maximum turnover of total protein through liver-blood-ovary axis was in the months of May and June when ovary recorded maximum vitellogenic growth. However, the turnover rate significantly decreased in May and June in Cd-treated specimens as was evident from data of ovary protein ($p < 0.001$, 0.005), GSI ($p < 0.0005$, 0.001) and oocytes diameter ($p < 0.005$).

Total Egg Count

As the experimental fishes were of approximately similar weight and size, the total egg count per fish provided an approximate data on the quantitative production of egg when exposed to Cd-toxicity. Data of total egg count in each month of experiment have been presented graphically in Fig. 28.3. Number of eggs during initial period (March) did not change significantly in Cd-treated specimens. However, in the following months there was noticeable reduction in the number of eggs per fish which was statistically significant as $p < 0.005$ (April), $p < 0.001$ (May) and $p < 0.025$ (June).

Energy Content

The data of fat, protein and ash content in ovary, liver and muscle as well as dry and wet weight of these tissues did not differ significantly ($p > 0.05$) in individual fishes under experiment. Hence, the data on these parameter were pooled and energy content was determined in terms of per gram tissues of both control group as well as Cd-treated specimens. Figs. 28.4-28.6 represent the per gram composition of wet tissues in control and Cd-treated specimens. The energy budget per gram tissue during experiment is given in Table 28.2 which also provides the total energy changes of all three tissues (liver, ovary and muscle) during the 4 months alongwith percent changes between March and June. Monthwise data indicate that while liver and ovary were affected by Cadmium Chloride, the muscle remained almost unaffected. However, the net energy changes between April and June indicate a different energy profile. In control group the mobilisation of energy in ovary is maximum (38.5%) between the period of experiments. But liver shows lesser mobilisation (8.2%) and muscle exhibited depletion (18.6%) in terms of

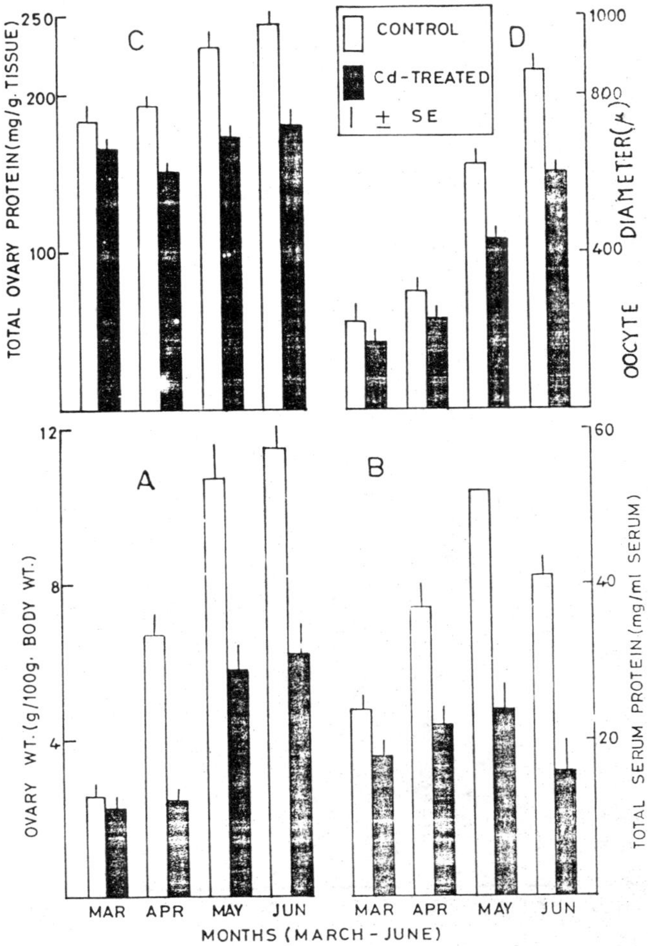

Fig. 28.2. Data of ovary and blood serum in control and Cd-treatd specimen between March and June.

A. Gonosomatic index (GSI)
March = Non-significant; April = $p < 0.001$;
May = $p < 0.005$; June = $p < 0.001$.

B. Total serum protein
March = $p < 0.05$; April = $p < 0.005$;
May-June = $p < 0.001$.

C. Total ovary protein
March = Non-significant; April-May = $p < 0.001$;
June = $p < 0.005$.

D. Oocyte diameter
March-April = Non-significicnat; May-June = $p < 0.005$

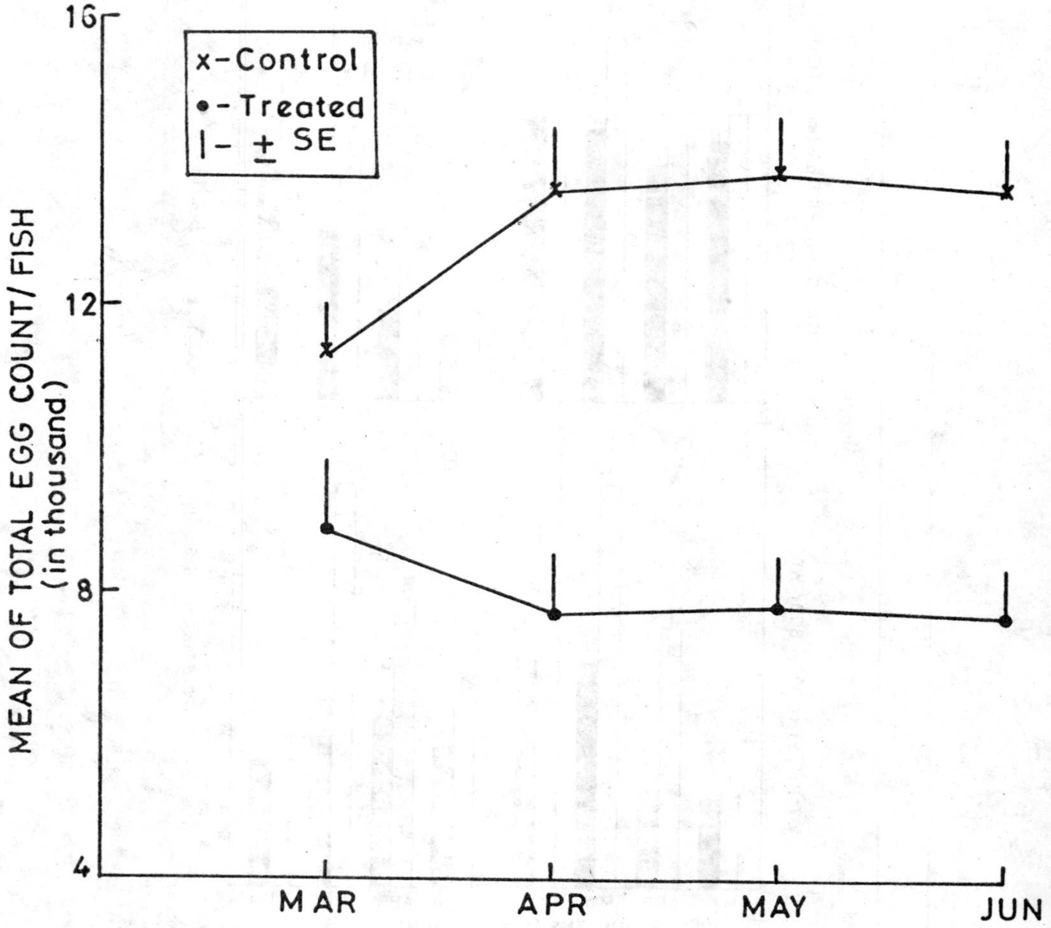

Fig. 28.3. Total egg count per fish of control and Cd-treated specimen between March and June. March = Non-Significant; April = $p < 0.005$; May = $p < 0.001$; June = $p < 0.025$.

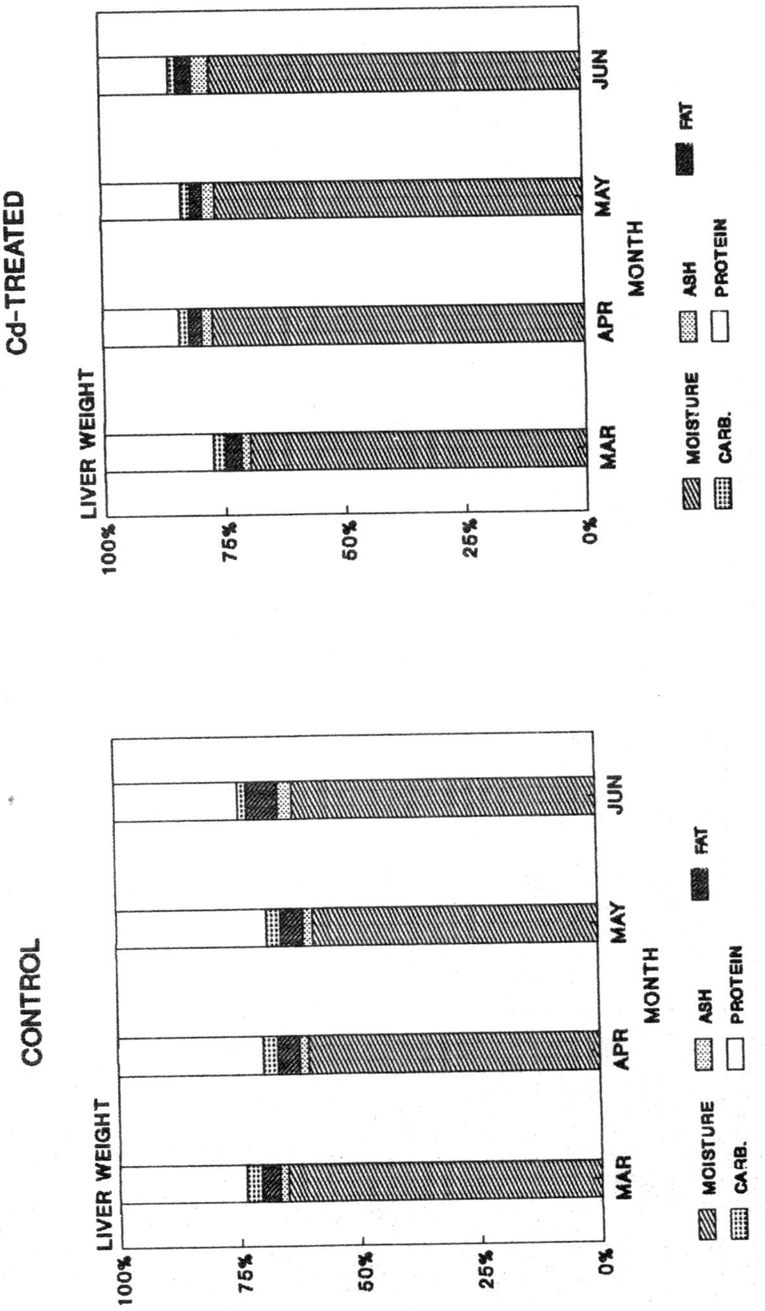

Fig. 28.4. Biochemical profile of per gram liver wet weight in control and Cd-treated fish.

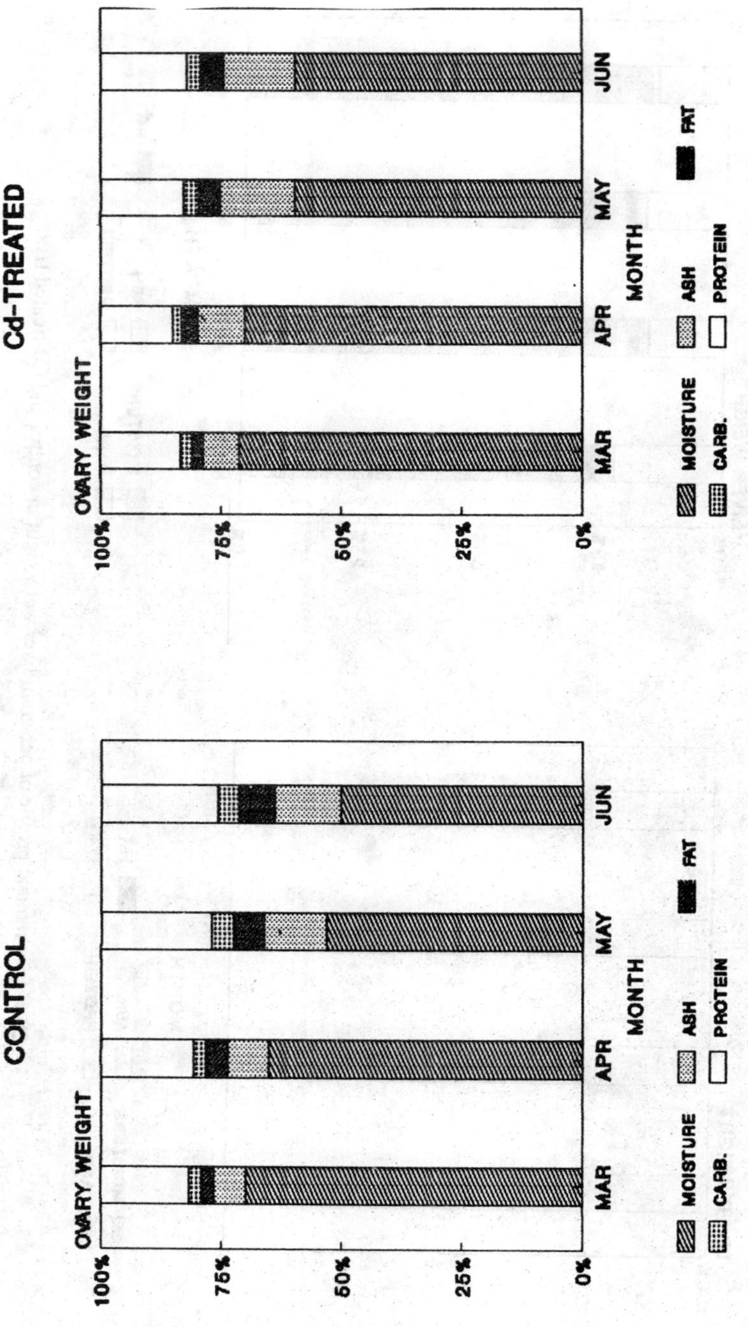

Fig. 28.5. Biochemical profile of per gram ovary wet weight in control and Cd-treated fish.

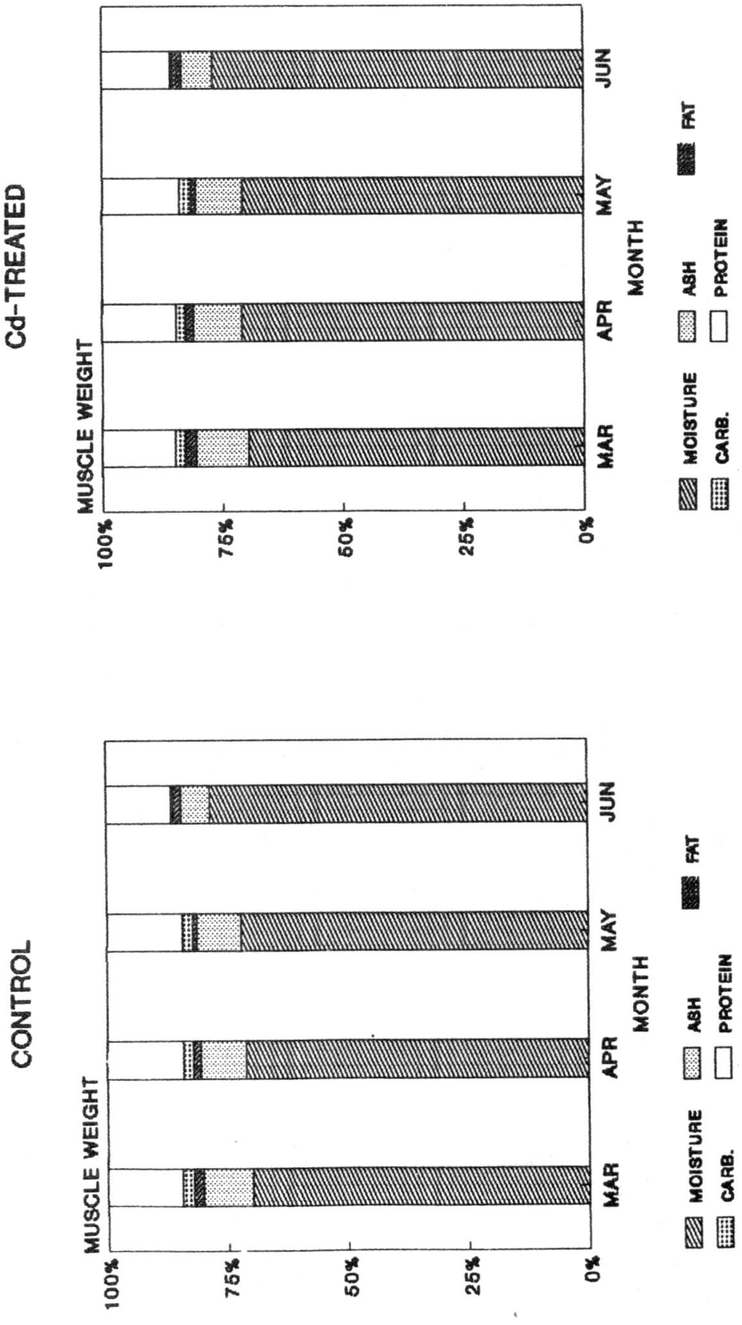

Fig. 28.6. Biochemical profile of per gram muscle wet weight in control and Cd-treated fish.

TABLE 28.2. *Calorific profile ($KCal^{-1}$) of the Liver, Ovary and Muscle and changes between March and June of control and Cd-treated fish. Differences indicate during three month period (April-June) as first analysis was done in March end.*

		March	April	May	June	Difference between March and June	Percent gain/loss between March and June	
Liver	C	1.982	2.237	2.335	2.161	+ 0.179	+ 8.2%	
	T	1.686	1.229	1.253	1.175	- 0.511	- 30.3%	- 38.5%*
Ovary	C	1.381	1.628	2.088	2.247	+ 0.866	+ 38.5%	
	T	1.248	1.246	1.536	1.589	+ 0.351	+ 22.0%	- 16.5%*
Muscle	C	1.159	1.121	1.061	0.944	- 0.215	- 18.6%	
	T	1.149	1.099	1.102	0.998	- 0.151	- 13.1%	- 5.6%*
Total	C	4.522	5.022	5.484	5.352	+ 0.830	+ 15.5%	
(L + O + M)	T	4.083	3.574	3.891	3.762	- 0.321	- 7.8%	- 23.3%*

- = Loss; L = Liver; O = Ovary; M = Muscle; + = Gain; C = Control; T = Cd-treated.
* Percent change in treated with respect to control.

calorific content. However, the energy profile obtained from the experiment reveal that the Cadmium causes significant fall in the calorific content of all the three tissues. In liver of Cd-treated fish the reduction of energy was 30.3 per cent which when compared to control group accounted for a net loss of 38.5 per cent between April and June. Though mobilisation by ovary was also evident in Cd-treated fish, it recorded 22.0 per cent fall in energy. When compared to control it accounted for net loss of 16.5 per cent in ovary. Similarly, depletion in energy content of muscle was evident both in control (18.6%) and Cd-treated (13.1%) specimens which indicated depletion in the muscle of Cd-treated by 5.6 per cent in comparison to control. As far as the total energy content of liver, ovary and muscle are concerned, there was a net gain of 15.5 per cent in control group between April and June while the Cd-treated specimens recorded a net loss of energy by 7.8 per cent during the same period when the total loss of energy by Cd-exposed fish was compared with control it accounted for a net loss of 23.3 per cent between April and June.

Discussion

In natural or experimental condition a sublethal concentration of a pollutant is most likely to produce sublethal effects so as to alter the morphological, physiological, histological and/or ethological condition of the fish (Rosenthal and Alderice, 1976), though it may not cause immediate death of the individual. Hence, the study of sublethal dose effect of a pollutant is comparatively more rationale than lethal or fatal dose, as in most of the polluted natural environment, sublethal concentration is met which may cause the alteration in the normal survival of organisms over a prolonged period of time. Accordingly, the present study was taken up in terms of sublethal dose response in the experimental fish.

Mortality percentage was very low (5.0%) during the course of experiment. However, as the mortality was noticed in both control and treated groups, it may not be attributed to Cd-toxicity. No visible morphological changes were noticed except that the pigmentation of skin became faint in Cd-treated fishes in comparison to controls. Similar changes in pigmentation have also been reported in Cd-exposed non-air breathing (*Tilapia mossumbica*) and air breathing (*Clarias batrachus*) fishes (Banerjee *et al.* 1978).

The reproductive phenomena in particular reference to vitellogenesis of teleosts is well studied and a generalized plan of activities of vitellogenesis during the reproductive phase of the year of the fish can be presented as follows to discuss our finding in required way.

Reproductive condition largely determines the body composition of the fish (Craik and Harvey, 1984; Strange and Pelton, 1987). Seasonal variation in the cellular constituents is a common observation of almost all fishes (Love, 1970). Cellular constituents of liver, muscle and gonad exhibit cyclical variation correlated with reproductive cycle of the fish, and it has been suggested to be the result of external (environmental) and internal (sex steroid) factors (Diana and Mackey, 1979; Emmersen and Emmersen, 1976; Medda *et al.* 1980; Nath and Sundararaj, 1981 a, b; Okuzawa *et al.* 1986; Quinitio *et al.* 1989).

Lapin and Basaanzhav (1989) also found the environmental conditions as one of the factors for metabolic activities related to the vitellogenesis. Various pollutants including heavy metals, therefore, may have substantial influence on oocyte development, maturation, hatching, growth and survival of different species of teleost fishes (Von Westernhagens, 1988). Cadmium is a non-essential heavy metal whose toxicity effects have been studied by different authors on a variety of fish species (Pickering and Gast, 1972; Hiltibran, 1971; Eaton, 1974; Von Westernhagen *et al.* 1975; Banerjee *et al.* 1978; Beattie and Pascoe, 1978; Alderdice *et al.* 1979 a, b; Wani and Latey, 1982, 1983). Rosenthal and Spiraling (1974) have reported the multifaceted effect on the development and survival of herring due to Cd-toxicity.

Onset of reproductive season is marked by the most remarkable metabolic activity that takes place in the liver where yolk protein (vitellogenin) is synthesized under the influence of estrogen (Ng and Idler, 1983; Mommsen and Walsh, 1988). The key product in the vitellogenesis is a multicomponent lipophosphoprotein called vitellogenin, synthesized in liver, which is transported by blood to ovary where it is cleaved into the final egg yolk protein (phosvitin and lipovitellin) and is incorporated in ovary as yolk granules (Tata, 1976; Nath and Sundararaj, 1981 a, b; Mommsen and Walsh, 1988; Quinitio *et al.* 1989). In normal fish the increased metabolic activity of liver during vitellogeninsynthesis causes proliferation and hypertrophy of liver resulting in the increased HSI as well as DNA, RNA and total peotein (Emmersen and Emmersen, 1976; Medda *et al.* 1980; Quinitio *et al.* 1989). As soon as a substantial amount of vitellogenin is synthesized in liver, it is cleared gradually by blood then incorporated in ovary by pinocytosis under the influence of pituitary gonadotropin (Nath and

Sundararaj, 1981 b). This phenomenon causes gradual rise of serum vitellogenin level (Emmersen and Emmersen, 1976; Nath and Sundararaj, 1981 a, b). Measurement of total protein in blood serum and liver thus provides, an overall indication of vitellogenin level (Emmersen and Emmersen, 1976; Medda *et al.* 1980). The entire process of vitellogenin and incorporation of yolk granules in ovary is thus carried out by hormone mediated liver-blood-ovary axis.

Garra mullya is a seasonal breeder and it enters the preparatory phase of gonadal development by February end (Khan and Mehrotra, 1991) at the site from where the fishes for present study were collected. Maturation of oocytes is completed by June end and spawning begins from mid-July. Hence, whatever impact cadmium may produce on vitellogenesis could be evident from the experiment between the beginning of March and June end. Guraya *et al.* (1975), while comparing the ovarian cycle of *Mystus tengara* in natural and confinded water has reported that the pattern of ovarian activity in confined water followed more or less same course as that in natural waters. Though, in temperate region some species like *Mugil cephallus* and *Mugil capito* exhibit rise in gonosomatic index mainly due to accumulation of athletic oocytes when confined inland locked habitat (Abraham *et al.* 1966), in *Mystus tengara*, a tropic species, the rise in GSI is mainly due to maturation of oocytes (Guraya *et al.* 1975). However, relatively slow growth of oocytes is noted in *Mystus tengara* when maintained in confined water. In the present study, the *Garra mullya* of control groups also exhibited relatively slow growth of oocytes in comparison to natural water (Khan and Mehrotra, 1991) as was evident from the data of total ovary protein, oocyte diameter and GSI (Fig. 28.2 A, C, D).

In control groups, a gradual rise of liver RNA was accompanied by corresponding rise in liver protein as well as HSI. Maximum level of liver protein, however, was recorded in April which indicates that protein mobilisation was at its peak in this month. Levels of RNA and HSI had maximum values in May and in June it decreased substantially (Fig. 28.1 B, C). The blood serum also began to increase from April and was maximum in May. However, the levels of ovary protein and GAS recorded sharp rise from May and were maximum in June (Fig. 28.2 A, C). The entire data of liver, serum and ovary showed well synchronisation of protein mobilisation in liver, its clearance by blood and incorporation in ovary. It is remarkable that large increase in liver weight and serum protein was found earlier than that in ovary. Thus, the highest metabolic activity of liver occurred when ovarian growth phase had advanced, and yolk deposition had started, as was evident from increasing ovarian protein, large oocyte diameter and higher GSI (Fig. 28.2 A, C, D). While ovarian vitellogenic growth was maximum in June, the total protein and RNA of liver as well as HSI and serum protein indicated a declining trend in June (Fig. 28.1 A, B; 28.2 B). These observations agree well with those made by Emmersen and Emmersen (1976) on *Platichthys flesus,* and by Quinitio *et al.* (1989) on *Cottus hangiongensis,* who found that liver weights in females increased during early phase of gonadal ripening, but then declined with progressive maturation of ovary upto the stage at which oocytes were ripe (June, in the present study). Khan and Mehrotra (1991) have also observed similar trends during the annual reproductive cycle of *Garra mullya.* Results indicate that the biochemical changes in liver, serum and ovary of control groups more or less followed the similar pattern as in natural waters.

Sublethal effects of pollutants on early developmental stages may be caused in two different ways-first by exposure of the parent fish and the second by the exposure of hatched larvae. Although, the complete suppression of ovarian egg development due to sublethal dose of heavy metals may be a rear case (Von Westernhagen, 1988), the inhibition of reproductive process in various stages has been reported due to copper toxicity in fathead minnow, *Pimephales promelas and Phoxinus phoxinus* (Bengtesson, 1974). Zinc, cadmium and copper at low concentration caused progressively decreasing spawning activity in fathead minnow (Eaton, 1973). Zinc, cadmium and copper produced significant effects on egg growth and production even at low concentration (parts per billion) in flagfish *Jordanella floridae* (Spehar *et al.* 1978). Heavy metals are also known to exert detrimental effects on the fertilization rates of spawned eggs (Ojaveer *et al.* 1980; Blaxter, 1977). Studies of the Cd-toxicity on Indian teleosts have revealed the inhibition of enzyme system in liver of *Clarias batrachus* and *Tilapia mossumbica* (Banerjee *et al.* 1978) and sever histological damage of liver and ovary of *Garra mullya* (Wani and Latey, 1982, 1983). Biochemical constituents of liver and ovary during the present investigation and also reported earlier, have also been found to fall significantly in *Garra mullya* (Khan *et al.* 1992 a), though the glycogen level remained unaltered in liver of *Garra mullya* under the influence of cadmium chloride (Khan *et al.* 1992 b). Banerjee *et al.* (1978) have also reported alteration in the haematological parameters in *Clarias batrachus* and *Tilapia mossumbica*.

The data of there present study indicate significant inhibitory influence of cadmium chloride at sublethal dose in the vitellogenin synthesizing system of liver. It is revealed by the noticeable decrease of total protein, RNA and DNA from the very first month of treatment in Cd-exposed fishes. Decreased rate of protein synthesis also resulted in fall of serum protein level which may be due to slower cleavness rate from liver as synthesized protein (vitellogenin) was available to blood in lesser amount in comparison to control group. Hiltibran (1971) noted considerable inhibition of phosphate metabolism in liver of vitellogenin, RNA and DNA, and inhibitory effect on its metabolism may cause retardation of protein synthesis in liver. Though phosphorus content was not estimated in the present study, it is possible that lower protein contents of liver and blood serum may have been caused due to inhibitory effect of liver phosphate metabolism (Hiltibran, 1971) and due to inhibition of liver enzyme system (Banerjee *et al.* 1978).

Ovary protein, GSI and oocyte diameter were found to exhibit non-significant changes during initial period of Cd-treatment in treated fishes. However, a gradual fall was noted in ovary protein, GSI and oocyte diameter with longer exposure time and decrease were statistically significant also. Major source of yolk protein is extra ovarian in the form of vitellogenin from liver though intra ovarian yolk synthesis has also been noticed in most teleosts (Droller and Roth, 1966; Anderson, 1974). Similarly in the present study, the protein, levels were found to exhibit significant fall in all three components – liver, serum and ovary of treated fish. It is indicative of the fact that low synthesis of protein in liver caused slower turn-over rate of yolk protein through liver-blood-ovary axis, resulting in decreased level of ovary protein, GSI and oocyte diameter which agrees well with the results of Brungs (1969); Hiltibran (1971); Benoit (1975, 1978); Banerjee *et al.* (1978); Wani and Latey (1982, 1983) and Khan *et al.* (1992 a). Indemnification of intra ovarian yolk was not done in the present study. Hence it is difficult to opine conclusively about the extent of influence of the cadmium salt that might have affected to intra ovarian process of yolk synthesis. However, Wani and Latey (1982) recorded severe histological damage of ovary in *Garra mullya* when exposed to cadmium and Powar and Katdare (1983) found drastic fall in GSI of the same species when exposed to sumithion, a commonly used insecticide.

Exposure of mature fish to sublethal levels of certain heavy metals may cause considerable reduction in number of eggs produced; in some cases even 80 per cent reduction has been noticed (review, Von Westernhagen, 1988). Copper administration in aquaria at 18-32 µg/l totally prevented egg deposition in fathead minnow *Pimephales promelas* (Mount, 1968; Mount and Stephan, 1969). However, the perusal of available literature indicates that complete suppression of egg production is not so common. A reduction upto 21 per cent in the number of eggs produced has been noted in *Phoxinus phoxinus* when exposed to 0.13 and 0.2 mg zinc/l (Bengtsson, 1974) and *Brachyrerio danio* also spawned lesser number of eggs when exposed to 5 mg Zn/l for nine days (Speranza *et al.* 1977). Eaton (1973) has also recorded progressive decrease in egg number per females when treated with cadmium and copper. Reduction in egg number due to different heavy metal toxicity at varied concentration have also been found in flag fish *Jordanella floridae* (Spehar *et al.* 1978), in guppy *Poecila reticulata* (Uviovo and Beatty, 1979) and Zebra fish *B. danio* (Kihlstran *et al.* 1971). In the present study, the progressive changes from month to month was recorded for number of eggs in both treated and control groups. A reduction in egg number was noticed from the very first month but it was statistically non-significant thus indicating that the cadmium toxicity at sublethal concentration takes more than a month in *Garra mullya* to produce detrimental effect on the numerical strength of the egg. In the following months the reduction was very significant resulting in a reduction of about 44 per cent in the June -- the last month of experiment when oocytes attain maturation. However, an average comparative reduction upto 31 per cent in the diameter of mature oocytes in June of Cd-treated specimens also indicate the possible non-viable nature of eggs produced which may result in either percentage failure of fertilization and/or detrimental effects on post-fertilization development (Blaxter, 1977; Alderdice *et al.* 1979 b; Ojaveer *et al.* 1980). Reduction in the number of eggs also suggests the possible inhibitory action on the germinal epithelial cells of ovary.

There is great paucity of information pertaining to the variation of calorific content in the tissues of fish maintained in heavy metal containing water. However, the biochemical constituents like protein, fat and carbohydrate, producing calorific energy have been found-to decrease on mexposure to heavy metals in several species of fish viz. blue gill (Hiltibran, 1971); fathead minnow *Pimephales promelas* (Banerjee *et al.*, 1978); *Clarias batrachus* and *Tilapia mossumbica* (Banerjee *et al.* 1978); flag fish *Jordanella floridae* (Spehar *et al.* 1978) and *Garra mullya* (Khan *et al.* 1992 a, b). The energy data of the present study in Table 28.2 indicate that cadmium toxicity significantly reduced the energy profile in liver, ovary and muscle. While in liver, the energy content was reduced by 38.5%, in ovary the reduction was 16.1% between April and June. Workers believe that the stored nutrient like protein, fat and carbohydrate in liver and body muscle are transferred to gonad and subsequently utilized as source of energy for metabolism as well as growth and maturation of gonad (Diana and Mackay, 1979; Blay Jr. and Eyeson, 1982; Strange and Pelton, 1987). In the present study maximum energy mobilisation was noticed in ovary of both control as well as treated groups. On the other hand energy depletion was also noticed from liver as well as muscle as the time of experiment advanced. Cadmium, however, was found to greatly influence the mobilisation-deposition process in all three tissues. Energy profile of control group indicates heavy depletion in muscle (18.6%) between April and June. However, the liver of control group mobilised lesser percentage (8.2%) energy indicating that liver too might have contributed substantially in ovarian calorific value where there was a gain of 38.5 per cent between the same period. However, the energy profile of Cd-treated specimens revealed that the depletion of energy content in muscle in this group was lesser (5.5%) with respect to control. Similarly, the liver of Cd-treated group recorded sharp decrease in the calorific content (38.5%) with respect to control between April and June which indicates that cadmium either caused large depletion from liver of riot inhibited energy mobilisation. However, latter possibility seems to be more likely as is evident from the month wise biochemical analysis of liver wet weight (Fig. 28.4). Reduced percentage depletion of energy from muscle and reduced mobilisation in liver of Cd-treated specimens is reflected well by a significant fall by 16.5 per cent in the energy content of ovary of treated fish. Data of total energy content of liver, ovary and muscle exhibited a gain of 15.5 per cent in control group between April and June where as in Cd-treated specimens there was a net loss of 7.8 per cent energy between the same period. Hence, it may be concluded that cadmium chloride inhibited energy mobilisation and depletion process of liver, ovary and muscle of *Garra mullya* which accounted for nearly 23.3 per cent over all loss of total energy in the tissue in question with respect to control group (Table 28.2).

As far as the mechanism of action of cadmium toxicity is concerned, the phenomenon is not fully understood (Nilsson, 1970; Rosenthal and Sperling, 1974). However, it is general agreement that cadmium ions first get deposited in tissues and eggs when exposed to cadmium or its compound (Pickering and Gast, 1972; Beattie and Pascoe, 1978; Benerjee *et al.* 1978; Rombough and Garside, 1982). Nilsson (1970) and Sax and Sax (1974) opine that mechanism of cadmium toxicity is probably due to high affinity of cadmium with sulphydryl and hydroxyl groups and ligands containing nitrogen. Therefore, the binding with such groups in chemical system make the control functions of the organisms vulnerable to cadmium even at low concentration. It is possible, therefore, that cadmium acts at cellular level to inhibit the enzyme system and other related biochemical processes of synthesis and breakdown. The reduced levels of protein, RNA, DNA and energy content of Cd-exposed fished during the period of investigation indicate that cadmium produces inhibitory effects on metabolic processes of liver, ovary and muscle which directly or indirectly contribute to the vitellogenic growth of the ovary.

Acknowledgement

One of the out0shors, E.A. Khan gratefully thank University Grants Commission, New Delhi for providing financial assistance in the form of a Senior Research Fellowship. Authors also thank Prof. P.N. Mehrotra, Head, Zoology Department, Ranchi University for providing Laboratory facilities of the experiment.

REFERENCES

Abalain, J.H., Jego, P. and Valotaire, Y. 1980. Effect of 17 β-Estradiol on the DNA, RNA, protein contents and on the DNA, RNA polymerase in the myllerian duct of the immature female Newt (*Pleurodeles Waltilii* Michah). *Gen. Comp. Endocrinol.* **40** : 402-408.

Abidin, A.Z. 1986. The reproductive biology of a tropical Cyprinid, *Hampala macrolepidota* (Van Hasselt), from Zoo Negara Lake, Kualalampur, Malaysia. *J. Fish. Biol.* **29** : 381-391.

Abrahim, M., Nelly, B. and Yashour, A. 1966. Oogenesis in five species of grey mullets (Teleostei : Mugilidae) from natural and land-locked habitats. *1st J. Zool.* **15** ; 155-172.

Alderdice, D.F., Rosenthal, H. and Velsen, F.P.J. 1979 a. Influence of salinity and cadmium on capsule strength of Pacific herring eggs. *Helgol. Wiss. Meeresunters.* **22** : 149-162.

Alderdice, D.F., Rosenthal, H. and Velsen, F.P.J. 1979 b. Influence of salinity and cadmium on the volume of Pacific herring eggs. *Helgol. Wiss. Merresunters.* **32** : 163-178.

Anderson, E. 1974. Comparative aspects of the ultrastructure of the female gamet. *Int. Rev. Cytol. Suppl.* **4** : 1-70.

Banerjee, A.K., Dastidar, G.G., Mukhopadhyay, P.K. and Dehadri, P.V. 1978. Toxicity of Cadmium. A comparative study in air breathing fish, *Clarias batrachus* (Linn.) and in non air breathing one, *Tilapia mossumbica* (Peters) *Ind. J. Exp. Biol.* **16** : 1274-1277.

Barannikova, I.A., Dyubin, V.P., Bukovskaya, O.S., Rebrova, L.V. and Travkin, V.G. 1989. Gonadotropic function of the hypophysics and the dynamics of gonadortropin and sex steroid in the blood of the autumn chum *Oncorhynchus keta* in the Amur River (Russian SFSR, USSR) after the completion of the sexual cycle. *Vopr. Ikhtiol.* **29** (5) : 823-830. (In Russian with English Summary).

Beattie, J.H. and Poscoe, D.P. 1978. Cadmium uptake by rainbow trout, *Salmo gairdneri* Richardson, eggs and alevins. *J. Fish Biol.* **13** : 631-637.

Bengtsson, B.E. 1974. The effects of zinc on the mortality and reproduction of the minnow, *Phoxinus phoxinus. Arch. Environ. Contam. Toxicol.* **2** : 342-355.

Benoit, D.A. 1975. Chronic effects of copper on survival, growth and reproduction of the blue gill (*Lepomis macrochirus*). *Trans. Am. Fish. Soc.* **104** : 353-358.

Benoit, D.A. and Holcombe, G.W. 1978. Toxic effect of zinc on fathead minnow (*Pimephales promelas*) in soft water. *J. Fish Biol.* **13** : 701-708.

Blaxter, J.H.S. 1977. The effect of copper on the eggs and larvae of placie and herring. *J. Mar. Biol. Assoc.* U.K. **57** : 849-858.

Blay, J. Jr. and Eyeson, K.N. 1982. Observations on the reproductive biology of the Shad *Ethmalosa fimbriata* (Bowdich) in the coastal waters of Cape Coast, Ghana. *J. Fish Biol.* **21** : 485-496.

Brungs, W.A. 1969. Chronic toxicity of zinc to the fathead minnow. *Pimephales promelas* Rafinesque. *Trans. Am. Fish. Soc.* **98** : 272-279.

Bruning, J.L. and Kintz, B.L. 1977. In : *Computational Handbook of Statistics.* Scott Foresman and Co. Palo, Alto, Calif.

Bulow, Frank, J. 1970. RNA-DNA ratios as indicators of recent growth rates of a fish. *J. Fish. Res Bd. Can.* *27* : 2343-2349.

Craik, J.C.A. and Harvey M. 1984. Biochemical changes occurring during final maturation of eggs of some marine and freshwater teleosts. *J. Fish. Biol.* **24** : 599-610.

Diana, J.S. and Mackay, W.C. 1979. Taming and magnitude of energy deposition and loss in the body, liver and gonad of northern pike (*Esox lucius*). *J. Fish. Res. Bd. Can.* **36** (5) : 481-487.

Droller, M.S. and Roth, T.F. 1966. An electron microscopic study of yolk formation during oogenesis in guppy *Lebistes reticulatus. J. Cell Biol* **28** : 209-232.

Eaton, J.G. 1973. Chronic toxicity of a copper, cadmium and zinc mixture to the fathead minnow (*pimephales promelas* Rafinesque). *Water Res.* **7** : 1723-173.

Eaton, J.G. 1974. Chronic cadmium toxicity to the blue gill (*Lepomis macrochirus* Rafinesque). *Trans. Am. Fish. Soc.* **103** : 729-735.

Eliassen, J.E. and Vahl, O. 1982 a. Seasonal variations in biochemical composition and energy content of liver, gonad and muscle of mature and immature Cod, *Gadus marhua* (L.) from Balsfjordan, Northern Norway. *J. Fish. Biol.* **20** : 707-716.

Emmersen, B.K. and Emmersen, J. 1976. Protein, RNA and DNA metabolism in relation to ovarian vitellogenic growth in the flounder *Platichthys flesus* (L.). *Comp. Biochem. Physiol.* **55** B : 315-321.

Emmersen, B.K. and Petersen, I.M. 1976. Natural occurrence and experiment induction by estradiol-17β of lipophospho-protein (vitellogenin) in flounder (*Plalichthys flesus* L.). Comp. Biochem. Physiol. **54** B : 443-446.

Finney, D.J. 1971. *Probit Analysis*. London : Cambridge University Press, p. 333.

Fisher, R.A. and Yates, C.B.E. 1974. *Statistical Tables.* Longman, p. 46.

Floch, J., Lees, M. and Stanly, G.H.S. 1957. A simple method for the isolation and purification of total lipids from animal tissues. *J. Biol. Chem.* **226** : 497-509.

Follet, B.K. and Redshaw, M.R. 1968. The physiology of vitellogenesis. pp. 219-308. In : *Physiology of Amphibia* (Ed. B. Lofts) Vol. 2, Academic Press, New York.

Guraya, S.S., Kaur, R. and Saxena, P.K. 1975. Morphology of ovarian changes during the reproductive cycle of the fish, *Mystus tengara* (Ham.). *Acta. Anat.* **91**: 222-260.

Hiltibran, R.C. 1971. Effects of cadmium, zinc manganese and calcium on oxygen and phosphate metabolism of blue gill liver mitochondria. *J. Water Pollut. Control Fed.* **43** : 818-823.

Jacob, Sheila Susan and Nair, N. Balakrishnan, 1983. Reproductive biology of the larvivorous fish *Macropodus cupanus* (Cuv. and Val.). *Proc. Ind. Acad. Sci. (Anim. Sci.).* **92** (2) : 159-170.

Khan, E.A. and Mehrotra, P.N. 1991. Variations of liver protein and RNA in relation to egg maturation in a hillstream teleost *Garra mullya* (Sykes). *J. Reprod. Biol. Comp. Endocrinol.* **3** (1) : 47-52.

Khan, E.A. Sinha, M.P., Saxena, N. and Mehrotra, P.N. 1992 a. Biochemical effects of Cadmium toxicity on a hillstream teleost *Garra mullya* (Sykes) during egg maturation. I. Total Protein, GSI, HSI. *Poll. Res.* **11** (3) : 157-161.

Khan, E.A., Sinha, M.P., Saxena, N. and Mehrotra, P.N. 1992 b. Biochemical effects of Cadmium toxicity on a hillstream teloest *Gara mullya* (Sykes) during egg maturation. II. Cholesterol and Glycogen. *Poll. Res.* **11** (3) : 163-167.

Kihlstrom, J.E., Lundberg, C. and Hulth, L. 1971. Number of eggs and young produced by zebra fishes (*Brachydanio rerio,* Hamilton-Buchanan) spawning in water contaning small amounts of phenyl mercuric acetate. *Environ. Res.* **4** : 355-359.

Kleiber, M. 1975. *The Fire of Life.* Robert T. Krieger Pub. Co., New York. p. 453.

Kuhnhold, W.W. 1972. The influence of crude oils on fish fry. pp. 315-318. In : *Marine Pollution and Sea Life* (Ed. M. Ruivo), Fishing News (Books), London.

Lapin, V.I. and Basaanzhav, G. 1989. Seasonal rhythm of physiological and biochemical processes in the Mongolian grayling *Thymallus brevirostris. Vopr. Ikhtiol.* **29** (5) : 831-841. (In Russian with English Summary).

Love, R.M. 1970. *The Chemical Biology of Fishes.* Academic Press, Inc. (London) Ltd.

Lowry, O.H., Rosenbrough, N.J., Farr, A.L. and Randall, R.J. 1951. Protein measurement with Folin phenol reagent. *J. Biol. Chem.* **193** : 265-275.

Mazmanidi, N.D. and Bazhasvili, T.R. 1975. Effects of dissolved petroleum products on the embryonic development of the black Sea flounder. *Hydrobiol. J.* **11** : 39-43.

Medda, A.K., Das Mahapatra, A.K. and Ray, A.K. 1980. Effect of Estrogen and Testosterone on the protein, nucleic acid contents of liver, muscle and gonad and plasma protein content of male and female (vitellogenic and non-vitellogenic) Singi fish, *Heteropneustes fossilis* Bloch. *Gen. Comp. Endocrinol.* **42** : 427-436.

Mommsen, T.P. and Walsh, P.J. 1988. Vitellogenesis and oocyte assembly. 347-406 pp. In : *Fish Physiology.* Vol. XI Part A. (Eds. W.S. Hoar and D.J. Randall), Academic Press, San Diego, New York, London, Tokyo.

Mount, D.I. 1968. cChronic toxicity of copper to fathead minnows (*Pimephales promelas,* Rafinesque). *Water. Res.* **2** : 215-233.

Mount, D.I. and Stephan, C.E. 1969. Chronic toxicity of copper to fathead minnow (*Pimephales promelas*) in soft water. *J.Fish. Res. Board Can.* **26** : 2449-2457.

Munro, H.N. and Fleck, A. 1966. The determination of nucleic acids. 113-176 pp. In : *Methods of Biochemical Analysis* (Ed. D. Glick) Vol. 14, Wiley-Interscience, New York.

Nath, P. and Sundararaj, B.I. 1981 a. Isolation and identification of femle specific serum lipophosphoprotein (vitellogenin) in the Cat fish, *Heteropneustes fossilis* (Bloch). *Gen. Comp. Endocrinol.* **43** : 184-190.

Nath,. P. and Sundararaj, B.I. 1981 b. Induction of vitellogenesis in the hypophysectomized Cat fish, *Heteropneustes fossilis* (Bloch) : Effects of piscine and mammalian hormones. *Gen. Comp. Endocrinol.* **43** :.191-200.

Ng, Bun T. and Idler, D.R. 1983. Yolk formation and differentiation in teleost fishes. 373-404 pp. In : *Fish Physiology* (Eds. W.S. Hoar, D.J. Randall and E.M. Donaldson). Academic Press, New York and London.

Nilsson, R. 1970. *Aspects on the toxicology of Cadmium and its Compounds*. Natl. Sci. Res. Coun. Stockholm.

Ojaveer, E., Annist, J., Jankowski, H., Palm, T. and Raid, T. 1980. On effects of copper, cadmium and zinc on the embryonic development of Baltic spring spawning herring. *Finn. Mar. Res.* **247** : 35-140.

Okuzawa, K. Furukawa, K., Aida, A. and Hanyu, I. 1986. Annual reproductive cycle of the homoronko *Gnathpogon elongatus caerulesens. Bull. Jap. Soc. Sci. Fish* **52** (11) : 1957-1960.

Pawar, K.R., and Katdare, M. 1983. Effect of sumithion on the ovaries of freshwater fish, *Garra mullya* (Sykes). *Curr. Sci.* **52** (16) : 784-785.

Pickering, Q.H. and Gast, M. 1972. Acute and chronic toxicity of cadmium to the fathead minnow (*Pimephales promelas*). *J. Fish. Res. Board Can.* **29** : 1099-1106.

Plack, P.A., Pitchard. D.J. and Fraser, N.W. 1971. Egg protein in Cod serum. Natural occurrence and induction by injection of estradiol 3-benzoate. *Biochem. J.* **121** : 847-856.

Quinitio, G.F., Takemura, A. and Goto, A. 1989. Ovarian development and changes in serum vitellogenin levels in the river sculpin, *gottus hangiongensis,* during an annual reproductive cycle. *Bull, Fac. Fish.* Hokkaido Univ. **40** (4) : 246-253.

Rombough, P.J. and Garside, E.T. 1982. Cadmium toxicity and accumulation in eggs and alevins of Altantic salmon *Salon salar. Can. J. Zool.* **60** : 2006-2014.

Rosenthal, H. and Alderdice, D.F. 1976. Sublethal effects of environmental stressors, natural and polutional, on marine fish eggs and larvae. *J. Fish. Res. Board Can.* **33** : 2047-2065.

Rosenthal, H. and Sperling, K.R. 1974. Effects of cadmium on development and survival of herring eggs. 383-396 pp. In : *The Early Life History of Fish* (Ed. J.H.S. Blaxter), Springer-Verlag, Berlin and New York.

Sax, N. and Sax, P.B. 1974. *Ind. Pollut.* Von Nostrand, New York.

Spehar, R.L., Leonard, E.N. and DeFoe, D.L. 1978. Chronic effects of cadmium and zinc mixtures on flagfish (*Jordanella floridae*). *Trans. Am. Fish. Soc.* **107** : 354-360.

Speranza, A.W., Seeley, R.J., Seeley, V.A. and Perlmutter, A. 1977. The effect of sublethal concentrations of zinc on reproduction in the zebra fish, *Brachydanio rerio* Hamilton-Buchanan. *Environ. Pollut.* **12** : 217-222.

Strange, R.J. and Pelton, J.C.P. 1987. Nutreint content of clupeid forage fishes. *Trans. Amer. Fish. Soc.* **116** : 60-66.

Tata, J.R. 1976. The expression of vitellogenin gene. *Cell* **9** : 1-14.

Uviovo, E.J. and Beatty, D.D. 1979. Effects of chronic exposure to zinc on reproduction in the guppy (*Poecilia reticulata*). *Bull. Environ. Contam. Toxicol.* **23** : 650-657.

von Westernhagen. H. 1968. Versuche zur Erbrutung der Eier des Schellfishches (*Melanogrammus aeglefinus* L.) unter kombinierten Salzgehalts-und Temperaturdingungen. *Ber. Dtsch. Wiss. Komm. Meeresforsch.* **19** : 270-287.

von Westernhagen, H. 1970. Erbrutung der Eier vom Dorsch (*Gadus morhua*), Flunder (*Pleuronectes flesus*) and Scholle (*Pleuronectes Platessa*) unter kombinierten Temperatur-und Salzgehaltsbedingungen. *Helgol. Wiss. Meeresunters.* **21** : 21-102.

von Westernhagen, H. 1988. Sublethal effects of pollutants on fish eggs and larvae. 253-346 pp. In : *Fish Physiology* Vol. XI Part A (Eds. W.S. Hoar and D.J. Randall), Academic Press, New York.

von Westernhagen, H. and Dethlefsen, V. 1975. Combined effects of cadmium and salinity on developments and survival of flounder eggs. *J. Mar. Biol. Assoc.* U.K. **55** : 945-957.

von Westernhagen H., Rosenthal, H. and Sperling, K.R. 1974. Combined effects of cadmium and salinity on development and survival of herring eggs. *Helgol. Wiss. Meeresunters.* **26**: 416-433.

von Westernhagen, H., Dethlefsen, V. and Rosenthal, H. 1975. Combined effects of cadmium and salinity on development and survival of garpike eggs. *Helgol. Wiss. Merresunters.* **27** : 268-282.

Wani, G.P. and Latey, A.N. 1982. Cadmium toxicity of gonads in a teleost fish, *Garra mullya* (Sykes). *Poll. Res.* **1** : 39-44.

Wani, G.P. and Latey, A.N. 1983. Toxic effects of Cadmium on the liver of a freshwater teleost, *Garra mullya* (Sykes). *Curr. Sci.* **52** (21) : 1034-1035.

Wilson, K.W. 1972. Toxicity of oil-spill dispersants to embryos and larvae of some marine fish. 318-322 pp. In : *Barine Polution and Sea Life* (Ed. M. Ruivo), Fishing News (Books), London.

Chapter 29

Environmental Epidemiology of Tirumala and Tirupati Pilgrim Towns

N.V. Nanda Kumar, D.M. Backiyavathy, E. Prabhakar AND M. Dhananjaya Naidu

Department of Zoology, Sri Venkateswara University, Tirupati-517 502, Andhra Pradesh

Abstract. The environmental epidemiology of Tirupati and Tirumala has been studied from 1984 to 1987 for the evaluation of impact of non-residents (pilgrims) population on the resident population of Tirupati and Tirumala. In the present review the health status with respect to diseases encountered by Tirupati and Tirumala subjects is compiled in relation to ground water pollution, storm water and seasonal change.

Key words : Epidemiology, Dysentery, Gastroenteritis, Encephalitis Cholera, Respiratory tuberculosis, Tirumala and Tirupati.

Introduction

Basis for environmental epidemiological studies

Epidemiological studies and results quickly become significant elements in public perceptions of health effects from environmental agents or factors (Radford, 1983). These epidemiological results help in regulating conditions which lead to health problems. Health morbidity status and its relation to environmental factors have been reported by a number of workers (Radford, 1983; Redmond, 1983). For example a number of lakes, reservoirs, soil and other terrestrial systems and groundwater are polluted due to domestic and industrial pollutants (Mishra and Trivedy, 1993; Mishra, 1989) in India and might cause water borne and other contagious diseases. In such situations epidemiological studies act as indices of environmental health hazards. In the present review the status of health with respect of diseases encountered by Tirupati and Tirumala subjects is compiled in relation to ground water pollution, storm water and seasonal change.

The epidemiological data was collected in the following major hospitals located in Tirupati and Tirumala.

1. Ramanarayana Ruya Government General Hospital,
2. Tirumala Tirupati Devasthanams, Ashwini Hospital at Tirumala.

Tirupati environmental epidemiology

A close surveillance was kept on the monthly-wise admission of patients of 34 communicable diseases as well as the mortalities recorded due to these diseases for four years. This is to account for–(1) The seasonal pilgrim influx Vs disease and post pilgrim influx effect on disease prevalence and health of resident population of Tirupati and Tirumala. (2) To investigate whether diseases occur also due to bad environment : water, drainage contamination etc., water borne or air borne disease. (3) To investigate the impact of Tirupati environment (Rainfall, seasonal changes etc.) on the seasonwise admission (monthlywise) of patients with various communicable disease. (4) S.V.R.R. Hospital is the biggest Government General Hospital in this district with medical faculty attached with teaching and clinic, at Tirupati. Health cards were analysed in this hospital for Tirupati epidemiology. (5) Ashwini Hospital, Tirumala a TTD (Tirumala Tirupati Devasthanam) managed Hospital at Tirumala hill is a well equipped and managed Hospital from where registered health records were analysed.

Results and Discussion

Epidemiological observation in Tirupati town

Diarrhoea : The incidence of diarrhoea in Tirupati was high in May, and July months of 1984 to 1987. The incidence show an increasing trend from April to July during 1984 to 1987 (Tables 29.1–29.3). The disease occurrence could not be correlated to rainfall (Table 29.5) and the subsequent rainfall wash into the drinking water wells, as the analysis of rainfall and disease, incidence did not show such trend. In Tirupati many wells are adjoining to drainage canals without cement lining.

However the disease association seems to be with pilgrim influx in Tirupati from April to June/July which are obviously the summer holidays, attracting the maximum number of pilgrims. The pilgrims influx combined with bad sanitation in Tirupati town may be the cause for diarrhoea. In old town open gutters carry night soil. The flies whose population is more during the above seasons and with arrival of mangoes attracting flies might be the cause for transmission and spreading the diarrhoea. Pilgrims from rural areas create bad sanitary conditions in open fields during the above season due to ignorance or sense of sanitation. This is the problem at Tirupati only and not at Tirumala. There were 54 deaths due to diarrhoea from 1984 to 1987 (Table 29.6). It is an example of pilgrim pressure on local residents.

Dysentery : Unlike diarrhoea, the dysentery incidence in Tirupati seems to be not associated with pilgrim influx season namely April-June/July (Tables 29.1–29.4). On the other hand the dysentery seem to be associated with rain fall, as shown in Tables 29.1–29.5 from 1984 to 1987. For example the high disease incidence appear correlated with the rainfall, wherein the high

TABLE 29.1. *Admission numbers for various communicable diseases at S.V.R.R. Govt. Hospital, Tirupati for 1984*

Sl. No.	Disease	Jan		Feb		Mar		April		May		June		July		Aug		Sep		Oct		Nov		Dec	
		A	D	A	D	A	D	A	D	A	D	A	D	A	D	A	D	A	D	A	D	A	D	A	D
1.	Diarrhoea	39	5	26	–	28	–	39	–	48	4	57	1	72	–	8	1			13	2			22	–
2.	Gastroenteritis	53	1	49	–	60	1	30	–	50	–	58	–	96	–	98	2			32	1			46	–
3.	Hepatitis	2	–	3	–	4	–	4	–	3	–	2	–	3	–	2	–			2	–			4	–
4.	Encephalitis	9	4	6	1	7	1	2	–	3	2	3	2	–	–	–	–			1	–			9	2
5.	Cholera	–	–	–	–	–	–	–	–	–	–	–	–	–	–	–	–			–	–			–	–
6.	Respiratory tuberculosis	81	2	76	2	94	4	99	6	90	3	89	4	78	3	–	–			103	–			58	6
7.	Dysentery	19	2	16	–	14	–	13	–	5	–	9	–	7	–	11	–			12	1			11	–

A = Number of admission D = Number of deaths

(a) Data from health records by the courtesy of Superintendent, S.V.R.R. Govt. Hospital, Tirupati.
For other 25 disease cases are not found or scanty or sporadic.

TABLE 29.2. *Admission number for various communicable diseases at S.V.R.R. Govt. Hospital, Tirupati for 1985*

Sl. No.	Disease	Jan		Feb		Mar		April		May		June		July		Aug		Sep		Oct		Nov		Dec	
		A	D	A	D	A	D	A	D	A	D	A	D	A	D	A	D	A	D	A	D	A	D	A	D
1.	Diarrhoea							31	4	37	3	39	--	63	4	39	5	31	--	21	3	26	1	52	4
2.	Gastroenteritis							37	3	52	1	65	--	42	--	54	1	34	1	26	--	37	--	45	1
3.	Hepatitis							6	--	6	--	6	--	4	--	3	--	2	--	4	--	1	--	6	1
4.	Encephalitis							2	1	--	--	5	4	3	3	5	3	3	--	2	--	1	--	2	2
5.	Cholera							--	--	--	--	--	--	--	--	--	--			--	--			--	--
6.	Respiratory Tuberculosis							85	8	72	4	85	10	54	3	96	5	79	2	92	2	63	3	102	3
7.	Dysentery							5	--	7	--	5	--	5	--	10	--	7	--	18	--	6	--	10	--

A = Number of admission D = Number of deaths

(a) Data from Health records by the courtesy of Superintendent, S.V.R.R. Govt. Hospital, Tirupati. For other 25 diseases cases are not found or scanty or sporadic.

TABLE 29.3. *Admission numbers for various communicable diseases at S.V.R.R. Govt. Hospital, Tirupati for 1986*

Sl. No.	Disease	Jan		Feb		Mar		April		May		June		July		Aug		Sep		Oct		Nov		Dec	
		A	D	A	D	A	D	A	D	A	D	A	D	A	D	A	D	A	D	A	D	A	D	A	D
1.	Diarrhoea					55	3	61	1	71	--	68	4	74	1	59	2	41	--	29	--	19	1	21	1
2.	Gastroenteritis					56	1	37	1	56	--	47	--	62	3	69	--	32	1	52	3	27	--	11	--
3.	Hepatitis					4	--	5	--	2	--	4	--	4	--	7	--	6	--	3	--	1	--	1	--
4.	Encephalitis					--	--	--	--	4	1	1	--	6	1	2	1	2	1	4	1	1	--	--	--
5.	Cholera					--	--	--	--	--	--	--	--	--	--	--	--			1658	82	--	--	--	--
6.	Respiratory Tuberculosis					94	1	93	6	75	8	83	20	69	5	59	6	73	4	57	2	16	--	9	1
7.	Dysentery					7	1	8	1	2	--	5	--	8	--	3	--	3	--	3	--	--	--	--	--

A = Number of admission D = Number of deaths

(a) Date from health records by the courtesy of Superintendent, S.V.R.R. Govt. Hospital, Tirupai.
For other 25 diseases cases are not found or scanty or sporadic.

TABLE 29.4. *Admission numbers for various communicable diseases at S.V.R.R. Govt. Hospital, Tirupati for 1987*

Sl. No.	Disease	Jan		Feb		Mar		April		May		June		July		Aug		Sep		Oct		Nov		Dec	
		A	D	A	D	A	D	A	D	A	D	A	D	A	D	A	D	A	D	A	D	A	D	A	D
1.	Diarrhoea	51	--	55	--	31	--	21	--	36	2	34	--	59	2										
2.	Gastroenteritis	33	2	27	1	21	--	21	2	39	--	40	--	57	--	57	--	56	2	43	2	64	2	54	1
3.	Hepatitis	6	--	2	--	3	--	--	--	3	1	4	--	4	--	5	--	7	--	6	--	2	--	9	--
4.	Encephalitis	--	--	1	--	3	1	2	--	2	--	3	--	1	1	4	--	6	1	3	--	6	4	5	--
5.	Cholera	--	--	--	--	--	--	--	--	--	--	--	--	--	--	--	--			1658	82	--	--	--	--
6.	Respiratory Tuberculosis	90	4	80	6	72	3	84	5	64	3	74	10	87	5	72	4	89	7	86	5	94	1	65	9
7.	Dysentery	5	--	3	--	3	--	3	--	3	--	.... Data not available													

A = Number of admission D = Number of deaths

(a) Data from Health records by the courtesy of Superintendent, S.V.R.R. Govt. Hosptial, Tirupati.
For other 25 diseases cases are not found or scanty or sporadic.

dysentery incidence after the rainfall. It is presumed that the rainfall-wash get into the drinking water wells, might result in dysentery which is mostly of amoebic origin. Hence, the disease showed an increase after rainfall or Post rain fall period. Four deaths due to dysentery were reported from 1984 to 1987 (Table 29.6) in Tirupati.

Gastroenteritis : The incidence of this disease was high during June, July August in all the years analysed viz., 1984 to 1987. This disease shows its association in the summer months and in the peak pilgrim season namely May, June and July (Tables 29.1–29.4). It is opined that gastroenteritis may be water borne and also spreads through the flies. Obviously during the summer months there is shortage of water, contamination of water due to shortage of water and fly nuisance. There were 33 deaths due to gastroenteritis from 1984 to 1987 (Table 29.6), thus, it warrants utmost caution in drinking water management and sanitation in Tirupati.

The association of the disease has been described as water borne previously. The analysis of rain fall data and gastroenteritis shows a close association (Table 29.1-29.5). It is opined that after rainfall, the rainfall wash along with drainage water get into the drinking water wells and might be responsible for the occurrence of the disease.

Hepatitis : The hepatitis admission rate did not show any monthly wise trend (Table 29.1–29.4). However, from August of December, 1987 (Table 29.4) from July to October 1986 (Table 29.3) from April to July 1985 (Table 29.2) and from March to May 1984 (Table 29.1) the incidence was high. The disease seems to show an association with rainfall (Table 29.5). For example in 1984 the rainfall was high in February and disease increased to maximum in March and again in July and month succeeding the November showed the association of disease with rainfall amount. There were four deaths from 1984 to 1987 (Table 29.6).

It is viewed that the hepatitis cases (Jaundice) are generally treated at home by indigenous Indian medicines or home remedies and Ayurvedic doctors. Hence, the admission rate is very very low (about four to five cases per month) in Tirupati in Government General Hospital.

Encephalitis : From 1984 to 1987 the number of deaths were forty four (Table 29.6) which is really a dangerous trend and an alarming number in pilgrim town. The number of admissions in each month did not show any relationship to season. The rainfall data and the disease incidence also did not show any appreciable relationship (Tables 29.1–29.5). The investigators opine from personal observations that there are a number of water pools in and around Tirupati which are

TABLE 29.5. *Monthly rainfall in Tirupati in the years of 1984, 1985 and 1986 (Rainfall in cm)*

Yera/Month	Jan	Feb	Mar	Apr	May	Jun	Jul	Aug	Sep	Oct	Nov	Dec
1984	0.0	12.0	1.6	2.1	2.5	4.2	20.4	2.0	15.2	6.0	20.1	2.3
1985	3.0	0.0	1.4	8.2	2.6	11.8	8.2	4.8	14.6	17.6	30.0	8.0
1986	6.8	0.5	0.0	0.0	1.0	2.5	4.6	6.5	8.5	12.8	19.0	1.4

mostly formed due to flow of drainage helping in mosquito breeding. There are no safe guards against regulating the stray pig population. The pig is the host for encephalitis virus. Hence, the regulation of stray pig population and water management together is recommended for prevention of encephalitis.

Cholera : The incidence of Cholera in 1984, 1985 and 1987 was nil. Whereas in 1986, in the month of October, there were 1955 cases of cholera with eighty two (82) deaths (Table 29.6). It is a unique and sporadic occurrence in October, 1986. It is corrclated to the Brahmostavam celebration at Tirumala during Dassarah, which attracts lakhs of pigrims from all over India. The Cholera outbreaks are very common during big festive days and congregations in India. However, the out-breaks was not reported in the years 1984, 1985 and 1987. The October 1986 incidence is an example of pilgrim pressure on lacal residents with reference to cholera incidence.

Respiratory Tuberculosis : The respiratory tuberculosis is a common and endemic disease in India. There is no association of the incidence of the disease with season from 1984 to 1987 (Table 29.1–29.4).The incidence of the disease however, was high in all the years and corroborate with Benjamin National atlas on tuberculosis. However, the investigators could deduce an interesting trend on the relationship of incidence of tuberculosis with amount of rainfall, in other words the incidence of tuberculosis was less with increase in the rainfall, and vice-cersa when analysed every month (Tables 29.1–29.5). It is presumed that during dry months

TABLE 29.6. *Number of admission and mortalities from 1984 to 1987 at S.V.R.R. Govt. Hospital, Tirupati*[(a)]

Sl. No.	Name of the disease	1984		1985		1986		1987	
		A	D	A	D	A	D	A	D
1.	Diarrhoea	352	13	338	24	498	13	287	4
2.	Gastroenteritis	572	6	392	7	449	9	512	12
3.	Hepatitis	29	–	38	3	37	–	51	1
4.	Encephalitis	40	13	23	13	20	5	36	9
5.	cholera	–	–		–	1655	82	–	–
6.	Respiratory tuberculosis	768	30	728	40	628	38	957	53
7.	Dysentery	117	3	73	–	39	1	19	–

A = Number of admissions D = Number of deaths

(a) Data collected from S.V.R.R. Govt. Hospital, by the courtesy of the Superintendent, S.V.R.R. Govt. Hospital, Tirupathi.

For other 25 diseases, cases were not found or scanty or sporadic.

in the absence of rainfall the emanation of dust due to vehicular traffic or wind action is more than in the rainy months. This dust might cause further aggravation or inflammation of the lungs of tuberculosis. It is suggested that dust containment by rain or water spray on the roads by water carrying automobiles might reduce the aggravation of the respiratory tuberculosis and the subsequent admission to hospitals.

Environmental Epidemiological Studies at Tirumala (UP Hills)

The Tirumala environmental epidemiological data was made possible due to the cooperation extended by medical officers particulary medical superintendent and other medical officers, T.T.D. Ashwini Hospital, Tirumala. The epidemiological data was collected from the Ashwini Hospital health records maintained by Ashwini Hospital. The environmental epidemiological data presented in this project was endorsed by the Superintendent, T.T.D. Ashwini Hospital, Tirumala. The Ashwini Hospital maintains the health records of residents and also the non-residents namely the pilgrims. The residents are mostly the employees of the T.T.D. administration of Tirumala and also the other residents who are residing in Tirumala belonging to various walks of life like business-men, fancy shop owners, street vendors, members of various religious institutions, hotelliers etc. However, a majority of health records pertains to local residents (1) employees and (2) non-employees of T.T.D. The non-employees are referred as "Native" in T.T.D. Health records and a very few health records for pilgrims. The number of pilgrims who visit Ashwini Hospital at Tirumala are about 5. to 10 per cent of the total resident visitors (Table 29.11). The epidemiological studies were made with a view to know whether the enormous pilgrim influx at Tirumala, do have effect on resident population of Tirumala, whether the peak pilgrim season, who's high admission rate of certain diseases and the incidence of diseases, has any relationship with environmental factors like season (monthly-wise), water shortages, pilgrim flux etc. The results are presented followed by discussion and conclusion.

A comparison is also made between Tirupati and Tirumala environmental epidemiological data in the subsequent sections.

The T.T.D. administration needs appreciation and more particularly the health wing of T.T.D. administration at Tirumala for maintaining the best sanitary conditions inside the living units (Cottages, choultries and offices) inside and outside the temple premises, closed drainage system in the T.T.D. administered living units. However, the residential localities which are not under the T.T.D. administration control and jurisdiction which are mostly age-old encroachments since decades and are living units of mostly non T.T.D. employees reflect bad environment or urban localities. In view of the good sanitary conditions and clean environment by T.T.D. administration, the health hazards or morbidity due to environmental conditions are the least and the disease incidence for various communicable diseases except one or two viral diseases is very low as compared to Tirupati environment.

Diarrhoea and Enteritis : The monthly-wise admissions for diarrhoea and enteritis in Tirumala is presented in (Tables 29.7–29.10) for the years 1984 to 1987. The incidence of diarrhoea and enteritis for the years 1984, 1985 and 1987 was very low when compared to number of visitors at Ashwini Hospital for respective years and the maximum number did not exceed six (6) in any given month (Table 29.7, 29.8 & 29.10), Whereas in 1986 the admission rate for diarrhoea and enteritis was unusually high (Table 29.9) as compared to other years. The maximum admission number was ninety two in May and was abver thirty cases in six other months and above five cases in rest of twelve months of 1986. A gradual decrease from May to December in 1986 was observed in the incidence of above disease. This trend is a clear correlation between the pilgrim influx and the incidence of the disease. Thus the diarrhoea and enteritis seems to associate with pilgrim influx. In other words, the incidence of the above disease is high among Tirumala residents during peak pilgrim influx in 1986. In Tirumala the pilgrim influx starts after January 15th and reaches its peak in May and gradually decreases after May (This increase of pilgrims depends upon various social and religious reasons. (1) One major cause is commencement of summer holiday from April onwards. (2) ''Uttarayana Sankramana'' commences from January 15th when the religious and social functions like marriages are celebrated). It is presumed that the increase in diarrhoea and enteritis cases in 1986 might be due to utilization of limited sources of water that is available in Tirumala. The limitation to water resource might have been caused due to shortage in rainfall. Accordingly, the rainfall data of Tirupati (Table 29.5) showed a low rainfall in 1986 as compared to 1984, 1985. For example, the total amount of rainfall from January to July in 1986 which are the dry months of the year recorded 15.4 cm. Whereas in 1984 the total rainfall from January to July was 42.8 cm and in 1985 it was 35.2 cm. In the same year (October) in Tirupati town also there was out break of cholera which is unique to 1986 only and can be well correlated to water shortage and contamination.

Infective Hepatitis : The incidence of infective hepatitis in Tirumala subjects is very low in all the years analysed in Tirumala and the number of admissions did not exceed four in any one month (Table 29.7–29.10). In 1984 from June to November, in 1985 from March to May and from September to December in 1987 there were no infective hepatitis cases. The data does not warrant any further health care due to good sanitation conditions at Tirumala.

Dysentery : Dysentery incidence pattern among Tirumala residents coincide with diarrhoea and enteritis in 1986. For example, from February to May in 1986 a gradual increase in incidence of dysentery and a decrease from June to December was observed (Table 29.9). Thus, the incidence of dysentery seem to associate with pilgrim influx. In other words, heavy inflow of pilgrims in summer coincide with incidence of dysentery among Tirumala residents in 1986 only. However, the number of dysentery cases did not exceed more than eleven in 1986 (Table 29.9) and did not exceed four in 1984, 1985 (Table 29.7 & 29.8) and no cases were reported in 1987 (Table 29.10) from March to December except one case in May. The dysentery is also water borne disease and also spread through flies through food. The shortage in water resource in Tirumala in 1986 which is discussed at length while discussing Diarrhoea and Enteritis at Tirumala. The investigators recommend a continuous minitoring of amoebae and other micoorganisms (bacteria) of tirumala water resources, namely at Papavinasanam and other water resources during peak pilgrim seasons only (February to June).

TABLE 29.7. *Seasonwise number of admissions for various diseases (including communicable) at T.T.D. Ashwini Hospital, Tirumala during the year 1984*

Sl. No. Name of the diseases	Jan	Feb	Mar	Apr	May	Jun	Jul	Aug	Sep	Oct	Nov	Dec	Total
I. Diseases of Alimentary system													
1. Peptic ulcer	2	–	1	1	–	–	–	1	–	–	–	2	7
2. Infective Hepatitis	–	2	1	2	–	–	–	–	–	–	–	1	8
3. Diarrhoea and Enteritis	3	1	2	5	1	3	1	6	–	2	2	1	27
4. Dysentery	2	3	–	4	1	2	2	2	1	4	1	–	22
5. Tumours	–	–	–	–	–	–	–	–	1	–	–	–	1
6. Anorectal disorders	1	2	–	–	–	–	–	–	–	–	–	–	1
7. Malaria	–	–	–	–	–	–	–	–	–	–	–	–	–
II. Diseases of Respiratory System													
(A) U.R.I, P.U.O. & V.F.	4	–	1	5	–	2	6	6	3	3	2	2	34
(B) L.R.I.													
1. Br. Asthma	1	2	2	4	1	1	3	4	3	--	--	--	21
2. Lobar pneumonia	--	--	1	--	--	--	--	2	--	--	--	--	3
3. Bronchitis	--	--	2	2	--	--	1	1	--	2	2	4	14
4. Bronchectaisis	--	--	--	--	--	--	--	--	--	--	--	--	--
5. Pneumothorax	--	--	--	--	--	--	--	--	--	--	--	--	1
6. Tuberculosis	3	2	6	4	1	5	4	4	3	1		2	37
III. Diseases of Genitourinary System	6	2	1	2	2	4	1	–	–	–	–	1	19
IV. Diseases of Gynae and Antinatal													
1. Abortions	–	4	–	3	2	–	4	2	–	2	2	3	22
2. Tumours	–	–	–	–	–	–	–	–	–	–	–	–	–
3. Pelvic infections	2	–	2	–	1	–	1	–	–	–	1	1	10
4. A.N.C.	2	4	–	–	–	–	2	–	1	2	–	–	11
5. Wound spesis	–	–	–	–	–	–	–	–	–	–	–	–	–
6. P.E.T. etc.	1	–	–	2	–	–	2	–	–	3	2	–	12
7. Prolapse	–	–	1	–	–	–	–	–	–	–	–	–	1

Data from Health records by the courtesy of Chief Medical Officer, T.T.D., Ashwini Hospital, Tirumala.
The Data is endorsed by Chief Medical Officer, Data for other diseases are not found or scanty or sporadic.

TABLE 29.8. *Seasonwise number of admissions for various diseases (including communicable) at T.T.D. Ashwini Hospital, Tirumala during the year 1985*

S.No. Name of the diseases	Jan	Feb	Mar	Apr	May	Jun	Jul	Aug	Sep	Oct	Nov	Dec	Total
I. Disease of Alimentary System													
1. Peptic ulcer	–	–	–	1	–	–	–	–	–	1	–	2	4
2. Infective diseases	–	–	–	–	–	–	–	–	–	–	–	–	–
3. Diarrhoea and Enteritis	–	1	–	2	1	5	2	4	1	1	2	2	21
4. Dysentery	–	–	–	4	1	1	2	2	4	3	3	1	21
5. Tumours	–	–	–	–	–	–	–	–	–	–	–	–	–
6. Infective Hepatitis	–	2	–	–	–	2	2	4	–	–	–	–	10
7. Malaria	–	–	–	1	1	–	–	4	2	2	–	–	10
II Diseases for Respiratory System													
(A) U.R.I., P.U.O. & V.F.	3	5	2	6	8	7	7	7	10	7	1	3	66
(B) L.R.I.													
1. Br. Asthma	--	4	2	1	--	--	3	3	3	3	3	3	25
2. Lobar pneumonia	--	--	--	1	--	--	--	--	--	1	--	--	2
3. Bronchitis	1	1	--	1	1	1	1	2	1	--	--	--	9
4. Pneumothorax	--	3	1	--	--	--	--	--	--	--	--	1	5
5. Tuberculosis	--	--	2	3	1	--	3	1	4	--	1	--	15
6. Broncho pneumonia	–	2	–	–	–	–	–	–	2	–	–	–	4
III. Diseases of Genitourinary System	3	–	–	–	1	5	2	–	1	2	–	–	14
IV. Diseases of Gynae and Antinatal													
1. Abortions	1	1	1	1	–	3	2	–	1	2	2	–	14
2. Tumours	–	–	–	–	–	–	–	–	–	–	–	–	–
3. Pelvic infections and wound sepsis	–	2	2	1	–	4	–	2	–	2	3	2	18
4. P.E.T. etc.	2	–	–	–	–	1	2	2	1	–	–	–	8
5. A.N.C.	1	–	–	–	1	1	–	–	–	1	–	1	5

Data from Health records by the courtesy of Chief Medical Officer, T.T.D., Ashwini Hospital, Tirumala.
The Data is endorsed by Chief Medical Officer, Data for other disease are not found or scanty or sporadic.

TABLE 29.9. *Seasonwise number of admissions for various diseases (including communicable) at T.T.D. Ashwini Hospital, Tirumala during the year 1986*

S.No.	Name of the diseases	Jan	Feb	Mar	Apr	May	Jun	Jul	Aug	Sep	Oct	Nov	Dec	Total
I.	Diseases of Alimentary System													
1.	Peptic ulcer	1	2	--	5	--	--	6	1	--	1	2	2	21
2.	Infective Hepatitis	--	2	--	--	2	3	3	2	1	--	1	1	15
3.	Diarrhoea and Enteritis	10	5	30	60	02	43	52	29	13	10	8	22	324
4.	Dysentry	4	5	6	10	11	6	4	3	1	7	4	1	62
5.	Tomours	--	--	--	--	--	--	--	--	--	--	--	1	1
6.	Anorectal Disorders	6	1	1	13	8	3	--	--	--	--	2	--	34
7.	Malaria	2	1	2	2	1	4	--	--	--	--	--	1	13
8.	Typhoid	3	2	4	5	1	--	--	--	4	--	--	--	15
II.	Diseases of Respiratory System													
	(A) U.R.I. P.U.O. & V.F.	10	3	30	39	43	35	27	14	13	16	15	15	260
	(B) L.R.I.													
1.	Br. Asthma	2	4	4	3	--	2	2	4	4	3	8	2	38
2.	Bronchitis	5	1	1	4	2	4	3	--	--	2	2	--	24
3.	Bronchectaisis	--	--	--	--	1	--	--	1	--	--	--	1	3
4.	Broncho pneumonia	--	--	2	1	--	--	--	1	--	3	--	--	7
5.	Pneumothorax	--	--	--	--	--	--	--	--	--	--	--	1	1
6.	Tuberculosis	4	--	3	2	--	2	1	--	--	3	4	6	25
III.	Diseases of Genitourinary System	1	1	--	--	2	--	1	1	--	1	2	2	11
IV.	Diseases of Gynae and Antinatal													
1.	Abortions	5	3	4	2	4	--	2	2	1	1	1	1	26
2.	Pelvic infection	3	--	--	--	--	--	--	--	--	--	--	--	3
3.	P.E.T. etc.	1	--	2	--	--	--	--	--	--	1	--	--	4
4.	A.N.C.	--	6	6	4	2	7	6	2	--	1	1	1	36
5.	Wound sepsis	--	--	--	1	--	1	2	2	2	--	--	1	9
6.	Prolapse	3	--	2	1	--	--	--	--	1	2	1	--	10

Data from Health records by the courtesy of Chief Medical Officer, T.T.D., Ashwini Hospital, Tirumala.
The Data is endorsed by Chief Medical Officer, Data for other disease are not found or scanty or sporadic.

Table 29.10. *Seasonwise number of admissions for various diseases (including communicable) at T.T.D. Ashwini Hospital, Tirumala during the year 1987*

S.No.	Name of the diseases	Jan	Feb	Mar	Apr	May	Jun	Jul	Aug	Sep	Oct	Nov	Dec	Total
I.	Diseases of Alimentary System													
1.	Peptic ulcer	--	1	5	3	4	1	--	2	1	--	--	2	19
2.	Infective Hepatitis	--	1	3	3	--	4	1	2	2	2	1	1	20
3.	Diarrhoea and Enteritis	2	--	2	2	--	1	1	3	1	2	1	--	15
4.	Dysentry	2	1	--	--	1	--	--	--	--	--	--	--	4
5.	Tumours	--	--	--	--	--	1	--	--	--	--	1	--	2
6.	Anorectal Disorders	1	--	--	--	--	--	--	--	--	--	1	--	2
7.	Malaria	1	2	1	1	2	1	1	--	1	2	--	1	13
8.	Typhoid	8	8	10	15	6	5	8	7	10	7	9	1	94
II.	Diseases of Respiratory System													
	(A) U.R.I.													
	P.U.O. & V.F.	--	1	4	--	2	1	--	3	--	--	3	--	14
	(B) L.R.I.													
1.	Br. Asthma	1	--	1	--	--	--	--	--	1	--	--	--	3
2.	Bronchitis	--	--	1	1	--	--	--	--	--	--	--	--	2
3.	Lobar pneumonia	--	--	--	2	--	--	--	1	--	--	--	--	3
4.	Bronchectaisis	1	4	1	2	2	4	2	2	2	1	--	1	23
5.	Tuberculosis	--	--	--	--	--	--	--	--	--	--	--	--	--
III.	Diseases of Genitourinary System	3	--	2	3	3	3	--	1	2	--	--	3	21
IV.	Diseases of Gynae and Antinatal													
1.	Abortions	1	1	5	1	1	1	--	1	1	--	1	--	13
2.	Pelvic infections	--	--	1	--	--	--	1	--	--	--	--	2	4
3.	P.E.T. etc.	--	--	--	--	--	3	--	--	--	--	--	--	3
4.	A.N.C.	--	--	--	1	--	--	--	--	3	2	--	--	6
5.	Wound sepsis	--	1	--	--	--	--	--	1	1	3	1	1	8
6.	Prolapse	--	--	--	--	--	--	--	--	--	--	--	--	--
7.	Tumors	--	--	--	--	--	--	--	--	--	--	--	--	--

Data from Health records by the courtesy of Chief Medical Officer, T.T.D., Ashwini Hospital, Tirumala.
The Data is endorsed by Chief Medical Officer, Data for other disease are not found or scanty or sporadic.

Malaria : The incidence of Malaria among Tirumala residents was very less and the number in any given month did not exceed four (Table 29.7–29.10). There were no cases in 1984. The authors are of the opinion that absence of malaria in Tirumala is due to absence of stagnant water pools which breed mosquitos. Since it is hilly area with mountain slope, the water could not stagnate and in the plains the closed drainage system help in prevention of mosquito breeding.

PUO & VF (Pyrexia due to unknown organism and viral fever) : The incidence of fevers caused by virus and other unknown organisms were unusually high among Tirumala residents in 1986 than in the other years. Thus the 1986 year showed high incidence of many communicable disease as described above. In 1986 the incidence of PUP & VF show a gradual increase from January to May and showed a decrease from June to December. The maximum number recorded in May was 43. Such trend was observed for the other communicable diseases also in the same year. The admission rates were fairly high in 1986 and in any given month (Table 29.9) and the maximum number of cases in any given month of 1984, 1985 and 1987 did and exceed ten (Tables 29.7, 29.8 and 29.10). It is presumed that the viral fevers are contagious and seem to associated with the influx of pilgrims. The heavy inflow of pilgrims in April, May and June might have caused high incidence of PUO and VF among Tirumala residents in 1986. It is a good example of pilgrim pressure on the incidence of communicabe diseases on Tirumala residents and particularly this disease in 1986 clsoely associates with pilgrim flux and is example of pilgrim-pressure on residents.

Typhoid : There were no reports of incidence of typhoid in 1984 and 1985 and very small number of cases in 1986 (Table 29.9). However, in 1987 on an average of eight cases per month were reported in the month of March, April and September (Table 29.10). No partricualr reason could be assigned for such an increase of typhoid cases in 1987 unlike the increase in the number of various communicable diseases in 1986. It is opined that the fly nuisance and food contamination might be cause for the number of typhoid cases in 1987.

Gynaec and Antinatal Dieases and Diseases of Genitourinary System : The incidence of these diseases did not show any partcular pattern or its association with any pilgrim influx among the residents of Tirumala.

Highlights of findings and Major Conclusions : The environmental epidemiology of Tirumala and Tirupati residents are studied from 1984 to 1987 for the evaluation of impact of non-residents (pilgrims) population on resident population of Tirupati and Tirumala.

There are two factors responsible for the incidence of diseases among the Tirupati and Tirumala residents (i) disease incidence caused due to pilgrims, (ii) the environmental factors for incidence of disease.

A factor called ''May factor of 1986'' of Tirumala has emerged from the epidemiological studies of Tirumala, wherein the communicable diseases like diarrhoea and enteritis, dysentery, viral fevers (PU & VF) disease of respiratory sysrtem, show maximum or peak incidence in the month of May which is also a peak pilgrim season. Though this pattern is observed in 1986 only, it reflects the impact of pilgrim pressure (non-resident pressure) on residents of Tirumala with respect to disease incidence.

Diarrhoea : The incidence of 54 deaths from 1984-87 at Tirupati warrants proper sanitary management system. The incidence of diarrhoea in Tirupati seem to associate with influx of pilgrims and the bad sanitary conditions of Tirupati (Please see discussion for details under Tirupati epidemiology) whereas in Tirumala, the disease incidence showed contrary to Tirupait, with negligible incidence of diarrhoea in 1984, 1985 and 1987 except May factor of 1986. This might be due to good sanitary conditions like closed drainage system at Tirumala. The May factor of 1986 is correlated to low rainfall and water shortage resulting in water contamination in Tirumala.

Dysentery : The incidence of dysentery in Tirupati town seem to associate with the environmental factors namely rainfall and rainfall wash into drinking water wells and ground water contamination and not associated with pilgrim influx like diarrhoea. Whereas in Tirumala, except the year of 1986, namely the May factor year, the other years showed the least incidence of dysentery for the years, 1984, 1985 and 1987. It is opined that there is not much amoebae and other microbial contamination causing dysentery.

Infective hepatitis : The incidence of infective hepatitis in Tirupati seem to associate with rainfall and not with the pilgrim influx. The incidence show a sort of relationship with post rainfall peaks. In Tirumala and Tirupati the incidence of the above disease is less as this disease (jaundice) is treated by Indian medicines and Indian medical practitioners with indigenous medicines. Hence, the number of admissions in hospitals following allopathi system is less.

PUO & VF : This particular disease incidence among Tirumala residents seem to associate with pilgrim influx in the year 1986 again following "May factor pattern". This disease is not associated with environmental factors.

Encephalitis : The incidence of encephalitis in Tirupati seems to have no relationship with pilgrim influx or season (environmental). However, the mortalities are alarming and the incidence is correlated to the formation of a number of water pools of waste drainge or rain water, and the presence of stray pig population which act as intermediate host.

In Tirumala there are no cases of encephalitis. Obviously, due to absence of pigs, clean and closed drainage system due to mountains slopes, and no water stagnation from drainage water.

Gastroenteritis : The incidence of this disease seems to associate with peak pilgrim influx and also the rainfall amount at Tirupati. The death rate were alarming in Tirupati. There were 33 deaths in Tirupati from 1984 to 1987. This warrants a cautious preventive method in ground water management and sanitation. The enteritis in Tirumala seem to associate with peak pilgrim influx.

Respiratory tuberculosis : The disease incidence was found high during dry months wherein the dust in Tirupati might have further aggravated the tuberculosis lungs. The dust containment

TABLE 29.11. *Number of visitors to S.V.R.R. Govt. General Hospital, Tirupati and T.T.D. Ashwini Hospital,* Tirumala during 1984-87.

Year	Jan	Feb	Mar	Apr	May	Jun	Jul	Aug	Sep	Oct	Nov	Dec
S.V.R.R.Govt. General Hospital												
1984	1466	1408	1476	1602	1529	1646	1574	1538	1514	1478	1320	1348
1985	1439	1499	1548	1496	1526	1548	998	1642	1453	1482	1189	1408
1986	1424	1307	1388	1526	1567	1638	1675	1429	1595	1655	strike	
1987	1552	1423	1461	1496	1071	1258	1650	1655	1899	2272	1599	1570
T.T.D. Ashwini Hospital												
1985	15044	17246	17776	20986	21259	19631	22620	23297	21485	23336	14765	16794
1986	15044	17246	17786	19433	19926	18521	19593	18140	18765	21624	19403	17283
1987	16338	16210	19366	19443	16587	15911	18563	16911	17666	18421	15024	12794

during rainy season brought down the admissions rate.

The comparative environmental epidemiology data reveals that the Tirumala environs are far superior in quality and in sanitation due to relatively excellent management of Tirumala by T.T.D. administration more particularly in the health care system, hospital management and sanitation than Tirupati Municipality. The Tirumala environmental management can be taken as a model for North Indian pilgrim towns like Banaras, Puri, Nasik, Gaya etc.

1. The May factor of 1986 described in the above sections namely the maximum incidence of communicable diseases among Tirumala residents during peak pilgrim months of May, 1986 particularly, warrants an integrated and combined efforts of environmentalists, T.T.D. administration and Medical personnel to take preventive measures from March to June to contain the incidence of the disease and to avoid May factor 1986 in future.
2. In Tirupati, since the communicable diseases are associated partly due to ground water pollution and partly due to pilgrim influx the municipality and concerned health department of Tirupati have to assure proper sanitation by preventing the possible contamination of drainage into drinking water well after rainwash. The extension areas with gutters without cement lining and seepage from gutters in old town due to weathering requires remedy.
3. The concerned authorities have to construct a number of paid and free lavatories and bath rooms to cope up with the non-residents and resident population.
4. Prevention of water pools formation from drainage in the new extension areas to avoid mosquito and fly nuisance.
5. To prevent drainage of faecal matter into open gutter in old town areas due to absence of modernised lavatory system.
6. The contaminents of dust by (a) extending the tar roads on the lateral sides (b) to temporarily contain the dust by wetting the roads should be attended to as the respiratory tuberculosis was more associated with dry months and the least during the rainy months.
7. An accurate and continuous monitoring from March to July for amoebae, bacterial and other microorganisms in water should be done to take preventive measures.
8. Street vendors should sell foods and fruits in low cost plastic bags or paper covers to avoid contamination form flies to prevent to diseases of gastroenteritis system.

Acknowledgemengs

Dr. N.V. Nanda Kumar, P.I., DOEn VII project thanks the Department of Environment (DOEn), Ministry of Environment & Forests, Government of India, New Delhi (Project No. 19/22/81-EN-I) for financial assistance.

The P.I. also thanks Medical Superintendent, S.V.R.R. Government Hospital, Tirupati and Medical Officer, S.V.U. Health Centre, Tirupati for health records and the Director, CFTRI for imparting training facilities for aflatoxin analysis at Discipline of Microbiology and Sanitation, CFTRI, Mysore.

We also thank Superintendent and Medical Officers, Ashwini Hospital, Tirumala for health records which are duly endorsed and for their valuable cooperation.

REFERENCES

Mishra, P.C. 1989. Soil pollution and Soil Organisms. Ashish Publishing House, New Delhi.

Mishra, P.C. and Trivedy, R.K. (Eds) 1993. Ecology and Pollution of Indian Lakes and Reservoirs. Ashish Publishing House, New Delhi 1-203 p.

Radford, P. Edward. 1983. Epidemiology of Radioation-induced cancer. *Environmental Health Perspectives* **52** : 45-50.

Redmond K. Carol. 1983. Cancer mortality among coke oven workers. *Environmental Health Perspectives.* **52** : 67-73.

Chapter 30

Biological Methods of Wastewater Treatment

P.C. Mishra

Department of Environmental Science Sambalpur University, Jyotivihar-768 019

Abstract. The review highlights the advantages of land treatment of wastewater and use of algae and aquatic macrophytes for reduction of pollution load. Integration of algal-bacterial symbiotic oxidation pond and water hyacinth treatment, with conventional primary treatment methods, have been suggested as a low cost efficient biological method of treatment for wastewater disposal.

Key words : Sewage, industrial wastewater, land treatment, Agriculture potentiality, Algae, water-hyacinth, wastewater treatment.

Introduction

Industrialisation has its unavoidable effect on pollution of air, water and soil based on the type of industry, nature of raw materials used and manufacturing process involved. Industrialisation as an index of modernisation, leads to alteration in the physical, chemical and biological properties of the environment. Since industrialisation and pollution are complementary to each other, measures are to be adopted to render pollution less harmful to the biosphere. The possible solutions to industrial pollution are treatment, waste recycling and disposal to the environment far off from the human habitation. But industries in general have one goal i.e. high production of their goods. They have as such neglected the proper treatment and disposal of wastewater. The authorities of central and state pollution prevention and control Boards have prescribed limit to various physico-chemical characteristics of industrial wastewater being discharged to open water bodies or land. But those limits are not generally adhered to due to many constraints.

The conventional chemo-mechanical methods for the treatment of wastewater is by far the best method of treatment. It employs several stages for the removal of pollution load. These are :

(i) Screening and/or macerating of flow to remove rags and coarse solids.

(ii) Grit removal chamber to protect down-stream equipments from damage by inert solids.

(iii) Primary sedimentation for the removal of gross solids.

(iv) Biological treatment (secondary treatment) to remove biodegradable organic matter.

(v) Tertiary treatment (Advanced chemical treatment) for nutrient removal.

(vi) Sludge treatment.

However, the design of water purification plants for the developing countries is very complex due to many constraints involved right from their conception. Some of the commonest constraints are :

1. Lack of funds for the initial investment.
2. Lack of trained personnel for operation and maintenance.
3. Lack of foreign exchange to import spare parts.
4. Lack of facilities for the repair of sophisticated equipments.

Thus a look at the conventional methods of wastewater treatment shows that not only are they cost prohibitive but also require regular and expensive maintenance and adequate technical manpower. So Developing countries have to look for an intermediate technology which may be less expensive and requires minimum technical trained personnel. The paper reviews some low-cost biological treatment methods with major emphasis on land treatment of wastewater, use in crop irrigation and use of aquatic macrophytes in the removal of pollution load.

Wastewater Generation in India

A survey carried out by Central Pollution Control Board (CPCB 1989) on the status of generation, collection, treatment and disposal of wastewater in 212 Class I cities covering 65 per cent of Indian population indicates a total generation of 12,145 million litres of wastewater per day (MLD) out of which only 2633 MLD (22%) are collected through sewerage system and rest are directly discharge to land or water without treatment (Table 30.1). From the total collection of 2633 MLD through sewerage system 859 MLD gets primary treatment alone and 1626 MLD gets both primary and secondary treatment. In the absence of tertiary treatment the released wastewater is capable of causing eutrophication in waterbodies.

In India, some 5000 large and medium industrial Units are responsible for pollution of our waterbodies. But their contribution to the total pollution load is less than 10 per cent. The rest is of domestic origin. For example, the industries of Bombay account for only 13 per cent of the total waste dumped into waterbodies and in Calcutta 11 per cent of wastes are of industrial origin. In Delhi the river Yamuna takes in 200 million litres of untreated human wastes everyday against 20 million litres of industrial wastewater. However, most of the times the industrial wastewater

TABLE 30.1. *Statewise Position of Wastewater Generation Collection and treatment in Class I Cities*

State/ Union Territory	Total No. of cities	Wastewater (MLD)		Treatment (MLD)	
		Generated	Collected	Primary only	Primary and secondary
Andhra Pradesh	17	781	163.40	--	172
Bihar	11	445	--	--	43
Gujarat	11	815	220.5	86	463
Haryana	9	130	38.53	--	--
Jammu & Kashmir	2	209	--	--	--
Karnataka	14	655	462	327	10
Kerala	6	223	24.08	–	5
Maharashtra	29	2838	190	293	–
Manipur	1	207	–	–	–
Madhya Pradesh	14	552	168	86	–
Orissa	6	207	22	–	–
Punjab	7	233	158	–	–
Rajasthan	11	318	–	–	–
Tamil Nadu	19	487	53	–	–
Tripura	1	13	–	–	–
Uttar Pardesh	29	1449	255	–	160
West Bengal	21	1107	27	–	20
Chandigarh	1	144	106	68	–
Delhi	1	1480	745	–	745
Pondicherry	1	38	–	–	–
Total	212	12145	2633	859	1626

Source - CPCB 1989.

pose a greater threat due to the presence of toxic chemicals and nonbiodegradable organic matter.

Characteristics of Wastewater

Composition and quality of sewage water from all sources differ from region to region and from season to season. Raw sewage effluents contain a large number of chemicals and organisms of various forms. While some of these are rich sources of nutrients to plants and fish, a host of them could cause toxicity or become detrimental when present in amounts larger than tolerable limits to bio-physical functioning of life.

The variation in total and available nutrients can be attributed to relative contribution from domestic and industrial units. A calculation on the average nutrient content of Indian sewage collected from 322 Class I cities and towns (CPCB 1989) reveals that the total quantity of plant nutrients generated per day from sewage comes to about 400 tonnes of Nitrogen, 136 tonnes of P_2O_5 and 480 tones of K_2O. This calculation is based on the average nitrogen, P_2O_5 and K_2O content of 50, 17 and 60 mg respectively per litre of sewage. Thus proper harvest of sewage could also be helpful in solving the nutrient scarcity of our farmers.

The industrial wastewater contains raw materials, processed chemicals, processed intermediates, processed by-products and impurities of the industries and the characteristics vary from industry to industry. The pollutants may be classified as :

(a) Organic substances that deplete the oxygen content and increase the Biochemical Oxygen Demand.

(b) In-organic substances like carbonate, chloride, nitrogen, phosphate etc. that render the waterbody unfit for use and encourages growth of microplants.

(c) Acids and alkalis which affect the growth of fish and other aquatic life.

(d) Toxic substances like cyanide, acetylene, phenol, heavy metals (mercury, lead, arsenic, cadmium etc.) which cause damage to the flora and fauna.

(e) Oil and other floating substances which interfere with self purification of waterbodies.

Thus no single tertiary or advanced treatment method is hundred percent efficient in removing the nutrients and other toxic chemicals from an industrial wastewater. Use of multiple methods has not been possible for economic consideration. Hence low cost nonconventional biological treatment methods have been tried by Scientists and have been proved to be successful on many cases.

Biological Methods of Disposal

Land Treatment

Land treatment is defined as the controlled and regulated application of partially clarified wastewater to the land surface to achieve a designed degree of treatment. Sewage has been used on land all over the world for several decades as a method of its final disposal. Recent studies have established the feasibility of wastewater reuse on land as a method of treatment for pollution control. The advantages of this system are :

(i) The water component and nutrient components of wastewater are recycled through crop irrigation,

(ii) Pollution of water sources can be eliminated, and

(iii) Increase in economic benefits in cases of high crop yield.

The prime objective of sewage farming is to achieve the purification of wastewater for control of water pollution simultaneously with the cultivation at low cost in comparison to the conventional treatment methods. The mechanism involved in the purification include complex physical, chemical and biological interactions operating together. The physical factors include filtration, absorption, adsorption, photo-oxidation, evapotranspiration, osmosis etc. The chemical factors are ion exchange, precipitation, solubilisation, oxidation, reduction etc. The biological factors include carbon assimilation, proteolysis, enzymatic hydrolysis, respiration, catalytic decomposition, ammonification, nitrification, denitrification, phosphate mobilisation etc. (Lochr 1979). The successful operation and efficient management of wastewater reuse system for crop irrigation depends upon the effluent quality (volume, organic load, nutrient content, total suspended solids, salinity, pH, pathogens etc.) soil properties (texture, permeability, sub-soil drainage, water table depth, cation exchange capacity; exchangeable sodium percentage etc.) and nature of crop species such as tolerance capacity, nutrient and oxygen uptake requirement, duration etc. (Mishra 1986).

It has been scientifically recognised that application of wastewater on land as a treatment process is very versatile, efficient and is considered as an advanced treatment process. This may also help in the improvement in physico-chemical and biochemical quality of soil in wasteland. A study conducted by the author shows a considerable reduction in suspended solid content (16-26%), BOD (27-42%) and COD (65-76%) of paper mills wastewater after land treatment. (Mishra *et al.*, 1992) and had no negative effect on ground water quality within a year (Table 30.2).

Research carried out at the Central Soil Salinity Research Institute, Karnal and in other places has shown that wastewater can be disposed off on land for growing suitable vegetation cover. This method is not only economical compared to conventional treatment plants but also help in greening the countryside and thereby improving the quality of life in the urban and rural areas (Das & Kaul 1992). Chaturvedi (1985, 1986) have found the suitability of a number of tree species in Uttar Pradesh when raised with tannery and dairy wastewater. AZRI, Jodhpur has used textile wastewater for growing trees (Aggarwal and Kumar, 1990). The Forest Department and local authorities in Haryana, Rajasthan and West Bengal have raised tree vegetation on a small scale using sewage and industrial effluents. Results of such research and scattered developmental efforts show that there is a considerable scope of utilising sewage and industrial wastewater for raising trees for greening wastelands and preventing their degradation.

In a study on survival and growth performance of 12 tree species to tannery wastewater

Table 30.2. *pH, BOD (ppm) and COD (ppm) and Papermills wastewater before and after land treatment*

Month	Before treatment			After treatment		
	pH	BOD	COD	pH	BOD	COD
Feb-Mar, 88	10.03	266	90	--	–	–
May-July	9.88	330	113	9.2	260 (22)	81 (29)
Aug-July	9.70	310	106	8.2	206 (33)	70 (34)
Nov, 88-June, 89	9.46	352	118	8.1	248 (30)	77 (35)
Feb-April	8.98	361	129	7.78	257 (29)	89 (31)
May-July	8.85	330	120	7.8	260 (22)	90 (25)
Aug-Oct	8.90	350	115	8.0	280 (20)	80 (295)

Figures in parentheses show % reduction.

Table 30.3. *Growth in diameter of twelve tree species as affected by application of Tannery wastewater at Kanpur*

Sl. No.	Species	Common name	DBH of treated plants (in cubic decimeter)	Control
1.	*Acacia nilotica* SBP *indica*	Babula	46.60	20.40
2.	*Terminalia arjuna*	Arjun	27.50	16.10
3.	*Ficus religiosa*	Pipal	12.20	8.06
4.	*Cordia dichotma*	Lasora	27.90	42.50
5.	*Haplophragma adenophlyllum*	Kathsagun	11.80	11.70
6.	*Azadirachta indica*	Neem	7.55	2.80
7.	*Pongamica Pinnata*	Papri	6.04	2.56
8.	*Syzygium cumini*	Jamun	8.60	6.35
9.	*Casia auriculata*	Awavam	3.30	1.62
10.	*Ficus bengalensis*	Bargad	3.07	4.61
11.	*Adhatoda vasica*	Arusa	0.40	0.55
12.	*Pinus roxburghii*	Chir	0.19	0.32

Source - Chaturvedi 1986, DBH - Diameter at breast height.

irrigation, Chaturvedi (1986) recorded high survival percentage in 11 species. In respect of biomass Babul, Arjun, Lasora, Awaram and Arus showed better growth of diameter and height when irrigated with tannery wastewater than with ordinary water (Table 30.3).

Similar studies at Central Arid Zone Research Institute (CAZRI) Jodhpur to test the tolerance in germination and seedling growth of 15 tree species with textile wastewater exhibited cent percent survival and good growth without any detrimental effect for the nine species i.e. *Eucalyptus camaldulensis, Azadirachta indica, Acacia nilotica* ssp. *indica, Acacia tortilis, colosphosperrmum mopane, Hardwickia binata, prosopis chilensis, prosopes cineraria and Tecomella undulate* (Agarwal and Kumar 1990). Chaturvedi (1985) has also recorded that *Syzygium cumini* attained a better height within one half year after planting when irrigated with wastewater from the diary plant compared to the plants irrigated with clean tubewell water. At Faridabad an area of 5.6 ha of wasteland has been developed into a city Park and Rose Garden by using the wastewater of Thermal Plant. The saplings of *Grewelia robusta* (Silver Oak), *Alstonia scholaris and Pongamin pinnata* alongwith shrubs like Bougenvillea have been successfully established. Similarly, at Murthal, 2000 Eucalyptus saplings have been established by irrigating with wastewater from breweries (NWDB, 1990).

Industrial wastewater has been tried by many scientists to see its effect on germination and crop growth. Spulnik (1949) reported that use of sulphite pulp mill wastewater had no adverse effect on the germination of sunflower seeds. Similar observations have been reported by Anon (1954) on crops like cotton, corn, grasses etc. Mckee (1957) have reported that wastewater from cheese factory and diary farm favoured growth of several crops. Irrigation of paper mills wastewater has shown encouraging results on oats and orchard grass (Hashimoto and Yokota, 1965) on paddy, wheat and groundnut (Khambatta and Kethar 1969, Mishra and Sahoo 1989), on germination of rice, block gram and tomatoes (Rajannan and Obliswamy 1979), on Populous (Hansen *et al.*, 1980) Mishra and Sahoo (1989) observed that diluted papermills wastewater (50%) showed germination efficiency and growth of paddy and maize comparable to that of control but pure wastewater inhibited the germination but favoured growth and crop production (Table 30.4 and 30.5).

As discussed earlier, sewage and associated sludge are rich in nutrients and their use has been found to improve the physical condition of the soil as well as increase in productivity. One known way of utilising the sewage and sludge has been to grow vegetables for growing demands of cities and towns. In Calcutta, a number of cooperative ventures have come up for raising vegetables, fish and crop. In Tamil Nadu, the scarcity of water coupled with shortage of fodder and forage has encouraged many Municipality corporations to establish sewage farms to produce forage and fodder grasses as well as a number of vegetables (Sree Ramulu 1990) sewage farms, due to the availability of sewage through out the year, can grow suitable varieties of Jowar, Maize, Lucern etc. for making highly nutritive poultry feeds. *Eucalyptus hybrid* has the highest capacity to transpire large amount of water through out the year and it has been therefore, found to be the most effective bio-drains and suitable for areas where large volume of sewage accumulates over small area. (Chhabra 1989). In USA study of wastewater reclamation and utilization revealed beneficial effects of sewage on grasses, shrubs and field crop like rye, wheat,

TABLE 30.4. *Effect of Paper Mills Wastewater on germination growth and crop production in Maize (X)*

Sl. No.	Parameters	Control	50%	100%
1.	Percent germination	73.33	70.0 (-4.54)	43.33 (-40.9)
2.	Plant height (cm)			
	(a) on 31st day	50.25	40.75 (−18.19)	63.5 (26.37)
	(b) on harvest day	142.5	149.8 (5.12)	156.3 (9.68)
3.	Total number of leaves on 31st day	30	27	34
4.	Pigment contents in leaves on 55th day (µg/g fresh weight)			
	Chlorophyll a	620.4	883.08 (+42.0)	1168.21 (+88.0)
	Chlorophyll b	452.6	723.2 (59.79)	739.33 (63.35)
	Carotenoid	0.27	0.48 (77.78)	0.61 (125.90)
5.	Dry weight of shoot (gm) after harvest	42.1	56.25 (33.5)	53.92 (27.98)
6.	Dry weight of ripe cobs (gm) after harvest	11.53	17.25 (49.6)	5.1 (−55.77)

Figures in parentheses show percent increase or decrease over control.

corn and alfalfa, (Anonymous, 1967), Bachmann (1954) found 52 per cent increase of hay yields due to sewage sprays. Similar observations were made by Bendixon *et al* (1968) in case of grasses when waste water was used. However Day *et al* (1963) in a study to compare the grain yield, grain quality and malt quality of two barley varieties irrigated with sewage and normal water found reduced weight and size of kernel on plots irrigated with sewage water because of high nitrogen content. Kardos (1967) found that hay yields were increased by 139 per cent, corn grain 78 per cent and oat grains 70 per cent due to use of sewage water. Similar results have also been obtained by Day and Tucker (1959) on barely and by Ball (1977) on eight varieties of perennial legumes.)

Sewage and industrial wastewater are rich in minerals and sometimes contains some toxic heavy metals. Raw sewage contain toxic elements, and heavy metals such as Fe, Zn, Mn, Pb, Cd, Ni, Co, Cr etc. Though many of these are essential plant nutrients, almost all become phyto-toxic at higher concentrations. The study on post Irrigation effect on the sewage sludge fed soils

Table 30.5. *Effect of Paper Mills Wastewater on Paddy*

Sl. No.	Parameter	Control	50%	100%
1.	Pigment content on 60th day (μg/g fresh wt.)			
	Chl. a	903.7	1256.78 (39.07)	1196.07 (32.35)
	Chl. b	589.88	833.48 (41.3)	688.18 (16.7)
	Carotenoid	0.432	0.626 (44.9)	0.512 (18.5)
2.	Number of tillers	31	31 (0)	30 (–32.)
3.	Weight of hundred grains (gm)	1.973	1.911 (–3.14)	1.879 (–4.76)
4.	Average weight of grains produced (gm)	23.15	34.12 (47.39)	34.4 (48.60)
5.	Average shoot length (cm)	67.83	66.5 (-1.96)	77.0 (2.17)
6.	Average dry shoot weight	30.87	30.97 (0.32)	34.76 (12.6)

Figures in parentheses show percent increase or decrease over control.

showed increased total Ni, Cr and Pb content (Sree Ramulu 1990). Studies carried out at Calcutta showed high concentration of Fe, Cu, Zn, Pb in parts of vegetables like gourd, mustard, spinach, cauliflower and radish. In most cases concentration of Pb was beyond the critical limits and thus unfit for human consumption. Although vegetables, crops and fish produced with sewage effluents around the Calcutta Metropolitan area for decades have not yet been connected with specific health hazards (Ghosh, 1990, referred by Das and Kaul, 1992) but the risk of heavy metals and other undesirable elements getting into the food chain cannot be ignored.

Somasekhar *et al* (1984) observed increased pH, organic carbon, Na, K and phosphate content but decreased waterholding capacity in soil after prolonged wastewater treatment. It is obvious that prolonged application of sewage and/or industrial wastewater which are rich in nutrients, will tend to increase the soil nutrient. The most important aspect is to utilize the wastewater after primary and secondary treatment thus removing suspended solids and organic matter load respectively. The wastewater containing high level of toxic heavy metals should not be used in irrigation.

The additional advantage of sewage irrigation is its capacity to sustain Blue green algae which are the nitrogen fixers. Growing algae over the sewage also decrease the concentration of toxic heavy metals and consequently reduce the soil and water pollution. However, reports are available concerning the spread of human pathogenic or toxin producing bacteria in soil through the use of sewage as an irrigation supplement or sludge as a fertilizer supplement (Gangawane 1989). Basu and Behera (1989) have shown that 50 per cent diluted paper mills effluent showed no significant change in microbial characteristics of soil as compared to control where as pure effluents decreased the composition, diversity and population of fungal species within 21 days of application. Thus diluted sewage and industrial wastewater application may not have any risk of nutrient enrichment or deterioration of soil quality provided it is free from toxic elements like heavy metals.

Use of Algae and Aquatic Macrophytes

In the secondary treatment process only bacteria and fungi are used to degrade the organic matter from wastewater aerobically. The use of algae in wastewater treatment could prove beneficial in different ways since they will bring about oxygenation and mineralization. Algae bacterial symbiosis has long been proved to be an inexpensive efficient method for reclamation of wastewater over bacterial action alone. Recently, scientists are busy in locating new photosynthetic organisms with high growth rate, higher biomass yields and high utilization potential which could be mass cultured in wastewater. Blue-green algae are ideally suited to perform these functions by virtue of their high flexibility to adopt to varied environments and their known nutritional and fertilizer value (Uma and Subramanian 1990).

Uma and Subramanian (1990) have been successful in removing 25 to 50 per cent of the total dissolved solids, calcium and chloride load from wastewater of ossein industry using *Halobacterium* sp. and *oscillatoria* sp., the blue green alga. *Oscillatoria Pseudogeminata* in combination with natural microbial population is efficient remover of nitrates, nitrites and other nutrients in paper mill effluent (Manoharan and Subramanian 1992). Reports are also available on the reduction of BOD, COD and high level of nitrogenous compounds by algal cultures in different industrial wastewater (Saxena *et al* 1974, Govindan 1984, 1985), Tam and Wong (1989). The capacity of cyanobacteria to remove large amount of phosphorous from wastewater has also been demonstrated by several workers (Chan *et al.* 1979, Tam and Wong, 1989).

Thus during secondary treatment phase with biological oxidation pond, algal-bacterial symbiosis is an ideal inexpensive method of wastewater reclamation where Algae provides oxygen, removes nutrients and other toxic elements and bacteria degrades organic matter aerobically.

Aquatic Hydrophytes

Recently there has been considerable interest in using aquatic plants in the treatment of wastewaters. The floating waterhyacinth (*Eichhornia crassipes*) has proven to be a significant economic burden to many tropical and subtropical regions (Holm *et al* 1969). It is a highly productive plant when grown in high nutrient waters and population can double their surface area coverage in two weeks (Penfound 1956, Penfound and Earle 1948). It adds organic matter to water after death and makes water anoxic.

The following characteristics of water hyacinth make it a suitable species for wastewater treatment :

(i) Easy availability

(ii) Easily cultiviable

(iii) Rapid growth rate

(iv) High mineral uptake and accumulation

(v) Easily harvested

(vi) Economic potential as feed, fertilizer and for generation of biogas

Mishra *et al.*, 1991 have recorded a decrease in conductivity (74%), total dissolved solids (57.2%), BOD (67%), COD (50%), Total N (25%), NO_3-N (67%) in fertilizer industry wastewater during 60 days of water hyacinth treatment. The reduction in conductivity and total dissolved solids in Aluminium and thermal wastewater was equally high. Fluoride reduction was 38 per cent in Aluminium industry wastewater (Table 30.6).

Use of water hyacinth has been successful in reducing BOD and COD significantly (BOD 94 per cent in sugar refinery BOD 84 per cent and COD 88 per cent in dairy waste, COD 44 per cent and BOD 58 per cent in Pulp and Paper mills, BOD 35 per cent and COD 70 per cent in tannery wastes, COD 92.97 per cent in textile industry waters). They are also efficient in removing nitrogen, phosphorous pheaols, pesticides etc. (Trivedy and Joag 1989).

Table 30.7 shows the percent reduction in metal content from wastewater over 60 days of water hyacinth treatment. The percent reduction was 78, 80 and 71.6 for aluminium, 100, 96, 99.6 for Iron, 75, 74, 88 for Zinc and 100, 76.7 and 100 for Copper in Aluminium, thermal and fertilizer wastewater respectively.

Removal of heavy metals like Cu, Cd, Hg, Mn, As, Cr, Ni, etc. by water hyacinth has been demonstrated by many workers. (Soutton and Blackburn 1971, Tokunage *et al.*, 1976, Nath *et al.*, 1984; Cooley *et al.*, 1978, Englande and Kaigate 1980, Yoeh 1979).

Thus integration of water hyacinth treatment plant associated with an oxidation pond having symbiotic action of algae and bacteria with primary treatment process may give a satisfactory result for removing most of the organic and inorganic pollutants from wastewater. In case, finance is not a constraint, the tertiary treatment method may also be adopted to bring the water quality to acceptable limit (Fig. 30.1).

TABLE 30.6. *Changes in Physiochemical Characteristics in wastewaters during water hyacinth treatment*

Parameter	Alluminium Industry				Thermal Power				Fertilizer Industry			
	Day				Day				Day			
	0	10	40	60	0	10	40	60	0	10	40	60
pH	9.2	8.0	83	81	9.0	8.8	8.7	8.9	8.0	8.4	8.4	80
Conductivity mho cm^{-1})	450	354	378	309	280	230	215	276	1460	1050	512	384
Total dissolved solids	131	122	97	87	220	148	120	109	210	211	205	90
Chemical oxygen demand	–	–	–	–	40	31	22	13	380	300	270	225
Biological oxygen demand	–	–	–	–	07	5	2	0	38	28	28	13
Flouride	15.6	15	9.4	10.5	–	–	–	–	–	–	–	–
Total Nitrogen	–	–	–	–	–	–	–	–	167	136	116	112
Total NO_3-N	–	–	–	–	–	–	–	–	14	9.7	6.5	5.9

TABLE 30.7. *Percentage reduction in some heavy metal content from waste water during water hyacinth treatment*

Days of treatment		Percentage reduction				
		Alluminium	Copper	Chromium	Iron	Zinc
0		--	--	--	--	--
3	AI	0	0	10	0	10
	TPP	0	67	0	3	11
	FI	--	--	--	17	0
10	AI	5.5	18	6	10	32.5
	TPP	12.5	67	40	12	56
	FI	--	100	12	17	20
17	AI	5.5	41	30.5	23	32.5
	TPP	50	67	58.0	67	48.0
	FI	--	100	30	16	70
25	AI	25	41	48.5	80	57.5
	TPP	42	67	66.0	88	63
	FI	--	100	6	54	--
32	AI	28.6	59	48.5	83	75.0
	TPP	44	67	66	90	63
	FI	--	100	12	92	74
40	AI	28.6	100	70	87	75.0
	TPP	73	67	74	95	74.0
	FI	--	100	41	93	75.0
48	AI	5.5	100	70	97	75.0
	TPP	75	67	80	95	74
	FI	--	100	53	96	87.5
55	AI	48	100	70	97	75.0
	TPP	75	77	80	95	74
	FI	--	100	70.5	96	87.5
60	'AI	48	100	70	97	75.0
	TPP	75	77	80	95	--
	FI	--	100	70.5	96	--

AI - Aluminium Industry TPP - Thermal Power Plant FI - Fertilizer industry.

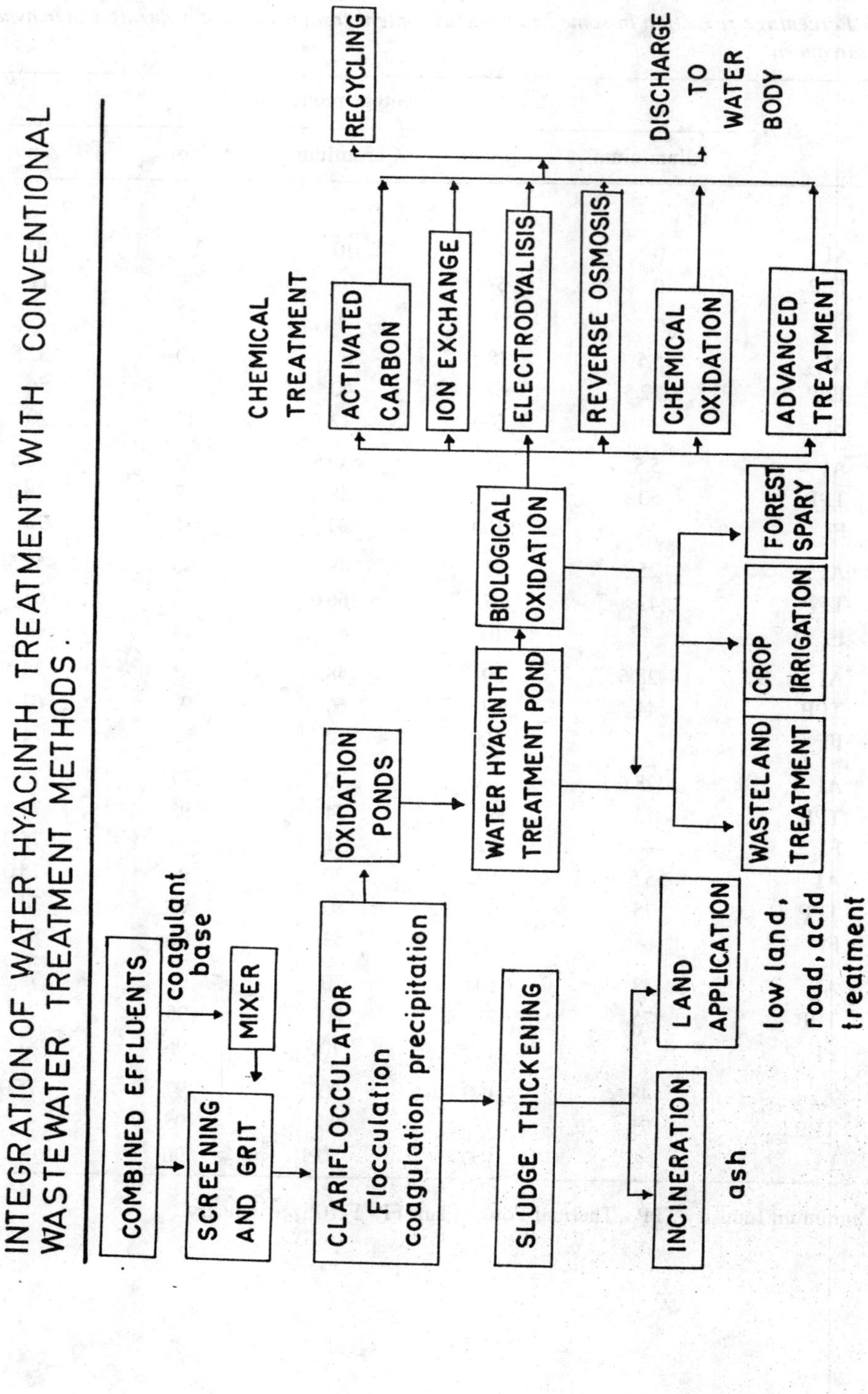
INTEGRATION OF WATER HYACINTH TREATMENT WITH CONVENTIONAL WASTEWATER TREATMENT METHODS.
COMBINED EFFLUENTS
coagulant base
MIXER
SCREENING AND GRIT
CLARIFLOCCULATOR Flocculation coagulation precipitation
OXIDATION PONDS
SLUDGE THICKENING
INCINERATION
ash
LAND APPLICATION
low land, road, acid treatment
WATER HYACINTH TREATMENT POND
BIOLOGICAL OXIDATION
WASTELAND TREATMENT
CROP IRRIGATION
FOREST SPARY
CHEMICAL TREATMENT
ACTIVATED CARBON
ION EXCHANGE
ELECTRODYALISIS
REVERSE OSMOSIS
CHEMICAL OXIDATION
ADVANCED TREATMENT
RECYCLING
DISCHARGE TO WATER BODY

Fig. 30.1

REFERENCES

Aggrawal, R.K. and Praveen Kumar, 1990, Textile industrial effluents. Implications and possible use for afforestation in Western Rajasthan *Annals* of arid Zone. **29** (4) : 295-302.

Anon, 1954. Mill Waste grows low and high corn. *Pulp and Paper*, **28** : 36.

Anonymous, 1967. *Study of Wastewater reclamation and utilization.* 3rd report of State Water Pollution Control Board. Seramento, California Publ, No. 18.

Bachmann, G., 1954. *Sewage Utilization Plant at Memmingen Wasserw.* Wass. Tech. **4** : 191-201, I., Ref. 1955, Yik : 157-344.

Ball, Robert, C. 1977. The Michigan State University water quality Management Project – A Facility for research in recycling treated wastewater. In *Waste Water renovation and Reuse* (Frank, M.D. ed.) : **7** : 205-253.

Basu, B. and Behera, N. 1989. Microbial Characteristics of Soil incubated with Paper Mills effluent. In *Soil Pollution and Soil Organisms.* (P.C. Mishra Ed.) Ashish Publishing House, New Delhi pp. 121-138.

Bendixon, T.W., Hill, R.D., Schwartz, W.A.; Robeck, G.G. 1968 Ridge and Furrow liquid waste disposal in Northern latitude, *J. Sanitary Dir.* ASCE, **94** : SAI : 147-157.

Central Pollution Control Board, 1989. Status of water supply and wastewater collection treatment and disposal in Class I cities. CPCB Report, Delhi.

Chan, K.Y.; Wong K.H.; and Wong, P.K., 1979. Nitrogen and Phosphorous removal from sewage effluent with high salinity by Chlorella Salina. Env. Poll. **18** : 139-146.

Chhabra, R.S. 1989. *Sewage Water Utilisation Through Forestry.* Central Soil Salinity Research Institute (CSSIR) ICAR, Karnal.

Chaturvedi, A.N. 1985. Jamun Plantations for Controlling Dairy effluent. *J. Trop. For.* **1** (3) : 210-274.

Chaurvedi, A.N. 1986. Trees and Shrubs for Control of Tannery Waste Water in India. *Environmental Conservation,* **13** (2) : 164-165.

Cooley, T.N. Ganzalez, M.H. and Martin, D.F. 1978. Radio Manganese, iron and phosphorous uptake by Water hyacinth and economic implications. *Eco. Bot.* **32** : 371-378.

Das, D.C. and Kaul, R.S. 1992. *Greening Waste lands through waste water.* National wastelands Development Board. Ministry of Environment and Forests, New Delhi-100 003; p. 64.

Day, A.D., Dickson, A.D., Tucker, T.C. 1963. Effect of City Sewage effluent on grain yield and grain molt quality of fullgrown irrigated Barley. *Agronomy Jour.*, **55** : 317-318.

Day, A.D. and Tucker, T.C. 1959. Production small grains pasture Forage using sewage effluent as a source of irrigation water and plant nutrients. *Agronomy Journal* **51** : 569-572.

Englande, A.J. and Kaigata, B. 1981. *Removal of Persistent heavy metais by vascular aquatic plant systems.* Ann. Meeting American Institute of Chemical Engineers, New Orleans, USA.

Gangawane, L.V. 1989. Sewage Farming and Micro organisms : a review. In : *Soil Pollution and Soil organisms* (Ed. P.C. Mishra), Ashish Publ. House New Delhi : 87-96.

Govindan, V.S. 1984. Studies on algae in relation to treatment of dairy wastewater *Ind. J. Environ. Health.* **26** : 261-263.

Govindan, V.S.1985. Treatment of tannery wastewaters by Stabilization pond method. *Ind. J. Env. Health,* **27** : 58-66.

Hansen, E.A.; Dawson, D.H. and Tolsted, D.N. 1980. Irrigation of intensively cultured plantations with papermill effluents. *Tappi,* **63** (11) : 139-143.

Hashimoto, T.and Yokota, H. 1965. Edaphological studies on the utilization of waste pulp liquor. *J. Sci. Soil Manure,* Tokyo, **36** : 231-234.

Holon, L.G.; Weloon, L.W.; and Blackburn, R.D. 1969. Aquatic Weeds, Science, **166** : 699-709.

Kardos, L.T. 1967. *Wastewater renovation by the land A living filter. Agriculture and the quality of our environment.* American Association for the advancement of science. Publ. No. 85, 1967. Washington D.C., pp. 241-250.

Khambatta, S.J. and Ketkar, C.M. 1969. *Present Scheme for treatment and utilisation of paper mill effluent on land for irrigation.* Presented in the symposium of problems connected with disposal of waste from various industries round about Poona; Nov. 29-30.

Loehr, R.C., 1979. *Land application of wastes* Vol I and II. Van Nostrad Reinhold Co. New York.

Manoharan, C. and Subramanian, G. 1992. Interaction between papermill effluent and the *cyanobacterium Oscillatoria pseudogeminata* var. unigranulata Poll. Res. **11** (2) : 73-84.

Mckee, F.J. 1957. Dairy waste disposal by spray irrigation *Sew. Industr. waste,* **29** : 1-57.

Mishra, R.P. 1986. Wastewater recycling for pollution control. *I.E. Journal.* Vol. XV (9) : 25-32.

Mishra, P.C., Panda, Madhumita, Patel, R.R. and Mohapatra, M.K. 1992. Physicochemical and Biochemical changes in ground water and soil during land treatment of paper mill wastewater. *J. Eco toxied Environ. monitor.* **2** (i) : 61-73.

Mishra, P.C. and Saha, S. 1989. Agropotentiality of papermills wastewater. In : *Soil Pollution and* Soil Organisms (Ed. P.C. Mishra). Ashish Publ. House, New Delhi, pp. 88-120.

Mishra, P.C., Patri Manorama and Panda, Modhumita 1991. Growth of water hyacinth and its efficiency in the removal of pollution load from industrial waste water J. Ecotoxicol Environ. monit **1** (3) : 218-224.

Nath, K.J., Rama Rao, S.V., Nair Sunanda, Gilman, R.H. and Mullick, D. 1984. Low cost wastewater treatment with water hyacinth pp. 655-663. In : G. Thyagarajan (Ed). *Proc. Int. Cong. Water Hyacinth* UNEP. Nairobi.

National, Wastelands Development Board, 1991. *Official Communications from various organisations. National Wasteland Development Board,* New Delhi.

Penfound, W.T. and Earle, T.T. 1948. The Biology of the water Hyacinth. *Ecol. Monogr.* **18** : 447-472.

Penfound, W.T., 1956. Primary production of vascular Aquatic plants. *Limnol. Oceanogr.* **1** : 92-101.

Rajannan, G. and Obliswamy, G. 1979. Effect of paper Factory effluents on soil and crop plants. *Ind. J. Environ. Health,* **2** : 110-130.

Saxena, P.N., Tiwari, A. and Khan, M.A. 1974. Effect of *Anacystis nidulans* on physicochemical and biological characteristics of raw sewage. *Proc. Ind. Acad. Sci,* **79** : 139-146.

Somashekar, R.K., Gowda, M.T.G., Shettigar, S.L.N. and Srinath, K.P. 1984. Effect of industrial effluents on crop plants. *Ind. J. Environ. Health* **26** (2) : 136-146.

Spulnik, J.B. 1949. Effect of waste sulphate liquor on soil properties and plant growth. *Soil Science* **49** : 37.

Sree Ramulu, U.S. 1990. *Studies on the Utilisation of Urban sewage and sludge for increasing crop production.* Proc. End. Sci Cong. Indore.

Sulton, D.L. and Blackburn 1971. Uptake of copper by water hyacinth. *Hyacinth Control Journal.* **9** : 18-20.

Tam, N.F.Y. and Wong, Y.S. 1989. Wastewater nutrient removal by *Chlorella pyrenoidosa and Scenedesmus* Sp. *Env. Poll.* **58** : 19-34.

Tokunaga, T.N. Furutas and Morimoto, M. 1976. Accumulation of cadmium in *Eichhornia crassipes Solms. Eisei Kagaku.* **22** : 234-239.

Trivedy, R.K. and Joag, G.A. 1989. Treatment of industrial waste water by water hyacinth application. In *Pollution Management in Industries* (Ed. R.K. Trivedy) Env. Publication, Karad. pp. 295-315.

Uma, L. and Subramanian, G. 1990. *Effective use of cyanobacteria in effluent treatment.* Nath Symp. on cyanobacterial nitrogen fixation IARI, New Delhi pp. 437-444.

Yoeh, B.G. 1979. *A. Preliminary study on the use of water hyacinth in the removal of heavymetals and B.O.D. from industrial waste waters.* Conf. on Chemical research in Malaysia, Kualalumpur.

Chapter 31

Our Life Style, Environment and Mutagenesis

R.K. Das

School of Life Sciences, Sambalpur University, Jyoti Vihar-768 019, Orissa

Abstract. In addition to disastrous effect on environment, the environmental pollutant may have 'direct' effect on our health, which may be again 'acute' or 'long term'. The 'long term' effect cannot be known immediately, it is expressed only long time after the exposure and thus of great human concern. In this communication attempts have been made to review long term health effects leading to induction of genetic damage and/or cancer caused by environmental pollutants associated with some of our modern life styles. Here mutagenic and carcinogenic efficiencies of indoor pollutants like cigarette smoke (both main stream and side stream), smoke released during burning of mosquito repellents and radiations emitted from our housing materials and consumer goods have been reviewed. The review has also included genotoxic and carcinogenic potentials of substances emitted from automobiles and of our food stuffs and beverages. The relation between mutation and cancer has also been discussed.

Key words : Life style, Environmental pollution, Mutation, Mutagenesis, Cancer, Carcinogenesis, Long-term health effect, Genotoxicity

Introduction

In an urge for providing sufficient food, comfortable life, more luxury and more leisure we have introduced into our environment thousands of chemicals of diverse nature, and machinery and equipments. Undoubtedly, all these have provided great comfort to human life and enormous benefit to the society; as a whole they have changed our life style greatly. As a result of these activities our environment is getting polluted. It is now realized that a price has to be paid in terms of quality of the environment and human health. It will not be amiss here to define environment. According to modern concept environment covers not only components like air, water, soil, etc., but refers to the one created by man's cultural habits and eating and drinking behaviour also (Ray, 1986).

A host of articles are being published in scientific as well as non-scientific journals and newspapers, and conferences, symposia, meetings, rallies, etc. are being organized to raise public awareness regarding the adverse effect of these environmental pollutants leading to environmental disaster and thereby causing human sufferings ("Indirect" effect : e.g. green house effect global warming, rising of the sea-level, ozone hole, etc.). But similar emphasis on their "direct" effect on human health is lacking. The direct effect may again be (a) acute : leading to instant death due to severe physiological disturbances (e.g. atomic bombardment in Hiroshima and Nagasaki, MIC gas tragedy in Bhopal, Chernobyl disaster, etc.), or (b) long term : causing irreparable damage to the cells and genetic make up of humans leading to the development of cancers and various hereditary disorders, which is of great concern (e.g. higher incidence of cancer among the survivors of Hiroshima, Nagasaki and Chernobyl, etc.).

Agents which are acutely toxic can be easily detected because the immediate effects allow rapid identification of their source. But mutations, cancers and tumours are only detected many years after exposure, they are not the immediate result of an exposure to mutagenic or carcinogenic agents. The latent period for tumour induction in humans may be as long as 20 years. Recessive mutations may take several generations to be detected. Obviously, a source of mutagenic or carcinogenic activity can only be traced after many persons have been exposed for many years. It is now high priority activity all over the world to screen mutagenic and carcinogenic agents in our environment - outdoor and indoor. The present review deals with the long term effects (i.e. mutagenic and carcinogenic) of environmental agents or factors associated with our modern life style.

Mutagenesis and Cancer

Today, cancer has become the biggest "killer" after heart disease. It is now well established that alteration in gene(s) of a normal cell, i.e. mutation, is responsible for initiation of cancer; thus cancer is genetic in origin and starts from somatic mutation (Weinberg, 1988). Earlier it was assumed that retroviruses (cancer causing viruses) were the culprit and caused cancer by transporting their cancer inducing v-oncogenes into the genome of normal persons on infection. But now a number of oncogenes (c-oncogenes) have been identified in the genome of a healthy person; however, they normally remain in inactive condition. Activation of those c-oncogenes results in the production of oncogene encoded protein(s) which initiates development of cancer. This activation is caused through gene mutation and chromosome aberrations, and physical and chemical carcinogens simply induce mutations and/or chromosome aberrations. Boveri (1914) was pioneer in postulating the "Mutation theory of cancer" as early as 1914 when genetics was in its infancy stage. Neoplasms like Chronic Myeloid Leukemia (CML), Burkitt's Lymphoma (BL), Multiple Myeloma (MM), Acute Lymphocytic Leukemia (ALL), etc. are associated with certain types of chromosomal abnormalities (Mitelman, 1990). There is also ample evidence that mutations in human somatic cells are initiators of cancer. The fact that all carcinogens are mutagens also supports the view of genetic origin of cancer.

It has turned out that mutagenicity tests can well be used as short term carcinogenicity test. Mutagenicity tests are less time consuming and much less costly than classical carcinogenicity tests.

Smoking and Mutagenesis

More than a century ago Dr. Samuel Johnson said, "A cigarete has fire at one end and a fool at the other." His comment is now found to be a naked truth. Cigarette, cigar, pipe, `bidi', `chutta', etc. are various types of tobacco preparations used for smoking; among them cigarette is the most popular and used world-wide. In India alone the tobacco industry produces 200 million sticks of cigarettes and 800 million bidis daily. In USA it is estimated that tobacco's contribution to all cancer death is 30 per cent; 85 per cent of all lung cancer cases are due to smoking; similarly 30-40 per cent bladder cancer death, 30 per cent pancreatic cancer death and 50 per cent oesophageal cancer death are smoking related (US Deptt. of Health & Human Services, 1982). Currently, tobacco use is estimated to account for 3 million death world over per year; in India alone in 1991 about 8,00,000 died due to tobacco (WHO, 1992).

Smoking generates two types of smokes : (a) mainstream (MS) smoke - the smoke inhaled by the smoker, and (b) sidestream (SS) smoke - the smoke coming out from the burning puff. Tobacco smoke, both MS and SS, contains more than 3800 compound (Fischer *et al.*, 1990). It is surprising that the amounts of most of them are 4-10 times more in the SS compared to the MS. Environmental tobacco smoke (ETS) which contains both SS smoke and MS smoke (exhaled by a smoker) is, therefore, qualitatively more harmful than MS itself. Hence passive smoking (inhaling ETS by non-smokers) is more dangerous than actual smoking. Smoking makes our environment particularly indoor environment polluted greatly. Among the compounds present in the cigarette smoke polynuclear aromatic hydrocarbon (PAH), polonium 210, aromatic amines and N-nitrosamines are strong carcinogens particularly the last group (IARC, 1986; Hecht and Hoffmann, 1988; Claxton *et al.*, 1989; Lofroth, 1989; Nair *et al.*, 1989; Fischer *et al.*, 1990). Epidemiological and experimental studies have confirmed that tobacco specific nitrosamines (TSNA) cause mutations and human cancer (Preussmann and Stewart, 1984; IARC, 1986). Recently contribution of aldehydes and oxy-radicals to tobacco carcinogenesis has been suggested because they cause DNA single strand break and DNA-protein cross link (Church and Pryor, 1985; Hsie *et al.*, 1986; Curren *et al.*, 1988). Recent investigation demonstrates that formation of endogenous nitrosamines in smokers is about twice as high as in non-smokers (Tsuda and Kurashima, 1991). Of the TSNAs nor-nicotine (NNN), anabasine (NAB), anatabine (NAT) and nitrosamineketone (NNK) are highly important (Hecht and Hoffmann, 1988).

Tobacco smoke has been studied by using various assays and conclusively been proved mutagenic in human cells in culture and in several other mammalian systems *in vivo* and *in vitro* (Vijayalaxmi and Evans, 1982; Obe *et al.*, 1984). Higher incidence of chromosome damage has been noted in the lymphocytes of smokers than of non-smokers (Vijayalaxmi and Evans, 1982). Significantly higher frequencies of micronuclei (MN) in buccal mucosa cells (Sarto *et al.*, 1987) and unscheduled DNA synthesis (Vijayalaxmi and Evans, 1982) have been recorded in smokers, compared to non-smokers, indicating more genetic damage in the former. Smokers produce more mis-shapen sperm than non-smokers (Evans *et al.*, 1981; Wyrobek *et al.*, 1983b). Incidence of spontaneous abortions and of having low birth-weight baby is also more in smoking mothers than non-smoking mothers (see Smoking and Health, 1979). Phillips *et al.* (1988) using autoradiographic analysis have demonstrated more DNA damage in smokers than non-smokers.

Not only the MS cigarette smoke but the ETS also is known to cause mutation. ETS particles collected from office buildings, restaurants, homes, railway compartments and planes have been demonstrated to be highly mutagenic in the Ames *Salmonella* test (Lofroth *et al.*, 1983; Husgafvel-Pursiainen *et al.*, 1986; Lofroth and Lazaridis, 1986; Lofroth *et al.*, 1988). A starting finding in Japan is that non-smoker wives who live with heavy smokers are twice as likely to die of lung cancer as wives of non-smoker husbands.

Genotoxicity has also been reported in whole mammals exposed to MS as well as SS cigarette smoke (Balansky *et al.*, 1988; Mohtashamipur *et al.*, 1987) and in new born mice the mothers of which were exposed to the smoke in pregnant condition (Balansky and Blagoeva, 1989). Thus, it seems the active agent(s) can cross the placental barrier. Lung cells of rats exposed chronically to cigarette smoke exhibit higher frequency of chromosome breakage (Rithidech *et al.*, 1989).

Urine of smokers, compared to that of non-smokers, induces more mutations even when both the groups are kept under same diet (Doolittle *et al.*, 1990). Urine of passive smokers even is known to contain significantly high quantity of mutagens (Mohtashamipur *et al.*, 1986).

Smokeless Tobacco and Mutagenesis

With the establishment of the fact that smoking causes irreparable damage to our genetic material and consequently the health, tobacco industries from different corners of the world have started making propaganda that use of smokeless (chewing) tobacco is less harmful. Probably this is at the root for increase in consumption of smokeless tobacco. But investigations in the recent past have proved that smokeless tobacco is as harmful as smoking, it is highly carcinogenic too (IARC, 1985). It is used in the form of snuff, snuff-dipping ('Khaini'), chewing, nass, 'gudakhu', etc. India has a dubious distinction of having the highest incidence of oral cancer in the world; of all the cancers in India 30-50 per cent suffer from oral cancer only. This is attributed to habit of taking smokeless tobacco. In India today 14.2 crore men and 7.3 crore women are slaves of the tobacco habit.

Aqueous and organic solvent extracts of tobacco have been reported to be genotoxic in bacterial and mammalian cells *in vivo* and *in vitro* (Shirname *et al.*, 1984; Whong *et al.*, 1987; Shirname More, 1991). Extracts of pyrolysed tobacco have been shown to induce chromosome aberrations in mice and 8-azaguanine resistant mutations in V79 Chinese hamster cells (Kulkarni *et al.*, 1987). Genotoxic effect of tobacco is also noted in habitual users. Stich and his co-workers (Stich *et al.*, 1982; Brunnemann *et al.*, 1987; Stich and Anders, 1989) have recorded higher incidence of micronuclei in buccal mucosa cells of habitual users of various tobacco preparations. Very recently we have noted 'gudakhu' induced genotoxicity in both mice and habitual users (Dash and Das, 1992, Das and Dash, 1992). The active principles for causing genotoxic effect are the same as mentioned in connection with smoking. Smokeless tobacco contains about 2500 compounds including TSNA, PAH, polonium 210, aromatic amines, etc. (IARC, 1985).

Mosquito Repellents and Mutagenesis

To combat mosquito menace in tropical and sub-tropical countries it is now a common practice to burn mosquito coil or mat in closed or semi-closed room. The burning of these mosquito repellents cause indoor air pollution greatly and persons are exposed to it for hours. The active ingredients of such mosquito repellents are various insecticides the quantity and types of which vary from brand to brand, and it is a trade secret. Insecticides belonging to the groups like pyrethroid, chlorinated hydrocarbon, organophosphorus and carbamates are usually used. These insecticides have already been proved to be highly genotoxic in mammalian and other test systems (Wild, 1975; Bhunya and Pati, 1990; Pati and Bhunya, 1989; Behera and Bhunya, 1989). Very recently we have reported significant increase of chromosome damages in pulmonary alveolar macrophages (PAMs) of rats exposed to different concentrations of a very widely used mosquito coil smoke @ 15 min/h, 8h/d (Das *et al.,* 1994). Moreover, the incidence of damage increases positively with the concentration of smoke. That mosquito coil smoke induces morphological changes to the cells of the respiratory passage has earlier been demonstrated by Liu and his coworkers (Liu and Sun, 1988; Liu *et al.,* 1989).

Our Dwellings and Mutagenesis

Urbanization goes hand in hand with our life style. Each of us wants a roof over his head - a permanent long lasting decent house to live in. However, recently we have come to know that the building materials like bricks, sand, concrete, stones, etc. emit a radioactive gas called radon, - a decay product of uranium, thorium, actinium and radium. It is also emitted by soils and rocks present under our feet. But in the open space it gets diluted too much and very little is absorbed in our body. On the other hand, inside the houses and buildings its concentration goes up to a dangerous level due to limited air circulation and thus our indoor gets polluted with radon - the latest health hazard. The Environmental Protection Agency (EPA), USA (1987) has estimated that 5000 out of 136000 lung cancer death per year in that country is due to indoor radon pollution, and that in 21 per cent homes, out of 15000 sampled, the level of radon radioactivity is more than the safety limit. Although no work has so far been done on possible genotoxic effect of radon the ionizing radiation is known to be highly mutagenic and damages DNA by inducing double strand breaks, single strand breaks, base damages and DNA-protein cross link (Natarajan, 1984). Besides, consumer products like TV and computer sets, smoke detectors, etc. which are now part and parcel of our modern life style also add to a considerable extent radiation to our indoor environment.

Automobiles and Mutagenesis

Our life style is greatly influenced by the use of automobiles. In today's world automobiles constitute one of the important sources of air pollution. The automobiles discharge obnoxious gases not only while `on' the road but while `off' the road also., though less in quantity. The major pollutants of a petrol engine are carbon monoxide (CO), nitrogen oxides, hydrocarbons, lead oxides and formaldehyde. Diesel engines discharge CO in traces but produce thick smoke. The emission from diesel engine is, therefore, cleaner than that of petrol engine so far obnoxious gases are concerned. Volumes of reports are now available on the occurrence of air pollution due to emission from automobiles and their acute effect on our health. However, we know little about their long term effect.

Lead (Pb) is highly poisonous and it is in the list of 'probable carcinogens' of IARC (1987). Pb is not mutagenic in bacterial assays, and both positive and negative reports are available from mammalian systems (Bhunya and Pal, 1986; Beyersman *et al.*, 1990). Chromosome analysis of workers occupationally exposed to Pb fails to exhibit higher incidence of aberrations; however, smokers among them do show elevated incidence of SCE (Maki-Paakkane *et al.*, 1980). Recently Rajah *et al.* (1990) have recorded higher incidence of chromosome damages in peripheral lymphocytes of printing press workers.

Polycyclic aromatic hydrocarbons (PAH) are highly carcinogenic and mutagenic and it has been mentioned earlier in connection with smoking (Lofroth, 1989; Husgafvel-Pursiainen *et al.*, 1986). Benzene has thoroughly been studied; its cacinogenic and aneuploidogenic activities have been tested in vertebrates and invertebrates (Liang *et al.*, 1983; Au *et al.*, 1988). The first case of leukemia reported in 1928 was attributed to benzene vapour. A very high incidence of chromosome aberrations has been noted in lymphocytes of man and PAMs of mice exposed occupationally and experimentally respectively to benzene (Au *et al.*, 1988). Methylbenzene mostly called toluene is an important volatile PAH discharged from automobile exhausts. Recently SCE frequency is found to be much higher in occupationally exposed persons than non-exposed ones (Harps *et al.*, 1990).

Organic solvent extract of diesel exhaust shoot has been reported to induce frame shift mutation in *Salmonella* (King *et al.*, 1981). Induction of mitotic anomaly and numerical chromosome aberrations has also been recorded with extracts of petroleum exhaust particulates in mammalian cells in culture by Hadnagy and Seemayer (1990).

Mutagens on Our Dining Table

Epidemiological studies have indicated diet related cancer in various organs of our body. Various food-stuffs are known to contain mutagens and carcinogens.

Fruits and vegetables : Flavonoids are common in vegetables, and every day we consume about 1 g of flavonoids. Among them quercetin, kaempferol, rhamnetin, wagonin, and nor-wogonin are very common and they have strong mutagenicity. The flavonoid quercetin has been studied extensively for its possible mutagenic and carcinogenic activity. It has been shown to induce mutation in *Salmonella* and in rodent cells *in vitro*. However, reports on its chromosome breaking capacity in mouse bone marrow cells are contradictory (See Nagao *et al.,* 1985).

Methanol extract of a Japanese vegetable - pickles has been reported to induce mutation in bacterial system and its active principle has been identified as kaempferol. High incidence of oesophagus cancer in China is attributed to consumption of Chinese pickle methanol extract of which has been found to induce revertant mutations in bacteria, and its mutagenic property is 6 times that of the Japanese pickle (Sugimura *et al.,* 1982). Bracken induced intestinal and bladder tumours is supposed to be caused by quercetin and kaempferol present in it.

Cooked Food : Cooking of foods particularly protein foods results into production of mutagens. Grilled fish and fried hamburger are found to be mutagenic in *Salmonella* and mammalian systems and mutagenic activity elevates with the increase of cooking time (Takahashi *et al.,* 1983; Takayama *et al.,* 1984b). Pyrolysed proteins and amino acids are later known to be responsible for such mutagenicity (Sugimura, 1985); pyrolysed starch, nucleic acid and vegetable oil have little or no mutagenicity. The most active mutagenic principles produced in pyrolysised proteins are the heterocyclic amines like IQ, MeIQ and MeIQx; and heated amino acids produce 7 mutagenic compounds - Trp-p-1, Trp-p-2, Glu-p-1, Glu-p-2, Orn-p-1, Phe-p-1 and Lys-p-1 (Sugimura *et al.,* 1981). Heated soyabean globulin also contains mutagens. All these substances are highly mutagenic to *Salmonella* and mammalian cells (Sugimura, 1985; Sugimura and Sato, 1983). Most of the substances are carcinogenic too (Takayama *et al.,* 1984a). High incidence of stomach and oesophagus cancer among Japanese fishermen is considered to be due to their habit of taking fish roasted on charcoal braisures. Again, excess nitrates supplied as fertilizer get accumulated in the plant body, upon intake of fruits and vegetables the nitrates are reduced to nitrites by oral flora of bacteria. Nitrites are also now widely used as food preservatives. They cause nitrosation of a number of promutagens present in the food stuffs in acidic stomach leading to formation of nitrosamines which are proven mutagens (discussed earlier).

Food Dyes : Dyes like Violet I, Red Dye 2, azo dyes (tetrazine, Brilliant Black BN and Orange GGN) and xanthene dye (Erythrosine B) have also been shown to be genotoxic in various mammalian and bacterial tests (Au and Hsu, 1979; Latt *et al.,* 1981; Fisher *et al.,* 1990; Price *et al.,* 1978). Sweetening agents like saccharin and cyclamate provide contradictory reports.

Alcoholic Beverages : Alcoholic drinks also contain nitrosamines, PAH and other Mutagenic agents, and chronic alcoholics show high degree of chromosome anomalies (Natarajan, 1984). Aldehyde is formed metabolically from ethanol in the human and it induces SCEs in human and CHO cells *in vitro* (Ristow and Obe, 1978). Evaporated residues of whisky, brandy and apple brandy have been reported to be mutagenic in *Salmonella* (Nagao *et al.,* 1981). Storage of alcohol is also related to the production of mutagens.

Coffee : Coffee is a common beverage throughout the world. Caffeine is the principal alkaloid of coffee; it is also available, though in less amount, in tea, cocoa; etc. Caffeine has been studied extensively for its chromosome damaging activity in plants. It inhibits repair of DNA damage and thus indirectly acts as a mutagen (Kihlman, 1977). Mutagenic efficiency of coffee has also been demonstrated in mammalian cells *in vitro* and in *Salmonella* and *E. coli*. Green coffee does not show mutagenicity. Mutagenic principles of coffee are diketones (Nagao *et al.*, 1985).

Papers used to pack food stuffs contain organo-chlorine which is highly toxic. Some amount of chlorine used in the paper industries to bleach pulp gets attached to the carbon atom of its pulp and constitutes organ-chlorine. A fraction of organo-chlorine is not washable and is of high concern. Chlorine is also used to disinfect water and during the process trichloromethane is produced. This organo-chlorine is highly poisonous. Western countries are now using ozone instead of chlorine to purify water.

Conclusion

Environmental pollutants arising out of our modern life style are numerous and it is beyond our scope to deal with all of them here, the present discussion represents just a cross sectional view. Among the pollutants mentioned above tobacco and tobacco smoke, pyrolysed protein-rich food, indoor ionizing radiations, insecticides, substances discharged by the automobiles, etc. have conclusively been proved to be genetically toxic, some of them are highly carcinogenic too. From the foregoing discussion it is clear how dangerous our modern life style is, and the geneticists are highly concerned about it. Our life-style is greatly responsible for increase in the incidence of cancer all over the world. Our future lies vulnerably in our hands. It is, therefore, of central importance to recognize mutagenic agents, to understand the mechanism of mutation induction, to limit the exposure to such mutagens, and to prevent careless introduction of new mutagens in our environment.

REFERENCES

Au, W.W., Bibins, P., Ward, J.B. and Legator, M.L. 1988. Development of a rodent lung macrophage chromosome aberration assay. *Mutat. Res.* **208** : 1-7.

Au, W. and Hsu, T.C. 1979. Studies on clastogenic effects of biological stains and dyes. *Environ. Mutagen.* **1** : 27-35.

Balansky, R.M. and Blagoeva, P.M. 1989. Tobacco smoke-induced clastogenicity in mouse fetuses and in newborn mice. *Mutat. Res.* **223** : 1-6.

Balansky, R.M., Blagoeva, P.M. and Mircheva, Z.I. 1988. The mutagenic and clastogenic activity of tobacco smoke. *Mutat. Res.* **208** : 237-241.

Behera, B.C. and Bhunya, S.P. 1989. Studies on the genotoxicity of asataf (acephate), an organophosphate insecticide, in a mammalian in vivo system. *Mutat. Res.,* **223** : 287-293.

Beyersmann, D., Hartwig, A. and Kortenkamp, A. 1990. Mutagenicity of carcinogenic metal compounds : Molecular mechanisms. pp. 57-60. In : *Environmental Hygiene* (Eds. N.H. Seemayer and W. Hadnagy) Springer-Verlag, Berlin.

Bhunya, S.P. and Pal, B.B. 1986. Cytogenetic effect of nickel fluoride in mouse. *Nucleus.* **29** : 132-136.

Bhunya, S.P. and Pati, P.C. 1990. Effect of deltamethrin, a synthetic pyrethroid, on the induction of chromosome aberrations, micronuclei and sperm abnormalities in mice. *Mutagenesis.* **5** : 229-232.

Boveri, T. 1914. Zur. Frage der Entstehung Maligner Tumoren. *Fischer,* Jena.

Brunnemann, K.D., Hornby, A.P. and Stich, H.F. 1987. Tobacco-specific nitrosamines in the saliva of inuit snuff dippers in the northwest territories of Canada. *Canc. Lett.* **37** : 7-16.

Church, D.F. and Pryor, W.A. 1985. Free-radical chemistry of cigarette smoke and its toxicological implications. *Environ. Health Perspec.* **64** : 111-126.

Claxton, L.D., Morin, R.S. Hughes T.J. and Lewtas, J. 1989. A genotoxic assessment of environmental tobacco smoke using bacterial bioassays. *Mutat. Res.* **222** : 81-99.

Curren, R.K., Yang, L.L., Conklin, P.M. Grafstrom, R.C. and Harris, C.C. 1988. Mutagenesis of xeroderma pigmentosum fibroblasts by aerolein. *Mutat. Res.* **209** : 17-22.

Das, R.K. and Dash, B.C. 1992. Genotoxicity of 'gudakhu', a tobacco preparation. II in habitual users. *Fd. Chem. Toxicol.* **30** : 1045-1049.

Das, R.K., Sahu, K. and Dash, B.C. 1994. Induction of chromosome aberrations and micronuclei in pulmonary alveolar macrophages of rats following inhalation of mosquito coil smoke. *Mutat. Res.* **320** : 285-292.

Dash, B.C. and Das, R.K. 1992. Genotoxicity of 'gudakhu', a tobacco preparation. I. In mice *in vivo. Mutat. Res.* **280** : 45-53.

Doolittle, d.J., Lee, C.K., Irett, J.L., Mirsalis, J.C. Riccio, E., Rudd., C.J., Burger, G.T. and Hayes, A.W. 1990. Comparative studies on the genotoxic activity of mainstream smoke condensate from cigarettes which burn or only heat tobacco. *Environ. Mol. Mutagen.* **15** : 93-105.

EPA (Environmental Protection Agency). 1987. *Federal Register,* USA.

Evans, H.J., Fletcher, J., Torrance, M. and Hargreave, T.B. 1981. Sperm abnormalities and cigarette smoking. *The Laceti* 627-629.

Fischer, A.B., Muller, D. and Wellhausen, F. 1990. Induction of sister chromaticd exchanges by food dyes. pp. 38-41. In : *Environmental Hygiene* II. (Eds. N.H. Seemayer and W. Hadnagy). Springer-Verlag, Berlin.

Fischer, S., Spiegelhalder, B., Eisenbarth, J. and Preussmann, R. 1990. Investigations on the origin of tobacco-specific nitrosamines in mainstream smoke of cigarettes. *Carcinogenesis.* **11** : 723-730.

Hadnagy, W. and Seemayer, N.H. 1990. The role of non-genotoxic mechanisms in carcinogenicity by airborne particulate pollutants. pp. 11-16. In : *Environmental Hygiene* II (Eds. N.H. Seemayer and W. Hadnagy) Springer-Verlag, Berlin.

Harps, E., Turhan, U., Kuhlmann, C., Mayer, N., Pfeiffer, E.H. 1990. Effect of chronical toluene exposure on sister chromatid exchanges and circulating immune complexes. pp. 157-160. In : *Environmental Hygiene* II (Eds. N.H. Seemayer and W. Hadnagy) Springer-Verlag, Berlin.

Hecht, S.S. and Hoffmann, D. 1988. Tobacco-specific nitrosamines, an important group of carcinogens in tobacco and tobacco smoke. *Carcinogenesis,* **9** : 875-884.

Hsie, A.W., Recio, L., Katz, D.S., Lee, C.Q., Wagner, M. and Schenley, R.L. 1986. Evidence for reactive oxygen species inducing mutations in mammalian cells. *Proc. Natl. Acad. Sci.* (USA) **83** : 9616-9620.

Husgafvel-Pursiainen, K., Sorsa, M., Moller, M. and Benestad, C. 1986. Genotoxicity and polynuclear aromatic hydrocarbon analysis of environmental tobacco smoke samples from restaurants. *Mutagenesis.* **1** : 287-292.

International Agency for Research on Cancer. 1985. Tobacco habits other than smoking; Betel-quid and Areca-nut chewing; and some related nitrosamines. *Monogr. 37, IARC,* Lyon.

International Agency for Research on Cancer. 1986. Tobacco smoking. *Monogr. 38, IARC,* Lyon.

International Agency for Research on Cancer, 1987. Evaluation of carcinogenic risk of chemicals to humans. *Monogr. Supply. 7,* IARC, Lyon.

Kihlman, B.A. 1977. Caffeine and Chromosomes. Elsevier, Amsterdam.

King, L.C. Kohan, M.J., Austin, A.C., Claxton, L.D. and Huisingh, J.L. 1981. Evaluation of the release of mutagens from diesel particles in the presence of physiological fluids. *Environ. Mutagen.* **3** : 109-121.

Kulkarni, J.R., Sarkar, S. and Bhide, S.V. 1987. Mutagenicity of extracts of brown and black masheri, pyrolysed products of tobacco using short-term test. *Mutagenesis.* **2** : 263-266.

Latt, S.A., Allen, J.W., Bloom, S.E., Carrano, A., Falke, E., Kram, D., Schneider, E., Schreck, R., Tice, R., Whitfield, B. and Wolff, S. 1981. Sister chromatid exchanges : a report of the gene-tox programme : *Mutat. Res.* **87** : 17-62.

Liang, J.C., Hsu, T.C. and Henry, J.E. 1983. Cytogenetic assays for mitotic poisons. The grasshopper embryo system for volatile liquids. *Mutat. Res.* **113** : 467-479.

Liu, W.K., Ng, T.B. and Wong, C.C. 1989. Biochemical and cellular changes in bronchoalveolar lavaged samples from rats after inhalation of mosquito coil smoke. *Toxicol. Lett.* **45** : 121-132.

Lit, W.K. and Sun, E.E. 1988. Ultra structural changes of tracheal epithelium and alveolar macrophages of rats exposed to mosquito coil smoke. *Toxicol. Lett.* **41** : 145-157.

Liu, W.K. and Wong, M.H. 1987. Toxic effects of mosquito coil (a mosquito repellent) smoke on rats. II. Morphological changes of the respiratory system. *Toxicol. Lett.* **39** : 231-239.

Lofroth, G. 1989. Environmental tobacco smoke : Overview of chemical composition and genotoxic components. *Mutat. Res. 222* : 73-80.

Lofroth, G. and Lazaridis, G. 1986. Environmental tobacco smoke (ETS) : Comparative characterization by mutagenicity assays of sidestream and mainstream cigarette smoke. *Environ. Mutagen.* **8** : 693-704.

Lofroth, G., Ling, P.I. and Agurell, E. 1988. Public exposure to environmental tobacco smoke. *Mutat. Res.* **202** : 103-110.

Lofroth, G., Nilsson, L. and Alfheim, I. 1983. Passive smoking and urban air pollution : *Salmonella/* microsome mutagenicity assay. pp. 515-525. In : Short-term Bioassays (eds. M.D. Waters, S.S. Sandhu *et al.*). Plenum, New York.

Maki-Pakkanen, J., Husgafvel-Pursiainen, K., Kalliomaki, P.L., Tuominen, J. and Sorsa, K. 1980. Toluene exposed workers and chromosome aberrations. *J. Toxicol. Environ. Health.* **6** : 775- 781.

Mitelman, F. 1990. Patterns of chromosome variation in neoplasia. pp. 86-100. In : Chromosomal Aberrations (eds. G.Obe and A.T. Natarajan) Springer-Verlag, Berlin.

Mohtashamipur, E., Muller, G., Norpoth, E. and Stucker, W. 1986. Urinary excretion of mutagens in passive smokers. *Toxicol. Lett.* **35** : 153-156.

Mohtashamipur, E., Norpoth, K. and Straeter, H. 1987. Clastogenic effect of passive smoking on bone marrow polychromatic erythrocytes of NMRI mice. *Toxicol. Lett.* **35** : 153-156.

Nagao, M., Takahashi, Y., Wakabayashi, K. and Sugimura, T. 1981. Mutagenicity of alcoholic beverages. *Mutat. Res.* **88** : 147-154.

Nagao, M., Wakabayashi, K. and Sugimura, T. 1985. Mutagens in food and drinks, and their carcinogenicity. pp. 69-85. In : Mutagenicity. Testing in Environmental Pollution Control (eds. F.K. Zimmermann and R.E. Taylor-Mayer). Ellis Horwood Ltd., Chichester.

Nair, J., Pakhale, S.S. and Bhide, S.V. 1989. Carcinogenic tobacco-specific nitrosamines in Indian tobacco products. *Fd. Chem. Toxicol.* **27** : 751-753.

Natarajan, A.T. 1984. Origin and significance of chromosomal alterations. pp. 156-176. In : *Mutations in Man* (ed. G.Obe) Springer-Verlag, Berlin.

Obe, G., Heller, W.D. and Vogt. H.J. 1984. Mutagenic activity of cigarette smoke pp. 223-246. In *Mutations in Man* (Ed. G. Obe) Springer-verlag, Berlin.

Pati, P.C. and Bhunya, S.P. 1989. Cytogenetic effects of fenvalerate in mammalian in vivo test system. *Mutat. Res.* **222** : 149-154.

Phillips, D.H., Hewer, A., Martin, C.N., Garner, R.C. King, M.M. 1988. Correlation of DNA adduct levels in human lung with cigarette smoking. Nature. **336** : 790-792.

Preussmann, R. and Stewart, B.W. 1984. N-Nitroso carcinogenesis. pp. 643-828. In : Chemical Carcinogenesis Vol. 2 (ed. C.E. Searles). American Chemical Society, Washington.

Price, P.J., Suk, W.A., Freeman, A.E., Lane, W.T., Peters, R.L., Vernon, M.L. and Huebner, R.J. 1978. In vitro and in vivo indications of the carcinogenicity and toxicity of food dye. *Int. J. Cancer.* **21** : 361-367.

Rajah, T.T., Chary, S.I. and Ahuja, Y.R. 1990. A study of chromosome aberrations and micronuclei in workers exposed to lead in a printing press. *Ad. Bios.* **9** : 69-73.

Ray, P.K. 1986. Environmental pollution and cancer. *J. Sci. Indus. Res.* **45** : 370-371.

Ristow, H. and Obe, G. 1978. Acetaldehyde induces cross-link in DNA and causes sister chromatic exchanges in human cells. *Mutat. Res.* **58** : 115-119.

Rithidech, K., Chem, B.T., Mauderly, J.L., Whorton, E.B. Jr. and Brooks, A.L. 1989. Cytogenetic effects of cigarette smoke on pulmonary alveolar macrophages of the rat. *Environ. Mol. Mutagen.* **14** : 27-33.

Sarto, F., Finotto, S., Giacomelli, L., Mazzotti, D., Tomania, R. and Levis, A.G. 1987. The micronucleus assay in exfoliated cells of the human buccal mucosa. *Mutagenesis.* **2** : 11-17.

Shirname-More, L. 1991. Forward mutation of *S. typhimurium* by smokeless tobacco extracts. *Mutat. Res.* **259** : 37-42.

Shirname, L.P., Menon, M.M. and Bhide, S.V. 1984. Mutagenicity of betel quid and its ingredients using mammalian test systems. *Carcinogenesis,* **5** : 501-503.

Smoking and Health, a report of the Surgeon General. 1979. U.S. Department of Health, Education and Welfare. Washington, D.C.

Stich, H.F. and Anders, F. 1989. The involvement of reactive oxygen species in oral cancers of betel quid/ tobacco chewers. *Mutat. Res.* **214** : 47-61.

Stich, H.F., Curtis, J.R. and Parida, B.B. 1982. Application of the micronucleus test to exfoliated cells of high cancer risk groups : tobacco chewers. *Int. J. Cancer.* **30** : 553-559.

Sugimura, T. 1985. Carcinogenicity of mutagenic and teratogenic amines formed during cooking process. *Mutat. Res.* **150** : 33-41.

Sugimura, T., Kondo, S. and Takebe, H. 1982. *Environmental mutagens and cancinogens.* University of Tokyo Press, Tokyo, Alan R. Liss. Inc., New York.

Sugimura, T., Nagao, M. and Wakabayashi, K. 1981. Mutagenic heterocyclic amines in cooked food. pp. 251-267. In : *Environmental Carcinogenesis* (eds. H. Egan, L. Fishbein, *et al.*) Vol. 4, IARC, Lyon.

Sugimura, T. and Sato, S. 1983. Mutagens and Carcinogens in food. *Cancer Res.* **43** : 24155-24215.

Takahashi, M., Fumio, F., Kasuke, N., Yoshitumi, M., Takeshi, K. and Yazo, H. 1983. Long term in vivo carcinogenicity test in fish meat pyrolysate in Syrian golden hamsters. *Gann.* **74** : 633-639.

Takayama, S., Matsuda, M., Mogani, M., Ohgaki, H., Sato, S. and Sugimura, T. 1984a. Induction of cancers in intestine, liver and various other organs of rats by feeding mutagens from glutamic acid pyrolysate. *Gann.* **75** : 205-213.

Takayama, S., Nakatsuru, Y., Matsuda, M., Ohgaki, H., Sato, S. and Sugimura, T. 1984b. Demonstration of carcinogenicity in F 344 rats of 2-amino-3-methyl-imidazo (4,5-1) quinoline from broiled sardine, fried and beef extract. *Gann.* **75** : 467-470.

Tsuda, M. and Kurashima, Y. 1991. Tobacco smoking, chewing, and snuff dipping : factors contributing to the endogenous formation of N-Nitroso compounds. *Toxicology*. **21** : 243-253.

Vijayalaxmi and Evans, H.J. 1982. *In vivo* and *in vitro* effects of cigarette smoke on chromosomal damage and sister-chromatid exchanger in human peripheral blood lymphocytes. *Mutat. Res.* **92** : 321-332.

Weinberg, R.A. 1988. Finding the antioncogenes. *Sci. Amer.* **259** : 44-53.

Wild, D. 1975, Mutagenicity studies on organophosphorus insecticides, *Mutat. Res.* **32** : 133-150.

WHO 1992. Women and tobacco. *Technical Report. World Health Organization,* Geneva. 1-122.

Whong, W.Z., Stewart, J.D., Wang, Y.K. and Ong, T. 1987. Acid-mediated mutagenicity of tobacco snuff : its possible mechanism. *Mutat. Res.* **177** : 241-246.

Wyrobek, A.J., Gordon, L.A., Burkhart, J.G., Francies, M.W., Kapp, R.W. Jr., Letz, G., Malling, H.V., Topham, J.C. and Whorton, M.D. 1983. An evaluation of human sperm as indicators of chemically induced alterations of spermatogenic function. *Mutat. Res.* **115** : 73-148.

Chapter 32

Species Composition and Seasonality in Soil-Surface Arthropod Populations in two Upland Agro-Ecosystems of Nagaland

M. Vikram Reddy[1] AND M. Alemla Ao

[1]*Kakatiya University, Environmental Biology Laboratory, Department of Zoology, Warangal, Andhra Pradesh-506 009, India*

Abstract. The present investigation reports on species composition and seasonal dynamics of surface arthropod populations in upland rice and maize ecosystems in the SASRD campus at Medziphema of Nagaland. The area, previously a Woodland, was slashed and burnt and used for 'Jhum' cultivation. A mean total number of 3791.7 surface dwelling arthropods belonging to 32 different taxa with 41.8 per cent Hymenoptera and 34.1 per cent Collembola were caught in rice ecosystem. In maize ecosystem, a mean total number of 3469 individuals belonging to 35 taxa with 32.68 per cent collembola and 31.1 per cent Hymenoptera were caught. The seasonal variations in the arthropod population were mostly related to climatic and soil factors, response of predators and abundance of prey population. The abiotic factors showed combined effects and influenced the temporal variation in population abundance of different arthropod taxa.

Key words : *Soil arthropods, seasonality, rice, maize, agroecosystems*

Introduction

Arthropods constitute a major component of the biotic communities inhabiting the soil surface. They are responsible for the litter breakdown and decomposition playing an active role in nutrient cycling (Seastedt, 1984), and help in soil aeration and turnover (Reddy, 1984). These arthropods of predator group can have a decisive influence on the structure and functioning of ecosystems (Paine, 1974, and Reise, 1977) by feeding on prey and preventing them from reaching a population size that exceeds the specific carrying capacity. Thus, together with other

[1]**Present address:** Resource Management Program, International Crops Research institute for the Semi-Arid Tropics (ICRISAT), Patancheru-502 324, Andhra Pradesh, India

factors, they contribute to the attainment and maintenance of the ecosystem equilibrium (Odum and Reichholf, 1980). The ground predator beetles may serve as natural control agents of agricultural pests. Several species of Carabidae feed on other sol insect pests (Frank, 1971, Best and Beegle, 1977), and others may aid in weed control through seed eating (Lund and Turpin, 1977). Finlayson and Campbell (1976) studied the Carabid and Staphylinid beetles from agricultural land. Different ecological aspects of the various species of ground beetle have been studied in several agricultural fields such as soybean (House and All, 1981, Price and Shephard, 1980, Ferguson and McPherson, 1985), wheat (Doane, 1981), and Alfalfa (Los and Allen, 1982). The increased interest in the use of pest management systems has emphasized the need for detailed knowledge of the ecology and population dynamics of the soil arthropods to bring them, particularly the pests, under control.

The soil surface dwelling arthropods may also indicate the vegetation and soil characteristics. This is possible, because most of them depend on vegetation and occupy specialized niches and, therefore, provide information about the area according to their presence or absence. Thiele (1977) reported that the Carabid populations have enormous potential as biological indicators through their response to environmental perturbations. Moreover, many investigators have explored on the possibility of soil surface dwelling arthropods as indicators of various factors (Majer, 1977, 1978; Reddy, 1986). Thus it is now realized that a consideration of these invertebrates is of crucial importance in environmental decision making.

Until the middle of the 20th century, even the western entomologists studying the surface dwelling arthropods had to go down on their knees with their pincers ready to pick up every fleeing and crawling insects that came their way. However, this method was time consuming, and the findings gave varying results mostly according to the collectors searching image and endurance, thus making the method unsuitable for quantitative investigation. The introduction of pitfall trap for the study of soil surface arthropods, therefore, came as a blessing to many of the entomologists and ecologists in the early 1930s (cf. Southwood, 1978) which formed the basis for the evolution of arthropod pitfallogy (Waage, 1985). Since then a variety of designs and modifications of pitfall traps had been used by various investigators depending with the choice of the individual investigator (Williams, 1958, Hutson, 1971, Wojeik *et al.*, 1972, Uetz and Unzicker, 1976, Thiele, 1977, Thomas and Sleeper, 1977, Smith, 1976 and Reddy and Venkataiah 1986).

Although, there are certain limitations of pit-fall traps used for studying particularly the quantitative composition of the soil surface inhabiting arthropods as the numbers of a species caught depend upon its abundance and activity, many entomologists and ecologists in the western countries have reported different activities and population densities of the surface active arthropods by using these traps. It has been used for gathering life history and distributional information of the arthropods (Mitchell, 1963, Rivard, 1964, Barlow, 1970, and Goulet, 1974). Nevertheless, it has been showed that continuous pit-fall trapping provides a reliable measure of soil arthropod populations. It was suggested that pitfall traps are most commonly used and are superior to many trapping devices available for studying the soil surface dwelling arthropods. (Thiele, 1977). Huffman and Harding (1980) reported that the method can indicate the location of species within various habitats and thus, record important parameter of the ecosystem. Besides, may author had studied the seasonality of the soil surface inhabiting invertebrates by pitfall traps. (McColl, 1975, Dunn, 1982, Reddy, 1984, Moeed and Meads, 1985).

In India, unfortunately the information on the soil surface inhabiting arthropods is very meager, may be because of the method which is still not known widely and has been used by very few researchers (Reddy, 1984). Oppenheimer and Tikader (1976) studied the seasonal changes in the activity of spiders and harvestman on the ground near Calcutta. Dutta (1983) studied the seasonal activity and density of the ground beetles. Reddy and Venkataiah (1986) reported on the summer abundance of soil surface inhabiting arthropods in different habitats of an urban ecosystem. During the present investigation the area of the woodland which was slashed and burnt and used for "Jhum" cultivation of upland rice and maize in the traditional manner was sampled as the crops grew in the "Jhum" cultivated field, with the help of pit-fall traps. The maize ecosystem was closer to the human habitation, and thus, it was more disturbed compared to the rice ecosystem. Besides, it has been cropped since last four years while the rice ecosystem was cropped since last one year. This report compares the species composition and seasonal population structure (relative density) of soil surface arthropods in relation to abiotic factors in the ecosystems.

Material and Methods

The present investigation was conducted in the SASRD (NEHU) campus at Medziphema (lat. 25°45'N; long. 93°53'E and altitude ca 456 m. MSL), Nagaland (North-East India). Pit-fall traps were used for sampling the soil-surface arthropods, details of which are described in Reddy and Venkataiah (1986). Abiotic factors such as air and soil temperatures were measures by simple thermometers, rainfall by rain guaze and the relative humidity by hygrometer. The soil moisture was measured gravimetrically. The other physico-Chemical factors such as Organic Carbon, pH, available phosphorus and potassium were measured as per method described in Jackson (1967). The numerical abundance of different taxa of soil surface arthropods were correlated with the abiotic environmental factors such as air temperature, rainfall, soil surface relative humidity, soil temperature, soil moisture, organic carbon, pH, available phosphorus and potassium being analyzed by step-down multiple regression analyses with the help of a personal computing system.

Results

Upland rice ecosystem

Qualitative and quantitative composition. A mean total number of 3791.7 surface dwelling arthropods belonging to 32 different taxa were caught from the upland rice system. The Formicidae (Hymenoptera) was caught in maximum number constituting 41.8 per cent of the total arthropods (Fig. 32.1) and being represented by ten species such as *Pheidole* sp. being (33.4%) followed by *Anoplolepis longipes* (2.5%), *Paratrechina* sp. (2.1%), *Meranoplus* sp. (1.62%), *Odontopnera transverse* (0.9%), *Camponotus* sp. (0.6%) and the other such as *Pachycondyla* sp., *Leptogenys* sp., *Anderomorpha* sp. and *Pheidologeton* sp. representing 0.4 per cent, 0.1 per cent and 0.1 per cent of the total arthropods, respectively. *Scelio* sp. (Scellonidae) comprised of 0.1 per cent of the total arthropods.

Collembola, the next dominant taxa, comprised of 34.1 per cent with Hypogastruridae constituting 20 per cent, Entomobryidae being mostly represented by *Dicranocentroides* cf. *salmoni* constituting 5.5 per cent, Isotomidae being mostly represented by *Isotoma trispinata* (MacGillivray) constituting 4.3 per cent, Onychiuridae constituting 4 per cent and Sminthuridae (*Sminthirus* sp) constituting 0.2 per cent of the total arthropods (Fig. 32.1). Acarina comprised of 9.1 per cent, with Cryptostigmata representing in maximum number (6.7%) followed by Prostigmata (1.4%), Mesostigmata (0.5%) and Astigmata (0.4%). Araneae, comprised of 1.1 per cent with Lycosidae being represented by *Trochosa* sp. and *Paradosa* sp. constituting 0.7 per cent, Gnaphosidae being represented by *Zelotes* sp. constituting 0.2 per cent and Clubionidae being represented by *Micaria* sp. accounting 0.1 per cent of the total arthropod (Fig. 32.1). Orthoptera comprised of 6.2 per cent with Gryllidae representing 6.0 per cent and Acrididae representing 0.2 per cent. Blattidae (Dictioptera) was represented by 1.0 per cent of the total arthropods. Homoptera constituted 0.4 per cent with Aphididae being represented by *Rhopalosiphum rufiabdominalis* comprising of 0.1 per cent and Lygaeidae being represented by *Oxycerenus* sp. and Horridipamera *neitneri* comprising 0.8 per cent of the total population (Fig. 32.1).

Coleoptera constituted 4.7 per cent of the total arthropods of which Scarabaeidae comprised of 2.8 per cent and being represented by *Onthophagus vaulogeri* (2.0%) and the other *Onthophagus* sp. (0.8%), Staphylinidae comprised of 0.6 per cent being represented by *Coenonica* sp. constituting 0.5 per cent and *Philonthus aenepennis* constituting 0.1 per cent. Elateridae comprised of 0.2 per cent being represented by *Agrypnus* sp. constituting 0.1 per cent and *Megapenthes* sp. constituting < 0.1 per cent. Carabidae comprised of 0.1 per cent being represented by *Scarites indus* and *Mastax laeviceps*, Chrysomelidae represented by *Chaetocnema basalis* constituted 0.1 per cent and *Monolepta signata* constituted < 0.1 per cent. Histeridae represented by *Neosantalus latitibius* constituted 0.1 per cent whereas Scolitidae being represented by *Coccotrypes carpophagus* constituted < 0.1 per cent, Curculionidae being represented by *Trochorhophalus sacchari* comprised of 0.04 per cent, Coccenilidae comprised of 0.6 per cent and the coleopterean larvae were represented by 0.2 per cent of the total arthropods (Fig. 32.1). The remaining arthropods belonging is Isoptera being represented by *Odontotermes* sp., Diptera, Lepidotera larvae, Dermaptera, Diplopods being represented by *Polydesmida* sp. Chilopoda being represented by *Lamyctes* sp. and Isopoda were represented by 0.8 per cent, 0.3 per cent, 0.1 per cent, 0.2 per cent, 0.1 per cent, and 0.2 per cent of the total arthropod, respectively (Fig. 32.1).

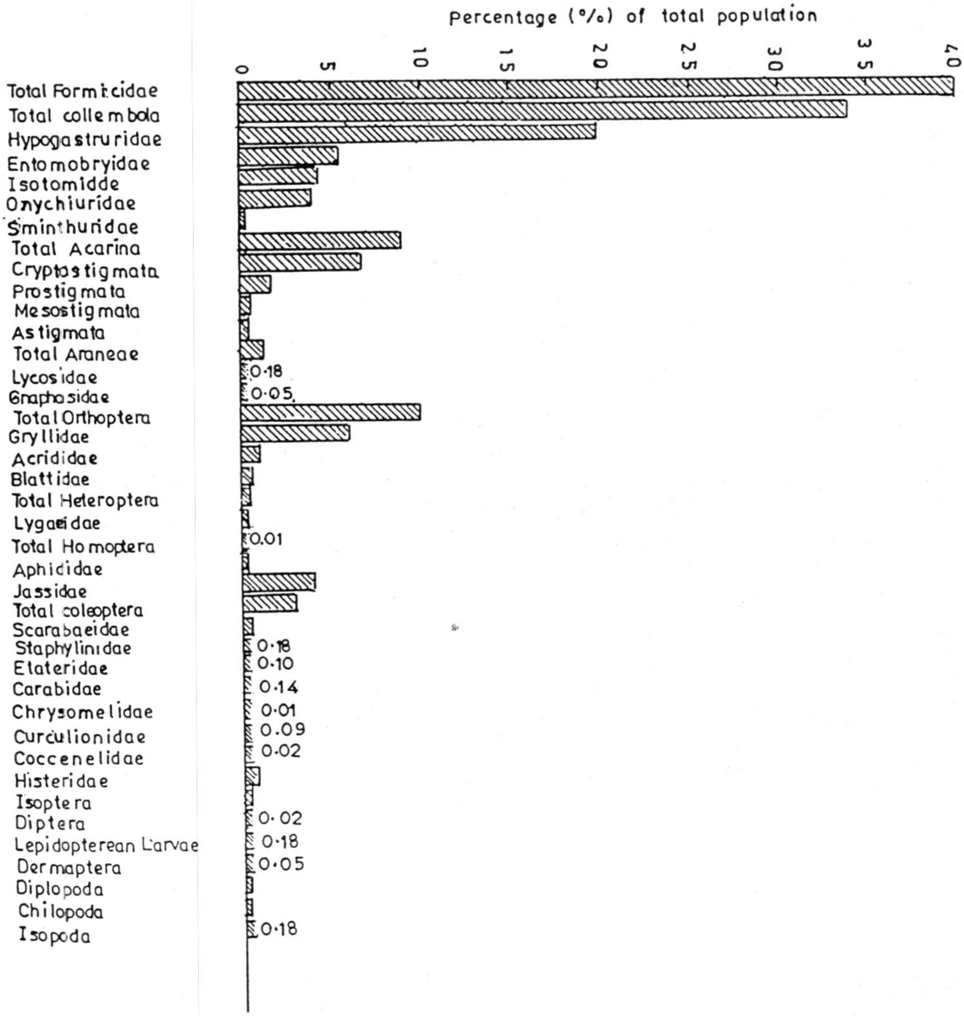

Fig. 32.1. Percentage composition of different taxa of soil surface arthropods inhabiting the soil-surface of the 'jhum' cultivated upland rice ecosystem.

Seasonality

Hymenoptera (Formicidae). Total Formicidae gradually decreased in abundance from November 1982 when it was higher in density and reached a low ebb during the second fortnight of April 1983 (Fig. 32.2). It increased to small peak of abundance during October and November, and again decreased in number in December 1983. However, it slowly increased and reached a small peak of increase in April 1984 *Pheidole* sp. was dominant throughout being recorded in higher number during the last fortnight of November 1982, followed by a smaller peak of increased during second fortnight of March 1983, while the number was low during the last fortnight of February 1984 (Fig. 32.2). *A. longipes*, the second dominant species, was higher in number during the last fortnight of December 1982, and was with a small peak of increase during first fortnight of February 1983. It decreased in number during the first fortnight of March 1983 (Fig. 32.2). *Paratrechina* sp. was low in number during November 1982, increased in December and decreased towards the end of the month. Its number reached a small peak of abundance during the second fortnight of March 1983, and decreased in April. It was absent till June 1983 when it reappeared, and reached the highest density in April 1984 (Fig. 2). *Meranoplus* sp. was irregular in its fluctuation. However, it was low in number in November 1982 and reached a small peak of increase in March 1983. It increased to higher number during 1983 and April 1984 (Fig. 32.2). *O. transverse* showed its peak of abundance during the first fortnight of December 1982, followed by a smaller peak of increase during second fortnight of May 1983. It was absent during July and August 1983 followed by irregular fluctuation (Fig. 32.2). *Camponotus* sp. was irregular in its fluctuation. However it was in higher number during the first fortnight of January 1984, and reduced in number during a number of times (Fig. 32.2). The other species viz., *Pachycondyla* sp. *Leptogenys* sp., *Pheidologeton* sp. and *Anderomorpha* sp., and *Scelio* sp. (*Scellonidae*) were caught in negligible number and were too few to show any distinct seasonal fluctuation.

Collembola. The Collembola was in low number in November 1982 which gradually increased reaching the peak during June 1983. It suddenly increased to a higher number during April 1984 (Fig. 32.3). The population densities of Hypogastruridae was maximum in number during the second fortnight of June 1983 followed by a smaller peak of increase during the first fortnight of February 1984 and a larger peak of increase during April 1984. Its density decreased during many times and was absent during January to first fortnight of April 1983, August to first fortnight of October 1983 and second fortnight of November to first fortnight of January 1984 (Fig. 32.3). The population density of Entomobryidae was irregular reaching highest number during the last fortnight of July 1983 followed by smaller peaks of abundance during the first fortnight of September and November 1983. It density was low during the second fortnight of August and November 1983 and during the first fortnight of January 1984 (Fig. 32.3). The population density of Isotomidae dwindled from November 1982 reaching higher in number during second fortnight of March 1983. It decreased to a smaller peak of increase in November 1983 and dwindled again. It was in low number many time and was absent during the first fortnight of December 1982, 17th May to 16th June 1983, 16th July to 15th August, first fortnight of September and 14th November to 14th December 1983 (Fig. 32.3). The population oscillation of Onychiuridae was irregular. Its

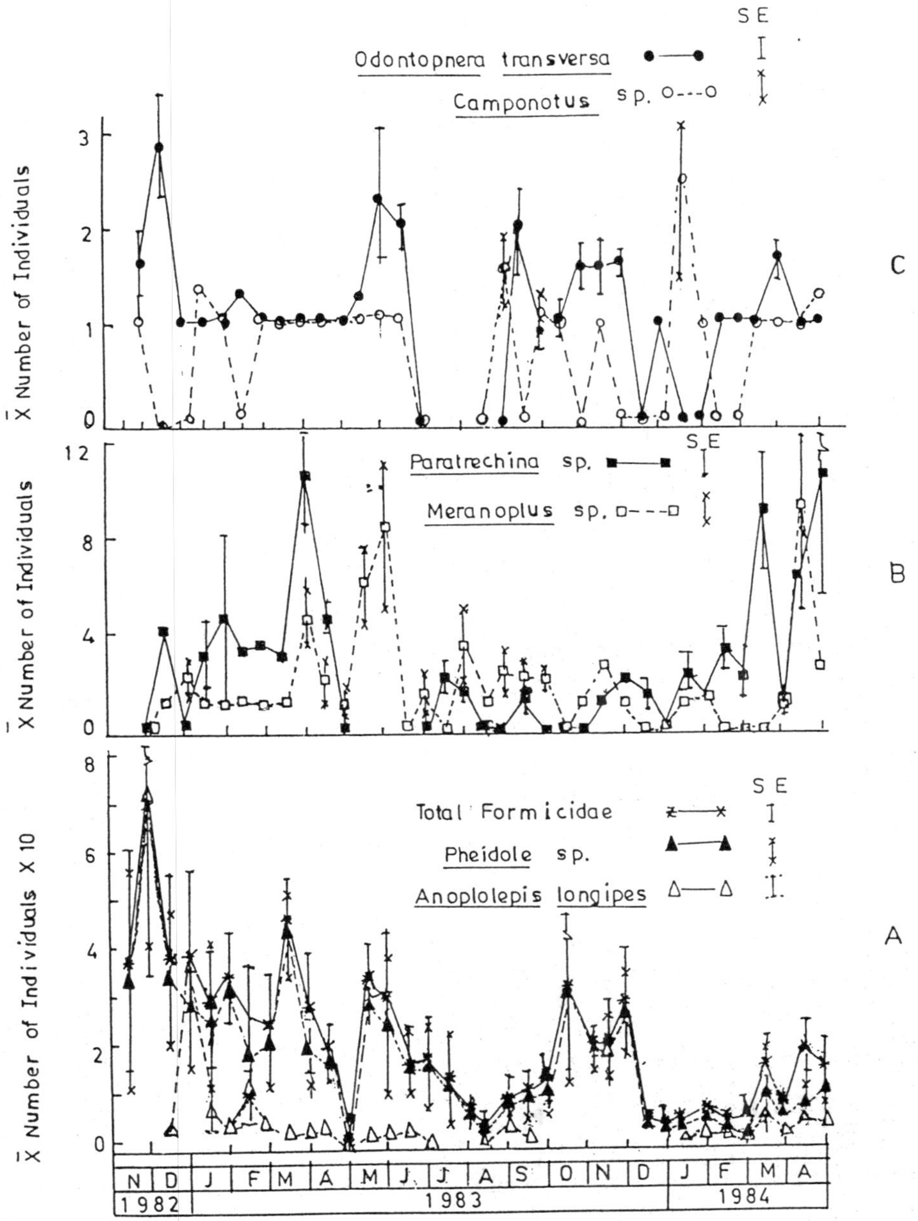

Fig. 32.2. Seasonal fluctuation in the relative population abundance of total Formicidae, *Pheidole* sp. and *Anoplolepis longipes* (A), *Paratrechina* sp. and *Meranoplus* sp, (B) and *Odontopnera transversa* and *Camponotus* sp. (C) inhabiting the soil surface of the 'Jhum' cultivated upland rice ecosystem.

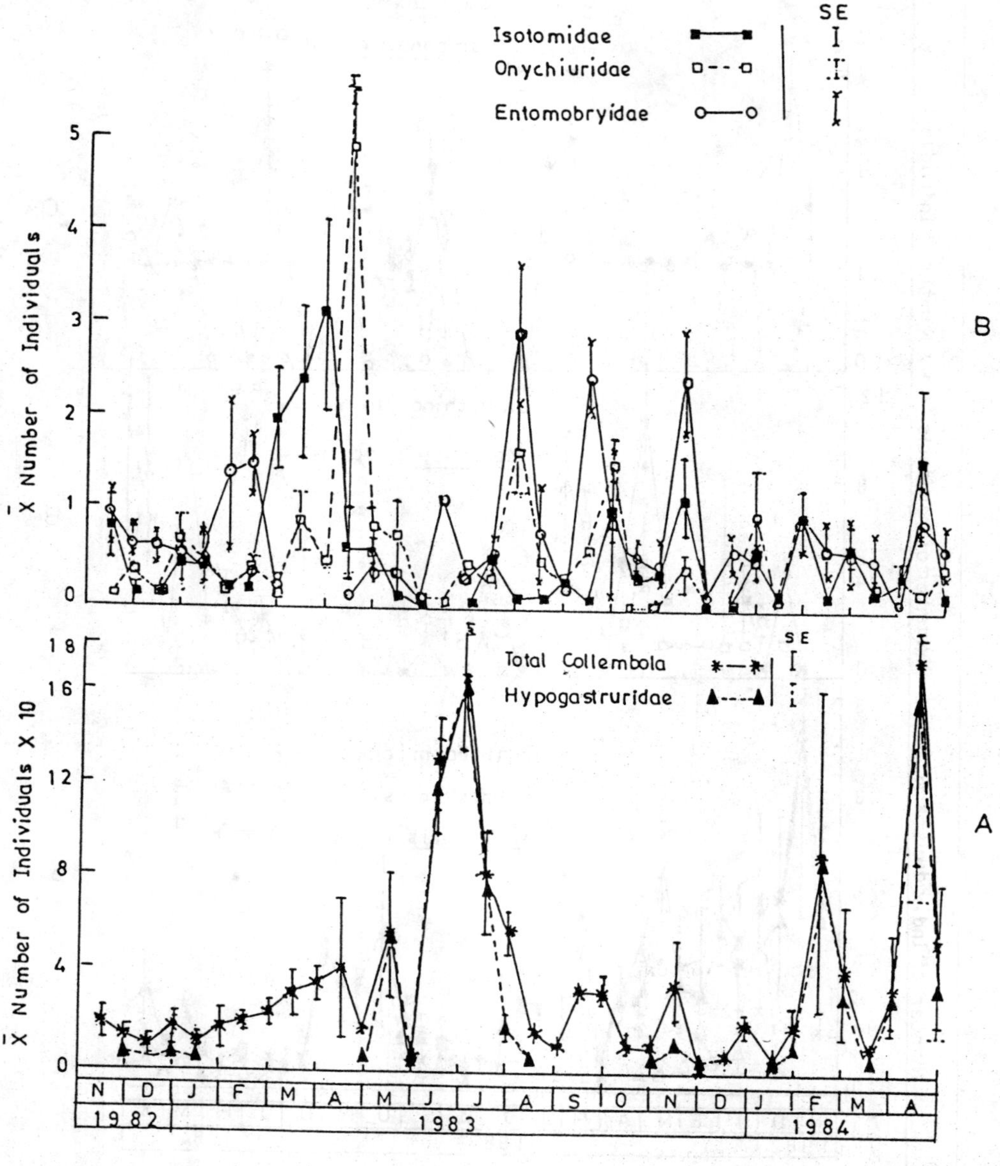

Fig. 32.3. Seasonal fluctuation in the relative population abundance of total Collembola and Hypogastruridae (A) and Isotomidae, Oncychiuridae and Entomobryidae (B) inhabiting the soil surface of the 'jhum' cultivated upland rice ecosystem.

density dwindled since November 1982 and increased to the highest peak of abundance during April 1983. It decreased to a low number during May followed by small peaks of increase in July and September 1983. Afterwards, the population dwindled again, and was absent during the first fortnight of November 1982 and first fortnight of December 1982, 17th May to 16th June 1983, October, 14th November to 14th December 1983 and January to first fortnight of March 1984 (Fig. 32.3). Sminthuridae was recorded seldom, only during first fortnights of August 1983 and April 1984 in negligible number.

Acarina. The population density increased to a higher peak of abundance during March followed by April 1983, and suddenly decreased to a low ebb during the following month (Fig. 32.4). It again increased to a smaller peak of abundance during May 1983, then, decreased and dwindled. It increased to a larger peak of abundance during April 1984. It was recorded minimum in number during second fortnights of November and December 1982 and February 1984. Cryptostigmata being the most abundant taxa among Acarina reflected the seasonal fluctuation similar to that of total Acarina (Fig. 32.4). The Prostigmata was recorded minimum in number during November-December 1982 which gradually reached the maximum during the second fortnight of March 1983 and suddenly decreased to a minimum during may 1983. They were absent during 2nd May to 16th June 1983 and reappeared in June 1983 and dwindled (Fig. 32.4). The astigmatid and Mesostigmatid mites were less abundant being recorded seldom, and in negligible numbers (Fig. 32.4).

Orthoptera. The total Orthoptera were recorded maximum during second fortnight of November 1982 which suddenly decreased in December 1982 (Fig. 32.5). With an increase during January 1983, their density decreased to a low ebb in April, and gradually increased reaching a smaller peak of abundance during July and gradually decreased to the minimum in August. They were absent during September and reappeared in October 1983 when they reached a larger peak of increase, then their abundance decreased and dwindled. Although the adult's of Gryllidae were recorded during November 1982 to January 1983 and April to November 1983 reaching the maximum during the second fortnight of June 1983, the Juvenile were recorded through out the study period except in September 1983 (Fig. 32.5). They were recorded in maximum number during the first fortnight of November 1982, followed by smaller peaks of increase during the first fortnight of January and second fortnight of October 1983, whereas they were minimum in number many time (Fig. 32.5). The Orthopterans belonging to Acrididae were recorded too few in number to show any distinct seasonal fluctuation. The seasonal fluctuation in the relative population abundance of Blattidae (Dictyoptera) was irregular. However, their number increased to a small peak of abundance in November 1982 followed by a larger peak of increase reaching the maximum un February 1983. They were minimum in number throughout the remaining period of study except in June, August to September and December 1983 to February 1984 when they were absent (Fig. 32.5).

Coleoptera. Coleoptera was absent during December 1982 and January 1983. Their density increased reaching a smaller peak of abundance during the second fortnight of May 1983 followed by a larger peak of increase reaching the maximum during second fortnight of August. Their density decreased to minimum during October 1983 and dwindled (Fig. 32.6). *O. vaulogeri*

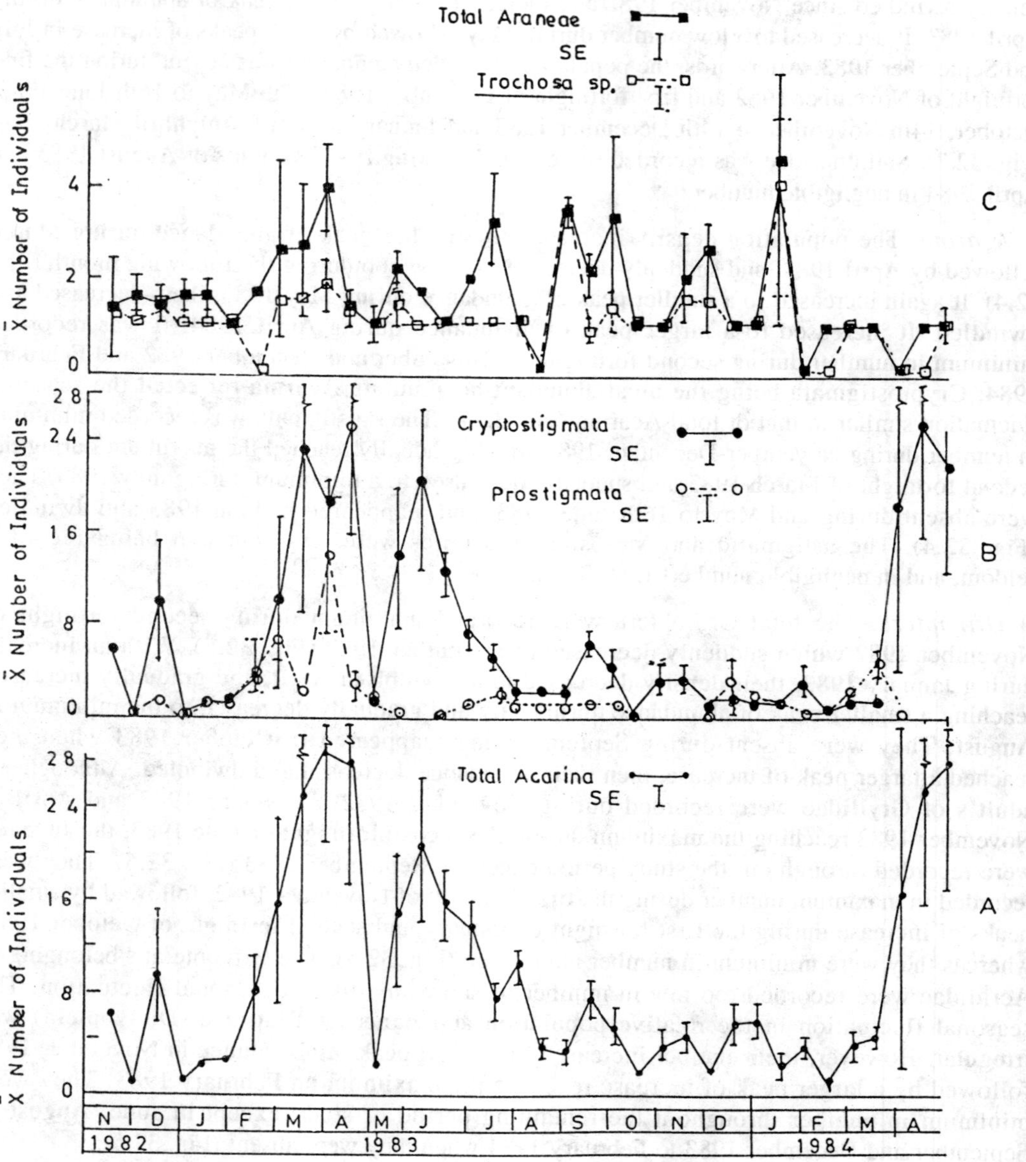

Fig. 32.4. Seasonal fluctuation in the relative population abundance of total Acarina (A), Cryptostigmata and Prostigmata (B), and total Araneae and *Trochosa* sp. (C) inhabiting the soil surface of the 'jhum' cultivated upland rice ecosystem.

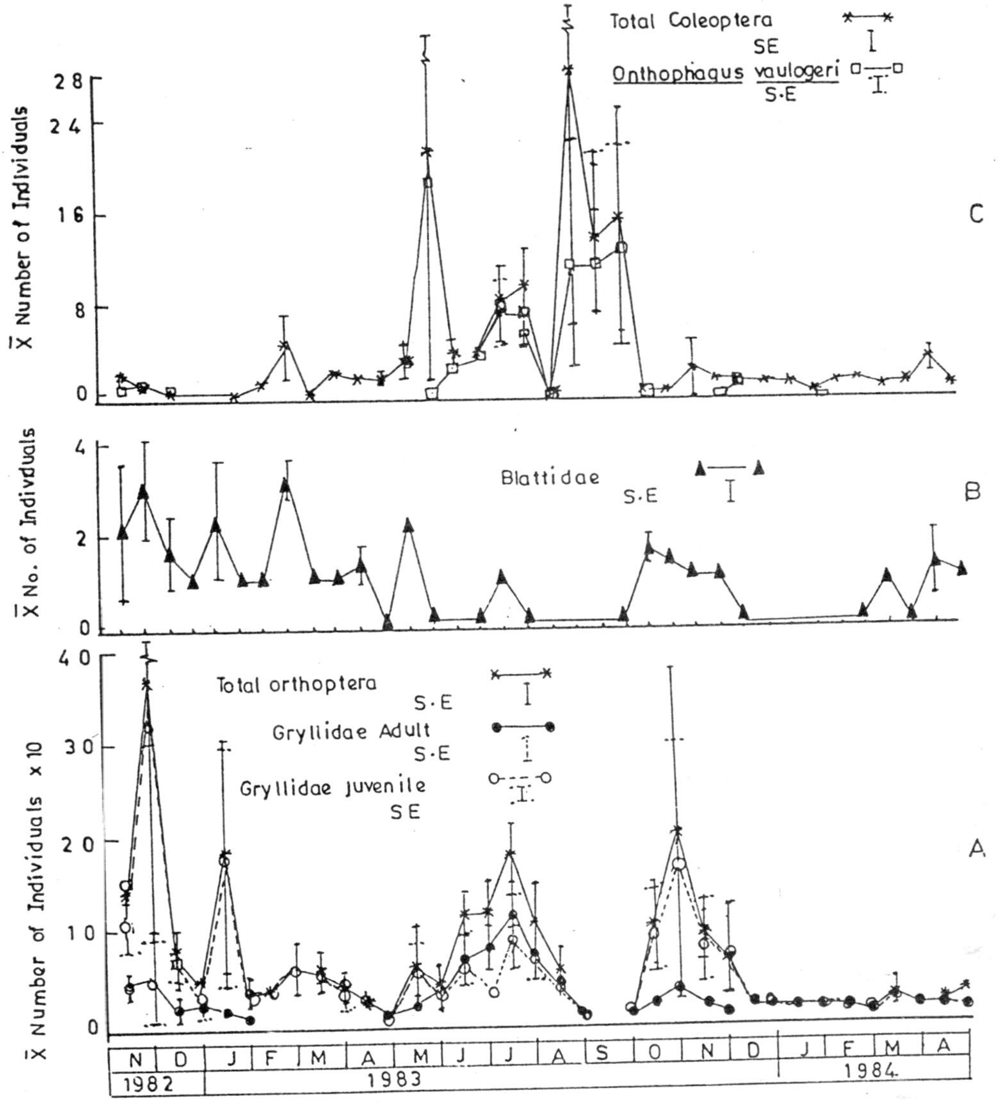

Fig. 32.5. Seasonal fluctuation in the relative population abundance of total Orthoptera and Gryllidae (A), Blattidae (B) and total Coleoptera and *Onthophagus* vaulogeri (C) inhabiting the soil surface of the 'jhum' cultivated upland rice ecosystem.

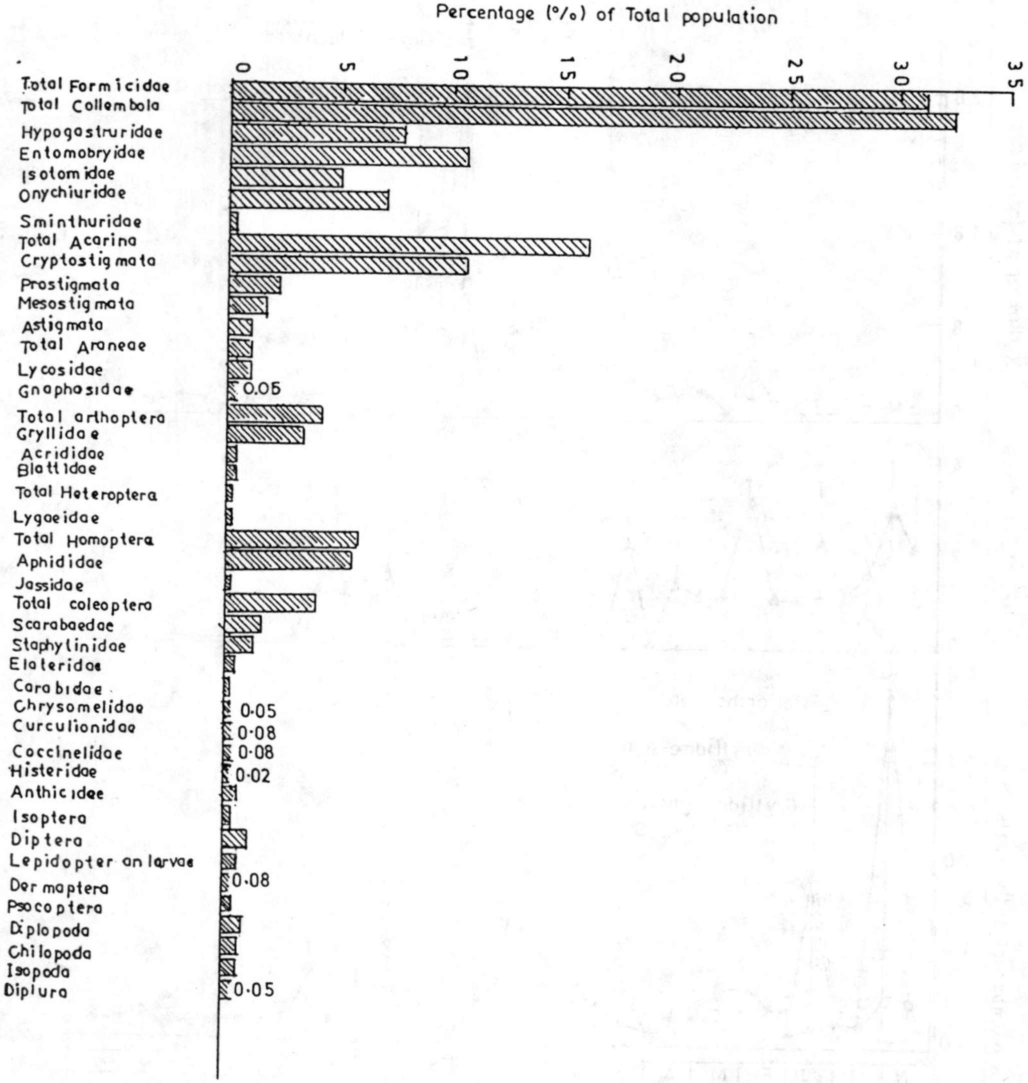

Fig. 32.6. Percentage composition of different taxa of soil surface arthropods inhabiting the soil-surface of the 'jhum' cultivated maize ecosystem.

was caught in minimum number in November 1982 and recorded highest in number during the second fortnight of May 1983, followed by a smaller peak of increase during the second fortnight of September 1983 (Fig. 32.6). The seasonal abundance of other species belonging to different families of Coleoptera were recorded irregularly and were too few in number to show any distinct seasonal fluctuation. The seasonal relative abundance of different arthropod belonging to Isoptera, Homoptera, Heteroptera, Diplopoda, Chilopoda, Diptera, Lepidopteran larvae, Dermaptera and Isopoda were numerically too low to show any distinct seasonality.

Maize ecosystem

Qualitative and quantitative composition. A mean total number of 3469 individuals in surface dwelling arthropods belonging to 35 different taxa were caught from the maize ecosystem. It showed that Collembola was recorded in maximum number constituting 32.68 per cent of the total arthropods and comprising of Entomobryidae being mostly represented by *D. salmon* (10.6%), Isotomidae being mostly represented by *I. trispinata* accounted for 7.2 per cent, Hypogastruridae accounted for 9.8 per cent, Onychiuridae represented for 5.1 per cent and Sminthuridae (*Sminthurus* sp.) represented for 0.2 per cent (Fig. 32.6). Formicidae (Hymenoptera) was the next dominant taxa comprising 31.1 per cent of the total arthropods. It was represented by eleven species, of which *Pheidole* sp. comprised of 17.0 per cent followed by *A. longipes* (8.5%), *Meranoplus* sp. (1.8%), *O. transverse* (1.3%), *Paratrechina* sp. (1.0%), *Pheidologeton* sp. (0.6%), *Camponotus* sp. (0.5), *Pachycondyla* sp. (0.46%), *Anderomorpha* sp. (0.3%), *Dorylus* sp. and *Leptogenys* sp. each comprising of 0.1 per cent and *Scelio* sp. (Scellonidae) comprising of 0.1 per cent of the total arthropods (Fig. 32.6). Acarina in total, comprised of 16 per cent with Cryptostigmata being numerically abundant constituting 10.7 per cent followed by Proștigmata (2.3%), Mesostigmata (1.8%) and Astigmata (1.2%). Araneae in total, comprised of 1.1 per cent with Lycosidae represented by *Trochosa* sp. constituting 1.17 per cent and Gnaphosidae being represented by *Zelotes* sp. constituting < 0.1 per cent of the total arthropods.

Orthoptera comprised of 4.2 per cent with Gryllidae representing 3.4 per cent and Acridiade representing 0.8 per cent of the total arthropod. Blattidae (Dictyoptera) was represented by 0.3 per cent and Lygaeidae (Heteroptera) being represented by *H. neitneri* constituted 0.2 per cent of the total arthropods. Homoptera, in total, constituted 5.8 per cent with Aphididae being represented by *R. rufiabdominalis* constituting 5.7 per cent and Jassidae Constituting 0.1 per cent of the total arthropods. Coleoptera in total constituted 4.3 per cent of the total arthropods, of which Scarabaeide comprised of 1.4 per cent being represented by *O. vaulogeri* constituting 0.2 per cent and the other *Onthophagus* sp. constituting 1.2 per cent, Staphylinidae comprised of 1.3 per cent of which *Coenonica* sp. constituted 1.1 per cent, *P. aenepennis* constituted 0.1 per cent and *Carpelimus* sp. constituted 0.1 per cent. Elateridae being represented by *Agrypnus* sp. comprised of 0.4 per cent. Carabidae being represented by *M. laeviceps* comprised of 0.2 per cent. Chrysomelidae comprised of < 0.1 per cent, being represented by *C. basalis* constituting 0.02 per cent and *M. signata* constituting 0.05 per cent. Histeridae being represented by *N.*

latitibius comprised of < 0.1 per cent. Curculionidae being represented by *T. sacchari* constituted 0.1 per cent. Coccinelidae comprised of 0.1 per cent and Anthicidae being represented by *Anthicus monstrosicornis* constituted 0.7 per cent of the total arthropods. The Coleopteran larvae were represented by 0.3 per cent of the total arthropods (Fig. 32.6). The remaining arthropods belonging to Isoptera being represented by *Odontotermes* sp. Diptera, Lepidopteran larvae, Dermaptera, Psocoptera, Diplopoda being represented by *Polydesmida* sp. and *Spirosteptus* sp., Chilopods being represented by *Lamyctes* sp. and *Rhysida* sp., Pseudoscorpions and Isopoda were represented by 0.2 per cent, 1.1 per cent, 0.6 per cent, 0.1 per cent, 0.4 per cent, 0.9 per cent, 0.3 per cent and 0.6 per cent of the total arthropods, respectively (Fig. 32.6).

Seasonality

Collembola. Collembola was at a low ebb during November 1982 and suddenly increased to a larger peak of abundance during the second fortnight of December 1982 (Fig. 32.7). Their relative abundance decreased suddenly during the second fortnight of January 1983, and fluctuated irregular by reaching higher in number during the first fortnight of September 1983 followed by smaller peaks of increase during second fortnights of January and April 1984 (Fig. 32.7). This population abundance of Entomobryidae was very irregular over the season dwindling from November 1982 till August 1983, and suddenly increased to maximum during the first fortnight of September 1983, then, it decreased and dwindled irregularly. They were absent during the first fortnight of November 1982, second fortnight of January 1983, 15th February to 17th March and 2nd April to 17th May 1983. The population abundance of Hypogastruridae was irregular during November 1982 to the first fortnight of February 1983 reaching a small peak during January 1983 and was absent till it increased in May 1983. However, its population reached a smaller peak of increase during second fortnight of January 1984 followed by a larger peak of increase being the maximum during second fortnight of March 1984. They were recorded minimum during the second fortnight of June 1983 and the first fortnight of February 1984 (Fig. 32.7). The population density of Isotomidae dwindled with smaller peaks of increase during the second fortnight of February and first fortnight of April 1983. However, the density reached the maximum suddenly during the first fortnight of November 1983 and then, decreased and dwindled. It was absent during 15th February 1983, July 1983, January 1984 and 12th February to 12th April 1984 (Fig. 32.7). The population abundance of Onychiuridae was irregular (Fig. 32.7). The population density of Sminthuridae was irregular and recorded seldom, and was too few in number to show any seasonality.

Hymenoptera (Formicidae). Formicidae was at low ebb from November 1982 till April 1983, and increased reaching the higher number during the second fortnight of May 1983 and suddenly decreased in August 1983 with a smaller peak of increased in July 1983. Their number dwindled till February 1984 when the density increased reaching a small peak of abundance in March 1984 and then, decreased. Their number was minimum during the second fortnight of August 1983 (Fig. 32.8). *Pheidole* sp. being the dominant depicted a seasonal fluctuation of its population abundance similar to that of the total Formicidae. However, its density was recorded minimum during the second fortnight of December 1982. The density of *A. longipes*

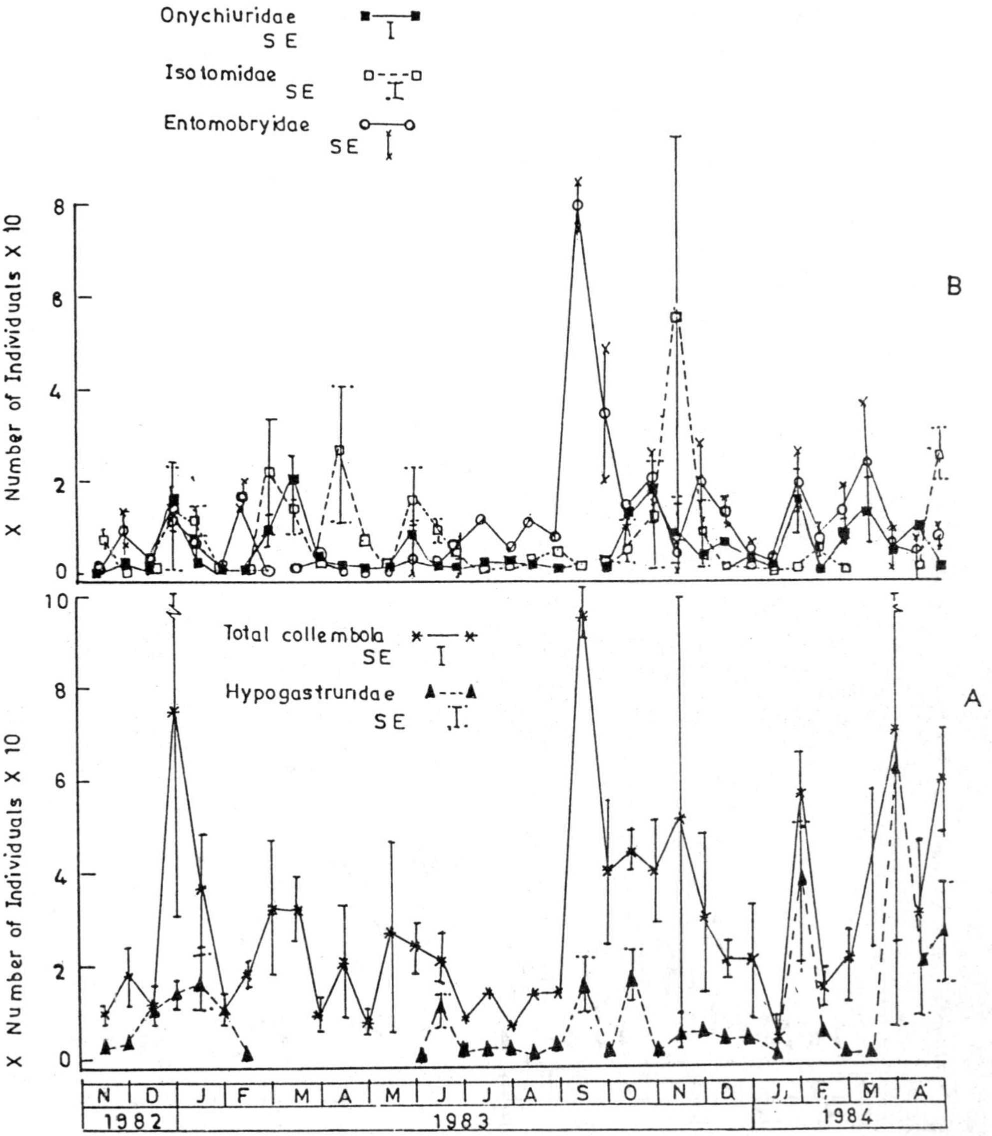

Fig. 32.7. Seasonal fluctuation in the relative population abundance of total Collembola and Hypodastruridae (A), and Onychiuridae, Isotomidae and Entomobryidae (B) inhabiting the soil surface of the 'jhum' cultivated maize ecosystem.

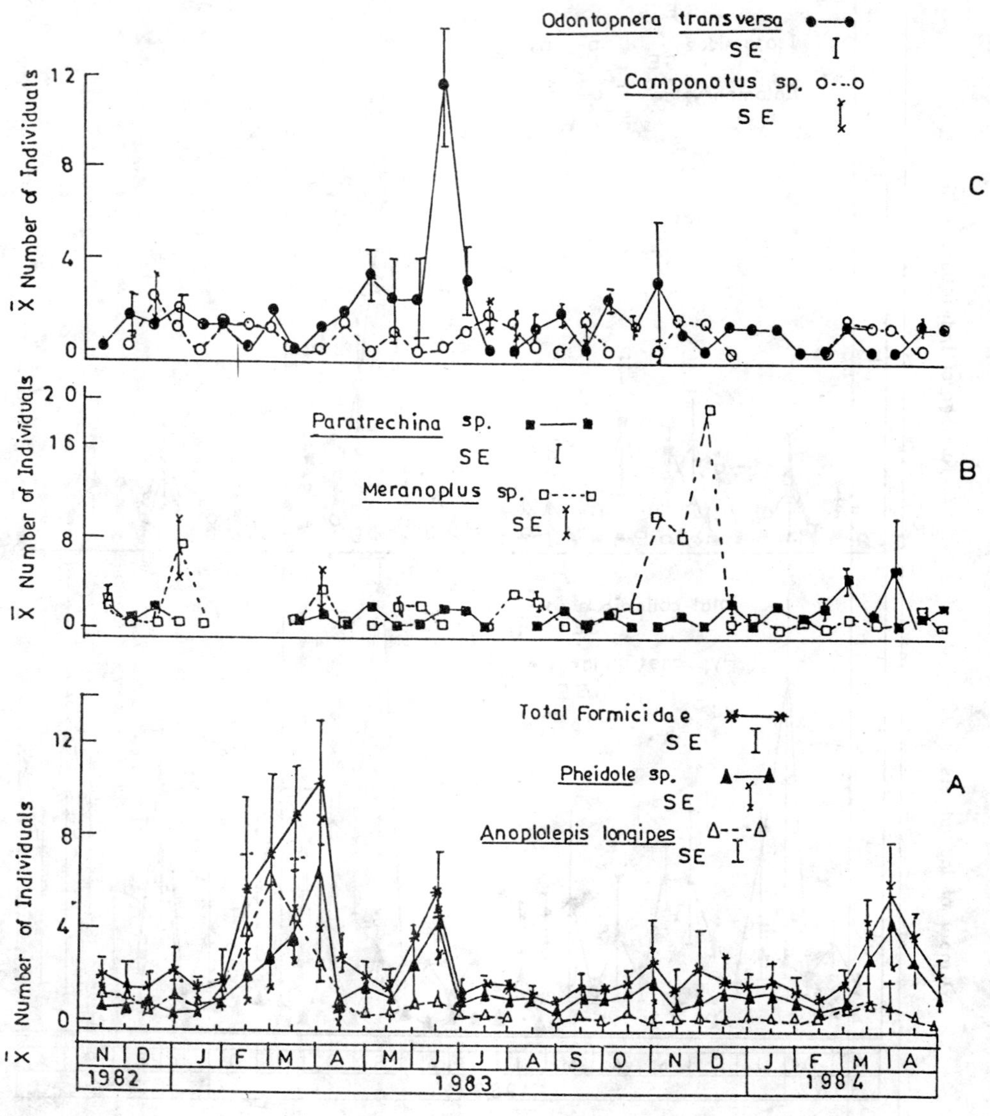

Fig. 32.8. Seasonal fluctuation in the relative population abundance of total Formicidae, *Pheidole* sp. and *Anoplolepis longipes* (A), *Paratrechina* sp. and *Meranoplus* sp. (B) and *Odontopnera transversa* and *Camponotus* sp. (C) inhabiting the soil surface of the 'jhum' cultivated maize ecosystem.

declined from November 1982 to the first fortnight of January 1983, and reached higher number during the second fortnight of February 1983 and gradually decreased followed by a small peak of increase during the first fortnight of June 1983 and during March 1984. However, it was absent during 16th July to 30th August 1983, 14th to 30th September, 15th October to 14th November, the first fortnight of December 1983 and January 1984 and during the second fortnight of April 1984 (Fig. 32.8). The population abundance of *Paratrechina* sp. dwindled being absent during the second fortnight of November 1982, 15th December 1982 to 17th March 1983, first fortnight of April to May, 1st July to 15th August, fire fortnight of September, October, second fortnight of November, first fortnight of December 1983 and the first fortnight of April 1984, However, its density was higher during the second fortnight of March 1984 (Fig. 32.8). *Meranoplus* sp. reached smaller peaks of increase during second fortnight of December 1982 and October 1983 followed by a larger peak of increase being maximum in number during the second fortnight of November 1983. It was absent during 15th November to 15th December 1982, 19th January to 17th March 1983, April, 1st June to 16th July, 15th August to 30th September, first fortnight of December, January and February 1984, March and the second fortnight of April 1984 (Fig. 32.8). The population abundance of *O. transversa* also dwindled through out the study period except during the first fortnight of June 1983 when it reached the maximum. *Camponotus* sp. was represented too few numerically to show any distinct seasonal fluctuation (Fig. 32.8). The seasonal population abundance of other formicidae viz., *Leptogenys* sp., *Pheidologetion* sp., *Anderomorpha* sp. and *Dorylus* sp. and *Scelio* sp. was irregular. They were recorded very less in number and did not show any distinct seasonal fluctuation.

Acarina. Total Acarina population dwindled from November 1982 to February 1983, their density reached a smaller peak of increase during the first fortnight of March 1983 followed by a larger one during April. Their number decreased with a small peak of increase during May and reached a low ebb. Their density reached another smaller peak of increase during the second fortnight of October 1983 and was minimum during the second fortnight of January 1984. It increased suddenly reaching a higher density during April 1984 (Fig. 32.9). Cryptostigmata was the most abundant taxa among Acarina. Its population density showed fluctuation similar to that of total acarina. The population density of Prostigmata fluctuated irregular with a smaller peak of increase during the second fortnight of November followed by a larger peak of increase during the second fortnight to February 1984. Their number was minimum during first fortnights of January and July 1983, and second fortnights of March and April.1984 and was absent during the first fortnight of November 1982, second fortnight of December 1982, second fortnight of October 1983, first fortnight of November 1983 and January 1984 (Fig. 32.9). The population density of Astigmata fluctuated irregularly reaching a higher number during the second fortnight of September 1983, followed by a smaller peak of increase during the second fortnight of October 1983. They were absent from November 1982 to 17th March 1983, 1st January to 12th February 1984 and 13th March to 12th April 1984. The population density of Mesostigmata reached a smaller peak of increase during the second fortnight of October 1983. The number reached a smaller peak of increase during first fortnight of March 1984, and was absent during 1st November 1982 to 15th February 1983, 17th March to

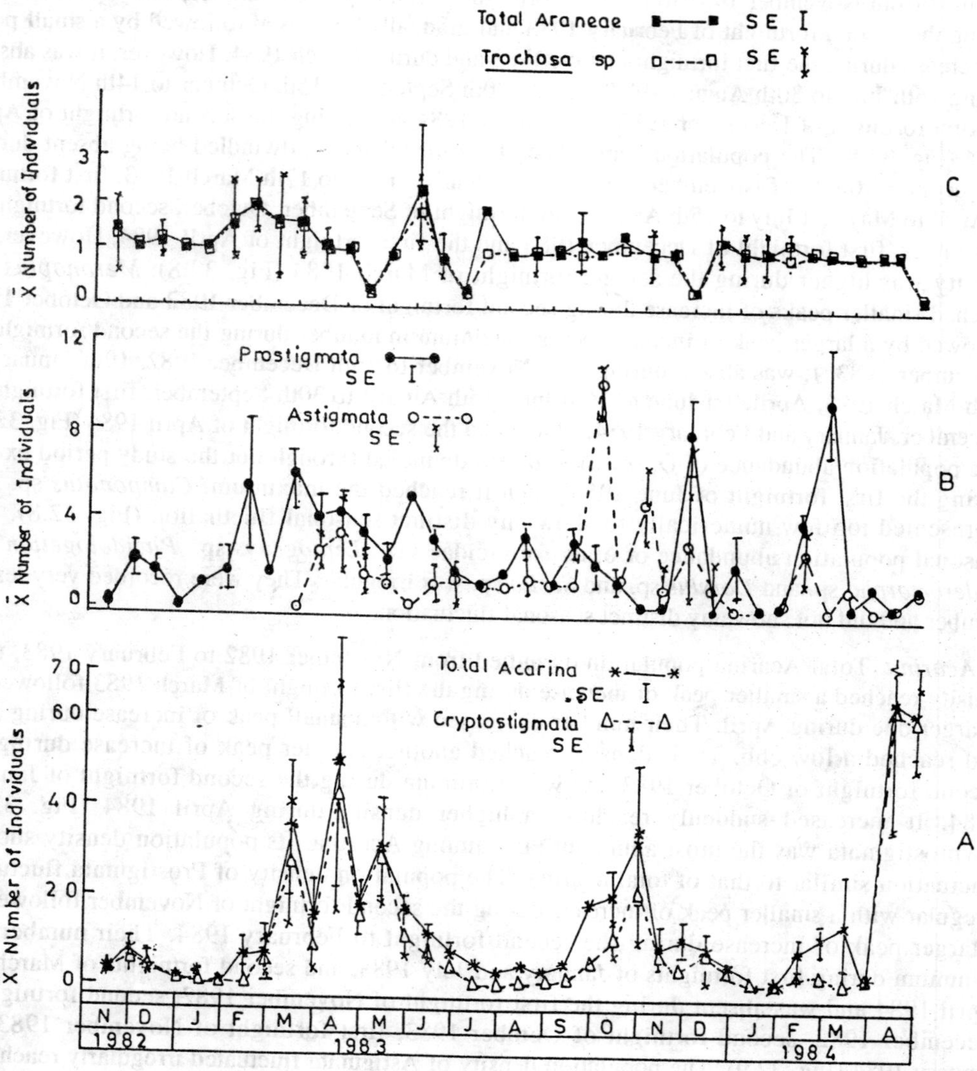

Fig. 32.9. Seasonal fluctuation in the relative population abundance of total Acarina, and Cryptostigmata (A), Prostigmata and Astigmata (B) and total Araneae and *Trochosa* sp. (C) inhabiting the soil surface of the 'jhum' cultivated maize ecosystem.

2nd May, June, 16th July to 14th September, second fortnights of November and December 1983, 12th January to 27th February 1984 and 13th March to 12th April 1984.

The population density of Araneae was irregular in their seasonal fluctuation. However, their number showed a smaller peak of increase during February 1983, followed a larger peak of increase being the maximum during the second fortnight of May. It showed another smaller peak of increase in July 1983 (Fig. 32.9). The *Trochosa* sp. being the dominant in Araneae showed a similar seasonal fluctuation in population as that of the total Araneae. It was absent during the second fortnights of April, June and November 1983 and April 1984. *Zelotes* sp. was represented too few in number to show any distinct seasonal fluctuation.

Orthoptera. Total Orthoptera recorded in a low abundance during November 1982 decreased to a low ebb during December 1982. Their density dwindled reaching the minimum during second fortnight of March 1983, and gradually increased thereafter reaching a higher number during the first fortnight followed by a decrease during the second fortnight of June 1983. It again increased to a small peak of abundance during the second fortnight of October 1983 and then decreased suddenly and dwindled. Gryllidae (adults) population dwindled from November 1982 reaching a higher number during the first fortnight of June 1983 followed by smaller peaks of increase during the second fortnights of July and October 1983. They were recorded in lower number during November and second fortnight of December 1982, second fortnight of April 1983, 16th June to 16th July, August and November 1983 and second fortnight of February 1984. Gryllidae (Juveniles) were recorded in higher maximum during the first fortnight of November 1982, followed by a smaller peak of abundance during first fortnight of July 1983 (Fig. 32.10). The abundance of Acrididae were too few to show any distinct seasonal fluctuation. Seasonal fluctuation of Blattidae was irregular being recorded in higher number during the first fortnight of August 1983 (Fig. 32.10).

Coleoptera. The population densities of Coleoptera dwindled from November 1982 till May, which suddenly decreased during June and dwindled thereafter reaching a smaller peak of increase during the second fortnight of April 1984 (Fig. 32.10). The population abundance of *Coenonica* sp. also dwindled from November 1982 to first fortnight of September 1983. Its density increased reaching the higher number in the second fortnight of October 1983, and suddenly decrease and dwindled. *O. vaulogeri* appeared only from February 10th to July 1983 during which its density was high in the second fortnight of May 1983. The seasonal fluctuation of the population density of other species belonging to different families of Coleoptera and that of the Coleopteran larvae were irregular and were in too less in number of show any distinct fluctuation. Besides, the densities of arthropods belonging to Isoptera, Diptera, Dermaptera, Psocoptera, Diplopoda, Chilopoda, Pseudoscorpionidae, Lepidopteran larvae and Isopoda were numerically very less, and did not show any distinct fluctuation.

Abiotic variables

Rainfall. The rainfall was scanty during November 1982 to January 1983, which gradually increased reaching the peak in July 1983. There was no rainfall in January to February 1984.

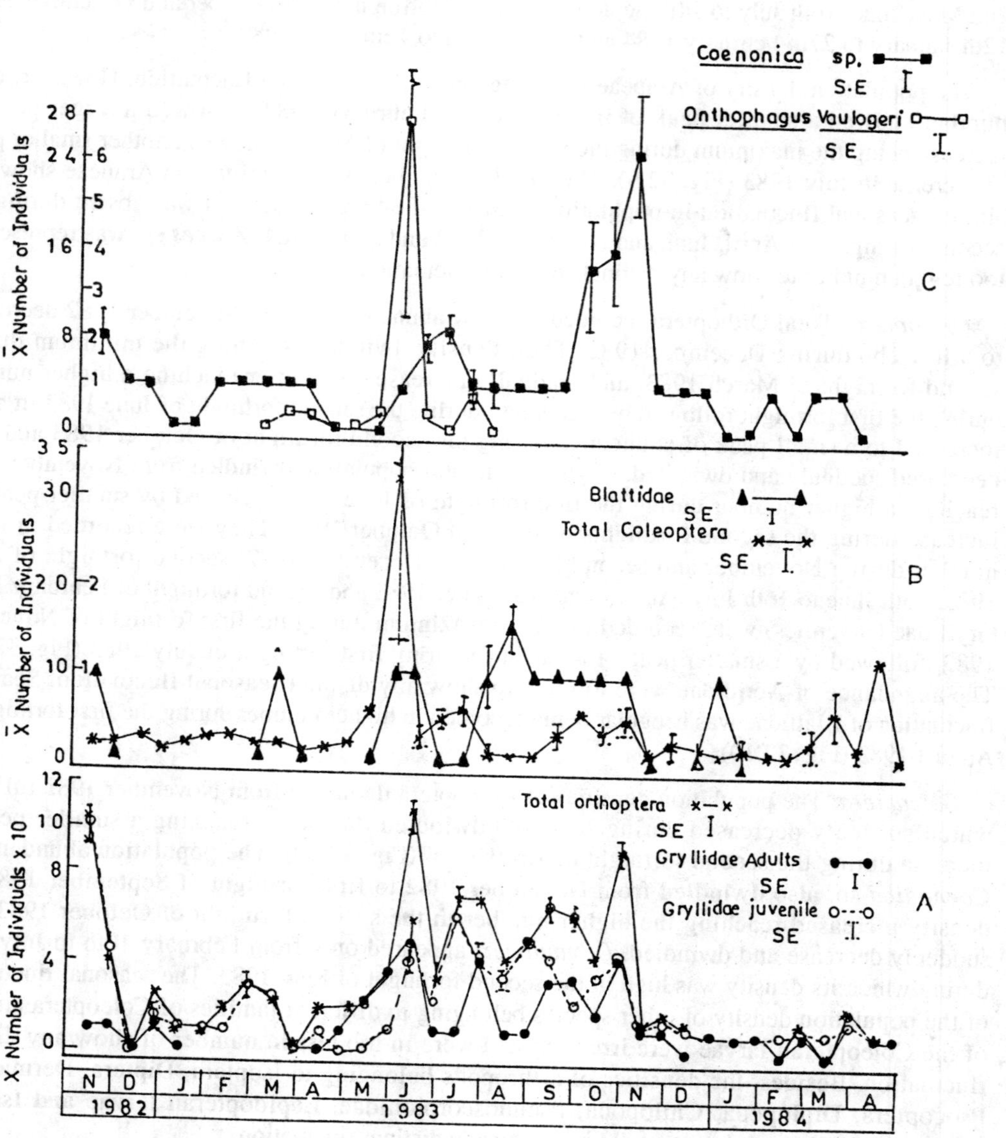

Fig. 32.10. Seasonal fluctuation in the relative population abundance of total Orthoptera and Gryllidae (A), Blattidae, total Coleoptera (B) and *Coenonica* sp. and *Onthophagus vaulogeri* (C) inhabiting the soil surface of the 'jhum' cultivated maize ecosystem.

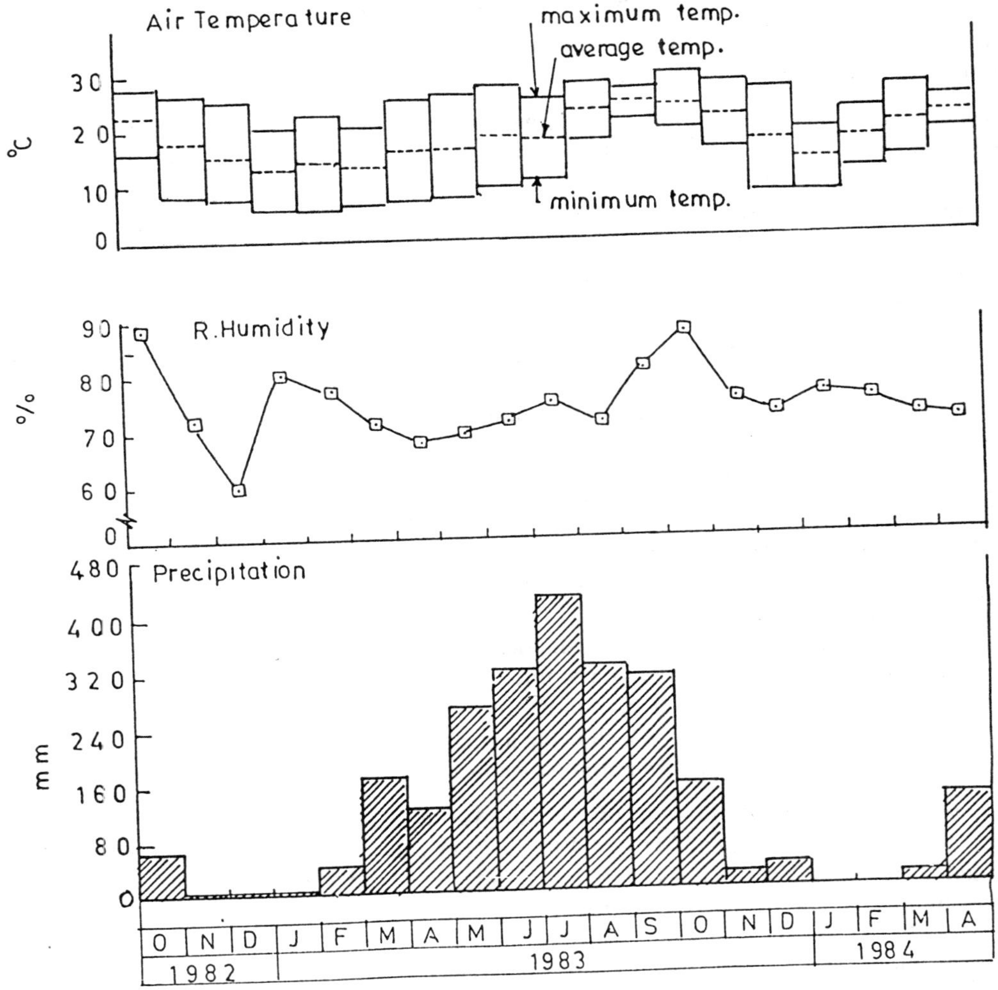

Fig. 32.11. Monthly variation in the total rainfall relative humidity and air temperature (at breast height) at the study site during the study period.

However, it was again recorded in March 1984 which increased in April 1984 (Fig. 32.3). The Number of rain days fluctuated as that of the total precipitation.

Relative Humidity. The seasonal Variation of relative humidity was irregular being recorded lowest during December 1982 and highest during October 1982 and 1983 (Fig. 32.11).

Temperatures. The maximum, minimum and average air temperatures ranged from 19.1 to 30.0°C, 5.1 to 21.7°C, and 13.0 to 25.0°C, respectively (Fig. 32.11). The maximum and average air temperatures were recorded highest during October 1983 whereas the minimum temperature was recorded highest during September 1983. However, the maximum temperature reached the low ebb during January 1984 and the average temperature was recorded lowest during January and March 1983, and the minimum temperature reached the minimum during February 1983.

The soil temperature at 5 cm depth of the rice ecosystem gradually decreased from October 1982 reaching a low ebb in January 1983, and again increased gradually reaching maximum in May 1983. It again decreased reaching the minimum in December 1983 and increased to the maximum in April 1984 (Fig. 32.12). In the maize agroecosystem it ranged from 15.5 to 29.0°C, showing a similar trend of fluctuation as that of rice, the maximum being recorded in August 1983 while minimum during December 1983 (Fig. 32.13).

pH. pH of both the surface and sub-soils of the upland rice (Fig. 32.12) and maize (Fig. 32.13) ecosystems ranged between 5.0 to 6.0 and were acidic its nature. In the rice ecosystem, the pH was recorded maximum in August 1983 and April 1984. In the sub-soil, it was recorded maximum in February 1983, followed by smaller peaks of increase during September 1983 and February 1984 and minimum being recorded during April and August 1983. In maize ecosystem, the pH and recorded maximum in December 1983 at the surface soil while minimum was recorded in April 1983, showing a smaller peak of increase in March 1983. In the sub-soil, the pH was recorded minimum in November 1982 and maximum in January 1984.

Soil moisture. The soil moisture of both the surface and sub-soil of the rice ecosystem ranged from 12 to 42.4 per cent and 14.4 to 39.7 per cent respectively, the maximum being recorded during May 1983 in the surface and during April and May 1983 in the Sub-soil while the minimum was recorded during February 1983 in the surface and during November 1982 in the sub-soil (Fig. 32.12). The soil moisture of both the surface and sub-soil of the maize ecosystem range between 12.3 to 47.0 per cent at surface and 14.7 to 39.0 per cent at the sub-soil being recorded maximum both at the surface and sub-soil during August 1983 and minimum during November 1983 (Fig. 32.13).

Potassium, Phosphorus and Organic Carbon. The monthly variations in available potassium (K_2O), phosphorus (P_2O_5) and organic carbon (OC) of surface and sub-soil of the rice agroecosystem is presented in Fig. 32.12. A perusal of which showed that in rice ecosystem the available K_2O at the surface and sub-soil ranged from 37.3 to 179.0 and 17.2 to 140.0 kg/ha respectively, the minimum being recorded in July 1983 and maximum during February 1984 followed by a smaller peak of increase during January 1983 at the former soil. In the sub-soil, it

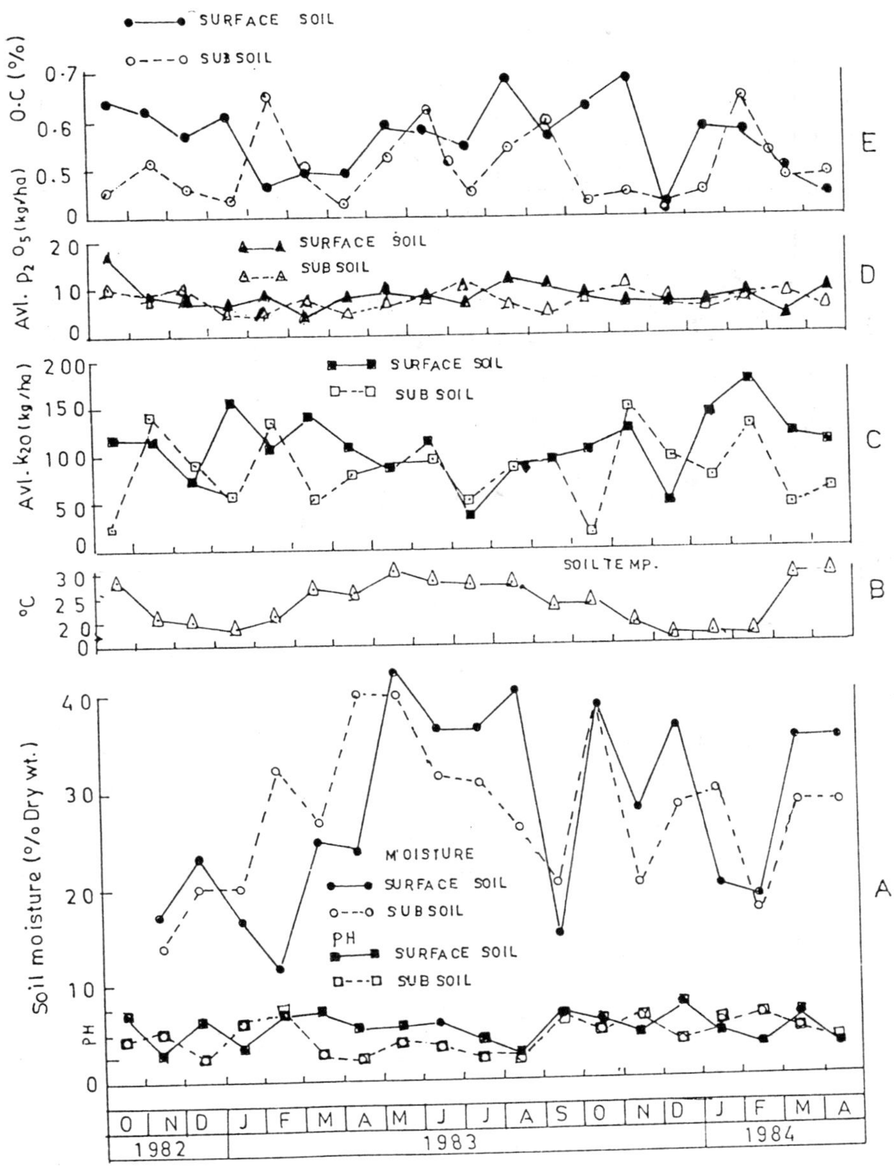

Fig. 32.12. Monthly variation in the soil pH and moisture (A), soil temperature (B), available K_2O (C), available P_2O_5 (D), and OC (E) of both the surface and sub-soil of the upland rice ecosystem during the study period.

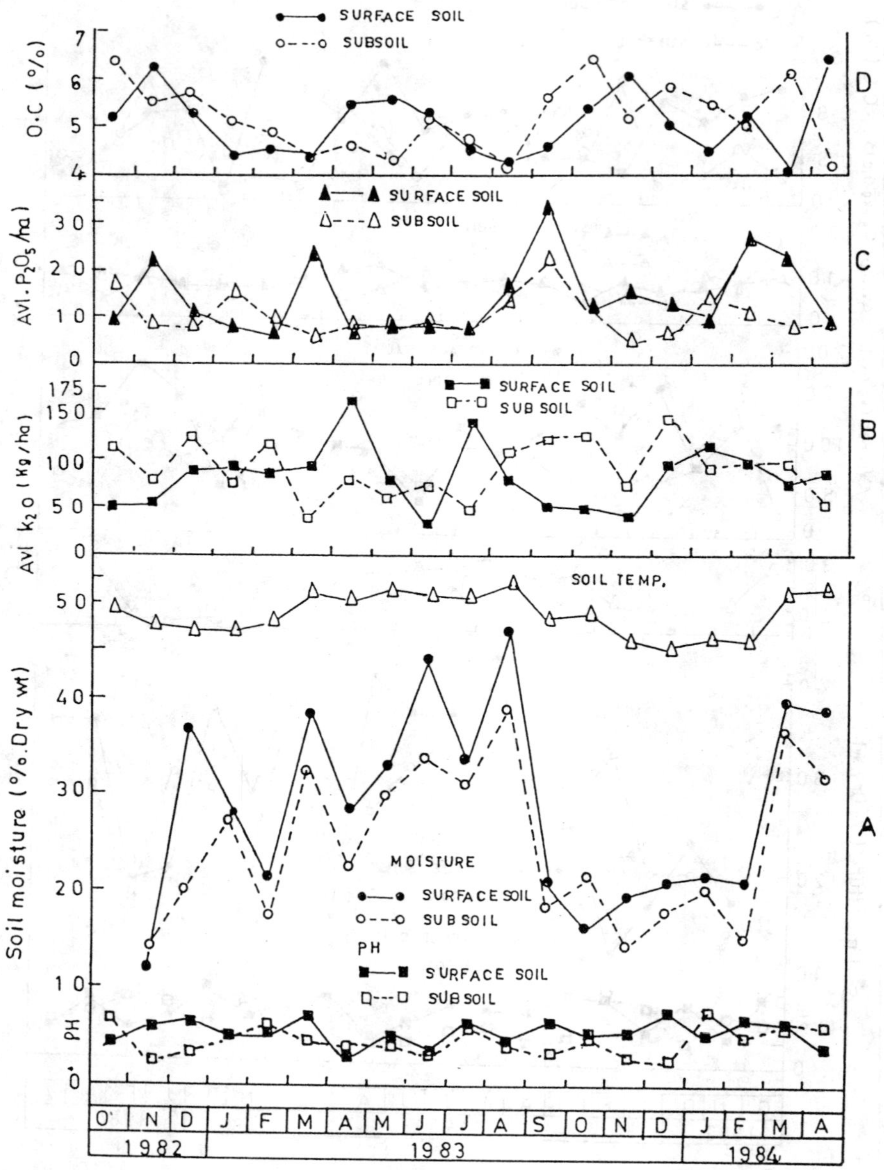

Fig. 32.13. Monthly variation in the soil pH, moisture and temperature (A), available K_2O (B), available P_2O_5 (C), and OC (D) of both the surface and sub-soil of the maize ecosystem during the study period.

was maximum in November 1983 and minimum during October 1983. However it increased in November 1982 (Fig. 32.12). Available P_2O_5 at the surface and sub-soil ranged from 5.0 to 17.5 and 5.5 to 11.5 kg/ha respectively, the maximum being recorded in October 1982 with a second peak of increase in August 1983 and the minimum in March 1984 at the surface soil, while in the sub-soil the maximum was recorded in July 1983 and the minimum in January 1983 (Fig. 32.12). The organic carbon content ranged from 0.42 to 0.65 per cent at both surface and sub-soil respectively (Fig. 3212), the maximum being recorded in August and November 1983 with a small peak of increase in October 1982 and minimum during December 1983 at the former soil. At the sub-soil the maximum percentage of organic carbon was recorded in February 1984 and minimum during October and December 1983, the variation being irregular in nature.

In maize ecosystem, available K_2O, available P_2O_5 and OC of surface soil ranged from 5.3 to 22.0 kg/ha, 6.0 to 34.3 kg/ha and 0.41 to 0.64 per cent, respectively (Fig. 32.13). In the sub-soil, these factors varied from 39.2 to 142 kg/ha, 5.3 to 22.0 kg/ha and 0.42 to 0.64 per cent, respectively. The available K_2O was recorded maximum in April 1983 and minimum in March 1983 and with a small peak of increase in October 1983 and reached the maximum, during December 1983. The available P_2O_5 was recorded maximum, both at the surface and sub-soil, during September 1983 whereas minimum was recorded in February 1983 at the surface soil and during November 1983 at the sub-soil. The organic carbon, in the surface soil, with smaller peaks of increase during November 1982 and 1983 and minimum in March 1984 was recorded maximum during April 1984, while in the sub-soil it was maximum during October 1982 and 1983 with a smaller peak of increase in March 1984, and minimum in August 1983 and April 1984.

Discussion

A mean total number of 3791.7 and 3468.0 belonging to more than 30 taxa of soil surface arthropods were recorded from the upland rice and maze ecosystems, respectively during the period of investigation. Formicidae were dominant followed by Collembola in the upland rice ecosystem whereas the Collembola were dominant followed by Formicidae in maize ecosystem. Moeed and Meads (1985) reported a total number of 93461 invertebrates belonging to 23 different groups from four types of habitats. Majer and Koch (1982) reported 12 groups of arthropods in three different habitats. Both the group of investigators found formicidae and Collembola as the major taxa in their studies. About ten species for Formicidae in each of the agroecosystems, *Pheidole* sp. being the most dominant were captured, Moeed and Meads (1985) reported six species of ants out of ten native species of Formicidae, However, Major and Koch (1982) reported a total of 39, 43, 32 spacies being trapped at three habitats respectively. Huffman and Harding (1980) reported seventeen species of Formicidae from various habitats.

Collembola was the most dominant group in maize ecosystem, while it was next of Formicidae in the upland rice system. In both the agroecosystems, individuals belonging to Entomobryidae were recorded in maximum number, which was most probably due to mainly their surface dwelling and being active on the soil surface. However, Moeed and Meads (1985) found Poduridae, as the dominant taxa. Majer and Koch (1982) collected individuals belonging Entomobryidae, Hypogastruridae, Isotomidae, Neauridae and Sminthuridae from three habitats. Among Acarina, Cryptostigmata was most abundant followed by Prostigmata, Astigmata and Mesostigmata. Moeed and Meads (1985) also reported Cyptostigmata and Prostigmata as the dominant taxa. Four species of Araneae belonging to three families were recorded from the upland rice agroecosystem, and two species belonging to two families were recorded from maize agroecosystem. Moeed and Meads (1985) collected individuals belonging eleven families of Araneae from the four habitats. Edwards *et al.* (1975) also reported about 50 species of spider. The capture of more Juveniles of Gryllidae than their adults, is in corroboration with the findings of Moeeds and Meads (1985), which suggested that the increase was following the breeding season.

In Coleoptera, fourteen species belonging to eight families were recorded of which *O. vaulogeri* (Scarabaeidae) followed by *Coenonica* Sp. (Staphylinidae) were dominant. However, the species composition is far less than findings of Moeed and Meads (1985) who reported 158 species of Coleoptera belonging to 36 families. This reduction in species composition may be due to the flush and burning practices, and other anthropogenic activities. Diptera, inspite of their greater mobility and power of dispersal were recorded in the pitfall traps. However, Moeed and Meads (1985) states that the catches of Diptera are believed to be related only approximately to their activity on the forest floor. They have reported different families and many of them were common in their collections. Huffman and Harding (1980) collected 28 species of Diptera belonging to 18 families. Diplopoda was represented by only one family and Chilopoda was represented by two families. However, Moeed and Meads (1985) recorded four families of Millipedes and two families of Centipedes from four habitats. This was probably due to the effect of previous 'Jhum' Cultivation. Particularly the burning and thereafter the arthropods could not recolonizes the habitat fully. However, Edwards *et al.*, (1975) recorded negligible number of Millipedes and Centipedes. The other arthropods belonging to Isopoda, Isoptera, Dermaptera, Psocoptera, Lepidopterean larvae and Pseudoscorpionidae were recorded in negligible numbers. Ostbye *et al.*, (1978) collected Plecoptera, Thysanoptera, Trichoptera, Lepidoptera, Brachycera and Symphyla in Poor numbers. Moeed and Meads (1980) reported Dermaptera and Pseudoscorpionidae in negligible numbers.

The seasonal changes in the population abundance of the soil surface arthropods during the present investigation was most probably due to the vatriations in the influence of several natural environmental or artificially induced factors. Phenology frequently regulates early or late season presence of arthropods and may explain some of the characteristics of their seasonal distribution. Moreover, being greatly affected by climatic factors the arthropods on reaching a developmental stage may prefer vertical movement. Thus, many of them such as beetles and others in the larval stage live within the litter of surface-soil whereas their adults live away from the soil and litter sub-system. Such temporary soil animals form a considerable percentage of the soil surface arthropods. Thus, the population of both the temperature and permanent soil surface dwelling arthropods may change in relation to the seasonal environmental factors, to the responses of predators, to the abundance of prey, and conversely, to the cannibalism in response to starvation leading to increase and decrease of the population densities (Price and Shepherd *et al*, 1977). Besides, the seasonal activity of these arthropods may be influenced by many other factors which may not be uniformly applicable to all the species. Nevertheless, the climatic factors such as temperature, humidity, precipitation, soil moisture and others effect the activity of these arthropods and cause increase or decrease in their population size to a great extent (Briggs, 1961; Duffey, 1962; Greenslade, 1964; and Mitchell, 1963). Vegetation of the site may also influence the arthropod population as most of them are plant dependent (Duffey, 1972). Even the absence or removal of the herbaceous layer near pitfall traps leads to either an increase or a decrease of the sample size (Greenslade, 1964).

It was observed that most of the surface dwelling arthropods were recorded more abundant during the rainy season although there was little correlation with the rainfall. However, the soil temperature showed a significant negative correlation with the total soil arthropod population in the upland rice ecosystem ($Y = 35.56 + 5.98 X$; $P < 0.05$). Edwards *et al.*, (1975) and Hutson (1978) stated that temperature is one of the most important factors in dealing with the activities of the arthropods. Almost all the arthropods were found to be affected by the fluctuation of temperature. Arthropod activity depend upon weather condition, especially temperature and soil moisture, and the general habitat surrounding the trap (Edwards *et al.*, 1975). In general, the ants were observed to be quite active throughout the year but were at low ebb during the winter. However, in u;land rice ecosystem, they were recorded highest in number during winter when temperature and soil moisture were low but in maize field its recorded in higher number during March 1983 when temperature and rainfall started increasing. Among the ants, *Pheidole* sp. increased during the summer and spring and decreased during the winter in the maize ecosystem when it was found in higher number in the upland rice ecosystem. However, the population abundance of ants being the predators seem to be less dependent on seasonal change in abiotic environmental factors. In the upland rice ecosystem, *A. longipes* showed a significant correlation with soil humidity ($Y = 36.24-0.46$; $P < 0.05$) whereas *Paratrechina* sp. was significant correlated with the organic carbon ($Y = 16.1-25.1 X$; $P < 0.01$). In maize field, the *Pheidole* sp. and *A. longipes* were significantly correlated with soil temperature ($Y = 12.7 + 1.23 X$; $P < 0.05$) and air temperature ($Y = 40.56-1.76$; $P < 0.05$), respectively. However, they did not show any correlation with any other abiotic factors. Majer and Koch (1982) observed the maximum activity of ants during spring and summer and low activity during the winter and concluded that the probable reason for the increase in ant population during spring and summer might be due the increase in the herbivore numbers. Nevertheless, observing the natural feeding rate of ants, Kajak *et al.*, (1972) stated that their activity was correlated with availability of food. Ants may, therefore, be a good general indicator of the seasonal trends in overall biological activity.

Majer (1981) reported that the increase in ants during spring and summer was connected with high temperature and availability of food such as herbivore invertebrates, plant sap and seeds, and the reason for their low activity during cooler months may be the reduced availability of such food sources. In the northerm hemisphere the seasonal activity of ant communities has been related to climatic factors with greatest summer activity occurring around June to August (Majer and Koch, 1982). McNeil *et al.*, (1978) observed that foraging of the red-wood ant *Formica Lugubris* Zett, was largely influenced by temperature, although excessive rain reduced foraging during cool periods. In a polish meadow, the ants exhibited grates activity in June-July, and were fond in significantly high numbers being positively correlated with temperature (Woyciechowski and Miszta, 1976). Whitford (1978) noted that increase in temperature and food served as threashold for ant population fluctuation and the increase in ant activity was correlated with the rainfall. However, Moeed and Meads (1985) and Majer and Koch (1982) reported that the increase in ant abundance was negatively correlated with rainfall. Edwards *et al.*(1975) reported peak numbers of ants being caught in June, and the reason for the ants being recorded highest in November in rice ecosystem may be their adaptability to xeric weather conditions and colonization habit.

In the upland rice ecosystem, the abiotic factors such as the air temperature, soil humidity, rainfall, soil temperature, soil moisture, pH, and organic carbon showed a significant combined effect on the seasonal population abundance of *Paratrechina* sp. and were responsible for 84 to 85 per cent of its seasonal variation. The rainfall, soil temperature, moisture and organic carbon were the important factors which regulated 84 per cent of its population fluctuation (Table 32.1). In the maize agroecosystem, the population fluctuation of *Pheidole* sp. was influenced by the combined effect of air temperature, soil humidity, rainfall, soil temperature, pH and available phosphorus, which were responsible for 62 to 69 per cent of its seasonal variation (Table 32.2).

There was higher abundance of Collembola in the upland rice ecosystem compared to the maize ecosystem. Such abundance may be attributable to the availability of relatively more vegetation with decaying plant matter on the soil surface in the rice ecosystem. The Collembola were most abundant during the summer seasons with intermittent rains when the temperature was high but the vegetation growth was flourishing. The population density decreased during the dry winter season. In the upland rice field, the population abundance of total Colembola showed a positive correlation with soil temperature ($Y = 56.56 + 3.97 X$; $P < 0.05$) while in the maize filed it has a negative correlationship with air temperature ($Y = -11.86 + 2.22 X$; $P < 0.05$). Wallace and Mackerras (1970) and McColl (1975) suggested that the decline in number of Collembola activity on the soil surface towards the end of summer (February) is attributable to the dry conditions. Joosse (1965) noted a positive correlation between the catch of *Orchesella cineta,* and negative correlation between the catches of *Entomobrya nivalis, Lepidocyrtus cyaneus* and *Tomocerus minor* and temperature. Besides, the total Collembola showed a positive correlationship with the rain fall in the upland rice ecosystem ($Y = 19.9 + 0.13 X$; $P < 0.05$) whereas Onychiuridae had a positive correlationship with the rainfall ($Y = 8.52 + 0.01 X$; $P < 0.05$) and Entomobryidae with soil humidity ($Y = 30.15 + 0.53 X$; $P < 0.05$) in the maize ecosystem. The rain season being prolonged from March to October, more Collembola were coptured during the season. In consistence to the present observation Moeed and Meads (1985) stated that the populations of Collembola were positively correlated with rainfall. It is evident that rainfall was most important climatic factor affecting the catches of Collembola (McColl, 1975).

TABLE 32.1. *Multiple regression (R^2) values of relationship between different abiotic environmental factors and the population densities of different taxa of soil-surface inhabiting arthropods of upland rice ecosystem.*

Arthropod taxa	R^2 for different Combination of abiotic factors					
Paratrechina sp.	AT+SH+RF+ST +SM+PH+OC+P +K R^2=0.852 F=5.14*	AT+RF+ST+ SM+pH+OC+ P+K+ R^2=0.852 F=6.48**	AT+RF+ST+ SM+pH+OC+ P R^2=0.851 F=8.18**	AT+RF+ST+ SM+pH+OC R^2=0.848 F=10.29**	RF+ST+SM+pH +OC R^2=0.845 F=13.16**	RF+ST+SM+ OC R^2=0.842 F=17.37*
Total Acarina	AT+SH+RF+ST +SM+PH+OC+P +K R^2=0.843 F=5.14*	AT+SH+ST+ SM+PH+OC+ K+P R^2=0.842 F=6.00*	AT+SH+ST+ SM+pH+OC+ K R^2=0.835 F=7.23**	SH+ST+SM+ PH+OC+K R^2=0.819 F=8.28**	SH+ST+SM+pH +OC R^2=782 F=8.6**	SH+ST+SM +OC R^2=0.73 F=8.8*
Cryptostigmata	AT+SH+ST+SM +OC+P+K R^2=0.714 F=3.64*	AT+SH+ST+ SH+OC+K R^2=0.712 F=4.53*	SH+ST+SM+ OC+K R^2=0.706 F=5.6*	SH+ST+SM+ OC R^2=0.694 F=.37*		
Total Coleoptera	AT+SH+RF+SM +P+K R^2=0.682 F=3.9*	AT+SH+RF+ SM+P R^2=0.68 F=5.12*	SH+RF+SM+ P R^2=0.676 F=6.79*	RF+SM+P R^2=0.673 F=9.6*		
Onthophagus vaulogeri	SH+RF+ST+SM +OC+P R^2=0.6898 F=4.07*	SH+RF+SM+ OC+P R^2=0.688 F=5.27*	SH+RF+OC+ P R^2=0.683 F=7.02	RF+OC+P R^2=679 F=9.9*		

*Significance at 5% level; **Significance at 1% level.

AT = Air Temperature; SH = Soil Humidity; RF = Rainfall; ST = Soil Temperature; SM = Soil Moisture; OC = Organic Carbon; P = Phosphorus; K = Postassium.

TABLE 32.2. *Multiple regression (R^2) values of relationship between different abiotic environmental factors and the population densities of different taxa of soil-surface inhabiting arthropods of maize ecosystem.*

Arthropod taxa	R^2 for different Combination of abiotic factors					
Pheidole sp.	AT+SH+RF+ ST+PH+P R^2=0.694 F=4.17*	AT+SH+RF+ ST+pH R^2=0.628 F=4.72*				
Entomobryidae	AT+SH+RF+ ST+SM+PH+ OC+P+K R^2=0.85 F=5.14*	AT+RF+ST+ SM+PH+OC+ P+K R^2=0.85 F=6.4**	AT+RF+SM+ PH+OC+P+K R^2=0.84 F=7.3**	AT+RF+SM+ OC+P+K R^2=0.83 F=8.66**	AT+RF+SM+OC +K R^2=0.81 F=10.44**	AT+RF+SM +OC R^2=0.78 F=11.5*
Total Acarina	AT+SH+RF+ ST+SM+pH+ OC+P+K R^2=0.797 F=3.5*	SH+RF+ST+ SM+pH+OC+ P+K R^2=0.795 F=4.37*	SH+RF+ST+ SM+pH+OC+K R^2=0.792 F=5.44*	SH+RF+ST+ SM+OC+K R^2=0.738 F=5.15*	SH+RF+ST+OC +K R^2=0.695 F=5.47*	SH+RF+OC +K R^2=0.642 F=5.86*
Cryptostigmata	AT+SH+RF+ ST+SM+PH+ OC+P+K R^2=0.829 F=4.29*	AT+SH+RF+ ST+SM+PH+ OC+K R^2=0.828 F=5.41*	SH+RF+ST+ SM+PH+OC+ K R^2=0.818 F=6.4*	SM+RF+ST+ SM+OC+K R^2=0.778 F=6.4*	SH+RF+SM+OC+ +K R^2=0.748 F=7.15*	RF+SM+OC +K R^2=0.719 F=8.3*
Total orthoptera	AT+SH+RF+ SM+pH+OC+ P+K R^2=0.781 F=4.01*	AT+SH+RF+ SM+PH+OC+ K R^2=0.752 F=4.34*	SH+RF+SM+ PH+OC+K R^2=0.707 F=4.4*	SH+RF+SM+ OC+K R^2=0.664 F=4.7*	SH+RF+SM+K R^2=0.641 F=5.86*	

*Significance at 5% level; **Significance at 1% level.

AT = Air Temperature; RF = Rainfall; SH = Soil Humidity; ST = Soil Temperature; SM = Soil Moisture; P = Phosphorus; OC = Organic Carbon; K = Potassium.

Although Collembola did not show any significant correlationship with pH they were recorded when the soil pH ranged 5.2 to 5.8. Since, very little work has been done on the effects of pH on the population of arthropods captured particularly in pitfall traps, it is difficult to comment on the effect of pH on soil surface arthropods. Moreover, from the field surveys, many authors did not find any correlation between soil pH and Collembola population and concluded that pH has little effect on them (Christiansen, 1964). However, some authors showed the preference of different species of Collembola of pH (Maclagen, 1932).

Organic carbon, and available phosphorus showed significant correlationship with the populations of Isotomidae ($Y = 13.86 + 36.3 X$; $P < 0.05$) and Entomobryidae ($Y = 15.63 + 1.54 X$; $P < 0.005$) respectively in the upland rice system. However, there was no correlationship with available potassium in either of the agroecosystems. Edwards *et al.*, (1975) reported that the potassium had little influence on the population abundance of Collembola. However, they have noted a positive correlation between *I. viridis* and the level of soil phosphorus. The population abundance of Entomobryidae was correlated with soil surface humidity ($Y = 30.15 + 0.53 X$; $P < 0.05$). Its population density in the maize field was significantly influenced by the combined effect of air temperature, soil surface humidity, rainfall, soil temperature, phosphorus and potassium, soil moisture, pH and organic carbon, that was responsible for 78 to 85 per cent of its seasonal variation (Table 32.2).

In upland rice ecosystem, the population abundance of Cryptostigmata was significantly correlated with soil humidity ($Y = 52.42–0.61X$; $P < 0.05$) and soil temperature ($Y = -12.98 + 0.87 X$; $P < 0.05$). The total Acarina also showed similar relationship with soil temperature ($Y = 12.84–0.94X$; $P < 0.01$). In the maize ecosystem, the Prostigmata was positively correlated with soil humidity ($Y = 12.81 + 0.21 X$; $P < 0.05$). The pH, available phosphorus and potassium did not show any correlationship with the Acarina population in either of the agroecosystems. However, the organic carbon showed significant correlation with Cryptostigmata (rice $Y = 49.97 + 119.23 X$; $P < 0.05$ and maize; $Y = 40.44–59.12 X$; $P < 0.05$), and total Acarina in both rice ($Y = 46.1–65.88 X$; $P < 0.05$) and maize ($Y = 43.4 + 113.63$; $P < 0.05$) ecosystems. Thus, their decrease in number during the winter was most probably due to the dry weather which might have caused them to migrate to the deeper layers of the soil. Moeed and Meads (1985) recorded high number of mites particularly belonging to Cryptostigmata and Prostigmata during spring and summer in broad leaf forest and found the correlation between the abundance of mites and temperature, but not with the rainfall. The abiotic factors showed combined influence on the seasonal population abundance of Cryptostigmata and that of total Acarina. In the upland rice ecosystem the abiotic factors such as soil surface humidity, air and soil temperatures, soil moisture, organic carbon and Potassium and Phosphorus showed a significant combined influence on the population density of Cryptostigmata, which ranged between 69 to 71 per cent of its seasonal variation. The combined effect of all the nine abiotic factors were responsible for 73 to 84 per cent of seasonal variation of total Acarina. The humidity, temperature, moisture and organic carbon being important one ear responsible for 73 per cent of the variation. In the maize ecosystem, the combined effect of the above nine factors were responsible for 71 to 82 per cent of seasonal variation of the Cryptostigmata and the rainfall, soil moisture, Organic Carbon and Potassium being responsible for 71 per cent of variation. Besides, the population abundance of total Acarina was significantly influenced by the combined effect of the above nine factors which were responsible for 64 to 79 per cent of the seasonal variation.

The abundance of Coleoptera was high during the rain season while they were absent during the winter. In the upland rice ecosystem, the rainfall ($Y = 0.51 + 0.25 X$; $P < 0.001$) and soil temperature ($Y = -7.44 + 0.43 X$; $P < 0.05$) showed significant correlation with *O. vaulogeri*. The beetle activity was recorded highest during the breeding season, and was correlated directly with temperature (Drift, 1959; Vlijm *et al.*, 1961; Mitchell, 1963; Greenslade, 1964 and Rickard and Haverfield, 1965). Ostbye *et al.*, (1978) reported high catches of Coleptera during June and July, and noted that this group, mostly predators specially of Carabidae and Staphylinidae tend to alternate in their predation activities. They reported Carbidae showing peak of its activity during spring while Staphylinidae being caught in higher number during summer. The Scarabid beetles were absent during winter-spring and autumn-winter seasons while they showed their higher activity from April to September. It is in consistence with the findings of King (1979) and King *et al.*, (1981) who recorded disappearance of the black beetles during autumn-winter and winter-spring seasons, which may also be due to the mortality and adult dispersal. They suggested that a significant number of adults migrate into favorable habitats. Besides, rainfall and soil moisture was highest during June-July which might have effected the activity of these beetles. Conversely, Moeed and Meads (1985) did not find any positive correlation between the Coleopteran abundance and rainfall. However, Mitchell (1963) and Greenslade (1964) reported the favorable effects of rainfall on the activity of beetles. In the upland rice field the available phosphorus showed negative correlationship with *O. vaulogeri* ($Y = -12.2 + 1.84 X$; $P < 0.05$) and the total Coleoptera ($Y = -11.02 + 17 X$; $P < 0.01$). Edwards *et al.* (1975) reported a positive correlation between available potash and beetle larvae and suggested that the effect may be direct as they spend most of their time in the soil. Although many of the abiotic environmental factors did not show any correlation with population abundance of *O. Vaulogeri* and total Coleoptera, the abiotic factors in combination effected their population densities significantly in the upland rice ecosystem. The soil humidity, rainfall, air temperature, soil moisture, and available phosphorus showed a combined effect n the population abundance of total Coleoptera whereas soil surface humidity, rainfall soil temperature, soil moisture Organic carbon and phosphorus affected combinedly on *O. vaulogeri*. These abiotic factions were responsible for 67 to 68 per cent of their seasonal variation ($P < 0.01$).

The adult Orthopterra were recorded in higher number during rain season and in low number during the winter. However, the Juveniles were caught in peak of abundance during November 1982 and during June-September 1983. But, they were less in number during winter and early summer. In the upland rice ecosystem, the Gryllidae (adults) ($Y = 0.18 + 0.12 X$; $P < 0.05$) and in the maize, the total Orthoptera ($Y = 1.67 + 0.01 X$; $P < 0.01$) showed significant positive correlationship with rainfall. In the maize field, the total Orthoptera showed a similar correlation with air temperature ($Y = 12.76 + 0.34 X$; $P < 0.05$). Moeed and Meads (1985) reported increase in number of Stenopelmatidae and Rhaphidophoridae during summer and autumn with large numbers of Juveniles. They observed the population to be positively correlated with temperature, and reported the ground Orthoptera, *Hemiandrus furcifer* to be most active from late spring to early autumn. Majer and Koch (1982) noted that the Orthoptera showed high levels of activity in spring, summer and in early autumn and were positively correlated with the temperatures of the current and the previous months, and negatively correlated with relative humidity and rainfall. They could not point out the main factor influencing the population as temperature was negatively correlated with relative humidity and rainfall. The total Orthoptera ($Y = -20.41 + 45.4 X$; $P < 0.05$) and Juvenile Gryllidae ($Y = 018.69 + 41.93 X$; $P < 0.05$) showed negative correlationship with organic carbon in the upland rice ecosystem. The abundance of juvenile Gryllidae in winter indicated that the lower temperature may have been favorable for their reproduction and multiplication. Besides, in the maze ecosystem the total Orthoptera were significantly influenced by the combined effect of air temperature, soil surface humidity, rainfall, soil moisture, pH, organic carbon, available phosphorus and postassium that was responsible for 64 to 78% of seasonal variation and the humidity, rainfall, moisture and potassium were the main factors (Table. 32.2).

Although there was no statistically significant correlationship between any of the abiotic factors and population abundance of Araneae, the *Trochosa* sp. being dominant among Araneae, was caught highest during rain and winter season in upland rice ecosystem and during summer in maize ecosystem. In corroboration, Hagvar *et al.*, (1978) reported the occurrence of peaks of increase of a few species of Araneae during July and in late August, suggesting that the difference in the pattern of catches may be attributable to different climatic factors. Moeed and Meads (1985) reported high catches of spiders in autumn from two habitats of froad leaf forests and stated that they were correlated with temperature but not with rainfall. Edwards *et al.*, (1975) reported more catches of spiders in August and lowest in December. They have reported population peaks of *Bathyphantes gracilis* and *Centromerita bicolor* in winter, the former being in January and the latter being in December and February. Locket (1975) recorded more or less similar findings. However, Edwards *et al.*, (1975) and Locket (1975) reported distinct peaks of *Trochosa terricola, Peachygnautha degeeri* and *Paradosa palustris* in May which is in consistence with the high catches of Araneae in maize ecosystem. Variation in curve shapes between habitats within the same year may partly be caused by different microclimate and weather conditions in the habitats (Hagvar *et al.*, 1978).

Polydesmida sp. and *Lamyctes* sp were recorded highest during summer with intermittent rains while *Rhysida* sp. showed highest peak in post-running seasons. It shows that high soil moisture and rainfall effected the population of the first two species while that of the latter was effected by high rainfall and temperature although there was no correlationship in between them. These findings are in consistence with those of Edwards *et al.*, (1975). Moeeds and Meads (1985) found more number of Chilopoda and Diplopoda during summer and autumn than in winter and spring. They found that the catches of Chilopoda and Diplopoda were both correlated with temperature, but not with rainfall.

Dipteran population were more abundant in the maize ecosystem compared to the upland rice and were highest early summer (February) in maize ecosystem while they showed the highest peak in March 1983 in upland rice ecosystem. There was no significant correlation between the abiotic factors and the Dipteran population in either of the ecosystems probably due to less abundance and irregular presence. Moeed and Meads (1985) recorded marked seasonal differences in the catches or Diptera between broad leaf and silver beech forest which showed the peaks during autumn, spring and summer. They recorded the catches of flies being correlated with temperature in one of the broad leaf forest and silver beech forest, and with rainfall in the other broad leaf forest. The population abundance of Homoptera and many other arthropods were too few to show any seasonal pattern. Hagvar *et al.*, (1978) stated that Homoptera was the only group among the different arthropods whose catches (differed greatly between habitats.

Thus the maize ecosystem, the one which was more closer to human habitation harboured less arthropod fauna and many of the arthropods present at the rice ecosystem were not recorded in the maize ecosystem probably because the human interference. Most of these arthropods were significantly influenced by the seasonal abiotic environmental factors and showed temporal variation. The abiotic factors showed combined effects and influenced the temporal variation in population abundance of different arthropod taxa.

Acknowledgement

This research was supported by funding by I.C.A.R. (New Delhi) in the form of a ad-hoc research project sanctioned to MVR, MAA is grateful to ICAR for the award of a Sr. Res. Fellowship. Scientists of the Commonwealth Institute of Entomology (London, UK) identified the arthropods.

REFERENCES

Barlow, C.A. 1970. Phenology and distribution of some *Pterostichus* (Coleoptera-Carabidae) of Eastern Canada. *N.Y. Ent. Soc.* **78** : 215-237.

Best, R.L. and Beegle, C.C., 1977. Food preference of five species of carabids commonly found in Iowa Cornfields. *Environ. Entomol.* **6** 9-12.

Briggs, J.B. 1961. A comparison of pitfall trapping and soil sampling in assessing populations of two species of ground beetle (Carabidae). *Rep. E. Malling Res. Sta.* 1960 108-12.

Christiansen, K. 1964. Bionomics of collembola. *Ann. Rev. Ent.* **9** : 147-178.

Doane, J.F. 1981. Seasonal captures and diversity of ground beetles (Coleoptera Carabidae) in wheat field and its grassy borders in Central Saskatchewan. *Quaest. Entomol.* **17** : 211-33.

Drift, J. Van Der 1959. Field studies on the surface fauna of forests. *Bijdr. Dierkde.* **29** : 70-103.

Duffey, E. 1962. A population study of spiders in Limestone Grassland. *J. Anim. Ecol.* **31** 571-599.

Duffey, E. 1972. Ecological survey and the arochonologist. *Bull. Brit. Arach. Soc.* **2** : 169-82.

Dunn, G.A. 1982. Ground beetles (Coleoptera : Carabidae). Collected by pitfall trapping in Michigan small grain fields. *The Great Lakes Entomologists.* **15** : 37-38.

Dutta, A.K. 1983. *Studies on ground beetles (Coleoptera : Insecta) of Dooars area, Jalpaiguri, West Bengal with special reference to their taxonomy, seasonal activity and density by pit-fall experiment.* Ph.D. Thesis, University of North-Bengal Darjeeling.

Edwards, C.A., Butler, C.G. and Lofty, J.R. 1975. The invertebrate fauna of the park grass plots II. Surface fauna. *Rothamsted Report for 1975.* Part 2. 63-89.

Ferguson, H.J. and McPherson, R.M. 1985. Abundance and diversity of adult carabidaein four soybean cropping systems in virginia. *J. Entomol. Sci.* **20** (2) : 163-171.

Finlayson, D.G. and Campbell, C.J. 1976. Carabid and staphylinid beetles from Agricultural land in the lower fraser valley. British Columbia. *J. Entomol. Soc. Brit. Columbia,* **73** : 10-20.

Frank, J.H. 1971. Carabidae (Colleoptera) as predators of the red backed cutworm (Lepidoptera-Noctuidae) in Central Alberta. *Can. Entomol.* **103** : 1039-44.

Greenslade, P.J.M. 1964. Pitfall trapping as a method for studying populations of Carabidae (Coleoptera). *J. Anim. Ecology.* **33** : 301-10.

Goulet, H. 1974. Biology and relationships of *Pterostichus adstricuts* Eschscholtz and Pterostichus pensylvanicus Leconte (Coleptera Carabidae). *Quaestiones Entomologicae.* **10** : 3-33.

Hagvar, S., Qstbye, E. and Melaen, J. 1978. Pitfall catches of surface active arthropods E.Qstbye in some high mountain habitats at finse, South-Norway. II. General results at emphasis on opiliones. Araneida, and Coleoptera. *Norw. J. Ent.* **25** : 195-205.

House, G.J. and All, J.N. 1981. Carabid beetles in soybean agroecosystems. *Environ. Entomol.* **10** : 194-6.

Huffman, F.R. and Harding, J.A. 1980. Pitfall collected insects from various Lower J.A. Harding Rio Grande Valley Habitats. *South-Western, Entomol.* **5** (1) : 33-46.

Hutson, K. 1971. A mechanical time sorting pitfall trap. *Ent. Mon. Mag.* **106** : 214-216.

Hutson, B.R. 1978. Influence of pH, temperature and salinity on the fecundity and longevity of four species of Collembola. *Pedobiologia.* **18** : 163-179.

Jackson, M.L. 1967. *Soil chemical analysis.* Prentice Hall of India Pvt. Ltd., New Delhi.

Joosse, E.N.G. 1965. Pitfall trapping as a method for studying surface dwelling Collembola *Z. Morph. Okol Tiere.* **55** : 587-596.

Kajak, A., Breymeryer, Petal, J. and Olechow, E. 1972. The influences of ants on the meadow invertebrates. *Ecol. Pol.* **20** : 163-171.

King, P.D. 1979. *Aspects of the ecology of black beetle, Heteronychus arator* (F) (Coleoptera Scarabaedae). Unpublished D. Phil. Thesis University of Waikato, New Zeland.

King, P.D., Mercer, C.F. and Meekings, J.S. 1981. Ecology of black beetle *Heteronychus arator* (Coleoptera Scarabaeidae) Population *N.Z J. of Agricultural Research.* **24** : 87-97.

Locket, G.H. 1975. *Spiders,* Leckford Record No., 3, p. 14.

Los, L.M. and Allen, W.A. 1982. Abundance and diversity of adult Carabidae in insecticide treated and untreated alfalfa fields. *Entomol.* **12** : 1068-72.

Lund, R.D. and Turpin, F.T. 1977. Carabid damage to weed seeds found in Indiana corn fields. *Environ. Entomol.* **6** : 695-8.

Maclagen, D.S. 1932. An ecological study of the Lucerne flea (*Sminthurus viridis* L.) *Bull. Ent. Res.* **23** : 101-145.

Majer, J.D. 1977. *The possible use of ant fauna as indicators of land use and of the success of revegetation attempts.* Proceedings, Western Australian Entomology Workshop, Perth, 8.

Majer, J.D. 1978. A preliminary survey of the epigaeic invertebrate fauna with particular reference to ants. In areas of different land use at dwelling up. Western Australia. Forest *Ecology and Management.* **1** : 321-24.

Majer, J.D. 1981. A flowering calendar for karagullen, a Northern Jarrah Forest Locality. *West Aust. Herbarium Research Notes.* **5** : 19-28.

Majer, J.D. and Koch, L.E. 1982. Seasonal activity of hexapods in woodland and forest leaf litter in the South-West of Western Australia. *J. Royal Soc. of Western Aust.* **65** : 37-45.

McColl, H.P. 1975. The invertebrate fauna of the litter surface of *Nothofagus truncata* forest Floor, and the effect of microclimate on activity. *New Zealand Journal of Zoology,* **2** : 15-34.

McNeil, J.N., Delisle, J. and Finnegan, R.J. 1978. Seasonal predatory activity of the introduced red wood ant. *Formica* lugubris (Hymenoptera. Formacidae) at Valcartier, Quebec in 1976. *Can. Ent.* **110** : 85-90.

Mitchell, B. 1963. Ecology of two carabid beetles. *Bembidion lampros* (Berbst) and Trechus guadristriatus (Schrank). 1. Life cycles and feeding behaviour. *J. Anim. Ecol.* **32** : 289-299.

Moeed, A and Meads, M.J. 1985. Seasonality of pitfall traped invertebrate M.J. Meads in three types of native forest, Orogorongo valley, *New Zealand. New Zealand Journal of Zoology.* **12** : 17-50.

Odum, E.P. and Reichholf, J. 1980. *Okologie, Grundbegriffe, Verknupfungen perspektiven.* BLV Munchen, 208 pp.

Oppenheimer, J.R. and Tikader, B.K. 1976. The ground activity of spiders (Araneae) and harvestmann (Phalangidae) in West Bengal, India. *J. Bombay Not. Hist. Soc.* **73** : 121-141.

Ostbye, E., Hagvar, S., Haande, K.M., Haande, P.S., Melaen, J. and Skartveit, A. 1978. Pit-fall catches of surface-active arthropod in some high mountain habitats at Finse, South Norway. 1. Aim. Methods and habitat descriptions. *Norw. J. Ent.* **25** : 189-193.

Paine, R.T. 1974. Intertidal community structure. *Oecologia,* **15** : 93-120.

Price, J.F. and Shepard, M. 1977. Striped earwig *Labidura riparian* Colonization of soybean fields and response to insecticides. *Environ. Entomol.* **6** : 679-83.

Price, J.F. and Shepard, M. 1980. Sampling ground predators in soybean fields. pp. 532-543. In. M. Kogan and D.C. Herzog [eds.], *Sampling methods in soybean entomology.* Springer-Verlag, N.Y.

Reddy, M.V. 1984. Ecology of soil and litter inhabiting arthropods. *Indian Rev. Life Sci.* **4** : 169-217.

Reddy, M.V. 1986. Soil-inhabiting arthropods as indicators of Environmental quality. *Acta Biologica Humgarica,* **37** : 79-84.

Reddy, M.V. and Venkataiah, B. 1986. Summer abundance of soil-surface arthropods in three vegetational covers an urban Ecosystem. *Environ. & Eco.* **4** : 259-262.

Rickard, W.H. and Haverfield, L.E. 1965. A pitfall trapping survey of darkling beetles in desert steppe vegetation. *Ecology.* **46** : 873-875.

Rivard, I. 1964. Carabid beetles (Coleoptera : Carabidae) from Agricultural lands near Beleville, *Ontario. Can entomol,* **96** : 517-520.

Seastedt, T.R. 1984. The role of microarthropods in decomposition and mineralization process. *Ann. Rev. Entomol.* **29** : 25-46.

Shepard, M., Carner, G.R. and Turnipseed, S.G. 1977. Colonization and resurgence of insect pests of soybeans in response to insecticides and field isolation. *Environ. Entomol.* **6** : 501-6.

Smith, B.J. 1976. A new application in the pitfall trapping of insects. *Trans. Ky. Acad. Sci.* **37** : 94-97.

Southwood, T.R.E. 1978. *Ecological methods with particular references to the study of insect populations.* John Whiley and Sons, N.Y. 524 pp.

Thiele Hana-ulrich 1977. *Carabid Beetles in their environments.* springer-verlag, Berlin, 369 pp.

Thomas, Donald, B. Jr. and Elbert, L. Seleeper 1977. The use of pitfall traps for estimating the abundance of arthropods with special reference to the Tenebrionidae (Coleoptera). *Ann. Ent. Soc. Amer.* **70** : 242-248.

Uetz, G.W. and Unzicker, J.D. 1976. Pitfall trapping in ecological studies of wandering spiders. *J. Arachnol.* **3** : 101-11.

Vlijm, L., Hartsuijker, L. and Richter, C.J.J. 1961. Ecological studies on carabid beetles. 1. Calathus melanocephalus (Linn) *Archive Neerlandises de Zoologie.* **14** : 410-422.

Waage, B.E. 1985. Trapping efficiency of carabid beetles in glass and plastic pitfall traps containing different solutions. *Fauna Norv. Ser. B.* **32** : 33-36.

Wallacel M.M.H., and Mackerras, I.M. 1970. The entognathous hexapods. In : *The insects of Australia.* The division of Entomology, Common Wealth Scientific and Industrial Research Organisation, Canberra, Mebourne University Press, Carlton, pp. 205-216.

Whitford, W.G. 1978. Structure and seasonal activity of Chihuahua desert and communities. *Insectes-soc.* **225** : 79-88.

Williams, G. 1958. Mechanical time-sorting of pitfall captures. *J. Anim. Ecol.* **27** : 27-35.

Wojeik, O.P., Banks, W.A. Hicks, D.M. and Plumley, J.K. 1972. A simple inexpensive pitfall trap. for collecting arthropods. *Florida Ent.* **55** : 115-116.

Woyciechowski, M. and Miszta, A. 1976. Spatial and seasonal structure of ant communities in a mountain meadow. *Ecol. Pol.* **24** : 577-592.

Chapter 33

Ecology of Some Fast Growing Nitrogen Fixing Trees in India

R.S. Ambasht, Alok K. Srivastava AND N.K. Ambasht

Ecology Research Laboratory Centre of Advanced Study in Botany Banaras Hindu University, Varanasi-221 005, India

Abstract. Nitrogen fixing and fast growing trees endowed with actinorhizal or rhizobial nodules in roots assume great importance in plantation forestry particularly on landslide affected, eroded or otherwise degraded lands. Among these, *Alnus nepalensis* in the Central and Eastern Himalayas, *Casuarina equisetifolia* on the Gangetic plains, Vindhyan plateau and coastal lands, *Leucaena leucocephala* in the social forestry programmes in most part of the country have been worked out for in situ seasonal and annual rates of biological nitrogen fixation, nitrogen release, uptake and retention, biomass production, litter fall and decomposition. There is a very efficient strategy evolved by these species to exploit available resources at high efficiency with least expenditure of energy. The paper discusses the application of these researches to the country's revegetation programme.

***Key words** : In situ N_2 fixation, Alness, Casuarina, Leucaena.*

Introduction

Nitrogen is a key element required by all plants and animals as it is an essential constituent of proteins which include all enzymes and DNA. Thus without it, life cannot exist. But the abundant stock of nitrogen present in the atmosphere is in inert form and cannot be readily used by organisms, unless it is first converted into usable forms like nitrates or ammonia. Biologically it is carried out by nitrogen fixing bacteria, acthinomycetes and blue-green algae which are largely found free living in nature and a few of these associate with roots of higher plants in form of nodules. Such tree species having potential of fixing nitrogen by *Rhizobium* (bacteria) (Dixon *et al*., 1981) and *Frankia* (actinomycete) (Johnsrud 1978) associations are referred as ''Nitrogen fixing trees''.

Among the important actinorhizal trees are *Alnus* (33 spp.), *Casuarina* (25 spp.), *Myrica* (26

Table 33.1. *Important Actinorhizal plant genera and their distributions*

Genera	No. of species	Family	Locality
Alnus	47	Betulaceae	N. America, S. America and Eurasia
Casuarina	16	Casuarinaceae	Australia and Oceania
Ceanothus	31	Rhamnaceae	North America
Comptonia	1	Myricaceae	North America
Coriaria	16	Coriariaceae	N. America, S. America and Oceania
Datisca	2	Datiscaceae	N. America and Asia
Discaria	5	Rhamnaceae	Oceania
Elaeagnus	38	Elaeagnaceae	N. America and Eurasia
Hippohae	2	Elaeagnaceae	Eurasia
Myrica	28	Myricaceae	N. America, Eurasia and Africa
Retanilla	2	Rhamnaceae	South America
Shepherdia	2	Elaeagnaceae	North America
Trevoa	2	Rhamnaceae	South America

spp.), *Elaeagnus* (17 spp.) and *Ceanothus* (31 spp.) and a few others. The name of important actinorhizal genera and their distribution are listed in Table 33.1. All such trees bear root nodules in which the actinomycetous micro-organisms *Frankia* are found. The *Rhizobial* nodulated higher plants are mostly legumes belonging to the subfamilies of family *Leguminosae*. More than 3200 leguminous species are reported to contain nitrogen fixing bacteria in their roots. Some of the common trees which are fast growing and N_2-fixing are *Acacia auriculiformis, A. nilotica, A. senegal, Aeschynomene* sp., *Albizia lebbeck, Leucaena leucocephala* and *Prosopis* spp. Many of these are not of Indian origin but are now acclimatized in the country. Table 33.2 shows the important N_2-fixing trees found in Indian tropics and subtropics, their origin and specific uses.

We can take up these two groups of N_2-fixing trees separately. However there is a third group of higher plants which associate symbiotic N_2-fixing cyanobacteria such as the coralloid roots of *Cycas* plants.

Non-Leguminous Nitrogen fixers

Symbiotic association of actinomycetes in the roots of some species belonging to the order Casuarinales, Myricales, Fagales, Rhamnales, Coriariales and Rosales are found mostly in the sub-tropical and temperate regions. *Frankia* genus involved in the N_2 fixation becomes more active in poor soils and in absence of H_2 and Co. *Frankia* infects the root hairs of the host plant

TABLE 33.2. *Important N_2-fixing trees found in Indian Tropics and Subtropics.*

Name of Species	Origin	Remarks
Leguminosae (Rhizobial)		
Acacia auriculiformis	NE Australia	Fuel Wood, Pulp wood
Acacia nilotica	Africa	Fire wood, Charcoal, tannin, gums, fodder
Acacia senegal	1Africa	Gum, Arabic, fire wood, Charcoal, Soil amelioration
Aeschynomene sp.	Tropical Africa	Light corky wood, fodder (leaves)
Alvizia lebbeck	Asia, Africa	Fuel wood, Furniture wood, medicine, green manure
Dalbergia sissoo	Indian sub continent	Timber, Furniture, fuel wood
Hardwickia binata	India	Timber, craft wood, fuel wood
Leucaena leucocephala	Central America, Mexico	Fuel wood, green manure, fodder
Prosopis cineraria	India	Fire wood, charcoal, green manure, posts
Prosopis juliflora	S W USA	Fuel wood, charcoal, green manure, posts
Betulaceae (Actinorhizal) *Alnus nepalensis*	Himalayas	Furniture, timber, fire wood
Casuarinaceae (Actinorhizal) *Casuarina equisetifolia*	Australia	Fuel wood, charcoal, windbreaks, coastal dune

and forms nodule or a coralloid structure, and it has three forms. Hyphae, Vesicles; where probably N_2-fixation takes place and bacteria like cells which primarily serve for the survival and dispersal of the organism. It is an obligate symbiont.

The search for more and more actinorhizal trees in going on and currently 24 genera with over 160 species of higher plants are recorded to possess *Frankia* bearing nodules. By and large, these species are capable of inhabiting nutrient poor soils and quickly establishing on landslide affected, eroded or otherwise degraded or stressed habitats. These species play important role in N_2 accretion and organic matter accumulation. In India the two most important actinorhizal N_2-fixing trees are *Alnus nepalensis* in the Himalayas and *Casuarina equisetifollia* in the plantation forest of the coastal areas and the degraded Vindhyan hills.

Alnus

Alnus nepalensis is the common alder species found in India throughout the central and eastern belts. This is a multiple use tree coming up commonly on the freshly eroded soils. It plays an important role in the N_2-economy of soil in the altitude of 1000-2800 metre in the Himalayas.

The common associated trees are *Cryptomeria japonica, Cupressus cashmiriana, Betula alnoides, Michelia champaca* and *Cinchona officinalis*. It is also planted as nurse tree for the *cordamum* plants grown profusely in the Darjeeling and Sikkim Himalayas. Another species *Alnus nitida* is found in the western Himalayas. The root nodules are coralloid in shape and pink to brown in colour. It occurs in natural forests as pioneers of freshly exposed lands and also as plantation forest. Sharma and Ambasht (1984, 1986, 1987, 1988, 1991) have studied the biomass, production, energetics, N_2-fixation rates, litterfall, decomposition, Nutrient release, nitrogen accretion and its energetics by different age classes of nodules in 7, 17, 30, 46 and 56 years old plantation stands of *A. nepalensis* in darjeeling Himalayas. This is also known as Himalayan alder and locally as *Uttis.* The associated forest floor vegetation is reported to change as the plantation grew old. Dominant understorey species in the age sequence of *A. nepalensis* plantation stands were : *Artemisia vulgaris* and *Polygenum Chinese* in 7-yr stand *Rebus reticulatus, Oxalis cerniculata* and *Solanum indicum* in 17-yr stand *Eupatorium adenophorum* in 30-yr stand, *Laportea terminalis, Girardiana palmota* and *Rumex nepalensis* in 46-yr stand and *Artemisia vulgaris* and *Rebus reticulatus* in 56-yr stand. The average tree height of the maximum girth class plants (upto 80 cm DBH) is about 35 meters.

Sharma and Ambasht (1991) have recorded that the standing crop biomass of the 5 different age class plantation gradually increased from 106 t ha^{-1} to 238, 350, 473 and 606 t ha^{-1} in 7, 17, 30, 46 and 56-yr old stands respectively. Among the plant components, the bole or trunk accounts for 30 per cent of total biomass in the youngest age and 66 per cent in the 56-yr old stands.

The leaf and twig production remains between 4.6 to 1.8 t ha^{-1} in different age groups. The share of understorey vegetation to total biomass was negligible. Litterfall was maximum during December to March and least in the rainy season during June to October. Litterfall ranged from 3.2 t ha^{-1} in 7-yr to 5.8 t ha^{-1} in 56-yr stand. But the forest floor littermass was much higher. This is because of the temperate nature of the habitat, where the decomposition is not as fact as in the tropics. The rate of primary production was fast in the younger age group accounting for 25 t ha^{-1} in 7-yr stand and decreased to 19.7 t ha^{-1} in 17-yr, 18.7 t ha^{-1} in 30-yr, 16.0. t ha^{-1} in 46-yr and 13 t ha^{-1} in 56-yr old stand. The above ground production accounts for 86 to 96 per cent of the total biomass whereas belowground net production for 4-14 per cent. The production efficiency of the younger leaf was about twice as much as in the old plantations.

The energy concentrationwise, the highest energy was in the range of 4000 cal g^{-1} in the fruits, about 4000 cal g^{-1} in branch, 3800 cal g^{-1} in trunk and 4300 cal g^{-1} in belowground parts. The annual net energy fixation of *Alnus nepalensis* was about 100×10^6 K cal ha^{-1} yr^{-1} in the 7-yr old stand to about 51×10^6 K cal ha^{-1} yr^{-1} in the 56-yr old stand. This is reasonably high energy fixation efficiency for the forest community accounting for 3.49 per cent of the total usable solar energy. Sharma and Ambasht (1984) measured the N_2-fixation, storage and release in this species using acetylene reduction method for nitrogenase activity. They have found that the nitrogen fixation is as higher as 102 kg ha^{-1} yr^{-1} in the 7-yr old stand and decreases to 49, 39, 32 and 25 kg ha^{-1} yr^{-1} in the 17, 30, 46 and 56-yr old stands respectively. In comparison to this, Bormann and Gordon (1980) have reported 50-106 kg ha^{-1} yr^{-1} in 5-yr old *Alnus rubra* and Binkley (1989) has found only 20 kg ha^{-1} yr^{-1} in *Alnus crispa* and as high as 130 kg ha^{-1} yr^{-1} in *A. rubra.* the maximum N_2-fixation was about 98 kg ha^{-1} in 7-yr stand during the monsoon from June to November while in the dry phase, only 4 kg/ha N was fixed in the other six months from December to May. The trends remain the same with the advancing age of trees but the overall nitrogen fixation quantity gradually decreased to half to one fourth. Young pink colour nodules were much more efficient fixers than the old brown nodules.

Seasonal litters and nodule decomposition were distinct with highest rate in warm rainy season followed by subsequent seasons of the year, with conditions gradually shifting from suitable to severe. The relative litter decomposition ratios were correlated with litter quality and environmental factors. Nodule decomposition is much faster process than litter decomposition. About 115 kg ha^{-1} yr^{-1} of N was released in 7-yr *Alnus* stand and 152 kg ha^{-1} yr^{-1} in 56-yr stand through litter decomposition process (Sharma and Ambasht 1987).

Casuarina

Casuarina (recently taxonomically revised into four separate genera) includes about 80 species of shrubs and trees that are primarily native to the southern hemisphere, mostly to Australia, where they occur in tropical, subtropical and temperate coastal regions as well as in arid inlands. A few species are native to Indo-pacific areas from peninsular Malaysia to Polynesia. *C. equisetifolia* and *C. cunnighamiana* are fairly widespread around the world (Doran and Hall 1981).

In India, *C. equisetifolia* was first introduced into Madras state in the 1860s for producting fuel for steam locomotives. Villagers then spread it along the coasts of the Indian peninsula, especially on the east coast. In due course, farmers brought it into the south Indian agricultural system. This tree is now one of the main firewood species in peninsular India. There are over 40,000 ha land under *Casuarina* plantation in India, normally harvested on a 7 to 15-years rotation cycle.

Casuarina is an actinorhizal nitrogen fixing tree. It has received the attention of ecologist because of its fast growing nature, wide ecological amplitude and its success in revegetation programme, restoring soil fertility and reconsiling fuel wood demand and supply. Srivastava (1992), Srivastava and Ambasht (1990, 1993 a, b, c and Ambasht and Srivastava 1991, 1992, 1993) have made extensive studies on *Casuarina equisetifolia* in the Sonbhadra district, in Shaktinagar forest unit of Renukoot forest division of Uttar Pradesh. Two stands of 5 and 8-yr age were studied for biomass and energy production, nitrogen storage in different plant components and the nitrogen transfer within soil-plant system, including biological N_2-fixation. The total biomass with increase in stand age ranged from 43.5 to 81.3 t ha^{-1} in 5-yr stand and 41.8 to 58.7 t ha^{-1} in 8-yr stand. Bole accounted for 55-67 and of the total biomass, followed by root (17-19%), foliage (6-13%) and branch (8-11%). The total annual litterfall ranged from 7 to 10 t ha^{-1} yr^{-1} in mature stands. Unlike *Alnus*, the maximum litterfall occurred in the month of may and minimum amount of litter fell during rainy months. Foliage litter accounted for 86 per cent to 97 per cent of the total litterfall. Net primary production and net energy fixation was calculated by adding the increment in component biomass and energy respectively, between two successive years. The net primary production ranged from 19.0 to 29.0 t ha^{-1} yr^{-1}. Bole accounted for 63-73 per cent of the total production followed by root branch and foliage. Fruits have very little share in total biomass and production. Likewise, the net energy fixation ranged from 365 × 10^6 to 567 × 10^6 kg ha^{-1} yr^{-1}. Computation of biomass accumulation ratio showed that *Casuarina* have potential of accumulating biomass. The energy conserving efficiency was found to be twice of 8-yr stand than in the 5-yr stand. Studies on seasonal and depth variation in total nitrogen, ammonium nitrogen and nitrate nitrogen in soil revealed that concentrations were generally highest in the top 10 cm depth and decreased with increased depth with little fluctuations. Litter decomposition rates were higher than in the temperate belts, and nearly 80 per cent of litter decomposed during one year period of study. In general decomposition rate is higher in tropics than the temperate and boreal biomass (Vitousek 1984, Ambasht and Srivastava 1992). Nitrogen fixation rates in *Casuarina* were higher in July-August period and gradually decreased to winter and minimum in summer months. However the nitrogen accretion to soil was maximum in the months of September-October. Nitrogenase activity was confined to only active root nodules and inactive root nodules did not show any activity. These findings confirmed the importance of *Casuarina* as pioneer species for poor and degraded soils.

Leguminous Nitrogen fixing trees

Leguminous trees are a potential source of fuel and hardwood (Brewbaker *et al.*, 1982). Now woody legumes are receiving interest for agroforestry and reafforestation programmes.

The family Leguminosae consists of 18,000 species of which nitrogen fixing trees are predominantly present.

Leucaena leucocephala (Lam. de wit.) a woody perennial belonging to the sub-family Mimosoidase has, in recent years, gained much attention for its fast growing nature and manifold uses. It is a native of Mexico. The giant varieties were discovered from central America. The tree has a high ecological amplitude. It thrives in naturalized stands in Hawaii from sea level upto 300 m and in Indonesia upto 500m. (Dijkman 1950). It can grown throughout the tropics and subtropics, with an annual rainfall in the range of 500 to 3000 mm and can thrive well in degraded soils, scrublands and even wastelands. *Leucaena* is known to have high quality forage. The wood is medium hardwood useful in lumber, timber, pulp and paper fuelwood and charcoal. It has been used as shade tree for coffee, cocao, paper and vanilla plantations (NRC 1984), and in alley cropping (Nagmbeki 1985).

The nitrogen fixing capability of the tree is derived from the mutual partnerships with bacteria of the genus *Rhizobium*. Recent researches or *L. leucocephala* have focussed the attention on its biomass, forage production and N cycling. In India Sandhu (1990), Sandhu *et al.* (1990). Ambasht *et al* (1992) have made extensive studies on *Leucaena* plantation in dry tropical region, with respect to biomass and energy production, N_2-fixation, litter decomposition, N release and soil nitrogen status. Total nitrogen content of the soil pool is reported to have increased by 290 kg ha^{-1} during one year period (Sandhu 1990). Nitrogen fixation rates were higher in July-August months but the maximum N accretion to soil was in the months of September-October. Annual litterfall was 10.0 t ha^{-1}, of which leaf litter accounted for 88 per cent of the total litterfall. Decomposition rate was fast and 81 per cent of the litter nitrogen was released during one year period of decomposition (Sandhu *et al.*, 1990). The total biomass was 157 t ha^{-1} at the 4-yr age of the plantation. The maximum biomass was accumulated in the bole component fallowed by branch leaf root.

Conclusions

Nitrogen, repeatedly needs to be fixed in order to become available to green plants. Greater attention is desired to fragile habitats which are subjected to heavy biotic pressures. There is a considerable controversy on the extent of natural ability of tropical soils to maintain their fertility against forces of anthropogenic overexploitation of natural vegetation. Here, actinorhizal and leguminous tree species assume importance. They are fast growing, good biomass producers and allow cutting to meet fuel energy demand. These options are very helpful especially for developing countries, where more than 80 per cent of the wood harvested is used as fuelwood. These species successfully cover the thin soil profiles and enrich soil fertility particularly nitrogen. The national commission on Agriculture (1976), has alarmed that the present requirement of fuelwood is 120-130 million tons and it will increase to 300-330 million tons in the year 2005, whereas the present supply from forests is only 50 million tons. This will necessitate fuelwood supply not only from forests but also from arable lands by developing appropriate energy plantations and agroforestry systems.

REFERENCES

Ambasht, N.K., Srivastava, A.K. and Agrawal, M. 1992. *Leucaena oryza* cropping interaction in wetland agriculture. Abs. INTECOLS IV International Wetlands Conference, Sec. V, 119, Ohio, USA.

Ambasht, R.S. and Srivastava, Alok K. 1991. Tropical litter decomposition, a holistic approach. In : Soil organisms and Litter Decomposition in Tropics (ed. M.V. Reddy), In press.

Ambasht, R.S. and Srivastava, Alok K. 1992. Ecology of nitrogen fixing trees in the tropics. Presented in ''International symposium on Rehabilitation of Tropical Rainforest Ecosystems : Research and Development Priorities'' held in Kuching, Sarawak, Malaysia (Proceedings in Press).

Ambasht, R.S. and Srivastava, Alok K. 1993. Nitrogen dynamics of actinorhizal *Casuarina* forest stands and its comparison with *Alnus* and *Leucaena* forests. In Current Science (Communicated).

Binkley, D. 1989. Interaction of site fertility and red alder on ecosystem production in douglas-fir plantation. *Forest Ecology and Management* **5** : 215-227.

Bormann, B.T. and Gordon, J.C. 1980. Stand density, leaf area, conopy structure and atmospheric nitrogen fixation in young red alder plantations. VI : North American Forest Biology Workshop, Aug. 1980. University of Alberta, Edmonton, Alta.

Brewbaker, J.L., Van Den Belt, R. and MacDicken, K. 1982. *Nitrogen fixing tree resources : Potentialities and Limitations.* Workshop on Biological Nitrogen Fixation Technology for Tropical Agriculture, Cali, Colombia.

Dijkman, M.J. 1950. *Leucaena* - a promising soil erosion control plant. *Economic Botany* **4** : 337-349.

Dixon, R.O.D., Berlier, Y.M. and Lespinat, P.A. 1981. Respiration and nitrogen fixation in nodulated roots of soyabean and pea. *Plant and Soil* **61** (1-2) : 135-142.

Doran, J.C. and Hall, N. 1981. Notes on fifteen Australian *Casuarina* species, Division of forest Research, CSIRO, Canberra, A.C.T., Australia, 34 pp.

Johnsrud, S.C. 1978. Nitrogen fixation by nodules of *Alnus incana* in a Norwegian forest ecosystem. *Oikos* **30** : 475-479.

National Research Council. 1984. *Leucaena* : Promising forage and tree crop for tropics. Second edition, National Academy Press, Washington, D.C.

Nagmbeki, D.S. 1985. Economic evaluation of alley cropping *Leucaena* with maize-maize, and maize-cowpea in southern Nigeria. *Journal of Agricultural Sciences* **107** (4).

Sandhu, J. 1990. Nitrogen dynamics in a plantation stand of *Leucaena leucocephala* (Lam.) de Wit. Ph. D. thesis, Banaras Hindu University, Varanasi.

Sandhu, J., Sinha, M. and Ambasht, R.S. 1990. Nitrogen release from decomposing litter of *Leucaena leucacephala* in the dry tropics. *Soil Biology and Biochemistry* **22** (6) : 859-863.

Sharma, E. and Ambasht, R.S. 1984. Seasonal variation in nitrogen fixation by different ages of root nodules of *Alnus nepalensis* plantations in the Eastern Himalayas. *Journal of Applied Ecology* **21** : 265-270.

Sharma, E. and Ambasht, R.S. 1986. Root nodule age class transition, production and decomposition in an age sequence of *Alnus nepalensis* plantation stands in the Eastern Himalayas. *Journal of Applied Ecology* **23** : 689-701.

Sharma, E. and Ambasht, R.S. 1987. Litterfall, decomposition and nutrient release in an age sequence of *Alnus nepalensis* plantation stands in the Eastern Himalaya. *Journal of Ecology* **75** : 997-1010.

Sharma, E. and Ambasht, R.S. 1988. Nitrogen accretion and its energetics in the Himalayan alder. Functional Ecology **2** : 229-235.

Sharma, E. and Ambasht, R.S. 1991. Biomass, productivity and Energetics in Himalayan Alder plantations. *Annals of Botany* **67** : 285-293.

Srivastava, Alok K. 1992. Biomass. Primary Production and Nitrogen dynamics in actinorhizal *Casuarina equisetifolia* (Forest.) plantations. Ph.D. thesis, Banaras Hindu University, Varanasi, India, 183 p.

Srivastava, Alok K. and Ambasht, R.S. 1990. Restoration of soil nitrogen status in an age sequence of *Casuarina equisetifolia* Plantation. IUFRO 19th world Forestry Congress, Report A, 570.

Srivastava, Alok K. and Ambasht, R.S. 1993 a. Spatial and Temporal variation of soil nitrogen in the actinorhizal nitrogen fixing *Casuarina equisetifolia* stands of different age. *Indian Journal of Forestry* **16** (2) : 115-122.

Srivastava, Alok K. and Ambasht, R.S. 1993 b. Root nodule biomass, production, decomposition and nitrogen release in *Casuarina* forest of Sonbhadea, India, Communicated in *Journal of Applied Ecology* U.K.

Srivastava, Alok K. and Ambasht, R.A. 1993 c. Soil moisture control of nitrogen fixation activity in dry tropical *Casuarina* plantation forest. *Journal of Environmental Management* (In press).

Vitousek, P.M. 1984. Litterfall, nutrient cycling and nutrient limitation in tropical forests. *Ecology* **65** : 285-298.

Chapter 34

Eco-Toxicological Effects of Mercury on Freshwater Fish

A.K. Panigrahi, A. Sahu AND N.B.K. Choudhury[1]

Laboratory of Environmental Toxicology, Department of Botany, Berhampur University, Berhampur-760 007, Orissa, India

Abstract. Acclimatized fishes (*Tilapia mossambica,* Peters) were exposed to a sub-lethal concentration (0.06 mg 1^{-1}) of phenyl mercuric acetate in aquaria under controlled laboratory conditions. Exposed fishes showed clinical symptoms, such as inappetence, ataxia, lateral paralysis of ventral fins, which appeared after 4 d of exposure. In addition involution was marked showing body bendings after 20 days of exposure. With increase in exposure period, exophthalmia, strained look, fixation of eye balls were observed. A significant decrease in brain somatic index, liver somatic index and gill somatic index was marked. Decrease in body weight of the PMA exposed fish was observed in contrast to the increase in body weight of the control fish. A significant depression in ventilation rate was marked in the exposed fishes. A partial recovery was marked when transferred to PMA free medium. The whole animal oxygen uptake of the exposed fish showed significant variation when compared to control fish. Significant recovery in respiration rate was marked at lower periods of exposure indicating a partial damage to the metabolic system. However, at higher exposure periods, only a partial recovery was observed, indicating vital damage to the metabolic system.

A striking difference in residual mercury accumulation was marked between the different organs of the exposed fish. Rate of accumulation on mercury in the brain of the exposed fish was the slowest, when compared to liver, muscle and gill. When the exposed fishes were transferred to PMA free medium, a decline in residual mercury concentration was marked, indicating excretion of mercury. Significant variation of haematological parameters in the exposed fish were observed, when compared to the control fish. A partial recovery was marked when the exposed fishes were transferred to PMA free medium.

Key words : Phenyl mercuric acetate, Toxicity, Fish, Respiration rate, Haematology, Residual mercury, Excretion, Body weight, Somatic index.

1. Secretary, Orissa Bigyan Academy Type VI A, 4/2, Unit-1, Bhubaneswar-751 009, Orissa.

Introduction

Pollution of the aquatic environment with mercury from industrial discharge, organic mercury and mercury based pesticides cause an accumulation of mercury in fish. It has been reported that environmental mercury can be dangerously toxic for living systems (Goldwater, 1971). Fishes have the ability to take up mercury from the aquatic environment through gills, skin and gastrointestinal tract and store it, in relatively high concentrations in different tissues (Panigrahi & Misra, 1978). Much work has been published pertaining to residual concentrations of mercury in fishes of freshwater lakes, offshore waters and also estuarine water (Scott and Armstrong, 1972; Forrester *et al.*, 1972; Greig *et al.*, 1975; Smith and Armstrong, 1975; Shaw *et al.*, 1986, 1988; Shaw & Panigrahi, 1986). Reports pertaining to experimental distribution studies of mercury have been studied (Olson *et al.*, 1973, Calabrease *et al.*, 1975; Gibilin and Massaro, 1973; Mc Kim *et al.*, 1976; Laaraman *et al.*, 1976 and Panigrahi & Misra, 1978). In all toxicological and pathological experiments, blood after shows significant changes before external signs of poisoning seen in the animals. The changes in the haematological parameters in fish, are therefore, serve as a better tool for investigating physiological changes caused by the toxicant. The literature available on the experimental toxic effects of heavy metal on blood of fish are of Panigrahi and Misra (1978), Panigrahi and Misra (1979), Panigrahi and Misra (1980), Pandey and Pandey (1977), Raizada and Singh (1980), Calabrese *et al.* (1975), Johansson-Sjobeck *et al.* (1975), Dawson (1979) and O'connor and Fromm (1975).

However, very little information is available relating to the effects of sub-lethal concentrations of organic mercury on the physiological and haematological functions of freshwater fish. The present investigation was undertaken to evaluate the extent of system damage caused by organic mercury on the physiological and haematological functions of freshwater fish. The present investigation was undertaken to evaluate the extent of system damage caused by organic mercury on *Tilapia mossambica*, Peters, in a laboratory conditions.

Materials and Method

Exposure

The test fish, *Tilapia mossambica,* Peters, 8-13 cm long, and weighing 10-14 g, were collected from the local nursery of the Fisheries Department of Berhampur (Ganjam, India), Orissa and were allowed to acclimatize in laboratory (controlled) reservoir for 15 days. The fishes were fed daily small slices of boiled egg during acclimatization holding and exposure period. Fifty fish each were maintained in two 100 l, fibre glass aquaria filled to 50 l, one exposed set where fishes were exposed to 0.06 mg l^{-1} of phenyl mercuric acetate and other the control set. Chlorine free tap water was used in both control and experimental aquaria. The exposure medium was changed daily to maintain the concentration of mercury at 0.06 mgl^{-1} of PMA. The experimental aquarium was thoroughly washed at 10 days interval to ensure the removal of any adhered mercury to the internal glass surface. The experimental aquaria were aerated for 8-10 h each day and kept under 10 h of illumination/day at 2,000 ± 200 Lux. The physico-chemical quality of the water of both control and the test waters were measured during experimental period. The dissolved oxygen was maintained at 95 per cent saturation. The following values were recorded in the control and exposure media : pH 7.2 ± 0.5, temperature 28 ± 2°C, optical density at 540 nm when compared to distilled water 0.015–0.025, conductivity 3.5 × 100 µmho, total hardness 74 ± 2.5 mg l^{-1} as calcium carbonate. Body weight, food consumption and oxygen uptake was recorded periodically.

Respiration

Tilapia fish were exposed to 0.06 mg l^{-1} (MAC = Maximum allowable concentration) of Phenyl mercuric acetate for 28 days and at a 7 days interval, the whole animal oxygen uptake was determined. At 7 days interval a set of fishes were taken out and kept in toxicant free, oxygen saturated water to test the recovery of the fish to this pre-test status. The whole animal oxygen uptake rates of the control and exposed fish were measured using five wide-mouth 21 capacity flasks. Each flask containing the test solution and a test fish, was hermetically sealed. A reference flask was kept with out fish to check any change of oxygen concentration, caused by the micro-organisms were computed with the final data. After 30 minutes, the dissolved oxygen of all experimental flasks were determined according to the modified Winkler's method (Ashby, 1973). The same procedure was repeated for the control fish. The reduction of dissolved oxygen concentration equals the amount of dissolved oxygen consumed by the fish in 30 minutes. The oxygen uptake was expressed as mg of O_2 g^{-1} h^{-1}.. The ventilation rates of *Tilapia* fish were counted by the number of opercular movements per minutes.

Haematology

Since unclotted blood samples are required for haematological studies and the anticoagulant should have no influence on the red cells or other cells, heparin, one of the most powerful anticoagulants was used for the collection of blood. Blood was collected from each fish by cardiac puncture. The sample was transferred gently into a glass vial containing the anticoagulant and frozen utile further use. This process was continued for all experimental purposes during exposure with the aid of a spectrophotometer by acid haematin method (Seivered, 1964). Red blood corpuscle (RBC) counts were made by drawing blood into a standard RBC diluting pipette diluting 1 : 200 with Hayem's fluid and the counting was conducted by the aid of a Neubauer's haemocytometer and a research microscope. The results were expressed as number of cells × 10^6 mm^{-3}. The haematocrit percentage or packed cell volume of blood was determined following Wintrobe's method (Dacie and Lewis, 1970).

Liver somatic index : Liver somatic index (LSI) was determined by the formula

$$\text{LSI} = \frac{\text{Liver (wet weight)} \times 100}{\text{Body weight}}$$

Brain somatic index : Brain somatic index (BSI) was determined by the formula

$$BSI = \frac{\text{Brain (wet weight)} \times 100}{\text{Body weight}}$$

Gill somatic index : Gill somatic index (GSI) was determined by the formula

$$GSI = \frac{\text{Gill (wet weight)} \times 100}{\text{Body weight}}$$

Residual accumulation of mercury and its excretion

Residual mercury accumulation inbrain, liver, muscle and gill tissues were estimated following the 'dithizone method' (Wanntorp and Dyfverman, 1955).

The respiration rate, ventilation rate, concentrations of residual mercury in brain, liver, muscle and gill, the haemoglobin percentage, erythrocyte count, haematocrite percentage, were determined after exposure periods of 7, 14, 21 and 28 days. After 28 days of exposure the fishes were transferred to mercury-free water and all parameters were recorded after further periods of 14 and 28 days. All and data were statistically analysed.

Results

All the exposed fish appeared lethargic after exposure to PMA. The Major clinical symptoms such as inappetence and ataxia appeared after 4 d exposure. After 21 days of exposure swelling of eyes followed by exophthalmia and blindness was noticed in 60% of the total *Tilapia* fish population. Towards the beginning of the 4th week, involution of the fish was observed. The exposed fish showed an initial lateral paralysis leading to total paralysis of the ventral fins. In addition to paralysis, periodic but erratic paralytic movements were observed in exposed fish. The affected fish most of the times settle to the bottom showing a downward head angularly placed body or bended body (Semi 'U' or 'V' shape) and fin floating irregularly. Exposed fish could not regain their pre-exposed status when transferred to toxicant free medium.

Autopsy studies revealed that the liver and brain of fish exposed to mercury after 28 day exposure was congested, wrinkled, flattened, pale and tender. The brain somatic index decreased from an initial 0.810 ± 0.042 to 0.526 ± 0.026, liver somatic index decreased from 0.574 ± 0.037 to 0.286 ± 0.043 and the gill somatic index declined from 2.283 ± 0.033 to 1.250 ± 0.085 within 28 days of exposure. The control fish remained clinically healthy through out the period of experiment (Fig. 34.1). the body weight of the test fish gradually decreased during the exposure period. A maximum of 18.4 ± 4.1 per cent decrease was observed on 28 d exposure. In contrast, a linear but slow increase of a maximum of 10.6 ± 3.2 per cent increase in body weight of the control fish was observed after 28 days (Fig. 34.2).

A striking difference of residual mercury accumulation was observed between the different

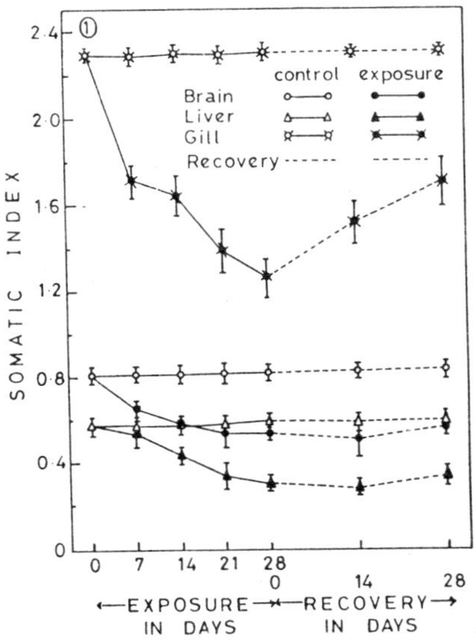

Fig. 34.1. Brain somatic index, liver sometic index and gill somatic index of control and PMA exposed fish, at different days of exposure and recovery. Bars indicate standard deviation.

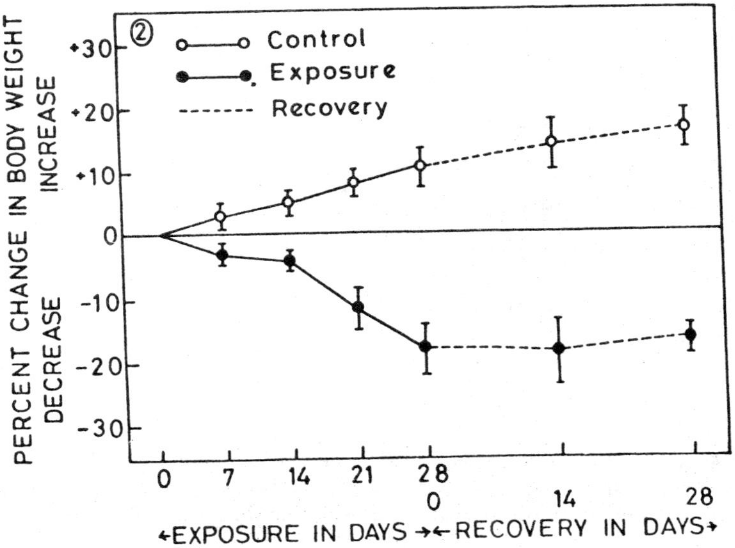

Fig. 34.2. Percent change in body weight of control and PMA exposed fish, at different days of exposure and recovery. Bars indicate standard deviation.

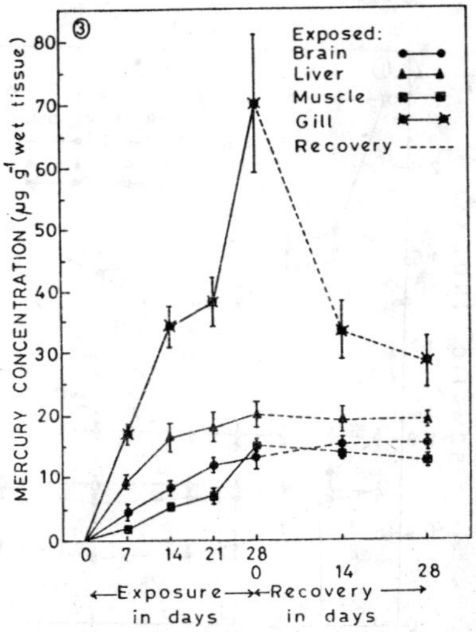

Fig. 34.3. Residual mercury concentration (μg g^{-1} wet tissue) in brain, liver, muscle and gill tissues of PMA exposed fish, at different days of exposure and recovery. Bars indicate standard deviation.

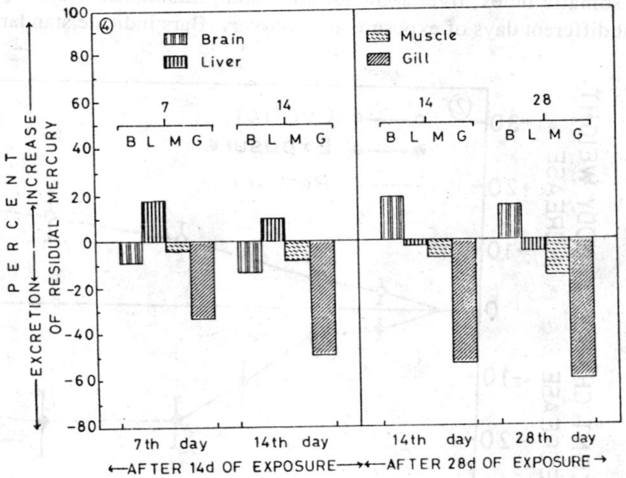

Fig. 34.4. Percent excrition/increase of residual mercury in brain, liver, muscle and gill tissues of PMA exposed fish, at 14 d and 28 d of exposure, when transferred to toxicant free medium for the same period of exposure. Data calculated from the mean of 5 samples.

organs of the same exposed fish. The residual mercury accumulation after 28 days of exposure was 12.66 ± 1.85 μg g^{-1} wet tissue in brain, 19.54 ± 1.85 μg g^{-1} wet tissue in liver, 14.69 ± 1.19 μg g^{-1} wet tissue in musical and 69.67 ± 11.22 μg g^{-1} wet tissue in gill tissues (Fig. 34.3). When the exposed fish was transferred to PMA free medium after 14 d and 28 d exposure interesting results were obtained. In brain after 14 days of exposure, decrease in mercury level was observed at 7 d and 14 d of recovery. However, after 28 days of exposure when the fish was transferred to recovery medium the residual level of mercury increased upto 15.00 ± 1.22 μg g^{-1} wet tissue after 14 days of recovery. On 28th day of recovery, a decline in residual level was marked when compared to 14 days recovery data but the value was still at higher level than 28 days of exposure mercury concentration value (Fig. 34.4). In liver after 14 days of exposure, there covery fish instead of showing any decrease in mercury level showed interesting increase of mercury level at all recovery periods. However, the 28 d exposed fish when transferred to recovery medium, a decrease in mercury level of liver was observed at 14 d and 28 d recovery period (Fig. 34.4). In both the sets of recovery, muscle tissue showed significant decline in mercury content. In case of gill tissues, significant decline was also observed in both the sets of recovery, indicating that gill was the most susceptible site of accumulation and most susceptible site for excretion (Fig. 34.4). The accumulation of mercury showed the existence of a positive correlation with the exposure period. Brain, liver, muscle and gill of the control fish was digested and analysed to determine the presence of background level of mercury. Several tests were conducted, the amount of mercury present in the sample were not traceable.

The ventilation rate in the test fish showed significant declining trend with the increase in the exposure period from 152 ± 09 strokes per minute to 65 ± 13 strokes per minute after 28 days exposure to PMA (Fig. 34.5). The data showed a partial recovery when the exposed fish was allowed to recover in PMA free medium for the same period of exposure. The ventilation rate of the fish has a direct relationship with that of respiration rate of the fish. The control fish did not show any significant variation in ventilation rate throughout the experimental period.

Initially the whole animals oxygen uptake was 0.843 ± 0.041 mg of O_2 g^{-1} h^{-1}, but when the fish was exposed to PMA, the whole animal oxygen uptake decreased linearly and significantly with the increase in exposure period (Fig. 34.6). After 7 d exposure, the whole animal oxygen uptake depleted to 0.542 ± 0.091 mg of O_2 g^{-1} h^{-1} and a significant depletion was observed on 28 d exposure (0.142 ± 0.038 mg of O_2 g^{-1} h^{-1}). The exposed fish were transferred to toxicant free medium after periodic exposure periods (7, 14, 21 and 28 days). The 7 d exposed fishes showed significant recovery and could reach to its pre-exposure level only after 22nd day of recovery. The 2nd batch of 14 d exposed fishes took 37 days to reach 100 per cent recovery. The 3rd batch of 21 d exposed fishes could not recover to its pre-exposure level. Even with the increase in recovery period the exposed fishes did not show any significant recovery. The final batch of 28th day exposed fishes could not recovery was observed, after 28 days of recovery. However, a partial recovery was observed, after 28 days of recovery. The value increased from 0.142 ± 0.038 to 0.292 ± 0.084 mg of O_2 g^{1} h^{-1} after 28 days of recovery (Fig. 34.6).

The haemoglobin concentration in the exposed fish decreased from 14.52 ± 0.76 g/100 ml to 9.16 ± 0.28 g/100 ml, after 28 days of exposure (Fig. 34.7). When the exposed fish was

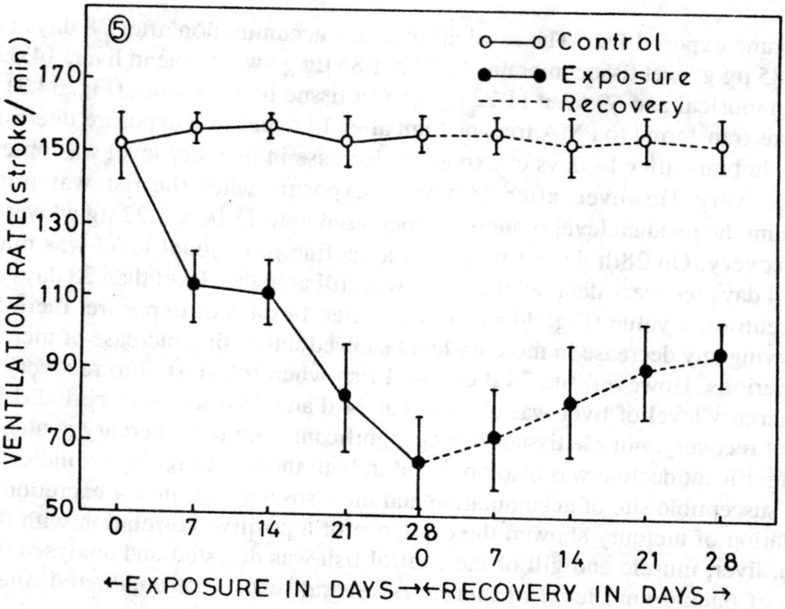

Fig. 34.5. Ventilation rate of PMA exposed fish at different days of exposure and recovery. Bars indicate standard deviation.

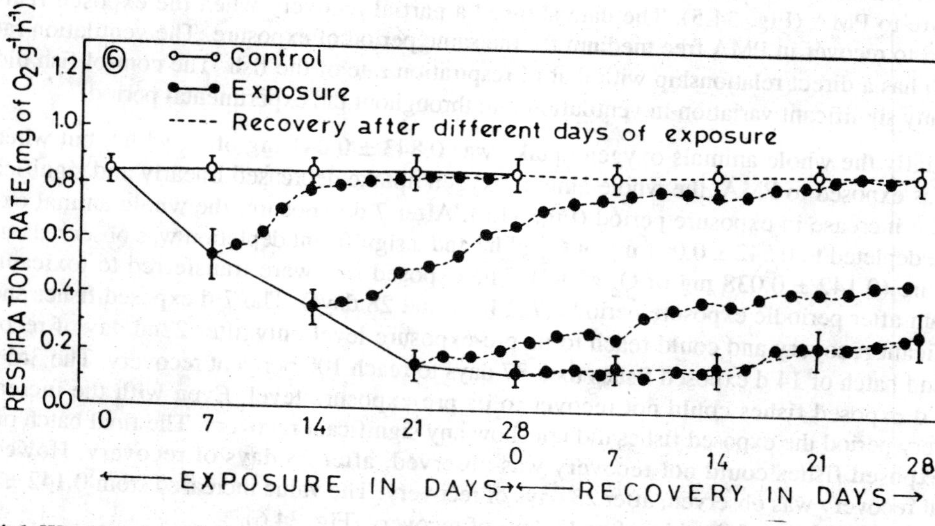

Fig. 34.6. Whole body oxygen uptake by control and PMA exposed fish at different days of exposure and recovery. Bars indicate standard deviation.

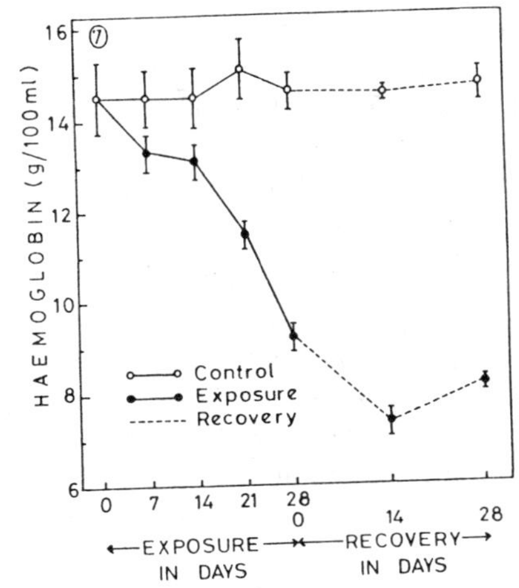

Fig. 34.7. Haemoglobin content of control and PMA exposed fish after different days of exposure and recovery. Bars indicate standard deviation.

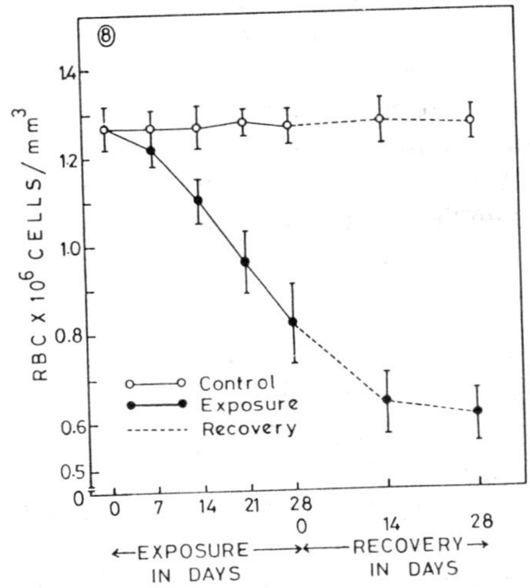

Fig. 34.8. Red blood count in control and PMA exposed fish after different days of exposure and recovery. Bars indicate standard deviation.

transferred to PMA free medium, the haemoglobin content further decreased upto 7.35 ± 0.31 g/100 ml at 14 d recovery. However, partial recovery was marked after 28 days of recovery. The haemoglobin content increased upto 8.15 ± 0.11 g/100 ml within a 28 d recovery period (Fig. 34.10). The red blood cell count also depleted from (1.27 ± 0.05) × 10^6 cells/mm^3 to (0.82 ± 0.09) × 10^6 cells/mm^3 on 28 d exposure (Fig. 34.8). When the exposed finish was transferred to PMA free medium, instead of showing any sign of recovery, the RBC count showed steady fall upto (0.61 ± 0.06) × 10^6 cells/mm^3 on 28 d recovery. The control fish showed no change in these parameters (Fig. 34.10).

The microhaematocrit percentage in exposed fishes showed significant declining trend (negative) at all exposure and recovery periods. The haematocrit value decreased from 35.20 ± 0.63 per cent to 22.25 ± 0.85 per cent at 28 days of exposure (Fig. 34.9). When the fish was transferred to PMA free medium microhaematocrit showed further deterioration. The value depleted to an extent of 17.38 ± 0.40 per cent at 28 d of recovery (Fig. 34.10).

Discussion

Long term sub-lethal effects, resulting from chronic exposure of organisms to low levels of toxicants, probably have the most important impact on the aquatic organisms. Biesinger *et al.* (1982) tested PMA added under renewed static conditions and reported that the probability of rapid breakdown of PMA in water (Matsumura *et al.*, 1971) made it impossible to resolve the relative toxicity of mercury as PMA. The same author reported loss of mercury by volatilization and adsorption, which has not often been reported by investigators and hence suggested that mercury concentrations given in the literature should be considered less than those reported, unless test solution measurements were included. Keeping the above in view, the test solutions were tested periodically to maintain the constancy of the medium and instead of a static test, test solutions were changed daily. If a difference, even marked during a day, PMA was added to maintain the constancy of the medium.

The rapid absorption of inorganic mercury through the gills, skin and gastro intestinal tract has been shown to occur in many species of fish (Gibilin and Massaro, 1973; Olson *et al.*, 1973; Panigrahi and Misra, 1978). The observed depression in active metabolism in mercury exposed fish are indicative of damage to nervous tissue, inhibition of enzymes or vital system was totally in agreement with Panigrahi & Misra (1978, 1980). It is possible that even a small reduction in the normal active metabolic rate may be indicative of stress in the animal (Panigrahi & Misra, 1980). Blindness, exophthalmia and body bending in fish were presumably caused by the effect of mercury on the brain and optic nerves. Berlin *et al.* (1975) have demonstrated that MeHg-induced visual impairment which is an early sign of pre-clinical changes. Panigrahi & Misra (1978 and 1980) reported exophthalmia and blindness in inorganic mercury poisoning and the bending of the body of the inorganic mercury exposed fish to be induced by only mercurial compounds (Panigrahi, 1980). Yamamura *et al.* (1987) reported no change in body weight in short form exposure in both control and MMC exposed rats. However, a decrease in body weight was reported after 8 week MeHg-exposed rats. Growth has often green used to assess the effect of

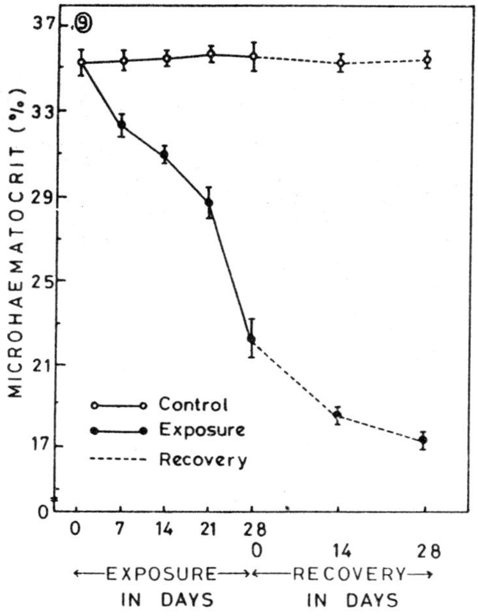

Fig. 34.9. Microhaematocrit value of control and PMA exposed fish after different days of exposure and recovery. Bars standard deviation.

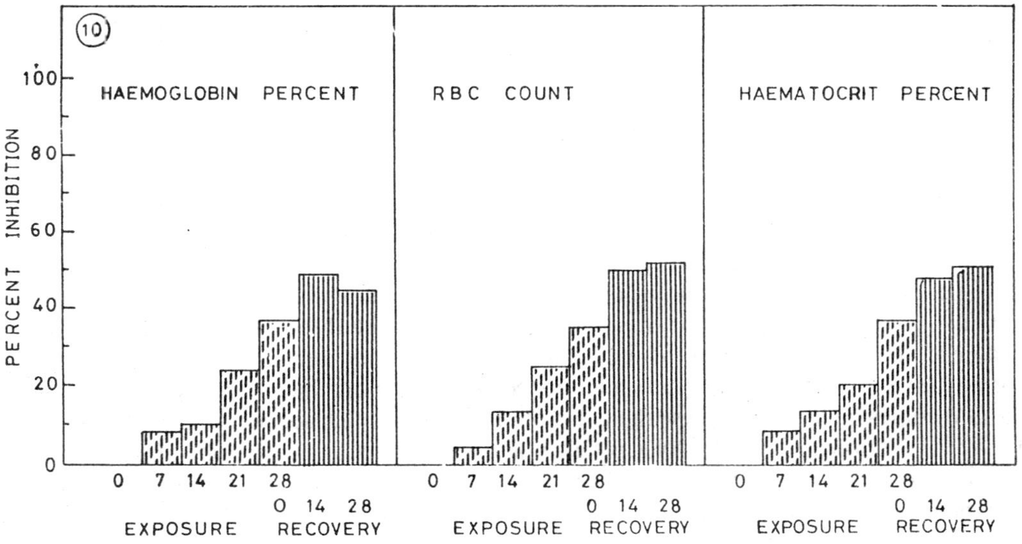

Fig. 34.10. Percent change in haemoglobin (%) RBC count and haematocrit (%) of PMA exposed fish after different days of exposure and recovery. Data computed from the mean of 5 samples. (REC =

toxicants on fish (Buikema *et al.* 1982). However, biochemical changes during toxicosis should precede reduction in growth because growth is the culmination of many biochemical processes; and physiological and biochemical effects are the underlying cause of measurable whole organism effects (Mehrle and Mayer, 1980). A decline in brain somatic index, liver somatic index and gill somatic index in exposed fish could be correlated with degeneration of cells and decrease in other biochemical variables.

The rate of accumulation of mercury is dependent on the type of species, its metabolism (Cutshall *et al.*, 1978) and feeding habit (Greig *et al.*, 1975). The rate of mercury uptake from water by *Tilapia* fish was quite rapid in all tissues. Mc Mik *et al.* (1976) reported the rapid rate of mercury uptake by the brook trout and the amount of mercury accumulated was directly proportional to the water MMC concentration. *Tilapia* fish could accumulate a higher concentration of mercury in gills when compared to other tissues like liver, muscle and brain. This is possible due to the fact that *Tilapia* fish totally depends on the gill for gaseous exchange. The pattern and distribution of mercury in different tissues of the fish reported by Mc Kim *et al.* (1976); Panigrahi and Misra (1978, 1980) are quite agreeable with our findings. The brain of *Tilapia* fish was found to accumulate mercury slowly than muscle, liver and gill and the excretion of mercury was at a much slower rate than other tissues. A maximum of 13.2 per cent excretion of mercury was calculated in 14 d exposed fish kept under recovery for 14 days in case of exposed brain tissue. No sign of excretion was observed at higher exposure periods. In case of exposed liver at lower exposure periods (14 days) no excretion of mercury was recorded. However, an insignificant amount to mercury (4.58%) was excreted after 28 days of recovery. Interestingly exposed muscle showed significant excretion of residual mercury (15.36%) at 28 d recovery of 28 d exposed fish. Most striking rate of excrertion (59.59%) was recorded in gill on 28 d recovery of the exposed fish (28 d). In recovery studies, the increase in mercury concentration in brain of the fish can be explained well. The route of entry of mercury to brain is mostly by blood vascular system. The fish was transferred to mercury free medium but mercury concentration in blood, liver, muscle and gill helps to build higher concentrations of mercury in brain. Similar finding were reported by Panigrahi & Misra (1980) in inorganic mercury intoxication, Gibilin and Massaro (1973) in rainbow trout, Hannerz (1968) in other species of fish. This is possible due to the retarding effect of blood brain barrier to a certain degree. The excretion rate of mercury in brain is very very slow as it lacks any special mechanism of excretion (Panigrahi, 1980). The data indicates that the rate of elimination of mercury is slow. The percentage of elimination from gills is higher than other tissues. It is possible that inorganic mercury biotransformed from methyl mercury (Omata *et al.*, 1980), phenyl mercury (present study) also disturbs cellular activities of the site of its accumulation in these tissues although qualitative and quantitative differences in the toxicity between these compounds remain to be elucidated. The data indicates a low concentration of mercury inmuscle than brain probably confirms the idea that the rate of elimination in the muscle is more than the rate of elimination in brain. Gill comes in direct contact with the exposure medium which accumulates more than brain and liver. The elimination rate of liver is faster than brain and muscle. The rate of accumulation is directly proportional to the period of exposure and to organic mercury. Reports that mercury reacts with complex biomolecules (Panigrahi, 1980) in different organs of the animals and forms complexes, and attachment of the compound to different specific membrane structures (Panigrahi, 1980) and other organelles of the tissues, explains the retention of mercury in the tissues for a longer period, instead of being excreted.

In contaminated waters fish eventually show signs of mercury poisoning and suffer blindness, nerve cell degeneration and muscular atrophy as observed by Takeuchi (1968) and Panigrahi & Misra (1978).

The exposure of the test fish to PMA was followed by loss of body weight, while the control fish increased their body weight. This was probably attributable to the pronounced decrease in the protein content of the muscle tissues of the exposed fish (Panigrahi & Misra, 1978, 1980 and Panigrahi, 1980). The stimulation of protein synthesis (Omata *et al.*, 1980) by MMC can not be fully ascribed to the stimulation of the synthesis of acute phase proteins in the liver since synthesis of all serum protein fractions was stimulated by MMC. Panigrahi & Misra (1978 & 1980) and Panigrahi (1980) reported drastic decline in protein content which contributes to the decline in protein content in inorganic mercury exposed fish. Hence it is really very difficult to say whether mercury (a heavy metal) has any direct role to play either for stimulation or for retardation in protein synthesis. Data were available on both the lines advocating both for stimulation and retardation. However, detailed work, addition of literature with proper interpretation is essential to conclude the effects of mercurials on synthesis of biomolecules.

The effect of organic mercury on the whole animals oxygen uptake and respiratory metabolism is as expected and followed the same trend of inorganic mercury poisoning (Panigrahi, 1980). Oxygen consumption decreased with the increase in exposure period. A drastic decline in respiration rate of the exposed fish, when compared to control fish, indicate the damage done by the toxicant. However, different periodic recovery clearly indicates the damage done by PMA to the respiratory system. In early exposure periods, it is very easy for the animals to recover. However, with the increase in exposure period recovery of the exposed fish becomes difficult, showing a damage to the respiratory system.

Heavy metals are known to effect the cytochrome systems when they attack directly the very first step of the electron transport system and the respiratory chain, as a result of which respiration is hibited partially and when the attack is on the subsequent steps, the respiration rate is reduced significantly and finally leads to death of the animal (Panigrahi, 1980). The interpretation of metal induced changes in respiration is complicated by the fact that such alterations differ from metal to metal, from species to species and from one experimental condition to another (Panigrahi, 1980).

Haematological tests are an important diagnostic tools and recent speculations have suggested that they may be equally valuable as indicator of disease or stress in fish (Calabrese *et al.*, 1975). Haematological observations have greater contribution to the pathological changes obtained during toxicological studies. This is the reason why the blood circulatory system in an animals is the most important system and circulation of blood is the most important function and at the same time before the onset of expression of an effect by any external chemical, blood as well as its constituents and characteristics, change to a significant level. The haemoglobin concentration decreased significantly with the increase in exposure period to PMA and also a drastic depletion in the number of RBC was noticed with the increase in exposure period. When the fishes were transferred to PMA free medium, no significant recovery was marked showing a permanent damage. The decrease in RBC was probably due to (1) reduction of erythrocyte life span due to metal poisoning (Panigrahi, 1980), (2) the destructive or suppresses effects (Panigrahi & Misra, 1978) of erythrogenic tissues, which is expressed as failure in red cell production (de Bruin, 1976), (3) haemoloysis causing haemolytic anemia (Webb, 1966; Johansson-Sjobeck *et al.*, 1975; Panigrahi & Misra, 1978, and Panigrahi, 1980). Mecury causes haemolysis by interacting with legands, constituents of the erythrocytic lipoprotein membrane, as the protein-SH group located in the membrane in the most favoured centre of attack (Panigrahi & Misra, 1979). Other legands including hydroxyl, phosphate and ammonia grouping may also precipitate forming complexes with metal ion. Such interactions while altering the permeability characteristic of the membrane, invariably results in derangement of the cation exchange function of red cell (de Bruin, 1976). The decrease in viability of red blood cells gives rise to accelerated destruction of the erythrocyte. A significant decrease in haematocrit percentage, with no sign of recovery was probably due to haemolysis of red blood cells caused by mercury (Webb, 1966, and Panigrahi, 1980). Larson and Lewander (1973) observed no significant change in the haematocrit values of the European eel, (*Anguilla anguilla*, L) after 145 days of starvation. Hence the decrease in haematocrit values observed in exposed fish can not be attributable as the effect of starvation but due to poisoning of mercury. The decrease in haemoatocrit value is also due to shrinkage of the blood cells (Sovio and Nyholm 1974). Our data are in agreement with those of Calabrese *et al.* (1975), Panigrahi and Misra (1978) and Johansson-Sjobeck *et al.* (1975). Drastic decline in haemoglobin content, RBC count and haematocrit values of the exposed fish is a direct indication of stress caused by mercury.

The difference between the control sets and exposed sets was only due to the addition of phenyl mercuric acetate (0.06 mg 1^{-1}) in the exposed sets. All the parameters studies has definite links, one with other parameters. Residual mercury accumulation in different organs of the exposed fish, clearly indicated the role of mercury played in the metabolic system. In addition, the changes observed in the exposed fish were correlated with the attributable to only residual mercury. If the system under observation, happens to be an important trophic level in the food chain, then biological magnification of the toxicant is inevitable and much higher concentration of the toxicant is predictable.

The results obtained in this study are only valid under the conditions of the experiments, since the interference of one of these factors of the environment will later the MAC values and might result in a different outcome. Such findings emphasize the need for broader research effort to obtain a comprehensive understanding of the consequences of mercury pollution in aquatic environments.

Acknowledgements

We express our gratitude to the Head, Dept. of Botany, Berhampur University, for giving us the necessary laboratory facilities, and to the CSIR (New Delhi) for financial assistance.

REFERENCES

Ashby, M. 1973, *Introduction to plant ecology.* Elbs, M. (ed.) pp. 27. London.

Berlin, M., Grant, C.A., Hellberg, J., Hellstrom, J., Schuta, A. 1975. Neurotoxicity of methyl mercury in sequirrel monkey. *Arch. Environ. Hlth.* **30** : 340-348.

Biesinger, K.E., Anderson, L.E., Eaton, J.G. 1982. Chronic effects of inorganic and organic mercury on *Daphnia magna,* toxicity, accumulation and loss. *Arch. Enciron. Contam. Toxicol.* **11** : 769-774.

Bruin, A. de 1976. Nucleic acid metabolism. In : *Biochemical toxicology of environmental agent.* Ed. Bruin, A. de, Elsevier/North Holland Biomedical Press, The Netherlands, pp. 612.

Buikema, A.L., Niederlehner, B.R., Cairns, Jr. J. 1982. Biological monitoring. Part IV toxicity testing. *Wat. Res.* **16** : 239-262.

Calabrese, A.L., Thurberg, F.P., Dawson, M.A., Wenzloff, D.R. 1975. Sub-lethal physiological stress induced by cadmium and mercury in white Hounder, *Pseudoplenoneectes americanus.* In : *Sub-lethal effects of toxic chemicals on aquatic animals.* Koeman, J.H., Strik, J.J.T.W.T. (eds.). Amsterdam : Elsevier.

Cutshall, N.H., Naidu, J.R., Pearcy, W.G. 1970. Mercury concentrations in Pacific hake, *Merluccivs productus* (Ayres) as a function of length and latitude. *Sci.* **200** : 1489-1491 (1978).

Dacie, J.V., Lewis, S.M. : *Practical haematology.* 4th Edn, J. and A. Churchil, London, pp. 19.

Dawson, M.A. 1979. Haematological effects of long term mercury exposure and subsequent period of recovery on the winter flounder, *Pseudopleuronactus americanus. Mar. Pollut. Func. Resp.*, pp. 171.

Forrester, C.R., Ketchen, K.S., Wong, C.C. 1972. Mercury content of spiny dogfish (*Squalus acanthias*) in the strait of Georgia, British Columbia. *J. Fish. Res. Bd. Can.* **29** : 1487-1490.

Gibilin, F.J., Massaro, E.J. 1973. Pharamacodynamics of methyl mercury in the rainbow trout (*Salmo gairdneri*) : Tissue uptake, distribution and excretion. *Toxicol. Appl. Pharmacol.* **24** : 81-91.

Goldwater, L.J. 1971. Mircury in the environment. *Sci. Am.* **224** : 15.

Greig, R.A., Wenzloff, D., Shelpuk, C. 1975. Mercury concentrations in fish, North Atlantic Offshore Waters-1971. *Pest. Monit J.* **9** : 15-20

Hannerz, L. 1968. *Experimental investigations on the accumulation of mercury in water organisms.* Rep. Inst. Freshwater Res. Drottingholm, **48** A : 120-126.

Johansson-Sjobeck, M.L., Dave, G., Lidman, U.A. 1975. Haematological effects of PCB in the European eel, *Anguilla anguilla* L. and rainbow trout *Salmo gairdneri,* Rich. *Sub-lethal effects of toxic chemicals on aquatic animals.* Koeman, J.H. Strik, J.J.T.W.A. (eds.). Amsterdam : Elsevier.

Laarman, P.W., Willford, W.A., Olson, J.R. 1976. Retention of mercury in the muscle of yellow Pearch (*Perca flavescens*) and rock bass (*Ambloplihes rupestris*) : *Trans. Amer. Fish. Soc.* **105** (2) : 296-300.

Larson, A. Lewander, K. 1973. Metabolic effects of starvation in the eel, *Anguilla anguilla,* L. *Comp. Biochem. Physiol.*, **44** : 367-374.

Matsumura, F., Gotoh, Y., Boush, G.M. 1971. Phenyl mercuric acetate : Metabolic conversion by micro-organisms. *Sci.* **173** : 49.

Mc Kim, J.M., Olson, G.F., Holcombe, G.W., Hunt, E.P., 1976. Long term effects of methyl mercuric chloride on three generations of brook trout (*Salvalinus frontainlis*) : Toxicity, accumulation, distribution and elimination. *J. Fish. Res. Bd. Can.* **33** : 2726.

Mehrle, P.M., Mayer, F.L. 1980. *Clinical tests in aquatic toxicology* : State of the art. *Environ. Hlth. Perspect.,* **34** : 139-143.

O'Connor, D.V., Fromm, P.C. 1975. The effect of methyl mercury on gill metabolism and blood parameters of rainbow trout. *Bull. Environ. Contam. Toxicol.* **13** : 406-41.

Olson, K.R., Bergman, H.L., Fromm, P.O. 1973. Uptake of methylmercuric chloride by trout : A study of uptake pathways into the whole animals and uptake by erythrocytes in vitro. *J. Fish. Res. Bd. Can.*, **30** : 1293-1299.

Omata, S., Sato, M., Sakimura, K., Sugano, H. 1980. Time-dependent accumulation of inorganic mercury in subcellular fractions of kidney, liver, and brain of rats exposed to methyl mercury. *Arch. Toxicol.* **44** : 231-241.

Pandey, K.C., Pandey, A.C. 1977. Haematology of a cat fish *Rita rita* (Ham.). *Proc. Ind. Acad. Sci.* **858** (6), 369.

Panigrahi, A.K. 1980. Eco-physiological effects of mercurial compound on some freshwater fishes. Ph. D. thesis, Berhampur University, Orissa, India.

Panigrahi, A.K., Misra, B.N. 1978. Toxicological effects of inorganic mercury on a freshwater fish, *Anabas scandens,* Cuv. & Val. and its ecological implications. *Environ. Pollut.*, **16** : 31-39.

Panigrahi, A.K., Misra, B.N. 1979. Effects of mercury on the morphology of erythrocytes in *Anabas scandens. Bull. Environ. Contam. Toxicol.* **23** : 784-787.

Panigrahi, A.K., Misra, B.N. 1980. Toxicological effects of a sub-lethal concentration of inorganic mercury on the freshwater fish, *Tilapia mossambica,* Peters. *Arch. Toxicol.* **44** : 269-278.

Raizada, M.N., Singh, C.P. 1980. Observations on haemotological values of freshwater fish, *Channa punctatus* (Bloch). *Comp. Physiol Ecol.* **5** (3) : 159.

Scott, D.P., Armstrong, F.A.J. 1972. Mercury concentration in relation to size in several species of freshwater fishes from Manitoba & North Western Ontario. *J. Fish. Res. Bd. Can.* **29** : 1685-1690.

Seivered, C.E. 1964. *Haematology for medical technologists.* Lea and Febiger, Philadelphia, pp. 334.

Shaw, B.P., Panigrahi, A.K. 1986. Uptake and tissue distribution of mercury in some plant species collected from a contaminated area in India : its ecological implications. *Arch. Environ. Contam. Toxicol* .**15** : 439-446.

Shaw, B.P., Sahu, A., Panigrahi, A.K. 1988. Mercury in plants, soil and water from a Caustic Chlorine industry. *Bull. Environ. Contam. Toxicol.* **36** : 299-305.

Soivio, A., Nyholm, K. 1974. The influence of changes in O_2 tension on the haemotocrit value of blood samples from asphysix rainbow trout, (*Salmo gairdneri*). *Aquacult.* **3** : 395.

Smith, T.G., Armstrong, F.A.J. 1975 Mercury in seals, terrestrial carnivores, and principal food items of the Inuit, from Holman, N.W.T. *J. Fish. Res. Bd. Can.* **32** : 795-801.

Takeuchi, T. 1968. *Pathology of Minamata disease*. Kutsuma, K. (ed.) Study group of Minamata Disease, Vol II, pp. 141-228. Japan.

Wanntorp, H,., Dyfverman, A. 1955. Identification and determination of mercury in biological materials. *Arkiv. Kemi* **9** : 7-27.

Webb, J.L. 1966. *Enzyme and metabolic inhibitors,* Vol. II. New York : Academic Press.

Yamamura, K., Maehara, N., Ohno, H., Ueno, N., Kohyama, A., Satoh, T., Shimoda, A., Kishi, R. 1987. Effects of MMC of Low concentration on the rat nervous system. *Bull. Environ. Contam. Toxicol.* **38** : 985-993.

Chapter 35

Ecotoxicological Studies on Aquatic Algae : A Review

S.C. Santra AND **S. Dan**

Department of Ecological Studies School of Environmental Sciences University of Kalyani, Nadia West Bengal, India

Abstract. Over past few decades, a considerable studies were carried out on the ecotoxicology on a variety of aquatic organisms, particularly on algae. In the present paper a comprehensive review on the subject were attempted.

Key words : *Aquatic algae, ecotoxicology, metal toxicity, Pesticides.*

Introduction

For a number of years there has been a series of investigations indicating that one or more algae could be used as organisms indicative of water quality. A variety of suggestions has been proposed, for instance Palmer (1955) demonstrated that algal assemblages could be used as indicators of clean water or polluted water. Later in 1969, Palmer published a composite rating of organisms such as *Euglena, Oscillatoria, Chlamydonomas, Scenedesmus, Chlorella, Stigeoclonium, Nitzschia* and *Navicula* etc. which could be used to indicate that the water was polluted. On the other hand different organisms such as *Lemanea, Stigeoclonium* and certain species of *Micrasterias, Staurastrum, Pinnularia, Meridion* and *Surirella* would indicate the one was sampling from a body of water considered clean water.

Identically Patrick (1972) also described a technique which could be useful in determining water quality. She prepared a normal species distribution curve for both clean and polluted water. Later two new assays, the algal bioassay compromise (ABC) and the sensitive algal assay procedure (SAAP) were often applied for a variety of ecotoxicological studies. Although the concept of combination assays might some day be a useful one, we have to carefully select the organism and the phenomenon we choose to manipulate. Certainly we need more concern for field conditions when we choose to manipulate. Certainly we need more concern for field conditions when we routinely study algal strains in culture. Otherwise, for some time in the future we will find extrapolation from one area to the other increasing difficult.

However, a basic problem with using an indicator assay concept is that we have to question whether we can use the abundant laboratory data to understand the functioning or organisms in nature. In some instances the conditions of culture which have been used for decades preclude any reasonable understanding of a field situation.

The direct, short-term effects of environmental contaminants on aquatic algae have been the subject of a myriad of different types of laboratory studies. However, the secondary or indirect effects of contaminants on algae, that is the change in the structure and function of algal communities resulting from toxic effects on other oganisms in the aquatic community, have received less attention. The role of algae in the disappearance and degradation of environmental contaminants has only recently been addressed.

Algal species offer useful test system for predictive ecotoxicology (Palmer, 1980; Walsh *et al*., 1982; Chawla *et al*., 1986, 1987), various protocols for ecotoxicity evaluation by algal growth studies have been suggested in the past by various agencies like EPA (1975), and OECD (1981).

In the classical treatise on ecotoxicology algae and other aquatic plant communities were used for biomonitoring purposes (Hunding and Lange, 1978; Moodey *et al*., 1983). They considered the parameters such as photosynthetic assimilation of carbon, liberation of oxygen, respiration, biomass, growth, uptake and transformation and accumulation of aquatic toxicants as possible indicators for both short term and long term exposure effects. The role of biological monitoring and ecotoxicology in the identification of problems and their prediction and prevention connected with fresh water pollution was fully realised, but still considerable in depth study is needed.

The presence of chemical contaminants has become a pervasive threat to many natural aquatic ecosystems. These contaminants can have toxic effects on many different types of organisms and affect biological processes at the cellular, population, community and ecosystem levels of organisations. These chemical contaminant are often called environmental contaminants which arises from a variety of sources that are usually classified as point source or non-point source pollutants. The effects of toxic chemicals have been measured on a wide variety of algal species by using different cultural methods and number of different biological responses. Algae commonly tested ranges from common, to rare and toxic fresh water, and marine species of various divisions. Cultural methods vary in composition of solutes in the nutrient medium, in nutrient concentration; in temperature and in intensity, periodicity, and quality of light. Various measurements of algal response to chemical contaminants omcide : (i) photosynthetic uptake of radiolabelled carbon dioxide, an indication of the functioning of photosystem I associated with chlorophyll-a; (ii) Evolution of oxygen, a measure of rate of the Hill reaction in photosystem II associated with chlorophyll-b; (iii) measures of relative population growth in time based on changes in cell numbers, amounts of chlorophyll-a extracted *in vitro*, turbidity of a cell suspension, and change in the dry weight of the culture; and

(iv) measures of critical physiological and biochemical rates, such as synthesis of lipids, protein, and nucleic acids, as well as the uptake of organic and inorganic nutrients from nutrients media.

However, the response of a particular species of algae to a potentially toxic contaminant is very dependent on cultural conditions. Many physical and chemical factors such as temperature, water hardness, pH and nutrient conditions may affect the toxicity of chemicals. To best understand the potential of direct effects of environmental contaminants on aquatic algae, it is important first to consider the examples of these effects on processes at the cellular level of organisation.

Indirect or secondary effects due to contaminant impact on organism other than algae are chiefly from alteration of aquatic macrophyte communities and reduction of zooplanition communities. Death of large masses of aquatic macrophytes may release large pulses of nutrients but only when accompained by severe deoxygenation of the water, complete suppression of macrophyte growth results in a reduction of nutrient concentrations in the water which results in less phytoplankton. Reduction in nutrients is due either to the loss of nutrient release of the macrophytes or to the increase of the oxidation-reduction potential of the mud-water interface which renders forms of nitrogen and phosphorus less soluble and reduces nutrient resuspension from the sediments.

The secondary effect on algae due to contaminant stress on the zooplankton community appears to be an increase in the algal standing biomass, photosynthetic rate and nitrogen fixation.

Algae take up and bioconcentrate a variety of environmental contaminants present in the water. The rate of chemical bioconcentration and the final amount are functions of the particular species of algae, a number of physical and chemical environmental factors, and the nature of the contaminant. Furthermore, aquatic algae have recently been implicated in the degradation of organic contaminants. Algae appear to act directly in the degradation of organic chemicals or mediate photolysis. Thus, algae may be an important factor in the mitigation of contaminant effects on other organisms by degradation to a harmless form.

Current Status on Ecotoxicological Studies on Algae

A. Toxic Metals

Metallic elements are an intrinsic component of the environment, their presence is considered unique in the sense that it is difficult to remove them completely from the environment once they enter in it. Erosion of surface deposits of minerals, forest fires and volcanic activity are some of the natural causes of environmental pollution with the toxic metals. The processes involved in the extraction of metals from the ores and their extensive use resulting from improved technology are also responsible for their dispersion in the biosphere. With the increasing use of a wide variety of metals in industry and in our daily life, problems arising from toxic metal pollution of the environment have assumed serious dimensions. Toxic metals, to a large extent, are dispersed in the biosphere through industrial effluents, organic wastes, refuse burning, transport and power generation. They can be carried to places many miles away from the soûrce by the winds depending on their physical state, viz., whether they are in the gaseous form or as particulates. Metallic pollutants are ultimately washed out of the air by rain onto land or the surface of the water ways.

During the past decade an increasing number of people concerned with environmental monitoring have gained access to sophisticated and rapid techniques for the analysis of heavy metals. At the same time there has been a relatively large number of accounts of biological methods for assessing the environmental impact of these metals, including some with freshwater algae (Table 35.1).

The literature concerning biological effects of heavy metal contamination in rivers and streams is extensive, with much of it directly or indirectly relevant to monitoring (Phillips, 1980; Whitton, 1984). In general the biological methods for monitoring heavy metal contamination fall mostly into four categories :

* Analysis of metal composition of organisms which are attached or have a restricted range of distribution;
* Toxicity assays;
* Analysis of species and community composition;
* Assessment of adaptation.

The subject of heavy metal toxicity to algae and cyanobacteria has received considerable attention because of the productivity of aquatics is maintained by a balance between photosynthesis and respiration of autotrophic/heterotrophic communities. Inhibition of either photosynthesis or respiration ultimately affects the productivity of the aquatic food chain (Rai *et al.*,1981; Prasad *et al.*, 1991).

Again among the algae, cyanobacteria are a fascinating group of prokaryotic photosynthetic N_2-fixers ecologically diverse habitats. With respect to metallic toxicity in blue green algal system, it is presumed that functional groups of many toxic metals were bound as ligands by various organic molecules of the BGA system.

Quite a considerable number of metal toxicity assay was made on BGA system alone. These include the effect of chromium, lead, cadmium, copper, and mercury on different species of *Anabaena* and *Nostoc* in particular (Rai and Raizada, 1988; Ray *et al.*, 1990, 1991; Husaini *et al.*, 1991; Prasad *et al.*, 1991).

In the past there were some review on the aspect of heavy metal toxicity on fresh water algae (Whitton, 1970; Hutchinson and Stokes, 1973; Rai *et al.*, 1981; Whitton, 1984). These reviews discussed the detailed accounts of algal bioassay for metal toxicity, as well as algaeas bioindicator of metal pollution too. Over past three decades, metallic effect on various algae were studied with respect to algicidal requirement or as trace metals required for growth and

TABLE 35.1. *Laboratory Bio-assay for Metal Toxicity*

Metals	Organisms	Observations	References
Cu	*Chlorella vulgaris*	Toxicity on growth	McBrien and Hassall (1967)
Cr, Zn	*Nitzschia linearis* (diatom)	Toxicity assay	Patrick and Crains (1968)
Cu	*Chlorella pyrenoidosa*	Photosynthesis	Steemann, Nielsen *et al.* (1969)
Polymetalic ores and concentrates (containing Cu, Zn, Sn, Pb)	*Microcystis aeruginosa*	Growth process	Dimitrieva (1969)
Cu	*Chlorella* sps.	Damage of cell	Gross *et al.* (1970)
Cu	Diatom	Toxicity of cell	Steemann, Nielsen *et al.* (1971)
Cu	Green and Blue green algae	Algicidal effect	Gibson (1972)
Al	*Chlorella pyrenoidosa*	Toxicity assay	Foy and Gerloff (1972)
Cd	*Scenedesmus quadracauda*	Growth	Klass *et al.* (1974)
Cr	*Chlorella pyrenoidosa*	Photosynthesis and Growth	Wium Anderson (1974)
Zn	*Chlorella vulgaris* and *Plectonema boryanum*	Toxicity assay	Rana and Kumar (1974)
Cu, Zn, & Cd	*Selenastrum capricornutun*	Toxicity assay	Bartlett *et al.* (1974)
Cu, Zn	*Hormidium rivulare*	Influence of P (H) on overall toxicity	Hargreaves & Whitton (1976)
Pb	*Selenastrum capricornutum*	Growth inhibition	Monahan (1976)
Hg & Zn	*Chlorella* sp.	Toxicity at sub-lethal concentration	DeFilippis & Pallaghy (1976)
Cn, Zn, Cd	*Selenastrum capricornutum*	Toxicity bioassay	Miller *et al.* (1976)
Zn	*Stigeoclonium tenue*	Resistance	Harding and Whitton (1976)
Hg	*Anacystis* sp.	Photosynthesis	Kleinen Hammans *et al.* (1976)
Zn	*Stigeochlonium tenue*	Toxicity alteration	Harding and Whitton (1977)
Cd	*Asterionella formosa*	Growth	Conway (1978)

(*Contd. Next Page*)

Metals	Organisms	Observations	References
Mn	Blue green algae	Toxicity assay	Singh and Kashyap (1978)
Cu	*Lungbya nigra*	Growth	Gupta and Arora (1978)
Cu	*Oscillatoria theibauti*	Toxicity effect	Ructer *et al.* (1978)
Hg	*Scenedesmus acutus*	Growth	Huisnan Ten Hooper (1978)
Pb	5 species of green algae	Toxicity assay	Weber *et al.* (1978)
Cd	*Euglena gracilis*	Growth	Nakano *et al.* (1978)
Pb	*Chlamydomonas* sp.	Toxicity assay	Schulze & Brand (1978)
Cr, Ni, Cu	*Scenedesmus quadricauda*	Toxicity assay	Wong *et al.* (1978)
Heavy metal compounds	*Chlorella cells*	Effect of cell permeability	De Filippis (1979)
Hg	*Anabaena ineqaualis*	Photosynthesis	Stratton *et al.* (1979)
Cr	*Chlorella pyrenoidosa*	Toxicity assay	Meisch and Schmitt Beckmann (1979)
Cu	21 algae	Extracellular complexing agent	Mcknight and Morel (1979)
Hg	*Chlorella vulgaris*	Toxicity reversal by ascorbic acid and glutathione	Rai (1979)
Cu, Pb	*Diatoma tenue,* var *elongatum*	Influence on polyphate bodies	Sicko-Goad & Stoerner (1979)
Cu	*Anacystis nidulans*	Resistance	Sarma (1979)
Cu	Filamentus green algae	Metal tolerance	Francke and Hillebrand (1980)
Cd	*Euqlena gracilis*	Cytotoxicity	Bonaly *et al.* (1980)
Cd & Cu	*Ankistrodesmus braunii* and *Anabaena* sp.	Ultrastructural changes	Massalski, Laub & Kusnner (1981)
Zn, Cd, Hg	*Euqlena gracilis*	Cell division and movement	De Filippis *et al.* (1981)
Zn	Blue green algae	Resistance	Shehata & Whitton (1981)
Cu, Zn	*Euqlena exigua*	Resistance	Whitton and Daiz (1981)
Zn	*Anacystis nidulans*	Resistance	Fahmi *et al.* (1982)
Heavy metals	*Plectonema boryanum*	Changes in cellular components	Jensen *et al.* (1982)

(*Contd. Next Page*)

Metals	Organisms	Observations	References
Co, Ni, Cu & Cd	*Anacystis nidulans*	Toxicity assay	Whitton and Shehata (1982)
Cd	*Nostoc calcicola*	Resistance	Singh and Pandey (1982)
Eight heavy metals (Mn, Zn, Hg, Pb, Cu, Cd, Co, Ni)	*Plectonema boryanum* and *Anabaena flosaquae*	Morphometric analysis and nitrogen fixing ability.	Rachlin *et al.* (1982)
Cd	*Anacystis nidulans*	Toxicity on enzymes	Singh and Yadava (1983)
Zn	*Chlorella vulgaris* and *Phormidium foveolarum*	Growth analysis	Pandey and Tripathi (1983)
Cu, Zn and CD	*Chronoccus pairs*	Toxicity and binding	Les and Walker (1984)
Cd	*Anacystis nidulans*	Toxicity of Enzymes	Singh and Yadava (1984, 1985)
Hg, Zn	*Spirulina platensis*	Toxicity assay	Kotangate *et al.* (1984)
Ni and Ag	*Nostoc muscorum*	Growth and other bio-chemical properties	Rai and Raizada (1985, 1986)
Cd	Some green algae	Toxicity	Vymazal (1987)
Cr, and Pb	*Nostoc muscorum*	Toxicity regulation	Rai and Raizada (1988)
Ag	Blue green algae	Toxicity	Rai (1989)
Cd	*Anabaena flos-aguae*	Toxicity monitoring	Rai *et al.* (1990)
Heavy metal	*Anabaena doliolum*	Toxicity assay	Dubey and Rai (1990)
Cu, Ni, Fe	*Anabaena doliolum*	Physiological and biochemical characteristtic of a copper tolerant and wild strain	Rai *et al.* (1991)
Pb	*Nostoc muscorum*	Metal induced inhibition of photosynthetic electron transpory chain	Prasad *et al.* (1991)
Cd	*Nostoc linckia*	Toxicity to photosynthetic and associated electron transporty system	Husaini *et al.* (1991)
Zn	*Chlorella vulgaris*	Survival and growth behaviour	Srivastava *et al.*

survival of algal population. In many cases, however, biochemical studies like photosynthetic activities, enzyme assay etc. were made with respect to metal toxicity. Identically considerable amount of investigations were made with respect to metal tolerance also. On the whole, majority of studies were made on unicellular green algae like *Chlorella* and *Scenedesmus* and also a number of blue green algae in *in vitro* system.

The earlier use of algae in laboratory assays were concerned mostly with algicides for swimming pools or blue green algal blooms in reservoirs. Assays on a range of possible algicides, including Cu and Ag were reported by G.P. Fitzgerald and his colleagues. Their studies used straight forward batch culture techniques, with liquid media (25 ml in 50 ml Flasks), but without shaking or aeration. The organisms tested were mostly one isolated specifically for the relevant sites. *Microcystis aeruginosa* and *Chlorella pyrenoidosa* were used mostly to assay both the algicidal and algistatic properties of copper supplied in various ways (Fitzgerald and Faust, 1963). However the most detailed studies of Cu as toxic substance or algicides were that of Elder and Horne (1978) and Kallqvist and Meadows (1978).

In 1969, Dimitrieva reported the effect of polymetallic ores and concentrates (containing Cu, Zn, Pb, Sn) on the growth of *Microcystis aeruginosa* in laboratory culture. Only 4 ppm metallic solution can able to reduce population in 10 days incubation period. Thereafter the effect of a number of heavy metals on algal growth and subsequently the interaction of various factors on toxicity of few selected metals were studied in separate experiments *in vitro* (Steemann *et al.*, 1969, 1971; Gross *et al.*, 1970; Foy and Gerloff, 1972; Gibson, 1972; Klaas *et al.*, 1974; Wium Anderson, 1974; Rana and Kumar, 1974; Hargreaves and Whitton, 1976; Monahan, 1976; De Filippis and Pallaghy, 1976; Miller *et al.*, 1976; Harding and Whitton, 1976; Kleinen Hammans *et al.*, 1976; Harding and Whitton, 1977; Conway, 1978; Singh and Kashyap, 1978; Gupta and Arora, 1978; Ructer *et al.*, 1978; Bonaly *et al.* 1980; De Filippis *et al.*, 1981; Massalski *et al.*, 1981; Shehata and Whitton, 1981; Whitton and Diaz, 1981; Jensen *et al.*, 1982; Whitton and Shehata, 1982; Singh and Pandey, 1982; Rachlin *et al.*, 1982; Singh and Yadava, 1983; Pandey and Tripathi, 1983; Rachlin *et al.*, 1984; Les and Walker, 1984; Singh and Yadava, 1984, 1985; Kotangate *et al.*, 1984; Rai and Raizada, 1985, 1986; Vymazal, 1987; Rai and Raizada, 1988; Dubey and Rai, 1990; Mallick and Rai *et al.*, 1990; Prasad *et al.* 1991; Husaini *et al.*, 1991; Srivastava *et al.*, 1991).

From the above studies, it appears that algae are a valuable tool for bioassay of metal toxicity. But most studies dealing with heavy metal toxicity to algae are confined to the laboratory microcosm under a defined set of culture media and environmental conditions (Whitton, 1984); using one species of algae and a single metal of bimetallic combinations. Although laboratory test data do give valuable information about toxicity, but they cannot be used directly to monitory changes in the algal communities of aquatic system in the field.

On careful examinations of different metal toxicity studies of algae, it appears that most of the studies were made to investigate the growth behavior, pigments, heterocyst differentiation, carbon and nitrogen fixation. A few studies however also made to understand the ultrastructural changes induced by toxic metals.

Although high concentrations of metals are known to be toxic to biota, reports on organisms surviving in extreme environments with elevated levels of heavy metals normally toxic to them are frequent. Tolerance to heavy metals in algae may develop either while growing in metal contaminated natural habitats or through successive cultivation at elevated doses in the laboratory. Cases of acquired resistance to various metals in algae growing in metal contaminated fresh water habitats are numerous (Massall, 1963; Gibson, 1972; Foy and Gerloff, 1972; Jensen *et al.*, 1982; Stokes *et al.*, 1977; Stokes, 1975 Mierle and stokes, 1976; Harding and Whitton, 1976; Foster, 1977; Singh and Kashyap, 1978; Hall *et al.*, 1979; Butler *et al.*, 1980; Whitton and Daiz, 1981; Harding and Whitton, 1981; Foster 1982; Whitton and Shehata, 1982; Whitton, 1984; Rai *et al.*, 1990). Despite a wealth of information on metal toxicity in algae, little attempt seems to have been made to furnish biochemical and physiological evidences to demonstrate metal effects and its tolerance in most algae particularly in cyanobacteria.

It is further interesting to note the fact that transport and subsequent binding of toxic metals to different parts of that algal components in another interesting area of investigation. There are a few studies carried out in the past on the aforesaid subject (Wang and Tischer, 1973; Wolk, 1973; Stokes, 1975; Murphy *et al.*, 1976; Davies, 1976; Hart and Scaife, 1977; Gowrinath and Rao 1990). In addition a few kinds of metal binding protein components were also isolated and characterised from metal tolerant strains of algae (Maclean *et al.*, 1972; Stokes *et al.*, 1977; Gingrich *et al.*, 1986 : Harwood-Sears and Gordon, 1990).

On the whole, algae have been employed as monitors (or indicators) of heavy metals in freshwaters using many different approaches, yet none of these approaches have been developed so well that it can be applied routinely for water management purposes. Thus there is an urgent need to develop a few standard methods which could lead to the use of algae becoming routine tools for monitoring metals in freshwaters.

B. Pesticides

Much literature has also been accumulated detailing the harmful or sometime accelerating effects of various pesticides and herbicides on algal system, more specifically on blue green algae (Table 35.2). The toxicity of different pesticides on Cyanobacteria in particular and factors affecting it received greater attention in past two decades or so. There were two interesting reviews published in the past on the said aspect (Wright, 1978; Padhy, 1985).

Among the pesticides, DDT (an organochlorine insecticide), is found to affect adversely many micro-algae (Cox, 1972). Growth of the Cyanobacteria *Anacystis nidulans* (Gregory *et al.*, 1969). *Microcystis aeruginosa* and *Anabaena cylindrica* (Vance and Drummond, 1969) were adversely affected at 1000 ppb (parts per billion) DDT. *Anabaena cylindrica* was found to be most sensitive. BHC, another organochlorine insectide was also tested on a number of blue green algae like *Microcystis aeruginosa, Aulosira fertilissima, Plectanema boryanum, Cylindrospermum* sp., *Anabeana spirodes, Microcystis flosaquae, Tolypothrix tenuis* and *Anabaenopsis raciborskii* etc. (Fitzgerald *et al.*, 1952; Palmer and Maloney, 1955; Ahmad and

TABLE 35.2. *Effects of Pesticides on Microalgae*

Pesticides	Organism	Observations	References
BHC and Dichlone	*Microcystis aeruginosa*	Growth inhibition	Fitzgerald *et al.* (1952)
BHC and Dichlone	*Microcystis aeruginosa*	Growth inhibition	Palmer and Meloney (1955)
Prathion	*Anabaena* sp.	Growth inhibition	Cook and Conners (1963)
Lindane	*Anacystis nidulans*	Growth inhibition	Lazaroff and Moore (1966)
Nabam and Captan	*Nostoc muscorum*	Growth inhibition	Morre (1967)
Captan	*Anacystis nidulans*	Growth inhibition	Moore, Dorword (1968)
DDT	*Anacystis nidulans*	Growth inhibition	Gregory *et al.* (1969)
DDT	*Microcystis aeruginosa and Anabaena cylindria*	Growth inhibition	Vance and Drummond (1969)
Dieldrin	*Anacystis nidulans*	Growth inhibition	Batterton *et al.* (1971)
DDT	Some microalgae	Growth and survibility	Cox (1972)
Zineb (dithame) and Ceresan	*Anabaena* sp. *Nostoc* sp. and *Tolypothrix tenuis*	Growth inhibition	Venkataraman and Rajyalakshmi (1971, 1972)
DDT	*Synechococcus elongatus*	Growth process	Worthen (1973)
BHC, Lindane, Endrin Carbaryl and Diazinon	*Aulosira fertilissima*	Growth process	Ahmad and Venkataraman (1973)
BHC, Endrin, and Diazinon	*Cylindrospermum* sp., *Plectonema boryanum and Aulosira fertilissima*	Growth process	Singh (1973)
Aldrin, Dieldrin and Endrin	*Microcystis aeruginosa*	Growth inhibition	Clegg and Kovening (1974)
Mexacarbate	*Oscillatoria terrebriformis*	Growth inhibition	Snyder and Sheriden (1974)
Panacid	*Aulosira* sp. and *Nostoc* sp.	Growth and cell damage	Gupta and Saxena (1974)
Aldrin and Dieldrin	*Anabaena cylindrica* and *Nostoc muscorum*	Growth inhibition	Schuberger and Wildman (1977)
BHC	*Anabaena spiroides* and *Microcystis aeruginosa*	Growth Process	Das (1977)

(*Contd. Next Page*)

Pesticides	Organism	Observations	References
BHC	*Tolypothrix tenuis*	Growth inhibition	Srinivasan and Emayavaramban (1977)
Captan	*Westiellopsis* sp., *Calothrix* sp., *Nostoc* sp. and *Tolypothrix* sp.	Growth inhibition	Gangawane and Saler (1979)
Dimecron, Ekalux	*Anabaena vaginicola*	Toxicity	Das and Singh (1978, 1979)
Blitox and Dithane	*Nostoc entophylum* and *Cylindrosprmum Muscicola*	Toxicity assay	Torar and Salpekar (1979), Anand and Veerappan (1980)
Thiram	*Nostoc* sp. and *Tolypothrix* sp.	Growth inhibition	Gangawane and Kulkarni (1979)
BHC	*Synechococcus aeruginosa*	Growth and survibility	Reddy and Israel (1983)
Zineb and Ziram	*Cylindrospermum* sp. and *Nostoc* sp.	Growth inhibition	Panigrahy and Padhy (1984)
Carbofuran	*Cylindrospermum* sp. and *Nostoc muscorum*	Growth inhibition	Panigrahy (1984)
Bosalin and feroxome	*Anabaena flos-aquae*	Growth and morphological features	Chinnaswamy and Patel (1986)
Some insecticides	*Nostoc linckia*	Growth and nitrogen fixation	Megharaj *et al.* (1988)

Venkatraman, 1973; Singh, 1973; Das, 1977; Srinivasan and Emayavaramban 1977; Reddy and Isreal, 1983, Pabbi and Vaishya, 1992). In most cases BHC reduces the growth of test algae from 10 to 50 per cent even at a concentration greater than 10 ppm.

The effect of other chlorinated pesticides like lindane dieldrin, endrin and aldrin camphochlor also studied on a number of blue green algal species viz., *Anacystis nidulaus* (Lazaroff and Moore, 1966; Batterton *et al.*, 1971; *Aulosira fertilissima* (Ahmad and Venkataraman, 1973; Singh, 1973); *Anabaena cylindrica, Anacystis nidulaus and Nostoc muscorum* (Schauberger and Wildman, 1977); *Microcystis aeruginosa* (Clegg and Kovening, 1974); *Plectonema boryanum* (Singh, 1973); and *Cylindrospermum licheriforme* (Palmer and Maloney, 1955). In most cases, it was reported that growth inhibition of test organisms were recorded in laboratory tests in between concentrations of pesticides about 2 to 5 ppm.

In contrast, the organophosphorus insecticides like parathion, methyl parathion, diazinon, malathion, phorate, (thimet) etc. affect the growth of test organisms viz., *Cylindrospermum* sp. *Aulosira fertilissima* and *Plectonema boryanum* (Singh, 1973), *Anacystis nidulaus* (Anonymous, 1977), *Chlorogloea fritschii* (Dasilva *et al*., 1975) and *Westiellopsis* sp., *Tolypothrix* sp. and *Aulosira* sp. (Srinivasan and Emayavaramban, 1977; Gangawane,. 1980); comparative at higher concentration (25 ppm).

Among the carbamate insecticides, carbaryl (sevin) at 10 ppm was found not to inhibit the growth of the cyanobacterium *Aulosira fertilissima* (Ahmed and Venkataraman, 1973), rather it had a growth enhancing effect as studied in details on *Cylindrospermum* sp. and *Nostoc muscorum* (Padhy and Singh, 1978).

Identically, carbofuran was found a growth enhancing effect on both cyanobacteria (*Cylindrospermum* and *Nostoc*) at 25 ppm. Both strains could not tolerate 300 ppm carbofuran on agar plate or in medium (Panigrahy, 1984). However, mexacarbate (zectran) beyond the 10 ppm level diminished the growth of *Synechococcus lividus* and *Oscillatoria terrebriformis* (Snyder and Sheridan, 1974) Recently Tandon *et al*., (1988) reported the interaction of endogulfan malathion with blue green algae, *Anabaena* and *Aulosira*.

Among the fungicides, the effect of dichlon on cyanobacteria was reported by Fitzgerald *et al*. (1952); Fitzgerald and Skoog (1954); Palmer and Maloney (1955); and Bisiach (1970). In most cases 5 to 55 ppm will be effective for killing the organisms. But Zineb (dithane) at 100 kg/ha or 50 ppm, did not inhibit the growth of species of *Anabaena, Nostoc, Aulosira, Tolypothrix* and *Anacystis* etc. (Venkataraman and Rajyalakshmi, (1972). However, it was reported that 30 species of algae (including blue green algae) were treated with Ziram between 0.25 to 2 ppm, diatoms were more affected than BGA and the green algae were least sensitive [Maloney and Palmer (1956)]. Paraschiv *et al*., (1972) reported that maneb inhibited the growth of two species of *Oscillatoria*, Mancozeb between 5 to 50 ppm levels permitted growth of 27 strains of Cyanobacteria (Venkataraman and Rajyalakshmi, 1972), identically Captan at lower concentration, 1 ppm suppressed the growth and development of *Nostoc muscorum* (Moore, 1967). *Anacystis nidulans* (Moore and Dorward, 1968); *Westiellopsis prolifica, Aulosira fertilissima, Tolypothrix tenuis* (Gangawane and Saler, 1979). Vapan at 1 ppm was not able to inhibit growth of *Nostoc muscorum* (Moore, 1967), but Copper dimethyl dithiocarbamate stopped growth of two species of *Oscillatorea* (Paraschiv *et al*., 1972). Ceresan did not inhibit growth of 8 strains of *Anabaena* and 11 strains of *Nostoc* and *Aulosira fertilissima* at 20 kg/ha and of *Tolypothrix tenuis* and *Anacystis nidulans* at 200 kg/ha (Venkataraman and Rajyalakshmi 1971), Panacide (dichlorophene) at 20 ppm was toxic to *Aulosira fertilissima* and *Nostoc* (Gupta and Saxena, 1974). MBC at 500 ppm level caused decrease in growth of *Nostoc* and *Calothrix* by 7.9 and 1.0 per cent respectively, while a 1,000 ppm level caused decrease in growth of *Aulosira* and *Tolypothrix* by 5.7 and 3.2 per cent respectively (Gangawane and Saler, 1979). The fungicide Topison-M at 50 ppm reduced 11.5 per cent growth of *Calothrix* at 1000 ppm. This fungicide reduced 31.6 per cent growth of *Westiellopsis* : at 300 ppm, it also reduced growth of *Aulosira* and *Tolypothrix* by 12.8 and 14.0 per cent respectively (Gangawane and Kulkarni, 1979). *Westiellopsis* sp. *Aulosira* sp., *Nostoc* sp., *Tolypothrix* sp.and *Calothrix* sp.

tolerated the highest concentration (1000 ppm) of brassicol. Bavistin was toxic to *Calothrix* at a 1 ppm level. Growth of *Nostoc* sp. and *Tolypothrix* sp. was reduced by 18.3 and 6.25 per cent by 500 level of fytolan (Gangawane, 1980).

C. Herbicides

Over past three decades, the effects of several herbicides like 2, 4-d, MCPA propazine, simazine, monuron, fluorneturon, phenyl carbamates, propanil, bipyridyl compounds, maleic hydrazide, dalapon, chloropicrin, and aminotrizole compound ect. were tested on number of micro-algae particularly the blue green algal species like *Microcystis aeruginosa, Anacystis nidulans, Oscillatoria* species, *Tolypothrix tenuis, Scytonema* sps., *Cylindrospermum licheniformis, Aulosira fertilissima, Anabaena* species, *Nostoc* species, *Phormidium autumnale, Gloeocapsa alpicola, Plectoneam boryanum, Chlorogloea fritischii,* and *Lyngbya* species etc. (Fitzgerald *et al.*, 1952; Palmer and Maloney, 1955; Maloney, 1958; Kim and Greulach, 1963; Dodge, 1964; Butler, 1965; Shilo 1965; Moutschen *et al.*, 1966; Platonova 1967; Singh and Subbramain 1970; Hamdi *et al.*, 1970; Arvick *et al.*, 1971; Venkataraman and Rajyalakshmi, 1971, 1972; Pillay and Tachan, 1972; Ahmad and Venkataraman, 1973; Voight and Lynch, 1974; Singh, 1974; Yamagishi and Hashizuma, 1974; Torres and O'-Flaherty, 1976; Sarma and Tripathi, 1976; Cullimore and McCann, 1977; Das, 1977; Hauxby *et al.*, 1977; Das and Singh, 1977; Mehta and Hauxby, 1977, 1979; Goulding and Badami, 1978; Wright *et al.*, 1977; Wright, 1978; Das and Singh, 1979; Anand and Veerappan, 1980; Tiwari *et al.*, 1981; Rao and Vidyavati, 1983; Sreenivas and Rana, 1990, 1991; Kolte and Goyal, 1992). In general the effect of different herbicides at concentration 0.03 ppm is stimulatory, but 0.1 ppm was mostly inhibitory. However, in most of the test organisms, prevention of photosynthesis or total growth inhibition were reported at 5 to 20 ppm of different kinds of herbicides. Sometime the said concentration may cause lysis of cells (Table 35.3).

In a study of the action of pesticides and herbicides on growth of the micro-algae, Shilo (1965) placed the toxicants in assay cups on agar plate culture of the algae. All the toxicants thus used either exhibits algistatic or algicidal effects. However, later an improved slide technique was developed by Pipe and Cullimore (1980) for this kind of bioassay using Cyanobacteria like *Oscillatoria, Lyngbya* and *Anabanea* species.

Particularly during last decades, several biochemical studies were carried out to investigate the mechanism of pesticide and herbicide actions on algal system (Singh *et al.*, 1979; Panigrahy, 1984; Tandon *et al.*, 1988; Sreenivas and Rana, 1990, 1991).

Pesticide resistance of micro-algae particularly Cyanobacteria are also frequently isolated by different workers with BHC, 2, 4-d, Blitox, Carbaryl, Zineb, etc. (Panigrahy, 1984).

Although the use of pesticides in rice fields and other crop fields is a common practice, the study of effects of pesticides on cyanobacteria is much concentrated in pure cultures initiated limited information are available on field studies. But on the whole it is well known that nitrogen fixing blue green algae were very much affected by the consistent use of pesticides and other toxic substances in crop fields.

TABLE 35.3. *Effects of Herbicides on Microalgae*

Herbicides	Organism	Observations	References
2, 4-D	*Microcystis aeruginosa*	Growth inhibition	Fitzgerald *et al.* (1952)
Maleic hydrazide	*Chlorella pyrenoides*	Effects on metabolism	Kim and Greulach (1963)
Maleic hydrazide	*Protocentrum micans*	Effect on metabolism	Dodge (1964)
Bipyridyl compound (Paraguat, diquat)	*Anabaena variabilis,* and *Nostoc* sp.	Growth process	Shilo (1965)
Maleic hydrazide	*Chara vulgaris*	Inhibition of metabolism	Moutsehen *et al.* (1966)
2, 4-D and Simazine	*Tolypothrix tenuis*	Growth process	Platonova (1967)
2, 4-D	*Cylindrospermum licheniformis*	Growth process	Arvik *et al.* (1971)
2, 4-D	*Nostoc and Anabena*	Growth process	Pillay and Tchan (1972)
BHC and MCPA	*Nostoc muscorum*	Algistatic and algicidal activity	Ahmad & Venkataraman (1973) and Singh (1973)
2, 4-D	*Anacystis nidulans*	Growth process	Voight and Lynch (1974)
2, 4-D	*Cylindrospermum* sp.	Growth process	Singh (1974)
Prometryne and Symetryne	*Tolypothrix tenuis*	Growth process	Yamigishi and Hashizume (1974)
2, 4-D	*Oscillatoria leutea*	Growth process	Torres and O'Flaherty (1976)
2, 4-D	*Lyngbya* sp. and *Oscillatoria* sp.	Growth and survibility	Cullimore and Mc Cann (1977)
2, 4-D	*Anabaena variabilis*	Growth process	Hayxby *et al.* (1977)
Maleic hydrazide	*Cosmerium impressulum*	Growth and cell division	Rama devi (1978)
Phenyl carbamate	Some cyanobacteria	Growth and metabolism	Wright (1977, 1978)
2, 4-D	*Anabaena doliolum*	Growth and sporulation	Tiwari *et al.* (1981)
Magnacid-H (2-propenal)	*Anabaena* sp.	Growth process	Fritz-Sheridan (1982)
Maleic hydrazide	*Cosmerium verrucosum*	Growth and cell division	Rao and Vidyavati (1983)
2, 4-D	*Anabaena doliolum*	Growth and sporulatum	Srivastava and Tiwari (1983)
Atrazine	*Nostoc* sp.	Growth & metabolism	Sreenivas and Rana (1990)
Triazine	*Nostoc* sp.	Detoxification	Sreenivas and Rana (1991)

Algal Bioassays

Organic pollution in inland water bodies is increasing day by day because of the rapid industrialization, disposal of municipal and other types of sewages and dumping of enormous amounts of refuge in these water bodies. These pollutants, not only bring about a change in the physico-chemical characteristics of water but also trigger a change in the biotic components of the ecosystem, resulting in the deleation of some, otherwise valuable species in the aquatic ecosystem. Addition of some undesirable and toxic components which destabilise the ecosystem render it useless for most of the purposes even for the biota living within. Recently the problem has assumed top priority and lots of studies have been conducted to highlight this menance and find out possible solutions. While a considerable volume of literature exists depicting the pollution by the changes in the physico-chemical parameters of water body brought about by the addition of pollutants by one or the other source, there are as yet only few studies in which algae have been exploited as the biological indicators of pollution (Vaidya, 1967; Fogg, 1960, 1961, 1965; Cholnoky, 1968; Round, 1965; Bick *et al.*, 1967; Palmer, 1969; Venkateswarlu, 1969, 1970; Rao, 1972, 1977; Moore 1974; Marathe and Nandkar, 1976; Besch *et al.*, 1972; Bahis, 1973; Schoeman, 1973, 1976; Whitton and Say, 1975; Prasad and Singh, 1982; Whitton, 1984; Somashekar and Ramaswamy, 1983, 1984; Pal *et al.*, 1986).

A large amount of evidences indicate that industrial and energy process effluents contain algal, animal toxicants and also algal growth stimulators. Algal laboratory tests are excellent indicators of pollution by these complex wastes. Like animals, algae in the laboratory respond to toxicants, but they also respond in an easily measurable way to substances that affect primary productivity. Algal population growth in the laboratory is rapid and may be measured directly in a variety of ways. In addition, rates of physiological processes within exposed populations may be estimated indirectly by measurement of evolution and uptake of oxygen.

Over past two decades, a large number of fresh water and marine algae were tested for assay of different kinds of effluents (Fisher *et al.,* 1973; Dunstan, 1975; Brack *et al.,* 1976; Tilman, 1976; Thomas *et al* ., 1980).

Algal tests have been used in the field and laboratory to determine biological potential of natural waters and of substances from industrial, municipal and agricultural sources dissolved in them. Normally laboratory assay was done by batch culture, for continuous culture techniques. The end points of toxicity assay tests was mostly measured by growth response curve mostly. Sometime chlorophyll content or photosynthetic rate was also estimated.

Vocke *et al.* (1980) tested scrubber ash slurry from the settling pond of a coal fired generating plant against *Microcoleus vaginatus, Ankistrodesmus falcatus, Selenastrum capricornutum* etc. In all tests growth was inhibited markedly.

Identically Walsh *et al.,*(1980) tested effects of 23 textile mill wastes on growth of *Selenastrum capricornutum* and *S. costatum.* They also found identical features. Litton (1970) also reported the growth inhibiton of *Selenastrum capricorutum* to limestone scrubber samples and synthetic fuels.

Walsh and Alexander (1980) and Walsh *et al.* (1982) also analysed the effects of different liquid wastes to a number of test algal species. Various reports however, suggests that different concentrations may act as inhibitors or stimulators of different kinds of algal cultures. As such it is highly essential to standardize a suitable test species for each kind of effluents. Walsh and Merrill (1984) made a comprehensive review on this subject and suggested a guideline for future research plan in this area.

REFERENCES

Ahmad, M.H. and G.S. Venkataraman 1973. Tolerance of *Aulosira fertilissima* to pesticides. Current Sci. **42** : 108.

Anand, N. and B. Veerappan 1980. Effect of pesticides and fungicides on blue green algae Phycos. **19** : 210.

Anonymous, 1977. Studies on biological nitrogen-fixation. In Nuclear Institute for agriculture and biology (five years of NIAB) : p 43.

Arvick, J.H., D.L. Willson and L.C. Darlington 1971. Response of soil algae to picloram 2, 4 - D mixtures. Weed science.**19** : 216.

*Bahis, L.L. 1973. J.Water Pollut Control Fed. **45** : 134.

*Bartlett, L., F.W. Rabe and W.H. Funk 1974. Water Res. **8** : 179.

Batterton, J.C., G.M.Bousch and F.Matsumura 1971. Growth responses of blue green algae to aldrin, dieldrin endrin, and their metabolites. Bull. Environ. Contam. toxicol. **6** : 589.

*Besch, W.K., M.Ricard and R.Cantin 1972. Int. Rev. Ges. Hydrobiol : **57** :39.

Bharati, S.G. and S.B. Angadi 1981. Short and long term effect of fungicides on nitrogen-fixing blue green algae. Phycos. **20** : 58.

*Bick, H., K.P. Krishnamurthi and J.S.S. Lakshminarayana 1967. WHOEBL, P-1.

*Bisiach, M. 1970 . Algal infestations in Italian soils. Riso. **19** : 129.

Bonaly, J., A. Bariand,S. Duret, and J.C. Mestre 1980. Cadmium cytotoxicity and variation in nuclear content of DNA in *Euglena gracilis*. Physiol Plant. **49** : 286.

Brack, G.S., A. Jensen and A. Mohus 1976 . J.Exp. Mar.Biol. Ecol. **25** : 37.

Buikema, A.L. Jr., B.R. Niederlehner and Jr. J. Crains 1982. Biological monitoring part IV toxicity testing. Water Res. **16** : 239.

Burton, K. 1968 Determination of DNA concentration with dipenylamine. In : Method in Enzymology (ed. kaplam *et al*, vol. 12B), 163.

Butler, P.A. 1965. Effects of herbicides on estuarine fauna. Proc.5th Weed Control. Conf. **18** : 576.

Butler, G.L. 1977. Algae and Pesticides. Residue Reviews **66** : 19.

Butler, M., A.E.J. Haskew and M.M. Young 1980. Copper tolerance in the green alga *chlorella vulgaris*. Plant Cell Environ **3** : 119.

Cairns, J.Jr., J.W. Hall, E.L. Morgan, R.E. Sparks, W.T. Waller and G.A. Westlake 1973. The Development of an automated biological monitoring system for water quality. Water Resource Research Center Bulletin. 59.

Cairns, J.Jr. and K.L. Dickson 1978. Field and laboratory protocols for evaluating the effects of chemical substances on aquatic life ; J.Test Eval . **6** : 81.

Carins, J.Jr., 1981. Biological monitoring part VI - Future needs Water Res. **15** : 941.

Chawla, G., P.N. Viswanathan and S.Devi. 1986. Effect of linear alkyl benzene sulphonate Environ. Exp. Bot. **26** :39.

Chawla, G., P.N. Viswanathan and S.Devi. 1987. Biochemical responses of *Scenedesmus quardicauda* to linear alkyl benzene sulphonate. Environ. Exp. Bot . **27** : 311.

Chlonoky, B.A. 1968. Die okologie der. Diatomeen in Binnengewassern, J.Cramer, Lehre, Germany, 699.

Chinnaswamy, R. and R.J. Patel, 1986. Bioassay to studey detoxification of 2-4D by BGA *Nostoc* sp. IBC . **3** :81.

Clegg, T.J. and J.L. Kovening 1974. The effect of four chlorinated hydrocarbon pesticides and one organo phosphate pesticides on ATP levels in three species of photosynthetic fresh water algae. Bot. Gaz. **135** :368.

*Conway, H.L. 1978 . J.Fish Res. Bd Canada **35** :286.

Cook, S.F., JR., and J.D. Conners 1962. The short term side effects of the insecticidal treatment of clear lake, Lake country, California. J.Econ . Entomol . **56** : 819.

Cox. J.L. 1972. DDT residues in marine phytoplankton. Residue Reviews **44** : 23.

Cullimore . D . R. and A.E. McCann 1977. Influence of four herbicides on the algal flora of a prairie soil. Plant and Soil **46** : 455.

Das, B. 1977 . Effect of herbicides and pesticides on the fresh water blue green algae. Utkal Univ., Vani Vihar (ph. D . dissert .)

Das, B. and P.K Singh 1977. The effect of 2-4-D on growth and nitrogen fixation of BGA *Anabaenopsis raciborskii*. Arch. Environ. Contam. Toxicol . **5** : 445.

Das, B. and P.K Singh 1978. Pesticides (hexachlorocyclohexane) inhibition of growth and nitrogen fixation in blue green algae *Anabaenopsis reciborskii* and *Anabaena aphanizomenoides*. Zeits. Allg . Microbiol . **18** : 161.

*Das, B. and P.K. Singh 1979 . Nova Hedwigia **63** : 161.

Dasilva, E.J., L.E. Henrickson, and E. Henrickson, 1975 . Effects of pesticides on blue green algae and nitrogen fixation. Arch. Environ. Contam. Toxicol. **3** : 193.

Davies, A.G. 1976. An assessment of the basis of mercury tolerence in *Dunaliella tertiolecta*. J. Mar Biol. Assoc. U.K. **56** : 39.

De Filippis, L.F. 1979. The effect of heavy metal compounds on the permeability of *Chlorella cells*. Z. Pflanzenphysiol. **92** : 39.

*De Filippis, L.F., R. Hampp and H. Ziegler, 1981. Arch Microbiol. **128** : 407.

De Filippis, L.F. and C.K. Pallaghy 1976. The effect of sublethal concentration of mercury and zinc on *Chlorella* III. Development and possible mechanisms of resistance to metals Z. Pflanzenphysiol. **79** : 323.

*Delmottle, A. 1980. J. Exp. Bot. **31** : 1107.

Demon, A., M. De Bruin, H. Th. Wolterbeek. 1988. The influence of pH on trace metal uptake by an algae (*Scenedesmus pannonicus* sub sp. Berlin) and Fungum (*Aureobasidium pullulans*) Environ Monitor and Assess. **10** : 165.

Dimitrieva, A.G. 1969. The effect of polymetallic ores and concentrates on the growth of cultures of *Microcystis aerugionsa* Kutz. In : L.F. Sirenko (ed) : Algals Blooms of Water. P 207 Kiev Naukova Dumka.

Dikshit, G. and G.L. Tiwari. 1990. Effect of pesticides on certain nitrogen fixing cyanobacteria. Proc. Not Symp. on Cyanobacterial Nitrogen fixation. (B.D. Kaushik ed). p-495.

*Dodge, J.D. 1964. Protoplasma **8** : 312.

Dubey, S.K. and L.C. Rai. 1990. Heavy metal toxicity in a nitrogen fixing cyanobacterium *Anabaena doliolum* : Regulation toxicity by certain environmental factor. Biomed. Environ. Sci. **3** (2) : 240.

*Dunstan, W.M. 1975. Environ. Sci. Technol. **9** : 635.

Elder, J.H. and A.J. Horne. 1978. Copper cycles and copper sulphate algicidal capacity in two California Lakes. Environ. Manag. **2** : 17.

EPA, 1975. Test methods for assessing the effects of chemical on plants. Final Report EPA. **560** : 575-008, office of toxic substances, Environmental Protection Agency; Washington, D.C.

Ferrante, G.M. and A. Battino-Viterbo 1974. Agar plate technique for potential algicide screening. Riso. **23** : 13.

*Fisher, N.S., L.B. Graham and E.J. Carpenter 1973. Nature **241** : 548.

Fisher, N.S., E.J. Carpenter, C.C. Remsen and C.F. Wurster 1974. Micub. Ecol. **1** : 39.

Fisher, N.S. and D. Frood, 1980. Heavy metals and marine diatoms influence of dissolved organic compounds on toxicity and selection for metal tolerance among four species. Mar. Biol. **59** : 85.

Fitzgerald, G.P., G.C. Gerloff, and F. Skoog, 1952. Studies on chemicals with selective toxicity to blue green algae. Sewage Ind. Wastes. **24** : 888.

Fitzgerald, G.P. and F. Skoog, 1954. Control of blue green algae blooms with 2-3, Dichloro-naptho quinone, Sewage Ind. Wastes. **26** : 1136.

*Fitzgerald, G.P. and S.L. Faust, 1963. Appl. Microbiol. **11** : 345.

Fogg, G.E. 1949. Growth and heterocyst production *Anabaena cylindrica* Lem. II. In relation to carbon and nitrogen metabolism. Ann. Bot. **13** : 241.

*Fogg, G.E. 1960. In. Proc. Symp. Algoloyg, ICAR, New Delhi p. 138.

Fogg, G.E. 1961. In physiology and Biochemistry of algae (Lewin, R.A. ed.) AP, NY, 449.

*Fogg, G.E., 1965. Proc. Royal Soc. London, **173** : 175.

Foster, P.L. 1977. Copper exclusion as a mechanism of heavy metal tolerance in a green alga. Nature. **269** : 322.

Foster, P.L. 1982. Metal resistances of chlorophyta from rivers polluted by heavy metals. Fresh Water Biol. **12** : 41.

Foy, C.D. and. G.C. Gerloff. 1972. Response of *Chlorella yrenoidosa* to alluminium and low pH. J. Phycol. **8** : 268.

*Francke, J.A. and H. Hillebrand, 1980. Aquatic Biol. **8** : 285.

Fritz-Sheridean, R.P. 1982. Impact of the herbicide magnacide-H (2-propenal) on algae. Bull. Environ. Contam. Toxicol. **28** : 245.

Gangawane, L.V. and L. Kulkarni. 1979. Tolerance of certain fungicides by nitrogen fixing blue-green algae and their side effects on rice cultivars. Pesticides **13** : 37.

Gangawane, L.V. and R.S. Saler, 1979. Tolerance of certain fungicides by nitrogen fixing blue green algae. Curr. Sci. **48** : 306.

Gangawane, L.V. 1980. Tolerance of N_2 fixing blue-green alge to brassicol, bavistan, fytolan. J. Indian. Bot. Sci. **59** : 150.

Gekler, W. E. Grill, E.L. Winnacker and M.H. Zenek. 1988. Algal sequester heavy metals via synthesis of phytochelalin complexes Arch. Microbiol. **150**: 197.

Gerloff, G.C., G.P. Fitzgerald, and F. Skoog. 1950. The isolation, purification and culture of blue green algae. Am. J. Bot. **37** : 216.

Gibson, C. E. 1972. The algicidal effect of Copper on a green and blue green alga and some ecological implications. J. Appl. Ecol. **9** : 513.

Gingrich, D.J., K.N. Weber, L.F. Shaw, J.S. Garvery and D.H. Petering 1986. Characterization of a highly negative and labile binding protein induced in *Euglena gracillis*. Environ Health Perspect. **65** : 77.

*Gonzalaves, E.U.A.A. and B.S. Vaidya, 1963. Hydrobiol. **21** : 188.

Gowrinath, K.P. and V.N. Rao, 1990. Detoxification of copper by *Nitzschia obtusa* Wm. Sm., a pennate Diatom. Bull. Environ. Contam Toxicol. **45** : 612.

Goulding, K. and N.V. Badami 1978. The algicidal activity of some substituted 1-3 diphenyl ureas, 1-phenyl-3 (2 pyridyl) ureas and the corresponding thioureas. Pest. Sci. **9** : 213.

Green, J.C., W.E. Miller, T. Shiroyama, R. A. Soltero and K. Pufman 1978. Mih. Inst. Verh. Theor. Angew. Limol. **21** : 372.

Gregory, W.W., J.K. Reed and L.E. Priester, 1969. Accumulation of parathion and DDT by some algae and Protozoa J. Protozool, **16** : 69.

Gross, R.E., P. Punso and W.M. Dugger, 1970. Observations on the mechanism of Copper damage in *Chlorella*. Plant Physiol. **46** : 183.

Gupta, G.S. and P.N. Saxena, 1974. Effect of panacide on some green and blue green algae. Current Sci. **43** : 492.

*Gupta, A.B. and A. Arora, 1978. Physiol. Plant. **44** : 215.

*Hall, A., A.H. Fielding and M. Butler 1979. Mar Biol. **54** : 195.

Hamdi, Y.A., E.I. Narway, and A.S. Tewfi. 1970. Effect of herbicides on growth and nitrogen fixation of alga *Tolypothrix Tenures*. Acta. Microbiol. Pol. **2** (Sc. B.) : 53.

Harding, J.C.P. and B.A. Whitton. 1976. Resistance to zinc of *stigeoclonium tenue* in the filed and the laboratory. Br. Phycol. J. **11** : 417.

Harding, J.C.P. and B.A. Whitton. 1977. Environmental factors reducing the toxicity of Zinc to *Stigeoclonium tenue*. Br. Phycol. J. **12** : 17.

*Harding, J.P.C. and B.A. Whilton, 1981. water Res. **15** : 301.

*Hargreaves, J.W. and B.A. Whilton, 1976. Oecologia. Besl. **26** : 235.

Harwood-Sears, V. and A.S. Gordon, 1990. Copper-induced production of copper binding supernatant proteins by the marine bacterium *Vibrio alginotyticus*. Appl. Environ. Microbiol. **56** : 1327.

Hassall, K.A. 1963. Uptake of copper and its physiological effects on *Chlorella vulgais*. Physiol. Plant. **16** : 323.

Hart, B.A. and B.D. Scaife, 1977. Toxicity and Bioaccumulation of cadmium in *Chlorella Pyrenoidosa*. Environ. Res. **18** : 327.

Hauxby, K., B. Tubea, J. Ownby, and E. Baseler, 1977. Effect of various herbicides on four species of algae Pest. Biochem. Physiol. **7** : 203.

*Huisman, J. and H.J.G. Ten Hoopen, 1978. Water Air Soil Pollut. **10** : 325.

Hunding, C. and R. Lange, 1978. In principle of ecotoxicology, (G.C. Butler ed) scope 12 scope 1cs, John Wiley and Sons, Chichesler. 234.

Husaini, Y., A.K. Singh and L.C. Rai, 1991. Cadmium toxicity to photosynthesis and associated electron transport system of *Nostoc linckia*. But Environ. Contam. Toxicol. **46** : 146.

Hutchinson, T.C. 1973. Comparative studies of the toxicity of heavy metals to phytoplankton and their synergestic interaction Water. Pollut. Res. Can. **8** : 68.

Jecobson, L. 1951. Maintenance of iron supply in nutrient solution by a single additiou of ferric potassium ethylene di-aminotetra acelate. Plant Physiol. **26** : 411.

Jensen, T.E., M. Baxter, J.W. Rachlin and V. Jain, 1982. Uptake of heavy metals by *Plactonema boryanum* (cyanophyceae) into cellular components, especially polyphosplate bodies. An X-ray energy dispersive study. Environ. Pollut. Ser A. **27** : 119.

Kallqvist, T. and B.S. Meadows, 1978. The toxic effect of copper on algae and rotifers from a soda lake (Lake Nukuru. East Ffrica) Water Res. **12** : 771.

Kaushik, B.D. and G.S. Venkatarama, 1983. Response of cyanobacterial nitrogen fixation to insecticides. Curr. Sci. **52** : 321.

*Kim, W.K. and V.A. Greulach. 1963. Phyton. Rev. Int. Bot. Exptl. **20** (2) : 127.

Klass, E., D.W. Rowe and E.J. Massaro, 1974. The effect of cadmium on population growth of the green algae, *Scenedesmus quadracanda* Bull. Environ. Contam. Toxicol **12** : 442.

Kleinen-Hammans, J.W., L.P.L.M. Rabou and H.Q. Pietersen. 1976. Participation of pigment complexes in uptake and incorporation of mercury ions by *Anacystis*. Photosynthetica **10** : 440.

Kotangate, L.R., R. Sarkar and K.P. Krishnamurthy, 1984. Toxicity of mercury and zinc to *Spirulina platensis*. Indian J. Environ. Health. **26** (1) : 41.

Kolte, S.O. and S.K. Goyal. 1992. Inhibition of growth and nitrogen fixation in *calothrix marchica* by herbicides *in vitro*. Proc. Natl. Symp. Cayanobacterial Nitrogen Fixation (Kaushik, B.D. ed) New-Delhi, p. 507.

Les, A. and R.W. Walker. 1984. Toxicity and binding of Copper, Zinc and Cadmium by the blue green alga *Chroococcus Paris*. Water Air Soil Pollut. **23** : 129.

Lee, C.M., H.L. Bergman, W.P. Wood and J.H. Rodger, Jr. 1982. Workshop summary and conclusion. In modeling the fate of chemicals in the Aquatic Environment. C. Dickson, K.L. Maki, A.W. Cairns, J.Jr. eds. Ann. Arbor. Science. **397**.

Lin, R.I. and O.A. Schjeide. 1969. Micro-estimation of RNA by the cupric ion catalysed orcinol reaction. Ann. Bio-chem. **27** : 473.

Litton, B. 1979. Final Report : Level I Bioassays on nine Conventional Limestone scrubbing samples with and without adipic acid additive, E.P.A. contract No-68-02-2681. Technical Directive No-105, Industrial Environmental Research Laboratory, Research Triangle Park, North Carolina p. 144.

Lowry, O.H., N.J. Rosebrough, A.L. Farr and R.J. Randall. 1951. Protein measurement with the Folin-phenol reagent. J. Biol. Chem. **193** : 265.

Mahapatra, P.K., P.K. Sethi and R.C. Mohanty. 1962. Effect of dimethoate on growth behaviour and nitrogen fixing ability of *Anabaena doliolum* under different environmental conditions. Proc. Natl. Symp. Cyanobacterial Nitrogen Fixation (Kaushik B.D. ed). p 465.

Maclean, F.I., O.J. Lucin, Z.A. Shaihk and E.R. Manz. 1972. The uptake subcellular distribution of Cd and Zn in Micro-organisms Fed. Proc. **31** : 699.

Mallick, N. and L.C. Rai, 1990. Effects of heavy metals on the biology of a nitrogen fixing cyanobacterium *Anabaena doliolum*. Toxic. Assess. **5** (3) : 207.

Maloney, T.E. and C.M. Palmer. 1956. Toxicity of six chemical compounds to thirty cultures of algae. Water Sewage Works **103** : 509.

Maloney, T.E. 1958. Control of algae with chlorophenyl dimethylurea. J. Amer. Water Works Assoc. **50** : 417.

Marathe, K.V. and B. Nandkar. 1976. Ind. J. Environ. Health **18** : 311.

Massalski, A. V.M. Laube and D.J. Kushner. 1981. Effect of Cadmium and copper on the ultrastructure of *Ankistrodesmus braunii* and *Anabaena* 7120. Microb. Ecol. **7** : 183.

McBrien, D.C.H. and K.A. Hassal. 1967. The effect of toxic doses of copper upon respiration. Photosysthesis and growth of *Chlorella vulgaris*. Physsiol. Plantarum. **20** : 114.

*McKnight, D.M. and F.M.M. Morel, 1979. Limnol. Oceanogr. **24** : 823.

Megharaj, M., K. Venkateswarlu and S. Rao, 1988. Effect of insecticides and phenolics on nitrogen fixation by *Nostoc linckia*. Bull Environ. Contam. Toxicol. **41** : 277.

Mehata, R.S. and K.Hauxby. 1977. Action of herbicides on blue green algae-ultrastructural observation on photosysthetic lamellat Mise. Ser. Bot. Soc. Amer. Pub. **154** : 51 (EN).

Mehata, R.S. and K. Hauxby, 1979. Effect of simazine on the blue green alga *Anacystis nidulans*. Bull. Environ. Contam. Toxicol. **23** : 319.

*Meisch, H.U. and I.S. Beckman, 1979. Z. Pflanzenphysiol. **94** : 231.

Mierle, G. and P.M. Stokes. 1976. Heavy Metal tolerance and metal accumulation by planktonic algae. In trace substances in Environmental Health (D.D. Hemphill ed.) Univ. Missouri. Columbia pp. 113.

Miller, W.E., J.C. Greene and T. Shiroyama. 1978. The *Selenastrum capricornutum* Printz. Algal Assay Bottle Test : Experimental design application, data interpretation protocal U.S. EPA Environmental; Research Laboratory, Corvallis Oregon. EPA-60019-78-018, p. 126.

*Monahan, T.H. 1976. J. Phycol. **12** : 358.

Moody, R.P., P. Wienberger and R. Green-Halgh. 1983 : In Aquatic Toxicology, Miragy (ed.) John Wiley and Soris, Chichester, pp. 503.

Moore, R.B. 1967. Algae as biological indications of pesticides. J. Phycol. **3** (suppl.) : 4.

Moore, R.B. and D.A. Dorward. 1968. Accumulation and metabolism of pesticides by algae. J. Phycol. **4** (suppl.) : 7.

*Moutschen, J., M.Dahmen and C. Gillet. 1966. La Cellulae. 58.

Murphy, T.P., D.R.S. Lean, C. Nalewajka, 1976. Blue green algae. Their excretion of iron selective chelators enable them to dominate other algae. Science NY. **192** : 900.

*Muller, K.W. and H.D. Payer, 1980. Physiol. Plant. **50** : 265.

Nakano. Y., K.O. Kamota, S. Toda and F. Fuwa, 1978. Toxic effects of cadmium on *Euglena gracilis* grown in zinc deficient and zinc sufficient media. Agric. and Biol. Chemis. **42** : 901.

OECD Guidelines for testing of chemicals (1981) section 2. Effect of biotic systems test 201. ''Alga growth inhibition test'' Adopted, 12 May 1981.

Pabbi, S., and A.K. Vaishya, 1992. Effect of insecticides on cyanobacterial growth and nitrogen fixation Proc. Natl. Symp Cyanobacterial Nitrogen Fixation. (Kaushik, B.D. ed.), p. 488.

Paddy, R.N. and P.K. Singh. 1978. Stabilizing effects of metallic ions in the blue green algal vrrus N-1. biochem. physiol. Pflanzen. **73** : 188.

Padhy, R.N. 1985. Cyanobacteria and Pesticides Residue Reviews. **94** : 1.

Pal, U.C., G. Bandyopadhyay and S.C. Santra. 1986. Importance of phytoplankton study in the assessment of water quality–a case study from Bhagiarathi Hoogly river basin Proc. All India Seminar On Water Quality Around Urban Ecosystem (K.S. Univ. Ed.), 1.

Palmer, C.N. and T.E. Maloney, 1955. Preliminary screening for potential algicides. Ohio F. Sci. **55** : 1.

Palmer, C.M. 1969. A. Composite rating of algae tolerating organic pollution. J. Phycol. **5** : 78.

Palmer, C.M. 1980. Algae and water Pollution. Castle Home Pulol., England; p-96.

Pandey, S.N. and A.K. Tripathi, 1983. Studies on the effect of Zinc on the growth on algae in all India, p. 206.

Panigrahy, K.C. 1984. Physiological and genetical effect of pesticides on blue green algae (effect of carbamate pesticides). Berham Univ. (Ph. D. dissetation).

Panigrahy, K.C. and R.N. Padhy, 1984. Toxicity of carbamate pesticides to cyanobacteria (in press).

Parachiv, M., E. Serbanescu, G.H. Popovici and C. Kjendow. 1972. Algicide action of chemical substances on blue green algae Rev. Roum. Biol. Ser. Bot. **17** : 195.

Parson, T.R. and Strickland, J.D.H. 1965. Particulate organic matter III. 1 : pigment analysis III 1.1 : determination of phytoplankton J. Fish Research BD Canada. **18** : 117.

Patrick, R. 1972. In : Indicators of Environmental Quality (Thomes; W.A. ed). Plenum Press, New York P1, 93.

*Patrick, R., J. Cairns Jr. and A. Scheier, 1968. Progressive Fish-Culturist. 137.

Phillips, D.J.H. 1980. Biological indicators and their use to monitor trace metal and organochlorine pollution. In : Quantitative Aquatic Biological Indicators (D.J.H. Phillips ed), Applied Science Publ., London.

Pillay, A.R. and Y.T. Tchan. 1972. Study of soil algae; Vii Adsorption of herbicides in soil and prediction of their rate of Application by algal method. Plant and Soil **36** : 571.

Pipe, A.E. and D.R. Cullimore, 1980. An implanted slide technique for examining the effects of the herbicide diuron on soil algae. Bull. Environ. Contam. Toxicol. **24** : 306.

Platonova, V.P. 1967. Effect of 2, 4-D, Simazine, Prometryne on Soil algae. Works Inter Universities Conf. in Kirosl. **20** : 215.

Prasad, B.N. and Y. Singh. 1982. On diatoms as indicators of water pollution. J. Ind. Bot. Soc. **61** : 326.

Prasad, S.M., F.B. Singh, L.C. Rai and H.D. Kumar. 1991. Metal induced inhibition of photosynthetic electron transport chain of the cyanobacterium *Nostoc muscorum*. FEMS Microbiology Letters. **82** : 95.

Rachlin, J.W., Jensen, T.E.M. Baxter, V. Jain, 1982. Utilisation of morphometric analysis in evaluating response of *Plactonema boryanum* (Cyanophyceae) to exposure to eight heavy metals. Arch. Environ. Contam. Toxicl. **11** : 323.

Rachlin, J.W. and B. Warkentine. 1984. The toxicological response of the algae *Anabaena flos-aquae* (cyanophyceae) to cadmium Arch. Environ. Contam. Toxicol. **13** : 143.

*Rai, L.C. 1979. Phycos. **18** : 105.

Rai, L.C., J.P. Gaur and H.D. Kumar, 1981. Phycology and heavy metal pollution. Biol. Rev. **56** : 99.

Rai, L.C. and M. Raizada, 1985. Effect of nickel and silver ions on survival, growth, carbon fixation and nitrogenase activity of *Nostoc muscorum* : regulation of toxicity by EDTA and calcium. J. Gen. Appl. Microbiol. **31** : 329.

Rai, L.C. and M. Raizada. 1986. Nickel induced stimulation of growth heterocyst differentation, CO_2 uptake and Nitrogenase activity in *Nostoc muscorum.* New Phytol. 104 : 111.

*Rai, L.C. and M. Raizada, 1988. Ecotoxicol. Environ. Saf. **15** : 195.

Rai, L.C. 1989. Sliver toxicity in a nitrogen-fixing cyanobacterium Interaction with chromium, nickel and lead. Biol. Metals. **2** : 122.

Rai, L.C., T.E. Jenson and J.W. Rechlin, 1990. A. morphometric and X-ray energy dispersive approach to monitoring pH altered Cadmium toxcity in *Anabaena flos-aquae*. Arch. Environ. contam. Toxicol. **19** : 479.

Rai, L.C., N. Mallick, J.B. Singh and H.D. Kumar. 1991. Physiological and biochemical characteristics of a copper tolerant and wild type strain of *Anabaena doliolum* under copper stress J. Plant Physiol. **138** : 68.

Rama Devi, K. 1978. Studies on the effects of mutagenic agents on *Cosmarium impressulum.* Ph. D. Thesis Osmania University, Hyderabad.

Rana, B.C. and H.D. Kumar. 1974. The toxicity of zinc to *Chlorella vulgaris* and *Plectonema boryanum* and its protection by phosphate. Physkos. **13** : 60

*Rao, V.S. 1972. Hydrobiol. **39** : 351.

*Rao, V.S. 1977. Hydrobiol. **53** : 13.

Rao, B.D. and Vidyabati. 1983. Maleic hydrazide induced variations in *Euastrum verrucosum.* In proc. Applied Phycological Cong., Advances in Applied Phycology (A.C. Sukla and S.N. Pandey eds). p. 274.

Reddy, T.R.K. and G. Isreal, 1983. Effect of BHC on blue green alga *Synechococcus aeruginosa* Proc. All India Applied Phycol. Cong., Kanpur (In Advance in Applied Phycology (A.C. Sukla and S.N. Pandey eds). pp. 291.

Richard, F.A. and T.G. Thompson, 1952. The estimation and characterisation of plankton population by pigment analysis. J. Mar. Res. **2** : 156.

Roe, J.H. 1955. The determination of sugar in blood and spinal fluid with authrone reagent. J. Biol. Chem. **212** : 335.

Round, F.E. 1965. In : Biol. Prob. Water Pollut. (Tarswell *et al.* ed) U.S. Dept. Health Edn and Walefare.

Ructer, J.G., J.J. McCarthy and E.J. Carpenter, 1978. The toxic effect of copper on *oscillatoria theirbautii.* Limnology and Oceanography. **24** : 558.

Sandman, G. and P. Boger. 1980 a. Copper Deficiency and toxicity in *Scenedesmus* Z. Pflanzenphysiol. **98** : 53.

Sandman, G. and P. Boger, 1980 b. Copper-induced exchange of plastocyanin and Cytochrome C-553 in cultures of *Anabaena variabilis* and *Plectonema boryanum.* Plant Sci. Lett. **17** : 412.

*Sarma, T.A. and A.K. Tripathi, 1976. Caryologia. **29** : 58.

Sarma, T.A. 1979. An attempt to find copper resistance in *Anacystis nidulans.* In : Advances in Cyanophyte Research (S.P. Singh. D.N. Tiwari. A.K. Kashyap and P.K. Yadawal eds). Dept. of Botany BHU India.

Sato, C., J.L. Schnoor, and D.B. McDonald, 1986. Characterisation of effects of Copper, Cadmium, Nickel on the growth of Nitrosomonas europaea. Environ. Toxicol Chem. **5** : 403.

Schauberger, C.W. and R.B. Wildman, 1977. Accumulation of aldrin and dieldrin by blue green algae and related effects on photo synthetic pigments. Bull. Environ. Contam. Toxicol. 17 : 534.

*Schoeman, F.R. 1973. Pretoria VI and 255 P-1.10.

*Schoeman, F.R. 1976. J. Limno. Soc. S. Afr. **2** : 21.

Schulze, H. and J.J. Brand, 1978. Lead toxicity and phosphate deficiency in *Chlamydomonas*. Plant Physiol. **62** : 727.

*Shehata, F.H.A. and B.A. Whitton, 1981. Verh. Int. Verein. Theor Angew. Limnol. **21** : 1466.

Shehata, F.H.A. and B.A. Whitton. 1982. Zinc tolerance in strain of the blue green algal *Anacystis nidulans,* Br. Phycol. J. **17** : 5.

Shilo, M. 1965. Study on the isolation and control of blue green algae from fish ponds, Bamidgjeh, **17** : 83.

*Sicko-Goad, L. and E.F. Stoermer, 1979. J. Phycol. **45** : 316.

Sicko-Goad, L. 1988. Effect of toxicants on phytoplanton with special reference to the Laureutian great lakes In : Toxic Contaminants and Ecosystem Health (M.S. Evans ed.) John Wiley and Sons, NY, p. 19.

Singh, R.N. and K. Subbramaiah. 1970. Effects of chemicals on *Fischerella Muscicola* (Thuret). Gom. Can. J. Microbiol. **16** : 19.

Singh, P.K. 1973. Effect of pesticides on blue green algae. Arch Microbiol. **89** : 317.

Singh, P.K. 1974. Algicidal effect of 2-4D on blue green algae *Cylindrospermusn* sp. Arch. Microbiol. **97** : 69.

Singh, S.P. and A.K. Kashyap. 1978. Manganese toxicity and mutagenesis in two blue green algae. Environ. Expt. Bot. **18** : 47.

Singh, H.N., H.R. Sing and A. Vaishampayan, 1979. Toxic and mutagenic action of herbicide Alachlor (Lasso) on various strains of the nitrogen-fixing blue green alga *Nostoc Muscorum* and characterisation of the herbicide induced mutant resistant to methylamine and L-methionine disulfoximine. Environ. Expt. Bot. **19** : 5.

Singh, S.P. and A.K. Pandey, 1981. Heavy metal interaction studies with a blue green alga. Acta. Bot. Indica. **9** : 290.

Singh, S.P. and A.K. Pandey, 1982. Cadmium mediated resistance to metals and antibiotics in a Cyanobacterium. Mol. Gen. Genet. **187** : 247.

Singh, S.P. and V. Yadava, 1983. Cadmium induced inhibition of nitrate uptake in *Anacystis Nidulans.* Interaction with other divalenet cations. J. Gen. Appl. Microbiol. **29** : 297.

Singh, S.P. and V. Yadava, 1984. Cadmium induced inhibition of ammonium and phosphate uptake in *Anacystis nidulans.* Interaction with other divalent cations. J. Gen. Appl. Microbiol. **30** : 79.

Singh, S.P. and V. Yadava 1985. Cadmium uptake in *Anacystis nidulans.* Effect of modifying factors. J. Gen. Appl. Microbiol. **31** : 39.

Snyder, C.E. and R.P. Sheridan. 1974. Toxicity of the pesticide Zectran on photosynthesis, respiration, growth in four algae. V. Phycol. **10** : 137.

Somashekar, R.K. and S.N. Ramaswamy, 1983. Algal indicators of paper mill waste waters. Phykos. **22** : 161.

Somashekar, R.K. and S.N. Ramaswamy. 1984. Biological assessment of water pollution : A case study of the river Kapila Intern. J. Environ. Studies, **23** : 261.

*Sorentino, C. 1979. Phykos. **18** : 149.

Sreenivas, S.S. and B.C. Rana, 1990. Growth and metabolic response of *Nostoc* sp. to atrazine, a triazine herbicide. Nat. Symp. Cyanobacterial Nitrogen Fixation. IARI.

Sreenivas, S.S. and B.C. Rana, 1991. Studies on detoxification of triazine herbicide by blue green algae *Nostoc* sp. Pollut. Res. **10** (1) : 47.

Sreenivasan, S. and N. Emayavaramban, 1977. Some observation on blue green algae. Aduthurai Rev. **1** : 98.

Srivastava, M., M.M.S. Rawat and R.K. Srivastava. 1991. Effect opf different concentrations of Zinc on the survival and growth behaviour of *Chlorella vulgaris*. Pollut. Res. **10** (1) : 49.

Stewart, W.D.P., G.P. Fitzgeraland and R.H. Burris, 1968. Acetylene reduction by nitrogen fixing blue green algae. Arch. Microbiol **62** : 336.

Steemann Nielsen, L., K. Nielsen and S. Wium-Anderson, 1969. The effect of deletorious concentration of copper in the photosynthesis of *Chlorella pyrenoidosa*. Physiol. Plant. **22** : 1121.

Steemann Nielsen, E. and S. Wium-Anderson, 1971.The influence of copper on photosynthesis and growth in diatoms. Physiol. Plant **24** : 480.

Stokes, P.M. 1975. Adaptation of green algae to high levels of copper and nickel in aquatic environments. International Conference on Heavy Metals in the Environment. Toronto, Canada, p. 137.

Stokes, P.M., T. Maler and J.R. Riordan, 1977. A low molecular weight Cu-binding protein in Cu-Tollerant strain of *Scenedesmus acuminatus*. In Trace Substances in Environmental Health XI (Hemphill D.D. ed.) Univ. of Missouri, Columbia, p. 146.

Stokes, P.M. 1986. Ecological effects of acidification on primary producers in aquatic systems. Water Air Soil Pollut. **30** : 421.

Stratton, G.W., A.L. Hubber and C.T. Corker, 1979. Effect of mercury ion on of *Anabaena inaqualis*. Applied Environ. Microbiol. **38** : 537.

Stratton, G.W. and C.T. Corke, 1979 a. The effect of cadmium ion on the growth, photosynthesis and nitrogenase activity of *Anabaena inaqualis*. Chemosphere. **5** : 277.

Strattion, G.W. and C.T. Corke. 1979 b. The effect of Cadmium ion on growth, photosynthesis and nitrogenase activity of *Anabaena inaqualis*. Chemosphere. **10** : 731.

*Stratton, G.W. and C.T. Corke, 1979 c. Can. J. MIcrobiol. **25** : 1094.

*Subramanian and Shanmugasundaram, 1986. Influence of herbicide 2, 4-D on nitrogen fixation and ammonia excretion by the Cyanobacterium *Anabaena*. Proc. Ind. Natl. Acad. B 52 **2** : 308.

Tandon, R.S., Ruplal and V.V.S. Narayana Rao. 1988. Interaction of endosulfan and malathion with blue green algae *Anabaena* sp. and *Aulosira* sp. Environ. Pollut. **52** (1) : 1.

Tarar, J.L. and T.H. Shewale, 1984. Studies on the effect of some fungicides on soil algae of paddy field. Phykos **23** (1 and 2) : 191.

*Thomas, W.H., J. T. Hollibaugh and D.L.R. Seibert, 1980. Phycologia, **19** : 202.

*Tilman, D. 1976. Science, **192** : 463.

Tiwari, D.N. A.K. Pandey and A.K. Mishra, 1981. Action of 2, 4-D Dichlorophenoxyacetic acid and rifampicin on heterocyat differentation in the blue green algae, *Nostoc Muscorum.* J. Biosci. **3** : 33.

Torar, J.L. and C.R. Salpekar, 1980. Relative tolerance of soil algae for selected insecticides. Science and Culture. **46** : 105.

Torres, A.M.R. and L.M.O-Flaherty, 1976. Influence of pesticides on *Chlorella, Chlorococcum, Stigoclonium* (chlorophyceae) *Tribonema Vancheri* (Xanthophyceae) and *Oscillatoria* (Cyanophyceae), Phycologia. **15** : 25.

*Vaidya, B.S. 1967. Hydrobiol. **29** : 256.

Vance, B.D. and W. Drummond, 1969. Biological concentration of pesticides by algae. J. Amer. Water Work Assoc. **16** : 360.

Venkateswarlu, V. 1969. An ecological study of the algae of the river Moosi, Hyderabad (India) with special reference to water pollution II. Factors influencing the distribution of algae Hydrobiol. **33** : 352.

Venkateswarlu, V. 1970. An ecological study of the algae of the river Moosi, Hyderabad (India) with special reference to water pollution IV. Periodicity of some common species of algae, Hydrobiol. **35** (1) : 45.

Venkataraman, G.S. and B. Rajyalakshmi, 1971. Tolerance of blue green algae to pesticides. Current Science. **40** : 143.

Venkataraman, G.S. and B. Rajyalakshmi, 1972. Relative tolerance of nirogen fixing blue green algae to pesticides Ind. J. Agri. Sci. **42** : 199.

Vocke, R.W., K.L. Sears, J.J.O. Toole and R.B. Wildman. 1980. Water Res. **14** : 141.

Voight, R.a. and D.L. Lynch, 1974. Effect of 2, 4-D and DMSO on Prokaryotic and eukaryotic cells. Bull. environ. Contam. Toxicol. **12** : 400.

Vymazal, J. 1987. Toxicity and accumulation of Cadmium with respect to algae and Cyanobacteria, a review. Toxicol. Assess. An. Int. Quart. **2** : 387.

*Walsh, G.E. and S.V. Alexander, 1980. Water Air and Soil Pollut. **13** : 45.

*Walsh, G.E., K.M. Duke and R.B. Foster, 1982. Water Res. **16** : 879.

Walsh, G.E. and R.G. Merrill, 1984. Algal bioassays of industrial and energy process effluents. In : algae as Ecological Indicators (L.E. Shubert ed.) academic Press, London, p. 330.

Wang, W.S. and R.S. Tischer, 1973. Study on the extracelluar Poly saccharides produced by a blue green algae *Anabaena floss-aquae* A. 37. Arch. Microbiol. **91** : 77.

*Weber, A., M. Melkonian, D.W. Lorch and M. Wettern, 1978. Mitt. Int. Verein. Limnol. Theor. Angew. **21** : 254.

Whitton, B.A. 1970. Toxicity of heavy metals to fresh water algae a review Phykos. **19** : 116.

Whitton, B.A. and P.J. Say, 1975. Heavy Metals in River Ecology (B.A. Whitton ed.) Blackwell, Oxford, p. 286.

Whitton, B.A. 1979. Plants as indicators of river water quality. In Biological Indicators of Water Quality (A. James and L. Evisor eds), Chichester, NY, John Wiley and Sons, p. 51.

*Whitton, B.A. and B.M. Diaz, 1981. Verh Int. Verein Theo, Angew Limnol. **21** : 459.

Whitton, B.A. and F.H.A. Shehata, 1982. Influence of Coblt, Nickel Copper and Cadmium on the blue green algae *Anacysits nidulans* Environ. Pollut. Ser. A. **27** : 275.

Whitton, B.A. 1984. Algae as monitors of heavy metals. In : Algae as Ecological Indicators (L.E. Shubert ed), London, Academic Press, p. 257.

*Wium-Anderson, S. 1974. Physiol. Plant, **32** : 308.

Wolk, C.P. 1973. Physiology and Cytological Chemistry of blue green algae Bacteriol. Rev. **37** : 32.

*Wong, P.T.S., Y.K. Chan and P.L. Luxon, 1978. Fish Res. B. Canada. **35** : 479.

Wong, S.L. and J.L. Beaver, 1980. Algal bioassays to determine toxicity of metal mixtures, Hydrobiol. **74** : 199.

Wong, P.T.S. 1980. Cadmium toxicity to phytoplankton and micro-organisms. In : Cadmium in the Environment pt. I (J.O. Nriagu ed) John Wiley and Sons. New York, p. 572.

Worthen, L.R. 1973. Interception and degradation of pesticides by aquatic algae, Rpt.PB-223/506, Prep for Office of Water Resources Research NIIS, US Dept. Commerce.

Wright, S.J.L. and A.F. Stainthorpe and J.D. Downs, 1977. Interactions of herbicides propanil and a metabolite, 3, 4-Dichloroaniline with blue green algae. Acta. Pat. Acad. Scient. Hung. **12** : 51.

Wright, S.J.L. 1978. Interactions of pesticides with micro-algae. In Pesticide Microbiology (I.R. Hill and S.J.L. Wright eds), London Academic Press, p. 535.

Yamagishi, A. and A. Hashizume, 1974. Ecology of green algae in paddy fields and their control with chemicals. Zasso Kenkyu. **18** : 39.

**Not seen in original.*

*Whitton, B.A. and B.M. Diaz. 1981. Verh Int. Verein Theor. Angew Limnol. 21 : 459.

Whitton, B.A. and P.H.A. Shehata. 1982. Influence of Cobalt, Nickel, Copper and Cadmium on the blue green algae *Anacystis nidulans* Environ. Pollut. Ser. A. 27 : 275.

Whitton, B.A. 1984. Algae as monitors of heavy metals. In : Algae as Ecological Indicators (L.E. Shubert ed). London, Academic Press, p. 257.

*Wium-Andersen, S. 1974. Physiol. Plant. 32 : 308.

Wolk, C.P. 1973. Physiology and Cytological Chemistry of blue green algae Bacteriol. Rev. 37 : 32.

*Wong, P.T.S., Y.K. Chan and P.L. Luxon. 1978. Fish Res. B. Canada. 35 : 479.

Wong, S.L. and J.L. Beaver. 1980. Algal bioassays to determine toxicity of metal mixtures. Hydrobiol. **74** : 199.

Wong, P.T.S. 1980. Cadmium toxicity to phytoplankton and micro-organisms. In : "Cadmium in the Environment pt. I (J.O. Nriagu ed) John Wiley and Sons: New York, p. 572.

Worthen, L.R. 1973. Interception and degradation of pesticides by aquatic algae. Rpt PB 223608, Prep. for Office of Water Resources Research NTIS. US Dept. Commerce.

Wright, S.J.L. and A.F. Stainthorpe and J.D. Downs. 1977. Interactions of herbicides propanil and its metabolite, 3, 4-Dichloroaniline with blue green algae. Acta. Phyt. Acad. Scient. Hung. 12 : 51.

Wright, S.J.L. 1978. Interactions of pesticides with micro-algae. In Pesticide Microbiology (I.R. Hill and S.J.L. Wright eds). London Academic Press, p. 535.

Yamagishi, A. and A. Hashizume. 1974. Ecology of green algae in paddy fields and their control with chemicals. Zasso Kenkyu. 18 : 39.

*Not seen in original.